Low Energy Electron Spectrometry

LOW ENERGY ELECTRON SPECTROMETRY

Kenneth D. Sevier

WILEY–INTERSCIENCE A Division of John Wiley & Sons, Inc.
New York · London · Sydney · Toronto

Library of Congress Catalog Card Number: 78-170684

ISBN 0-471-77850-8

Printed in the United States of America.

10 9 8 7 6 5 4 3 2 1

Preface

In 1965 I inquired of several publishers about the feasibility of a book on the topic of electron detection at low energies. In response I received a stimulating and unique suggestion from the late editor of Interscience Publishers, Robert A. Schetty. The suggestion was that perhaps a larger work, for example, one embracing the entire field of low energy electron spectroscopy, might be a more suitable goal.

The merit of this suggestion was immediately clear: experimental work concerned with the measurement of low energy electron spectra, in particular below about 50 keV, had expanded and intensified immensely during the preceding 10 to 15 years. New techniques were making it possible to overcome previous difficulties related to electron sources and their preparation. Electron spectrum measurements using much improved magnetic spectrometers and the more recently developed and improved electrostatic spectrometers were becoming highly accurate. New and improved methods of electron detection at low energies had been advanced to facilitate electron spectrometric studies at very low energies.

A treatment of the topic of low energy electron spectrometry at first suggested stressing the actual lowering of the low energy limit for electron spectrometry investigations and the instrumental and experimental refinements that have made this possible. Chapter 2 deals, therefore, with instrumentation and detection techniques; electron sources as a topic of specific treatment has been omitted because of time limitations.

However, by lowering that energy limit, new topics are bared and become subject to intensified investigation; older, previously diffuse topics (e.g., atomic level widths, effects of atomic recoil after alpha-decay) come into focus. In addition, since X-ray spectroscopy is already at home in this energy region, great strength is realized by combining what is learned through each of the two fields in the investigation of topics of common interest in atomic physics. Some of these topics have also been investigated through research in widely differing fields (e.g., neutron scattering techniques supplement electron scattering and infrared absorption excitation of lattice vibrations in crystalline solids), and such multifield considerations of the topics are rich with information. A very striking example of a topic where several fields of research are applied is the study and determination of atomic electron binding energies (see Appendix F).

In view of these several considerations, topics in atomic physics are treated in general in Chapters 3 through 8, with emphasis on their experimental investigation by using methods of electron spectrometry. Other relevant information originating from work in different fields is also integrated into the treatments.

The selection of the chapter subjects was made to embrace the subject of atomic physics in the broadest feasible manner. Chapter 3, treating the Auger and Coster–Kronig effects and autoionization, deals with electron emission rearrangement transitions in initially excited atoms. Also included, as is often done in treatments of this subject, is a consideration of the alternative radiative deexcitation of these atoms so as to complement the main treatment. Chapter 4 concerns internal conversion. Although this phenomenon is both nuclear and atomic and has usually been examined from the nuclear standpoint, it is treated here in such a way as to focus attention on its atomic aspects (e.g., as a method to study the distribution of the various orbital electrons throughout the atom).

In Chapter 5 are treated nuclear recoil after alpha-decay, its effects on internal conversion electron line structures, and applications of its effect in studying changes in binding energies due to multiple ionization. Chapter 6 surveys the subject of atomic level widths as measured by electron spectrometric and other methods of investigation. Chapter 7 attempts to give a broad treatment and review of the subject of atomic electron binding energies as measured for atoms in both the gaseous and solid states, with methods of electron spectrometric methods, X-ray and ultraviolet spectroscopy, electron scattering measurements, and so on. Also discussed here are alterations of atomic electron binding energies as resulting from changes in the chemical or physical states of the element of interest.

Chapter 8 treats electron scattering in solids. Here, high resolution electron energy loss measurements may be combined with electromagnetic radiation absorption or neutron scattering methods to learn of the nature of plasmon, exciton, and phonon excitations.

Finally, there are appendices containing information of special value to the electron spectrometrist, as well as to workers in other fields. The table of atomic electron binding energies has been extended to include results not only from X-ray and photoelectron methods, but also from Auger, electron scattering, and internal conversion methods. Appendix H provides the derivation of the two forms of the Ritchie formula which are most used in electron scattering investigations.

A topic of science may be divided up into its several aspects as follows: theoretical treatment of the subject; experimental investigation of the subject; and the application of knowledge in the subject to investigate other, perhaps unrelated, topics or even for industrial or practical use. The three subdivisions are actually strongly interrelated (see, e.g., the introduction to Chapter 8). In this book emphasis is placed on the experimentation moment of the topics, with enough theory included to give, hopefully, a unifying character. This makes it possible to refer to unexpected or yet unexplained experimental results in the text without a strong break of continuity, rather than having to refer to exceptions of a theoretical interpretation of a phenomenon. At the end of most of the chapters are brief citations of practical or investigational applications of knowledge of the topic.

The aim of this book is to provide a single, unique source of descriptive material and references about the many uses of electron spectrometric techniques in a variety of fields. Hence this work is essentially "historical-descriptive." Some of the chapters are reviews of research done in the several separate subjects treated (e.g., the Auger effect, internal conversion). Certain chapters, however, consider topics also of interest in fields other than electron spectrometry (e.g., atomic binding energies, natural widths of atomic levels). The treatments are arranged to inform a specialist what has been done with electron spectrometry and other techniques in a variety of fields other than his own. The unifying factor throughout is the emphasis on methods of electron spectrometry used for the research performed. The information obtained by these methods is complemented by that obtained through applying investigational techniques of other fields.

A continuous effort has been made to keep the treatment objective. The main purpose is to bring a review of work on the several topics under one cover. A high degree of organization among the chapter topics, which is so correctly desirable, has not been accomplished because of the previously mentioned time limitations.

Also, no attempts have been made to unify extensively in topics where the unification or critical analysis of theoretical and experimental data has not yet been carried out by specialized workers. The treatments of these topics are at times merely statements or descriptions of completed work. Extension and/or reiteration of earlier, successful attempts at unification have, however, been included. Further coordination in and of the various subjects must be done by future workers.

The title of this book is descriptive of the orientation of the effort that has gone into it. "Low energy" is, of course, a relative term and embraces different ranges in different fields: in nuclear reactions low energy may be anything above some megaelectron volts; in nuclear physics, above perhaps 100 to 200 keV; in neutron physics, either several megaelectron volts or in the millielectron-volt region, depending on the nature of the research. As stated previously, electron spectrometric research activities at energies below about 50 keV have been strongly expanded recently, and, except for the Auger effect and internal conversion, this book deals with work at energies below 50 keV. When this approximate upper limit was cited in response to a nuclear physicist's query: "What do you mean by low energy?," he responded, "That's not low energy, that's too low energy." And indeed this, too, is low energy.

The last two words of the title, "electron spectrometry," refer to the spectrometric analysis of electron spectra, regardless of the source—whether by an

Auger transition, internal conversion, photoelectric effect, or electron scattering. It will be noted that the term "spectrometry" has been chosen over the older term spectroscopy. This has been done following a suggestion by Dr. Manfred O. Krause. In essence, he holds, and I agree, that as a field develops to a point at which accurate quantitative measurements are possible, the field has become transformed from one of "-scopy" to one of "-metry." This has been true of mass separation, which matured from "mass spectroscopy" to "mass spectrometry." The same maturing has occurred in "electron spectral analysis." Hence the title *Low Energy Electron Spectrometry.*

A charm of a personal nature has also been lent to this work. While in Paris, in 1969, I had the pleasure of meeting and having a most pleasant and stimulating visit with Professor Pierre Auger. In response to my invitation that he introduce Chapter 3 with an idea or sentiment, he gave that which stands at the fore of that chapter—an observation which, indeed, provokes one's thought and more.

KENNETH D. SEVIER

Paris, France
May 1971

Acknowledgments

The scope of this book is exceedingly great. Indeed, it would no doubt have been impossible to accomplish its writing between October 1966 and March 1970 without the inspiring and willing aid of many experts working with the individual topics.

It was most fortuitous that Professor M. Eugene Rudd, Lincoln, Nebraska, consented to provide a very singular treatment of electrostatic instruments and also a review of the topic of earth magnetic field annulment (Sections 3 and 4 of Chapter 2).

Experts in the several fields provided ready aid in checking and criticizing the various chapters. They are: M. Eugene Rudd, Chapter 2, instrumentation and electron detection; Werner Mehlhorn, Münster, Germany, now at Freiburg, Chapter 3, the Auger, Coster–Kronig, and autoionization effects; Otakar Dragoun, Heidelberg, Germany, now at Prague, Czechoslovakia, Chapter 4, internal conversion; G. A. Korolev, Leningrad, Chapter 5, atomic recoil after alpha-decay and consequences and applications; Geoffrey Kaye, formerly at Liverpool and now with I.B.M. in England, Chapter 6, natural widths of atomic levels; M. O. Krause, Oak Ridge, Tennessee, Chapter 7, atomic electron binding energies; J. B. Swan, Nedlands, Australia, and A. Otto, Munich, Germany, Chapter 8, electron losses by scattering in solids; and C. Béguin, Grenoble, France, and D. M. Hercules, Athens, Georgia, Appendix D, accuracy comparison of chemical analysis techniques.

In addition, it was possible to contact many workers, some by mail, but most personally, through study trips organized for the purpose of discussing research problems, interests, and activities. In this connection I extend my special gratitude to the following persons for their kind and inspiring help and cooperation: R. G. Albridge, W. N. Asaad, Y. Baer, J. Berkowitz, F. Boehm, H. Boersch, A. Bohg, B. Brehm, C. Briançon, R. E. Burge, P. G. Burke, G. Busch, E. J. Callan, M. Campagna, T. A. Carlson, W. A. Chupka, B. Cleff, K. Codling, F. J. Comes, M. A. Coulthard, A. V. Crewe, M. Creuzburg, C. J. Danby, J. Daniels, F. J. de Heer, W. de Wieclawik, J. Sant 'Ana Dionisio, H. Ehrhardt, J. Eland, G. T. Ewan, A. Faessler, R. W. Fink, S. Fraga, M. S. Freedman, H. U. Freund, D. C. Frost, J. S. Geiger, (Chalk River), J. Geiger, (Berlin), E. Gilberg, H. B. Gilbody, A. Gizon, J. Gizon, R. L. Graham, E. M. Gunnersen, S. Hansen, S. K. Haynes, E. Holøien, F. Illés, P. Jahn, E. Karttunen, C. E. Kuyatt, J. Lafoucrière, R. E. LaVilla, P. Longe, G. Malli, G. V. Marr, J. W. Mayer, E. J. McGuire, A. Meisel, A. H. Muir, D. Nagel, H. Niedrig, N. N. Perrin, F. T. Porter, W. C. Price, H. Raether, T. N. Rhodin, V. Schmidt, C. Sébille, H. C. Siegmann, D. W. Turner, M. Valadares, L. Vriens, R. J. Walen, G. Weich, K. Zeppenfeld, and T. Åberg.

I was resident in Uppsala, Sweden, during the first three years of working on this book, having also lived there the seven previous years. It was there, and in Stockholm, that I received much aid and inspiration from friends and associates. I express my appreciation to W. Seibt, P. Kleinheinz, R. Vukanavič, Ml. Župančič, R. Nordberg, A. Fahlman, A. Nigavekar, S. Bergvall, E. Källne, B. Ekstig, L. Samuelsson, H. Slätis, P. Erman, B. Santesson, and E. Söderberg. I am especially grateful to R. Spohr for his constant good interest and aid on this project during the past four years.

It was in Uppsala, at the Institute of Physics and at

the University's Carolina Bibliotek, that most of the reference work was done. I thank the representatives of the University of Uppsala. Also, I thank Professor Fröman of the Institute of Theoretical Physics for permission to use the library of that institute.

Appreciation is due also to Ruth Almegaard of the Niels Bohr Institute in Copenhagen for permission to use the Institute's library. I have also made use of the library of the physics division, Argonne National Laboratories, and the physics libraries of the University of California at Berkeley and the University of Nevada at Reno.

I express my gratitude to the many able typists who put the text into its first, readable form. I am especially indebted to Barbro Kinnwall of Uppsala, who not only typed at least one-half the total text, but improved the text in the process.

I am indebted to the following typists in Uppsala for their aid and good work in typing: Gunilla Carlsson, Maud Gustavsson, Gunvor Jacobsson, Barbro Oldberg, and Britt Palmborg. I am also indebted to Marie Johnson, New York, Sylvia Gottlieb, Indiana, and Diane Nowell and Caroline Kelly, Nevada, for their quality typing.

Finally, I thank my mother, Myrtle Sevier, for many hours of typing much of the text and for her devoted interest in this project.

K. D. S.

May 1971

Contents

*By M. E. Rudd, University of Nebraska, Lincoln.

. . . and this, the jagged edge of the subject's frontier, must be puzzled over into the future.

K. D. S.

Chapter One

Introduction

For the purposes of this book, the subject of low energy electron spectrometry includes those scientific research topics which are concerned with the measurement and analysis of electron spectra. The nature of the sources of these electrons—in particular the physical phenomena giving rise to these spectra—may vary greatly from one to another.

Several of these phenomena have been under investigation for many years (e.g., the Auger effect, Chapter 3, internal conversion, Chapter 4, electron scattering in solids, Chapter 8). Others have gained in interest or come to be investigated seriously for the first time only in recent years (e.g., autoionization, Chapter 3, effects of nuclear recoil after alpha decay, Chapter 5). Again, various atomic properties that have been investigated earlier via X-ray spectroscopic and other techniques have only recently gained interest for the application of electron spectrometric methods (e.g., atomic electron binding energies and their alterations with chemical state, Chapter 7; atomic level widths, Chapter 6).

The several topics just named differ in the atomic phenomena of their concern and the information they provide the investigator about the phenomena themselves or about physical properties of the atom. However, all these subjects share one point in common: they may be, or in some cases must be, investigated using methods of spectrometric analysis of electron spectra.

Up until the late 1950s, electron spectrometric investigations were confined mainly to higher energy electron spectra ($E > 30$–50 keV). This was due mostly to the unavailability of precise instrumentation for lower electron energies (i.e., energy analysis and electron detection techniques). Another handicap, which was especially strong for the investigation of low energy electron spectra from radioactive substances, was represented by difficulties with electron source preparation. More recently, improvements in magnetic spectrometer design and the very rapid development and improvement of electrostatic spectrometers have made it possible to investigate electron spectra at much lower energies—down to a few electron volts. Simultaneously, there were improvements in methods of low energy electron detection, and new methods were developed. These instrumental developments for both electron spectral analysis and electron detection are treated in Chapter 2.

Spectra from gaseous electron sources can be investigated at very low energies ($\lesssim 500$ eV) more suitably than can those from solid sources. This may be partially because of surface contamination effects on the solid source surface resulting from insufficiently low vacuum. It is significant to note that many researchers who are now experimenting in the $E \lesssim 10$ keV energy region, having worked earlier at much higher energies, often continue to employ vacuums of $> 10^{-6}$ torr. On the other hand, solid state physicists and workers who are concerned through their fields of interest with surface purity use $\lesssim 10^{-9}$ torr when carrying out electron spectrometric investigations.

Figure 1 shows the energy ranges over which the topics of interest to the several chapters span. It is interesting to note that this span includes seven decades below ~ 50 keV, which has been a frequent lower limit in previous works dealing with electron spectrometry. It should also be noted that electron spectrometric investigations can be carried out down to energies approaching 1 eV. The several listed topics that span energies lower than this are concerned with changes or differences in higher electron energies (e.g., atomic level widths, Chapter 6; and inelastic electron scattering, Chapter 8).

The energy region shown in Figure 1 is one that also interests workers in other fields of atomic physics: X-ray, ultraviolet, and infrared spectroscopy, solid state physics, and so on. For this reason, in many of the chapters experimental results concerning the same or similar physical quantities as obtained by methods other than electron spectrometry are included and/or compared. In addition, several of the chapters end with short descriptions serving to refer the reader to work in general related to the chapter topics, although

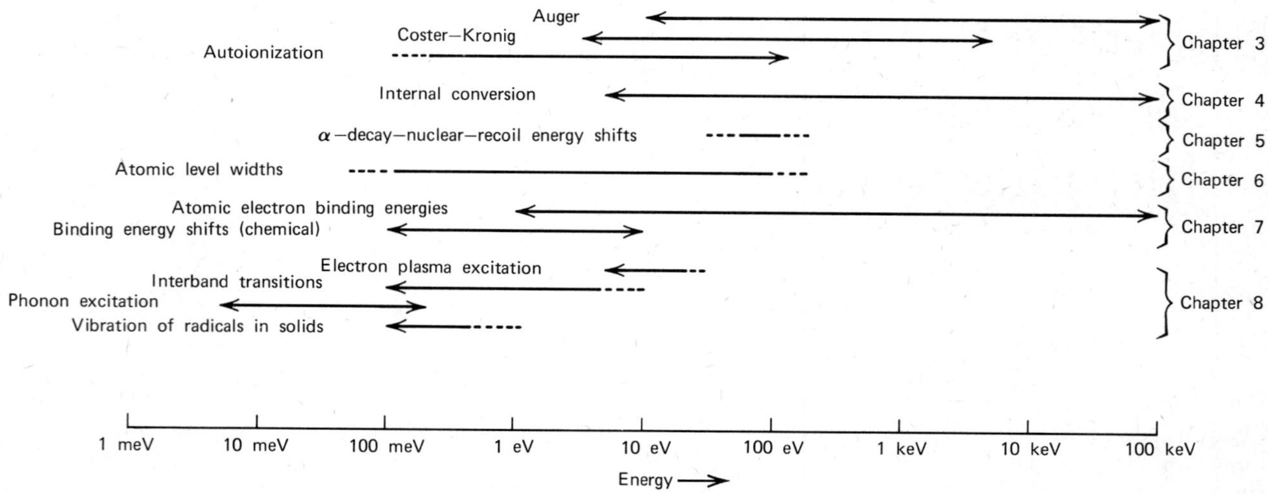

Figure 1. Energy ranges of interest in this book.

not close enough to be included in its own right. Also are cited technical or other applications of knowledge of the chapter subject.

Aside from the topics included in the chapters that follow, there are several others that might also be collected under the title of this book. Some of these, however, have been treated extensively elsewhere and the reader is referred to these works for further information: electron diffraction (including low energy electron diffraction) and its applications,[1] atomic excitation and ionization processes in gases by electron impact,[2] and proton, ion, and atomic bombardment.[3] In addition, two other topics have been omitted here: a general treatment of electron sources and electron scattering in gases (to complement Chapter 8). However, there are limited treatments available[4] of the second topic.

Supplementary tables and discussions have been included in Appendices A through H. The tables are those which are of constant use to the worker engaged in electron spectrometry. In addition, certain useful evaluations and derivations are included.

Finally, a convention adopted to represent error limits of numerical values should be explained. For the sake of simplification, error limits relating to quoted values have been represented in a more convenient manner. For example, the value 3.420 ± 0.015 is written 3.420(15).

REFERENCES

1. P. B. Hirsch, *Electron Microscopy of Thin Crystals*, Butterworths, London, 1965; K. W. Andrews, D. J. Dyson, and S. R. Keown, *Interpretation of Electron Diffraction Patterns*, Plenum Press, New York, 1967; R. E. Weber and W. T. Peria, *J. Appl. Phys.* **38** (1967) 4355; P. W. Palmberg, T. N. Rhodin, and C. J. Todd, *Appl. Phys. Letters* **11** (1967) 33; *Appl. Phys. Letters* **10** (1967) 122; P. W. Palmberg, C. J. Todd, and T. N. Rhodin *J. Appl. Phys.* **39** (1968) 4650; P. W. Palmberg and T. N. Rhodin, *J. Phys. Chem. Solids* **29** (1968) 1917; *J. Chem. Phys.* **49** (1968) 134; *J. Chem. Phys.* **49** (1968) 147; R. L. Gerlach and T. N. Rhodin, *Surface Sci.* **8** (1967) 1; *Surface Sci.* **10** (1968) 446.
2. B. L. Moiseiwitsch and S. J. Smith, *Rev. Mod. Phys.* **40** (1968) 238; F. H. Field and J. L. Franklin, *Electron Impact Phenomena and the Properties of Gaseous Ions*, Academic Press, New York, 1957; G. F. Drukarev, *The Theory of Electron-Atom Collisions*, Transl., Ed., J. B. Hasted, Academic Press, New York, 1965.
3. W. Lichten, "Resonant Charge Exchange," *Advan. Chem. Phys.*, Vol. VIII, I. Prigogine (Ed.), Wiley-Interscience, New York, 1968; H. H. Fleischmann and R. A. Young, *Phys. Rev. Letters* **19** (1967) 941; E. W. Thomas and G. D. Bent, *Phys. Rev.* **164** (1967) 143, 151; F. W. Byron, Jr., and C. J. Joachain, *Phys. Rev.* **164** (1967) 1; *Phys. Letters* **24A** (1967) 616; D. Jaecks, F. J. de Heer, and A. Salop, *Physica* **36** (1967) 606; J. W. R. Fennema, *Physica* **39** (1968) 13.
4. See Chapter 7, Section 3.6.2.

Chapter Two

Instrumental Methods of Electron Spectrometry and Methods of Electron Detection

1. INTRODUCTION

The production and investigation of electron spectra may be divided into three moments: electron sources, electron energy or momentum spectrum analysis, and electron detection. Each of these is to a large extent independent of the other two, and the tripartite subdivision is at once both convenient and natural.

Influences of electron energy straggling in the source and difficulties of detector window penetration and electron detection have mostly negligible adverse influence on the quality of the measurement results for higher energies (in hundreds of keV); however, these effects become increasingly magnified with decreasing electron kinetic energies. As a consequence, methods of electron source preparation and electron detection must be greatly improved, and the nature of these adverse effects and their influence on the quality of experimental information must be probed and understood. At the same time, the question of how these may be reduced or eliminated must be constantly considered.

As the name indicates, the subject of spectrometry is primarily concerned with the analysis of electron spectra. For this reason methods of electron energy analysis and detection are treated here. For a review of electron sources, other literature must be consulted.[1]

In this chapter the various types of instrumentation that give high energy or momentum resolving power ($\gtrsim 0.1\%$) are reviewed. These comprise *two* main subdivisions: magnetic (Section 2) and electrostatic spectrometers (Section 3). The iron-free magnetic instruments and the electrostatic spectrometers require the annulment of the earth's magnetic field in the region where energy analysis takes place. For this reason Section 4 deals with the annulment of the earth's magnetic field within the spectrometer volume. Section 5 treats briefly alterations in electron microscopy instrumentation, which make electron energy analysis possible. This is done either through the employment of electrostatic fields of special form or the superposition of both magnetic and electrostatic fields.

Over the past ten years the advances and improvements made regarding instruments in the second subdivision are strongly overshadowed by those enjoyed by spectrometers of the first subdivision. Magnetic spectrometers are therefore reviewed with only the naming of the general properties and performance of those most commonly used—the specifics may be found in the original articles and in other reviews.[1–5] Information with significant bearing on electron spectrometry at low energies, however, is reported here.

In the second subdivision are included the various types of electrostatic spectrometers of large acceptance angle, whereas an additional subgroup would include those electrostatic (e.g., the Möllenstedt type) or combined electrostatic and magnetic field instruments (e.g., the Wien filter), both of which have small acceptance angle and high resolving power. (See Section 5.)

The different methods of electron detection, for low kinetic energies, are reviewed in Section 6.

1.1 Basic Formulas in Electron Spectrometry

Electron spectra may be presented in terms of number of electrons per unit momentum value versus momentum or per unit energy value versus energy. The first of these is convenient when a magnetic spectrometer is employed and the second when one of the electrostatic type is used.

In the first case, momentum is measured in terms of $B\rho$ (magnetic field value in Gauss times the radius of trajectory curvature in that field for the electron energy of interest). By setting force equal to the product of mass and acceleration of an electron traveling a circular trajectory of radius ρ, one obtains

$$e\mathbf{v} \times \mathbf{B} = m\mathbf{a} \qquad \text{or} \qquad evB = \frac{mv^2}{\rho}, \tag{1}$$

where e is the electron charge, v the electron velocity, and the relativistic electron mass

$$m = m_0\left(1 - \frac{v^2}{c^2}\right)^{-1/2}. \tag{2}$$

In equation 2, m_0 is the electron rest mass. Also, momentum $= p = mv = eB\rho$.

When measuring spectrum intensity in a variable field magnetic spectrometer of fixed radius, since the instrumental resolution $\Delta(B\rho)/B\rho$ is constant, the plot of intensity versus momentum will be that intensity in the momentum band $\Delta(B\rho)$.

The velocity of an electron of a given energy is given by the expression

$$v = \frac{2E}{eB\rho}. \tag{3}$$

This relationship is plotted in Appendix B, with the function m versus E for the electron. It is seen that for $E \approx 50$ keV and ≈ 5 keV the relativistic effects have increased the electron mass by 10 and 1%, respectively.

Turning now to the electrostatic spectrometer, here we measure the electron intensity as a function of electron kinetic energy. Now, in order to transform an electron intensity spectrum as a function of momentum into one as a function of energy, the momentum intensity distribution must be divided by

$$\frac{dE}{dp} = \frac{p}{m}. \tag{4}$$

Because the momentum scale is used with magnetic spectrometers and the energy scale with electrostatic spectrometers, and since it is convenient to correlate kinetic energies with binding energies, and so on, it is very useful to have an accurate table available for converting energy to momentum values and the reverse.

The relativistic equation for the kinetic energy of an electron is

$$E = mc^2 - m_0 c^2. \tag{5}$$

Using now equation 2,

$$(mc^2)^2 = [(m_0 c^2)^2 + e^2(B\rho)^2c^2] \tag{6}$$

and

$$E = m_0 c^2\left[\left(\left(\frac{e}{m_0 c}\right)^2 B\rho^2 + 1\right)^{1/2} - 1\right]. \tag{7}$$

From Appendix A, the 1965 fundamental constants,[6] one obtains the values of $m_0 c^2$ and $e/m_0 c$. A table giving $B\rho$ (in G-cm) versus E (in keV) in steps of 0.1 G-cm for energies from 0 to near 50 keV is given in Appendix C.

Setting $m_0 c^2 = \mu$,

$$\frac{\Delta E}{E} = \left(1 + \frac{\mu}{\mu + E}\right)\frac{\Delta(B\rho)}{B\rho}, \tag{8}$$

which expresses the energy-dependent relaionship between the resolution expressed in terms of energy and that in terms of momentum. The ratio $(\Delta E/E)/[\Delta(B\rho)/B\rho]$ varies but little with E for low energies: this ratio has the asymptotic value of 2.0 for $E \to 0$, and it is within 2% of 2.0 for $E < 10$ keV, and equal to 1.837 for $E = 100$ keV. The resolutions in the two units are, therefore, related approximately as $R(E) \approx 2R(B\rho)$.

1.2 Nomenclature

The description and comparison of the wide variety of electron spectrometer types necessitates the definition of a general set of properties or merit scales. The functional qualities or figures of merit of these spectrometers (e.g., the resolving power, transmission, etc.) may then be compared to rate the spectrometric quality of the instruments.

The terms in Section 1.2.1 refer to spectrometers that have neither extended sources, detectors, nor combinations of these; such instruments are treated in Section 1.2.2.

1.2.1. Single or Limited Sources and Detectors

Consider now a source of electrons from which a certain fraction of emitted electrons of all energies is allowed to pass the entrance baffle system. This is called the *solid angle* Ω of the instrument and is expressed in percent of 4π rad. The product of the source area (in cm^2) and the solid angle is called the *luminosity* $\Lambda = \Omega\sigma$.

The electrons admitted into the spectrometer are separated relative to their momenta (magnetic instrument) or energies (electrostatic instrument) and electrons lying between $B\rho$ and $B\rho + \Delta(B\rho)$ or E and $E + \Delta E$, respectively, are allowed to pass the defining slit and enter the detector. However, the *transmission* T of a spectrometer is the fraction of monoenergetic electrons passing into the detector. The product of the transmission and the source area is called the *overall luminosity*, $L = T\sigma$.

Table I.

Resolutions and electron line parameters for magnetic and electrostatic spectrometers.

	Magnetic	Electrostatic
FWHM	$\Delta(B\rho)$	ΔE
Base width	$\Delta^\circ(B\rho)$	$\Delta^\circ E$
Resolution	$R = \frac{\Delta(B\rho)}{B\rho}$	$R = \frac{\Delta E}{E}$
Basal resolution	$R^\circ = \frac{\Delta^\circ(B\rho)}{B\rho}$	$R^\circ = \frac{\Delta^\circ E}{E}$

The resolutions are related as in equation (8).

The electron intensity passing the detector slit is dependent on the size of that slit. If the detector slit size is equal to the image of the source at the detector, the *factor of acceptance* F will be equal to unity. As the slit is reduced in size, F is reduced and the transmission is reduced by the same factor. F usually takes values between 0.75 and 1.0.

The spectral electron line profile is usually given either in terms of its *full width at half-maximum intensity* (FWHM), also called halfwidth, or in terms of its total base width, the former being approximately half of the latter. Although the FWHM is usually the quantity measured and employed in spectrometry (see Chapter 6), the base width is more conveniently calculated theoretically. The *resolution* R of a spectrometer of fixed electron trajectories will be the FWHM of an electron line divided by the energy or momentum value of the line. These terms for electrostatic and magnetic spectrometer measurements are collected in Table I.

The *electron-optical quality* of a spectrometer is well represented by the ratio of its transmission to its resolving power T/R. Account of source extension can be taken by replacing T/R with L/R.

The dispersion D of a magnetic spectrometer is the change in electron-optical image position per unit change in ρ,

$$D = B\frac{dx}{d(B\rho)} = \frac{2}{1-\alpha}, \tag{9}$$

where α is the coefficient of the second term in the binomial expansion of the field with respect to the radial distance. For a homogeneous field instrument, $\alpha = 0$.

For the electrostatic spectrometer the dispersion is the position change in terms of electron energy.

1.2.2. Extended Sources and Detectors

The luminosity of an experimental arrangement may be increased in two manners, and each one increases thereby the intensity of monoenergetic electrons striking on the detector.

1. The luminosity may be increased by placing several narrow sources side by side, each one at a unique electrostatic potential, so that electrons of the same initial energy will be focused at the same position in the focal plane. The total luminosity for a multistrip source will be, then, $L_m = m_m L$ cm^2, where L is that for one of the strips and m_m is the number of strips.

Extended sources are usually multistrip. Alternatively, a converter may be exposed to an X-ray spectrum so that photon energy varies with position on the converter and photoelectrons of the same initial binding energy will focus at the same position in the focal plane (cf. Section 2.1.3.5).

2. The luminosity may also be increased by introducing an orbit corrector arrangement (electrostatic or magnetic) between the source and the detector. Here the luminosity, as well as the transmission, will become augmented by a factor of m_c. A spectrometer employing both the first and second methods will have a final luminosity $L_{m,c} = m_m m_c L$ cm^2, where L is the luminosity of the unaltered instrument with but one source strip.

The information per unit time obtained in a measurement can be greatly improved by increasing the number of single detectors n placed in the focal plane. The *information* obtained is defined then in general as $I = m_m m_c n L$ cm^2, assuming that the other modifications of multistrip source and electrostatic corrector are also made.

When a single extended position-sensitive detector is used, and assuming that $m_m = m_c = 1$, and letting L be a function of $B\rho$, the information may be given by

$$I = \int_{B\rho_{\min}}^{B\rho_{\max}} \frac{d(B\rho)L(B\rho)}{B\,\Delta\rho} = \int_{B\rho_{\min}}^{B\rho_{\max}} \frac{d(B\rho)L(B\rho)}{DRB\rho}, \quad (10a)$$

where $\Delta\rho = DR\rho$ is the channel in radius increment subtended by 1 FWHM and D is the dispersion of the instrument. For a single-channel detector and with L a constant, it is seen that $I = L\ \text{cm}^2$.

For electrostatic analyzers, we have the analogous expression

$$I = \int_{E_{\min}}^{E_{\max}} \frac{dEL(E)}{\Delta E} = \int_{E_{\min}}^{E_{\max}} \frac{dEL(E)}{DRE}, \quad (10b)$$

where $\Delta E = DRE$.

2. MAGNETIC SPECTROMETERS

2.1. Uniform Magnetic Field Spectrometers

Among the wide variety of magnetic spectrometer types, the one that has been employed with great success for the longest period of time is the homogeneous 180° (semicircular) magnetic spectrometer. The first work employing an instrument of this type was carried out during the years just following 1910[7] (see Refs. 5, 8). Although it boasts only single focusing of the first order, high resolving powers are easily obtainable; the reduction in instrumental transmission resulting from resolution enhancement is strongly compensated for when the entire wide energy range available for analysis is recorded. This advantage is lost, however, when a single narrow energy band detector such as a Geiger–Müller (G–M) counter is employed.

This type of instrument is most versatile. It has been used to study the spectra of both high and low energy conversion electrons, beta and Auger electron spectra (e.g., Refs. 8–14) and photoelectron spectra.[15,16] The effects of nuclear recoil on conversion electron line shape, following alpha-decay, have been examined with a homogeneous 180° magnetic spectrometer (Chapter 5, Figures 8, 9, 14, 15); electron coincidence measurements have also been studied[17] and for electron bombardment measurements the electron energy has been selected with such a spectrometer to investigate for example, semiconductors.[18] The combination of its simplicity of construction and use and its high resolution capability has made this instrument of unique value to atomic and nuclear physicists.

The source of the focusing principle of the semicircular spectrometer is the fact that an electron of a given energy will trace, in a homogeneous field, a circular path of a radius proportional to the strength of the field. If a bundle of such electrons were to originate from a given point, but diverges slightly within an angle 2ϕ, their paths will be most concentrated after having traversed a mean value of 180° along their circular paths. Figure 1 illustrates the focusing principle and shows two alternative set-ups; (*a*) where the electrons selected by the defining slit, width a, are focused in a plane including the source,[19–22] and (*b*) in a plane including the aperture slit.[23] In the case represented by Figure 1*a*, the electron spectrum may be recorded by placing a photographic plate or other detector in the focal plane p, which includes the source, and in Figure 1*b* along p' in the plane including the aperture slit. Here, the slit width defines the resolving power of the instrument through defining the aperture angle 2ϕ. It can be shown that the aperture angle $2\phi = a/d$. In each of these cases the contribution to line width due to the source width must be added in order to obtain the total width of the final profile.

An alternative arrangement, used in the work of Robinson,[24] and also by Ference,[25] is shown in Figure 1*c*. Here the source is large, and the collimating slit is narrow, and it is mainly the source width s, which determines the resolving power of the instru-

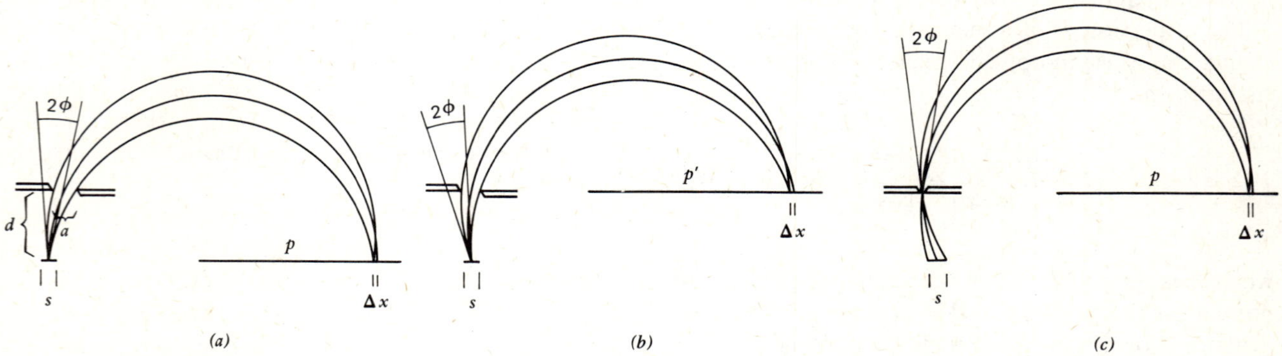

Figure 1. (*a*) and (*b*) Two possible geometries for a semicircular spectrograph: s, the source width, d, the source-slit distance, a, the slit width subtending an angle ϕ, p, the focal plane; (*c*) the arrangement using a wide source and a narrow slit.

ment. Here also the setup may be altered to cause the focal plane to coincide with that of the source rather than with that of the defining slit.

Still other arrangements, with other relative positioning of the source, slit, and focal plane have also been considered.[26–28] Although each has its merits, only slight actual gain is made over the arrangement shown in Figure 1*b*, which is considered in detail in the next paragraph. Note, however, that those arrangements in Figure 1 allow the employment of the extended focal plane p or p', usually through the use of a continuous (photographic) or multiple detector arrangement. It is also possible to use a single detector (e.g., G–M counter) placed at some position in the focal plane. For single-channel detection, the aperture slit may be placed elsewhere along the electron trajectory (see, e.g., Ref. 26) without adversely effecting the quality of the detected spectrum.

Returning now to Figure 1*b*, it is seen that spherical aberration has caused monoenergetic electrons to spread out to form an intensity profile, or line, of base width Δx in the focal plane. Assuming that the source has the width s, then

$$\Delta x = s + 2\rho(1 - \cos\phi), \tag{11}$$

where ρ is the radius of curvature for the electrons of interest. This expression gives the line width at the base. Assuming ϕ to be small and expanding $\cos\phi$, one obtains for the base resolution of the spectrometer

$$R^\circ = \frac{\Delta^\circ(B\rho)}{B\rho} = \frac{s}{2\rho} + \frac{\phi^2}{2}. \tag{12}$$

It is seen that the first term is proportional to the source width and can be reduced in practice to whatever minimum will still allow sufficient source strength; the resolution is dependent on the second power of the half-acceptance angle ϕ, and first-order focusing is thereby obtained. However, for a given source strength, the narrower the source, the greater the electron straggling effects owing to increased source thickness. For radioactive samples, usually 0.5 to 0.1 mm in width, this is dependent on the specific activity of the deposit. For thorium (B + C + C″) sources, however, where effectively only the atoms of interest are collected, a source wire may be made as narrow as ~0.005 mm without appreciable straggling effects on 25-keV electrons.[29]

The basal resolution expressed in equation 12 can be minimized by setting its two terms equal, so that they contribute equally to the total resolution. For this optimal condition one obtains the relationship

$$\phi = \sqrt{s/\rho}. \tag{13}$$

This relationship between ϕ, s, and ρ holds in theory, but in practice, for very small resolutions, inhomogeneities in the magnetic field will be expected to dictate an absolute minimum in resolution for decreasing ϕ and s.[30]

The electron source assumed here has only a width s and no length; the resulting electron line intensity profile can be calculated without great difficulty, assuming also detection in one dimension. This calculation is reported in Ref. 5. From such calculations we obtain the profile shown in Figure 2*a*. Note that for small ϕ, which is replaced according to equation (13), the base of the profile is reduced to $2s$ and takes the "optimal" shape shown in Figure 2*b*. The width of the line at half-maximum becomes then s, and $R = \frac{1}{2}R^\circ$.

A more complete calculation of the line shape requires inclusion of the finite length of the source, as

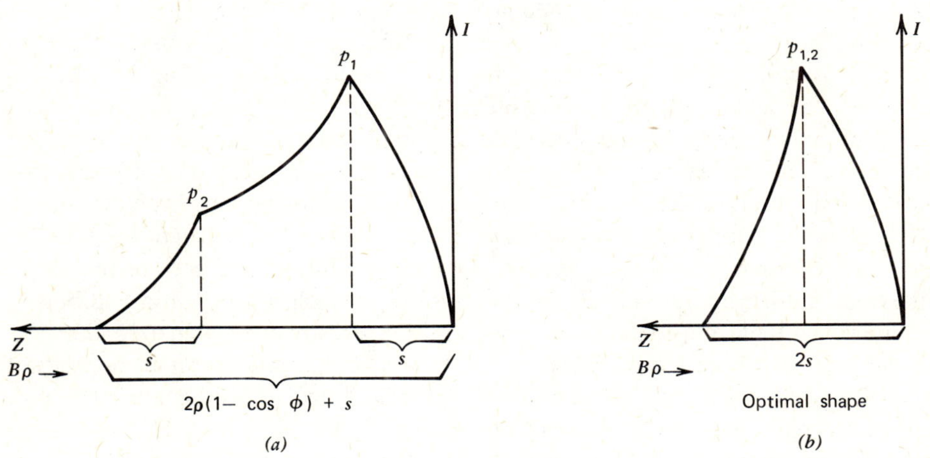

Figure 2. (*a*) The intensity distribution of a focused monoenergetic electron line from a source of width s. (*b*) The line intensity profile for optimum slit adjustment for a given source width.

has been done by Kerberlé;[31] here the intensity was calculated only as a function of the distance from the source and assuming a detector of zero width in the focal plane.

Considering the effect of source length l, we let ψ denote the angle between two electrons of the same energy, one from each of the extremities of the source wire, as they strike the center of the focal plane. The trajectory will have had a radius $\rho \cos \psi$ in the plane perpendicular to the magnetic field, and the distance less than 2ρ from the source at which it will strike the detector will be

$$\Delta x' = 2\rho - 2\rho \cos \psi. \quad (14)$$

Since ψ is small, equation 14 will give

$$\Delta x' = \frac{l^2}{\pi^2 \rho}. \quad (15)$$

As before, the increase in electron line width owing to source length should not exceed the source width contribution, or s. We obtain for the optimum condition

$$l \leq \pi \sqrt{s\rho}. \quad (16)$$

Assuming that a counter with a slit width w is used for electron detection, the total base resolution becomes

$$R^{\circ}_{\text{tot}} = \frac{1}{2\rho} [s + w + \rho(\phi^2 + \psi^2)]. \quad (17)$$

However, if photographic detection of low energy electrons ($E < 40$ keV) is employed, the electron tracks in the emulsion are quite short ($\gtrsim 10\ \mu$m). If electron track counting methods are used for obtaining the spectral intensity distribution from a plate, given sufficiently narrow counting area, the term w in equation 17 becomes negligible.

From equation 17 it is noted that as ρ is increased, the resolving power may be held constant by a suitable increase in s and/or an increase in ψ through a lengthening of the source (i.e., an increase in l). The two ρ-dependent terms in equation 17 are the first, $s/2\rho$, and the last, $\psi^2 = 2l^2/\pi^2\rho^2$. Optimally s and l may both be increased with the increase in ρ, to keep R° constant, as well as all the other figures of merit for the instrument. Note, however, that for a constant electron source strength, the number of electrons emitted per unit source area is decreased; that is, for a radioactive source its thickness is reduced. For photoelectron sources, the converter area is increased, and also, in proportion to this, the source strength. Therefore large spectrometers are seen to be advantageous regarding source conditions and signal strength; disadvantages include effects caused by field inhomogeneities and difficulties with spectrometer temperature control.

In the past several years a number of large spectrometers have been constructed, the orbital radius extending to as large as 0.75 m. Several of these are cited here for reference, with their maximum radii: 75 cm,[20,21,32] 42 cm,[33,34] 40 cm,[35] 22 cm,[15] 15 cm,[36,8] 12 cm,[37] 10 cm.[38]

Turning now to the evaluation of the momentum values of spectral lines on a photographic plate, for the arrangement in Figure 1*b*, these may be calculated using

$$a^2 + x^2 = \frac{4(B\rho)^2}{B^2}, \quad (18a)$$

where x is the distance from the slit to the spectral line of interest and the field is assumed to be truly homogeneous (see, e.g., Ref. 38). If the electron source under investigation is accurately interchanged with one emitting conversion electrons of several accurately known $B\rho$ values, the instrument may be calibrated by solving for a and B the overdetermined system of equations 18a, obtained by inserting the x_i and ρ_i values for the several calibration lines. The momenta of the unknown lines may then be easily determined from equation 18a. This method may also be programmed and carried out by a computer.[39]

Unfortunately, "uniform field" magnetic spectrometers can actually often show field nonuniformities up to 1% for fields between 50 and 100 G. Such nonuniformities can be greatly reduced by utilizing shims and several thick parallel iron plates separated by air gaps;[35] proper choice of the magnetization procedure applied to the plates will also help.

In the magnetization procedure, current is passed through coils about the spectrometer pole pieces alternatingly greater and smaller than and converging to the value corresponding to the desired final field strength.[40] The best procedure, which may be determined through experimentation, will give the most desirable final field intensity distribution throughout the spectrometer gap. Even desired nonuniform distributions may be created by such a procedure: for example, the pole pieces of the large spectrograph at Orsay, which form the top and bottom of the vacuum chamber, are magnetized so that the field is slightly lower in the center than at the edges. When the chamber is closed and evacuated, the pole surfaces curve slightly into the tank and the field becomes very uniform over the entire volume.[41]

Field nonuniformity will be expected to effect the shapes of individual spectral lines but slightly, since monoenergetic rays experience very nearly the same magnetic field values along their closely spaced trajectories. However, electrons of increasingly differing energies will experience increasingly different average B-values along their paths. Equation 18a can be generalized to account for these variations in average B-value as a function of electron momentum for a given instrumental field setting. This may be done by expanding

$$B = B_0 + B_1 x + B_2 x^2 + \cdots \tag{18b}$$

and evaluating the coefficients using a standard calibration source. Alternatively, one may construct a plot of B versus ρ by the measurement of a large number of standard conversion lines. It is possible to determine this relationship with an accuracy of better than 0.01%.[33, 34]

2.1.1. Figures of Merit of the Semicircular Spectrometer.

The solid, or acceptance, angle for the 180° spectrometer, its luminosity, and its transmission are easily derived:

$$\Omega = \frac{\phi\psi}{2\pi}, \tag{19}$$

$$\Lambda = \frac{sl\phi\psi}{2\pi}\ \text{cm}^2, \tag{20}$$

$$T = \frac{l}{4\pi^2\rho}\,\phi\ \text{cm}^2. \tag{21}$$

The ratio of the transmission to the resolution, the electron-optical quality $T/R°$, can be determined by use of equations 17 and 21, assuming that $R \simeq \frac{1}{2}R°$;

$$\frac{T}{R} = \frac{l\phi}{\pi^2}[s + w + \rho(\phi^2 + \psi^2)]^{-1}; \tag{22}$$

and the overall information obtained will be the product of the overall luminosity and $d(B\rho)/DRB\rho$ integrated over the momentum region of detection. A single G–M counter will give an information value of

$$I \equiv L\ \text{cm}^2, \tag{23}$$

so that $d(B\rho)/DRB\rho = 1$. Actually R may vary strongly over the $B\rho$ region of interest, and this must be taken into account.

One can estimate that if detection is over the region of $B\rho_0 \pm 20\%(B\rho_0)$ and the momentum resolution is a constant 0.05% then, using equation 10a,

$$I \approx 4 \times 10^3 L(B\rho_0)\ \text{cm}^2.$$

2.1.2. Special Applications of Postacceleration Techniques

The several pre- and postacceleration techniques described in other sections of this chapter may be employed for the semicircular spectrometer having a single detector, but special mention should be made here of a unique technique for the extended detector.

The blackening efficiency of emulsion plates by low energy electrons is low and varies strongly with energy. Standard techniques of accelerating all electrons striking the plate will enhance the blackening of the plate, but correction for the variation of the blackening efficiency with energy must still be made to obtain a measure of spectral electron intensity versus $B\rho$. Slätis[42] has developed a postacceleration arrangement that will accelerate all electrons in a region near the focal plane, so that they will both strike the plate with the same energy and show no appreciable alteration of spectral line shape. A low preacceleration voltage was also applied to the source wire, but this does not alter the focusing properties. The thorium (B + C + C″) electron spectrum from about 3 to 20 keV was investigated applying a 6.5-kV post- and 0.66-kV preacceleration voltage.

2.1.3. Other Uniform Field Spectrometers

In the foregoing sections the characteristics of the 180° spectrometer have been considered by assuming a geometry in which the detector is either in the focal plane including the source or in that including the aperture slit. Over the years, however, several other geometries that exhibit unique focusing properties with associated resolution or transmission enhancements have been presented in the literature. A few of these are described briefly here.

2.1.3.1. 109° SPECTROMETER. The 109° spectrometer was described by Paris[43] in the early 1960s. Here both the source and detector are placed outside of the magnetic field, as shown in Figure 3*a*. Electrons leaving the source S, which is situated in the field free region, enter the homogeneous field at an angle as illustrated. After traveling in the field along a circular path, they leave the field region to become focused on a point along the focal plane $S'F$. (S and S' are the same distance on either side of the field boundary.) In this geometry the line width is independent of the source width, which is not so in the 180° case. In addition, two other advantages over the 180° geometry are reported: the focusing here is second order and an increase in luminosity by 3/2 is realized. Noteworthy disadvantages are that the quality of recorded spectra is very sensitive to source positioning and to fringing effects at the field boundary.

2.1.3.2. SECTOR FIELDS. The fringing field at the edge of a uniform field may be utilized to give a focusing effect in the direction parallel to the field. This manner of obtaining two-directional focusing for the uniform magnetic field instrument has been treated by Cross.[44] He showed that vertical focusing may be realized at the position of horizontal focusing by the proper adjustment of the pole boundary curvatures and the electron trajectory entrance and exit angles to those boundaries. Under special circumstances, both second order and double focusing are theoretically realizable. The instrument by Paris just described is actually a special case of this general sector-shaped field.

Green et al.[45] have constructed an instrument of this type, a 55° instrument with circular and straight field boundaries for entering into and exiting from the field, respectively (Figure 3*b*). Momentum resolutions were better than 0.1% for $\alpha \gtrsim 5°$, $D \simeq 4.5$, and $\Omega = 10^{-2}$, with the energy range simultaneously detected being from 0.5 E_0 to 2.5 E_0, where E_0 is a chosen energy.

An earlier 90°-sector uniform field spectrograph was constructed by Browne and Buechner.[46] The instrument, which is diagrammatically shown in Figure 3*c*, employs a circular magnet field boundary of 50-cm radius; although the instrument was employed to measure the high energy spectra of protons, deuterons, and alpha-particles, the principle may also be applied to the analysis of electron spectra. From geometrical considerations, this 90° deflection instrument had an aberration that was nearly proportional to ϕ^3 (i.e., approximately second-order focusing in the plane normal to the magnetic field). The energy range simultaneously focused in the focal surface was 0.49 E_0 to 1.2 E_0, where E_0 is the energy for the "central" trajectory. The focusing due to fringing field effects here was negligible; therefore, the transmission was reduced greatly after a total particle path length of between 1 and 2 m.

Resolution and transmission have been shown by Yagi[47] to be improved simultaneously by employing a homogeneous field sector spectrometer in tandem with a preceding quadrupole-magnet doublet. Again, this design was intended for proton beam momentum analysis, but it could in principle be applied for electron energy analysis. Yagi's article discusses this design and compares its figures of merit with those of other designs.

2.1.3.3. IRON-FREE SPECTROMETER. An iron-free 180° spectrometer has been reported by Henrikson.[37] The field is produced by two iron-free coils spaced to give a nearly homogeneous field; the slight inhomogeneity, however, causes some second-order focusing. Figure 3*d* illustrates the principle of the instrument's focusing properties. The central ray travels in a circle where the magnetic field value is constant and maximum, corresponding to the center of the field strength profile plotted in Figure 3*e*. Clearly an electron traveling to either side of the central ray will be in a field of intensity less than that along the central ray path, and the radii of path curvature will be larger than that of the central ray. Therefore, after a mean angle of 180°, these noncentral rays will approach the central ray more closely than for the uniform field instrument. Although the instrument employed exhibited only weak second-order focusing, one can see that the proper magnetic field strength profile (Figure 3*e*), will allow all electron trajectories within the 2ϕ aperture angle to coincide at the detector. The resolution is seen to be still dependent on the source width; also, only a single channel detector at D can be employed. Several L-Auger spectra recorded using this instrument are shown in Figure 17, Chapter 3.

2.1.3.4. ELECTRONS IN A HOMOGENEOUS FIELD. In a homogeneous magnetic field, electrons of any energy will impinge on their points of origin after they have traveled 360°.[48,49] If the electrons are allowed to have a small velocity component in the direction of the magnetic field, monoenergetic electrons of momentum $B\rho_0$ will strike the source axis at a distance of $D = 2\pi\rho_0 \cos\psi \sin\psi$ above the point of origin. For small angles ψ, $D = 2\pi\rho_0\psi$. This geometry is illustrated in Figure 3*f*, with a side view of the same arrangement appearing in Figure 3*g*. However, although electrons will strike the source axis at a distance D above the source, which is proportional to the electron momentum, this separation is too small ot allow special spectrum analysis along the axis. Instead, we can select the energy of interest by placing a baffle at the 180° focusing position. Several such paths can be so defined around the source, and the electrons focused at the source axis may be detected with a single detector.

The construction of such an instrument has been reported:[48,50] the radioactive source and the silicon detector were separated by a block of gold; the maximum trajectory diameter, about 12 cm, was defined by aluminum diaphragms. The quality of spectrometer performance was in agreement with predictions. The instrument was used together with a gamma-detector for $\gamma - e^-$ coincidence measurements.

A second and similar possible geometry employing

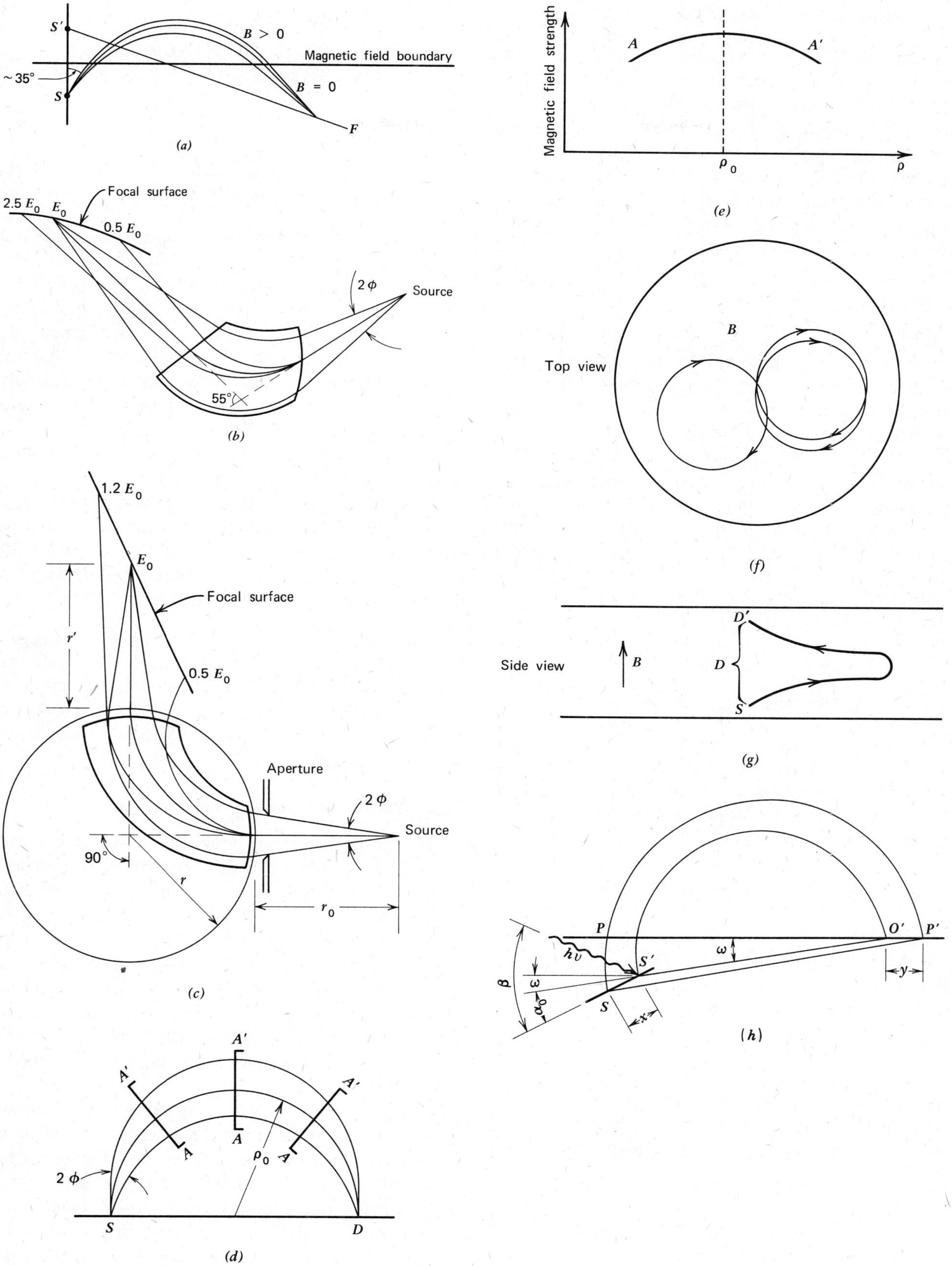

Figure 3. (*a*)–(*c*), (*f*)–(*h*) Various instrumental arrangements using a homogeneous magnetic field. (*d*) Focusing properties of a magnetic field of the strength profile in (*e*).

an annular homogeneous magnetic field, with the source and detector both situated in the central field free region, was also suggested. The main drawback of this type of spectrometer is again that the source and detector are too close to each other to allow effective shielding of the latter from the former. This is especially critical if radioactive sources are to be used from which are emitted β, γ, or X-radiations of high energy. Even if the electron source is one of photoelectrons, associated X-radiation will give a high radiation background at the detector.

2.1.3.5. COMPENSATING CONVERTER. A very interesting manner of compensating for the inherent energy width of X-radiation employed to obtain photoelectrons for spectrum analysis in a semicircular spectrometer has been advanced by Siegbahn et al.[51] By subjecting an extended converter to X-radiation, whose energy varies slowly over the converter's surface, it is possible to orient the converter so that contributions to the electron spectral line width due to the width of the converter can be compensated for at any given energy.

The arrangement is shown in Figure 3*h*. It is seen that

$$(2\rho)^2 = [2\rho_0 \sin \omega + x \sin (\omega + \alpha_0)]^2 + [x \cos (\omega + \alpha_0) + 2\rho_0 \cos \omega + y]^2, \quad (24)$$

where ρ_0 and ρ are the radii of trajectory curvature for two electrons of the energies E_0 and $E_0 + \Delta E$, respectively. Assuming x and y to be much smaller than ρ_0, one may neglect terms higher than first order in x and y. One then obtains

$$\left(\frac{\rho}{\rho_0}\right)^2 = 1 + \frac{x \cos \alpha_0 + y \cos \omega}{\rho_0}. \quad (25)$$

If $\Delta E \ll E_0$, then for nonrelativistic energies one obtains

$$E_0 + \Delta E = E_0 \left(\frac{\rho}{\rho_0}\right)^2, \quad (26)$$

Now, due to the crystal dispersion $D = dx/d(h\nu)$, we can allow the energy difference between X-rays striking the converter at S and S' to be just ΔE by adjusting β:

$$x = \frac{D\Delta E}{\sin \beta}, \quad (27)$$

and the distance y between the points O and P' where the electrons strike the detector becomes zero:

$$y = 0 = \frac{\rho_0 \Delta E}{E_0 \cos \omega}\left(1 - \frac{DE_0 \cos \alpha}{\rho_0 \sin \beta}\right) \quad (28)$$

or

$$E_0 = \frac{\rho_0 \sin \beta}{D \cos \alpha}. \quad (29)$$

The test of this theoretical treatment was described in Appendix 10 of Ref. 51 and its validity was thereby established: The chlorine K photoelectron lines from sodium chloride using the X-radiation energy range including both copper $K\alpha_1$ and copper $K\alpha_2$ were shown to coalesce for an angle of $\beta = 12.5°$. It should be noted that this coalescence is realized only at one energy value and that, for photoelectron lines situated at other energies, β must be altered according to equation 29. This is because E_0 is not proportional to $B\rho_0$ in a magnetic spectrometer.

This experimental arrangement is in effect similar to that in which a multistrip radioactive source is employed where potentials are applied to the strips so that electrons of identical initial energy are detected at the same detection point (see Section 2.3.1).

2.1.3.6. ANOTHER IRON-FREE SPECTROMETER. An iron-free homogeneous field helical trajectory spectrometer was reported in 1957 by Du Mond.[52] In the report of the construction and performance of this instrument, an extensive investigation of its electron optical properties was made.

2.2. The Double-Focusing Spectrometers

The principle of the double focusing of charged particles as employed in the cyclotron and the betatron[53] was first applied to the analysis of electron spectra in 1946 by Siegbahn and Svartholm.[54,55] A magnetic field of rotational symmetry and with radial and azimuthal field gradients will cause an electron to oscillate in both the radial and azimuthal directions when its initial direction is not normal to these directions. These first spectrometers employed suitably shaped iron pole pieces to create the desired magnetic field form.

Iron-core instruments have given an electron momentum measurement precision of approximately one part in 10^4, and a momentum resolution of down to $\sim 3 \times 10^{-4}$—a figure that is limited by imperfections in the magnetic field. Attempts to overcome this limitation were later made by producing a magnetic field of a desired shape using an array of iron-free coils. Both the resolution and the momentum reasurement precision were thereby bettered by a factor of 10.

Defining the electron optical circle as the path followed by electrons traveling perpendicularly to both ρ and z and of a momentum $B\rho_0$, then electrons departing initially by the angles ϕ_z and ϕ_ρ from this path will return to the optic circle after the "focusing angles" ψ_ρ and ψ_z (the angles about the z-axis at which the optic circle is intersected). For small values of ϕ_z and ϕ_ρ these focusing angles are

$$\psi_\rho = \pi\left[1 + \frac{\rho_0 B'(\rho_0)}{B(\rho_0)}\right]^{-1/2}, \qquad (30)$$

and

$$\psi_z = \pi\left[-\rho_0 \frac{B'(\rho_0)}{B(\rho_0)}\right]^{-1/2}, \qquad (31)$$

where

$$B'(\rho_0) = \left[\frac{dB(\rho)}{d\rho}\right]_{\rho=\rho_0}.$$

The focusing angles are related by

$$\psi_\rho^{-2} + \psi_z^{-2} = \pi^{-2}, \qquad (32)$$

which may be solved in general to give

$$\psi_\rho = \frac{\pi}{\sqrt{1-\alpha}} \quad \text{and} \quad \psi_z = \frac{\pi}{\sqrt{\alpha}}. \qquad (33)$$

The dispersion is given in general by equation 9, where α is the same as that used here.

2.2.1. *Iron-Core $\pi\sqrt{2}$ Spectrometer*

The simplest solution of equation 33 is given by $\alpha = \frac{1}{2}$, or $\psi_\rho = \psi_z = \pi\sqrt{2} = 254.56°$. Then simultaneous focusing in both directions ($\theta_\rho = \theta_z = \theta = \pi$ i.e., the electron returns to the optical circle for the first time; see below) will occur at this common focusing angle. Setting equations 30 and 31 equal, it is clear that

$$B'(\rho) = -\frac{1}{2\rho}B(\rho), \qquad (34)$$

and the magnetic field form in the vicinity of the optic axis will vary as $1/\sqrt{\rho}$.

The preceding treatment refers to small values of ϕ_r and ϕ_z; as these are increased, however, we expect increasing aberrations in the two focusing directions. Specifically, an electron of proper momentum leaving the source plane (Figure 4) in the trajectory direction t_A, which is perpendicular to both ρ_0 and z, will travel along the optic circle A and will intersect the image plane at $\pi\sqrt{2}$ rad, where $\theta = \pi$. Its coordinates will be $\eta(\theta) = 0$ and $\tau(\theta) = 0$ for all values of θ along the path. An electron leaving the source plane along path C with an axial angle ϕ_z will reach its maximum displacement from the optic circle ($\theta = \pi/2$) at $\frac{1}{2}\pi\sqrt{2}$ rad and return to it after $\pi\sqrt{2}$ rad ($\theta = \pi$). Similarly,

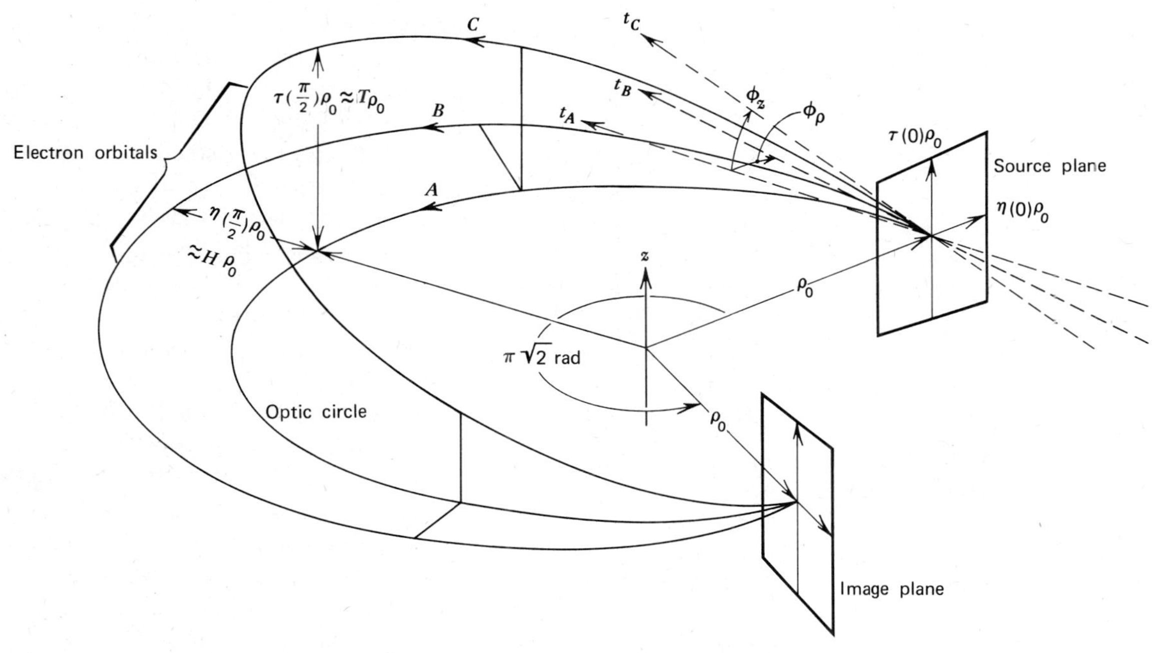

Figure 4. The trajectories of electrons departing with angles ϕ_ρ and ϕ_z to the normal of the source plane and from its center in a $\pi\sqrt{2}$ spectrometer. (From Ref. 74.)

an electron starting on trajectory t_B at a radial departure angle ϕ_ρ will become displaced the greatest distance from A at $\frac{1}{2}\pi\sqrt{2}$ rad ($\theta = \pi/2$) and return to the optic circle at the image plane. Finite values of $\eta(\pi)$ and $\tau(\pi)$ are termed radial and axial aberrations.

We may generalize the treatment, which will allow a detailed investigation of aberrations and also of the suitability of other field forms, by expressing the z-component of the field at $z = 0$ as

$$B_z(\rho, 0) = B_0\left[1 + \alpha\left(\frac{\rho - \rho_0}{\rho_0}\right) + \beta\left(\frac{\rho - \rho_0}{\rho_0}\right)^2 + \cdots\right]$$
$$= B_0(1 + \alpha\eta + \beta\eta^2 + \cdots), \qquad (35)$$

where $\eta = (\rho - \rho_0)/\rho_0$ and $\alpha, \beta, \ldots$ are constants to be adjusted to minimize aberration. Comparison with equation 34 shows that a choice of $\alpha = -\frac{1}{2}$ will give first-order axial, as well as radial, focusing. The second coefficient may take the values $\beta = \frac{1}{8}, \frac{1}{4}$, or $\frac{3}{8}$. Lee-Whiting and Taylor[56] have shown that $\frac{1}{8}$ and $\frac{3}{8}$ are better than $\frac{1}{4}$ when a rectangular aperture is being used. Calculations of $\eta(\pi)$ and $\tau(\pi)$ were carried out for an electron leaving the source at the coordinates $\eta(0) = \delta\rho/\rho_0$ and $\tau(0) = \delta z/\rho_0$ and in the direction ϕ_ρ and ϕ_z to the optic circle for $\alpha = -\frac{1}{2}$ ($1/\sqrt{\rho}$ field form). The base resolution takes the form

$$R^\circ = \frac{s}{4\rho_0} + \frac{\omega}{4\rho_0} + \frac{4\beta - 3}{48\rho_0^2}h^2 + \frac{2 - 16\beta}{12}\phi_\rho^2 + \frac{16\beta - 6}{12}\phi_z^2. \qquad (36)$$

The second-order dependence on ϕ_z will disappear when $\beta = \frac{3}{8}$;[57,58] if $\beta = \frac{1}{8}$ is chosen, the resolution is independent of ϕ_ρ to the second order. (The value of $\beta = \frac{3}{8}$ has been the most extensively used.) Alternatively, Rosenblum[59] responds that the minimum width of the image of a rectangular source is obtained when $\beta = \frac{1}{4}$,[60] on the basis of a study by Shull and Dennison.[61] If $\phi_\rho = \phi_z$, in equation 36, the resolution is independent of β. Also, from equation 36 it is seen that the dispersion for the $\pi\sqrt{2}$ spectrometer is 4 or twice as large as for the semicircular instrument.[54]

The determination of the coefficients in equation 35 to optimize the instrument's resolving power has been carried out to the sixth order by Lee-Whiting and Taylor.[56]

The instrument constructed by Bartlett et al.[62] ($\rho_0 = 30$ cm), with one set of coils with inner radius larger than the maximum radius of the vacuum tank and a second set with outer radius smaller than the tank's minimum radius, is capable of producing any field form required. It is therefore possible to optimize the performance of the instrument, which results in very good figures of merit, as seen in Figure 6. A similar ($\rho_0 = 30$ cm) instrument was more recently constructed in Fribourg by Chen et al.[63], also using "inner" and "outer" coils. The luminosity versus resolution for the instrument is plotted in Figure 6.

There have been attempts to optimize the performance of certain $\pi\sqrt{2}$ iron-core spectrometers. Bartlett and Keith[64] investigated the effect of altering the focusing angle from $-9.5°$ to $+21.5°$ about $\pi\sqrt{2}$. They showed that the operation of the "high aperture," $\beta = \frac{3}{8}$, spectrometer is not adversely affected by small deviations of the focusing angle about $\pi\sqrt{2}$ when readjustment of the field was made so as to optimize the focusing properties. Bobykin and Babenkov,[65] however, found that they could increase the resolving power of their instrument to a maximum value by increasing the focusing angle to ~273° or $\pi\sqrt{2} + 19°$.

2.2.2. *Iron-Free* $\pi\sqrt{2}$ *Spectrometer.*

The theoretical treatment of the $\pi\sqrt{2}$ spectrometer's focusing properties will give the optimal magnetic field shape for a desired order of focusing. The first double-focusing spectrometers were iron instruments which allowed the field shape to be formed by proper shaping of the iron pole pieces; however, it is also possible to produce the desired magnetic field form with a suitable arrangement of current coils. The coils are placed about a spectrometer tank of a nonmagnetic material in such a way as to produce the proper field form within the tank. Advantages of this arrangement are that the field strength is accurately known, being proportional to the current that is passing through the coils; one also avoids possible hysteresis effects in magnet iron, which may cause small divergences of the field form from that desired. This last point is especially important for low electron energies. On the other hand, there will no longer be screening of the earth's magnetic field, which must therefore be compensated for with high accuracy. This may be done by placing an array of Helmholz coils about the space where the spectrometer is located; by proper adjustment of the current through these coils, the earth's field—and most other disturbing magnetic fields—may be nullified. This question is treated further in Section 4.

The construction of several iron-free $\pi\sqrt{2}$ instruments has been reported since that of the first one ($\rho_0 = 30$ cm) in 1956[66] (see Figure 4). Large versions of this device ($\rho_0 = 50$ cm) have more recently been constructed in Berkeley[67] and in Uppsala.[68] Both were improved over the earlier instrument by carefully determining the proper baffle shapes in order to optimize the transmission for a given resolution. The addition of field correction coils to improve focusing has also been carried out.[69]

Different high-resolution iron-free instruments of the $\pi\sqrt{2}$ type have been reported in other laboratories: $\rho_0 = 30$ cm,[70,72] $\rho_0 = 50$ cm,[73] $\rho_0 = 100$ cm.[74] The figures of merit of these are included in Figure 6.

2.2.3. *Other Iron-Free Double-Focusing Instruments*

Double-focusing spectrometers with $\alpha > \frac{1}{2}$ will, by equation 9, have larger dispersions, but they may also, by equations 30 and 31, have the drawback that double focusing will be realized after an angle of $\psi_\rho = \psi_z > 360°$. This will necessitate either the electron beam passing the source position before detection (>360°) or the source and detector being in close proximity of each other.

Reviews of such instruments and their performances have been presented by Mladjenovič.[75,76] See also the articles by Lee-Whiting.[77,78] Here only one such device is considered by way of example, namely, the $\pi\sqrt{13}/2$ ($\rho_0 = 30$ cm) spectrometer designed[79–81] and constructed[82,83] by Daniel, Jahn, and co-workers. The focusing characteristics of this instrument are shown in Figure 5. Here $\alpha = \frac{9}{13}$ and the first radial focus is at $\psi_\rho = \pi\sqrt{13}/2$, and the first z focus is at $\psi_z = \pi\sqrt{13}/3$. There is not double focusing for any single angle of ψ_ρ and ψ_z. In order to have the slit position at as large an angle of ψ as possible and yet less than 2π, the focal plane was chosen at the larger of the two above-mentioned single focusing positions, giving thereby the larger dispersion, or at $\psi_\rho = \pi\sqrt{13}/2 = 324.5°$. The detection slit must have a curved form in order to accept efficiently the electrons focused there.[81] The dispersion of the instrument is $D = 6.5$. (See also Figure 6.)

A second instrument that might be mentioned is the trochoidal double-focusing spectrometer, constructed by Mugnier and Lafoucrière.[84,85] This instrument, with a magnetic field intensity falling off as ρ^{-1}, has for a point source extremely good optical quality: 6% transmission for a basal resolution of 0.1%. The dispersion is $D = 2.56$.[78] Since the usable source area is strongly reduced, the luminosity also exhibits great reduction. The luminosity of such an instrument the size of the Chalk River, Ontario, $\pi\sqrt{2}$ spectrometer would be only 20% as great[78] (see Figure 6).

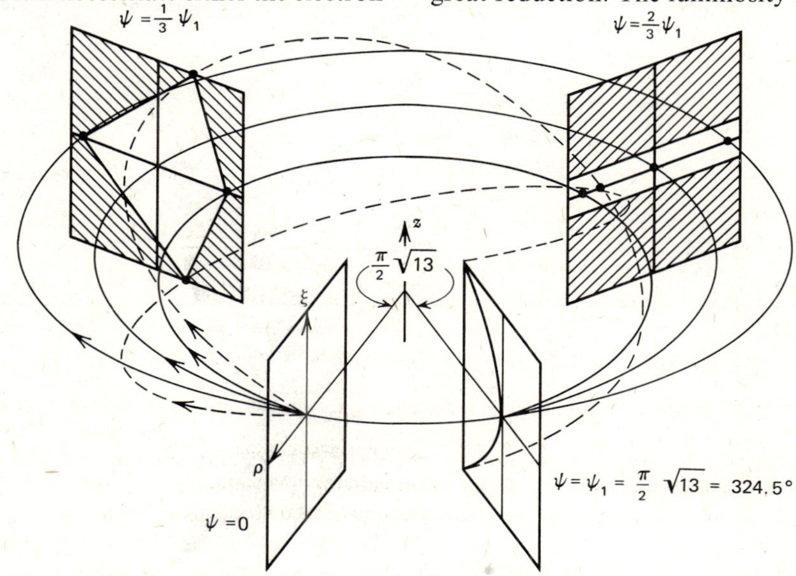

Figure 5. The baffle system and limit electron trajectories for the $\pi/2\ \sqrt{13}$ iron-free spectrometer. (From Ref. 83.)

Another possibility for improving the focusing properties of a double-focusing spectrometer is to allow the field to vary with azimuthal angle. Theoretical investigations of the properties of such instruments have shown them to have both high resolution and high luminosity.[86–89]

2.3. Methods for Increasing the Luminosity and the Information for a Given Resolution

2.3.1. *Multistrip Sources*

The resolution and luminosity of a spectrometer are related so that an increase in one will reduce the other; for example, an increase in source area will result in a decrease in resolution. However, it has been shown

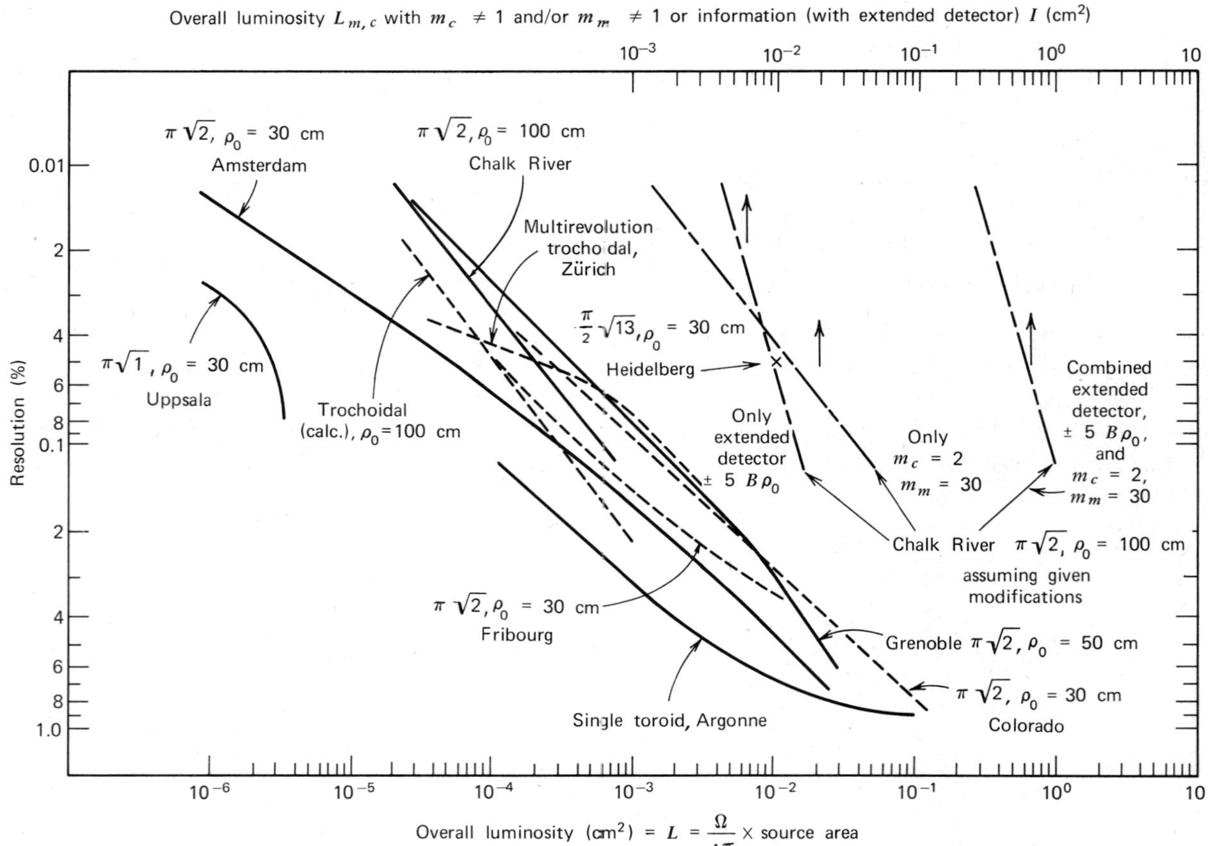

Figure 6. Comparison of figures of merit for several double-focusing spectrometers. Solid lines relate to iron-core instruments, and dashed lines to iron-free ones. The upper scale refers only to the three curves to the right, which are the figure-ofmeritcurves expected for the Chalk River instrument employing multisource, electrostatic corrector, and extended detector combinations. Instruments of the following institutes are represented: Uppsala,[66] Amsterdam,[70,72] Fribourg,[63] Chalk River,[74] Zurich,[89] Grenoble,[73] Colarado,[62] Heidelberg,[82,83] Argonne.[71] Also given for comparison is the theoretical curve for the trochoidal instrument.[78]

by Bergkvist[90] that if the activity is placed on a multistrip substrate and the strips are electrically insulated from each other, electrons of the same initial kinetic energy can be made to strike the detector slit of a spectrometer by applying the proper electrostatic potential to each of these strips. In addition, inclining the source plane 45° to the radial direction will give optimum focusing conditions. In this way, it is possible to increase the luminosity by a factor m_m, the number of strips. Ideally the resolution will remain that for a single strip of the single strip width. In his initial arrangement, Bergkvist used 30 source strips. He used this apparatus in combination with an electrostatic aperture defect corrector that allowed an additional increase in transmission (see Section 2.3.2).

Subsequent uses of multistrip radioactive sources have been reported. Gelletly et al.[91] have applied such a source arrangement in the investigation of internal conversion electron spectra. In their set-up the multistrip source plane, composed of 40 elements, was placed at an angle of 17° to the spectrometer radius for optimal focusing. For a given total source activity, the reduced activity source density on the multistrip source resulted in a noticeable reduction of the low energy tail intensities on the conversion lines, and it became possible to achieve more accurate relative intensity determinations.

A similar arrangement was employed by Nilsson et al.,[92] where 15 strips were used and the source plane angle to the spectrometer radius was 43°. The arrangement was used in combination with a multidetector system (see Section 2.3.3).

2.3.2. *Field Correction*

An electrostatic corrector, consisting of grids of the proper shapes placed at the angular position $\pi\sqrt{2}/2$ in a $\pi\sqrt{2}$ spectrometer ($\rho_0 = 50$ cm) will reduce the aperture defect in the instrument's focusing properties. This arrangement was used by Bergkvist[90] in combination with the multistrip souce to obtain a total increase in luminosity by a factor of 120. With 30 source strips employed, the electron optical quality was increased a factor of 4.

Calculations of the shape of a Bergkvist-type electrostatic detector have been carried out for the Chalk River $\pi\sqrt{2}$ spectrometer by Lee-Whiting.[93] It was estimated that such a corrector should increase the transmission of that instrument by a factor of about 2. Assuming this to be the exact increase in transmission, considering the employment of a 30-strip source, and assuming that the resolution thereby obtained is equal to that for a single strip, one would expect an increase in luminosity by a factor of 60. The resulting resolution versus luminosity ($L_{m,c}$) curve is shown in Figure 6.

2.3.3. *Multidetector Systems*

The information I obtained for a spectrum will be a function of how much of the focal surface is employed for a given resolution. Several detectors may be placed side by side in this plane and the spectrum swept; the obtained individual spectra may then be superimposed to give the final composite measurement. Nilsson et al.[92] tested such an array of ten $0.8 \times 15\ \text{mm}^2$ gold anodes deposited on an n-type silicon crystal slice with a 0.2-mm separation between them. Used in combination with a multistrip source (15 strips), the total arrangement gave an expected increase in information of a factor of 150. Bartlett and Spenny,[94] on the other hand, have constructed a detector with 14 proportional counter units for their iron-core spectrometer.

The possibility of constructing a double-focusing iron-core spectrograph ($\rho_0 = 30$ cm) has been reported by Wild and Huber[95] and Huber et al.[96] The momentum range measurable during one exposure was $+8$ and -8% about $B\rho_0$.

In an attempt to estimate crudely the information obtainable with such an arrangement, assume that the luminosity $L(B\rho)$ and the resolution R are constant over the detectable momentum range. According to equation 10a and using data for the Chalk River $\pi\sqrt{2}$, $\rho_0 = 100$ cm instrument ($D = 4$) set for a resolution of 0.012% and a luminosity of $\sim 2 \times 10^{-5}$ cm^2, and letting detection take place $\pm 5\%$ about ρ_0 over which region L, R, and B are assumed to be constant

$$I = \frac{L}{DR} \ln \left[\frac{(B\rho)_{\max}}{(B\rho)_{\min}}\right] \cong 4 \times 10^{-3},$$

or an increase by a factor of 200. For $R = 10^{-3}$ and $L = 6 \times 10^{-4}$ cm^2, we obtain $I \cong 1.5 \times 10^{-2}$, or an increase of 250. This curve is shown in Figure 6.

Combining the extended detector arrangement with the extended source and the corrector of Section 2.3.2 gives an overall information figure of merit of about 12,000 over that obtainable with the unmodified Chalk River instrument, as shown in Figure 6.

3. ELECTROSTATIC ANALYZERS*

3.1. Retarding Field Analyzers

One of the earliest electrostatic analysis methods, and also one of the simplest, is that of placing a retarding potential on the detector or on an aperture or screen before the detector. Increasing this potential by a small increment cuts out the incoming charged particles with energies within that increment. A plot of detector signal versus retarding potential is an integral energy distribution curve. Various methods of differentiation have been devised to yield a true energy distribution.

Several configurations of retarding fields are used. In the Lozier tube[97] a cylindrical field is used. Ions from a section of the beam pass through the space between collimating plates and on through a screen to a cylindrical collector. Wei and Kuppermann[98] have recently pointed out some difficulties associated with the use of the screen, namely elastic and inelastic scattering and secondary emission. Spherical retarding fields are also used, generally with grids as in the recent work of Wei, Cho, and Caldwell,[99] where again spurious effects attributable to the presence of grids are noted.

An important application of the spherical geometry is the use of the LEED (low energy electron diffraction) systems for analyzing Auger electron spectra from surfaces bombarded by electron beams. Such systems are described by Weber and Peria,[99a] Harris,[99b] Palmberg and Rhodin,[99c] and by Taylor,[99d] and are enjoying increasing usage as analytical tools particularly for detection of surface conditions of solids.

Fields with plane symmetry are used to analyze parallel beams of charged particles. This idea was used by Fox, Hickam, Kjeldaas, and Grove[100] in 1951 to obtain in effect a nearly monoenergetic beam of electrons for ionization studies. This was done by taking differences in the signals produced by electron beams using two slightly different retarding potentials. This method, which yields the signal due only to the narrow slice of the energy distribution between the two potentials chosen, has come to be known as the

* Sections 3 and 4 have been contributed by Prof. M. E. Rudd, Dept. of Physics, University of Nebraska, Lincoln, Nebraska.

RPD method. Chantry[101] has recently employed a multiscaler with this technique to automatically reduce the data to a difference form. Figure, 7*a* and 7*b* illustrate the method. As the retarding potential difference (RPD) was alternately swept over the

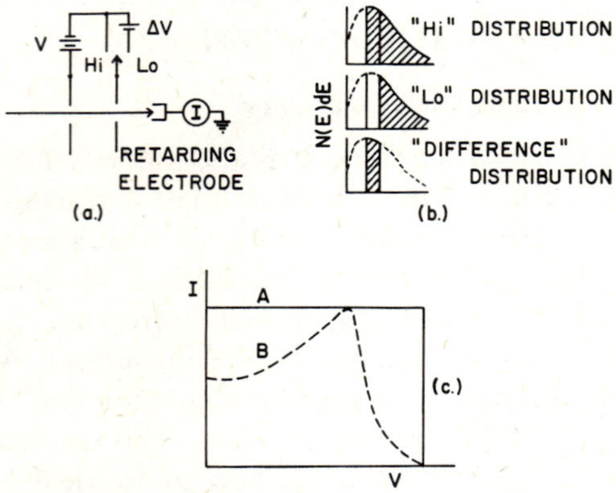

Figure 7 The retarding potential difference (RPD) technique. (*a*) Incoming beam is retarded by potentials V and $V + \Delta V$ in "Hi" and in "Lo" positions, respectively. Only the particles with energy great enough to overcome the potential reach the detector. (Adapted from Ref. 101.) (*b*) By taking the difference between the readings at "Hi" and "Lo," the number of particles in a narrow energy interval can be determined. (From Ref. 101.) (*c*) Current versus retarding potential for a monoenergetic beam in the ideal (A) and actual (B) cases. (Adapted from Ref. 103.)

required range of energies with and without a small incremental voltage added, the ions resulting at each step were recorded in the multiscaler channels. During these sweeps the multiscaler was alternately switched between the add and subtract modes to yield the difference distribution. Although considered here as a monochromator, a similar system could be used as an energy analyzer. Chantry uses 0.1-eV resolution with his apparatus. Using a similar principle and special care to ensure that the electron paths are accurately parallel to the retarding field, Golden[102] reports resolution as good as 8 meV.

In an excellent article on RPD analyzers, Simpson[103] demonstrates why a simple parallel-plate retarding field analyzer generally yields a curve such as B in Figure 7*c* instead of the ideal A. The tail is due to particles with velocity components not parallel to the field. The dip at low values of V is caused by reflections from the collection plate. The common use of an axial magnetic field to help collimate the beam has been shown to give potentially erroneous results, since it allows through the collimation apertures particles that could have large transverse velocity components. Simpson examines spherical capacitor RPD analyzers also and shows that the maximum energy spread in the analysis of particles emitted from a central sphere of radius a and falling on the outer sphere of radius b is $\Delta E/E = (a/b)^2$. In the same article Simpson discusses the use of a Faraday cage (to reduce the reflection of electrons) and also the filter lens, which employs electron optical means to convey electrons to a retarding field at low energy and thence to the collector at high energy. A number of advantages are demonstrated, but it is pointed out that this device suffers from a serious fault shared by all retarding field analyzers; namely, that an integral spectrum is obtained with its inherently poor signal-to-noise ratio when differentiated. Deflection analyzers, discussed in the next section, do not have this fault because they produce a differential energy spectrum directly. Taylor [99d] has discussed the conditions under which the signal-to-noise ratio favors deflection analyzers over that for retarding field analyzers.

3.2. Electrostatic Deflection Analyzers

Several analyzers have been devised in which an electrostatic field deflects particles according to their energies. Those within a narrow range of deflections (and thus a narrow range of energies) are selected by a slit or other aperture and allowed to pass to the detector. These analyzers have been utilized in a wide variety of systems over energy ranges from less than 1 eV to tens of kiloelectron volts and higher.

We first describe briefly four of the most common types of analyzers and show that there are basic similarities among them. A derivation for the transmission function is given both with and without aberration terms. Criteria are given next for choosing slit widths and collimation apertures to optimize the overall transmission. The use of analyzers for measurement of absolute values of cross sections differential in energy of the detected particle is discussed and evaluated. A simple method for making analyzers direct reading is presented, and finally consideration is given to the distortion caused in the energy spectrum by sweeping an analyzer over energies while using a detector with a finite time constant.

In a series of articles,[104–106] Paolini and Theodoridis have obtained the transmission characteristics of particular analyzers with given dimensions and operating conditions. In their treatment, the transmission is limited by rays striking inner surfaces of the analyzer. Here we treat the more easily controlled case of transmitted rays limited by entrance and exit slits, and it is assumed that the analyzer dimensions are such that the walls of the analyzer do not limit the rays.

In general an analyzer is asked to accept a beam of charged particles which is diverging in two dimensions at the entrance slit. Angles α in the plane of the analyzer and β in the perpendicular plane are measured from the direction of the central ray of the bundle. The limiting values of these angles are $\pm\alpha_m$ and $\pm\beta_m$. For the central ray the ratio of the analyzer potential V_p to the energy E_0 passed by the analyzer is a constant depending on the geometry. Define the analyzer constant as $C = qV_p/E_0$.

To be useful, an analyzer should focus the diverging beam in at least one plane and the higher the order of focusing the greater the detector current will be for a given resolution. As with any analyzer, high resolution can be obtained only at the expense of intensity, but better focusing properties will improve the quality of the inevitable compromise. For perfect focusing, the image position would have to be independent of the angles α and β. In the equation giving the image position in an actual analyzer, the terms containing these angles are called the aberration terms. If the lowest power of either angle in this equation is n, the focusing is said to be of $(n - 1)$st order in that angle.

Two useful terms are the dispersion D, defined as the displacement of the image point per unit fractional change of energy, and the trace width w_t, which is the width of the image for a point monoenergetic source for a given range of the angle α or β. The trace width is a measure of the aberrations. Large values of D and small values of w_t are obviously desirable.

3.2.1. *Four Types of Deflection Analyzers*

3.2.1.1. PARALLEL-PLATE ANALYZER OR $\pi/4$ ANALYZER. The simplest analyzer to construct and also to treat theoretically consists of two plane parallel plates held at different potentials. A slit in one plate allows particles to enter at an angle ϕ. They traverse a parabolic path and leave via a second slit. Referring to Figure 8*a*, when a particle of energy E and charge q enters at an angle ϕ, the component of the range in the plane of the paper is

$$l = (2sE \sin 2\phi/qV_p)(1 - \sin^2\beta/\sin^2\phi)^{1/2},$$

where β is the angle the incoming ray makes with the plane of the paper. The maximum range occurs for $\phi = 45°$ if β is small, and therefore focusing takes place for this angle. If we let α represent the (small) deviation of ϕ from 45° and also assume that β is small, the equation reduces to

$$l \simeq \frac{2sE}{qV_p} \cos 2\alpha \cos^2\beta \simeq \frac{l_0 E}{E_0}(1 - 2\alpha^2 - \beta^2) \quad (37)$$

to second order. Since l_0 is to be the range for $\alpha = \beta = 0$ and $E = E_0$, we have $l_0 = 2sE/qV_p$; so the analyzer constant is $C = qV_p/E_0 = 2s/l_0$.

The dispersion is seen from equation 37 to be $D = E_0(\partial l/\partial E)_{E=E_0, \alpha=\beta=0} = l_0$ if aberration terms are neglected. The trace width is $w_t = l_0(2\alpha^2 + \beta^2)$.

Yarnold and Bolton[107] were the first to describe this analyzer. A further analysis was given by Harrower,[108] who derived some of the basic equations. Eland and Danby[109] have used such an analyzer with a resolution of 25 meV. Volz[110] has derived the shape of the resolution curve neglecting the aberrations.

Recently, it has been shown that adding a field free section to the normal parallel-plate analyzer will yield second order focusing. In this case the focusing

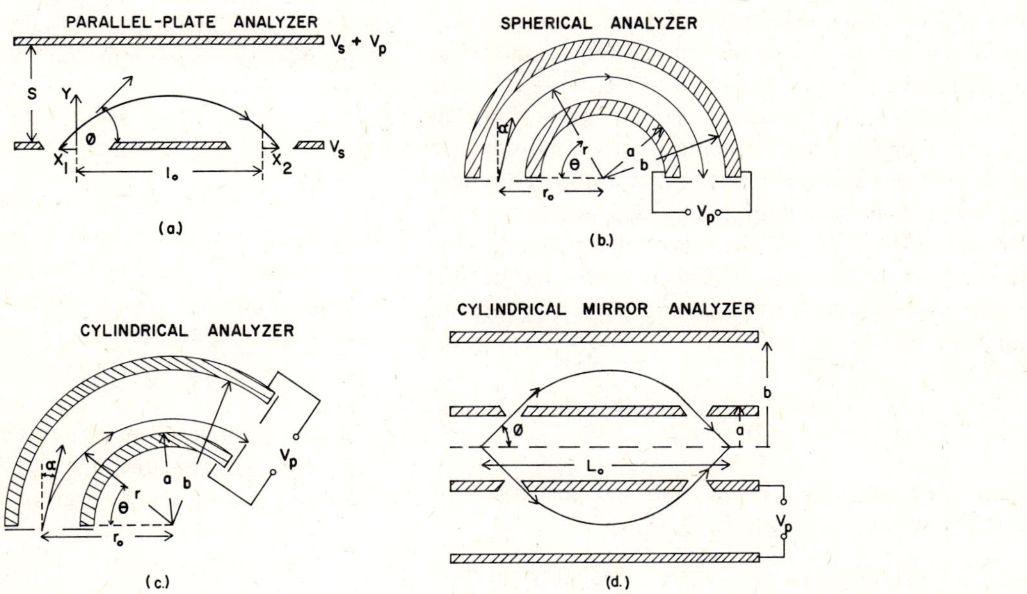

Figure 8. Schematic diagrams of four electrostatic analyzers: (*a*) Parallel-plate analyzer, (*b*) spherical analyzer, (*c*) cylindrical analyzer, (*d*) cylindrical mirror analyzer.

occurs for an entrance angle of $\phi = \pi/6 = 30°$. For treatments of the $\pi/4$-, $\pi/6$-, and modified $\pi/6$-analyzers, see the articles by Green and Proca and by Schmitz and Mehlhorn[110a] and Section 3.2.1.5.

3.2.1.2. SPHERICAL ANALYZER. Purcell,[111] following a suggestion of Aston,[112] introduced an analyzer with focusing in two dimensions in the form of the spherical capacitor; he also gave an analysis of the orbits, including relativistic effects. Hafner, Simpson, and Kuyatt[113] developed the theory further in their comparison of this analyzer with the cylindrical mirror type. Ashby[114], Rogers,[114a] and Kessler and Weichert,[114b] have also given relativistic treatments of the spherical analyzer. Browne, Craig, and Williamson[130] have presented design and construction information.

Figure 8*b* shows the main features of this analyzer. Since the field is radial and equal to $(V_p/r^2)ab/(b-a)$, it follows that the relation between the energy E_0 of the central ray and the analyzer potential is such that the analyzer constant is $C = (b/a) - (a/b)$, provided the central ray has an initial value of r given by $r_0 = \bar{r} = \frac{1}{2}(a+b)$ (i.e., that it enters midway between the two spherical plates). The differential equation of the path of a charged particle entering at arbitrary r_0 is $d^2u/d\theta^2 + u = \bar{r}E_0/r_0 E\cos^2\alpha$ where $u = r_0/r$. It is obvious from the spherical symmetry that this analyzer yields perfect focusing for the angle β. The solution of the differential equation is $r_0/r = (\bar{r}E_0/r_0 E\cos^2\alpha) \times (1-\cos\theta) + \cos\theta - \tan\alpha\sin\theta$. Expanding the factors containing α, we find all powers from the first present. However, at the angle $\theta = 180°$, the last term vanishes and with it the only first power of α. So for that angle we have first-order focusing. Choosing that value of θ and expanding the equation to second order in α, we have $r_0/r_{180} = (2E_0\bar{r}/Er_0)(1+\alpha^2) - 1$.

The analyzer is usually operated with $V = 0$ at $r = \bar{r}$ so a charged particle entering the analyzer at a value of $r_0 \neq \bar{r}$ will enter a region at a nonzero potential and will lose or gain energy depending on what part of the slit it enters. The loss of energy is $qV(x_1) = 2E_0 x_1/\bar{r}$, where $x_1 = r_0 - \bar{r}$. This is subtracted from E in the previous equation. Letting $x_2 = r_{180} - \bar{r}$, we can write finally

$$\frac{x_2}{2\bar{r}} = -\frac{x_1}{2\bar{r}} + \frac{\Delta E}{E_0} - \alpha^2, \tag{38}$$

where $\Delta E = E - E_0$. From this the dispersion is $D = 2\bar{r}$ and the trace width is $w_t = 2\bar{r}\alpha^2$.

3.2.1.3. CYLINDRICAL ANALYZER. Probably the first practical electrostatic deflection analyzer was developed by Hughes and Rojansky[115] in 1929, when they showed that focusing of electron beams takes place every 127° 17′ in a radial, inverse first-power electric field. Hughes and McMillen[116] built such a cylindrical analyzer and showed that it could be used as an energy analyzer. It has since found wide acceptance in a variety of applications. Marmet and Kerwin[117] improved it by adding secondary electron suppression grids and used it as an energy selector or monochromator. Rudd[118] derived an expression for its transmission curve, and Kuyatt and Rudd[119] proved that it could be used for measurement of absolute cross sections. Transmission functions for a specific analyzer operated under conditions similar to those used in satellite work are calculated by Theodoridis and Paolini.[105] Rogers[120] has extended the theory of the cylindrical analyzer to include relativistic effects.

Referring to Figure 8*c*, the differential equation of a charged particle in a cylindrical field is

$$\frac{d^2u}{d\theta^2} + u = \frac{E_0}{uE\cos^2\alpha},$$

where $u = r/r_0$. A first-order solution given by Hughes and Rojansky[115] is

$$\frac{r_0}{r} = \sqrt{\frac{E_0}{E\cos^2\alpha}}\,(1-\cos 2^{1/2}\theta) + \cos 2^{1/2}\theta - 2^{-1/2}\tan\alpha\sin 2^{1/2}\theta.$$

As with the spherical analyzer, the last term (which contains the largest aberration terms) vanishes for a certain choice of angle θ—namely, $\theta = 2^{-1/2}\pi = 127°17'$—so this is the angle at which a first-order focus is obtained. A solution correct to second order in α which holds at the 127° focus angle is obtained from the paper of Hughes and Rojansky:[115]

$$\frac{r_0}{r_{127}} = \frac{2E_0}{E}\left(1 + \frac{4\alpha^2}{3}\right) - 1. \tag{39}$$

The effect of the angle β can readily be included. Particles of velocity v_0 (where $\frac{1}{2}mv_0^2 = E_0$) for which $\beta \neq 0$ have a velocity component in the plane of the central orbit of $v_0\cos\beta$. This reduces the effective energy from E_0 to

$$E_0\cos^2\beta \simeq E_0(1-\beta^2). \tag{40}$$

Again, particles lose (or gain) energy upon entering the field between the plates. Assuming $V = 0$ at $r = \bar{r}$ this loss of energy is given by

$$qV(x_1) = 2E_0\ln\left(\frac{r_0}{\bar{r}}\right) \simeq \frac{2E_0 x_1}{\bar{r}}, \tag{41}$$

where $x_1 = r_0 - \bar{r}$.

* $\frac{r_0}{r_{127}} = \left(2\sqrt{\frac{E}{E_0}} - 1\right)\left(1 + \frac{4}{3}\alpha^2\right)$ (39)

Table II.
Summary of information about three electrostatic analyzers

	Parallel-plate	Spherical	Cylindrical
Electric field	$\varepsilon_y = -V_p/s$ $\varepsilon_x = \varepsilon_z = 0$	$\varepsilon_r = -V_p ab/r^2(b - a)$ $\varepsilon_\theta = \varepsilon_\phi = 0$	$\varepsilon_r = -V_p/r\ln(b/a)$ $\varepsilon_\theta = \varepsilon_z = 0$
Energy of central ray	$E_o = qV_p \ell_o/2s$	$E_o = qV_p/(b/a - a/b)$	$E_o = qV_p/2\ln(b/a)$
Analyzer constant C ($=qV_p/E_o$)	$2s/\ell_o$	$b/a - a/b$	$2\ln(b/a)$
Differential equation	$d^2y/dt^2 = -2E_o/m\ell_o$	$d^2u/d\theta^2 + u = \bar{r}E_o/r_oE\cos^2\alpha$ where $u = r_o/r$ and $\bar{r} = (1/2)(a + b)$	$d^2u/d\theta^2 + u = E_o/uE\cos^2\alpha$ where $u = r_o/r$
Solution of differential equation	$\ell = \ell_o(E/E_o)\sin 2\phi$ [handwritten: $\cos^2\beta$] $(1 - \sin^2\beta/\sin^2\phi)^{1/2}$ [struck through]	$r_o/r = \frac{\bar{r}E_o(1 - \cos\theta)}{r_oE\cos^2\alpha} + \cos\theta - \tan\alpha \sin\theta$	$\frac{r_o}{r} \approx \sqrt{\frac{E_o(1 - \cos 2^{1/2}\theta)}{E\cos^2\alpha}} + \cos 2^{1/2}\theta - 2^{-1/2}\tan\alpha \sin 2^{1/2}\theta$
Focusing condition	$\phi = \pi/4$	$\theta = \pi$	$\theta = \pi/2^{1/2} = 127°17'$
2nd order sol'n under focusing condition	$\ell \simeq \ell_o(E/E_o)(1 - 2\alpha^2 - \beta^2)$ $\alpha = \phi - \pi/4$	$r_o/r_{180} \simeq 2E_o\bar{r}(1 + \alpha^2)/Er_o - 1$	$r_o/r_{127} = \frac{2E_o}{E}(1 + \frac{4}{3}\alpha^2) - 1$ [struck through; handwritten: $r_o/r_{127} = \left(2\sqrt{\frac{E_o}{E}} - 1\right)\left(1 + \frac{4}{3}\alpha^2\right)$]
Potential	$V(y) = V_a + 2E_oy/q\ell_o$	$V(r) = (2E_o\bar{r}/q)(1/\bar{r} - 1/r)$ if $V = 0$ at $r = \bar{r}$	$V(r) = (2E_o/q)\ln(r/\bar{r})$ if $V = 0$ at $r = \bar{r}$
Loss of energy on entering analyzer	V_a	$qV \simeq 2E_ox_1/\bar{r}$	$qV = 2E_o\ln(r_o/\bar{r}) \simeq 2E_ox_1/\bar{r}$
Basic equation	$(x_1 + x_2)/\ell_o = \Delta E/E_o - 2\alpha^2 - \beta^2$	$\frac{x_1 + x_2}{2\bar{r}} = \frac{\Delta E}{E_o} - \alpha^2$	$\frac{x_1 + x_2}{\bar{r}} = \frac{\Delta E}{E_o} - \frac{4}{3}\alpha^2 - \beta^2$
Dispersion D	ℓ_o	$2\bar{r}$	$r_o\ (\simeq \bar{r})$
Trace width	$\ell_o(2\alpha^2 + \beta^2)$	$2\bar{r}\alpha^2$	$\bar{r}(4\alpha^2/3 + \beta^2)$
Path length of central ray	1.148 D	1.571 D	2.222 D
A	2	1	4/3
B	1	0	1

Combining equations 39 through 41 and setting $x_2 = r_{127} - \bar{r}$ and $x_1 = r_0 - \bar{r}$, we get the result

$$\frac{x_2}{\bar{r}} = -\frac{x_1}{\bar{r}} + \frac{\Delta E}{E_0} - \frac{4\alpha^2}{3} - \beta^2, \tag{42}$$

and from this we read off $D = \bar{r}$ and $w_t = \bar{r}(4\alpha^2/3 + \beta^2)$. Table II summarizes some of the important information about the three analyzers discussed.

3.2.1.4. CYLINDRICAL MIRROR ANALYZER. Two coaxial cylinders (Figure 8*d*) with the source and detector of charged particles on the axis form the cylindrical mirror analyzer. The entrance and exit apertures are two annular slits in the inner cylinder. Blauth[121] seems to be the first to have published a description of this analyzer, but he attributes it to someone else. (According to Dr. W. Mehlhorn, Prof. Chr. Gerthsen first proposed this analyzer and Dr. R. Gremmelmaier was the first to construct and use one.) Mehlhorn[122] has used it extensively in Auger spectroscopy. Others using various versions of it are Edwards et al.,[123] Ogurtsov et al.,[124] and Toburen.[125] Zashkvara et al.[126] gave the first analysis of its operation, and Sar-el,[127] proceeding independently, gave both a theoretical description and an experimental verification. He further suggested the use of only a portion of the cylinders. This modification is useful for measuring angular distributions, but one of its most attractive features is partly lost, namely, its large azimuthal acceptance angle. Nevertheless, it may still be superior to the parallel-plate analyzer, similarly positioned, because of its focusing action in two dimensions. Hafner, Simpson, and Kuyatt[113] made a comparison of the trace-width/dispersion characteristics for this analyzer and the spherical analyzer and concluded that the cylindrical mirror is superior in performance to any other previously available analyzer; but they also point out the difficulty in designing lenses to be used with it.

It has been shown[126,127] that for a point object on the axis to be imaged back on the axis after analysis, the rays must be confined to a range around an angle of 42° 18.5′. From the symmetry, one sees that there is exact focusing relative to varying azimuthal angles, which permits a very wide acceptance angle for a given resolution.

The analysis of the cylindrical mirror instrument is somewhat different from the previous ones we have considered and only the result given by Zashkvara et al.[126] and quoted by Hafner, Simpson, and Kuyatt[113] is given here for the basic equation

$$\frac{\Delta L}{L_0} = \frac{x_1 + x_2}{D} = \frac{\Delta E}{E_0}(1 + 1.84\alpha) - 2.85\alpha^3,$$

where L_0 is the distance between the object and image as measured along the axis and ΔL is the change in axial position of the image due to a shift of energy by ΔE and a variation of angle ϕ from the optimum 42° 18.5′. The distances x_1 and x_2 are distances along the axis measured from the intersections of the axis and the central ray; the dispersion D is here equal to L_0. The trace width is 2.85 $L_0 \alpha^3$.

3.2.1.5. VARIATIONS. Four basic types of deflection analyzers have been discussed, but a great many variations of these exist. It is possible, for example, to put the object and image points at some distance from the actual entrance and exit plane of an analyzer. In this situation the entire 127° for the cylindrical analyzer and the 180° for the spherical analyzer are not needed and a smaller sector angle can be used depending on the positions of object and image. This is discussed by Dempster,[128] by Herzog,[129] by Siegbahn et al.,[51] by Browne et al.,[130] and by Wollnik.[130a]. Siegbahn concludes that the sector angle should not be reduced too much if operation is to remain optimum. An analogous situation is encountered with the cylindrical mirror when it is desired to put the source and/or detector at the slits instead of on the axis. In such a situation the optimum angle is somewhat different from the 42° 18.5′ given previously.

Some have added additional electrodes to "standard" analyzers to improve the focusing. An example is the use by Matsuda[130b] of top and bottom plates on a 90° cylindrical analyzer to vary the focal length electrically.

The so-called fountain spectrometers of Edelmann and Ulmer[131] and Kuyatt[132] extended the idea of the parallel-plate analyzer to allow a 2π aximuthal angle of acceptance. However, small detectors (such as electron multipliers) cannot be used with these, and the observation angle is fixed as well (see also Section 3.2.1.1.)

A variation of the spherical analyzer called the Keplertron attains a very high transmission by placing the source on the surface of the inner sphere. This is discussed by a number of authors, notable Zashkvara, et al. [126] Birkhoff and Ritchie and others.[133, 134]

If the entire 2π azimuthal angle of the cylindrical mirror is used, one is ordinarily restricted to viewing particles coming from a beam at the one angle 42° 18.5′. To measure angular distributions Ogurtsov et al.[124] have recently reported some success in using an additional cylindrical grid electrode between the scattering center and the first slit. With the proper potential on it, particles coming from the beam at different angles are deflected into the entrance slit and analyzed in the usual way from there. Angles from 30° to 75° are attainable. The major difficulty is in the calculation of the trajectory to determine the angle and the geometrical quantities needed for calculation of cross sections.

A general treatment of toroidal analyzers (which include the spherical and cylindrical analyzers as special cases) is given by Wollnik.[130a]

3.2.2. *Analyzer Transmission Function without Aberrations.*

Equations 37, 38, and 42 for the parallel-plate, spherical, and cylindrical analyzers all have the form

$$\frac{x_1 + x_2}{D} = \frac{\Delta E}{E_0} - A\alpha^2 - B\beta^2, \qquad (43)$$

where x_1 and x_2 are the displacements of the object and image points, respectively. A, B, and D have different values for the various analyzers and are given in Table 1. The cylindrical mirror analyzer equation cannot be expressed in this form and thus the part of the following development, which involves the aberration terms, does not apply to that type of analyzer.

All four of the analyzers just listed can be considered as imaging devices that focus an image of the entrance slit in the image plane; this plane is, for example, at $\theta = 127°$ for the cylindrical analyzer. In each device, the magnification is unity, since x_1 and x_2 enter the basic equation in the same way. This also means that the analyzer would work equally well if the entrance and exit slits were interchanged. Additionally, all four analyzers exhibit linear dispersion; that is, the position x_2 of the image is a linear function of ΔE.

If all particles enter the analyzer with $\alpha = \beta = 0$, or if $A = B = 0$, the system would be aberration free. We consider this first, and derive the corresponding transmission function, which is a graph of the fraction of the particles at the entrance slit that are transmitted by the analyzer versus their energy at a given setting of the analyzer voltage V_p. As the energy varies, the image of the entrance slit sweeps across the exit slit. Assuming the latter slit is long enough so that no particles with the proper value of x_2 are blocked by it, and also assuming a uniform illumination of the entrance slit, the shape of the transmission function depends only on the two slit widths w_1 and w_2 and the dispersion D. The basic equation 43 proves that when the angle terms are zero there is a 1 : 1 correspondence between object position x_1 and image position x_2 for a monoenergetic source. Then the entrance slit will be exactly imaged in the plane of the exit slit. Figure 9 shows, for $A\alpha_m{}^2 = 0$, the entrance slit

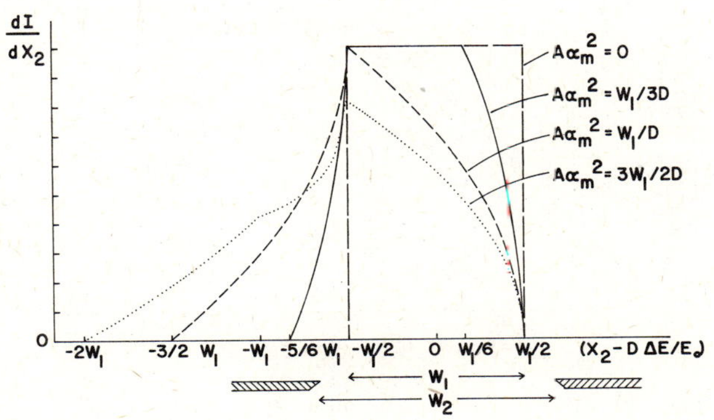

Figure 9. Distribution of intensity in the image plane coming from a slit of width w_1 in the object plane and having an initial angular spread of $\pm\, \alpha_m$ for various values of α_m. As the energy (or analyzer potential) is varied, this distribution sweeps over the exit slit of width w_2.

image of width w_1 superimposed on the exit slit of width w_2 for the case $w_1 < w_2$. For $E = E_0$ their centers will coincide, and clearly the fraction transmitted will be unity. But for $w_2 < w_1$ the fraction is $f = w_2/w_1$. This fraction will stay constant as the energy is varied until $\Delta E/E_0 = \pm\frac{1}{2}|w_1 - w_2|/D$ and

Table III. Constants for an aberration-free analyzer.

Quantity	$w_1 > w_2$	$w_2 > w_1$	$w_1 = w_2 = w$
Base width $\Delta E/E_0$	$(w_1 + w_2)/D$	$(w_1 + w_2)/D$	$2w/D$
Top width $\Delta E/E_0$	$(w_1 - w_2)/D$	$(w_2 - w_1)/D$	0
FWHM/E_0	w_1/D	w_2/D	w/D
Height f_m	w_2/w_1	1	1
Area $\int f\, dE$	$E_0 w_2/D$	$E_0 w_2/D$	$E_0 w/D$

then will decrease linearly to reach zero at $\Delta E/E_0 = \pm\frac{1}{2}(w_1 + w_2)/D$. Therefore, the transmission function is a trapezoid with constants given in Table III. These may be summarized for all three cases as follows:

base width = (sum of slit widths)/D,
top width = (difference of slit widths)/D,
FWHM = (larger slit width)/D,
area = E_0(exit slit width)/D.

These results are graphed in Figure 10. An especially important result is that the area under the curve is independent of the size of the entrance slit width.

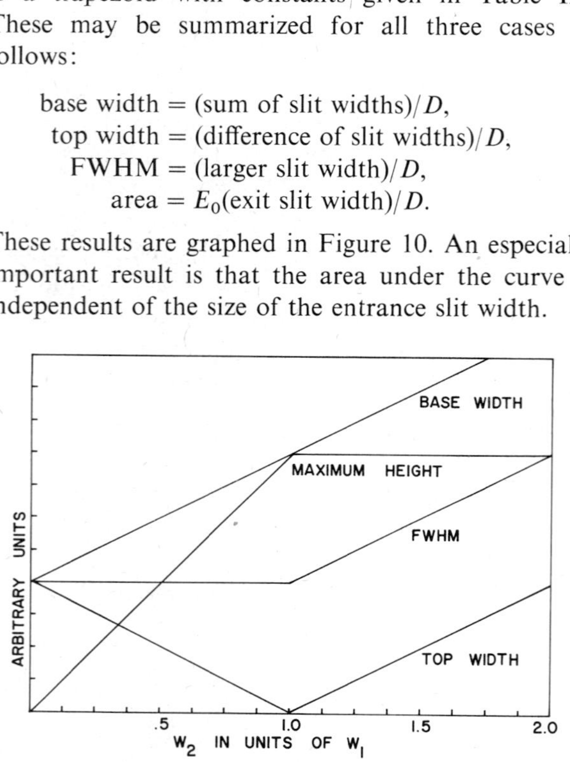

Figure 10. Plot of base width, full width at half-maximum, top width, and maximum height of the analyzer resolution curve for an analyzer without aberrations that is uniformly illuminated at its entrance slit. The width of the exit slit in units of the width of the entrance slit is plotted on the abscissa.

3.2.3. *Analyzer Transmission Function with Aberrations*.

We now take up the more realistic case where the angle terms in equation 43 are not zero. This problem is quite complicated if done in a completely general way, since the shape of the transmission function depends on the distribution of current over the angles α and β, on the relation of this to the distribution over the entrance slit width, on the relation between the maximum sizes of the two aberration terms, and finally on the relative sizes of the two slits.

To simplify matters, assume again that the current is uniformly distributed over the width of the entrance slit and over all values of α from $-\alpha_m$ to $+\alpha_m$. Further assume that the term in β is zero, which is the case for the spherical analyzer. For the other analyzers we note that α and β enter the basic equation 43 in the same way, so that finding the effect of finite values of β is done in precisely the same way as for α. If both terms are present, the shape of the transmission function will, of course, be more complicated than the one we calculate here (except in the special case $\alpha_m/\beta_m = (B/A)^{1/2}$, in which case the two angle terms merge into one). Nevertheless, we can deduce certain effects of the aberration terms which are simply additive when both terms are included.

With these assumptions the basic equation becomes $\Delta E/E_0 = (x_1 + x_2)/D + A\alpha^2$. From this equation we may derive expressions for the intensity distribution in the image plane and then the transmission function of the analyzer.

Consider first a monoenergetic line source at the entrance slit. Then $dx_1 = d(\Delta E) = 0$. Let k' be the intensity (or current) per unit angle α. Then if I_0 is the total intensity, we have $k' = I_0/2\alpha_m$. The intensity in the image plane from x_2 to $x_2 + dx_2$ is twice that at the entrance slit from α to $\alpha + d\alpha$, since α is squared in the basic equation and thus equal positive and negative angles contribute intensity to the same image position. Therefore the image intensity is $dI = 2k'd\alpha$ and the intensity per unit width x_2 is $dI/dx_2 = (dI/d\alpha)(d\alpha/dx_2)$. Then, since the basic equation gives $dx_2 = 2DA\alpha d\alpha$, we have $dI/dx_2 = k'/DA\alpha$. Now spread out the line source over the entire entrance slit and define k to be the entering intensity per unit angle α and per unit width x_1. Then $k = I_0/2\alpha_m w_1 = k'/w_1$, and the element of image intensity coming from dx_1 in the slit is $d(dI/dx_2) = kdx_1/DA\alpha$. Using the basic equation to substitute for α we get

$$\frac{dI}{dx_2} = \frac{k}{DA^{1/2}} \int_a^b \frac{dx_1}{[\Delta E/E_0 - (x_1 + x_2)/D]^{1/2}},$$

or

$$\frac{dI}{dx_2} = \frac{2k}{(AD)^{1/2}} \left| \left(D\frac{\Delta E}{E_0} - x_1 - x_2 \right)^{1/2} \right|_a^b.$$

Choosing the limits of integration a and b is not trivial, since a given image point may be illuminated by rays starting from a range of values of x_1 with correspondingly varying values of α. The results are given in Table IV for the case $AD\alpha_m{}^2 \le w_1$. This gives us the intensity distribution in the image plane or the plane of the exit slit. Figure 9 shows how this distribution looks for the case $\alpha_m = 0$ (no aberrations) and for the cases $A\alpha_m{}^2$ equal to $w_1/3D$ and w_1/D. Also shown are the results calculated from a different set of equations for the case $A\alpha_m{}^2 = 3w_1/2D$.

As the energy is changed, this distribution sweeps past the exit slit, and the fraction falling within the limits of that slit as a function of energy is the transmission function. Obviously the final shape of this function depends on the interrelation of w_1, w_2, and α_m. For the special case of equal slit widths and small

Table IV.

Intensity distribution at the exit slit for $AD\alpha_m^2 \leqq w_1$

Range	a	b	dI/dx_2
$-\frac{w_1}{2} - AD\alpha_m^2 + D\Delta E/E_o \le x_2 \le -\frac{w_1}{2} + D\Delta E/E_o$	$\frac{w_1}{2}$	$-x_2 + D\Delta E/E_o - AD\alpha_m^2$	$2k\alpha_m - 2k(-x_2 + D\Delta E/E_o - \frac{w_1}{2})^{1/2}/(AD)^{1/2}$
$\frac{w_1}{2} + D\Delta E/E_o \le x_2 \le \frac{w_1}{2} + D\Delta E/E_o - AD\alpha_m^2$	$-x_2 + D\Delta E/E_o$	$-x_2 + D\Delta E/E_o - AD\alpha_m^2$	$2k\alpha_m$
$\frac{w_1}{2} + D\Delta E/E_o - AD\alpha_m^2 \le x_2 \le \frac{w_1}{2} + D\Delta E/E_o$	$-x_2 + D\Delta E/E_o$	$-\frac{w_1}{2}$	$2k(\frac{w_1}{2} - x_2 + D\Delta E/E_o)^{1/2}/(AD)^{1/2}$

Table V. Transmission function for $w_1 = w_2 = w$ and $A\alpha_m^2 = w/D$.

Range	Fraction transmitted
$-w \leq D\Delta E/E_0 \leq -w + AD\alpha_m^2$	$\frac{1}{w}\frac{2(w + D\Delta E/E_0)^{3/2}}{3\alpha_m (AD)^{1/2}}$
$-w + AD\alpha_m^2 \leq D\Delta E/E_0 \leq 0$	$\frac{1}{w}\left[D\Delta E/E_0 + w - \frac{1}{3}AD\alpha_m^2\right]$
$0 \leq D\Delta E/E_0 \leq AD\alpha_m^2$	$\frac{1}{w}\left[D\Delta E/E_0 + w - \frac{1}{3}AD\alpha_m^2 - \frac{4}{3\alpha_m (AD)^{1/2}}(D\Delta E/E_0)^{3/2}\right]$
$AD\alpha_m^2 \leq D\Delta E/E_0 \leq w$	$\frac{1}{w}\left[-D\Delta E/E_0 + w + \frac{1}{3}AD\alpha_m^2\right]$
$w \leq D\Delta E/E_0 \leq w + AD\alpha_m^2$	$\frac{1}{w}\left[-D\Delta E/E_0 + w + \frac{1}{3}AD\alpha_m^2 + \frac{2}{3\alpha_m (AD)^{1/2}}(D\Delta E/E_0 - w)^{3/2}\right]$

α_m, there are five stages of overlap of the image and the exit slit, yielding the results given in Table V. A graph of the function is shown in Figure 11 for $A\alpha_m^2$ equal to 0, $w/3D$, and w/D.

By differentiation we find that the maximum of the transmission function is $f_{max} = 1 - AD\alpha_m^2/4w$ and that

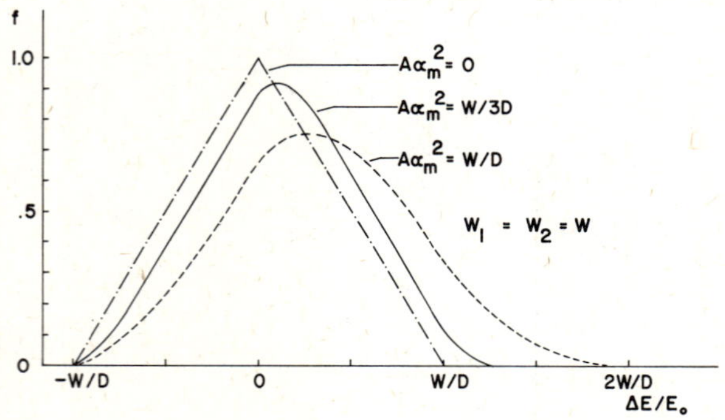

Figure 11. Shape of the transmission function of an electrostatic analyzer with equal slit widths ($w_1 = w_2 = w$) with no aberrations (dashed line) and with $A\alpha_m^2 = w/3D$.

it occurs at $\Delta E_m/E_0 = \frac{1}{4}A\alpha_m^2$. From this the transmission function half-maximum points can be found and the full width at half-maximum may be calculated, as well as the shift S of the center due to the aberration. The results are $\text{FWHM}/E_0 = w/D + \frac{1}{4}A\alpha_m^2$ and $S/E_0 = A\alpha_m^2/3$. Since the α and β terms enter the basic equation in the same way, the effect on the FWHM and S are the same. The contribution to the FWHM due to the two angular aberrations is not strictly additive, but for equal slits one may assume that this gives, at most, a small percentage of error. Then approximately, including the contribution due to β,

$$\frac{FWHM}{E_0} = \frac{w}{D} + \frac{1}{4}A\alpha_m^2 + \frac{1}{4}B\beta_m^2 \qquad (44)$$

and

$$\frac{S}{E_0} = \frac{A\alpha_m^2}{3} + \frac{B\beta_m^2}{3}.$$

A very important result is that, in spite of the presence of the aberration terms, the area under the transmission function curve is the same as what we calculated for the case of no aberrations, namely, area $= E_0 w_2/D$. This can be shown by actual integration under the curve, section by section, or by a more general method.[119]

3.2.4. *Optimizing the Analyzer*

Since electrostatic analyzers are used in a wide variety of situations, it is impossible to give operating conditions that will provide optimum results in every case. We present here a calculation to optimize the construction of the analyzer and collimating slits for a specific but fairly common set of circumstances, namely, the analysis of particles emitted from collisions of a beam with a gas target. A typical arrangement is shown in Figure 12. A pair of slits of widths s_1 and s_2

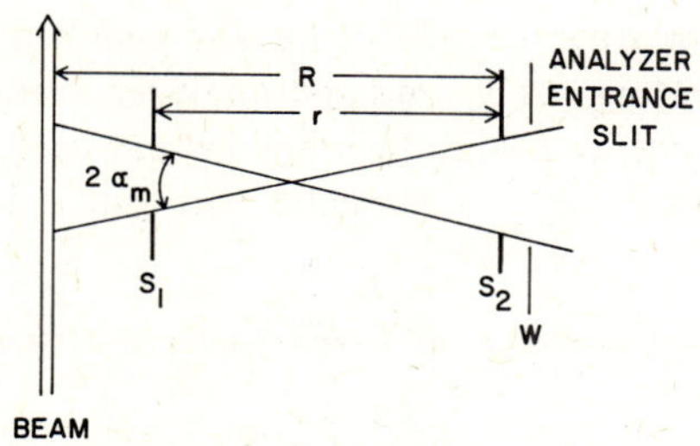

Figure 12. Arrangement of apparatus to analyze charged particles formed by collision of beam with gas molecules. First and second defining slits, which are of widths s_1 and s_2, respectively, determine the angle α_m. The height h_2 of the second slit largely determines the angle β_m in the plane perpendicular to the plane of the diagram.

define the angles α_m and β_m within which particles enter the analyzer. Sometimes the entrance slit w serves as the second defining slit, in which case $s_2 = w$. At any rate, we will assume that

$$s_2 = cw_1, \tag{45a}$$

where c is a constant of order of magnitude unity. It is well known (see, e.g., Ref. 132, p. 11) that the current I_{in} entering the analyzer is proportional to $(l d\Omega)_{eff}$ where l is the length of beam path viewed and $d\Omega$ is the solid angle subtended by slit 2 at the collision center. The quantity $(l d\Omega)_{eff}$ is proportional to $s_1 s_2 h_2$, where h_2 is the length of slit 2. Using equation (45a) we have

$$I_{in} \propto s_1 w_1 h_2 . \tag{45b}$$

Now consider two situations involving narrow and wide analyzer resolution functions.

3.2.4.1. NARROW ANALYZER RESOLUTION FUNCTION. If the FWHM of the analyzer is small compared to the range over which a significant variation in the energy distribution to be measured takes place (such as in a continuous spectrum or even in a line spectrum under very high resolution), then the output current from the analyzer is proportional to the area under the resolution curve, which is $E_0 w_2/D$ times the input current. Then the output current is

$$I_{out} \propto s_1 w_1 w_2 h_2 . \tag{45c}$$

For an analyzer *without* aberrations, the FWHM of the analyzer is $w_g E_0/D$, where w_g is the greater of w_1 and w_2. Then for a fixed FWHM, the greatest value of I_{out} occurs when $w_1 = w_2 = w_g = w$. Choosing equal slit widths, then, consider next the optimization of an analyzer *with* aberrations.

For equal slit widths equation 44 holds. Denoting FWHM/E_0 by F,

$$F = \frac{w}{D} + \frac{1}{4} A\alpha_m^2 + \frac{1}{4} B\beta_m^2. \tag{46}$$

We now inquire into the optimum relation among the three contributions to F in equation 46. To find this we vary α and β, keeping r, R, E_0, and D fixed, and we find the values that maximize I_{out} subject to the condition that F have a given value. We note that $\alpha_m = (s_1 + s_2)/2r$ and we make the approximation that $\beta_m \simeq h_2/2r$. Using these and equation 45a, equation 45c becomes

$$I_{out} \propto (2r\alpha_m - cw)w^2\beta_m,$$

or using equation 46 to eliminate w,

$$I_{out} \propto (2r\alpha_m - cDF + \tfrac{1}{4}AcD\alpha_m^2 + \tfrac{1}{4}BcD\beta_m^2) \times (F - \tfrac{1}{4}A\alpha_m^2 - \tfrac{1}{4}B\beta_m^2)^2\beta_m . \tag{47}$$

Taking the partial derivative of I_{out} with respect to α_m and equating to zero for a maximum of I_{out}, we get

$$(\tfrac{5}{4}A\alpha_m^2 + \tfrac{1}{2}B\beta_m^2 - 2F)r + 3AcD(\tfrac{1}{8}B\beta_m^2 + \tfrac{1}{8}A\alpha_m^2 - \tfrac{1}{2}F)\alpha_m = 0.$$

Remembering that F is of the order α_m^2 or β_m^2, we drop the terms in the second set of parentheses, since they are of higher order than those in the first. Then the condition on α_m for maximum detector current is

$$\frac{1}{4} A\alpha_m^2 = \frac{F - \frac{1}{4}B\beta_m^2}{5}. \tag{48a}$$

Differentiating equation 47 with respect to β_m, equating to zero, and dropping higher order terms, we find the condition on β_m is

$$\frac{1}{4} B\beta_m^2 = \frac{F - \frac{1}{4}A\alpha_m^2}{5}. \tag{48b}$$

Both equations 48 are satisfied if

$$\frac{w}{D} = \frac{2F}{3}$$

$$\frac{1}{4} A\alpha_m^2 = \frac{1}{4} B\beta_m^2 = \frac{F}{6}$$

or

$$A\alpha_m^2 = B\beta_m^2 = \frac{w}{D}.$$

In other words, the angle terms in equation 46 should be equal and each should be one-fourth the slit term.

Putting in values of A and B for the parallel-plate analyzer, one finds that the angles need to be restricted to about 3 to 4° for optimum conditions at 1% resolution and to 1 to 2° for 0.1% resolution.

3.2.4.2. WIDE ANALYZER RESOLUTION FUNCTION. At the opposite extreme, we consider the case in which the

Max I_0:
each angle term = $\frac{1}{6}F$ ie $\frac{1}{4}A\alpha_m^2 = \frac{1}{6}F$
slit term = $\frac{2}{3}F$ ie $\frac{w}{d} = \frac{2}{3}F$
or each angle term = $\frac{1}{4}$ slit term ie $\frac{1}{4}A\alpha_m^2 = \frac{1}{4}\frac{w}{D}$

energy distribution varies considerably over the energy range represented by the FWHM of the analyzer. Such a case is encountered when a spectrum with sharp peaks (i.e., a line spectrum) is analyzed under low or medium resolution.

We may choose to maximize the *area* under the peak as we sweep over it with the analyzer. As shown in the next section, this area is proportional to w_2 and, since the input current is still given by equation 45b, the area under the peak in the output current is proportional to $s_1 w_1 w_2 h_2$. Thus the analysis of Section 3.2.4.1 holds.

Alternatively, we might wish to maximize the *height* of the measured peak. If the peak in the energy distribution is very narrow compared to the FWHM of the analyzer, the height in the output peak is approximately the height of the input peak multiplied by the maximum of the transmission function. Since for equal slits the transmission function maximum was shown in Section 3.2.3 to be

$$f_{\max} = 1 - \frac{AD\alpha_m^2}{4w} - \frac{BD\beta_m^2}{4w},$$

the height of the maximum in the output current peak is

$$I_{0\,\max} \propto (2r\alpha - cw)\left(\frac{w}{D} - \frac{1}{4}A\alpha_m^2 - \frac{1}{4}B\beta_m^2\right).$$

Again using equation 46 to eliminate w, differentiating with respect to α_m, and equating to zero, we get $3A\alpha_m^2/2 = F - \frac{1}{2}B\beta_m^2$ when we discard higher order terms. Similarly, the condition on β_m comes out $3B\beta_m^2/2 = F - \frac{1}{2}A\alpha_m^2$. These two conditions are satisfied by choosing

$$\frac{w}{D} = \frac{3F}{4},$$

$$\frac{1}{4}A\alpha_m^2 = \frac{1}{4}B\beta_m^2 = \frac{F}{8},$$

or therefore $\frac{1}{4}A\alpha_m^2 = \frac{1}{4}B\beta_m^2 = w/6D$. So in this case the angle terms should be one-sixth the slit term.

It should also be noted that if only the quantity $(l d\Omega)_{\text{eff}}$ is maximized (without multiplying by the area under the resolution curve as in Section 3.2.4.1 or by $f_{\max}$ as in Section 3.2.4.2) by the same method, again for a fixed FWHM, the result is that each angle term should be one-half the slit term. Under a somewhat different set of conditions, Kuyatt and Simpson[135] have suggested making the angle term one-half the slit term in a spherical analyzer to minimize tailing of the transmission function. The present analysis indicates that the angle terms should be even smaller than this, at least under the conditions specified.

In this treatment no focusing was assumed between the collimating slit s_2 and the analyzer entrance slit. However, the analysis still holds in that case *provided* equation 45a still remains valid and that c is not too large.

3.2.5. *Measurements of Cross Sections Differential in Energy*

It was noted in a previous section that the area under the resolution curve (transmission function) of most electrostatic analyzers is given simply by $E_0 w_2/D$. This result, which is independent of the size of the entrance slit, holds regardless of whether there is a spread of directions of the incoming particles, provided they are distributed uniformly over the slit and over all possible angles. This result was first demonstrated by Rudd[118] for the cylindrical analyzer, but obviously it may be extended to other analyzers with the same basic equation. Kuyatt and Rudd[119] showed in a more general and elegant manner that the area under the curve retains the above-mentioned value even when the entering particles are distributed over the entrance slit and over the angles α and β in any arbitrary manner whatsoever. This remarkable result is characteristic of all analyzers with linear dispersion, the only condition being that the distribution over the entrance slit be independent of energy. This result has important implications for the use of electrostatic analyzers for the measurement of absolute values of cross sections differential in energy of the detected particles, i.e. cross sections per unit energy range of some emitted charged particle.

An analyzer that had 100% transmission over a narrow energy range dE and 0% transmission outside this range would be ideal for such measurements. Referring to Figure 12, if dI is the observed signal current in that energy range from a length l of the primary beam of current I_0 through a medium of density of particles n, the differential cross section is $\sigma(E) = (I_0 nl)^{-1} dI/dE$. But because analyzers do not have this ideal-square transmission function, what is actually measured is a value of dI/dE averaged over the resolution curve, the average being weighted by the transmission function.

Consider first a slowly varying energy distribution with dI/dE the current per unit energy interval incident on the entrance slit of the analyzer. Following Kuyatt and Rudd,[119] the output current of the analyzer when set for a center energy E_0 is

$$I_{\text{out}} = \int \frac{dI}{dE} f\left(\frac{E - E_0}{E_0}\right) dE,$$

where the integral extends over the resolution curve of the analyzer. But if dI/dE does not change appreciably over the energy range of the resolution curve, it can be removed from the integral. The remaining integral is simply the area under the transmission function $E_0 w_2/D$, so

$$\left(\frac{dI}{dE}\right)_{av} = \frac{I_{out} D}{E_0 w_2}.$$

Since the quantities on the right are easily measurable, the average dI/dE is determined. The average is equal to its exact value if it is varying linearly and if the transmission function is symmetric. Otherwise it represents a close approximation if the variation is small over the resolution curve.

It has also been shown[119] that in a distribution containing a sharp peak represented by $I_{in} g(E)$, where $\int g(E) dE = 1$, the current I_{in} can be determined by measurements made on the output of the analyzer. The result is

$$I_{in} = \frac{SD}{w_2 E'},$$

where

$$S = \int I_{out}(E_0) dE_0$$

represents the measured area under the peak and E' is the average energy of the peak. The only restriction here is that the width of the peak be small compared to E'.

3.2.6. *Direct Energy Reading Analyzers*

It is sometimes desirable to so construct or adjust electrostatic analyzers that some potential (in volts) applied to the analyzer is equal to the energy (in electron volts) of the particles being passed by the analyzer. This can be done quite precisely, and the arrangement is often worthwhile. Conceptually, the simplest way of making a direct reading analyzer is to adjust the dimensions of the device so that the analyzer constant is equal to unity. Conditions for this state are given in Table VI.

However, it is difficult to machine dimensions as precisely as needed and sometimes the dimensions prove to be impractical. If, instead, one makes the needed adjustments electrically, at the same time the operation is made more flexible. This was suggested by Rudd[136] for a special case. A more general treatment is given here. Kuyatt and Simpson,[136a] Chamberlain,[136b] and Park and Schowengerdt,[136c] have developed similar circuits for energy loss measurements.

We distinguish two modes of operation. First, there is the *constant-acceleration* mode in which the energy of the incoming particles is increased or decreased by a constant amount while the analyzer is varied over its energy range. This also includes the special case of zero preacceleration. Second is the *constant-energy* mode. Here the incoming particles are accelerated or decelerated before analysis by an amount that varies with the energy to be observed so that analysis always takes place at the same energy. Since the resolution is directly dependent on the energy, the constant-energy mode ensures a constant resolution over the energy range. This is a definite advantage in some cases, such as the study of line spectra.

Table VI.

Conditions on analyzer dimensions to make it direct reading.

Type of analyzer	Analyzer constant $C = qV_p/E_0$	Condition for $C = 1$
Parallel plate	$2s/\ell_0$	$\ell_0 = 2s$
Spherical	$b/a - a/b$	$b/a = 1.618$
Cylindrical	$2 \ln b/a$	$b/a = 1.649$
Cylindrical mirror	$0.77 \ln b/a$	$b/a = 3.67$

The various analyzers can be made direct reading in either mode with the proper electrical connections. Two voltage sources are required; one a fixed voltage V_f and the other variable in accordance with the relation

$$V_v = rV' + V_c, \tag{49}$$

where r is an adjustable parameter, V_c is a constant voltage, and V' is the voltage that is varied to carry the analyzer through the desired energy range. For example, such a voltage source could be conveniently obtained from a simple circuit resembling that of Figure 13, where r is simply the voltage divider ratio. We distinguish next between spherical and cylindrical analyzers on the one hand and parallel-plate and cylindrical mirror analyzers on the other. In the latter type, incoming particles must pass through one of the electrodes and are thus changed in energy by the voltage on that electrode *before* entering the analyzer.

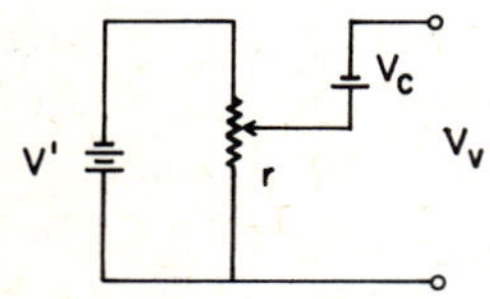

Figure 13. Simple circuit to give a potential varying in accordance with equation 49.

3.2.6.1 SPHERICAL AND CYLINDRICAL ANALYZERS. The electrodes of spherical and cylindrical analyzers are usually operated symmetrically in voltage about the potential V_s of the entrance slit, so that charged particles enter approximately along an equipotential at the same potential as the entrance slit. The connections are shown in Figure 14*a*.

The equation for the central ray for any analyzer is $qV_p = CE$, where C is the constant of the analyzer, and E is the energy of the entering particles. Hence for spherical or cylindrical analyzers with the entrance slit at V_s we have

$$qV_p = C(E - qV_s), \tag{50}$$

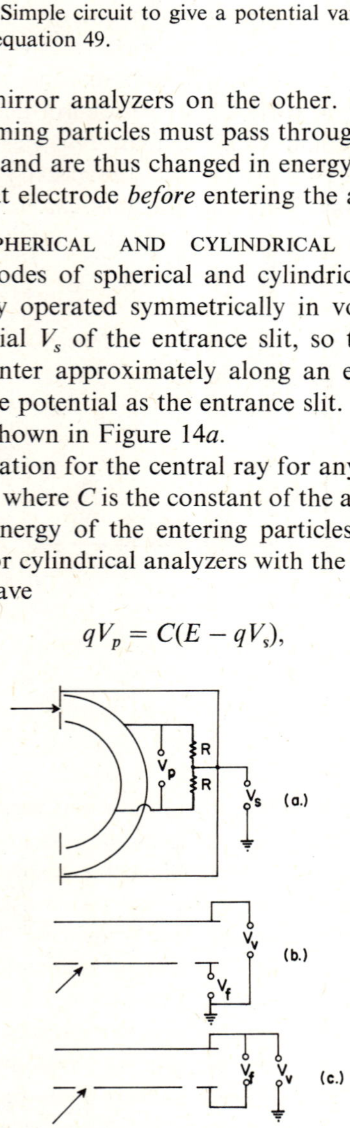

Figure 14. (*a*) Connections of a spherical or cylindrical analyzer to ensure that the curved plate electrodes are operated symmetrically in voltage about the entrance and exit slit voltage. (*b*) Connections for a parallel-plate analyzer for use in the constant-acceleration mode. (*c*) Connections for a parallel-plate analyzer for use in the constant-energy mode.

where E is the energy of the particles when they are in a region at zero potential. For a direct reading analyzer, $E = qV'$. Combining this with equation (49) we have $V_v = rE/q + V_c$. For the constant-acceleration mode, V_p is supplied by V_v and the slit voltage V_s by V_f. Then from equation 50, $q(rE/q + V_c) = C(E - qV_f)$. Equating coefficients of the variable terms, we obtain $r = c$, and equating the constants we have $V_c = -CV_f$.

For the constant-energy mode, V_p is instead supplied by V_f and V_s by V_v. Rewriting the equation and equating coefficients we now have $r = 1$ and $V_c = -V_f/C$.

3.2.6.2. PARALLEL-PLATE AND CYLINDRICAL MIRROR ANALYZERS. There are several ways of connecting the voltage supplies to the two plates for parallel-plate and cylindrical mirror analyzers, but the ones that have been found to work out best are shown in Figures 14*b* and 14*c*. For the constant-acceleration mode, the connections are the same as those in Figure 14*b* with $r = C$ and $V_c = (1 - C)V_f$. The connections of Figure 14*c* are for the constant-energy mode and $r = 1$ and $V_c = V_f(C - 1)/C$.

It should be noted that when preacceleration (or deceleration) is required for either mode, suitable electrostatic lenses must be provided, but this subject is adequately treated elsewhere[136d] and is not considered here.

3.2.7. *Effect of Sweep Speed on a System with a Time Constant*

In some analyzing systems a continuous sweep is made over the energy range to be covered by the analyzer. If the detection system has a time constant, the effective shape of the analyzer transmission function is distorted, since the detector cannot follow instantaneous changes. An example of such a situation is the use of a count rate meter with its time constant as part of the detection system.[136] In such a case it is important to know how the distortion may be minimized or corrected for. We present here an analysis for two shapes of analyzer function—triangular and square. The effects for trapezoidal and most other functions should fall between these. The important effects are a reduction of the height of the maximum, a shift of the center, and an increase in the FWHM.

The differential equation for the output voltage V of the detector as a function of time for an input voltage function V_i and a time constant T is $dV/dt = V_i/t - V/T$. Consider first a triangular analyzer function as an input. This would result from sweeping at constant speed over a monoenergetic line with an analyzer with equal slit widths and no aberrations. Referring to Figure 15, the exact solutions of the equation are

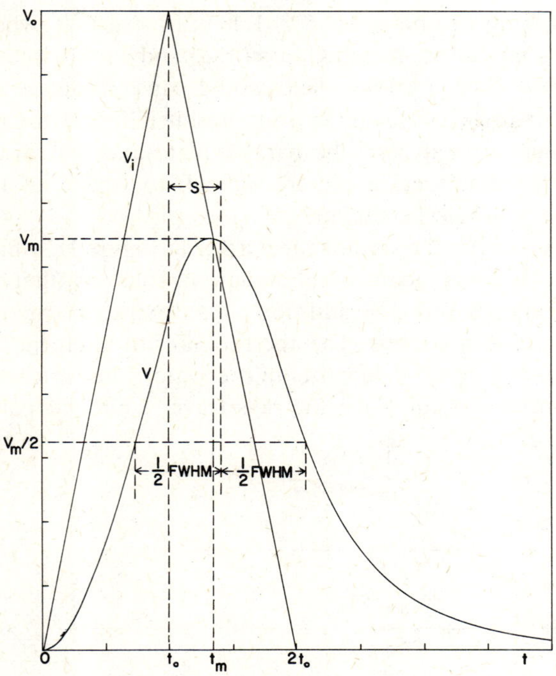

Figure 15. Effect of sweep speed on a triangular analyzer function. Triangle V_i is the input voltage function to a detection system with a time constant T; V is the output voltage function when $T = \frac{1}{2}t_0$.

$$\frac{V}{V_0} = \begin{cases} \dfrac{t}{t_0} - \dfrac{T}{t_0}(1 - e^{-t/T}) & \text{for } 0 \le t \le t_0, \\ \dfrac{T}{t_0}(1 - 2e^{t_0/T})e^{-t/T} - \dfrac{t}{t_0} + 2 + \dfrac{T}{t_0} & \\ & \text{for } t_0 \le t \le 2t_0, \\ \dfrac{T}{t_0}(1 - 2e^{t_0/T} + e^{2t_0/T})e^{-t/T} & \text{for } t \ge 2t_0. \end{cases}$$

The maximum of the curve always occurs in the region between t_0 and $2t_0$ and in fact must lie on the undistorted straight line function. It is given by

$$t_m = t_0 + T \ln (2 - e^{-t_0/T})$$

$$\frac{V_m}{V_0} = 1 - \frac{T}{t_0} \ln (2 - e^{-t_0/T}).$$

The area under the distorted curve is exactly the same as that under the undistorted curve, and thus determinations of absolute cross sections can be made without having to correct for sweep speed distortion, provided that measurements are made of areas.

Expressions for the shift S of the center and the FWHM cannot be obtained explicitly but must be obtained by approximations or numerically. An approximation good for small time constants gives

$$S \cong T - T(2 - e^{-t_0/T})e^{-(t_0/2T+1)},$$

$$\text{FWHM} \cong t_0 + T \ln 2.$$

A more exact calculation yields the results graphed in Figure 16.

For the square resolution curve of width t_0, the solution is

$$\frac{V}{V_0} = \begin{cases} 1 - e^{-t/T} & \text{for } 0 \le t \le t_0, \\ e^{-t/T}(e^{t_0/T} - 1) & \text{for } t \ge t_0. \end{cases}$$

The maximum here always comes at $t = t_0$ and is of height $V_m/V_0 = 1 - e^{-t_0/T}$. For the square resolution curve it is possible to find exact expressions for S and FWHM. They are

$$S = T \ln 2 - \tfrac{1}{2}T \ln (1 + e^{-t_0/T}),$$

$$\text{FWHM} = t_0 + T \ln (1 + e^{-t_0/T}).$$

Figure 16 also shows a plot of these results.

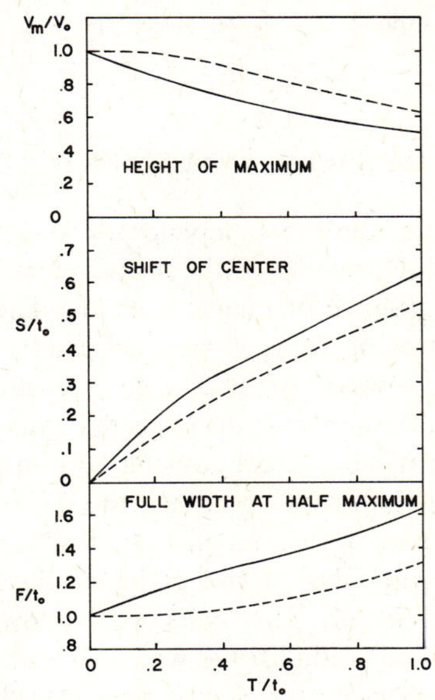

Figure 16. Effect of sweep speed and detector time constant on the height of the maximum, shift of the center, and the full width at half-maximum: full curves, triangular analyzer function; dashed curves, square analyzer function.

In general, the shorter the time constant or the slower the sweep speed, the more faithful the reproduction of the resolution curve. In practice a fairly good approximation is obtained when the time constant is less than about one-fifth of t_0 (or of the FWHM). Figure 17 gives two actual runs over the helium autoionization electron spectrum with time constants of 10 sec and 1 sec. These correspond to $T/t_0 \simeq 1.4$ and 0.14, respectively. The loss of resolution with the larger time constant is evident.

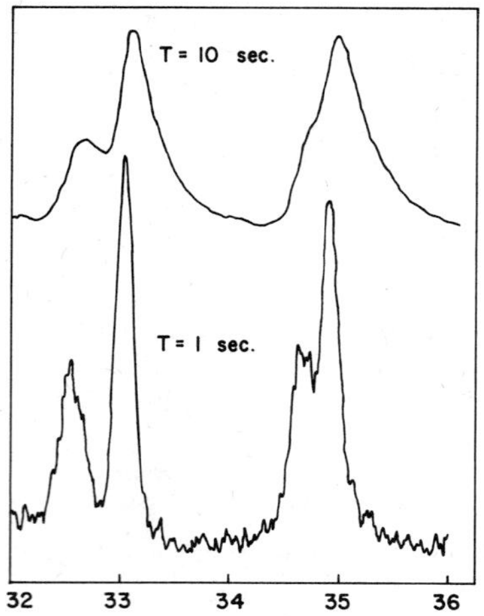

Figure 17. Helium autoionization electron spectrum using parallel-plate analyzer with approximately a triangular transmission function. Both curves traced at the same sweep speed (0.025 V/sec) but with different time constants as shown. Note apparent loss of resolution when long time constant is used without a correspondingly slower sweep speed. Energy scale is uncorrected.

4. MAGNETIC FIELD ANNULMENT*

Electrostatic analyzers depend for their operation not only on the presence of the proper electric field but also on the absence of magnetic fields. The magnetic force on a moving charged particle causes difficulties in at least two ways; (*a*) it changes the effective constant of the analyzer from the expected value, and (*b*) it deflects the particles before, during, and after analysis so that fewer than expected reach the detector.

As an example of the first effect, consider a magnetic field B perpendicular to the plane of the orbit in a cylindrical analyzer. This causes an additional outward (or inward) radial force which must be compensated by an extra inward electric field in order that the particles still reach the exit slit. The increment of electric field is equated to the magnetic force per unit charge, giving

$$\frac{\Delta V_p}{r}\left(\ln\frac{b}{a}\right)^{-1} = B\left(\frac{2E_0}{m}\right)^{1/2}.$$

Then since $V_p = (2E_0/q)\ln(b/a)$, we can write

$$\frac{\Delta V_p}{V_p} = \frac{Bqr}{(2mE_0)^{1/2}}.$$

The fractional change in the plate potential V_p is the same as the fractional change in the analyzer constant C. As an example, for electrons in an analyzer with $r = 5$ cm we obtain $\Delta C/C = 1.48 \times 10^{-3}\, B/E^{1/2}$, where B is measured in milliGauss (mG) and E_0 in electron volts. Then a 10-mG field would cause about a 1% error in C at 2 eV. The same result is found for the spherical analyzer. The parallel-plate and cylindrical mirror analyzers are more difficult to treat, but the results would be similar.

To clarify the second effect, refer to Figure 18, which illustrates a geometrical situation such as that in Figure 12 with the addition of an extra obstructing aperture of width s. The aperture fails to intercept the beam when no field exists in the region; but interception does occur when the rays traverse circular paths

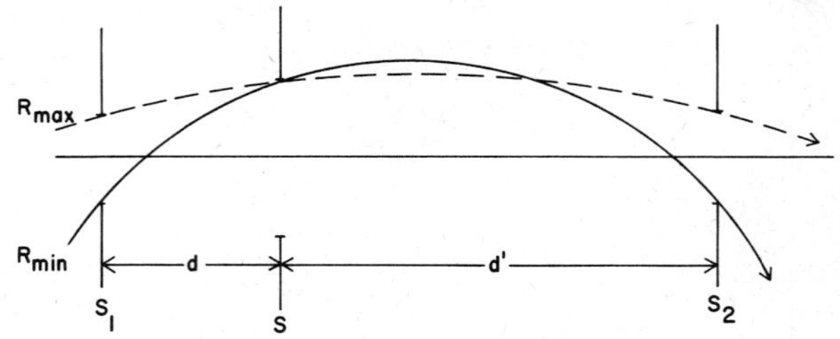

Figure 18. The effect of a magnetic field on the transmission of charged particles by collimating slits s_1 and s_2. If the radii of curvature of the orbits are smaller than R_{min}, obstructing aperture s prevents the transmission of particles. For radii greater than R_{max}, all particles are transmitted.

as they do in a magnetic field. Even in the absence of such apertures, comparable obstructions always exist in the form of walls of tubes, and so on. Shown in Figure 18 are two extreme paths. The dashed ray has the greatest curvature for which all the beam can be passed through the aperture s_2. The solid ray has the smallest curvature for which the entire beam is cut off. The radii of curvature of these extreme rays can be calculated by an application of the sagitta formula. For the case $s_1 = s_2$ we find $R_{min} = dd'/(s + s_1)$ and $R_{max} = dd'/(s - s_1)$. The relation between the radius of curvature R, energy E_0, and magnetic field is $B = (2mE_0)^{1/2}/R$. For electrons this becomes $B = 3370\, E_0^{1/2}/R$ where B is in milliGauss, E_0 in electron volts, R in centimeters, and m is taken to be equal to m_0 which is the electron rest mass. The transmission through this slit system is 100% until the magnetic field is great enough to give a radius of curvature equal to R_{max} and then it drops to reach zero at the field corresponding to R_{min}.

In the United States the horizontal component of the earth's magnetic field is approximately 200 mG and the total field is 500 to 600 mG. For most work involving analysis of ions at low energies and electrons at energies below several hundred electron volts, the earth's field must be eliminated or reduced. This can be done by placing the apparatus inside a system of

* The author (M. E. R) thanks R. Spohr for the use of his bibliography on magnetic field compensation.

current-carrying coils designed to provide a field equal to the earth's field but in the opposite direction. The same result can be accomplished by enclosing the apparatus in a high-permeability magnetic shield. The latter method is not discussed here; the reader should refer to the excellent article by Wadey[137] and the further references given there.

At the center of a single circular coil of wire is a field that is relatively uniform, since the first derivative of the field with respect to axial displacement vanishes there. However, the region of uniformity in this simple system is not great enough for most applications. The next most complicated coil system consists of a pair of identical coaxial coils separated by a distance equal to their radii. A coil pair such as this, usually attributed to Helmholtz, has vanishing first, second, and third derivatives on the axis halfway between the coils. This configuration, combining simplicity, reasonably good uniformity of fields, and easy access to the working region inside, has been widely used to annul the earth's field in the region occupied by various types of experimental apparatus, including electrostatic analyzers. In some cases a single pair of coils is tilted so that the axis is aligned with the earth's field. Since it is rather difficult to adjust two angular degrees of freedom and the current in the coils simultaneously, it is generally better to use three pairs of coils with their axes mutually perpendicular. Each then annuls one component of the earth's field and is independent of the action of the others, at least at the center point.

Bock[138] has discussed the uniformity of the field both on and off axis for a Helmholtz pair. Firester,[139] who suggested that square coils are easier to construct and allow better access to the interior, showed that the spacing yielding the best uniformity at the center (which we will call the Helmholtz spacing) for square coils is 0.5445 times the length of one side. His article lists contours for constant deviation of the axial component of the field both on and off axis. Ference et al.[140] have used circular coil pairs to demonstrate that the uniformity for points off the axis is better when the spacing of the coils is somewhat less than the Helmholtz spacing. But for a particular pair of "thick" coils, Crownfield[141] has shown that a given uniformity on the axis can be obtained over a larger distance if the spacing is increased above the Helmholtz spacing. Rudd and Craig[142] confirmed this for "thin" coils both round and square and showed that, if the separation is 20% greater than the Helmholtz spacing, the field is uniform to within 1.5% over the entire axis from one coil plane to the other. Cacak and Craig[143] have recently calculated total field deviation contours for single round and square coil pairs at various spacings and also for three pairs of perpendicular coil pairs. In addition, they give the maximum spherical volume for 0.1, 1, and 5% uniformity as a function of coil spacing and find that these curves usually peak quite close to the Helmholtz spacing (see Figure 19).

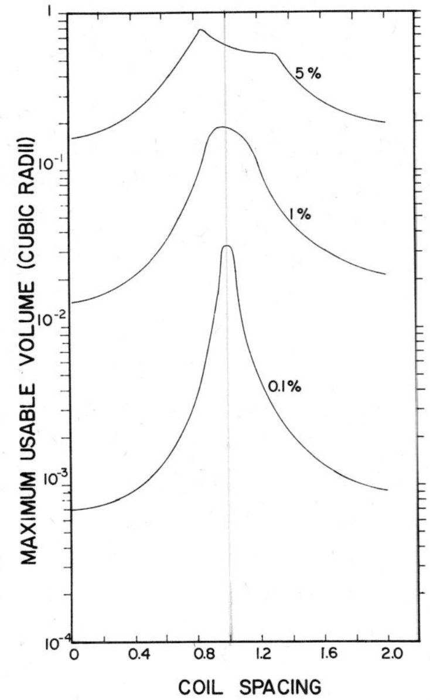

Figure 19. Maximum spherical volume of a given uniformity in a system of three mutually perpendicular round coil pairs as a function of their spacing. (From Ref. 143.)

Additional coils can be added to the Helmholtz pair for better uniformity over a larger region. This, of course, increases the complexity and further limits the accessibility. A system of three coaxial coils, the center one having half the number of turns of the outer two, is claimed by Barker[144] to have twice the size of the region of uniformity as a Helmholtz pair. Braunbeck[145] devised a system of two pairs of coaxial coils, which gives a considerable increase in uniformity of field over a single pair. Such an arrangement is often used in cases where very high uniformity of field is desired (see e.g., Scott.[146])

Fanselau[147] has treated systems of one, two, and three pairs of coils, all coaxial, and a four-coil system has been worked out by Lee-Whiting[147a] for the Chalk River iron-free instrument. Bloom et al.[148] suggested the use of octagonal coils which, they point out, are not more difficult to build than square ones, since corner bracing is needed in the latter case. They prove an interesting theorem:

> In any system of N coils designed to cancel all field derivatives up to the $2N$th, if the coils are to be in the shape of regular polygons, then the polygons must have not less than $2N$ sides.

This would indicate, for example, that not much is to be gained in going from one square coil pair to two. A very general treatment of axially symmetric fields is given by Garret,[149] who treats the field in general without reference to its source. Then, when an optimum field that is consistent with Laplace's equation has been designed, a suitable set of currents is sought to provide the field. After such an analysis he suggests two systems of high uniformity. One is an elongated Helmholtz pair which is rather like a solenoid with a gap at the center. The other is a short solenoid with double windings on the ends. Rubens[150] has proposed a set of five equally spaced square coils with unequal numbers of turns which form a cube. Perhaps the ultimate in uniformity is a system described by Clark.[151] He notes that the field everywhere inside an ellipsoid is uniform if coils are wound on the outside in such a way that each unit of length measured along the axis of the coils contains the same number of turns. This is true even if the axis of the coils does not coincide with an axis of the ellipsoid. He has mounted an experimental apparatus inside a sphere wound in such a manner.

Generally speaking, improvements in field uniformity are bought at the price of greater complexity, inaccessibility of the exterior volume, or both.

5. ELECTRON ENERGY ANALYSIS IN ELECTRON MICROSCOPY

The electron microscope is an instrument with which one can investigate objects and materials of interest by scattering a parallel beam of high energy electrons, instead of photons, and recording the image in the focal plane of the instrument. It is possible to obtain magnifications as high as 10^5 to 10^6, depending on the quality of the optics and the energy of the electron beam.

In the scattering process some of the electrons will suffer energy losses and at the same time be scattered through some angle (see Chapter 8). The resulting energy spectrum of the electrons may then be analyzed using electron optical methods. In this way the electron microscope may be modified to become an energy analyzer of extremely high energy resolution. Such instruments have thereby come to play an important role in the investigation of these scattering phenomena.

The energy analysis may be made after the electrons have been focused in the image plane[152–154], and therefore without altering the microscope focusing properties. The alterations of interest to this section, however, are those where an energy analyzing element is introduced into the microscope itself (e.g., an electrostatic or a superimposed electrostatic and magnetic field analyzer).

Section 5.1 mentions briefly the retardation methods of electron spectral analysis. Section 5.2 is devoted to the properties of the Möllenstedt analyzer, and Section 5.3 covers the characteristics of the superimposed electrostatic and magnetic field analyzer, such as the Wien filter.

5.1. Retarding Field Analyzers

Retarding field analyzers of large solid angle were mentioned in Section 3.1. Here, however, it is of interest to discuss those techniques employed to analyze the energy spectrum of a nearly parallel beam of electrons.

The instrumental arrangement in Figure 20 is that used by Boersch and Schweda.[155] The electrons are passed through the slit B, through the reflection chamber R, and into the retardation chamber G, which is placed at a certain potential $U_0 + U_G$: U_0 is the cathode potential, and U_G is a variable potential applied to the chamber. Electrons of energy $E \geq eU_0 + eU_G$ will be collected in G (giving a current I_G), and those of energy $E < eU_0 + eU_G$ will be repelled and collected in the reflection chamber R (giving a current I_R). Obviously the sum of the currents for the two chambers will be constant.

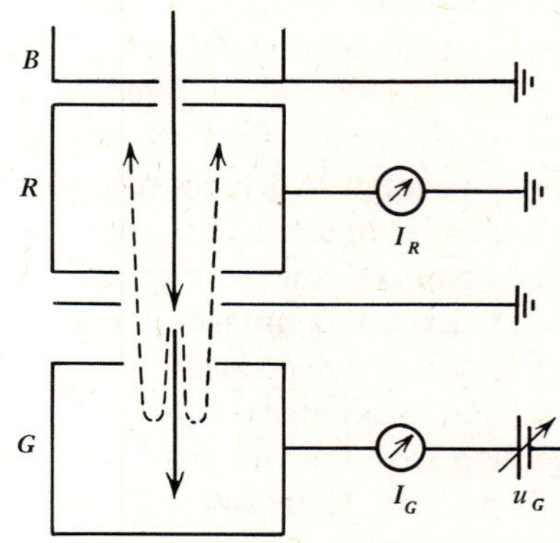

Figure 20. Retarding field analyzer as used by Boersch and Schweda.[155]

This instrument will measure the integrated currents for electron energies above as well as below the set energy $eU_0 + eU_G$. Differentiation of these with respect to energy will give the electron energy spectrum.

Chamber R may be removed and only the retardation chamber G used, as has been done by Boersch and Miessner[156] and Horstmann et al.[157] Horstmann and Meyer[158] have used such an arrangement to separate inelastically and elastically scattered electrons. The obtained resolving power for the instrument gave $\Delta E \sim 0.03$ eV.

5.2. The Möllenstedt Analyzer

The electrostatic analyzer introduced by Möllenstedt in 1949[159] was the first cylindrical electrostatic lens of this type. The resolutions of the first instruments were very high ($\Delta E/E \sim 3 \times 10^{-5}$); subsequent improvements allowed an additional improvement of a factor of two in this figure.[160] More recent arrangements[161] have boosted this to 2×10^{-6}.

The Möllenstedt analyzer has been used by many workers to investigate electron energy loss spectra.[162–164] It is also reported that modified versions of the instrument have yielded improvements on the resolution or the intensity of the detected electron beam.[165,166]

Figure 21 shows diagrammatically the trajectories of electrons of two slightly differing energies. The variation of position at which a beam of a given energy will be focused as a function of displacement x of the entrance beam from the plane of symmetry appears at the base of the figure. It is seen that there is a distance x at which the change dy/dx, and therefore also dy/dE, is

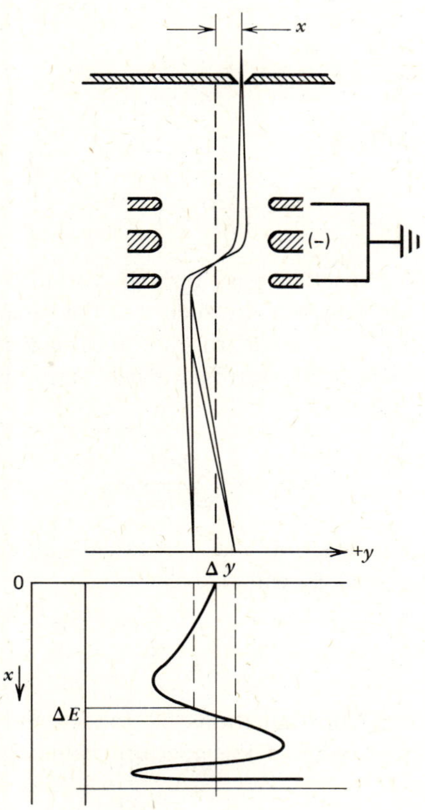

Figure 21. Möllenstedt analyzer.

moving smoothly through the plane of symmetry on the y-axis. Therefore, a small change in electron energy from a given energy E_0 will result in a slight change in y-position in the image plane.

Theoretical treatments of the characteristics of the Möllenstedt analyzer have been presented by several authors.[159,160,167–170] These articles should be referred to for more information.

The Möllenstedt analyzer analyzes the electron spectrum in only one direction in the focal plane; the perpendicular direction will show the scattering probability as a function of scattering angle.[171] The recording of the signal in this focal plane will have the coordinates of energy loss and scattering angle.

The ultimate energy resolution for a Möllenstedt analyzer will be limited by the energy distribution of the electrons from the electron source. Thermal emission electrons have been shown to have an energy distribution of ~ 0.2 to 0.6 eV in width.[156,172–174] For this reason, an analyzer must be used before the object to reduce this width.[168] In this way it is possible to reduce the width to between 10 and 20 meV and perhaps even less.

5.3. Superimposed Electrostatic and Magnetic Field Analyzers

The possibility of constructing electron energy analyzers employing electrostatic and magnetic fields has been considered from time to time.[175] However, the very successful Wien filter, which is of this type, was first reported in 1897.[176] It has served in investigational work since then (e.g., Ref. 177), but it has recently been further developed and employed with particular success.[178–180] For this reason this instrument merits special consideration in this section; a second, more recent instrument is also referred to.

Other instruments using the superposition of electrostatic and magnetic fields employ the focusing properties of one field to correct for aberrations in those of the other. This is the case in Section 3.2, where an electrostatic corrector is used in a magnetic spectrometer. Another application of this type is the correction of aberrations in an electron mirror by employing the two kinds of fields simultaneously.[181]

5.3.1. *Wien Filter*

The Wien filter consists of a magnetic and an electrostatic field which are placed perpendicular to each other. Energy analysis of an electron beam may occur when the beam passes into the crossed field volume perpendicular to the two fields.

The electron energy analyzing characteristics of a Wien filter are shown in Figure 22. Following the

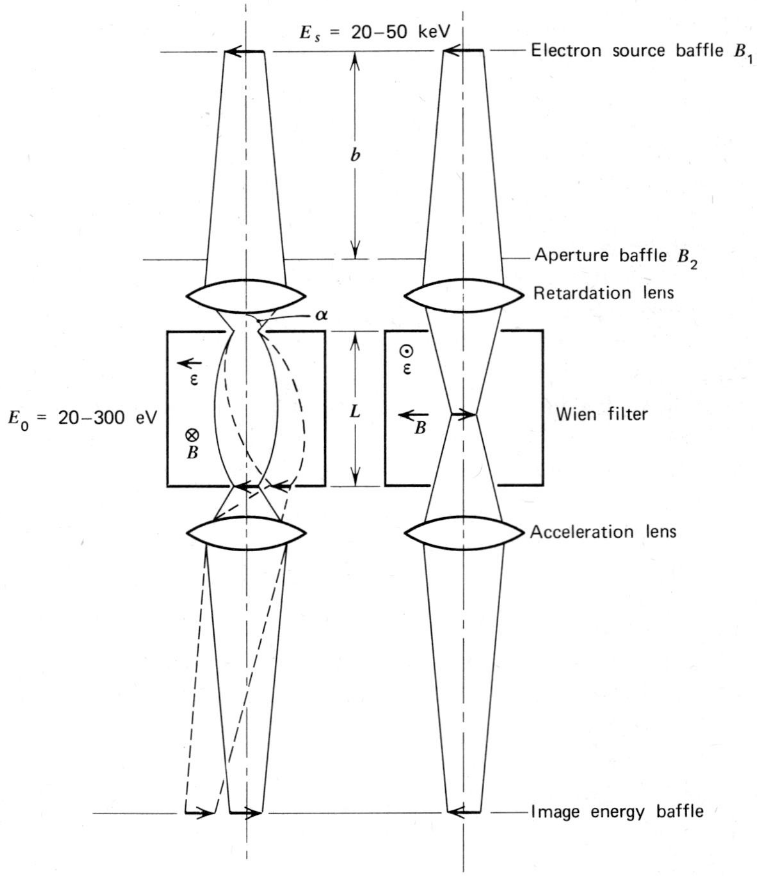

Figure 22. Wien filter analyzer used by Boersch et al.[182]

analysis procedure by Boersch et al.,[182] electrons of an initial energy between $E_s = 20$ and 50 keV are retarded in the retarding lens to an energy between $E_0 = 20$ and 300 eV, whereafter they enter the Wien filter. The lateral magnification of the filter becomes unity when the following relationship is realized:

$$\frac{L}{E_0} = \frac{2\pi}{\varepsilon}, \tag{51}$$

where ε is the electric field strength and L is the filter length. The electron source image size at the filter exit is the baffle B_1 diameter $2r_1$ times the magnification M_1 of the retardation lens:

$$\Delta y_p = 2r_1 M_1. \tag{52}$$

The filter entrance angle α will cause a width

$$\Delta y_\alpha = \frac{3L\alpha^2}{\pi}. \tag{53}$$

With r_2 the radius of baffle B_2, then

$$\alpha = \frac{r_1 M_1 + r_2}{bM_1}\sqrt{\frac{E_s}{E_0}},$$

or, since $r_1 M_1 \ll r_2$,

$$\alpha = \frac{r_2}{bM_1}\sqrt{\frac{E_s}{E_0}}. \tag{54}$$

The electrons that pass through the filter will be those whose velocities are such that the forces due to interactions with the electrostatic and magnetic fields cancel, or when the electrostatic field strength has the following value:

$$\varepsilon = -\sqrt{\frac{2eE_0}{mB}}, \tag{55}$$

m being the electron mass. A small difference in energy ΔE from E_0 will give

$$\Delta y_D = \frac{L}{\pi E_0}\Delta E. \tag{56}$$

Therefore from equations 52 to 54, and 56, the maximum opening of the aperture baffle B_2 will be

$$r_{2,\max} = bM_1\sqrt{\frac{\Delta E}{3E_s}} \tag{57}$$

for a given resolution ΔE, and the maximum extent of the electron source B_1 will be

$$r_{1,\max} = \frac{L\,\Delta E}{2\pi M_1 E_0}. \tag{58}$$

Entrance and chromatic errors due to the retardation and acceleration lenses, considered to be small, are neglected. Influence of the opening errors on the resolution can be decreased by reducing the opening B_2. The limit is given by $\vartheta \approx \lambda/2r_2$ ($\lambda = h/\sqrt{2meE_s}$), so

$$\Delta y_B = \frac{hM_1 b}{E_s r_2}\sqrt{\frac{E_0}{2me}}. \tag{59}$$

Eliminating r_2 using equation 57, we obtain the theoretical resolution of the analyzer:

$$\Delta E_{\text{theor}} = \sqrt[3]{3}\,E_0\left(\frac{\lambda\pi}{L}\right)^{2/3}. \tag{60}$$

For example, if $E_s = 25$ keV, $E_0 = 20$ eV, $L = 1$ cm, then $\Delta E_{\text{theor}} = 5.2 \times 10^{-5}$ eV. In addition, the aperture baffle opening must be $r_2 = 1.6 \times 10^{-4}$ cm, and the size of the electron source $r_{1,\max} \approx 10^{-6}$ cm.

The dispersion of the instrument may be increased by a factor of up to 3 by dividing the potential across the electrodes unevenly.

The resolving power of the Wien filter will depend on the energy width of primary electrons, resulting from thermal and anomalous energy distribution contributions.[172,173] In electron energy loss experiments, for example, this primary energy width may be improved by passing the electron beam through a Wien filter monochromator before the scattering site. This will increase the resolving power by a factor of as high as 50 or more.[183] For an analyzer with $\alpha = 6 \times 10^{-3}$ rad and $E_0 = 20$ eV, the theoretical resolution of 13 meV compares well with the experimental value of 17 meV.[184]

5.3.2. *Antiparallel Field Analyzer*

A second energy analyzer is being worked on in London.[185] Here the analyzer consists of the magnetic and electrostatic fields oriented antiparallel to each other and both perpendicular to the electron beam. Deceleration from the initial kinetic energy of 75 keV to about 1 eV is caused before entrance into the analyzer and acceleration to 75 keV again takes place after exit. Preliminary measurements show a resolution down to ~10 meV.

6. ELECTRON DETECTION DEVICES

Electron detection at energies below 50 keV, and especially below 20 keV, has often posed a problem to workers in electron spectrometry. The methods employed with advantage at higher energies may be used favorably when detector construction refinements are applied (e.g., the G–M counter) or with difficulty when the detector is of another type, even if the instrument is precision constructed and the associated electronics accurately made (e.g., solid state detectors, scintillation counters). Of the two groups, the former is the more commonly used. The characteristics of these are enumerated in Section 6.1. For comparison, the suitabilities of semiconductor counters and scintillation counters for low energy electron detection are given in Sections 6.2 and 6.3.

Another type of detector that has proved to be useful in the detection of low energy electrons is the open-ended multidynode electron multiplier. With this type of device, which can be used either as a dc current amplifier or as a single electron counter, one can detect electrons of energies as low as 100 eV. A general survey of these instruments is contained in the article by Kuyatt,[132] which also treats the Faraday cup electron collector. However, the continuous channel electron multiplier, which may be considered in this group, is discussed in Section 6.4, because of its increased use as a detector at very low electron energies.

Finally, the use of photographic emulsions in electron detection at low energies is presented in Section 6.5. This treatment is divided into the integral (film blackening indicates electron intensity) and differential (the individual electron track density indicates intensity) electron detection techniques.

In the remainder of this chapter the application of pre- or postacceleration techniques to improve detection efficiency and special modifications of the detectors to enhance performance (e.g., to give position sensitive detection) are cited.

Reviews of charged particle detectors have been presented in the past, but their use for higher energies, $E \gtrsim 30$ keV, has usually been stressed. These articles may be consulted for further reference.[186]

Reports of actual applications of various methods of low energy electron detection are to be found in articles dealing with low energy electron spectrum analysis, such as Auger and Coster–Kronig spectrum investigations. The review of such work in Chapter 3. Sections 5.2 and 5.3, includes information about methods of electron energy analysis and detection. The reader is referred to that chapter for further references.

6.1. Gaseous Ionization Counters

An electron passing through a gas will dissipate its energy primarily through ionization of the gaseous atoms or molecules. If this is allowed to take place in an imposed electrostatic field, the positive and negative charges in the ionization track will become separated and collected at the appropriate electrodes. A high field may be used to accelerate the ionized gas particles or electrons, to produce additional ionization in the counter gas through collision. In this way a proportional counter may give an amplification of an initial pulse by a factor of up to 10^5 with little loss of proportionality. The resolution of this detector is of the order of a few keV,[187–192] depending on the amplification and the nature of the counter gas. Also, there is a small shift in gain with counting rate,[193] which can, however, be effectively eliminated.[194] For additional treatments of the characteristics of these counters, see Refs. 195, 196, and 189.

Even though this counter is invaluable in energy discrimination of both electrons and soft X-rays, its resolution is not high enough to allow it to function as other than an electron detector in high resolution spectrometry.

When the acceleration field is further increased in magnitude, the pulses are no longer amplified proportionally, a pulse from a G–M counter will then indicate the detection of an electron, but not its energy.

The G–M counter is the most commonly used device in electron spectrometry above 5 to 10 keV when thinnest windows are employed. Below this energy, counter window penetration difficulties may be reduced or eliminated with pre- or postacceleration techniques. The construction of several such detectors has been reported,[197,198] and their characteristics and properties have also been considered by a number of researchers: the angular dependence of detector sensitivity,[199] the avalanche effect and secondary electron production for different geometries,[200,201] counter background,[202] counting characteristics at high intensities,[203] characteristics of counter dead time,[204] resolving time,[205] and the suitabilities of various filling gases.[189,206]

The electron detection sensitivity of a gaseous counter is mainly limited by the thickness of the window used. Electron detection efficiencies are usually determined experimentally by measuring the change in counting rate of a known flux of electrons as a function of electron energy. This efficiency is in fact the product of the window penetration and the detection efficiency of the counter for the penetration electrons of whatever energy distribution they may have after traversing the window. This sensitivity falls off rapidly when the electron energy gives a penetration range that is comparable to the window thickness. Stopping powers of films measured in this way will actually reflect the efficiency of the window-film–counter-gas combination as a counter. The cut-off energy of a window will be the initial electron energy below which no electrons are detected.

The possibility of using no window over the detector has been investigated by Langer et al.[207] It was found that the gas escaping through the detector slit acted as an insensitive region and the cut-off was equivalent to that of a film of $\sim$1-keV thickness.

6.1.1. Counter Windows

The properties of thin films as counter windows have been treated by several authors.[208] The investigation of the transmission properties of thin films for electrons of energies 0 to 50 keV presented by Lane and Zaffarano,[209] provides a very complete treatment of the topic. In this work, the transmission curves for films of a given material were found for various thicknesses. A curve thus derived could then be used as an intensity correction curve. It was demonstrated by these authors that the transmission curves for a nonconducting film measured as a function of increasing or decreasing electron energy were different. This "hysteresis" curve (see Figure 23) resulted from charging of the foil by the nontransmission of low energy electrons when using increasing energy for the measurement. This difficulty can be overcome by evaporating a thin conducting coat of metal on the window or by using a fine mesh support grid. The latter is often used for very thin, nonself-supporting windows.

An energy dependent transmission curve is, of course, closely related to the mean range-energy relationship for the material in question. Many experimental investigations of this relationship have resulted in subsequent empirical or semiempirical expressions for the range-energy relationship (e.g., the article by Feldman[210] for 1 to 10 keV energies and also Refs. 211, 212). It is even possible to relate a known mean range value for a given window material and electron energy through a definitive equation, thus obtaining a fairly accurate intensity correction curve for the film. This alternative has been considered for $E > 10$ keV by Subba Rao.[213]

Films with very low cut-off energies (especially $\gtrsim 20$ keV) are not self-supporting when the detector window is too large. Here, then, a conducting grid must be placed under the film for the necessary support. Such grids are available, constructed of exceedingly thin wire, and their transmissions can reach as high as 95%.

Backscattering of electrons from increasingly thin films becomes proportional to the film thickness, and

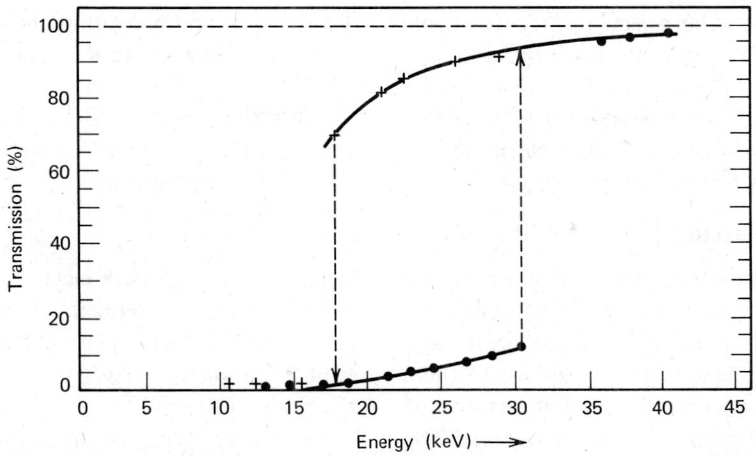

Figure 23. Transmission curve for a formvar film, 0.3 mg/cm², with no aluminum coating, showing a hysteresis effect: dots, increasing energy; pluses, decreasing energy. (From Ref. 209.)

Just et al.[214] have determined film thickness from backscattered electron intensity measurements by exploiting this property. However, the backscattering probability is usually small. (See Archard's[215] treatment.) In any case, if the experimentally determined transmission function of a given counter window is determined uniquely, the effect of backscattering will be included.

Vobecký and Dragoun[216] have compiled a very useful description of the preparation and properties of some seven types of thin films appropriate as counter windows as well as for electron source backings.

6.1.2. Pre- and Postacceleration Techniques and Position Sensitive Detection

Detector window transmission correction at low energies may be overcome by pre- or postaccelerating the electrons to be detected. With preacceleration, the electrons are accelerated through an electrostatic field before energy analysis takes place. Postacceleration is the use of acceleration after analysis and just before detection.

Preacceleration has been employed, for example, by Dionisio, Albridge, and others.[217,218] Since a semicircular spectrograph was used for the electron energy analysis, pre-, rather than postacceleration was advisable. The electric field vector in the vicinity of the source must parallel as closely as possible the electron trajectories so that no spurious focusing effects will be introduced. Such effects were found by Novakov and Hollander (see Section 4.2.2 of Chapter 7) when a potential of several kilovolts was placed on the source.

Techniques of post acceleration instituted just prior to electron detection, have been employed by several workers,[118,219–223] (see also Chapter 3, Sections 5.2 and 5.3). Mehlhorn and Albridge[198] used a ~100-V deceleration voltage before their 8.6-kV acceleration voltage in order to reject extremely low energy electrons and reduce the background counting rate. Although the background was strongly reduced, there remained a ~15% excess intensity over the transmission curve for the window used. This was due either to electron emission within the acceleration chamber, or, more probably, to the focusing into the counter of some of the electrons diverging after passing the focal plane. (See also Ref. 223.) Similar effects were found by Krisciokaitis and Haynes.[219]

From Section 6.1 it is seen that the pulse size for the cylindrical geometry detector is dependent on the intensity of the field gradient and therefore also on the radius of the central wire of a proportional counter. Grennberg[224] has exploited this effect to make a position sensitive radiation detector. Although his counter was used for X-ray detection, it may also be applied for electron detection. A central wire was used whose radius varied several orders of magnitude over its length, and the detector pulses were lead to a multichannel analyzer. The strongest limiting factor here is the statistical spread of pulse amplification for any one given energy. For copper K X-radiation, the resolution due solely to a statistical contribution was estimated to be about 12%. The experimentally obtained resolution was ≤ 20%.

A second position sensitive proportional counter, the performance of which is not dependent on pulse heights, has been developed by Morgue and Chery.[225,226] Here an incoming ionizing particle, alpha-particle or electron, will create a path of ion-electron pairs. Pulses from the electron cascade to the counter's central electrode will reach the two electrode ends with a time difference depending on the position of particle entry into the counter. A time-to-pulse height spectrum, giving the relative pulse delay spectrum, recorded in a multichannel analyzer, will give the position (and thus also energy) spectrum of the detected particles. For the electronic and detector arrangement employed, alpha-particle detection positions could be determined to about $\pm$ 1 mm over its utilized 75-mm length.

6.2. Semiconductor Detectors

The semiconductor radiation detectors have achieved increasing importance in recent years. In some laboratories they have largely replaced scintillation and gaseous detectors in nuclear reaction studies. In these studies, alpha- beta-, and gamma-radiations in the 2 to 50 MeV range, however, were of interest. The properties of such detectors have been reviewed by several authors, and their articles should be referred to for further information.[227,228]

The energy required to produce an ion pair in a semiconductor is dependent on the electronic structure of the material used, namely, on the energy gap value for the material. This parameter depends in turn on other factors, such as the crystalline structure, the purity, the amount of doping (if employed), the temperature, and so on.

Pure germanium has an energy gap of some 0.67 eV, and 2.94 eV of electron or gamma-ray energy is required to create one electron-hole pair. The corresponding energies for silicon are 1.12 eV energy gap and 3.75 eV for pair-hole production.[229,227] These figures for electron-hole pair production energy are on the average almost 10 and 100 times smaller than the equivalent quantity for gaseous and scintillation detectors, respectively. Therefore, statistical fluctuations in the charge production process are small. Also, pulse amplifier noise here is reduced by a factor 10 over the gaseous counter, since charge flow is larger by the same factor. However, the unfortunate difficulties with collection of ionization in solids necessitate rigid requirements on the character of the solid.[227]

The average hole-electron pair production energy should be as small as possible. In a given detector material, silicon or germanium, 5 valency atoms (donor) may replace some of the atoms in the lattice, the resulting material being called *n*-type. If 3 valency atoms are used, acceptor holes are produced, and the material is referred to as *p*-type. The effect of donors is to introduce permitted discrete levels in the impurity atom vicinity and, in an energy band model, in the forbidden gap close to the valence band. Such donor and acceptor atoms are easily ionized, the ionization energies being factors of $\sim$20 and $\sim$60 less than for pure silicon and germanium, respectively. The donor lithium, interstitial rather than replacement, is particularly suited for drifting and gives the ionization energies of 0.033 eV and 0.0093 eV in silicon and germanium, respectively.[227]

Electrons down to 30 keV energies may be detected with a peak width of $\sim$ 2 keV with Ge(Li) counters. The Si(Li) detector will apparently function satisfactorily down to $\sim$ 50 keV as an electron detector. Hollander[230] has, however, used a Ge(Li) gamma-detector and a Si(Li) electron detector for $(e^- - \gamma)$ coincidence measurements. The detected electron energies were $\gtrsim$80 keV. The efficiency of electron detection by a Si(Li) detector is $\sim$50% at 60 keV.[231] (See also Ref. 232.)

Of course, temperature is very important in producing the sensitive lithium drifted detectors.[233] The drifting is carried out over extended periods of time at temperatures of several hundred degrees centigrade. When the detector is finally prepared, it must be stored at low temperatures to reduce further diffusion to a minimum. The lithium density at the surface of a detector is especially sensitive to temperature and other perturbing parameters. Reduced concentration can result in an effectively insensitive region over the first microns of the surface depth, and the presence of such a region will have a strong effect on low energy particle detection efficiency.

Another problem is the attachment of an electrically conducting layer to the detector surface. These are often too thick to allow effective penetration and detection of lowest energy electrons.

As a partial remedy for both these difficulties, the depletion region may be reduced and the contact layer produced by implanting ions of an appropriate element into the surface. Gunnersen[234] has obtained 0.22(2)-μm thick windows (i.e., contacts) on Ge(Li) detectors through implantations of 80-keV ^{69}Ga. Work of this nature has also been carried out in Karlsruhe.[235] (See also Ref. 237.)

It should also be mentioned that the Ge(Li) counter itself contributes 0.1 keV to the resolution; electronics adds a constant noise equivalent to $\sim$0.5 keV. Thus, for 10-keV electrons, the overall resolution is $\sim$5%, and for 100-keV electrons, 0.5%.[234] Noise has been shown to drown out signals for electrons of energies below $\sim$50 keV.[236] See Ref. 238 for an early review of limitations of various radiation detectors.

As in the case of gaseous counters, such problems as fluctuations in ionization[189,239] and leakage and current noise[240] limit the use of semiconductor detectors for low energy electron detection, but the limitations are less serious for the semiconductor devices. References 227, 228, 188 are cited for a coverage of such drawbacks.

In recent years reports have been made of a semiconductor radiation detector that gives internal amplification through the application of a high electrostatic field.[241,242] For short-range particles, (stopped mainly in the thin acceptor region of a detector), amplification factors of up to 10^3 are obtained, whereas minimum ionizing particles give factors of 20 to 50.

Several position-sensitive semiconductor devices have been reported. These are mostly for particles of higher energies. They have functioned as alpha- and heavy-particle detectors employing time delay coincidence between pulses coming from the back and the face of the detector.[243–245]

An array of ten semiconductor detectors has been tested for electron detection in a $\pi\sqrt{2}$ iron-free spectrometer by Nilsson et al.[92] Each detector, of an array positioned in the spectrometer focal plane, could be used independently. Also tested was a multistrip source with each strip placed at an appropriate potential to cause electrons of the same original energy to strike at the same position in the detector plane.

Semiconductors, or solid state detectors, are of great value and use as heavy-particle, X-ray, gamma-ray, and high energy electron detectors. The research carried out to determine how the applications may be made most efficient has also contributed much to the understanding of the nature of semiconductor materials. However, this knowledge has not yet been focused on the problem of electron detection for energies below some 50 keV.

6.3. Scintillation Counters

The response of sodium iodide, anthracene, and plastic scintillators to electrons of low energy has only recently been investigated in any detail. As detectors of individual electrons, scintillation counters give mean particle energy dissipation in steps of ~300 eV per energy loss, which is some 100 times larger than that for single crystal semiconductor detectors, as mentioned in the previous section. Hence the statistical fluctuations in total light output in the scintillators will be appreciably greater. On the other hand, this is partially counterbalanced by the response (light output/keV) showing a maximum in the vicinity of 15 to 20 keV,[245] where it is some 20% larger than that at 1 MeV. Also, comparing NaI(Tl), anthracene, and plastic scintillators, relative responses to 500-keV electrons are reported to be 1.6, 1.0 and 0.38, whereas those to 20-keV electrons are 2.5, 1.0, and 0.41, respectively.[246]

The reader is referred to the work by Porter et al.[246] (5–1000 keV) and Faber[247] (2–40 keV), who treated NaI(Tl), anthracene, and plastic scintillators; by Schmidt and Wallek,[248] (1–29 keV) who investigated anthracene, and by Pleitz[249] (3–11 keV) for plastic scintillators. See also Refs. 250, 251, and the articles just cited for information on backscattering and noise effects, as well as for further references.

6.4. Continuous-Channel Detectors

The recent introduction and improvement of the so-called continuous-channel electron detector, or channeltron, has been welcomed by many workers in low energy electron spectrometry. Such windowless detectors have shown detection efficiencies of above 50% over an appreciable low energy range (see Figure 24).

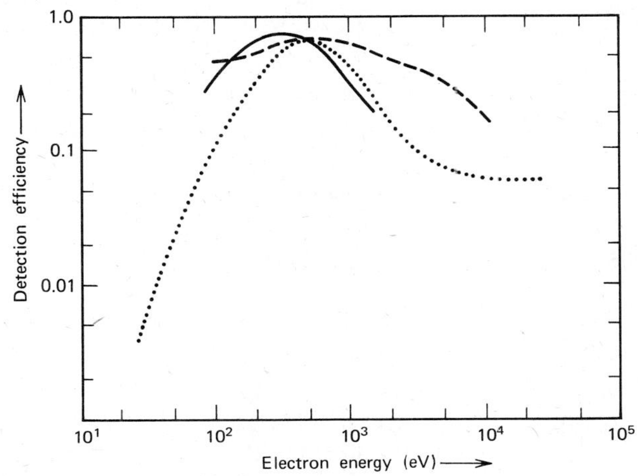

Figure 24. Electron detection efficiency as a function of electron energy for the Bendix channeltron as measured by Bosqued[258] (solid curve), Frank[260] (dotted curve), and Sharber et al.[257] (dashed curve).

The continuous-channel planar electron multiplier was first reported in 1934 by Farnsworth.[252] In 1961, Hamisch[253] investigated further the characteristics of such an instrument. Later designs employed linear cylindrical tubes with a semiconducting material coating the inside.[254–256] In these reports were considered and tested the advantages of a curved cylinder geometry, as employed by the Bendix Corp. and Mullard, Ltd., in their commercially available detectors. An extensive report of the characteristics of these channeltrons has been recently presented by Sharber et al.,[257] Bosqued,[258] Bosqued and Reme,[259] and Frank.[260] Egidi et al.[261] have investigated the variation of gain with time and the effects of fatigue on detection efficiency for electrons and protons.

The electron detection efficiency for the Bendix channeltron, as a function of impact energy, is shown in Figure 24, as measured and reported by three investigators.

Finally, the "focused mesh" electron multiplier, produced by Johnston Laboratories, Ltd., should be mentioned, because of its high detection efficiency, although it is actually a multidynode instrument. Each dynode (20 in all) is made up of a raft of specially formed, raised electron sensitive surfaces. These devices, having >90% electron detection sensitivity at 200 eV, have been favorably employed in several laboratories working with low energy electron spectrometry.

The electron detectors described must, of course, be placed under high vacuum when employed. Although the electron energy dependent detection efficiency of the detector can, in effect, be determined at a given time, this function can change a great deal with time because of fatigue effects and alteration of the semiconductor surface by exposure to the residual gases in the "vacuum." Therefore, the detector is at present an accurate device for determining the relative intensities of spectral components only with simultaneous energy dependent detection efficiency calibration.

6.5. Photographic Electron Detection

One of the most common electron detectors is the photographic emulsion plate, which has been used almost exclusively in combination with the semicircular spectrograph. The existence of photographic emulsion made it possible to take advantage of the high resolving power of this instrument in electron spectrum investigations even before 1930, when other types of high resolution spectrometers had not yet been developed.

An emulsion exposed to an electron spectrum will record the spectrum as bands of blackening which darken as a function of increased electron intensity. Determining spectral component intensities from the caused blackening may be regarded as the "integral" method of measurment. Alternatively, an electron may enter an emulsion giving rise to an individual track, which can be seen under a microscope. The counting of the individual electron track density as a function of position on the photographic plate, which will give the electron intensity spectrum directly, may be referred to as the "differential" method of measurement. These two alternatives are discussed in the following sections.

The photographic emulsion has been used advantageously as an energy-discriminating high energy alpha-, beta-, and so on, particle detector. This is made possible by measuring the particle track length in the emulsion and comparing it with an accurately known energy versus range function for the emulsion. The properties of photographic, or nuclear research, emulsions have been investigated in detail in connection with this work. Treatments dealing with this field should be referred to for information on the general properties of these emulsions.[262–265]

This method has also been applied for electrons of relative low energies (~50–~200 keV and above) in beta-spectrum measurements.[266]

6.5.1. *Integral Method*

Emulsion blackening as a measure of electron intensity has been employed by many workers as a means of electron detection for a semicircular spectrograph.[8,12,24,34,217,267–269] From these photographic spectrum recordings accurate momentum or energy determinations of spectral components can be obtained. It is more difficult, however, to obtain accurate relative intensity determinations between components of widely differing energies. There are two reasons for this. First, electrons of differing energies have different track lengths in emulsion, and hence different numbers of developed grains. Second, the mean number of grains made developable, per increment path length, by the passing electron is a function of the electron energy; hence the number of grains per unit electron track length will vary with energy.

However, determination of the mean blackening per electron as a function of initial electron energy will give relative spectral component intensities from a measured electron spectrum. The problem of relating the net blackening of an emulsion, measured with a densitometer as a function of position on the plate, to the electron intensity for the corresponding energy has been treated extensively by Slätis.[8,267,268,270] Here it was found that, for Ilford Selochrome 120 photographic emulsion, the relation between electron density ε, and the blackening density D, was of the form $D = 2.58 \log_{10} (\varepsilon + 1)$. The blackening density, per constant number of electrons, determined as a function of electron energy had a maximum at about 60 keV, the energy for which the range is approximately equal to the emulsion thickness. Also, the function did not go through zero because there was an insensitive protective coating on the emulsion. Smith and Hollander[269] employed Eastman X-ray film on a glass backing for their measurements. They constructed a blackening efficiency versus electron energy curve from data of earlier investigators (see also Ref. 271). For investigations of these and other emulsions, see the articles by Burge, Garrard, and Browne[272] and Klein and Moisar.[273]

For energies less than ~12 keV, Sujkowski and Slätis[274] found that the electron detection sensitivity, or the average blackening caused by a single electron, was enhanced if the emulsion was acquired in gel form, melted by heating, and spread onto the glass plate by the investigator. A second important point was noted: for low energy electrons the emulsion must be thin enough to allow fixing to occur in a matter of less than ~4 min, which means that the emulsion thickness should be less than 10 to 15 μm.[275] The emulsion surface, containing the entire tracks of electrons of low energy, becomes dissolved by the fixing solution during prolonged submersion.

The ranges of a low energy electron in Ilford G5 emulsion are about 5.6, 4.5, 3, and 2 μm for the energies 25, 20, 15, and 10 keV, respectively.[276] From Ross and Zajac's[277] measurements, the grain density along an electron track approaches ~1.5 μm^{-1} toward the track's terminus. Therefore, for energies less than ~10 keV, the number of grains per track will be few indeed. The mean deviation of this number will allow electrons to come to rest without producing a track with increasing probability for decreasing energies. By comparing lines of known relative intensity and different energies, it was estimated that electron detection efficiency is effectively 100% for energies ≳12 keV and falls off slowly toward lower energies to ~85% at 7 keV and then more rapidly to ~15% at 3 keV.[278]

In this connection, the results of Wittekindt[279] indicate that the individual electron detection sensitivity of 50-μm thick Ilford G5 emulsion is about 40% at the energy of 30 keV. The reason for this disagreement with the results cited previously is not obvious, but the thickness of the emulsion used may be responsible. It was found in the work related to Refs. 275 and 280 that for Ilford K5 emulsion of thickness 50 μm, the *M*-conversion lines of the 39.85 keV transition in ^{208}Tl (energy ~36 keV) showed quite clearly. On the same emulsion plate, however, the *L*-conversion lines appeared but weakly. Further investigation revealed that the surface of the emulsion becomes dissolved in the fixing process. The fixing time required for the 50-μm emulsion was sufficient to allow the shallow *L*-conversion line tracks to be removed with the dissolving emulsion. The ends of the more deeply penetrating 36-keV electron tracks, however, remained. It was for this reason that thinner emulsion (10–15 μm) was later used for the lower energy investigations.[280,281] It is therefore possible that Wittekindt's measured low detection sensitivities resulted from emulsion dissolution during fixing.

Returning to lower energy electron detection, consider the example of Ilford K5 emulsion, having an undeveloped grain diameter of ~0.21 μm, which will be expected to have a factor of at least three more grains threaded per micron than actually become developed. It is assumed that the emulsion is 50% gel by volume. The problem of increasing the emulsion's electron detection sensitivity becomes one of sensitizing the emulsion grains so that all threaded grains become developable.

There have been reports on a number of ways in which emulsions may be sensitized,[282–289] mainly through chemical treatments. Another possibility is to make extra-fine grain emulsions with a higher grain density.[290]

A very important consideration is the removal of latent tracks and developable grains in the emulsion prior to exposure in an experiment. The development of these features will cause an undesirable background darkening in the final plate. Eradication of latent tracks and grains can be effected with varying degrees of success by heating the emulsion, either in a humid atmosphere or dry,[275,291] using oxidizing agents,[292] or applying a pulsed electric field.[293]

It has been found that the emulsion plates prepared by heating and spreading emulsion in gel form onto glass plates have an advantage over the commercial emulsion plates, for the heating of the gel tends to destroy latent image grains to a large extent. Also, the plates can be made of any predesired thinness, and one is assured of a sensitive emulsion surface.

Blackening per impinging electron can also be increased by pre- or postacceleration of the electrons before impact. This has been employed by several investigators[294] (see also other sections in this chapter). Slätis[42] has constructed a postacceleration arrangement that allows all electrons to strike the emulsion plate with the same energy (see Section 2.1.2). Slätis[295] has also demonstrated that very weak lines can be made more obvious by making contact copies of the plates twice in succession. The second copy shows enhanced intensity of weak, previously unseen lines.

6.5.2. *Differential Method*

It is also possible to determine an electron spectrum by counting the electron tracks in an emulsion plate. This was first done by Antonova[296] and Kleinheins[9] for higher energies. Faler[297] used Ilford G5 emulsion to investigate spectral lines at ~280-keV and at ~30-keV energies.

The method was taken up again later[298] and subsequently developed further[275] and employed in various investigations.[29,280,299] The counting was carried out manually using a microscope of 1000 X magnification. The number of electron tracks per unit area was plotted as a function of position along the plate, and the spectrum was thus obtained. (Such spectra are shown in Figures 8 and 9 of Chapter 5.)

The differential method was later taken over by

other workers who built a precision semicircular spectrograph, and an investigation of the photoelectron lines of copper, L-subshells, using copper $K\alpha_1$ X-rays was carried out shortly thereafter.[281] A similar experiment on germanium was reported the following year.[300]

The technique was developed further by this group. The electron track counting method was automatized,[301] with a controlled movement of the emulsion plate.[302] The track or grain counting method has also been used in obtaining optical spectra from emulsion plates.[303] There have been several such automatic grain or track counting arrangements reported in association with high energy physics (e.g., Refs. 304, 265).

REFERENCES

1. K. T. Bainbridge, in *Experimental Nuclear Physics*, E. Segré (Ed.), Vol. I, John Wiley & Sons, Inc., New York, 1953, p. 559.
2. T. R. Gerholm, *Handbuch der Physik* (*Ed. Flügge*) **33** (1956), 609.
3. M. Deutsch and O. Kofoed-Hansen, in *Experimental Nuclear Physics*, E. Segré (Ed.), Vol. III, John Wiley & Sons, Inc., New York, 1959, p. 427.
4. C. S. Wu and C. Geoffrion, in *Nuclear Spectroscopy*, Ajzenberg-Selove (Ed.), pt. 9A., Academic Press, New York, 1960, p. 70; W. W. Buechner, *ibid*. p. 55.
5. K. Siegbahn, *Alpha-, Beta-, and Gamma-Ray Spectroscopy*, K. Siegbahn (Ed.), Vol. I, North Holland Publishing Company, Amsterdam, 1965, p. 79.
6. E. R. Cohen and J. W. M. Du Mond, *Rev. Mod. Phys*. **37** (1965) 537.
7. J. Danysy, *Le Radium* **9** (1912) 1; *Le Radium* **10** (1913) 4.
8. H. Slätis, *Ark. Fys*. **6** (1953) 415.
9. P. Kleinheins, *Z. Naturforsch*. **11a** (1956) 252.
10. A. H. Muir, Jr., thesis, California Institute of Technology, 1960.
11. J. M. Hollander, *Phys. Rev*. **103** (1956) 1590.
12. R. G. Albridge, J. M. Hollander, C. J. Gallagher, and J. H. Hamilton, *Nucl. Phys*. **27** (1961) 529.
13. J. Valentin, D. J. Horen, and J. M. Hollander, *Nucl. Phys*. **31** (1962) 353; *Nucl. Phys*. **31** (1962) 373.
14. R. G. Albridge, *Nucl. Instr. Meth*. **29** (1964) 61.
15. A. Fahlman and K. Siegbahn, *Ark. Fys*. **32** (1966) 111.
16. A. D. Baker, D. P. May, and D. W. Turner, *J. Chem. Soc*. (*B*) **(1968)** 22.
17. J. Kyles, C. G. Campbell, and W. J. Henderson, *Proc. Phys. Soc*. (*London*) **A66** (1953) 519.
18. V. Krynchko and V. F. Synorov, *Sov. Phys.-Instr. Exp. Tech*. no. 6 (1966) 1346.
19. K. Siegbahn, *Ark. Mat. Astr. Fys*. **30A** (1944) 20.
20. D. Berényi and F. Illés, *At. Kozl*. **3** (1961) 83.
21. F. Illés and D. Berényi, *At. Kozl*. **3** (1961) 91.
22. S. János, D. Berényi, and F. Illés, *At. Kozl*. **3** (1961) 105.
23. E. Rutherford and H. R. Robinson, *Phil. Mag*. **26** (1913) 717.
24. H. Robinson, *Proc. Roy. Soc*. (*London*) **A104** (1923) 455.
25. M. Ference, Jr., *Phys. Rev*. **51** (1937) 720.
26. C. Geoffrion and G. Giroux, *Can. J. Phys*. **34** (1956) 920.
27. C. Geoffrion, *Rev. Sci. Instr*. **20** (1949) 638.
28. G. Lindström, *Ark. Fys*. **4** (1952) 1.
29. K. D. Sevier, *Nucl. Phys*. **61** (1965) 601.
30. K. Yagi, *Nucl. Instr. Meth*. **39** (1966) 88.
31. E. Kerberlé, *J. Rech. Cent. Nat. Rech. Sci*. **49** (1959) 283.
32. F. Illés and D. Berényi, *Acta Phys. Hung*. **18** (1965) 265.
33. J. Kormicki, H. Niewodniezański, Z. Stachura, K. Zuber, and A. Budziak, *Acta Phys. Polon*. **31** (1967)317.
34. J. Kormicki, H. Niewodniczański, and Z. Stachura, *Inst. Nucl. Phys*., Cracow, Int. Rept. No. 476/PL, 1966.
35. C. Sébille, thesis, Orsay, 1968.
36. H. Slätis, *Nucl. Instr. Meth*. **2** (1958) 332.
37. H. E. Henrikson, California Institute of Technology, Spec. Tech. Rept. 24, October, 1956.
38. E. Karlsson and K. Siegbahn, *Nucl. Instr. Meth*. **7** (1960) 113.
39. W. Schneider and T. Lindqvist, *Nucl. Instr. Meth*. **13** (1961) 21.
40. P. Paris and G. Serge, *Nucl. Instr. Meth*. **35** (1965) 242.
41. Private communication from C. Sébille, 1969.

42. H. Slätis, *Ark. Fys.* **29** (1965) 485.

43. P. Paris, *J. Rech. Cent. Nat. Rech. Sci.* no. 54 (1961) 65.

44. W. G. Cross, *Rev. Sci. Instr.* **22** (1951) 717.

45. A. E. S. Green, R. J. Berkley, C. E. Watson, and C. F. Moore, *Rev. Sci. Instr.* **37** (1966) 415.

46. C. P. Browne and W. W. Buechner, *Rev. Sci. Instr.* **27** (1956) 899.

47. K. Yagi, *Nucl. Instr. Meth.* **31** (1964) 173.

48. P. Paris, thesis, Orsay, 1966.

49. B. Elbek, *Nucl. Instr. Meth.* **48** (1967) 353.

50. P. Paris and J. Treherne, *Rev. App. Phys.* **4** (1969) 291.

51. K. Siegbahn, C. Nordling, A. Fahlman, R. Nordberg, K. Hamrin, J. Hedman, G. Johansson, T. Bergmark, S.-E. Karlsson, I. Lindgren, and B. Lindberg, *Nova Acta Reg. Soc. Sci. Upsal. Ser. IV* **20** (1967), Uppsala; Tech. Rept. AFML-TR-68-189, Wright-Patterson Air Force Base, Ohio, Oct., 1968.

52. J. W. M. Du Mond, *Ann. Phys.* (*Paris*) **2** (1957) 283.

53. D. W. Kerst and R. Serber, *Phys. Rev.* **60** (1941) 53.

54. N. Svartholm and K. Siegbahn, *Ark. Mat. Astr. Fys.* **33A** (1946) 21.

55. K. Siegbahn and N. Svartholm, *Nature* **157** (1946) 872.

56. G. E. Lee-Whiting and E. A. Taylor, *Can. J. Phys.* **35** (1957) 1.

57. F. N. D. Kurie, J. S. Osoba, and L. Slack, *Rev. Sci. Instr.* **19** (1948) 771.

58. A. A. Bartlett and K. T. Bainbridge, *Rev. Sci. Instr.* **22** (1951) 517.

59. E. S. Rosenblum, *Phys. Rev.* **72** (1947) 731.

60. E. Arbman and N. Svartholm, *Ark. Fys.* **10** (1955) 1.

61. F. B. Shull and D. M. Dennison, *Phys. Rev.* **71** (1947) 681; *Phys. Rev.* **72** (1947) 256.

62. A. A. Bartlett, R. A. Ristinen, and R. P. Bird, *Nucl. Instr. Meth.* **17** (1962) 188.

63. T. Y. Chen, O. Huber, J. Kern, L. Schellenberg, B. Walthard, and H. Wild, *Nucl. Instr. Meth.* **29** (1964) 181.

64. A. A. Bartlett and J. Keith, *Nucl. Instr. Meth.* **36** (1965) 119.

65. B. V. Bobykin and M. I. Babenkov, *Zhr. Tekh. Fiz.* **36** (1966) 931 [*Sov. Phys.—Tech. Phys.* **11** (1966) 685].

66. K. Siegbahn and K. Edvarson, *Nucl. Phys.* **1** (1956) 137.

67. K. Siegbahn, C. Nordling, and J. M. Hollander, UCRL-Rept. 10023, Berkeley, 1962.

68. K. Siegbahn, C. Nordling, S.-E. Karlsson, S. Hagström, A. Fahlman, and I. Andersson, *Nucl. Instr. Meth.* **27** (1964) 173.

69. K.-E. Bergkvist and J. M. Hollander, *Nucl. Instr. Meth.* **53** (1967) 25.

70. C. de Vries and A. H. Wapstra, *Nucl. Instr. Meth.* **8** (1960) 121.

71. M. S. Freedman, F. Wagner, Jr., F. T. Porter, J. Terandy, and P. P. Day, *Nucl. Instr. Meth.* **8** (1960) 255.

72. C. de Vries, thesis, Amsterdam, 1960.

73. A. Baudry and A. Moussa, *Nucl. Instr. Meth.* **72** (1969) 167.

74. R. L. Graham, G. T. Ewan, and J. S. Geiger, *Nucl. Instr. Meth.* **9** (1960) 245.

75. M. Mladjenovič, *Nucl. Instr. Meth.* **7** (1960) 11.

76. M. Mladjenovič, Rept. at Int. Conf. Radioactivity in Nuclear Spectroscopy, Vanderbilt University, Nashville, Tenn. (August, 1969).

77. G. E. Lee-Whiting, *Can. J. Phys.* **35** (1957) 570.

78. G. E. Lee-Whiting, *Nucl. Instr. Meth.* **43** (1966) 182; *Can. J. Phys.* **41** (1963) 496.

79. H. Daniel, *Rev. Sci. Instr.* **31** (1960) 249.

80. H. Daniel and L. J. Laslett, *Rev. Sci. Instr.* **31** (1960) 1225.

81. H. Daniel and L. J. Laslett, *Nucl. Instr. Meth.* **10** (1961) 48.

82. H. Daniel and P. Jahn, *Nucl. Instr. Meth.* **14** (1961) 353.

83. H. Daniel, P. Jahn, M. Kuntze, and G. Spannagel, *Nucl. Instr. Meth.* **35** (1965) 171.

84. D. Mugnier and J. Lafoucrière, *Nucl. Instr. Meth.* **45** (1966) 45.

85. D. Mugnier, P. Beuzit, and J. Lafoucrière, *Nucl. Instr. Meth.* **50** (1967) 77.

86. K.-E. Bergkvist and A. M. Sessler, *Nucl. Instr. Meth.* **46** (1967) 317; J. A. Bruner and F. R. Scott, *Rev. Sci. Instr.* **21** (1950) 545.

87. A. Hofman, *Nucl. Instr. Meth.* **40** (1966) 13.

88. F. Schmutzler, H. Daniel, and L. Tauscher, *Nucl. Instr. Meth.* **35** (1965) 176.

89. R. Balzer, D. S. Bharucha, F. Heinrich, and A. Hofmann, *Nucl. Instr. Meth.* **57** (1967) 277.

90. K.-E. Bergkvist, *Ark. Fys.* **27** (1964) 383, 439; *Internal Conversion Processes*, Academic Press, New York, 1966, p. 107; *Nucl. Instr. Meth.* **43** (1966) 170.

91. W. Gelletly, J. S. Geiger, and R. L. Graham, *Phys. Rev.* **157** (1967) 1043.
92. Ö. Nilsson, S.-E. Karlsson, I. Andersson, C. Nordling, and K. Siegbahn, *Nucl. Instr. Meth.* **47** (1967) 13.
93. G. E. Lee-Whiting, *Nucl. Instr. Meth.* **43** (1966) 194.
94. A. A. Bartlett and D. L. Spenny, *Nucl. Instr. Meth.* **50** (1967) 125.
95. H. Wild and O. Huber, *Helv. Phys. Acta* **30** (1957) 3.
96. O. Huber, L. Schellenberg, and H. Wild, *Helv. Phys. Acta* **33** (1960) 534.
97. W. W. Lozier, *Phys. Rev.* **36** (1930) 1285.
98. P. S. P. Wei and A. Kuppermann, *Rev. Sci. Instr.* **40** (1969) 783.
99. P. S. P. Wei, A. Y. Cho, and C. W. Caldwell, *Rev. Sci. Instr.* **40** (1969) 1075.
99a. R. E. Weber and W. T. Peria, *J. Appl. Phys.* **38** (1967) 4355.
99b. L. A. Harris, *J. Appl. Phys.* **39** (1968) 1419.
99c. P. W. Palmberg and T. N. Rhodin, *J. Appl. Phys.* **39** (1968) 2425.
99d. N. J. Taylor, *Rev. Sci. Instr.* **40**, (1969) 792.
100. R. E. Fox, W. M. Hickam, T. Kjeldaas, Jr., and D. J. Grove, *Phys. Rev.* **84** (1951) 859.
101. P. J. Chantry, *Rev. Sci. Instr.* **40** (1969) 884.
102. D. Golden, and A. Zecca, *Rev. Sci. Instr.* **42** (1971) 210.
103. J. Arol Simpson, *Rev. Sci. Instr.* **32** (1961) 1283.
104. F. R. Paolini and G. C. Theodoridis, *Rev. Sci. Instr.* **38** (1967) 579.
105. G. C. Theodoridis and F. R. Paolini, *Rev. Sci. Instr.* **39** (1968) 326.
106. G. C. Theodoridis and F. R. Paolini, *Rev. Sci. Instr.* **40** (1969) 621.
107. G. D. Yarnold and H. C. Bolton, *J. Sci. Instr.* **26** (1949) 38.
108. G. A. Harrower, *Rev. Sci. Instr.* **26** (1955) 850.
109. J. D. H. Eland and C. J. Danby, *J. Phys. E.* (*J. Sci. Instr.*) **1** (1968) 406.
110. D. J. Volz, Ph.D. thesis, University of Nebraska, 1968.
110a. T. S. Green and G. A. Proca, *Rev. Sci. Instr.* **41** (1970) 1409; G. A. Proca and T. S. Green, *Rev. Sci. Instr.* **41** (1970) 1778; W. Schmitz and W. Mehlhorn, (to be published).
111. E. M. Purcell, *Phys. Rev.* **54** (1938) 818.
112. F. W. Aston, *Phil. Mag.* **38** (1919) 710.
113. H. Hafner, J. Arol Simpson, and C. E. Kuyatt, *Rev. Sci. Instr.* **39** (1968) 33.
114. N. Ashby, *Nucl. Instr. Meth.* **3** (1958) 90.
114a. F. T. Rogers, *Rev. Sci. Instr.* **22** (1951) 723.
114b. J. Kessler and N. Weichert, *Nucl. Instr. Meth.* **29** (1964) 245.
115. A. L. Hughes and V. Rojansky, *Phys. Rev.* **34** (1929) 284.
116. A. L. Hughes and J. H. McMillen, *Phys. Rev.* **34** (1929) 291.
117. P. Marmet and L. Kerwin, *Can. J. Phys.* **38** (1960) 787.
118. M. E. Rudd, Ph.D. thesis, University of Nebraska, 1962.
119. C. E. Kuyatt and M. E. Rudd, *Bull. Amer. Phys. Soc.* **8**, No. 4 (1963) 336.
120. F. T. Rogers, Jr., *Rev. Sci. Instr.* **8** (1937) 22; *Rev. Sci. Instr.* **11** (1939) 19; *Rev. Sci. Instr.* **14** (1943) 216.
121. E. Blauth, *Z. Phys.* **147** (1957) 228.
122. W. Mehlhorn, e.g., *Z. Phys.* **160** (1960) 247 and *Z. Phys.* **208** (1966) 1.
123. A. K. Edwards, J. S. Risley, and R. Geballe, *Abstr. 6th Int. Conf. Physics of Electronic and Atomic Collisions*, Cambridge, 1969, p. 563.
124. G. N. Ogurtsov, I. P. Flaks, and S. V. Avakyan, *Abstr. 6th Int. Conf. Physics of Electronic and Atomic Collisions*, Cambridge, 1969, p. 274, and *Zh. Tekhn. Fiz.* **39** (1969) 1293.
125. L. Toburen, *Phys. Rev.* **3** (1971) 216.
126. V. V. Zashkvara, M. I. Korsunskii, and O. S. Kosmachev, *Zh. Tekhn. Fiz.* **36** (1966)132. [*Sov. Phys.—Tech. Phys.* **11** (1966) 96.]
127. H. Sar-el, *Rev. Sci. Instr.* **38** (1967) 1210 and *Rev. Sci. Instr.* **39** (1968) 533.
128. A. J. Dempster, *Phys. Rev.* **51** (1937) 67.
129. R. Herzog, *Z. Phys.* **89** (1934) 596.
130. C. P. Browne, D. S. Craig, and R. M. Williamson, *Rev. Sci. Instr.* **22** (1951) 952.
130a. H. Wollnik, "Electrostatic Prisms," in *Focusing of Charged Particles*, Vol. II, A. Septier (Ed.), Academic Press, New York, 1967, p. 163.
130b. H. Matsuda, *Rev. Sci. Instr.* **32** (1961) 850.

131. F. Edelmann and K. Ulmer, *Z. Angew. Phys.* **18** (1965) 308.
132. C. E. Kuyatt, in "Methods of Experimental Physics," Vol. 7A, *Atomic Interactions*, L. Marton (Ed.), Academic Press, New York, 1968.
133. R. D. Birkhoff, J. M. Kohn, H. B. Eldridge, and R. H. Ritchie, *Nucl. Instr. Meth.* **6** (1960)313.
134. R. H. Ritchie, J. S. Cheka, and R. D. Birkhoff, *Nucl. Instr. Meth.* **6** (1960) 157.
135. C. E. Kuyatt and J. Arol Simpson, *Rev. Sci. Instr.* **38** (1967) 103.
136. M. E. Rudd, *Rev. Sci. Instr.* **37** (1966) 971.
136a. C. E. Kuyatt and J. A. Simpson, in *Atomic Collision Processes*, M. R. C. McDowell (Ed.), North-Holland Publ. Co., 1964, p. 191.
136b. G. E. Chamberlain, *Phys. Rev.* **155** (1967) 46.
136c. J. T. Park and F. D. Schowengerdt, *Rev. Sci. Instr.* **40** (1969) 753.
136d. See, e.g., *Vacuum Tubes*, by K. R. Spangenberg, McGraw-Hill, New York, 1948.
137. W. G. Wadey, *Rev. Sci. Instr.* **27** (1956) 910.
138. R. Bock, *Z. Phys.* **54** (1929) 257.
139. A. H. Firester, *Rev. Sci. Instr.* **37** (1966) 1264.
140. M. Ference, A. E. Shaw, and R. J. Stephenson, *Rev. Sci. Instr.* **11** (1940) 57.
141. F. R. Crownfield, *Rev. Sci. Instr.* **35** (1964) 240.
142. M. E. Rudd and J. R. Craig, *Rev. Sci. Instr.* **39** (1968) 1372.
143. R. K. Cacak and J. R. Craig, *Rev. Sci. Instr.* **40** (1969) 1468.
144. J. R. Barker, *J. Sci. Instr.* **27** (1950) 197.
145. W. Braunbeck, *Z. Phys.* **88**, (1934) 399.
146. C. G. Scott, *Rev. Sci. Instr.* **28** (1957) 270.
147. G. Fanselau, *Z. Phys.* **54** (1929) 260.
147a. G. E. Lee-Whiting, Chalk River Report (Ontario) CRT-673 (1957).
148. A. L. Bloom, D. J. Innes, R. C. Rempel, and K. A. Ruddock, *J. Appl. Phys.* **36** (1965) 2560.
149. M. W. Garret, *J. Appl. Phys.* **22** (1951) 1091.
150. S. M. Rubens, *Rev. Sci. Instr.* **16** (1957) 243.
151. J. W. Clark, *Rev. Sci. Instr.* **9** (1938) 320.
152. A. V. Crewe, *6th Int. Cong. Electron Microscopy*, Kyoto, 1966, pp. 625, 631; *Science* **154** (1966) 729.
153. J. Hillier and R. F. Baker, *J. Appl. Phys.* **15** (1944) 663.
154. L. Marton, J. A. Simpson, and T. F. McCraw, *Rev. Sci. Instr.* **26** (1955) 855.
155. H. Boersch and S. Schweda, *Z. Phys.* **167** (1962) 1.
156. H. Boersch and H. Miessner, *Z. Phys.* **168** (1962) 298.
157. M. Horstmann, G. Meyer, and H. Raether, *Z. Phys.* **154** (1959) 633.
158. M. Horstmann and G. Meyer, *Z. Phys.* **159** (1960) 563; *Z. Phys.* **164** (1961) 21.
159. G. Möllenstedt, *Optik* **5** (1949) 499.
160. G. Möllenstedt, *Optik* **9** (1952) 473.
161. K.-H. Gaukler, *Naturwiss.* **50** (1963) 438.
162. W. Lippert, *Optik* **12** (1955) 467.
163. J. Geiger, *Z. Phys.* **175** (1963) 530.
164. A. Carillon and B. Gauthé, *Compt. Rend.* **263** (1966) 1242.
165. A. Carillon and B. Gauthé, *3rd Euro. Reg. Conf. Electron Microscopy*, Prague, 1964, p. 51.
166. W. A. M. Hartl, *Z. Phys.* **191** (1966) 487.
167. F. Leonhard, *Z. Naturforsch.* **9a** (1954) 727.
168. K.-H. Gaukler, *Z. Phys.* **196** (1966) 85.
169. C. A. Williamson, M.Sc. thesis, University of London, 1966; I am indebted to Prof. R. E. Burge for kindly sending me a copy of this thesis.
170. A. J. F. Metherell and M. J. Whelan, *J. Appl. Phys.* **37** (1966) 1737.
171. J. Geiger, *Z. Phys.* **161** (1961) 243.
172. H. Boersch, *Z. Phys.* **139** (1954) 115.
173. W. Dietrich, *Z. Phys.* **151** (1958) 519.
174. W. Dietrich, *Z. Phys.* **152** (1958) 306.
175. R. Herzog, *Z. Phys.* **97** (1935) 596; *Physik Z.* **41** (1940) 18.
176. W. Wien, *Verhandl. Deut. Phys. Ges.* **16** (1897) 165.

177. W. Henneberg, *Ann. Phys.* **19** (1934) 335; R. Herzog, *Z. Phys.* **89** (1934) 447.
178. P. S. Farago, *Proc. Roy. Soc.* (*Edinburgh*) **A66** (1962) 1.
179. W. Legler, *Z. Phys.* **171** (1963) 424.
180. T. M. El Sherbini, *FOM*—Instituut voor Atoom en Molecuulfysica, Amsterdam (February, 1969).
181. E. Kasper, *Optik* **28** (1968) 54.
182. H. Boersch, J. Geiger, and W. Stickel, *Z. Phys.* **180** (1964) 415.
183. H. Boersch, J. Geiger, and H. Hellwig, *Phys. Letters* **3** (1962) 64.
184. H. Boersch, J. Geiger, and W. Stickel, *3rd Eur. Reg. Conf. Electron Microscopy*, Prague, 1964, p. 23.
185. Private communication, M. T. Browne, Queen Elizabeth College, London, 1969.
186. Articles in *Alpha-, Beta-, and Gamma-Ray Spectroscopy*, K. Siegbahn (Ed.), North Holland Publishing Company, Amsterdam, 1965.
187. S. Forsén and J. Rydberg, *Nucl. Instr. Meth.* **7** (1960) 204.
188. K. Matsuda, I. Nonaka, K. Omata, K. Yagi, and M. Koike, *Nucl. Instr. Meth.* **53** (1967) 82.
189. G. D. Alkhazov, A. P. Komar, and A. A. Vorob'ev, *Nucl. Instr. Meth.* **48** (1967) 1.
190. J. H. Carver, and P. Mitchell, *Nucl. Instr. Meth.* **52** (1967) 130.
191. M. W. Charles and B. A. Cooke, *Nucl. Instr. Meth.* **61** (1968) 31.
192. P. W. Benjamin, C. D. Kemshall, and J. Redfearn, *Nucl. Instr. Meth.* **59** (1968) 77.
193. L. Fergason, *Rev. Sci. Instr.* **37** (1966) 964.
194. N. Spielberg, *Rev. Sci. Instr.* **37** (1966) 1268.
195. G. E. Kocharov and G. A. Korolev, *Bull. Acad. Sci., Phys. Ser., USSR* **27** (1964) 308.
196. R. Gold and E. F. Bennett, *Phys. Rev.* **147** (1965) 201.
197. C. J. Andreen, W. Parker, and H. Slätis, *Nucl. Instr. Meth.* **15** (1962) 205.
198. W. Mehlhorn and R. G. Albridge, *Nucl. Instr. Meth.* **26** (1964) 37.
199. P. Polly, *Nucl. Instr. Meth.* **49** (1967) 346.
200. J. P. Budliger and A. Laisne, *Nucl. Instr. Meth.* **61** (1968) 253.
201. R. C. Sastri and S. D. Chatterjee, *Ind. J. Phys.* **41** (1967) 617.
202. B. L. Cohen, *J. Sci. Instr.* **37** (1960) 475.
203. H. Lindeman, *Nucl. Instr. Meth.* **40** (1966) 159.
204. L. Fava and V. Sangiust, *Nucl. Instr. Meth.* **42** (1966) 325.
205. H. Lindeman and E. Mornel, *Physica* **33** (1967) 691.
206. A. J. L. Collinson, I. C. Demetsopoullos, and J. M. Zarzycki, *Proc. Phys. Soc.* (*London*) **78** (1961) 785.
207. L. M. Langer, J. W. Motz, and H. C. Price, Jr., *Phys. Rev.* **77** (1950) 798.
208. W. Paul and H. Steinwedel, *Beta- and Gamma-Ray Spectroscopy*, K. Siegbahn (Ed.), North Holland Publishing Company, Amsterdam, 1955 pg.1; H. Slätis, *ibid* pg. 259; R. D. Birkhoff, *Handb. Physik.* **34** (1958) 53; H. Slätis, and W. C. Parker, Ref. 5.
209. R. O. Lane and D. J. Zaffarano, US-At. Energ. Comm. Rept., ISC-439 (December, 1953) Government Printing Office, Washington, D.C., 20225.
210. C. Feldman, *Phys. Rev.* **117** (1960) 455.
211. G. Stybe, *Z. Phys.* **159** (1960) 237.
212. S. P. Khare and Y. P. Varshni, *Ann. Phys.* (*Paris*) **7** (1961) 220; K. H. Weber, *Nucl. Instr. Meth.* **25** (1964) 261; A. R. Shul'man, I. R. Kanicheva, and B. F. Barzdo, *Fiz. Tver. Tela* **5** (1963) 3318 [*Solid State* **5** (1964) 2430]; L. Morbitzer and A. Scharmann, *Z. Phys.* **181** (1964) 67.
213. B. N. Subba Rao, *Nucl. Instr. Meth.* **44** (1966) 155.
214. T. Just, H. Niedrig, and H. Yersin, *Z. Angew. Phys.* **25** (1968) 89.
215. G. D. Archard, *J. Appl. Phys.* **32** (1961) 1505.
216. M. Vobecký and O. Dragoun, *Jad. Energ.* (*Czech.*) **3** (1957) 410. This article, now translated to English, is available from the authors as Nucl. Res. Inst. Rept.–2148 (1969), Ustav Jaderného Vyzkumu, ČSAV, Řež, Prague, Czechoslovakia.
217. J. S. Dionisio, *Ann. Phys.* (*Paris*) **8** (1963) 747; *Ann. Phys.* (*Paris*) **9** (1964) 29.
218. R. G. Albridge, *Nucl. Instr. Meth.* **29** (1964) 61; R. G. Albridge, thesis, University of California, UCRL–Rept. 8642 (1960); H. Slätis, *Ark. Fys.* **29** (1965) 485; M. I. Margues and M. T. Ramos, *Compt. Rend.* **265** (1967) 1209; and Ref. 12.
219. R. J. Krisciokaitis and S. K. Haynes, *Nucl. Instr. Meth.* **58** (1968) 309.
220. S. Charalambus and A. Theordossiou, *Nucl. Instr. Meth.* **36** (1965) 331.
221. S. C. Curran and J. M. Reid, *Rev. Sci. Instr.* **19** (1948) 67; K. Gubernator, *Z. Phys.* **152** (1958) 183; B. Schonland, *Proc. Roy. Soc.* (*London*) **104A** (1923) 235; D. Butt, *Proc. Phys. Soc.* (*London*) **63A** (1950) 986.
222. A. Juillard, thesis, University of Grenoble, 1961.

223. M. J. Zender, W. Pou, and R. G. Albridge, *Z. Phys.* **218** (1969) 245.

224. B. Grennberg, University of Uppsala, Institute of Physics Rept., UUIP–481, 1966; *J. Sci. Instr.* **44** (1967) 203.

225. M. Morgue, thesis, University of Lyon, 1969.

226. M. Morgue and R. Chery, *Congr. Int. Electronique Nucléaire*, Versailles (September, 1968).

227. S. Goulding, University of California Rept., UCRL–16231 (July 1965); *Nucl. Instr. Meth.* **43** (1966) 1.

228. J. W. Mayer, *Nucl. Instr. Meth.* **43** (1966) 55.

229. R. H. Pehl, F. S. Goulding, D. A. Landis, and M. Lenzlinger, *Nucl. Instr. Meth.* **59** (1968) 45.

230. J. M. Hollander, *Nucl. Instr. Meth.* **43** (1966) 65.

231. H. T. Eastman, A. J. Haverfield, and J. M. Hollander, *Nucl. Instr. Meth.* **32** (1965) 333.

232. T. Tamura and T. Kuroyanagi, *Nucl. Instr. Meth.* **67** (1969) 38.

233. J. Brownridge and D. McLoughlin, *Nucl. Instr. Meth.* **60** (1968) 116.

234. Private communication from E. M. Gunnersen, Aldermaston, 1967.

235. O. Meyer, *Nucl. Instr. Meth.* **70** (1969) 279, 285.

236. H. J. West, Jr., F. P. Burns, L. Wang, and F. P. Zumba, *Rev. Sci. Instr.* **33** (1962) 380.

237. G. Dearnaley, A. G. Hardarce, B. D. Rogers, *Nucl. Instr. Meth.* **71** (1969) 86.

238. F. S. Goulding, *IEEE Trans. Nucl. Sci.* **NS-11** (1964) 117.

239. I. Ogawa, *Nucl. Instr. Meth.* **49** (1967) 325.

240. M. Masuda, T. Takeda, and H. Chisaka, *Nucl. Instr. Meth.* **49** (1967) 77.

241. G. C. Huth, *IEEE Trans. Nucl. Sci.* **NS-13** (1966) 36.

242. G. C. Huth, J. B. Trice, and J. A. Shannon, *IEEE Trans. Nucl. Sci.* **NS-12** (1964) 275.

243. E. Laegsgaard, F. W. Martin, and W. M. Gibson, *Nucl. Instr. Meth.* **60** (1968) 24.

244. S. Kalbitzer and W. Melzer, *Nucl. Instr. Meth.* **56** (1967) 301.

245. W. Melzer and F. Pühlhofer, *Nucl. Instr. Meth.* **60** (1968) 201.

246. F. T. Porter, M. S. Freedman, F. Wagner, Jr., and I. S. Shuman, *Nucl. Instr. Meth.* **39** (1966) 35.

247. K. Faber, *Z. Angew. Phys.* **23** (1967) 240.

248. V. Schmidt and L. Wallek, *Z. Phys.* **210** (1968) 164.

249. W. Pleitz, *Z. Phys.* **190** (1966) 446.

250. D. K. Butt, *Physica* **18** (1952) 1142.

251. H. Paul and I. Hofmann, *Nucl. Instr. Meth.* **22** (1963) 141.

252. P. T. Farnsworth, *J. Franklin Inst.* **218** (1934) 411.

253. H. Hamisch, *Z. Angew. Phys.* **14** (1962) 322.

254. D. S. Evans, *IEEE Trans. Nucl. Sci.* **NS-12** (1964) 34.

255. D. S. Evans, *Rev. Sci. Instr.* **36** (1965) 375.

256. J. Adams and B. W. Manley, *Electron. Eng.* March, 1965.

257. J. R. Sharber, J. D. Winningham, and W. R. Sheldon, *IEEE Trans. Nucl. Sci.* NS–15(1968) 536.

258. J. M. Bosqued, thesis, University of Toulouse, 1967.

259. J. M. Bosqued and H. Reme, *Nucl. Instr. Meth.* **57** (1967) 6.

260. L. A. Frank, University of Iowa Rept. 65–22 (1965); *Rev. Sci. Instr.* **40** (1969) 685.

261. A. Egidi, R. Marconero, G. Pizzella, and F. Sperli, *Rev. Sci. Instr.* **40** (1969) 88.

262. *Particle Photography*, P. Demers (Ed.), Vol. 3, Montreal, 1964.

263. G. Sprague, *J. Appl. Phys.* **32** (1961) 2410.

264. C. E. K. Mees, *The Theory of the Photographic Process*, Macmillan Company, New York, 1954.

265. W. H. Barkas, *Nuclear Research Emulsions*, Vol. 1: *Techniques and Theory*, Academic Press, New York, 1963.

266. A. Bonetti, "Applications of Nuclear Emulsions in Measurements of Beta-Ray Spectra," *Bull. Centre Phys. Nucl. Univ. Libre Bruxelles*, No. 19, May, 1950 (unpublished).

267. H. Slätis, *Ark. Fys.* **8** (1954) 441.

268. M. Mladjenovič and H. Slätis, *Ark. Fys.* **8** (1954) 65.

269. W. G. Smith and J. M. Hollander, *Phys. Rev.* **101** (1956) 746.

270. H. Slätis, *Nucl. Instr. Meth.* **16** (1962) 51; *Ark. Fys.* **22** (1962) 517.

271. R. Glocker, *Z. Phys.* **160** (1960) 568.

272. R. E. Burge, D. F. Garrard, and M. T. Browne, *J. Phys. E.* (*J. Sci. Instr.*) **1** (1968) 707; R. E. Burge and D. F. Garrard, *J. Phys. E.* (*J. Sci. Instr.*) **1** (1968) 715.

273. E. Klein and E. Moisar, *Naturwiss.* **48** (1961) 23.

274. Z. Sujkowski and H. Slätis, *Ark. Fys.* **14** (1959) 101.
275. K. D. Sevier, *Nucl. Instr. Meth.* **22** (1963) 345.
276. J. M. Blum, *J. Phys. Rad.* **12** (1951) 860.
277. M. A. S. Ross and B. Zajac, *Nature* **162** (1948) 923.
278. Work done in connection with Refs. 280, 29, 281.
279. R. Wittekindt, *Z. Phys.* **158** (1960) 572.
280. K. D. Sevier, *Ark. Fys.* **25** (1963) 87.
281. K. D. Sevier, A. Fahlman, and K. Siegbahn, *Ark. Fys.* **29** (1965) 367.
282. J. W. Mitchell, *J. Phys. Chem.* **66** (1962) 2359.
283. E. Klein, *J. Phys. Chem.* **66** (1962) 2407.
284. M. Debeauvais-Wack, *Ann. Phys.* **8** (1963) 41.
285. H. Faraggi and A. Garin, *Nuovo Cim.* **17** (1960) 830.
286. E. S. Baker, *J. Opt. Soc. Amer.* **58** (1969) 1378.
287. R. Rechenmann, *Ann. Phys.* **6** (1961) 1001.
288. L. Avan and C. Dubois, *Compt. Rend.* **251** (1960) 1280.
289. G. Rasigni, *Compt. Rend.* **261** (1965) 2868.
290. N. A. Perfilov, N. R. Novikova, and E. I. Prokove'na, *At. Energ.* **4** (1958) 45 [*J. Nucl. Energ.* **9** (1959) 90].
291. E. Dahl-Jensen and E. Skrzypczak, *Nuovo Cim. Suppl.* **23** (1962) 150.
292. H. Ackermann and H. Faissner, *Nucl. Instr. Meth.* **10** (1961) 339.
293. D. M. Samoilovich and I. V. Ardashev, *Dokl. Akad. Nauk SSSR* **179** (1968) 59 [*Doklady* **13** (1968) 211]; *Dokl. Akad. Nauk SSSR* **184** (1969) 68 [*Doklady* **14** (1969) 327].
294. F. Gautier, *Compt. Rend.* **262B** (1966) 1628; *Compt. Rend.* **263B** (1966) 1411.
295. H. Slätis, *Ark. Fys.* **37** (1968) 25.
296. I. A. Antonova, *J. Eksp. Teoret Fiz. SSSR*, **30** (1956) 571 [*J. Exptl. Theoret. Phys., USSR*, **3** (1956) 461.].
297. K. T. Faler, thesis, University of California, UCRL Rept.–8664, 1959 (unpublished).
298. K. D. Sevier, University of California, UCRL Rept.–9809, 1961 (unpublished).
299. K. D. Sevier, thesis, University of Uppsala, 1965.
300. R. Carlsson, A. Fahlman, and K. Siegbahn, *Ark. Fys.* **32** (1966) 103.
301. R. Carlsson, A. Fahlman, and K. Siegbahn, University of Uppsala, Institute of Physics, UUIP Rept.–485, 1966.
302. T. Berglund, R. Carlsson, A. Fahlman, and K. Siegbahn, University of Uppsala, Institute of Physics, UUIP Rept.–544, 1967.
303. R. Carlsson, A. Fahlman, R. Hallin, and K. Siegbahn, *Ark. Fys.* **32** (1966) 99.
304. A. Bernheim, M. Combes, J. C. Fayolle, H. de Surville, and M. Morand, *Rev. Phys. Appl.* **3** (1968) 59.

Chapter Three

The Auger, Coster-Kronig, and Autoionization Effects

Sans doute un principe plus subtil que celui de l'entropie nous donnera-t-il un jour des vues assez profondes pour satisfaire notre soif de "raison suffisante." Peut-être sera-t-il basé sur la notion de Forme, plus raffinée que celle d'Ordre, sur laquelle reposaient jusqu'à présent tous nos espoirs.

Pierre Auger
Paris, July 6, 1969

1. INTRODUCTION

The radiationless de-excitation of atoms was predicted by Rosseland,[1] and first experimentally established by Auger[2,3] in 1923. Photographs of the interior of an X-ray-irradiated cloud chamber clearly showed paired electron tracks originating from the X-ray paths. One of these tracks was that of a photoelectron; the other was due to an Auger electron emitted in the filling of the created atomic vacancy.

The establishment of this effect made it possible to explain a number of previously unclarified experimental observations, for example, the extra lines in photoelectron spectra recorded by Robinson and coworkers[4,5] (see also Burhop[6]).

In the event of an initial vacancy occurring in the K-shell, the filling of the vacancy will result in the emission of either an X-ray or an Auger electron. In the former case, the energy of the X-radiation is the change in the atomic binding energy of the transition electron; in the latter case, the energy released in the filling of the initial K-shell vacancy is imparted to a second atomic electron, which is expelled from the atom.

An L-subshell vacancy may be filled by either an X-ray or Auger transition or, for the L_1 and L_2-subshells in certain regions of the periodic table, by a Coster–Kronig transition process. In such regions it is energetically possible for an initial L_1- or L_2-subshell vacancy to filled by an electron from a higher L-subshell and the released energy to be carried away by an outer shell electron. This third possibility was described in 1935 by Coster and Kronig.[7] The Coster–Kronig process plays a dominant role in the excitation of satellite lines in L X-ray spectra.

Clearly an important fundamental process of atomic de-excitation, the Auger effect is of interest in the fields of atomic and solid state physics. But despite this, it has been omitted from most textbooks dealing with these fields.

On the other hand, the Auger effect has often been treated in connection with the field of nuclear physics. Since the necessary initial atomic vacancy may be created, for example, by internal conversion processes and by orbital electron capture, the Auger effect can be observed in nuclear decay. Therefore, certain knowledge of the intensities and energies of Auger spectral components will aid greatly in the interpretation of closely lying or superimposed electron conversion lines (see Section 5.3.2.). Also, an inner shell vacancy creation can, through a cascade of Auger transitions, leave an atom in a state of high ionization; and this state will persist if the atom is in a gaseous or vapor state or, to a lesser extent, in an electrically insulating environment. The state of high ionization is expected to effect the decay rate of nuclear transitions, as shown in Chapter 4, Section 5.2.1. Such states of high ionization may also cause perturbations in angular correlation measurements.[8]

Until the late 1950s there was little interest in the Auger effect and little theoretical or experimental work done in the field. In 1952 Burhop[6] presented an extensive review of work carried out to that date. A shorter treatment by Bergström[9] was to follow shortly after, in 1955.

Interest in experimental and theoretical work on the Auger effect increased, however, after about 1958. This development was mainly as a result of advances in theory and improvements in high resolution electron spectrometry, and it led to a greatly enhanced understanding of the effect.

Advances have been rapid since 1958. An extensive review article covering the field was published in 1960;[10] this was followed by three shorter articles treating progress in the field through about 1963.[11–13] An up-to-date review of the theoretical treatment of the Auger effect has been recently given by Mehlhorn.[14] The reader should refer to Mehlhorn's article for a lucid and concise theoretical treatment of the effect.

This chapter reviews the results of work done in the field up to the middle of 1969. Here more emphasis has been layed on the lower energy Auger and Coster–Kronig effects than is found in other articles.

The subject material in Chapter 3 is subdivided into the following main sections. A description of the possible sources of Auger electrons included in the next section is followed by a description of the classification of Auger electrons. Section 4 deals with the *K*-series Auger component energies and relative intensities. Section 5 includes treatments of the energies and intensities of the *L*, *M*, and so on, Auger series and of the Coster–Kronig effect. Section 6 deals with autoionizing levels in atoms, for their nonradiative decay is equivalent to the Auger process after the creation of an inner shell vacancy; and Section 7 treats recent investigations of the angular distribution of Auger electrons. Section 8 discusses the fluorescence yields for *K*-, *L*-, and *M*-shells and the techniques by which they may be measured. Section 9 covers several special topics involving or relating to the Auger effect. Finally, the applications of knowledge of the Auger transition process in determining nuclear and atomic properties, and to industrial and technical ends, are listed in Section 10.

2. SOURCES OF AUGER ELECTRONS

As has already been stated, an Auger electron may be emitted as a result of the filling of an ionized internal atomic shell. There are a number of possible ways of forming the initial vacancy, each with its own particular advantages and detriments to the investigation of resulting Auger spectra. They may be enumerated as follows.

2.1. Orbital Electron Capture and/or Internal Conversion in a Radioactive Element

Whether orbital electron capture, internal conversion in a radioactive element, or both, are involved, the relative probabilities of a vacancy occurring in the various atomic shells and subshells (K, L_1, L_2, etc.) are dependent on the mode of the nuclear transition as well as its energy. Of course, if the energy of the transition is lower than the *K*-shell electron binding energy, for example, no initial *K*-shell vacancies can be formed.

Radioactive materials have often been used as Auger electron sources. The strength of such sources can be adjusted as desired, and their compactness and ease of handling and mounting for measurement are of great advantage. For high resolution studies, however, the methods of source preparation must be refined in order to obtain satisfactorily thin sources and the necessary reduction of electron energy losses due to straggling, all the while keeping the total activity at a satisfactory level. In addition, electrons of higher initial energy can become scattered in the spectrometer, thus increasing the general background. Electrons having suffered energy losses while being backscattered from the source substrate material will also enhance the general background as well as increase the spectral line low energy tails.

2.2. The Photoelectric Effect

The initial vacancy is caused through the ionization of an atomic electron by an X-ray or gamma quantum. All electrons of atomic binding energies less than the X-ray or gamma energy may be ejected, the relative ionization probability for each atomic electron depending on the characteristics of its bound atomic state and the X-ray or gamma energy. If the entire continuous X-ray spectrum irradiates the converter material, the resulting electron spectrum will be the superposition of three electron spectra: (*a*) photoelectron lines resulting from external conversion of the characteristic X-rays in all the atomic shells of the converter material having binding energies less than the X-ray energies, (*b*) the continuous underlying electron spectrum due to the external conversion of the continuous X-radiation, and (*c*) the Auger electron spectrum arising through the de-excitation of the atom.

However, a single strong X-ray line can be selected out of the X-ray emission spectrum by reflection from a crystal monochromator, thus greatly reducing the underlying electron continuum. Care should be exercised in choosing the energy of the line, in order to avoid overlapping of photoelectron lines with the Auger lines of interest. Unfortunately, only a small amount of the chosen X-radiation is reflected from the monochromator.

2.3. Electron Bombardment

An inner atomic shell electron may be excited to a positive energy state by an electrostatic interaction with an impinging high velocity electron. This ionization process has a greater cross section than the photoelectric effect and has been advantageously employed in the study of *KLL*-Auger spectra of gaseous elements of low *Z*-value.

The impinging electron, however, will most probably lose only a part of its energy in the ionization of a bound atomic electron. The ionized electron spectrum then will be in effect a continuum upon which the Auger spectrum is superimposed. In addition, the impinging electron can cause multiple atomic ionization, or excitation, and the resulting complete Auger spectrum will become the superposition of spectra originating from atoms in a variety of atomic excitation and ionization states. The ensuing complexity of the Auger spectrum may hinder its analysis.
tion, and the resulting complete Auger spectrum will become the superposition of spectra originating from atoms in a variety of atomic ionization states. The ensuing complexity of the Auger spectrum may hinder its analysis.

The electron-bombarded material is usually in the gaseous state and electron straggling effects can be made negligible. In using this ionization technique, one has the advantage that the spectral line shapes will depend solely on the effective source size and the focusing properties of the spectrometer.

Electron bombardment is often used for exitation of autoionizing states in atoms in the gaseous state (see Section 6).

2.4. Ion Bombardment

The study of the Auger electron spectrum arising from ion impact of solid surfaces can supplement other methods of obtaining information about the purity of the surface or the thickness of a thin covering surface layer of another material. The applications are primarily technical (see Section 10).

However, the study of ion or atomic collisions in a vacuum may be carried out—for example, by investigating the ejected electron spectrum. The presence of Auger electrons on the otherwise fairly continuous electron background will give information on the degree and nature of the atomic excitation and ionization upon impact. The measurement of electron spectra arising from ion and atomic impact, which can also provide valuable data on the nature of autoionization levels, has been used to great advantage in the study of these phenomena (see Section 6).

Ion bombardment, including proton bombardment, of gaseous atoms will cause the formation of multi-ionized multi-excited atomic states more effectively than in the case of electron bombardment (see Section 4.1.3).

3. CLASSIFICATION

A single initial vacancy in the *K*-shell or an *L*-, *M*-, and so on, subshell of an atom may be filled by an electron from an outer subshell with the released energy being transferred to a second outer shell electron, which is ejected. If, for instance, an initial vacancy in the *K*-shell is filled by an L_2-electron with the expulsion of an M_3-electron, the emitted electron is referred to as a KL_2M_3-Auger electron. All *K*-series Auger electrons are of the general type KX_pY_q, where *X* and *Y* refer to electron shells lying outside the *K*-shell, and *p* and *q* are subshell indices. When one does not wish to distinquish between subshells, the indices are dropped, and reference is made to *KLL* or *KMN*, and so on, Auger groups, or, more generally, *KXY*-Auger groups. Sometimes in the literature, however, *X* and *Y* may refer to the atomic shells lying outside of the *L*-shell, the *L*-shell being named specifically. This is done, for example, in dealing with the ratios of *KLX* and *KXY* Auger group intensities to that of the *KLL* group.

The *L*-Auger electrons arise through processes of the type $L_iX_pY_q$, where *X*- and *Y*-shells lie outside the *L*-shell. Coster–Kronig transitions are of the types $L_iL_pX_q$, $M_iM_pX_q$, and so on, and involve an electron transition between two *L*-, *M*-, and so on, subshells, the released energy being carried away by an electron from an outer shell.

The Auger process was once considered to be a two-step process—internal production and photoeffect of X-rays—rather than a one-step nonradiative atomic de-excitation process. Theoretically, in the nonrelativistic limit, both viewpoints provide the same radiationless transition rate. However, for the effect to be a two-step process, the KL_1L_1 Auger electron would have to arise as a result of the internal conversion of a K–L_1 X-ray, and this X-ray transition is always forbidden by the selection rule $\Delta l = \pm 1$ for radiative dipole transitions. Therefore, the Auger process must be one of radiationless reorganization of an excited atom.

After an Auger transition takes place in an initially singly ionized atom, the atom is in a state of double ionization. The final energy state of the atom is dependent on the atomic electron subshells in which the two atomic vacancies are situated and on the residual interaction between the remaining atomic electrons. The energy of the emitted Auger electron will therefore be directly related to the energy of the final atomic state.

The manner in which the possible states of the final ionized atom are characterized determines the notation to be used for the various components of each Auger group. In general, different quantum numbers must be used for different Z-value regions of the periodic table; the choice of numbers will be governed by the necessity of taking into account (*a*) relativistic effects and (*b*) the residual interaction between the individual final state atomic electrons.

In the simplest model that may be used to classify Auger electrons, the various lines are denoted by symbols of the type KL_1L_3, $L_2M_2N_6$, and so on. This classification results from the assumption that there is extreme *j–j* coupling between the electrons in the nonclosed shells of the atom. In the *j–j* coupling limit, account is taken of relativistic effects but not of interelectron interactions. In this extreme coupling case the spin s and orbital momentum for each electron l combine to form a total angular momentum j. Hence, the L_1, L_2, and L_3 electron states of an atom may be described by the l, j values of the states: 0, $\frac{1}{2}$; 1, $\frac{1}{2}$; and 1, $\frac{3}{2}$, respectively. The final atomic states after an Auger transition can thus be designated by labels signifying the l, j values of the original atomic electron states of the two electrons taking part in the transition (i.e., L_1L_1, L_1L_2, etc.), as predicted by relativistic theory.

The relativistic effects may be neglected by assuming that there is no $L_{2,3}$ subshell splitting. This approximation leads, for example, to a triple-component KLL-group: KL_1L_1, $K_1L_{2,3}$, and $KL_{2,3}L_{2,3}$. These represent Auger transitions to final $2s^02p^6$, $2s^12p^5$, and $2s^22p^4$ L-shell electron configurations, respectively. In practice, this approximation is useful when the KLL-Auger spectra of elements of low Z-value are investigated at low resolution.

For many years it was assumed that the Auger effect could be satisfactorily described in terms of the *j–j* coupling model just described; this was primarily because instrumental resolution was not sufficiently high to resolve more than six KLL-components. In addition, most of the investigations were carried out for the heavier elements, where *j–j* coupling predominates, to avoid difficulties with electron detection and source preparation. In 1957, however, seven lines were reported in the KLL-Auger group of copper.[15] The following year, in a theoretical treatment of the K-Auger spectrum presented by Asaad and Burhop,[16] the final state of the atom was expressed in terms of intermediate coupling; that is, the interelectron interaction lies somewhere between the limits of pure *j–j* and pure L–S (Russell-Saunders) coupling. In this coupling, nine KLL-Auger lines are predicted. As it happens, however, the reported seventh line for copper does not correspond in energy position to any transition predicted on the basis of the intermediate coupling scheme.

Evidence of a satellite of the KL_1L_3 line was found soon after[17] in the KLL Auger spectrum for plutonium ($Z = 94$), showing that even for high atomic numbers the coupling is not pure *j–j*. Later the same satellite line was reported by Graham and Merritt[18] for $Z = 66$, by Ewan and Merritt[19] for $Z = 64$, by Graham and others[20–22] for $Z = 63$, and by Ewan et al.[23] for $Z = 62$. Eight lines were reported in 1961[24] for cesium ($Z = 55$), which included two satellite lines that were different from the one previously reported. The two last named satellites were convincingly established shortly[20] thereafter for iodine ($Z = 53$). Also in 1962 Hörnfeldt et al.[25] first observed a total of nine KLL-Auger lines for $Z = 38$, 40, and 42.

In the theoretical treatment of Asaad and Burhop,[16] which dealt with both the KLL- and KLM-Auger groups, two additional possible bases for classification of Auger lines were presented, both of which had been considered long before by Ference.[26]

3.1. Pure *L–S* Coupling

Relativistic effects are small and often not taken into account where pure L–S coupling is concerned—whereas the electrostatic interaction between the electrons in the nonclosed shells is included. This classification can be applied for small Z-values as well as for atomic shells where the electron binding energy is small (i.e., outer lying shells) and the electron-electron interaction is not negligible. In this limit the total angular momentum L and the total spin S are conserved

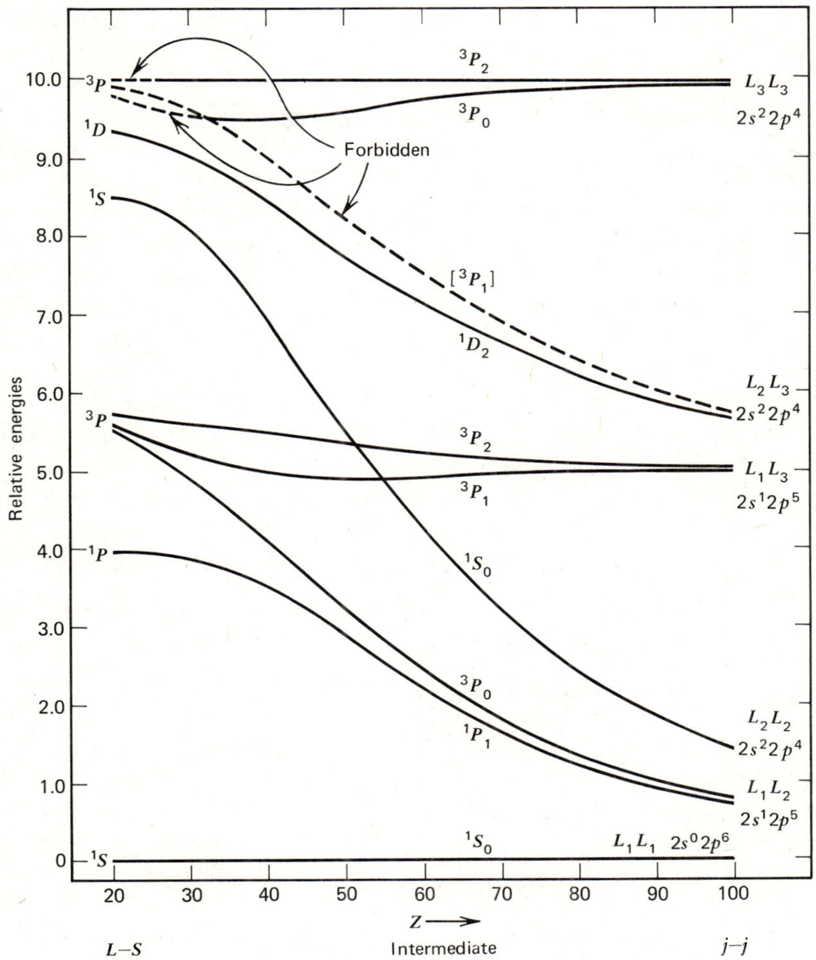

Figure 1. Coupling, as a function of atomic number, between final atomic vacancy states following a *KLL*-Auger transition.

separately, and there are six possible energy levels of an atom doubly ionized in the *L*-shell. These are qualitatively illustrated in the atomic level energy diagram in Figure 1. Auger transitions to the $L_{2,3}L_{2,3}(^3P)$ configuration toward the limit of *L–S* coupling are forbidden by the requirement of conservation of angular momentum and parity.

3.2. Intermediate Coupling

For the interaction lying between the extreme limits of pure *j–j* and pure *L–S* coupling, both the electron-electron interaction and the relativistic effects must be considered in order to obtain a complete description of the final atomic states. In the intermediate interaction region, the energy of the atom depends in general on the total angular momentum *J* of the final electron system. This results in a splitting of each line in the *j–j* or *L–S* coupling schemes into a multiplet of components of different *J*-values. This leads in principle to ten final atomic states; however, transitions to one of these, $L_2L_3(^3P_1)$, is forbidden theoretically for any coupling.[16] The transitions to the final $^3P_0$ and $^3P_2$ states, which were forbidden in the extreme *L–S* coupling limit, are allowed in the intermediate coupling region. More specifically, an Auger transition from the $K(^2S_{1/2})$ state (an initial *K*-hole) to any of the ^3P states in the final $2s^22p^4$ configuration of the *L*-shell is forbidden in the pure *L–S* coupling limit by the nonconservation of parity, and five components are contained in the *KLL* spectrum. However, assuming intermediate coupling, a mixing will arise between pure *L–S* states having the same total angular momentum (e.g., $^3P_0$ and $^1S_0$, $^3P_2$ and $^1D_2$) and being of the same $2s^22p^4$ configuration. It will be possible then for transitions to take place to these final mixed states without violation of parity conservation.

Returning to the $^3P_1$ state, note that there are no other states of the $2s^22p^4$ configuration having the same total angular momentum that can mix with it.

A possibility of configuration mixing between this $^3P_1$ state and the $^1P_1$ states of the $2s^12p^5$ configuration has been suggested.[13] This would make possible transitions to the $^3P_1$ state, which would give rise to a tenth line in the KLL-Auger group. However, the $2s^22p^4$ configuration has parity +1, whereas the $2s^12p^5$ configuration has parity −1. Since there is no atomic interaction that can change parity (see Section 6.3.3), the tenth line is truly forbidden.

In the intermediate coupling region, the components of the Auger spectrum may be classified with equal validity in terms of either j–j or L–S coupling. The former can be used when the experimental resolution is not high enough to allow the less intense lines to be observed. When the satellite lines are distinguishable, however, it is necessary to refer unambiguously to the components of the spectrum; thus each line is labeled according to (*a*) the limit to which it tends in the extreme j–j coupling scheme and (*b*) the symbols describing it in the intermediate coupling scheme. The nine lines of the KLL-Auger spectrum are then $KL_1L_1(^1S_0)$, $KL_1L_2(^1P_1)$, and so on, in accordance with the nomenclature included in Figure 1. Three of the nine lines—namely, $KL_1L_2(^3P_0)$, $KL_1L_3(^3P_2)$, and KL_3L_3-$(^3P_0)$—are quite weak and have been termed "Auger-satellite lines" because they are so close to the more intense $KL_1L_2(^1P_1)$, $KL_1L_3(^3P_1)$, and $KL_3L_3(^3P_2)$ lines, respectively, in medium and large Z-value regions.

An Auger transition is symbolized then by a sequence of letters designating the shell or subshell vacancies in the atom in the order of their occurrence. In a like manner, if a given Auger transition takes place in an atom that is in an initial state of double ionization, the transition may again by symbolized by a designation of the order of occurrence of the atomic vacancy states. A transition taking place from a KL_3 double vacancy state to an $L_1L_1L_3$ triple vacancy state can be symbolized as either KL_3–$L_1L_1L_3$ or $KL_1L_1(L_3)$; such transitions give rise to "Auger-vacancy-satellite lines." (see Section 5.1.1). The KL_3–$L_1L_1L_3$ form is to be preferred when a more complete designation of the initial and final states is desired [e.g., one may wish to specify the $KL_1(^3S)$–$L_1L_{2,3}L_{2,3}(^2P)$ transition in neon]. For the heavier elements, where j–j coupling is predominant, the alternative symbolism may be employed to designate the Auger-vacancy-satellite lines.

4. THE *K*-AUGER SERIES

Auger electrons emitted from an atom with an initial vacancy in the K-shell can be divided into distinct groups according to their energies. The KLL-group, which is the most intense group and completely isolated from the other K-Auger groups, is always the primary subject of K-Auger electron spectrum studies. Investigations of the KLM, KMM, and so on, groups have until recently been carried out primarily for the determination of KLX/KLL and KXY/KLL group intensity ratios. Lately, with improvements in spectrometric techniques related to resolution and source preparation, there has been more interest in the analysis of these subordinate groups.

The KLX-group is so isolated and intense, relative to the KLL-group, that its investigation may be carried out at high resolutions and employing quality sources. The KLM-group is the most intense, and although there is a disturbing amount of overlapping between its component subgroups, making analysis difficult, the $KL_{1,2}M_{1,2,3}$ subgroups together stand fairly isolated for intermediate and high atomic numbers. The KXY-group is comprised of comparatively weak and extremely numerous Auger transition lines.

4.1. The *KLL*-Auger Group

For convenience, the treatment of the KLL-Auger group is divided into two parts: the component energies and the component intensities.

In the first of these (see Section 4.1.1), the development of the theoretical treatment of the Auger component energies is reviewed and compared with experimentally determined values for these. The theoretical calculation of the component intensities is reviewed in Section 4.1.2, together with pertinent experimental results. Comprehensive tables of experimental data regarding both energies and intensities are presented in the latter section.

4.1.1. Energies of KLL-Auger Electrons.

The energy of a KLL-Auger electron can be estimated by considering the Auger process as an internal X-ray photoeffect. The approximate energy of a KL_pL_q Auger electron will then be $E(KL_pL_q) = [E(K) - E(L_p)] - E(L_q)$, where $E(K)$, $E(L_p)$, and $E(L_q)$ are the atomic binding energies of the K, L_p, and L_q electrons, L_q being the ejected electron.

However, it was noticed at an early date[27,28] that the KLL-Auger electron kinetic energies are not simply linear combinations of the binding energies of the atomic electrons taking part in the transition. The energies estimated in this way and those obtained from experimental measurements differ by an amount $\Delta E(L_pL_q)$, and the equation for the energy of a KL_pL_q Auger electron is then

$$E(KL_pL_q) = E(K) - E(L_p) - E(L_q) - \Delta E(L_pL_q). \tag{1a}$$

The last term may be considered as the increase of the L_q electron binding energy caused by a vacancy in the L_p-shell.

Bergström and Hill[29] suggested that, since the energy $E(L_q) + \Delta E(L_p L_q)$ lies between $E^Z(L_q)$ and $E^{Z+1}(L_q)$ (the L_q electron binding energies for the atoms of atomic numbers Z and $Z+1$), the binding energy discrepancy $\Delta E(L_p L_q)$ can be assumed to be linearly dependent on an increase in Z due to a change in screening. This effective incremental charge, ΔZ is expressed as

$$\Delta Z = \frac{\Delta E^Z(L_p L_q)}{E^{Z+1}(L_q) - E^Z(L_q)}. \tag{1b}$$

In their measurements of the *KLL*-Auger spectrum of mercury ($Z = 80$), Bergström and Hill found $\Delta Z = 0.55$ and 0.76 for the $L_{1,2}$-and the L_3-subshell electrons, respectively, being ejected. It was clear that ΔZ for their experiments was less than unity. (See also Ref. 6.) It has been shown that each of the two values of ΔZ is of approximately the same size for $Z \sim 55$ to ~ 83.[30]

Albridge and Hollander[28] pointed out some time later that the two values ΔZ for $L_p = L_1, L_2$, or L_3 and $L_q = L_1$ or L_2 and $\Delta Z'$ for $L_p = L_1, L_2$, or L_3 and $L_q = L_3$ are not entirely independent parameters. This is because KL_pL_q and KL_qL_p are equivalent expressions, and it is arbitrary whether L_p or L_q is adjusted. One obtains the relation

$$\frac{\Delta Z'}{\Delta Z} = \frac{E^{Z+1}(L_p) - E^Z(L_p)}{E^{Z+1}(L_q) - E^Z(L_q)}.$$

Also, it was pointed out, a single average correction factor, ΔZ_{av}, defined by

$$E(KL_pL_q) = E(K) - E(L_p) - E(L_q) - \tfrac{1}{2}\Delta Z_{av}[E^{Z+1}(L_p) - E^Z(L_p) + E^{Z+1}(L_q) - E^Z(L_q)],$$

is sufficient to calculate all *KLL*-Auger electron energies for uranium to an accuracy of about $1:10^3$.

In 1957 Geoffrion and Nadeau[30] suggested an alternative method for calculating $\Delta E(L_pL_q)$. Their proposed formula has the form

$$\left[\frac{1}{\Delta E^Z(L_pL_q)}\right]^n = \left[\frac{1}{E^{Z+1}(L_p) - E^Z(L_p)}\right]^n + \left[\frac{1}{E^{Z+1}(L_q) - E^Z(L_q)}\right]^n$$

or, more simply,

$$\Delta E^Z(L_pL_q) = \frac{E^{Z+1}(L_q) - E^Z(L_q)}{\left\{1 + \left[\dfrac{E^{Z+1}(L_q) - E^Z(L_q)}{E^{Z+1}(L_p) - E^Z(L_p)}\right]^n\right\}^{1/n}}. \tag{2}$$

As n approaches infinity, we obtain $\Delta E^Z(L_pL_q) = E^{Z+1}(L_q) - E^Z(L_q)$, which is equivalent to setting $\Delta Z = 1$ in equation 1b. A comparison was made between experimental energy values for the *KLL*-Auger components for elements of $Z = 83$, 80, 78, and 47, and the values predicted using equation 2 for $n = 1$ and $n = 2$. It was found that the exponent chosen depends on which subshells are involved in the Auger transition.

Aside from the empirical treatments of *KLL*-group component energies just described, only a limited amount of work has been done on the theoretical calculations of these energies. In 1958 Asaad and Burhop[16] published their nonrelativistic calculations assuming intermediate coupling between the final state atomic vacancies. The relative transition probabilities and the energies of the nine lines of the *KLL*-group were calculated. The transition probabilities were also calculated for the *KLM*-group, but only the energy splitting of the states for each possible final state configuration was given and not the relative energies of the configurations.

In their calculations of the *KLL*-Auger component energies, Asaad and Burhop expressed the nearly closed shell configurations $2s^02p^6$, $2s^12p^5$, and $2s^22p^4$, as completely closed shell configurations, together with the two-electron configurations $2s^2$, $2s^12p^1$, and $2p^2$, respectively. The states of the two systems are equivalent after the sign of the spin-orbit coupling term has been reversed and certain terms have been added to the diagonal elements of the electrostatic interaction matrix.[31] The energies of the electrons in the *KLL*-Auger spectrum can be calculated in terms of the electrostatic interaction between these two electrons, that is, the Slater–Condon integrals:

$$F^\nu(nl, n'l') = e^2 \iint_{r_1, r_2 = 0}^{\infty} \gamma_\nu R_{nl}^2(r_1) R_{n'l'}^2(r_2) r_1^2 r_2^2 \, dr_1 \, dr_2$$

$$G^\nu(nl, n'l') = e^2 \iint_{r_1, r_2 = 0}^{\infty} \gamma_\nu R_{nl}(r_1) R_{n'l'}(r_2) \times R_{nl}(r_2) R_{n'l'}(r_1) r_1^2 r_2^2 \, dr_1 \, dr_2, \tag{3}$$

where $R_{nl}(r)$ and $R_{n'l'}(r)$ are the radial parts of the wave functions of electrons in the nl and $n'l'$ bound states, respectively, and

$$\gamma_\nu = \begin{cases} \dfrac{r_1^{\nu}}{r_2^{\nu+1}} & \text{for } r_1 < r_2), \\[2ex] \dfrac{r_2^{\nu}}{r_1^{\nu+1}} & \text{for } r_2 < r_1. \end{cases}$$

The Slater–Condon parameters have been evaluated by Fisk and Fraga from analytical Hartree–Foch functions and collected in tabular form[32] (see also Ref. 33). The energies of the nine *KLL* lines become, after utilizing the identity $E(L_2) - \zeta_{21} = E(L_3) + \frac{1}{2}\zeta_{21}$, where $\zeta_{21} = \frac{2}{3}[E(L_2) - E(L_3)]$:

$$KL_1L_1(^1S_0): \quad E(K) - 2E(L_1) - F^0(20, 20)$$

$$KL_1L_2(^1P_1): \quad E(K) - E(L_1) - E(L_2) - F^0(21, 20) + \tfrac{3}{4}\zeta_{21} - \{[\tfrac{1}{3}G^1(21, 20) - \tfrac{1}{4}\zeta_{21}]^2 + \tfrac{1}{2}\zeta_{21}{}^2\}^{1/2}$$

$$KL_1L_3(^3P_1): \quad E(K) - E(L_1) - E(L_3) - F^0(21, 20) - \tfrac{3}{4}\zeta_{21} + \{[\tfrac{1}{3}G^1(21, 20) - \tfrac{1}{4}\zeta_{21}]^2 + \tfrac{1}{2}\zeta_{21}{}^2\}^{1/2}$$

$$KL_1L_2(^3P_0): \quad E(K) - E(L_1) - E(L_2) - F^0(21, 20) + \tfrac{1}{3}G^1(21, 20)$$

$$KL_1L_3(^3P_2): \quad E(K) - E(L_1) - E(L_3) - F^0(21, 20) + \tfrac{1}{3}G^1(21, 20)$$

$$KL_2L_3(^1D_2): \quad E(K) - E(L_2) - E(L_3) - F^0(21, 21) + \tfrac{2}{25}F^2(21, 21) + \tfrac{3}{4}\zeta_{21} - \{[\tfrac{3}{25}F^2(21, 21) + \tfrac{1}{4}\zeta_{21}]^2 + \tfrac{1}{2}\zeta_{21}{}^2\}^{1/2}$$

$$KL_3L_3(^3P_2): \quad E(K) - 2E(L_3) - F^0(21, 21) + \tfrac{2}{25}F^2(21, 21) - \tfrac{3}{4}\zeta_{21} + \{[\tfrac{3}{25}F^2(21, 21) + \tfrac{1}{4}\zeta_{21}]^2 + \tfrac{1}{2}\zeta_{21}{}^2\}^{1/2}$$

$$KL_3L_3(^3P_0): \quad E(K) - 2E(L_3) - F^0(21, 21) - \tfrac{1}{10}F^2(21, 21) - \tfrac{3}{2}\zeta_{21} + \{[\tfrac{3}{10}F^2(21, 21) - \tfrac{1}{2}\zeta_{21}]^2 + 2\zeta_{21}{}^2\}^{1/2}$$

$$KL_2L_2(^1S_0): \quad E(K) - 2E(L_2) - F^0(21, 21) - \tfrac{1}{10}F^2(21, 21) + \tfrac{3}{2}\zeta_{21} - \{[\tfrac{3}{10}F^2(21, 21) - \tfrac{1}{2}\zeta_{21}]^2 + 2\zeta_{21}{}^2\}^{1/2}. \quad (4)$$

In equation 4 $E(K)$, $E(L_1)$, $E(L_2)$, $E(L_3)$ are the electron binding energies of the designated shell or subshell. The values of the $F^\nu(nl, n'l')$ and $G^\nu(nl, n'l')$ terms were determined by using the experimentally obtained *KLL*-Auger electron energies for $Z = 83$ as reported by Mladjenovič and Slätis.[34] They were found to be appreciably smaller than those obtained from Hartree's self-consistent-field calculations[35]—presumably the discrepancy resulted from the neglect of relativity. Accounting for the effect of relativity by using Burhop's[36] expression $(1 - \alpha Z^2)$, the G's and F's were approximated by

$$A(Z - Z_s)(1 - \alpha Z^2), \quad (5)$$

where Z_s is the screening constant. Using this equation, the values of A, α, and Z_s required to yield values of each of the F's and G's were deduced using the results of the measurements given in Ref. 34 for $Z = 83$. A comparison of the Auger energy values given by these calculations and the few Auger energies measured up to that time showed fair agreement.

The predicted values agreed well with those obtained experimentally by Johnson and Foster[37] for silver ($Z = 47$), but they were about 40 eV below those obtained for copper ($Z = 29$) by Bellicard et al.[38] Geoffrion and Nadeau[30] have since pointed out, however, that the measured energies for the silver *KLL*-Auger spectrum were about 40 eV too low because of an error in the energy of the calibration standard. The corrected experimental values for silver are therefore also in disagreement with the theoretically predicated energy values.

In 1959 Asaad[39] presented the results of relativistic calculations, assuming *j–j* coupling, for the *KLL*-Auger spectrum for mercury ($Z = 80$), based on Møller's semiclassical treatment for the interaction of radiation with electrons. Calculation results for both energies and relative intensities of the *KLL*-Auger components for mercury were presented and are in good agreement with experiment. These values are included in Table IV of Section 4.1.2.

It was shown some time later,[25,40] in a report of newly measured *KLL*-Auger line energies, that expression 5 does not account with sufficient accuracy for the variation of *KLL*-Auger electron energies with *Z*-values. A new set of values for A was calculated to agree with the new results on the measurement of the nine component Auger spectra for strontium ($Z = 38$), zirconium ($Z = 40$), and molybdenum ($Z = 42$). It was found that this adjustment gave close agreement between the observed and calculated energy values over an extensive *Z*-region. However, the energy values given by the expression increased too quickly with Z; that is, simultaneous agreement with experiment in both the heavy and light regions could not be obtained with any single set of constants.

Hörnfeldt[41] showed that better agreement could be obtained with experimental data if $(1 - \alpha Z^2)$ were replaced by $(1 + \beta Z^3)$. Calculating a least-mean-squares adjustment of constants A and β, using published experimental Auger energy data, a table of *KLL*-Auger energies was obtained. The screening constants Z_s were taken from Ref. 16. The obtained values of the constants are included in Table I.

Using these values for A and β, the *KLL*-transition energies were calculated for the elements from calcium ($Z = 20$) to fermium ($Z = 100$). Agreement between the experimental and the calculated energy values of the $KL_2L_3(^1D_2)$ line was better than 0.05% in most cases. The relative differences between the various *KLL*-line energies in the table are also in very good agreement with experimental values.

Table II shows the *KLL*-Auger line energies as calculated by Hörnfeldt[41] for $Z \geq 40$. A few small

Table I.

F and G function parameters.

Integral	Z_s [16) 25) 41)]	α [16) 23)] $\times 10^6$	A(keV) [16)] $\times 10^{-3}$	A(keV) [25)] $\times 10^{-3}$	A(keV) [41)] $\times 10^{-3}$	β [41)] $\times 10^{-6}$	Relativistic calculations [45)] (no screening) A(keV) $\times 10^{-3}$	a $\times 10^6$	Integral
$F^\circ(20,20)$	3.8	17.6	4.08	2.54	2.10	1.40	4.092	19.830	$\mathcal{F}^\circ(20,20)$
$F^\circ(21,20)$	4.0	32.5	4.41	2.64	2.30	1.10	4.411	21.327 14.451	$\mathcal{F}^\circ(20,21)$ $\mathcal{F}^\circ(20,21)$
$G^1(21,20)$	2.6	--	2.39	2.66	2.66	--	2.392	10.364 18.312	$\mathcal{G}^1(20,21)$ $\mathcal{G}^1(20,21)$
$F^\circ(21,21)$	4.5	25.0	4.94	3.02	2.70	0.80	4.942	23.775 14.628 5.481	$\mathcal{F}^\circ(2\bar{1},2\bar{1})$ $\mathcal{F}^\circ(2\bar{1},21)$ $\mathcal{F}^\circ(21,21)$
$F^2(21,21)$ $G^2(21,21)$	5.1	--	2.39	2.75	2.75	--	2.392	4.036 -0.232	$\mathcal{F}^2(2\bar{1},21)$ $\mathcal{G}^2(2\bar{1},21)$

Table II.

The energies for the KLL Auger transitions (in keV) obtained from semi-empirical calculations. Values for Z less than 40 are taken from ref.[42] and Z greater than 39 from ref.[41]

Z	Element	KL_1L_1 $(^1S_0)$	KL_1L_2 $(^1P_1)$	KL_1L_2 $(^3P_0)$	KL_1L_3 $(^3P_1)$	KL_1L_3 $(^3P_2)$	KL_2L_2 $(^1S_0)$	KL_2L_3 $(^1D_2)$	KL_3L_3 $(^3P_0)$	KL_3L_3 $(^3P_2)$
6	C	0.243	0.252	0.258	0.258	0.258	0.265	0.266	0.267	0.267
7	N	0.356	0.362	0.369	0.369	0.369	0.373	0.375	0.377	0.377
8	O	0.474	0.486	0.495	0.495	0.495	0.504	0.507	0.509	0.509
9	F	0.610	0.627	0.638	0.638	0.638	0.650	0.654	0.657	0.657
10	Ne	0.761	0.781	0.794	0.794	0.794	0.808	0.813	0.816	0.816
11	Na	0.928	0.952	0.967	0.967	0.967	0.984	0.989	0.993	0.993
12	Mg	1.105	1.135	1.151	1.151	1.151	1.172	1.179	1.183	1.183
13	Al	1.301	1.336	1.354	1.354	1.354	1.379	1.387	1.392	1.392
14	Si	1.516	1.554	1.574	1.574	1.575	1.602	1.611	1.616	1.617
15	P	1.742	1.784	1.805	1.806	1.806	1.835	1.845	1.851	1.852
16	S	1.982	2.034	2.057	2.058	2.059	2.096	2.107	2.114	2.115
17	Cl	2.249	2.305	2.329	2.330	2.331	2.370	2.382	2.389	2.391
18	Ar	2.527	2.586	2.612	2.613	2.614	2.656	2.669	2.677	2.679
19	K	2.815	2.881	2.909	2.910	2.912	2.959	2.973	2.981	2.984
20	Ca	3.122	3.195	3.224	3.225	3.227	3.279	3.294	3.303	3.306
21	Sc	3.456	3.533	3.563	3.564	3.567	3.622	3.638	3.647	3.651
22	Ti	3.799	3.886	3.916	3.919	3.922	3.985	4.002	4.011	4.016
23	V	4.168	4.259	4.290	4.293	4.298	4.362	4.381	4.391	4.397
24	Cr	4.557	4.651	4.683	4.687	4.692	4.757	4.778	4.788	4.795
25	Mn	4.956	5.056	5.089	5.094	5.100	5.169	5.191	5.202	5.211
26	Fe	5.374	5.480	5.514	5.519	5.527	5.598	5.622	5.634	5.644
27	Co	5.808	5.923	5.957	5.964	5.972	6.049	6.075	6.088	6.099
28	Ni	6.264	6.384	6.419	6.426	6.436	6.514	6.542	6.556	6.568
29	Cu	6.732	6.861	6.896	6.905	6.916	7.000	7.030	7.045	7.059
30	Zn	7.214	7.348	7.384	7.394	7.407	7.493	7.526	7.543	7.558
31	Ga	7.712	7.852	7.888	7.900	7.915	8.000	8.037	8.057	8.073
32	Ge	8.216	8.365	8.401	8.416	8.433	8.523	8.563	8.586	8.603
33	As	8.749	8.903	8.939	8.957	8.975	9.063	9.107	9.133	9.152
34	Se	9.283	9.447	9.483	9.504	9.524	9.616	9.665	9.695	9.715
35	Br	9.840	10.014	10.049	10.074	10.096	10.189	10.244	10.279	10.300
36	Kr	10.412	10.594	10.630	10.658	10.682	10.777	10.837	10.877	10.899

Table II. (cont.)

Z	Element	KL_1L_1 ($^1S_0$)	KL_1L_2 ($^1P_1$)	($^3P_0$)	K_1L_3 ($^3P_1$)	($^3P_2$)	KL_2L_2 ($^1S_0$)	KL_2L_3 ($^1D_2$)	KL_3L_3 ($^3P_0$)	($^3P_2$)
37	Rb	10.995	11.186	11.221	11.255	11.280	11.376	11.442	11.487	11.511
38	Sr	11.595	11.795	11.830	11.870	11.897	11.992	12.066	12.118	12.143
39	Y	12.213	12.422	12.457	12.503	12.532	12.626	12.708	12.767	12.793
40	Zr	12.850	13.067	13.103	13.156	13.187	13.277	13.369	13.436	13.463
41	Nb	13.504	13.730	13.765	13.826	13.859	13.948	14.049	14.125	14.153
42	Mo	14.178	14.414	14.448	14.518	14.553	14.638	14.750	14.836	14.865
43	Te	14.868	15.112	15.147	15.227	15.263	15.343	15.467	15.563	15.594
44	Ru	15.574	15.827	15.862	15.953	15.991	16.066	16.202	16.310	16.341
45	Rh	16.297	16.559	16.594	16.697	16.737	16.806	16.955	17.076	17.109
46	Pd	17.040	17.311	17.346	17.461	17.503	17.565	17.729	17.863	17.897
47	Ag	17.797	18.078	18.113	18.242	18.286	18.339	18.519	18.668	18.702
48	Cd	18.567	18.857	18.892	19.037	19.082	19.125	19.322	19.488	19.523
49	In	19.353	19.653	19.688	19.849	19.896	19.929	20.144	20.327	20.363
50	Sn	20.157	20.465	20.501	20.680	20.728	20.749	20.984	21.135	21.223
51	Sb	20.977	21.295	21.331	21.529	21.579	21.588	21.843	22.065	22.104
52	Te	21.813	22.142	22.178	22.397	22.449	22.444	22.722	22.965	23.004
53	I	22.666	23.004	23.041	23.282	23.336	23.314	23.616	23.883	23.923
54	Xe	23.530$^+$	23.878$^+$	23.918$^+$	24.185$^+$	24.240$^+$	24.205$^+$	24.530$^+$	24.820$^+$	24.860$^+$
55	Cs	24.409	24.775	24.812	25.104	25.160	25.109	25.464	25.782	25.825
56	Ba	25.316	25.690	25.728	26.047	26.105	26.032	26.416	26.763	26.806
57	La	26.250	26.630	26.668	27.017	27.076	26.977	27.392	27.769	27.813
58	Ce	27.200	27.590	27.628	28.008	28.068	27.945	28.393	28.801	28.846
59	Pr	28.170	28.571	28.609	29.024	29.085	28.936	29.419	29.862	29.908
60	Nd	29.163	29.573	29.612	30.062	30.126	29.947	30.468	30.948	30.995
61	Pm	30.173	30.594	30.633	31.122	31.187	30.977	31.538	32.058	32.106
62	Sm	31.197	31.630	31.669	32.198	32.264	32.023	32.626	33.186	33.234
63	Eu	32.247	32.690	32.730	33.303	33.371	33.092	33.740	34.345	34.395
64	Gd	33.315	33.769	33.809	34.428	34.497	34.181	34.877	35.527	35.578
65	Tb	34.401	34.867	34.908	35.575	35.645	35.290	36.035	36.734	36.786
66	Dy	35.510	35.986	36.028	36.747	36.819	36.419	37.218	37.971	38.023
67	Ho	36.639	37.127	37.168	37.943	38.016	37.569	38.425	39.233	39.287
68	Er	37.787	38.286	38.328	39.161	39.235	38.740	39.654	40.521	40.575

Table II. (cont.)

Z	Element	KL_1L_1 ($^1S_0$)	KL_1L_2 ($^1P_1$)	KL_1L_2 ($^3P_0$)	KL_1L_3 ($^3P_1$)	KL_1L_3 ($^3P_2$)	KL_2L_2 ($^1S_0$)	KL_2L_3 ($^1D_2$)	KL_3L_3 ($^3P_0$)	KL_3L_3 ($^3P_2$)
69	Tm	38.957	39.469	39.511	40.405	40.480	39.933	40.910	41.839	41.894
70	Yb	40.150	40.673	40.716	41.675	41.752	41.148	42.191	43.186	43.242
71	Lu	41.361	41.896	41.940	42.967	43.045	42.383	43.496	44.559	44.616
72	Hf	42.584	43.132	43.176	44.275	44.354	43.630	44.816	45.952	46.010
73	Ta	43.831	44.392	44.437	45.610	45.691	44.903	46.166	47.376	47.435
74	W	45.104	45.676	45.721	46.976	47.058	46.196	47.541	48.834	48.894
75	Re	46.387	46.972	47.017	48.358	48.441	47.503	48.936	50.315	50.376
76	Os	47.693	48.297+	48.337+	49.769	49.854	48.829+	50.357+	51.831	51.893
77	Ir	49.032	49.641	49.687	51.211	51.296	50.194	51.812	53.375	53.438
78	Pt	50.388	51.005	51.052	52.679	52.766	51.566	53.289	54.956	55.020
79	Au	51.745	52.388	52.436	54.161	54.249	52.975	57.797	56.563	56.628
80	Hg	53.143	53.799	53.847	55.682	55.771	54.398	56.331	58.206	58.272
81	Tl	54.544	55.221	55.270	57.218	57.309	55.840	57.888	59.878	59.944
82	Pb	55.968	56.672	56.721	58.787	58.879	57.317	59.483	61.590	61.658
83	Bi	57.436	58.155	58.204	60.394	60.487	58.813	61.105	63.338	63.406
84	Po	58.92	59.65	59.70	62.02	62.12	60.31	62.74	65.11	65.18
85	At	60.44	61.17	61.22	63.70	63.79	61.84	64.41	66.93	67.00
86	Rn	61.96	62.71	62.76	65.39	65.48	63.40	66.13	68.79	68.86
87	Fr	63.52	64.29	64.35	67.12	67.22	65.00	67.88	70.70	70.77
88	Ra	65.11	65.90	65.95	68.89	68.99	66.61	69.66	72.65	72.72
89	Ac	66.70	67.51	67.56	70.67	70.77	68.24	71.46	74.62	74.69
90	Th	68.36	69.16	69.22	72.51	72.61	69.90	73.30	76.64	76.71
91	Pa	70.01	70.84	70.89	74.37	74.47	71.60	75.19	78.71	78.79
92	U	71.71	72.55	72.61	76.28	76.39	73.33	77.12	80.84	80.92
93	Np	73.43	74.29	74.34	78.23	78.33	75.07	79.07	83.01	83.08
94	Pu	75.17	76.04	76.10	80.20	80.31	76.84	81.06	85.21	85.29
95	Am	76.93	77.82	77.88	82.21	82.32	78.64	83.09	87.47	87.55
96	Cm	78.73	79.64	79.70	84.23	84.34	80.48	85.13	89.71	89.79
97	Bk	80.54	81.48	81.54	86.29	86.41	82.34	87.22	92.03	92.11
98	Cf	82.37	83.33	83.38	88.39	88.50	84.21	89.34	94.40	94.48
99	Es	84.24	85.21	85.26	90.53	90.64	86.10	91.49	96.81	96.89
100	Fm	86.17+	87.15+	87.21	92.73	92.85	88.05	93.71	99.28	99.37

+ Re-evaluated by interpolation between adjacent values (see text).

changes (denoted by crosses) have been made in regard to the energies of some of the lines for xenon ($Z = 54$), osmium ($Z = 76$), and fermium ($Z = 100$). In studying the original table it was found that the KL_2L_3 ($^1D_2$) energy was a continuously varying function of Z except for discontinuities at $Z = 54$ and 76, where the magnitudes were larger than the quoted ~0.05% accuracy of the values. These two energy values were then replaced by values obtained by interpolation. The difference between the energies tabulated for the $KL_2L_3(^1D_2)$ line and for the other Auger lines were determined for each Z-value. Any irregularities greater than 0.05% of the Auger energy in these energy differences expressed as a function of Z were redetermined by interpolation.

For $Z < 40$, the energy values listed in Ref. 42 are entered. These were calculated using a separate set of parameters as listed in Ref. 25 (see Table I), where a Z^2 dependence was found to be more appropriate for this Z-value region.

Calculations have also been carried out using the relativistic self-consistent-field method for the elements magnesium ($Z = 12$), potassium ($Z = 19$), and copper ($Z = 29$), and compared with experimental values.[42] The disagreement (ca. 10 eV for $Z = 12$) becomes less with increasing atomic number. The Auger energies were calculated for the free atom and the measurements were carried out for atoms in a chemical compound in the solid form. This discrepancy is apparently due to a solid state effect. (One has seen similar discrepancies and agreements between theoretical binding energy calculations compared with experimental values obtained from elements in solid chemical compounds and in the gaseous state, respectively.) This view is supported further by a comparison of the measured values of the KLL-Auger spectrum for neon with theoretical values and the semiempirical values listed in Table III. Here (see Table III) the measured values are in much better agreement with the theoretical values (to ~4 eV) than with the semiempirical values (to ~13 eV).

Recently Hillig[42a] measured the KLL- and KLM-Auger spectra for sodium in the vaporized state. He found that all of the KLL components were of energies 13–15 eV lower than the values measured for the metallic oxide (Na_2O) by Fahlman et al.[49] Since in Na_2O the bonding is ~82% ionic, it is also of interest to compare the KM_1–LLM_1 transition energies for the sodium vapor with Fahlman's values: The KM_1–LLM_1 component energies lay at 18–20 eV below the corresponding KLL component energies for the oxide;[42a] that is, they were in worse agreement with the oxide

Table III.
Comparison of experimental[42)51)] and theoretical values (relativistic Hartree-Fock-Slater[42)] and non-relativistic Hartree-Fock[44)]) for the KLL Auger lines for neon.

Transition	Energy (eV)				
	Experiment		Theory		Semi-empirical
	ref.[51)]	ref.[42)]	ref.[44)]	ref.[42)]	Table II.
$KL_1L_1(^1S)$	748.05(42)	748.1(1)	746.91	748.13	761
$KL_1L_{2,3}(^1P)$	771.45(42)	771.6(1)	770.81	775.81	781
$KL_1L_{2,3}(^3P)$	781.95(42)	781.9(1)	782.95	786.36	794
$KL_{2,3}L_{2,3}(^1S)$	800.45(42)	800.5(1)	800.90	804.47	808
$KL_{2,3}L_{2,3}(^1D)$	804.15(40)$^+$	804.15^{++}	805.98	808.52	813

$^+M_5N_1N_{2,3}(^1P_1)$ and $M_5N_{2,3}N_{2,3}(^1D_2)$ Auger lines of krypton and $L_3M_1M_{2,3}(^1P_1)$ Auger line of argon used for energy calibration. Energy differences from the $KL_{2,3}L_{2,3}(^1D)$ line accurate to ±0.1 eV.

$^{++}$Energies measured only; the $KL_{2,3}L_{2,3}(^1D)$ energy was taken from ref.[51)] (private communication from T. Bergmark).

values than were the sodium vapor *KLL* component energies.

The *KLL*-Auger spectra for sodium and magnesium in both the solid metallic and oxide states has been measured by Weichert.[42b] He finds that in both cases there are shifts of several electron volts in component energies between the two chemical states for each element.

As yet there are no detailed measurements carried out on the *KLL*-Auger spectrum for argon. A similar trend in agreement between the various corresponding sets of values might be expected there as well.

Recently Asaad[45] calculated the Slater's integrals for the *K*- and *L*-shells using relativistic hydrogenic radial wave functions (effects of screening were not considered). It was shown that the relativistic Slater integrals can be expressed in the form $AZ(1 + aZ^2)$, where a is positive for all integrals excepting $\mathcal{G}^2(21, 21)$. The values obtained for A and a are given in Table I. It was thus established that the factor in the Slater's integrals takes the form $(1 + aZ^2)$ and not $(1 + \beta Z^3)$ as suggested by Hörnfeldt.[41]

The calculations by Gelius and Nordqvist[46] of the binding and *K*-Auger energies for elements from sodium ($Z = 11$) through copper ($Z = 29$) were carried out using both the relativistic and nonrelativistic Hartree–Fock–Slater self-consistent-field method. The difference between the *K*-Auger energies calculated using the two methods varied from ~2 eV for magnesium ($Z = 12$) to ~20 eV for calcium ($Z = 20$). Nestor et al.[44] have compared the importance of the influence of relativity with that of exchange on the calculation of the *ns*-shell electron binding energies for a number of elements from $Z = 10$ to $Z = 79$. It was seen that, when the electron binding energy is above ~2 keV, the relativistic effect becomes the more important one. Furthermore, it was demonstrated that the influence of the relativistic effect does not become negligible even for the outer shells.

These results indicate, then, that it is preferable to use relativistic wave functions in calculating Auger transition probabilities and energies even for the lightest elements.

It is seen that the semiempirical energy values in Table II agree well with experimental values for elements in the solid form for atomic numbers ranging from 100 down to about 10. Figure 2 shows the differences between the experimentally determined energies of the $KL_1L_1(^1S_0)$ line and the other Auger lines plotted as a function of Z (see Table VIIIa). Also included are the theoretical values obtained for $Z = 10$ and $Z = 80$.

The $KL_1L_1(^1S_0)$ to $KL_3L_3(^3P_2)$ energy difference is normalized to unity. Curves indicating energy differences of 0.1, 0.5, 1.0, 5.0, and 10.0 keV from the $KL_1L_1(^1S_0)$ energy values are also inserted. At the top of Figure 2 is a plot of the variation of the approximate KL_1L_1 and $KL_{2,3}L_{2,3}$ Auger line half-widths with Z-value; the vertical scale is made identical to that for the relative Auger line positions below. For comparison, a separation of 0.1% in $B\rho$-value from the $KL_1L_1(^1S_0)$ line is also plotted as a function of Z in the upper part of the figure. Obviously the combined effects of the natural Auger line widths (see Chapter 6, Section 4.1) and the resolution of the spectrometer used, would act with any line broadening effects due to electron energy losses in the source to make it impossible for higher Z-values to resolve clearly some of the Auger line doublets. It has been possible, however, to employ graphic analysis to establish the presence of some of the weak satellite lines in the spectra for elements of high Z-values. It can also be seen that for intermediate Z-values (from approximately $Z = 30$ to $Z = 65$) it is possible to resolve the total nine-line *KLL*-spectrum using an instrumental resolution of ~0.06% and sufficiently thin sources,[47, 25] the $50 \leq Z \leq 56$ region being excepted because of overlapping of some of the lines.[20] Values obtained from theoretical energy calculations for $Z = 80$ are also included[39] (see Table VI), as are the values, taken from Ref. 44, for $Z = 10$ (see Table III).

At the bottom of Figure 2 the relative differences between the experimental and semiempirical calculated energies of the $KL_2L_3(^1D_2)$ Auger transitions appear.

It is important to remember that absolute energies of the electron lines are directly dependent on the accuracy of the calibration energy. The error in the relative energy difference between two electron lines, however, may be quite low in comparison. Graham et al.[20] indicated these electron line energy accuracies by writing, for example, 21.78_7 keV for the energy of the $KL_1L_1(^1S_0)$ Auger line for tellurium ($Z = 52$). This figure indicated that the absolute line energy is 21.787(10) keV, whereas the error of energy difference between this line and any other in the *K*-Auger spectrum will probably be much less than 10 eV.

Figure 3*a* shows a *KLL*-Auger spectrum in ^{125}I ($Z = 53$) obtained by Graham et al.[20] The *K*-vacancies in ^{125}I result from *K*-capture in 18-hr ^{125}Xe and internal conversion in ^{125}I. In order to obtain as thin a source as possible with high specific activity, the gaseous ^{125}Xe was mass separated and allowed to impinge on an aluminum foil backing at a retarded energy of 1.5 keV. As can be seen, the $KL_2L_2(^1S_0)$ and $KL_1L_3(^3P_1)$ lines have very nearly the same energy

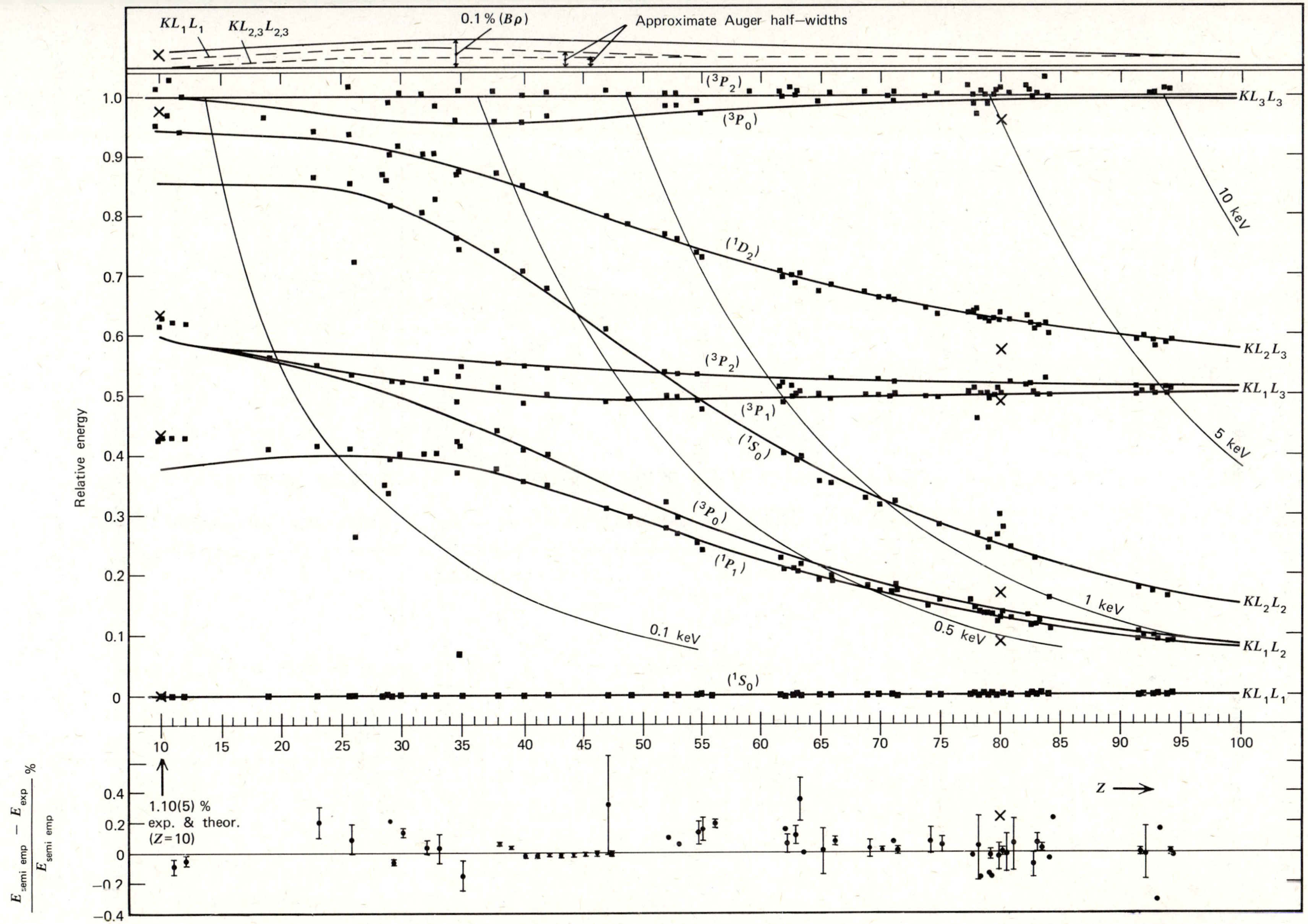

Figure 2. Relative energy separations of the *KLL*-Auger line components from the $KL_1L_1(^1S_0)$ line in units of the energy $E[KL_3L_3(^3P_2)] - E[KL_1L_1(^1S_0)]$ taken from the semiempirical values in Table II. Entered are also the energy differences of 0.1, 0.5, 1.0, and so on. keV from the $KL_1L_1(^1S_0)$ line position, as a function of atomic number. At the top of the figure the KL_1L_1 and $KL_{2,3}L_{2,3}$ natural half-widths are compared to a 0.1 % electron *momentum* difference as a function of Z-value. The upper vertical scale, in relative energy units, is identical to that below.

At the bottom of the figure are shown the percentage deviations of experimental energy values of the strongest $KL_2L_3(^1D_2)$ Auger line from semiempirical values in Table II. These values are based on the energy measurements of Auger spectra from atoms in the solid state. For $Z = 10$, however, where the neon atoms are in the gaseous state, the relative Auger energies are in agreement with the general trend, whereas the absolute values, both experimental and theoretical, are much lower than the semiempirical values (see text).

The corresponding relative and absolute energy values for $Z = 10$ (see Table III) and $Z = 80$ (see Table VI), from theoretical calculations, are entered for comparison at the points marked with crosses.

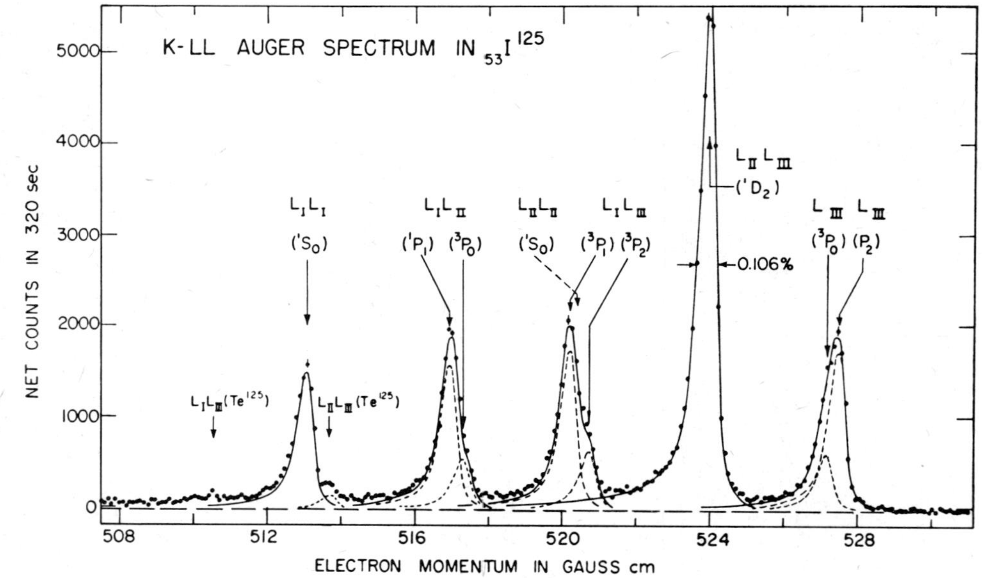

(a)

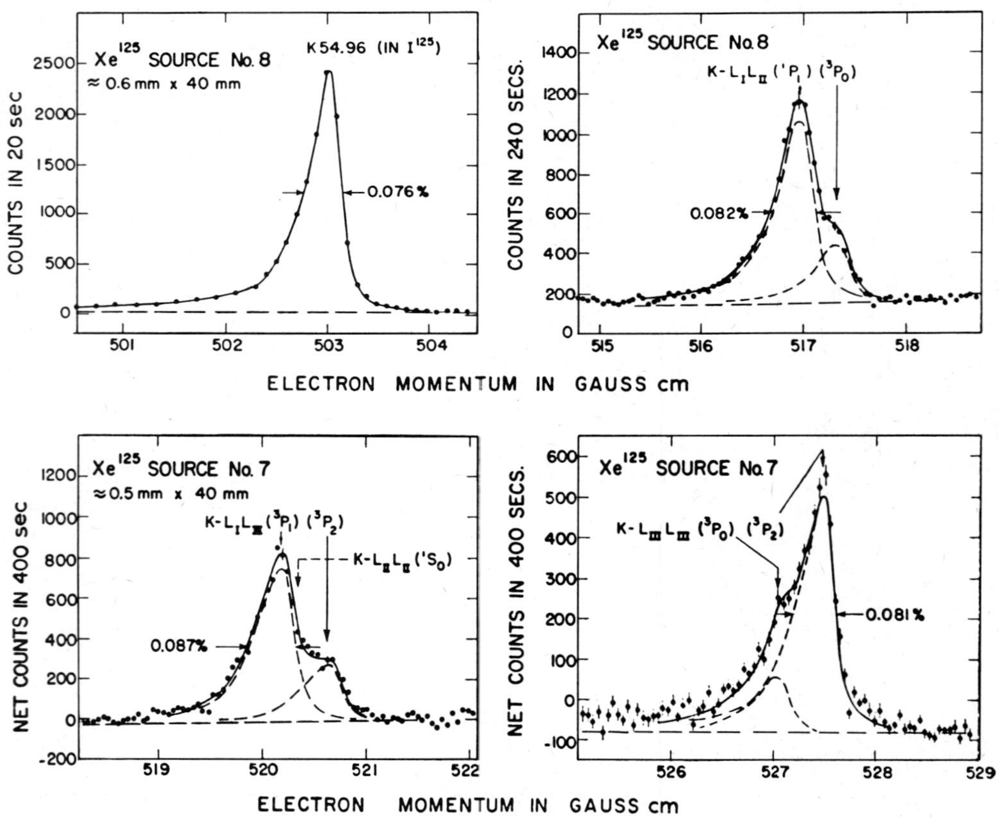

(b)

Figure 3. (*a*) The *KLL*-Auger spectrum in ^{125}I ($Z = 53$) measured by Graham et al.[20] (*b*) Study of the complex line groups using a higher resolution.

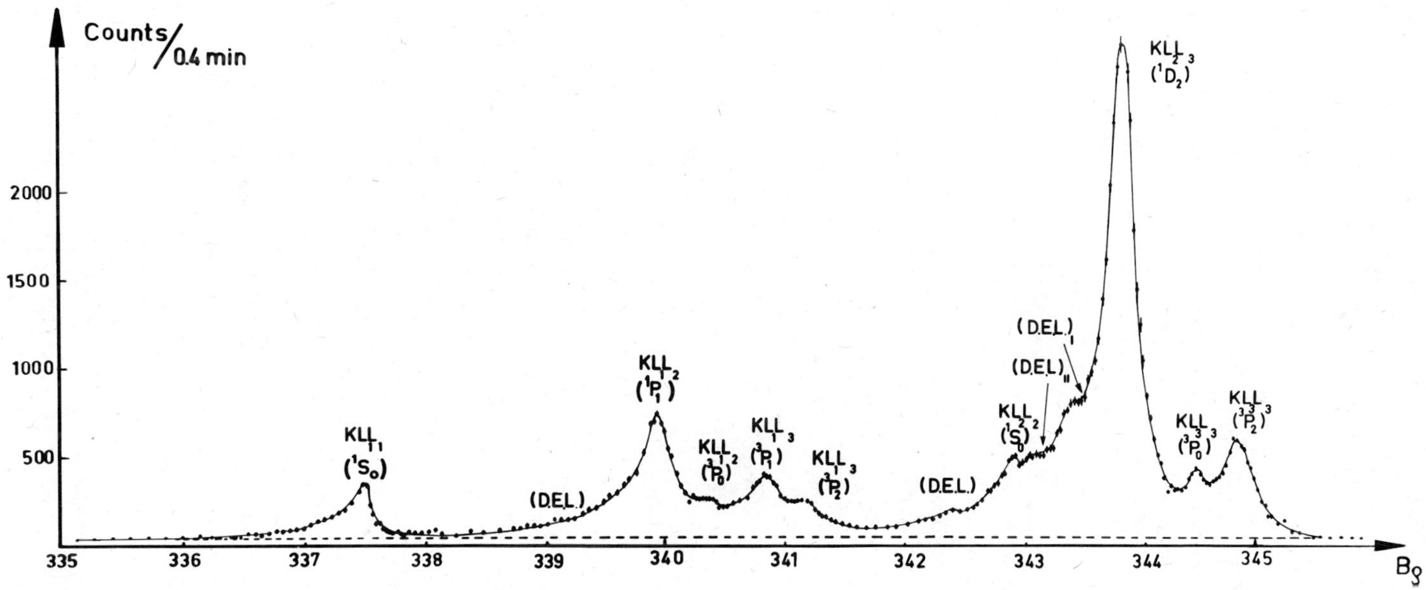

Figure 4. The *KLL*-Auger spectrum in ^{79}Br ($Z = 35$) obtained by Erman et al.[47] Structures interpreted as owing to discrete energy losses (D.E.L.) suffered in the solid radioactive source are denoted.

value for this Z-value as shown in Figure 2, and only eight KLL-components can be proved to be present. The complex line groups in the spectrum were then scanned in detail at an increased instrumental resolution to demonstrate the multiplet character. The results are shown in Figure 3*b*.

The KLL-Auger spectrum in ^{79}Br ($Z = 35$) was calculated by Erman et al.,[47] who employed source preparation techniques similar to those used by Graham et al.[20] The measured spectrum clearly showing the nine-component spectrum is reproduced in Figure 4. When the relative line positions are entered into Figure 2, all the spectral components except the $KL_1L_1(^1S_0)$ line agree with the predicted relative energy values, if the entire spectrum is shifted upward so that the $KL_3L_3(^3P_2)$ component position falls at the top of the figure. Asaad[48] investigated the effects of configuration interaction on the $KL_1L_1(^1S_0)$–$KL_2L_2(^1S_0)$ and $KL_1L_1(^1S_0)$–$KL_3L_3(^3P_0)$ transition energy differences in an attempt to explain this discrepancy, but it became obvious that this was not the solution. The $KL_1L_1(^1S_0)$–$KL_2L_3(^1D_2)$ component energy difference for bromine was later remeasured by Fahlman et al.,[49] employing X-ray photoionization. They obtained a value within 5 eV of the value from Hörnfeldt's table as compared with the discrepancy of 47 eV obtained by Erman et al.[47]

A number of investigations of the KLL-Auger spectrum for elements having $Z < 13$ have been reported. The initial K-shell vacancy is most advantageously produced by employing photoionization or electron bombardment methods. At the low energy values involved, difficulties arising from electron straggling in a solid converter material are greatly reduced (see Chapter 8, Section 7).

Körber and Mehlhorn have twice investigated the KLL-Auger spectrum for gaseous neon ($Z = 10$) using an electrostatic spectrometer: initially at an energy resolution of 1.2% employing X-radiation[50] to produce atomic ionization, and later at the much higher energy resolution of 0.12% using electron bombardment.[51] (See table III.) The spectral electrons were post-accelerated to ensure penetration of the counter window. By using this arrangement, the effects on the spectral line shapes of electron energy losses due to straggling were effectively negligible. Figure 5 shows the measured electron spectrum for neon, using an energy resolution of 0.5% in the energy region of the KLL-Auger components. The expected five KLL-Auger components, together with a number of additional electron lines, were observed. The latter were assigned to KL–LLL type Auger transitions. This effect is treated further in Section 4.1.3.

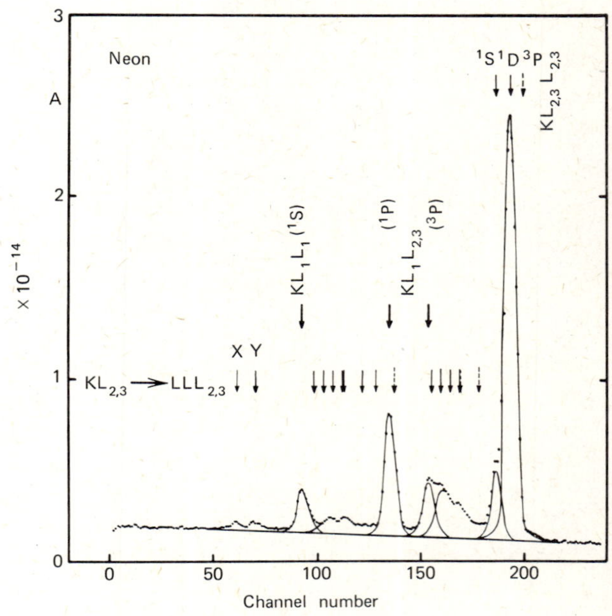

Figure 5. The KLL-Auger spectrum observed in neon ($Z = 10$) by Körber and Mehlhorn[51] following initial ionization by electron bombardment. A number of KL–LLL Auger-vacancy-satellite lines are also visible.

Albridge et al.[54] investigated the KLL-Auger spectrum from fluorine ($Z = 9$) in the solid chemical forms LiF, NaF, and MgF_2. The observed variation among the energies of the Auger spectral components was interpreted as a result of the differences between the three cations (see Section 4.1.3).

As can be seen in Table VIIIa, there have been only a few high resolution investigations of the KLL-Auger spectrum for elements having atomic numbers lying between 12 and 30. Mehlhorn and Albridge[55] found six components in the spectrum for iron ($Z = 25$) and five in that for vanadium ($Z = 23$). The radioactive isotopes ^{57}Co and ^{51}Cr were used as electron sources; these decay by electron capture to the stable isotopes ^{57}Fe and ^{51}V, respectively. Source thickness effects on the experimental spectral line shapes were observed to be strong, and the KLL-Auger spectrum for iron was analyzed into its components by using the K-conversion line shape of the 14.4-keV transition in ^{57}Fe. In the case of vanadium, the ^{51}Cr and ^{57}Co activities were electroplated simultaneously onto the same source backing and the 14.4-K conversion line shape in ^{57}Fe was used to make possible a qualitative analysis of vanadium. The presence of the KL_3L_3 Auger line was established in the spectrum for iron but not in that for vanadium, which exhibited larger spectral line width values resulting from the increased source thickness.

Measured energy values of the *KLL*-components, as reported in the literature, are collected in Table VIIIa together with their measured relative intensities. Also included is information on the type of electron source, instrumentation, calibration, and, for low Z-values, the chemical and physical state of the substance investigated. These tabulated energy values were used to obtain the data included in Figure 2.

4.1.2. *KLL-Auger Line Intensities*

In 1927 Wentzel[56] demonstrated that transition probabilities for the Auger effect, calculated with the nonrelativistic approximation, were nearly independent of atomic number Z, and equal to $\sim 10^{15}$ sec^{-1}. This independence of atomic number has been confirmed in later nonrelativistic treatments (see, e.g., Refs. 57 and 16), whereas relativistic effects appear to account for an increase in the Auger transition probability by ~83% for gold ($Z = 79$) and by ~84% for mercury ($Z = 80$), with radiative transitions being decreased by ~75%.[39]

In calculations based on the nonrelativistic approximation, consider the two electrons moving in their initial states in both the field due to the nucleus and the average field due to the other atomic electrons. The Coulomb interaction between the two electrons involved in the Auger transition is considered to be the perturbation that causes one electron to undergo a transition from its initial bound state to a final state in the positive energy continuum, whereas the second electron fills the initial atomic vacancy. Letting $\varphi_i(\mathbf{r}_1)$ and $\psi_i(\mathbf{r}_2)$ represent the initial single-electron wave functions of the two electrons in the atom having an inner level vacancy and $\varphi_f(\mathbf{r}_1)$ and $\psi_f(\mathbf{r}_2)$ the final states of these two electrons, one can write according to first-order perturbation theory

$$W_a = \frac{2\pi}{\hbar}\left|\iint \varphi_f^*(\mathbf{r}_1)\psi_f^*(\mathbf{r}_2)\frac{e^2}{|\mathbf{r}_1 - \mathbf{r}_2|} \times \varphi_i(\mathbf{r}_1)\psi_i(\mathbf{r}_2)\,d\mathbf{r}_1\,d\mathbf{r}_2\right|^2,$$

where W_a is the Auger transition probability and $e^2/|\mathbf{r}_1 - \mathbf{r}_2|$ is the perturbation energy. In addition, one must consider that the initial and final wave functions must be antisymmetrical in their coordinates, since the two electrons are indistinguishable. This is done by replacing the product $\varphi_i(\mathbf{r}_1)\psi_i(\mathbf{r}_2)$ by $2^{-1/2}[\varphi_i(\mathbf{r}_1)\psi_i(\mathbf{r}_2) - \varphi_i(\mathbf{r}_2)\psi_i(\mathbf{r}_1)]$ and making a similar replacement of the product $\varphi_f^*(\mathbf{r}_1)\psi_f^*(\mathbf{r}_2)$. The wave functions for the bound states are normalized to unity, and the continuum wave function is normalized to represent one ejected electron per unit time.

In the relativistic theory, the Auger transition rate cannot be calculated in a manner analogous to that used for the nonrelativistic approximation. Instead, we assume that the transition of one electron from its initial state to the positive energy continuum is caused by the interaction of that electron with the electromagnetic radiation emitted by the other electron when it fills the atomic vacancy. The Auger transition process is thereby considered to be one of internal absorption of radiation.

In 1935 Burhop[58] calculated the transition probabilities for Auger transitions of the K-series in silver ($Z = 47$) using a nonrelativistic treatment and assuming j–j coupling between final atomic states. Hydrogen-like wave functions were employed and screening was taken into account by use of Slater's screening constants The transition amplitudes used in these calculations were later shown to be in error and were eventually recalculated by Callan.[59] (See also Ref. 60 for a treatment of the computation of hydrogenic functions in calculating bound-free matrix elements.) At about the same time, calculations of Auger transition probabilities for atomic shells and subshells up through $4p$, using unscreened hydrogen wave functions, were published by Pincherle[57] (see Section 4.3). The following year Massey and Burhop[61] calculated the KL_1L_1:KL_1L_2:KL_1L_3 transition probability ratios for gold ($Z = 79$) using a relativistic calculation and employed Dirac wave functions (see Table IV). The calculations were only exploratory, and their object was to estimate the magnitude of the relativistic effect.[39] It was later pointed out that these calculations included a number of errors.[10]

Ramberg and Richtmeyer[62] carried out nonrelativistic calculations on the Auger transition probabilities for gold ($Z = 79$) assuming j–j coupling. Screening was according to the Thomas–Fermi statistical model. Their results are included in Table IV.

In 1955 Rubenstein and Snyder[63,64] assumed extreme L–S coupling in their investigation of the Auger spectra for argon ($Z = 18$), krypton ($Z = 36$), and silver ($Z = 47$). They used wave functions for a self-consistent field. The transition probability values obtained for KL_1L_1, $KL_1L_{2,3}$, and $KL_{2,3}L_{2,3}$ are listed in Table IV. (See also Table VII.)

In 1958 Asaad and Burhop[16] presented nonrelativistic calculations, on the basis of intermediate coupling, for the energies and relative intensities of the components of the *KLL*- and *KLM*-Auger groups for five atomic number values from $Z = 25$ to $Z = 80$ (see also Section 4.1.1). For the *KLL*-spectrum, the transition amplitudes of Burhop[58] were used, whereas for the *KLM*-spectrum those of Rubenstein and

Table IV.

Comparison of theoretical KLL Auger transition probabilities (10^{-3} atomic units).

	Non-relativistic								Relativistic				
	Pincherle[57]	Burhop[58]	Ramberg & Richtmyer[62]	Asaad[39]	Krause et al[69]	Rubenstein and Snyder[63][64]			Massey & Burhop[61]	Asaad[39]	Listengarten[65][66]		
	All Z	Z=47	Z=79	Z=80	Z=10	Z=18	Z=36	Z=47	Z=79	Z=80	Z=65	Z=81	Z=92
	H,u	H,s	T-F	HSC	H-F	HSC			D	HSC	T-F-D		
KL_1L_1	2.5	2.54	2.02	2.04	0.577	1.15	1.91	1.92	1.74	9.001	6.88	10.12	18.54
KL_1L_2	--	--	--	2.41	--	--	--	--	9.52	12.956	10.33	17.18	35.59
KL_1L_3	--	--	--	4.75	--	--	--	--	9.21	7.367	9.11	9.18	13.48
KL_2L_2	--	--	--	0.51	--	--	--	--	--	0.778	0.59	0.56	0.77
KL_2L_3	--	--	--	10.89	--	--	--	--	--	13.177	12.96	13.27	16.59
KL_3L_3	--	--	--	6.18	--	--	--	--	--	5.902	7.61	5.58	6.93
$KL_1L_{2,3}$	8.83	8.65	7.10	7.16	2.11	3.74	6.79	6.86	18.73	20.32	19.44	26.36	49.07
$KL_{2,3}L_{2,3}$	24.3	17.11	19.26	17.58	12.20	9.38	18.57	18.50	--	19.86	21.16	19.41	24.29
ΣKLL	35.6	28.30	28.38	26.78	14.89	14.27	27.27	27.28	--	49.19	47.48	55.89	91.90

H: using hydrogenic wave functions; u: assuming an unscreened atom; s: using Slater screening constants; T-F: using wave functions for a Thomas-Fermi statistical potential; HSC: using wave functions for a Hartree self-consistent field; H-F: using wave functions from Hartree-Fock solutions; D: using Dirac wave functions; T-F-D: using wave functions for a Thomas-Fermi-Dirac statistical potential.

Snyder[63,64] were employed. Table V gives the results of their calculations of the transition probabilities for the components of the two Auger groups.

This treatment, assuming intermediate coupling, gave relative *K*-Auger intensities in much improved agreement with experimental values. Some discrepancies remained, however, particularly for high atomic numbers where extreme *j–j* coupling is approached. The neglect of relativity was responsible for these.

Soon after his nonrelativistic calculations Asaad[39] published results for the relative intensities and energies of the *KLL*-Auger lines for mercury ($Z = 80$). The results of these relativistic calculations were in very favorable agreement with experimentally determined values. Also included were the results of nonrelativistic limiting values calculated for both *j–j* and intermediate coupling. The agreement between calculations for these two couplings shows that, for $Z = 80$, intermediate coupling approaches almost exactly the limit of *j–j* coupling. The calculated relative intensities of the *KLL*-Auger lines for the two types of coupling and their predicted energies are listed in Table VI.

Some years later Callan,[59] using screened nonrelativistic hydrogenic wave functions and assuming *L–S* coupling, calculated the absolute and relative intensities of the *KLL*-components for 40 elements from $Z = 12$ through $Z = 80$. He noticed the influence of relativistic effects on Auger transition rates in heavier atoms, although the calculations were nonrelativistic. This was attributed to the derivation of the Auger energies from atomic energy level values, which are, of course, subject to relativistic effects. Callan also included a recalculation of Burhop's[58] Auger transition amplitudes for silver ($Z = 47$).

In 1961 Listengarten[65] published relativistic calculations of Auger transition rates for thallium ($Z = 81$) assuming *j–j* coupling; screening was accounted for by using the Thomas–Fermi–Dirac statistical model of the atom. The following year the same author[66] extended this relativistic treatment to include the Auger transition probabilities for terbium ($Z = 65$) and uranium ($Z = 92$). The obtained values for the three elements are entered in Table IV. These theoretical values give relative Auger transition probabilities in good agreement with experimental values (see Figures 6 and 7).

In 1963 Asaad[67] published calculations of the intensities of the nine lines of the *KLL*-Auger spectrum based on intermediate coupling theory. The transition amplitudes computed by Callan[59] were used, and the calculations were carried out for the same elements treated by the latter. Good agreement was obtained between the calculation results and experiment for elements of medium atomic numbers, relativistic effects becoming important for high *Z*-values. Deviations from pure *L–S* coupling were observed from $Z = 25$ upward, although extreme *j–j* coupling was not fully approached even at $Z = 80$. Computed values were also included for the Auger line strengths in the *j–j* coupling limit. In addition, a comparison was made between the two ways of treating the final atomic electron system: as one of two electrons or, the correct way, as one of two holes. At the same time Asaad presented in another article[68] a general expression for Auger transition probabilities from any single initial vacancy to any final configuration of the doubly ionized atom. Extreme *j–j* coupling was assumed in the final and initial configurations. This expression was in general terms of radial integrals, Clebsch-Gordon coefficients, and Racah coefficients. Tables were included giving the Auger transition probabilities in terms of radial integrals for initial $s_{1/2}$, $p_{1/2}$, and $p_{3/2}$ vacancies and final vacancies through $d_{5/2}d_{5/2}$.

In their study of the de-excitation of neon atoms initially ionized in the *K*-shell by X-rays, Krause et al.[69] calculated the KL_1L_1, $KL_1L_{2,3}$, and $KL_{2,3}L_{2,3}$ transition probabilities for this element. Pure *L–S* coupling was assumed. Their results are entered in Table IV.

Chattarji and Talukdar[70] carried out relativistic calculations of the KL_1L_1 Auger transition probability for elements lying between $Z = 72$ and 80. It was seen that relativistic effects cause this probability to increase from 3.741 to 6.035 ($\times 10^{-3}$ in atomic units) over this region of atomic number. These values may be compared with those in Table IV.

The *KLL*-Auger spectra for the heaviest elements ($Z \geq 85$) have been investigated only twice with high enough resolution to allow measurement of the satellite lines of the KL_1L_2 and KL_1L_3. Ewan et al.[17] were able to show that the KL_1L_3 line for $Z = 94$ had a satellite of low intensity. Later Briançon[122] measured a satellite structure for both the KL_1L_2 and KL_1L_3 transitions in uranium ($Z = 92$). The relative satellite intensity for the latter transition disagreed with that of Ewan et al. by a factor of 5. In view of this disagreement with earlier results, Briançon carried out several measurements of the satellite structure, all of which supported her initial result. Subsequently when the KL_1L_2 line plus its satellite was remeasured in Chalk River, the intensity ratio between the two components agreed with Briançon's measurements. The earlier Chalk River value for the satellite intensity was in error due to the use on an erroneous line profile for analysis.[70a]

Up to this time the intermediate coupling theory had

Table V.

Relative intensities of KLL and KLM Auger lines calculated by Asaad and Burhop[16] using the intermediate coupling approximation.

Z=	25	30	37	47	80
$L_1L_1(^1S_0)$	1.00	1.00	1.00	1.00	1.00
$L_1L_2(^1P_1)$	2.99	2.80	2.40	1.77	1.15
$L_1L_2(^3P_0)$	0.047	0.047	0.047	0.047	0.047
$L_1L_3(^3P_1)$	0.24	0.40	0.80	1.42	2.03
$L_1L_3(^3P_2)$	0.24	0.24	0.24	0.24	0.24
$L_2L_2(^1S_0)$	0.41	0.38	0.32	0.24	0.15
$L_2L_3(^1D_2)$	6.08	5.54	5.08	4.71	4.32
$L_3L_3(^3P_0)$	0.012	0.038	0.096	0.185	0.28
$L_3L_3(^3P_2)$	0.47	0.86	1.30	1.72	2.12
$L_1M_1(^3S_1)$	0.001	0.001	0.001	0.002	0.002
$L_1M_1(^1S_0)$	0.17	0.19	0.20	0.29	0.31
$L_1M_2(^3P_0)$	0.005	0.006	0.007	0.011	0.013
$L_1M_2(^1P_1)$	0.19	0.18	0.17	0.19	0.18
$L_1M_3(^3P_1)$	0.035	0.059	0.114	0.25	0.31
$L_1M_3(^3P_2)$	0.027	0.031	0.037	0.057	0.062
$L_2M_1(^3P_0)$	0.005	0.005	0.005	0.006	0.006
$L_2M_1(^1P_1)$	0.146	0.138	0.118	0.150	0.152
$L_3M_1(^3P_1)$	0.100	0.115	0.170	0.25	0.28
$L_3M_1(^3P_2)$	0.023	0.024	0.024	0.033	0.035

Table V. (cont.)

Z=	25	30	37	47	80
$L_1M_4(^3D_1)$	--	0.001	0.002	0.010	0.010
$L_1M_4(^1D_2)$	--	--	0.001	0.004	0.009
$L_1M_5(^3D_2)$	--	0.001	0.003	0.015	0.010
$L_1M_5(^3D_3)$	--	0.002	0.005	0.022	0.022
$L_2M_2(^1S_0)$	0.080	0.077	0.059	0.087	0.071
$L_3M_2(^3S_1)$	0.001	0.001	0.001	0.001	0.001
$L_3M_3(^1P_1)$	--	--	0.0004	0.001	0.001
$L_3M_3(^3P_0)$	0.020	0.029	0.058	0.125	0.142
$L_2M_3(^3P_1)$	--	0.0002	0.0004	0.0009	0.0010
$L_3M_2(^3P_2)$	0.087	0.34	0.39	0.75	0.80
$L_2M_3(^1D_2)$	1.08	0.93	0.88	1.035	0.93
$L_2M_2(^3D_1)$	0.001	0.001	0.001	0.002	0.002
$L_3M_3(^3D_2)$	0.015	0.016	0.088	0.60	0.70
$L_3M_3(^3D_2)$	--	0.002	0.003	0.005	0.005
$L_3M_4(^3P_0)$	--	--	0.0002	0.0007	0.0014
$L_3M_4(^3P_1)$	--	--	0.0003	0.0012	0.0030
$L_2M_5(^3P_2)$	--	--	0.0006	0.0026	0.0058
$L_2M_4(^1P_1)$	--	--	0.0004	0.0014	0.0023
$L_3M_5(^3D_1)$	--	--	0.0004	0.0019	0.0041
$L_2M_4(^3D_2)$	--	--	0.0049	0.028	0.053
$L_3M_4(^3D_3)$	--	--	0.0099	0.138	0.315
$L_3M_5(^1D_2)$	--	--	0.0014	0.0055	0.0057
$L_3M_4(^3F_2)$	--	--	0.0015	0.0057	0.0146
$L_3M_5(^3F_3)$	--	--	0.0084	0.0225	0.040
$L_3M_5(^3F_4)$	--	--	0.0135	0.070	0.137
$L_2M_5(^1F_3)$	--	--	0.049	0.158	0.263
$\frac{KLM}{KLL}$ =	0.173	0.189	0.215	0.382	0.432

Table VI.

KLL Auger line intensities and energies for mercury (Z=80)[39]

Transition	Energy (keV)	Relative intensities	
		Relativistic	Non-relativistic
KL_1L_1	53.293	1.00	1.00
KL_1L_2	53.713	1.44	1.18 (1.27)
KL_1L_3	55.780	0.82	2.33 (2.25)
KL_2L_2	54.133	0.09	0.25 (0.25)
KL_2L_3	56.199	1.46	5.34 (5.37)
KL_3L_3	58.266	0.66	3.03 (3.00)

The relativistic values are for j-j coupling where the total KLL transition probability is 49.19×10^{-3} atomic units; the non-relativistic values are for j-j coupling and intermediate (in parentheses) coupling, the total KLL transition probability being 26.78×10^{-3} atomic units.

given good agreement with experimental values for intermediate atomic numbers. Discrepancies between theoretical and experimental values for large Z-values are removed by taking into consideration the effects of relativity. As more experimental data were collected on the relative intensities of KLL-Auger transitions for low Z-values, it became evident that there was disagreement with theory ($Z = 23$, 26, Ref. 55; $Z = 29$, 32, Ref. 71; $Z = 35$, Ref. 47; $Z = 38$, 40, 42, Ref. 25). As Figure 6 illustrates, the experimental values for the relative intensities of the $KL_1L_{2,3}$ and $KL_{2,3}L_{2,3}$ transitions increase with decreasing Z-value, whereas the theoretical values of Callan[59] and Rubenstein and Snyder[63,64] show the opposite tendency. The problem, as described by Assad,[48] became one of not merely improving wave functions used in the calculations, but also of finding a "mechanism" that greatly diminishes the KL_1L_1 line intensity with decreasing Z-value and/or greatly increases the total transition probability for the final $2s^12p^5$ and $2s^22p^4$ configurations. However, Asaad's calculations[45] showed that the interaction between the configurations $2s^02p^6$ and $2s^22p^4$ gives rise to a strong variation of the KL_1L_1 transition probability with Z.

The calculation of the KLL-line intensities, determined using the transition probability amplitudes by Archard,[72] Rubenstein and Snyder,[63,64] Burhop[58] (see also Ref. 59), and Callan,[59] were made by using two different methods. For "method A," use was made of eigenvectors and mixing parameters determined from tabulated electron binding energies[73] and Hörnfeldt's[41] semiempirical expressions for the terms $F^0(20, 20)$, $G^1(20, 21)$, $F^0(21, 21)$, and $F^2(21, 21)$ (see Section 4.1.1). For "method B," analytical wave functions were used in the calculations. The transition probabilities, obtained by method A, to the $2s^12p^5$ and $2s^22p^4$ configurations relative to that to the $2s^02p^6$ configuration are included in Table VII. The results of the calculations using method A are included in Figures 6 through 8. The electron binding energies of several levels (especially of the L_1 level) for the elements sodium ($Z = 11$) through copper ($Z = 29$) have since been remeasured, and a number of errors in previous tabulated values for elements of low atomic number have been corrected[74] (see below).

The practice of relating the intensities of the various KLL-Auger lines to that of the KL_1L_1 line has been exercised for the following reasons: (*a*) this line has the same strength regardless of whether the calculations were based on extreme L–S, extreme j–j, or intermediate coupling; (*b*) it has no satellite lines; and (*c*) the variation of its strength with Z-value is small. However,

Figure 6. The comparison of experimental and theoretical $KL_{2,3}L_{2,3}/KL_1L_1$ and $KL_1L_{2,3}/KL_1L_1$ Auger component intensity ratios Open and solid circles are experimental values (Table VIIIa), and the latter were obtained by graphical analysis of overlapping components. The following symbols and initials are used: L, Listengarten's relativistic calculations;[65,66] A, Asaad's relativistic calculations;[39] A & B, Asaad and Burhop's calculations.[16] The remaining symbols refer to sets of numerical values listed in Table VII and are clearly designated there.

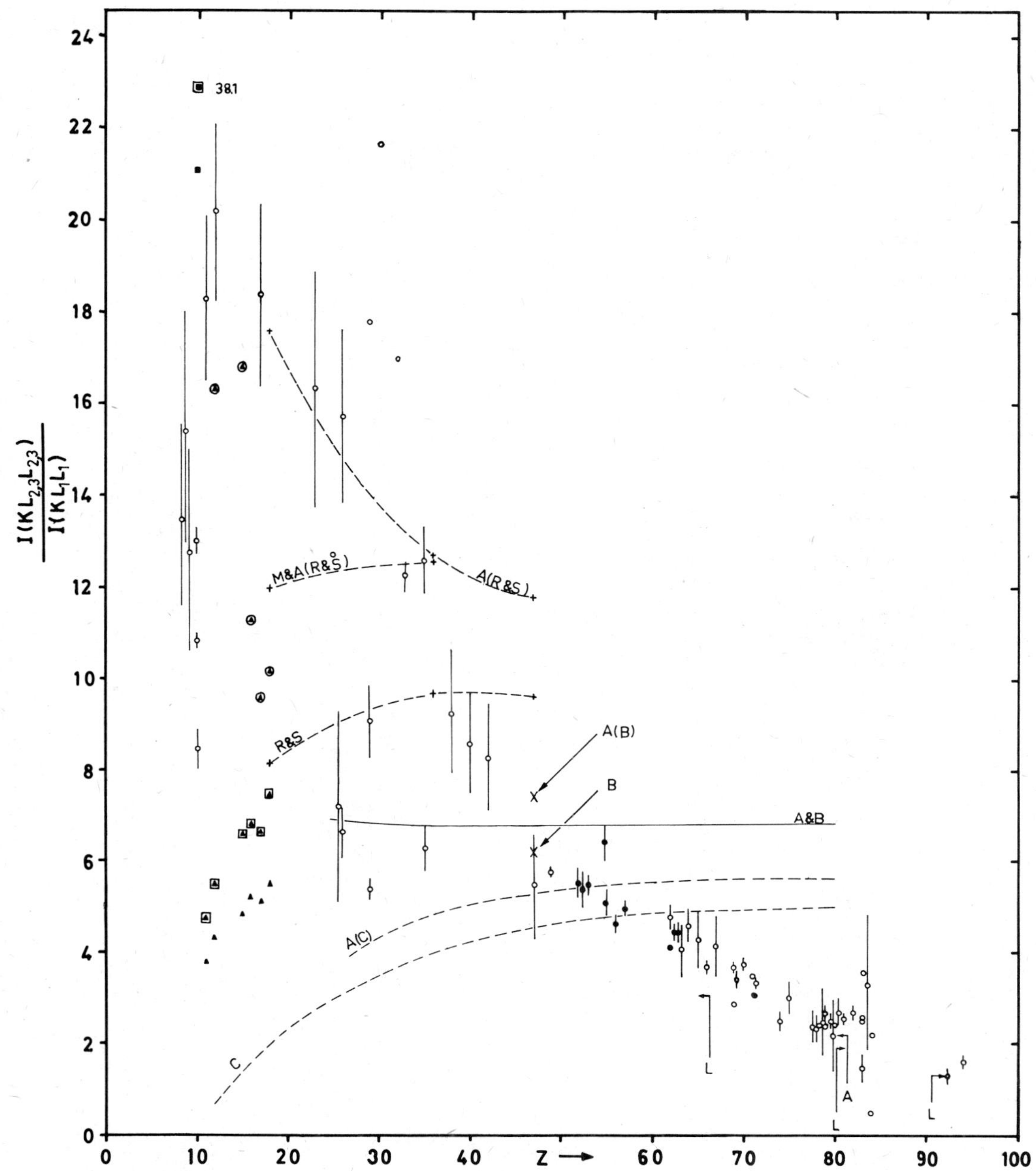

$\frac{I(KL_{23}L_{23})}{I(KL_1L_1)}$
38.1
M&A(R&S)
A(R&S)
R&S
A(B)
B
A&B
A(C)
C
L
A
Z →

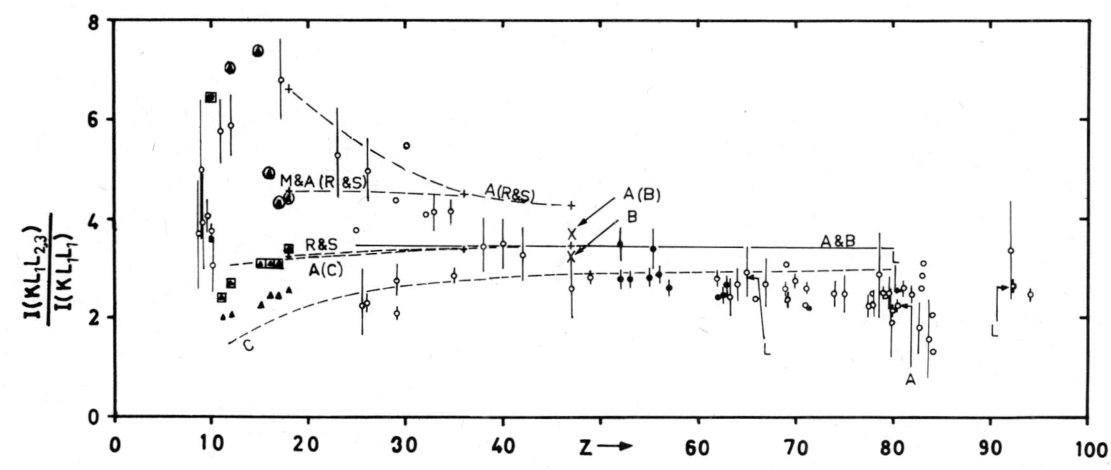

$\frac{I(KL_1L_{23})}{I(KL_1L_1)}$
M&A (R&S)
A(R&S)
A (B)
B
R&S
A(C)
A&B
C
L
A
Z →

Table VII.

Relative transition probabilities calculated using various authors' transition amplitudes.

Z	$I(KL_1L_{2,3})/I(KL_1L_1)$ without configuration interaction	with configuration interaction: Asaad[48)	with configuration interaction: Mehlhorn and Asaad[75)		$I(KL_{2,3}L_{2,3})/I(KL_1L_1)$ without configuration interaction	with configuration interaction: Asaad[48)	with configuration interaction: Mehlhorn and Asaad[75)	
		"Meth. A"	I	II		"Meth. A"	I	II
			Rubinstein and Snyder[63) 64)					
	"R&S"	"A(R&S)"	"M&A(R&S)"		"R&S"	"A(R&S)"	"M&A(R&S)"	
18	3.25	6.62	4.60	4.57	8.16	17.6	11.98	11.87
36	3.55	4.54	4.49	4.48	9.72	12.7	12.57	12.52
47	3.57	4.30	--	--	9.64	11.8	--	--
			Archard[72)					
	"▲"	"(▲)"	"[▲]"		"▲"	"(▲)"	"[▲]"	
11	2.04	--	2.46	2.45	3.78	--	4.78	4.75
12	2.15	7.03	2.73	2.72	4.30	16.3	5.50	5.47
15	2.42	7.39	3.15	3.12	4.84	16.8	6.60	6.52
16	2.51	4.99	3.16	3.14	5.21	11.3	6.82	6.77
17	2.52	4.37	3.16	3.13	5.11	9.61	6.67	6.60
18	2.60	4.50	3.40	3.38	5.48	10.2	7.48	7.43
			Burhop[58) recalculated by Callan[59)					
	"B"	"A(B)"			"B"	"A(B)"		
47	3.24	3.76	--	--	6.28	7.45	--	--
			Krause et al.[69)					
	"■"		"[■]"		"■"		"[■]"	
10	3.66	--	6.48	--	21.13	--	38.1	--
			Callan[59) 67)					
	"C"	"A(C)"			"C"	"A(C)"		
12	1.50	3.06	1.64	--	0.734	2.54	0.921	--
13	1.66	3.28	--	--	0.956	2.86	--	--
14	1.79	3.21	--	--	1.18	2.91	--	--
15	1.90	3.17^+	2.24	--	1.38	2.96^+	1.80	--
16	1.99	3.19	--	--	1.56	3.09	--	--

Table VII. (cont.)

	"C"	"A(C)"			"C"	"A(C)"		
17	2.09	3.06	--	--	1.77	3.05	--	--
18	2.17	3.18	2.61	--	1.97	3.35	2.57	--
19	2.24	3.25	--	--	2.15	3.58	--	--
20	2.30	3.26	--	--	2.32	3.70	--	--
21	2.35	3.28	--	--	2.47	3.85	--	--
22	2.39	3.11	--	--	2.60	3.68	--	--
23	2.43	3.06	--	--	2.72	3.69	--	--
24	2.47	3.02	--	--	2.85	3.71	--	--
25	2.50	3.08	3.05	--	2.98	3.89	3.85	--
26	2.53	3.03	--	--	3.08	3.88	--	--
27	2.56	3.02	--	--	3.19	3.94	--	--
28	2.59	3.06	--	--	3.29	4.08	--	--
29	2.62	3.09	--	--	3.40	4.19	--	--
30	2.65	3.14	3.11	--	3.52	4.35	4.32	--
31	2.66	3.13	--	--	3.60	4.43	--	--
32	2.69	3.17	--	--	3.69	4.53	--	--
33	2.71	3.20	--	--	3.78	4.64	--	--
34	2.73	3.22	--	--	3.86	4.73	--	--
35	2.75	3.23	--	--	3.94	4.80	--	--
36	2.77	3.24	3.23	--	4.02	4.88	4.86	--
37	2.78	3.25	--	--	4.09	4.94	--	--
38	2.80	3.26	--	--	4.15	5.00	--	--
39	2.81	3.27	--	--	4.20	5.04	--	--
40	2.82	3.26	--	--	4.25	5.07	--	--
41	2.84	3.28	--	--	4.30	5.12	--	--
44	2.87	3.29	--	--	4.45	5.25	--	--
45	2.88	3.29	--	--	4.50	5.28	--	--
46	2.89	3.29	--	--	4.54	5.31	--	--
47	2.90	3.29	--	--	4.58	5.34	--	--
50	2.93	3.30	--	--	4.68	5.41	--	--
55	2.97	3.30	--	--	4.85	5.51	--	--
57	2.98	3.30	--	--	4.91	5.55	--	--
70	3.03	3.28	--	--	5.14	5.65	--	--
72	3.04	3.28	--	--	5.16	5.65	--	--
80	3.06	3.27	--	--	5.22	5.64	--	--

[+]Error in ref.[48] corrected in ref.[75]

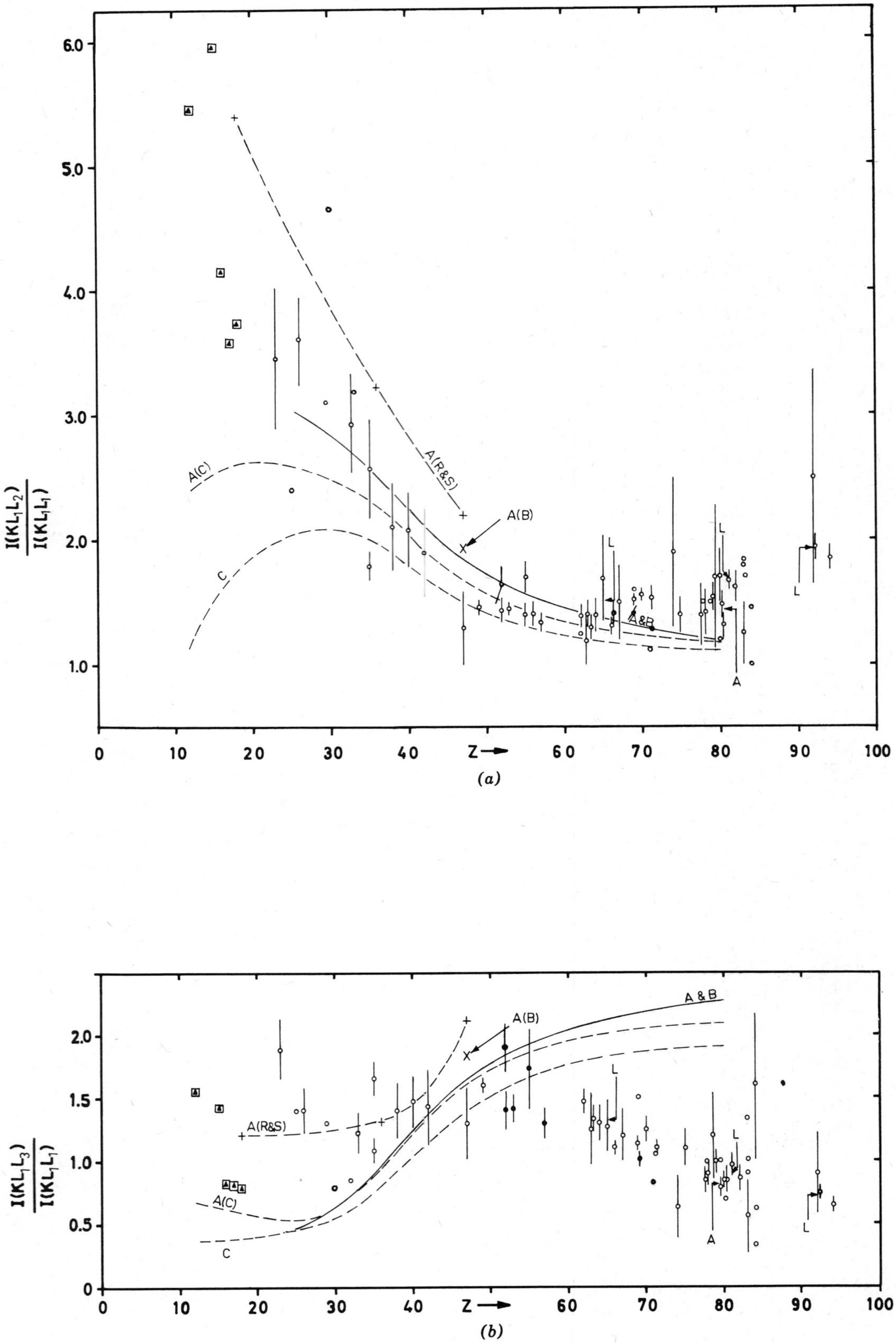

Figure 7. The experimental intensities of the KL_1L_2, the KL_1L_3, the KL_2L_2, the KL_2L_3, and the KL_3L_3 Auger lines relative to that of the KL_1L_1 line. The symbols and initials referring to theoretical values are the same as in Figure 6.

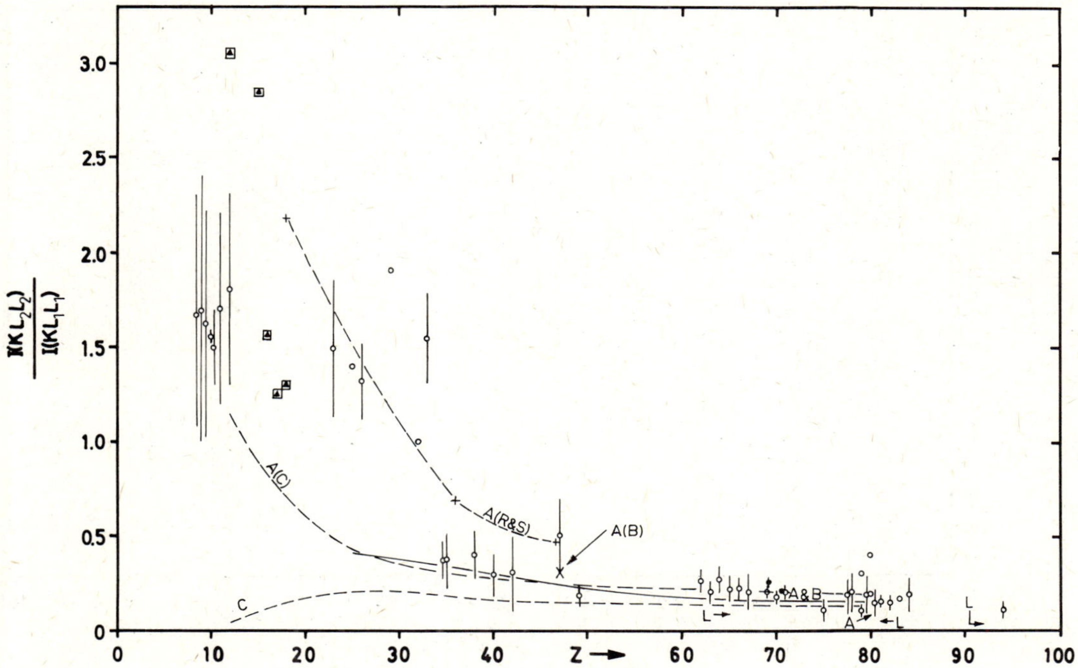

3.0
2.5
2.0
1.5
1.0
0.5
0
I(KL2L2) / I(KL1L1)
A(C)
A(R&S)
A(B)
A&B
C
L
A
0
10
20
30
40
Z
60
70
80
90
100

(c)

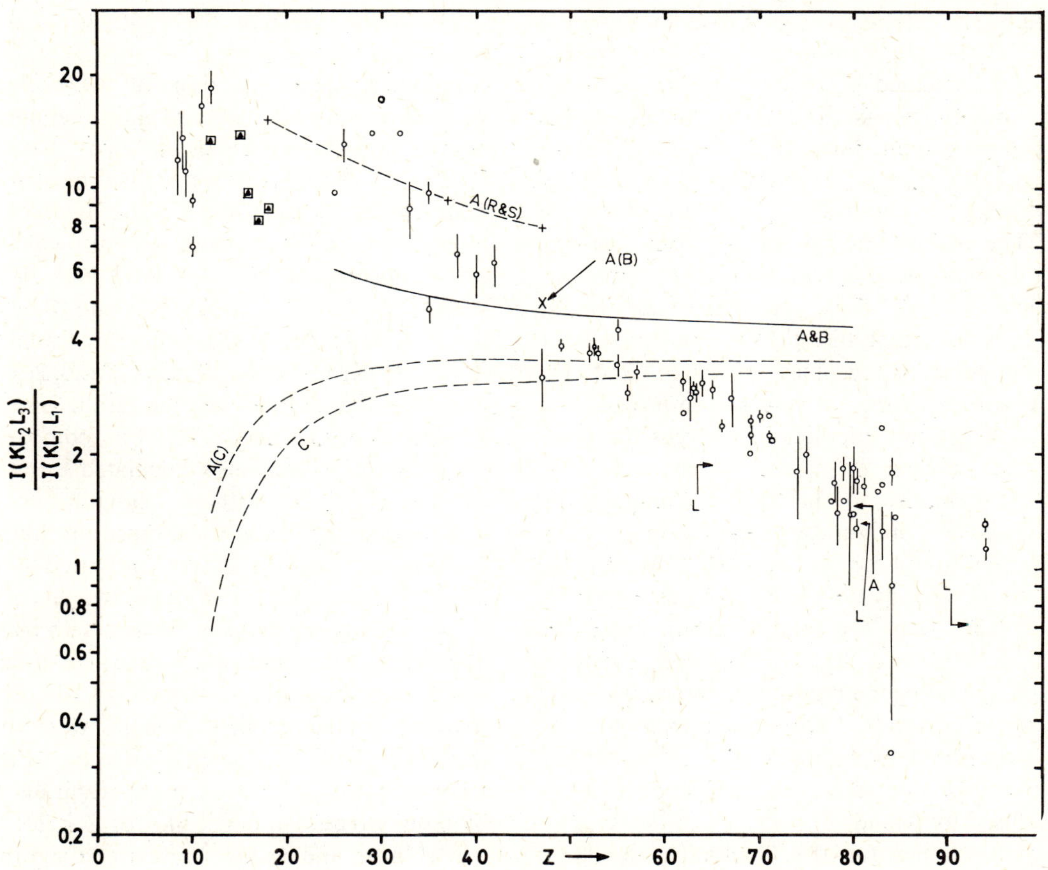

20
10
8
6
4
2
1
0.8
0.6
0.4
0.2
I(KL2L3) / I(KL1L1)
A(R&S)
A(B)
A&B
A(C)
C
L
A
0
10
20
30
40
Z
60
70
80
90

(d)

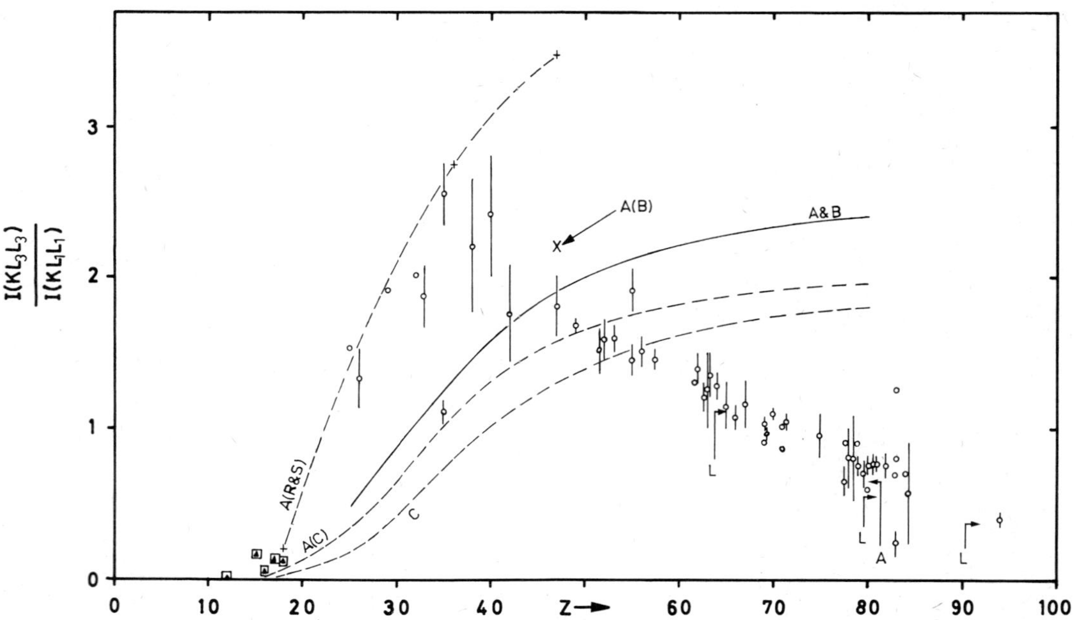

(e)

Asaad's calculations[48] showed that, because of configuration mixing, the interaction of this configuration with the $KL_2L_2(^1S_0)$ and $KL_3L_3(^3P_0)$ configurations must be considered when the KL_1L_1 line intensity is determined. This interaction gives rise to a strong variation of the KL_1L_1 transition probability with Z.

The values calculated by Asaad[48] using Callan's[59] transition amplitudes, method A, for the fractional intensities of the satellite lines, $^3P_0/(^3P_1 + {}^3P_0)$ for the KL_1L_2 line, $^3P_2/(^3P_1 + {}^3P_2)$ for the KL_1L_3 line, and $^3P_0/(^3P_2 + {}^3P_0)$ for the KL_3L_3 line, are shown in Figure 8. The new values for the fractional satellite intensities for the KL_1L_3 and the KL_1L_2 configurations coincide with those calculated earlier by Asaad.[67] It is seen that the satellite intensity of the KL_3L_3 configuration becomes increased and the agreement with experimental values is improved as a result of configuration interaction, whereas the other two fractional satellite intensities, which depend on transition probabilities to the $2s^12p^5$ configuration, are not affected by configuration interaction.

Early in 1966 Mehlhorn and Asaad[75] presented the results of recalculations on the *KLL*-Auger spectrum for $10 \leq Z \leq 36$. Here the relative group intensities $I(KL_1L_{2,3})/(KL_1L_1)$ and $I(KL_{2,3}L_{2,3})/I(KL_1L_1)$ were recalculated including configuration interaction and using more accurate L_1 atomic electron binding energies, remeasured by Fahlman et al.,[74] and more accurate Coulomb and spin-orbit interaction energies than those used by Asaad.[48]

Because the values for the Coulomb interaction matrix elements F and G obtained by extrapolation of Hörnfeldt's formulas[41] did not agree with those calculated using the experimental energies of the *KLL*-Auger components (equations 4) for $Z = 10$ (see Ref. 51), new expressions for $F^0(21, 21)$, $F^2(21, 21)$, $G^1(20, 21)$ linear in Z-value were deduced for the low Z-region.

The matrix element $F^0(20, 20)$ was calculated via the experimental energy of the $KL_1L_1(^1S_0)$ Auger component (first equation of 4) both with and without consideration of configuration interaction. The relative group intensities $I(KL_1L_{2,3})/I(KL_1L_1)$ and $I(KL_{2,3}L_{2,3})/I(KL_1L_1)$ were then recalculated using matrix elements, including $F^0(20, 20)$ with configuration interaction. The same ratios were also calculated using transition amplitudes for $Z = 18$ and 36 by Rubenstein,[63] for $Z = 11$, 12, 15, 16, 17, and 18 by Archard,[72] for $Z = 12$, 15, 18, 25, 30, and 36 by Callan,[59] and for $Z = 10$ by Krause et al.,[69] using $F^0(20, 20)$ determined with consideration of configuration interaction (indicated as I in Table VII). For comparison, these intensity ratios were also calculated using the matrix element $F^0(20, 20)$ without configuration interaction and employing only the transition amplitudes of Rubenstein[63] and Archard[72] (indicated as II in Table VII). The agreement between the two sets of values was within 1%; hence it is not necessary to include effects of configuration interaction in calculating $F^0(20, 20)$ when only group intensities are of interest. The fact that the relative group intensities were smaller than those calculated by Asaad,[48] as seen in Figure 7, was attributed to the use of more accurate L_1-electron binding energies, in particular for $Z < 24$.

Mehlhorn and Asaad[75] prepared diagrams showing the percentage change that possible errors in the electron binding energy values and the matrix elements would have on the relative group intensity values.

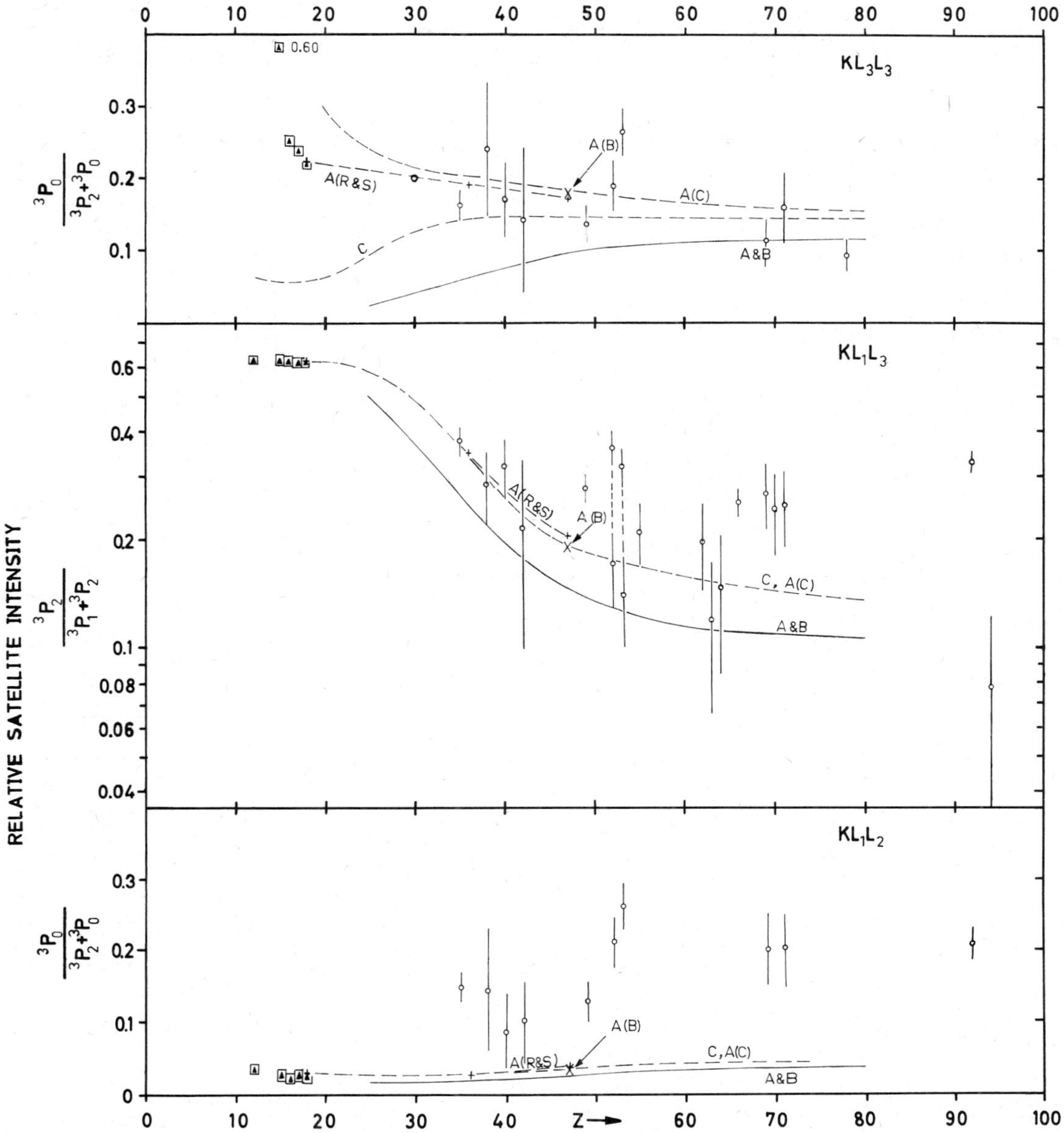

Figure 8. The experimental relative *KLL*-Auger satellite component intensities for the KL_1L_2, KL_1L_3, and KL_3L_3 transitions. Symbols and initials are the same as in Figure 6.

These were calculated for $Z = 12$ using the transition amplitudes by Archard[72] and for $Z = 18$ and 36 using those by Rubenstein.[63] From those calculations and from the experimental results on the relative group intensities of the *KLL*-Auger spectrum for neon ($Z = 10$),[50,51] it was evident that the main difference between theoretical and experimental values is owing to incorrect wave functions used in the theoretical treatment at low Z-values. (The *KLL*-component intensities of Ref. 51 are to be superseded by later measurements[52] where an earlier error source has been eliminated.) The same tendencies in relative group intensities were later found in investigations of other low Z-value elements ($Z = 9$, $Z = 11$, 12, and $Z = 17$ in Refs. 54, 49, and 76, respectively).

For comparison, the experimental results of the $I(KL_1L_{2,3})/I(KL_1L_1)$ and $I(KL_{2,3}L_{2,3})/I(KL_1L_1)$ intensity ratios are shown in Figure 6, together with the results of the various theoretical calculations referred to previously. The experimental intensities of the individual *KLL*-Auger components relative to that of the KL_1L_1 line are gathered and presented in Figure 7 along with theoretical values. The experimentally determined fractional intensities of the satellite lines are compared with values from theoretical treatments in Figure 8.

Table VIIIa.

Experimental energies (in keV) and relative

Ref.	Z	KL_1L_1 ($^1S_0$)	KL_1L_2 ($^1P_1$)	KL_1L_2 ($^3P_0$)	KL_1L_3 ($^3P_1$)	KL_1L_3 ($^3P_2$)
85	7	--	—	0.3360(15)	—	—
		--	—	0.36(3)	—	—
85	8	--	—	0.4630(37)	—	—
		--	—	0.32(4)	—	—
54	9^+	0.6091(4)	0.6281(4)	—	0.6365(4)	—
		1.00(15)	3.92(63)	—	1.11(25)	—
54	9^+	0.6096(4)	0.6287(4)	—	0.6370(4)	—
		1.00(14)	2.68(32)	—	1.05(24)	—
54	9^+	0.6083(4)	0.6271(4)	—	0.6352(4)	—
		1.00(15)	3.01(53)	—	0.94(27)	—
42	10	(-56.0 eV)[§]	(-32.6 eV)[§]	—	(-22.2 eV)[§]	—
		--	--	—	--	—
51	10	0.74805(42)	0.77145(42)	—	0.78195(42)	—
52		1.00	2.73(4)	—	1.01(2)	—
53	10	--	--	—	--	—
		1.00	2.5(4)	—	0.6(4)	—
50	10	0.7476(10)	0.7712(6)	—	0.7822(10)	—
		1.00	2.9(3)	—	1.2(2)	—
49	11	0.9228(4)	0.9508(4)	—	0.9634(4)	—
		1.00	4.3(5)	—	1.5(4)	—
49	12	1.1013(4)	1.1347(4)	—	1.1496(4)	—
		1.00	4.6(5)	—	1.3(3)	—
76	17	--	--	—	--	—
		1.00	4.6(6)	—	2.2(5)	—
42	19	2.809	2.878	—	2.904	—
		--	--	—	--	—
55	23^+	4.159(6)	—	4.254(4)	—	4.284(6)
		1.00(15)	—	3.45(31)	—	1.89(28)
86	25^+	4.958(8)	—	5.065(7)	—	5.102(7)
		1.00	—	2.4	—	1.4
86	25	4.962(2)	—	5.068(2)	—	--
		--	—	--	—	--
55	26^+	5.365(7)	—	5.475(5)	—	5.508(7)
		1.00(11)	—	3.60(16)	—	1.40(16)
77	26^+	5.376	—	—	5.446	—
78		1.00	—	—	2.33(20)	—
79	26^+	5.376	—	—	5.446	—
		1.00(29)	—	—	2.36(22)	—
77	29	6.732	—	—	6.846	—
		1.00	—	—	2.13(9)	—

intensities of KLL Auger lines

KL_2L_2 ($^1S_0$)	KL_2L_3 ($^1D_2$)	KL_3L_3 ($^3P_0$) … ($^3P_2$)	Method --- En. calibr.	Instr. --- Chem. form
——	0.35990(75)	——	X - C	ES, Poa
——	1.00	——	E1C	N_2, gas
——	0.4902(12)	——	X - C	ES, Poa
——	1.00	——	E1C	O_2, gas
0.6515(5)	0.6544(4)	—— -- ——	X - C	IFDF, Poa
1.69(62)	13.7(6)	—— -- ——	XPE	LiF, solid
0.6517(5)	0.6546(4)	—— -- ——	X - C	IFDF, Poa
1.62(54)	11.9(5)	—— -- ——	XPE	NaF, solid
0.6504(5)	0.6533(4)	--	X - C	IFDF, Poa
1.67(51)	11.1(5)	--	XPE	MgF_2, solid
(-3.7 eV)§	(0 eV)§	—— -- ——	EI - C	ES, Poa
--	--	—— -- ——	--	Ne, gas
0.80045(42)	0.80415(40)	—— -- ——	EI - C	ES, Poa
1.55(3)	9.31(15)	—— -- ——	Auger lines	Ne, gas
--	--	—— -- ——	PI - C	ES
1.5(2)	7.0(4)	—— -- ——	--	Ne, gas
——	0.8035(6)	——	X - C	ES, Poa
——	13.0(13)	——	E1C	Ne, gas
0.9856(4)	0.9898(4)	—— -- ——	X - C	IFDF, Poa
1.7(5)	16.6(17)	—— -- ——	XPE	Na_2O, solid
1.1748(4)	1.1797(4)	—— -- ——	X - C	IFDF, Poa
1.8(5)	18.4(18)	—— -- ——	XPE	MgO, solid
——	--	——	EI - C	ES, Poa
——	18.2(2.0)	——	--	CCl_4, gas
--	2.972	—— -- ——	X - C	IFDF, Poa
--	--	—— -- ——	Auger lines	K salt
4.357(6)	4.374(4)	—— -- ——	RS - C	IFDF, Poa
1.49(36)	——	—— 14.90(40) ——	CL	Cr, metal
5.191(8)	5.205(6)	—— 5.225(8) ——	RS - C	IFDF, Poa
1.4	9.75	—— 1.52 ——	CL	Fe, metal
--	5.204(2)	—— 5.227(2) ——	X - C	IFDF, Poa
--	--	—— -- ——	--	?
5.595(7)	5.618(5)	—— 5.639(7) ——	RS - C	IFDF, Poa
1.32(20)	13.10(40)	—— 1.32(20) ——	CL	Co, metal
——	5.570	——	RS - C	IFDF
——	6.65(60)	——	CL	Co, metal
——	5.570	——	RS - C	IFDF
——	7.2	——	CL	Co, metal
——	7.016	——	RS - C	IFDF
——	5.41(24)	——	CL	--

<table>
<tr><th rowspan="2">Ref.</th><th rowspan="2">Z</th><th>KL_1L_1</th><th colspan="2">KL_1L_2</th><th colspan="2">KL_1L_3</th></tr>
<tr><th>$(^1S_0)$</th><th>$(^1P_1)$</th><th>$(^3P_0)$</th><th>$(^3P_1)$</th><th>$(^3P_2)$</th></tr>
<tr><td>38</td><td>29^+</td><td>6.735
1.00(9)</td><td colspan="2">——
——</td><td>6.845
2.8(2)</td><td>——
——</td></tr>
<tr><td>71</td><td>29^+</td><td>6.739(6)
1.00</td><td colspan="2">6.867(4)
3.1</td><td colspan="2">6.909(6)
1.3</td></tr>
<tr><td>87</td><td>30^+</td><td>7.220(4)
$1.0^{\times\times\times}$</td><td colspan="2">7.358(2)
4.7</td><td colspan="2">7.399(3)
0.8</td></tr>
<tr><td>71</td><td>32^+</td><td>8.212(6)
1.00</td><td colspan="2">8.367(4)
3.3</td><td colspan="2">8.415(6)
0.85</td></tr>
<tr><td>88</td><td>33</td><td>8.742(10)
1.00</td><td colspan="2">8.902(10)
2.93(40)</td><td colspan="2">8.958(10)
1.22(15)</td></tr>
<tr><td>80</td><td>35</td><td>9.86(1)
1.00</td><td colspan="2">10.05(1)
1.79(12)</td><td colspan="2">10.11(1)
1.09(10)</td></tr>
<tr><td>47</td><td>35</td><td>(-369(5) eV)§
1.00</td><td>(-229(3) eV)§
2.19(20)</td><td>(-206(5) eV)§
0.38(5)</td><td>(-176(3) eV)§
1.02(10)</td><td>(-155(5) eV)§
0.62(10)</td></tr>
<tr><td>25</td><td>38^+</td><td>11.5844(16)
1.00(13)</td><td>11.7891(15)
1.8(2)</td><td>11.823(4)
0.3(2)</td><td>11.8628(18)
1.00(13)</td><td>11.885(4)
0.40(13)</td></tr>
<tr><td>25</td><td>40^+</td><td>12.8518(15)
1.00(12)</td><td>13.0702(13)
1.9(2)</td><td>13.102(4)
0.18(12)</td><td>13.1502(15)
1.00(12)</td><td>13.187(4)
0.47(12)</td></tr>
<tr><td>25</td><td>42^+</td><td>14.1761(13)
1.00(13)</td><td>14.4161(12)
1.70(19)</td><td>14.452(4)
0.19(12)</td><td>14.5213(13)
1.12(19)</td><td>14.549(4)
0.31(19)</td></tr>
<tr><td>37</td><td>47^{++}</td><td>17.74(6)
1.00(20)</td><td colspan="2">18.02(6)
1.3(2)</td><td colspan="2">18.18(6)
1.3(2)</td></tr>
<tr><td>89</td><td>49</td><td>19.352(1)
1.00</td><td>19.651(1)
1.28(3)</td><td>--
0.19(4)</td><td>19.849(1)
1.15(3)</td><td>--
0.44(4)</td></tr>
<tr><td>20</td><td>52</td><td>21.787(10)×
1.00(5)</td><td>22.116(10)
1.13(7)</td><td>22.168(10)
0.30(5)</td><td>22.380(10)
1.16(8)</td><td>22.423(10)
0.51(6)</td></tr>
<tr><td>90</td><td>52</td><td>--
1.00(6)</td><td colspan="2">--
1.63(15)</td><td colspan="2">--
1.90(14)</td></tr>
<tr><td>20</td><td>53</td><td>22.652(10)×
1.00(2)</td><td>22.990(10)
1.07(2)</td><td>23.019(10)
0.38(4)</td><td>23.272(10)
1.22(2)</td><td>23.317(10)
0.46(4)</td></tr>
<tr><td>24</td><td>55</td><td>24.39(2)
1.00(6)</td><td>24.75(2)
——</td><td>--
1.71(7)</td><td>25.08(2)
1.58(19)</td><td>25.14(3)
0.42(19)</td></tr>
<tr><td>43</td><td>55</td><td>24.40(2)
1.00</td><td colspan="2">24.74(2)
1.41(10)</td><td colspan="2">25.07(3)
1.73(15)</td></tr>
<tr><td>91</td><td>56</td><td>25.251(6)
1.00</td><td colspan="2">25.657
1.4(1)</td><td colspan="2">26.011(6)
1.8(1)</td></tr>
<tr><td>92</td><td>57</td><td>--
1.00</td><td colspan="2">--
1.34(8)</td><td colspan="2">--
1.56(9)</td></tr>
<tr><td>93</td><td>62</td><td>31.15
1.00</td><td colspan="2">31.61
1.25</td><td colspan="2">32.19
1.50</td></tr>
<tr><td>23</td><td>62</td><td>31.20(2)
1.00(5)</td><td colspan="2">31.62(2)
1.39(6)</td><td>32.19(2)
1.18(6)</td><td>32.25(2)
0.29(9)</td></tr>
</table>

KL_2L_2 ($^1S_0$)	KL_2L_3 ($^1D_2$)	KL_3L_3 ($^3P_0$)	($^3P_2$)	Method / En. calibr.	Instr.
———	——— 7.015	———	———	RS - C	IFDF
———	——— 9.1	———	———	CL	
7.0060(25)	7.0345(10)	—— 7.0630(25)	———	X - C	IFDF
1.9	14	—— 1.9	———	XPE	
--	7.5354(10)	7.550(3)	7.5663(10)	RS - C	MLS
--	17.3	0.9	3.5	CL	
8.524(4)	8.5610(25)	—— 8.6010(25)	———	X - C	IFDF
1.0	14	—— 2.0	———	XPE	
9.074(10)	9.105(9)	—— 9.139(9)	———	RS - C	IFDF, Poa
1.55(24)	8.8(15)	—— 1.86(21)	———	CL	
10.20(1)	10.26(1)	—— 10.33(1)	———	RS - C	ICDF
0.38(15)	4.8(4)	—— 1.10(8)	———	CL	
(-49(5) eV)§	(0 eV)§	(+41(5) eV)§	(+61(3) eV)§	RS - C	ICDF
0.37(10)	9.72(70)	0.41(5)	2.13(20)	--	
11.989(4)	12.0599(11)	12.109(4)	12.1352(12)	X - C	IFDF
0.40(13)	6.67(33)	0.53(27)	1.67(2)	XPE	
13.285(4)	13.3719(12)	13.437(4)	13.4644(13)	X - C	IFDF
0.29(12)	5.90(29)	0.41(20)	2.0(2)	XPE	
14.641(4)	14.7509(13)	14.841(4)	14.8665(13)	X - C	IFDF
0.31(19)	6.25(31)	0.25(19)	1.5(2)	XPE	
18.29(6)	18.46(6)	—— 18.65(6)	———	RS - Ph	SCS
0.5(2)	3.2(4)	—— 1.8(2)	———	CL	
--	20.143(1)	--	20.362(1)	RS - C	IFDF
0.18(6)	3.90(4)	0.20(4)	1.48(3)	Auger lines	
--	22.699(10)	22.945(10)	22.983(10)	RS - C	IFDF
with KL_1L_3	3.69(6)	0.30(7)	1.28(7)	CL	
--	--	—— --	———	RS - C	IFDF
--	3.78(5)	—— 1.50(13)	———	--	
--	23.602(10)	23.884(10)	23.912(10)	RS - C	IFDF
with KL_1L_3	3.66(5)	0.42(7)	1.17(2)	CL	
--	25.43(2)	--	25.79(2)	RS - C	IFDF
with KL_1L_3	4.23(7)	—— 1.91(6)	———	CL	
--	25.43(2)	—— 25.77(2)	———	RS - C	ICDF
———	3.39(25)	—— 1.44(10)	———	CL	
———	26.366(6)	—— 26.737(6)	———	RS - C	IFDF
———	2.90(15)	—— 1.5(1)	———	CL	
--	--	—— --	———	RS - C	IFDF
with KL_1L_3	3.28(10)	—— 1.45(7)	———	--	
———	32.58	—— 33.19	———	RS - C	ICDF
———	2.56	—— 1.31	———	CL	
32.01(2)	32.61(2)	—— 33.22(2)	———	RS - C	IFDF
0.26(6)	3.14(13)	—— 1.39(6)	———	CL	

Table VIIIa (cont.)

<table>
<tr><th rowspan="2">Ref.</th><th rowspan="2">Z</th><th>KL_1L_1</th><th colspan="2">KL_1L_2</th><th colspan="2">KL_1L_3</th></tr>
<tr><th>$(^1S_0)$</th><th>$(^1P_1)$</th><th>$(^3P_0)$</th><th>$(^3P_1)$</th><th>$(^3P_2)$</th></tr>
<tr><td rowspan="2">94</td><td rowspan="2">63</td><td>32.208(20)</td><td colspan="2">32.656(20)</td><td colspan="2">33.308(20)</td></tr>
<tr><td>1.00(13)</td><td colspan="2">1.19(19)</td><td colspan="2">1.25(25)</td></tr>
<tr><td rowspan="2">95</td><td rowspan="2">63</td><td>32.15(10)</td><td colspan="2">32.59(5)</td><td colspan="2">33.22(5)</td></tr>
<tr><td>1.00</td><td colspan="2">1.3(1)</td><td colspan="2">1.5(1)</td></tr>
<tr><td rowspan="2">20,21, 22</td><td rowspan="2">63</td><td>32.24</td><td colspan="2">32.70</td><td colspan="2">33.32</td></tr>
<tr><td>1.00(5)</td><td colspan="2">1.41(7)</td><td>1.17(8)</td><td>0.16(8)</td></tr>
<tr><td rowspan="2">19</td><td rowspan="2">64</td><td>--</td><td colspan="2">--</td><td colspan="2">--</td></tr>
<tr><td>1.00(7)</td><td colspan="2">1.40(10)</td><td>1.11(9)</td><td>0.19(9)</td></tr>
<tr><td rowspan="2">96</td><td rowspan="2">65</td><td>34.43(5)</td><td colspan="2">34.88(5)</td><td colspan="2">35.62(5)</td></tr>
<tr><td>1.00(15)</td><td colspan="2">1.69(20)</td><td colspan="2">1.27(13)</td></tr>
<tr><td rowspan="2">18</td><td rowspan="2">66</td><td>35.492(6)</td><td colspan="2">35.963(7)</td><td>36.732(7)</td><td>36.806(10)</td></tr>
<tr><td>1.00(3)</td><td colspan="2">1.31(5)</td><td>0.83(5)</td><td>0.28(3)</td></tr>
<tr><td rowspan="2">97</td><td rowspan="2">66</td><td>35.54(2)</td><td colspan="2">36.03(4)</td><td colspan="2">--</td></tr>
<tr><td>1.00</td><td colspan="2">1.40(10)</td><td colspan="2">--</td></tr>
<tr><td rowspan="2">97</td><td rowspan="2">67</td><td>--</td><td colspan="2">--</td><td colspan="2">--</td></tr>
<tr><td>--</td><td colspan="2">--</td><td colspan="2">--</td></tr>
<tr><td rowspan="2">98</td><td rowspan="2">67</td><td>--</td><td colspan="2">--</td><td colspan="2">--</td></tr>
<tr><td>1.00(13)</td><td colspan="2">1.51(20)</td><td colspan="2">1.21(15)</td></tr>
<tr><td rowspan="2">99</td><td rowspan="2">69</td><td>--</td><td colspan="2">--</td><td colspan="2">--</td></tr>
<tr><td>1.00</td><td colspan="2">1.6</td><td colspan="2">1.5</td></tr>
<tr><td rowspan="2">100</td><td rowspan="2">69</td><td>--</td><td>--</td><td>--</td><td>--</td><td>--</td></tr>
<tr><td>1.00(3)</td><td>1.23(7)</td><td>0.30(6)</td><td>0.83(5)</td><td>0.31(5)</td></tr>
<tr><td rowspan="2">101</td><td rowspan="2">69+</td><td>38.958(25)</td><td colspan="2">40.420(35)</td><td colspan="2">40.420(35)</td></tr>
<tr><td>1.00(4)</td><td colspan="2">1.40(6)</td><td colspan="2">1.02(4)</td></tr>
<tr><td rowspan="2">102</td><td rowspan="2">70</td><td>40.149(4)</td><td colspan="2">40.672(4)</td><td>41.681(4)</td><td>41.762(10)</td></tr>
<tr><td>1.00(2)</td><td colspan="2">1.56(3)</td><td>0.93(5)</td><td>0.30(8)</td></tr>
<tr><td rowspan="2">103</td><td rowspan="2">71</td><td>41.368(10)</td><td>41.929(12)</td><td>41.959(20)</td><td>42.989(12)</td><td>43.059(20)</td></tr>
<tr><td>1.00</td><td>1.22(8)</td><td>0.30(6)</td><td>0.82(5)</td><td>0.28(6)</td></tr>
<tr><td rowspan="2">101</td><td rowspan="2">71</td><td>--</td><td colspan="2">--</td><td colspan="2">--</td></tr>
<tr><td>1.00</td><td colspan="2">1.39</td><td colspan="2">0.89</td></tr>
<tr><td rowspan="2">104</td><td rowspan="2">71</td><td>41.334</td><td colspan="2">41.881</td><td colspan="2">42.936</td></tr>
<tr><td>1.00</td><td colspan="2">1.25</td><td colspan="2">1.05</td></tr>
<tr><td rowspan="2">105</td><td rowspan="2">74+</td><td>45.08(4)</td><td colspan="2">45.63(5)</td><td colspan="2">46.95(4)</td></tr>
<tr><td>1.00(20)</td><td colspan="2">1.9(4)</td><td colspan="2">0.64(14)</td></tr>
<tr><td rowspan="2">106</td><td rowspan="2">75</td><td>46.400(25)</td><td colspan="2">47.010(25)</td><td colspan="2">48.370(25)</td></tr>
<tr><td>1.00(10)</td><td colspan="2">1.40(14)</td><td colspan="2">1.10(11)</td></tr>
<tr><td rowspan="2">20
107</td><td rowspan="2">78</td><td>50.35</td><td colspan="2">51.06</td><td colspan="2">52.68</td></tr>
<tr><td>1.00</td><td colspan="2">1.40(25)</td><td colspan="2">0.90(10)</td></tr>
<tr><td rowspan="2">108</td><td rowspan="2">78</td><td>50.32(10)</td><td colspan="2">50.97(11)</td><td colspan="2">52.67(11)</td></tr>
<tr><td>1.00</td><td colspan="2">1.42(18)</td><td colspan="2">0.85(12)</td></tr>
<tr><td rowspan="2">12
109</td><td rowspan="2">78</td><td>50.44</td><td>51.07</td><td>--</td><td>52.54</td><td>--</td></tr>
<tr><td>1.00</td><td colspan="2">1.5</td><td colspan="2">1.0</td></tr>
</table>

KL_2L_2 $(^1S_0)$	KL_2L_3 $(^1D_2)$	KL_3L_3 $(^3P_0)$	KL_3L_3 $(^3P_2)$	Method --- En. calibr.	Instr.
--	33.702(20)	—— 34.386(20)	——	RS - C	ICDF
--	2.82(19)	—— 1.25(19)	——	CL	
32.98(5)	33.62(5)	—— 34.29(5)	——	RS - C	IFDF
	3.0(1)	—— 1.2(1)	——	CL	
33.09	33.74	—— 34.39	——	RS - C	IFDF
0.20(5)	2.92(16)	—— 1.35(7)	——	CL	
--	--	—— --	——	RS - C	IFDF
0.27(7)	3.08(18)	—— 1.28(9)	——	--	
35.27(8)	36.03(5)	—— 36.78(5)	——	RS - C	ICDF
0.22(7)	2.96(15)	—— 1.14(6)	——	CL	
36.369(7)	37.195(6)	—— 38.002(9)	——	RS - C	IFDF
0.22(5)	2.39(5)	—— 1.07(8)	——	CL	
--	--	—— --	——	RS - C	ICDF
--	--	—— --	——	CL	
37.60(2) ××	38.45(2) ××	—— 39.29(2)	——	RS - C	ICDF
1.00(50) ××	8.15(45) ××	—— 3.15(20) ××	——	CL	
--	--	—— --	——	RS - C	ICDF
0.20(10)	2.80(15)	—— 1.16(5)	——	--	
--	--	—— --	——	RS - Ph	SCS
--	2.0	—— 0.9	——	--	
--	--	--	--	RS - C	IFDF
0.20(2)	2.44(4)	0.12(3)	0.92(5)	--	
39.911(40)	40.916(25)	—— 41.889(23)	——	RS - C	ICDF
0.25(2)	2.23(9)	—— 0.95(2)	——	CL	
41.120(10)	42.184(4)	—— 43.234(4)	——	RS - C	IFDF
0.17(1)	2.53(4)	—— 1.09(2)	——	CL	
42.376(15)	43.493(10)	44.567(20)	44.632(12)	RS - C	IFDF
0.20(3)	2.24(4)	0.17(4)	0.87(5)	CL	
--	--	—— --	——	RS - C	ICDF
0.22	2.15	—— 0.92	——	CL	
--	43.466	—— 44.571	——	RS - Ph	SCS
--	2.52	—— 1.00	——	CL	
--	47.505(40)	—— 48.855(40)	——	RS - Ph	SCS
--	1.8(3)	—— 0.715	——	CL	
47.510(25)	48.910(25)	—— 50.370(25)	——	RS - Ph	SCS
0.10(5)	2.0(2)	—— 0.95(10)	——	CL	
--	53.28	—— 55.03	——	RS - C, Ph	MLS, SCS
0.20(10)	1.40(25)	—— 0.80(20)	——	CL	
51.55(10)	53.27(10)	54.87(11)	55.04(11)	RS - C	IFDF
0.19(10)	1.69(22)	0.062(13)	0.59(9)	CL	
--	53.39	--	54.90	RS - C	IFDF
--	1.5	—— 0.9	——	CL	

Ref.	Z	KL_1L_1 ($^1S_0$)	KL_1L_2 ($^1P_1$)	KL_1L_2 ($^3P_0$)	KL_1L_3 ($^3P_1$)	KL_1L_3 ($^3P_2$)
109a	79	51.78(2)	52.42(2)		54.20(2)	
		1.00	1.54(11)		0.99(9)	
110	79	51.87	52.51		54.26	
		1.00(25)	1.7(4)		1.2(3)	
12	79	51.83	52.48	--	54.26	--
111		1.00	1.5		1.0	
112	80	53.118(35)	53.761(36)		55.672(37)	
		1.00			1.11(10) (also KL_2L_3)	
29	80	53.18(4)	53.79(4)		55.71(4)	
		1.00	1.2		0.7	
113	80	53.56	54.19		56.15	
		1.00	1.32(10)		0.85(6)	
114	80	--	--		--	
		1.00(10)	1.68(13)		0.85(10)	
115	80	53.12(7)	53.80(7)		55.67(14)	
		1.00	1.48(11)		0.79(8)	
116	81	54.51(10)	55.18(10)		57.28(10)	
		1.00	1.67(9)		0.97(10)	
117	82	--	--		--	
		1.00	1.62(13)		0.87(10)	
34	83	57.466(18)	58.186(18)		60.417(18)	
118		1.00	1.8		1.1	
119	83	57.458(35)	58.147(35)		60.423(35)	
		1.00	1.72		0.91	
120	83+	57.49ø	58.15ø		60.55ø	
		1.00(16)	1.24(19)		0.57(24)	
121	84+	--	--		--	
		1.00	1.45		0.62	
120	84+	58.92ø	--		62.17ø	
		1.00(40)	--		1.6(6)	
28	92	71.69(10)	72.54(10)		76.29(16)	
		1.00(25)	2.5(6)		0.90(22)	
122	92	71.74(2)	72.55(2)	72.61(2)	76.30(2)	76.42(2)
		1.00	1.54(10)	0.40(6)	0.50(5)	0.25(3)
123	93	73.7	74.6		78.52	
		1.00			0.59	
124	93	73.41	74.25		78.21	
		--	--		--	
125	94	75.19(22)	76.07(23)		80.29(24)	
		--	--		--	
17	94	75.180(15)	76.050(15)		80.240(15)	80.40(6)
		1.00	1.86(10)		0.60(6)	0.05(3)

KL_2L_2 ($^1S_0$)	KL_2L_3 ($^1D_2$)	KL_3L_3 ($^3P_0$) ($^3P_2$)	Method / En. calibr.	Instr.
52.95(3)	54.81(2)	—— 56.63(2) ——	RS - C	IFDF
0.11(3)	1.83(13)	—— 0.75(7) ——	CL	
53.10	54.89	—— 56.64 ——	RS - Ph	SCS
∿0.3	1.4(4)	—— 0.8(2) ——	CL	
--	54.87	-- 56.70	RS - Ph	SCS
--	1.5	—— 0.9 ——	CL	
--	56.328(37)	—— 58.223(73) ——	RS - C	IFDF
--	with KL_1L_3	—— 0.27(7) ——	CL	
54.52	56.35(5)	—— 58.27(5) ——	RS - Ph	SCS
∿0.2	1.4	—— 0.6 ——	CL	
55.07	56.73	—— 58.70 ——	RS - C	IFDF
0.40(3)	1.28(8)	—— 0.76(5) ——	CL	
--	--	—— -- ——	RS - C	ICDF
0.14(7)	1.84(13)	—— 0.76(8) ——	--	
54.5(2)	56.34(7)	—— 58.27(7) ——	RS - C	ICDF
0.19(10)	1.69(16)	—— 0.69(7) ——	CL	
55.80(12)	57.86(10)	—— 59.89(10) ——	RS - C	ICDF
0.156(25)	1.66(9)	—— 0.76(5) ——	CL	
--	--	—— -- ——	RS - C	ICDF
0.15(4)	1.80(15)	—— 0.75(8) ——	--	
--	61.090(18)	—— 63.404(18) ——	RS - Ph	SCS
<0.2	1.6	—— 0.8 ——	CL	
58.783(35)	61.070(35)	—— 63.367(35) ——	RS - C	ICDF
0.17	1.64	—— 0.69 ——	CL	
--	61.16$^{\phi}$	—— 63.48$^{\phi}$ ——	RS - Ph	SCS
--	1.25(19)	—— 0.23(9) ——	CL	
--	--	—— -- ——	RS - C	ICDF
<0.15	1.36	—— 0.70 ——	--	
59.92$^{\phi}$	62.77$^{\phi}$	—— 65.33$^{\phi}$ ——	RS - Ph	SCS
1.9(8)	0.89(36)	—— 0.57(24) ——	CL	
--	77.13(13)	—— -- ——	RS - Ph	SCS
--	1.4	—— -- ——	CL	
--	77.12(1)	—— -- ——	RS - Ph	SCS
--	1.30(6)	—— -- ——	CL	
75.3	79.32	—— 83.3 ——	RS - Ph, C	SCS
<0.07	with KL_1L_3	—— 0.15 ——	CL	
--	78.95	—— 82.99 ——	RS - Ph	SCS
--	--	—— -- ——	CL	
--	81.08(24)	—— 85.29(25) ——	RS - Ph	SCS
--	--	—— -- ——	CL	
76.78(4)	81.060(15)	—— 85.300(15) ——	RS - C	IFDF
0.11(4)	1.11(7)	—— 0.40(5) ——	CL	

Table VIIIa (cont.)

Symbols and abbreviations:

+ Intensities renormalized so that that of the $KL_1L_1(^1S_0)$ line is unity.

++ Energies corrected according to ref.[30)]

§ Energy displacements from the $KL_2L_3(^1D_2)$ line position.

ϕ Error in energy 0.01 - 0.05 keV.

× Error limits according to a private communication from R. L. Graham, (see also text).

×× Intensities normalized to $I(KL_2L_2) = 1.00$.

××× KL_1L_1 intensity in error as published by factor 10 (private communication from M. S. Freedman).

Method:

RS = Atomic vacancy produced by electron capture and/or internal conversion in a radioactive source.

X = Atomic vacancy produced by X-ray bombardment.

EI = Atomic vacancy produced by electron bombardment.

PI = Atomic vacancy produced by proton bombardment.

C = Electrical counting of electrons.

Ph = Photographic emulsion plates for electron detection.

Instrument:

ES = Electrostatic spectrometer.

ICDF = Iron core double focusing spectrometer.

IFDF = Iron free double focusing spectrometer.

MLS = Magnetic lens spectrometer.

Poa = Post-acceleration techniques applied.

Pra = Pre-acceleration techniques applied.

Energy calibration:

ElC = Electron cannon used.

XPE = X-ray photoelectron line energies used.

CL = Energies of electron lines due to internal conversion of gamma-rays.

Chemical form.

Here the chemical form, and sometimes physical form, of the element of interest is given. This is given only for the lower Z-values where the chemical or physical states can influence the energies or intensities of the K-Auger spectral components most strongly.

Table VIIIb.

Additional experimentally determined Auger electron line energies.

$KL_2L_3(^1D_2)$ line[25)127)]

Z	Energy (keV)	Z	Energy (keV)	Z	Energy (keV)
38	12.0599(11)	41	14.0519(13)	44	16.2039(14)
39	12.7044(12)	42	14.7509(13)	45	16.9569(15)
40	13.3719(12)	43	15.4683(14)	46	17.7294(15)
				47	18.5163(16)

Values from ref.[129)].

Z = 29 $KL_2L_3(^1D_2)$ line: 7.03463(5) keV$^+$

Z = 32 " " " : 8.56205 keV

$^+$Evaluation based on 1952 values of the fundamental constants.

It will be noticed in Figure 6 that some of the results for $Z = 26$ and $Z = 29$ (Refs. 77–79 and 77, 38, respectively) fall below the general trend of the other data. This can be attributed to value errors arising from difficulties in analyzing the component intensities when low instrumental resolution is used, and when it is difficult to correct for the low energy tails of spectral lines. The earlier measurement[80] of the bromine spectrum ($Z = 35$), made using low resolving power, gives relative intensity values lying far below the values obtained later using higher resolution.[47] These discrepancies result, no doubt, from the same analysis difficulties.

Slätis has attempted to correlate the intensities of the several *KLL*, *KLM*, and so on,[81] and *L*-Auger[82] spectral components in the electron spectrum following the decay of thorium B by defining "selection rules" for the transitions, determined by quantum numbers of the levels involved in the transitions. Surprisingly good qualitative agreement was obtained between the predicted relative intensities and the experimental. The reader is referred to these articles for details of the treatment.

The experimental data from which the experimental relative and absolute energies are taken are presented in Table VIII, which also provides the experimentally determined relative intensities of the *KLL*-Auger spectral components. The experimental relative *KLL*-Auger component intensities plotted in Figures 6 through 8 are derived from these tabulated values.

The reader is also referred to recent calculations of Auger transition rates and fluorescence yields by McGuire.[82a]

4.1.3. *Atomic and Molecular Effects on KLL-Auger Spectra*

4.1.3.1. *KL–LLL* AUGER SPECTRUM. Körber and Mehlhorn[51] assigned the additional lines observed in their *KLL*-Auger spectrum for neon to transitions from an initial atomic state with one vacancy in each of the *K*- and *L*-shells, to a final atomic state of triple ionization in the *L*-shell. Figures 5 and 10*a* and 10*b* show the spectrum measured at two different instrumental resolutions. Assuming pure *L–S* coupling, the possible initial states are $KL_1(^1S, {}^3S)$ and $KL_{2,3}(^1P, {}^3P)$, and the possible final states are $L_{2,3}L_{2,3}L_{2,3}(^4S, {}^2D, {}^2P)$, $L_1L_{2,3}L_{2,3}(^4P, {}^2D, {}^2S, {}^2P)$, and $L_1L_1L_{2,3}(^2P)$. Certain of the transitions are not allowed by the selection rules—$\Delta L = \Delta S = \Delta J = 0$ and no parity change.

The energies of all transitions, except the $KL_1(^1S)$ to *LLL* transitions, which were not observed in the electron spectra, were calculated from the difference between the energies of the initial *KL*-states and, using the values known from optical spectroscopy, those of the final *LLL*-states. Table IX gives the experimental relative energies and intensities, together with the calculated relative energies. The intensities are relative to that of the KL_1L_1 Auger transition. A similar tabular comparison was presented in Ref. 42.

Körber and Mehlhorn[51] used electron bombardment to cause the initial *K*-shell atomic ionization in neon. The ionization method gives rise to a significant

Table IX.

Relative energies and intensities of the KL-LLL Auger transitions in neon following electron[51)] and proton[53)] impact ionization.

Transition		Relative energy[+] in eV[51)]		Relative intensities[++]		
Initial state	Final state	Experiment	Theory	Electron bombardment	Proton bomb.	Results regarding electron impact spectra[51)]
$KL_{2,3}(^3P)$	$L^3_{2,3}(^4S)$	--	-13.5(1.0)	--	--	forb.; Line E, fig. 10a
	(^2D)	-18.4(2)	-18.6(1.0)	0.84(5)	1.9(3)	alwd.; Line B, fig. 10a
	(^2P)	-21.0(3)	-21.2(1.0)	0.41(4)[+++]	1.6(4)	alwd.; Line A, fig. 10a
	$L_1L^2_{2,3}(^4P)$	--	-36.3(1.0)	--	--	alwd.; Line T, fig. 10b
	(^2D)	-44.9(3)	-45.9(1.0)	0.19(4)	0.8(3)	alwd.; Line R, fig. 10b
	(^2S)	--	-50.6(1.0)	--	--	alwd.; Line N, fig. 10b
	(^2P)	-52.8(4)	-53.1(1.0)	0.16(4)	0.7(2)	alwd.; Line M, fig. 10b
	$L^2_1L_{2,3}(^2P)$	-73.6(5)	-73.6(1.0)	0.21(5)	1.1(2)	alwd.; Line X, fig. 5
$KL_{2,3}(^1P)$	$L^3_{2,3}(^4S)$	--	-8.7(1.0)	--	--	forb.; Line F, fig. 10a
	(^2D)	-13.8(2)	-13.8(1.0)	0.50(4)	0.8(2)	alwd.; Line D, fig. 10a
	(^2P)	-16.6(3)	-16.4(1.0)	0.41(4)	0.9(2)[a]	alwd.; Line C, fig. 10a
	$L_1L^2_{2,3}(^4P)$	--	-31.5(1.0)	--	--	forb.; Line U, fig. 10b
	(^2D)	--	-40.2(1.0)	--	0.2(2)[a]	alwd.; Line S, fig. 10b
	(^2S)	--	-45.8(1.0)	--	--	alwd.; Line P, fig. 10b
	(^2P)	-48.4(3)	-48.3(1.0)	0.22(4)	0.5(2)	alwd.; Line O, fig. 10b
	$L^2_1L_{2,3}(^2P)$	-68.6(5)	-68.8(1.0)	0.20(5)	0.3(2)	alwd.; Line Y, fig. 5
$KL_1(^3S)$	$L_1L^2_{2,3}(^4P)$	--	-3.7(1.0)	--	--	forb.; Line L, fig. 10a
	(^2D)	(-11.8(4))	-12.4(1.0)	0.16(5)	--	alwd.; Line K, fig. 10a
	(^2S)	--	-18.0(1.0)	--	--	forb.; Line H, fig. 10a
	(^2P)	--	-20.5(1.0)	--	--	alwd.; Line G, fig. 10a
	$L^2_1L_{2,3}(^2P)$	--	-41.0(1.0)	--	0.2(2)[a]	alwd.; Line V, fig. 10b

Table IX (cont.)

$KL_1(^1S)$	$L_1L_{2,3}^2(^4P)$				--	forbidden
	(^2D)				0.6(3)	forbidden[b]
	(^2S)				--	allowed[b]
	(^2P)				--	allowed
	$L_1^2L_{2,3}(^2P)$				0.5(2)	allowed

$^+$The energies are given relative to the $KL_{2,3}L_{2,3}(^1D_2)$ transition.

$^{++}$The intensities are given relative to the intensity of the $KL_1L_1(^1S_0)$ transition.

$^{+++}$Together with the $KL_1(^3S)$--$L_1L_{2,3}^2(^2P)$ transition.

[a]Lines not well resolved.

[b]Interchanged erroneously in ref.[51] (Private communication from W. Mehlhorn, 1969)

multiple ionization cross section in the target atoms, which results in the appearance of Auger-vacancy-satellite lines. In the photoelectric or electron bombardment production of an initial K-hole, the probability for an additional vacancy being simultaneously produced is also appreciable.[83] According to the sudden approximation, multiple ionization should be independent of excitation mode (impact or photoionization)[83] if correlation between orbital electrons is negligible and if the incident particle energies are sufficiently above the ionization threshold. For electron impact ionization of neon, the ratio of double to single initial ionization reaches the sudden approximation value at and above $E_e/[E(K) + E'(L)] \cong 1.6$, where E_e is the impact energy, $E(K)$ is the K-shell electron binding energy, and $E'(L)$ is the binding energy of an L-shell electron when a K-vacancy exists.[84] Recent measurements[84a] show that for neon the ratio of the KL-LLL Auger vacancy satellite intensity to the KLL-Auger intensity is the same for X-ray ionization (1.5-keV aluminum $K\alpha$ X-rays) and for electron impact ionization (3–10 keV impact energy).

The preceding data refer to atoms in the gaseous state. However, and unexpectedly, the KLL-spectra for $Z = 11$ and 12 in the form of a solid compound show no Auger-satellite lines of appreciable strength following photoionization vacancy production in the K-shell.[49]

When electron, proton, or ion impact is used to produce the initial K-shell atomic vacancy, the probability of double initial ionization, relative to single ionization, increases strongly. Auger-vacancy-satellite lines of intensities comparable to the KLL-spectral components may be observed. In addition, triply and higher initial ionized states are produced more strongly in particle impact excitation, causing the Auger spectrum to become even more complex. For a comparison, the relative intensities of the KL–LLL Auger-vacancy-satellite components for neon obtained by proton bombardment ionization, as reported by Edwards and Rudd,[53] are entered in Table IX. The electron spectrum obtained by these workers is shown in Figure 9 and can be compared with Figures 5 and 10.

Since the cross section for ionization increases with proton energy, the maximum bombardment energy of 300 keV was used for neon. Table X compares experimentally determined relative excitation probabilities of K and KL vacancy states in neon, as caused by electron and proton bombardment. It is noticed that the cross section for KL double vacancy production relative to that for K single vacancy production is in general greater for proton bombardment than it is for electron bombardment. This proportion is not expected on the basis of the sudden approximation limit of the shake-off process, which should be independent of the primary particle.[126] However, the ionization cross section for direct collision processes has a dependence on the inverse square of the particle velocity and has been shown to be negligible for beta-decay down to energies of about 50 keV.[128] In the case just cited, for electron energies of 5 keV and proton energies of 300 keV, the direct collision process may play a more important role in atomic ionization. In addition, the electron velocity is about five times greater than that of the protons.

As can be seen, the cross section for excitation of a given state by electrons and by protons is different. However, upon excitation of that state one would expect the relative probabilities of decay through the various modes to the various final states to be independent of the method of excitation. It can be seen in Table IX that there are some rather large departures from this expectation. Whether the departure reflects experimental inaccuracies or something more fundamental is not yet clear.[53]

Stalherm[130] has measured the relative intensities of the $L_{2,3}MM$ and $L_{2,3}M$ to MMM Auger spectral components for argon using electron bombardment excitation; the spectrum has also been measured by Rudd et al.,[131] who used proton bombardment excitation. Again the spectrum initiated by proton bombardment shows stronger satellite lines than that obtained through the electron impact excitation.

4.1.3.2. EXCITATION LINES IN AUGER SPECTRA. Fahlman et al.[132] reported the presence of a strong spectral line lying 5.0 eV in energy below the strong $KL_2L_3(^1D_2)$ Auger line from potassium in potassium chloride. The line was also noticed at 4.1 eV below the $KL_2L_3(^1D_2)$ Auger line of potassium in potassium sulfate. A hump 9 eV below the corresponding Auger line for chlorine in potassium chloride was considered to be of the same origin.

Various possible origins of the line were examined and rejected upon testing: characteristic energy loss suffered by the $KL_2L_3(^1D_2)$ Auger electrons, another Auger transition of the KLL-group, a photoelectron line, a chemical shift effect. It was finally decided that the line arises from the simultaneous occurrence of a $KL_2L_3(^1D_2)$ Auger transition and an excitation in the periphery of the atomic electron structure.

For the sulfur there were found two lines, at 7 and 14 eV below the $KL_2L_3(^1D_2)$ Auger line. The $L_{2,3}$ photoelectron spectrum of sulfur in potassium sulfate exhibited a characteristic energy loss at about 13.5 eV below the main peak, so that the former line at 7 eV

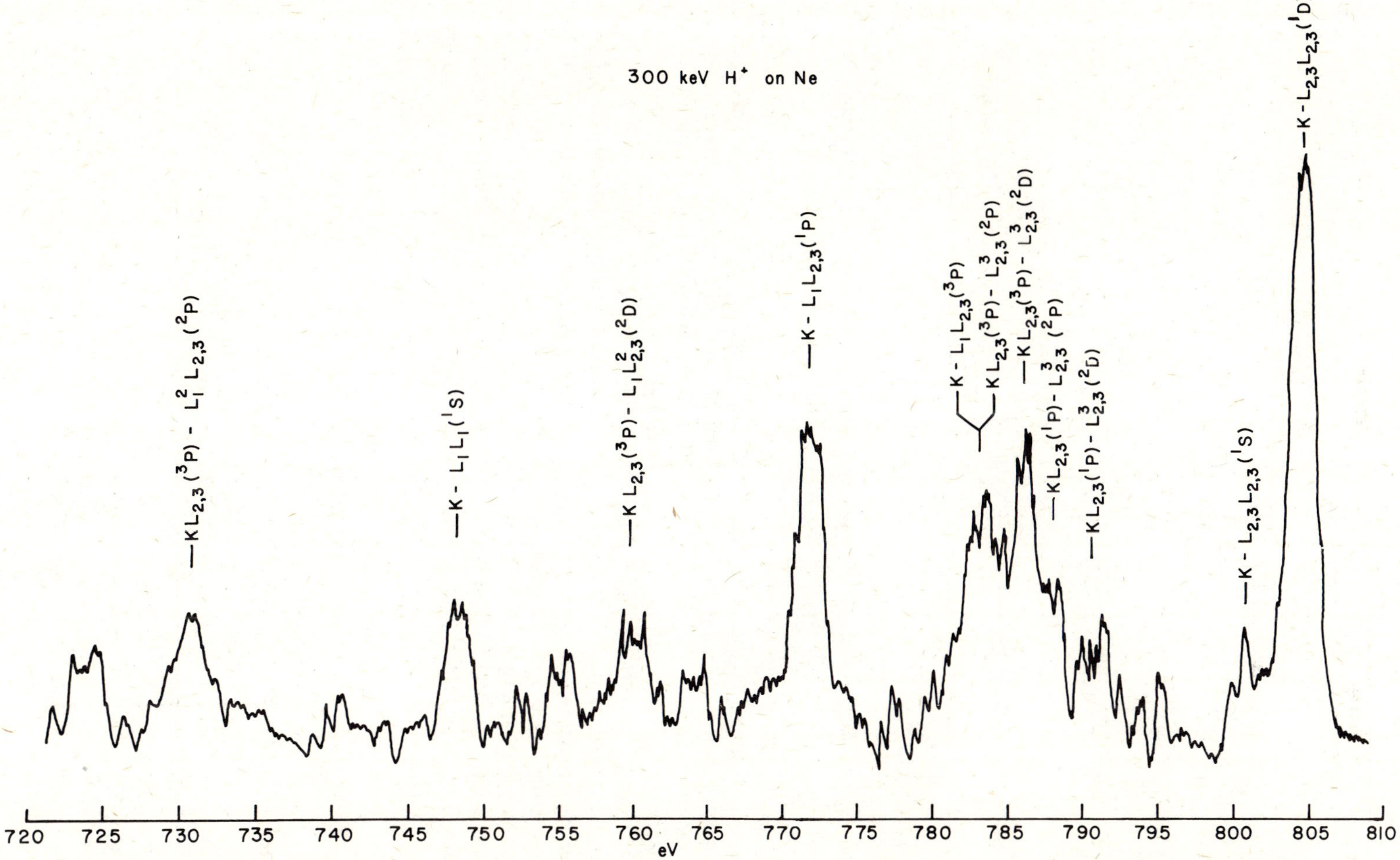

Figure 9. The electron energy spectrum from neon produced by 300-keV H^+ bombardment as measured by Edwards and Rudd.[53] The peaks are due to Auger electrons from initial *K*-vacancy states and *KL*-vacancy states. Above 795.5 eV, the vertical scale has been reduced by a factor of 2. It is interesting to compare this spectrum with those in Figures 5, 10*a*, and 10*b* obtained by electron bombardment ionization.

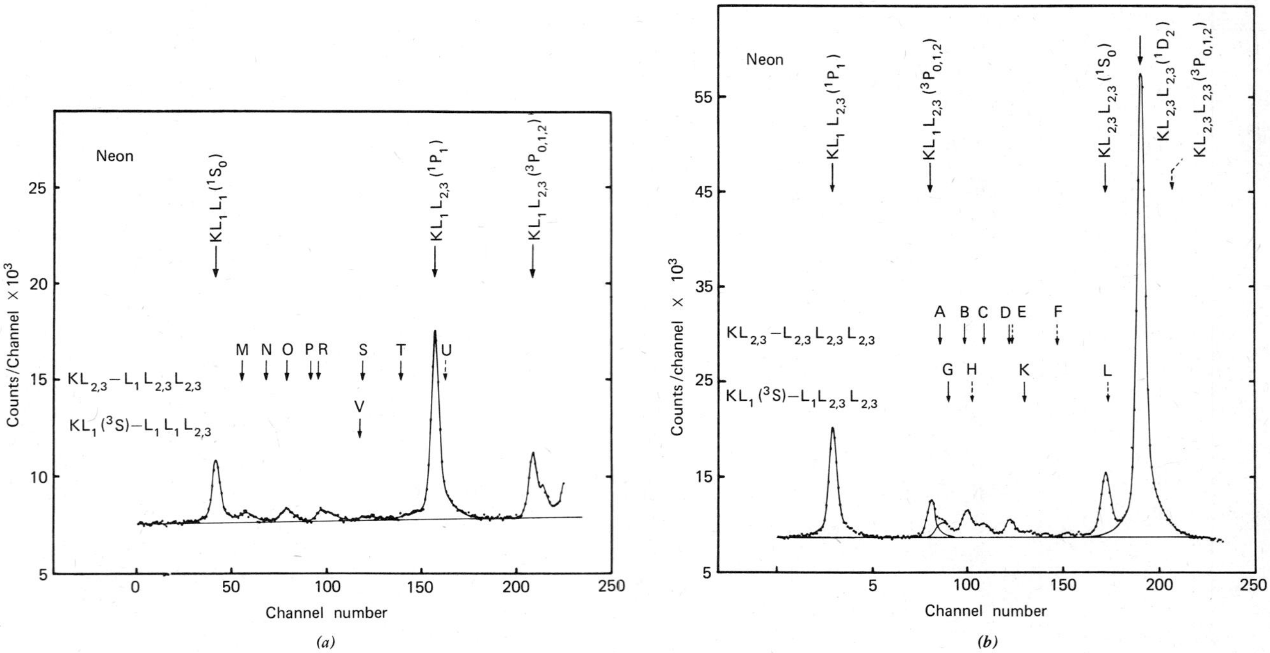

Figure 10. The *KLL* and *KL–LLL* Auger spectra of Figure 5, by Körber and Mehlhorn,[51] in greater detail. (See also Section 4.1.3.1).

may also be emitted in the same type of nonradiation transition as the lines found in the potassium and chloride Auger spectra from potassium chloride. (See also Chapter 8, section 3.2.3.)

4.1.3.3. EFFECT OF CHEMICAL OXIDATION STATE AND CHEMICAL BONDING ON THE AUGER TRANSITION ENERGY. The influence of the chemical state of an atom on the binding energies of the atomic electrons is treated in Chapter 7. In sodium thiosulfate the two sulfur atoms are in different oxidation states: 2^- and 6^+. The *KLL*-Auger electron spectra for sulfur in this compound will then be the superposition of the spectra, differing slightly in energy from the two types of atoms.[133,134] Coulson and Gianturco[135] have made theoretical Hartree–Fock calculations of the effects of chemical bonding of sulfur, silicon, and aluminum chemical compounds on *KLL*-Auger and *K* X-ray energies. Poor agreement with experimental values for sulfur show that refinement of the theoretical treatment is necessary.

The differences in energy of the copper KL_2L_3 Auger line for a copper atom in the metallic and the bivalent oxide forms were investigated by Nordling

Table X.

Relative cross-sections for excitation of K and KL vacancy states in neon (from ref.[53]).

State	Proton impact ref.[53] (E_p = 300 keV)	Electron impact ref.[51] (E_e = 5700 eV)	Electron impact ref.[52]
$K(^1S)$	12.6	19.81	15.60
$KL_{2,3}(^3P)$	6.1	1.81	
$KL_{2,3}(^1P)$	2.7	1.33	
$KL_1(^3S)$	0.9	0.16	
$KL_1(^1S)$	1.1	--	

et al.[136] (See also Chapter 6, Section 4.2.) The energy shift of this line by 1.0(3) eV toward lower energies for the oxide was obtained. This value compared favorably with a roughly estimated value obtained from a comparison of the *K*-, L_2-, and L_3-electron binding energies between the two cases.

In their investigation of the *KLL*-Auger spectrum for fluorine in the fluorides of lithium, sodium, and magnesium, Albridge et al.[54] noted a dependence of relative component intensities on the chemical compound; the component energies also appeared to alter, but accuracy limitations made it impossible to investigate this phenomenon further. It was observed that the component energies increased an average of approximately 0.4 eV and decreased an average of approximately 1.1 eV for going from lithium fluoride to sodium fluoride and to magnesium fluoride, respectively.

The sodium fluoride and lithium fluoride bonds were estimated to be ~90% ionic and that of magnesium fluoride ~85% ionic; therefore, the fluorine *L*-shell was assumed to be completely filled. The *KLL*-Auger spectrum for fluorine in these compounds should then be analyzable in terms of transitions to the usual five final states predicted by *L–S* coupling. However, there is to be expected a small dependence of the *KLL*-component energies on the chemical bonding of the fluorine; the ionic character is largest for the sodium fluoride bond and will decrease through lithium fluoride to the smallest value for magnesium fluoride. Hence the amount of negative charge in the vicinity of the fluorine atom will decrease in the order NaF, LiF, MgF_2; this will in effect cause a rise in the atomic binding energies of fluorine, whereupon a decrease of the Auger transition energies will occur. Not only was the energy shift expected to be about the same for all Auger components, but it was anticipated that the relative energies of these components for the three compounds would agree; and indeed, such agreement was shown by measurement.

4.1.3.4. *K*-AUGER EFFECT IN GASEOUS MOLECULES. Until 1968 only a few Auger spectra have been recorded for gaseous molecules. Mehlhorn[85] and Siegbahn et al.[42] investigated carbon in CH_4 and Spohr[137] studied carbon in CH_4, CH_3Br, CH_2Br_2, $CHBr_3$, and CBr_4. These spectra show complicated structure, caused by the coupling complexity of the final state double vacancy in the carbon *L*-shell. One must analyze the final state in terms of molecular orbitals. The *M*-Auger spectra for bromine were also investigated.[137]

Stalherm et al.[138] have also carried out an investigation of the *KLL*-Auger spectrum from nitrogen with an extensive treatment of the electronic states in the final N_2^{+2} ion. The complex structure observed could be extensively analyzed on this basis. Although no evidence of vibrational structure could be found, its absence was not proved, since the instrumental resolution did not allow detailed enough investigation. Details of the treatment are provided in Ref. 138 and in Ref. 14. Carlson et al.[138a] have also investigated the *KLL*-Auger spectrum of N_2 from both electron impact and photo-ionization in the *K*-shell. The reader should also see Moddeman's[138b] thesis which reports the study of *KLL*-Auger spectra for a wide variety of diatomic through polyatomic gaseous molecules (e.g., N_2, O_2, CO, NO, H_2O, CO_2, halide, and derivatives of methane and silane, and the like).

4.2. The *KLM*-Auger Group

Unlike the *KLL*-Auger group, the *KLM*-Auger group has been the subject of relatively few investigations. The group is complex, being composed of the KL_1M, KL_2M, and KL_3M subgroups, which partially overlap each other. The entire KL_1M subgroup is situated free from the others for $Z \gtrsim 30$, and the $KL_1M_{1,2,3}$ lines are free of the other subgroups for for $Z \gtrsim 70$. The $KL_1M_{1,2}$ lines stand alone for all *Z*-values. In addition, the $KL_{1,2,3}N,O$ subgroups may also overlap portions of the former subgroups, increasing the difficulties to be met in analyzing the *KLM*-groups.

A summary of experimental values of some of the *KLM*-Auger component energies and intensities are collected in Table XI.

4.2.1. *Energies of KLM-Auger Electrons*

Theoretical calculations of the energy splittings of the states within the various possible atomic electron configurations after the emission of a *KLM*-Auger electron have been calculated for several atomic numbers by Asaad and Burhop,[16] who assumed intermediate coupling for the final atomic configuration. The calculated energy splittings for each final configuration were plotted as a function of the effective nuclear charge. However, since the authors do not give the relative energies of the configurations, it is impossible to compare their calculations with experimental results where the fine structure of the lines is not resolved.

Mladjenovič and Slätis,[34] applying the procedure of Bergström and Hill[29] (see Section 4.1.1), calculated for $Z = 83$ an effective incremental charge:

$$\Delta Z = \frac{\Delta E^Z(L_p M_q)}{E^{Z+1}(M_q) - E^Z(M_q)}, \qquad (6)$$

which is correlated to the energy discrepancy

Table XI.

Experimentally determined energies (in keV)
the components of the

Energies

Ref.	Line / Z	KL_1M_1	KL_1M_2	KL_1M_3	KL_2M_1	KL_1M_4	KL_1M_5	KL_2M_2	KL_2M_3	$KL_2M_{4,5}$	KL_3M_1
37)	47[+]	20.92 (6)	21.06(6)		21.21 (6)			21.17(6)			
102)	70	48.36 (2)	48.60 (2)	48.80 (2)	48.86 (3)				49.34 (2)	49.76 (2)	49.92 (2)
103)	71	49.875	50.118	50.374		50.950				51.29	51.512
104)	71	49.845	50.101	50.350					50.846		51.492
105)	74	55.51 (5)	54.74 (5)	55.03(5)					55.53 (5)		56.37 (5)
106)	75	56.190 (28)	56.390 (28)	56.680(28)				56.910 (28)	57.300 (29)	57.710 (29) (KL_2M_4 only)	58.110 (29)
108)	78	61.093	61.365	61.731 (without $KL_1M_{4,5}$)					62.326 ($KL_1M_{4,5}$)[+]	62.839	63.406
12) 111)	79	62.83	63.05								
144)	83		69.85	70.25					71.08		
29)	83	70.02 (4)	70.39 (4)		70.74 (4)				71.55 (4)		
120)	83	70.19[§]	70.41		70.80						
120)	84	71.74[§]			72.82				73.60		
121)	84	72.01	72.32	72.91				(with KL_1M_3)	75.46 (KL_1M_5)[+]		

Intensities

Ref.	Z	KL_1M_1	KL_1M_2	KL_1M_3	KL_2M_1	KL_1M_4	KL_1M_5	KL_2M_2	KL_2M_3	$KL_2M_{4,5}$	KL_3M_1
47)	35	1.0[×]	1.78(20)		0.84(10)			3.85(40) (+KL_1N)		0.35 (10)	1.16 (25)
80)	35	0.20 (5)	0.21 (5)	0.22 (5)	0.17 (10)				0.85 (15)		0.26 (15)
37)	47	0.8(2)	0.8(2)		0.2(1)			1.6(3)			

and relative intensities of
KLX Auger groups.

KL_3M_2	KL_3M_3	KL_1N_1	$KL_1N_{2,3}$	$KL_3M_{4,5}$	KL_2N_1	$KL_2N_{2,3}$	KL_3N_1	$KL_3N_{2,3}$	KL_3O	Other
21.51(6)				21.72 (6)		21.89 (6)		22.06 (6)		
50.13 (3)	50.33 (3)			50.73 (3)	50.99(3)		51.98(2)		52.35 (3)	
51.734	51.966		52.06	52.380 ($KL_1N_{4,5}$ + KL_1O_1)		52.611	53.650		53.99 ($KL_3N_{4,5}$ +)	
51.739	51.975			52.421 (KL_3M_4 only)			53.573			53.827 (KL_3N_4 only)
56.64 (5)	56.96(5)				57.43(5)		58.82(5)			
58.350 (29)	58.650 (29)			59.150 (29)			60.660 (30)		61.080 (30)	59.600 (39) (KL_2O)
63.694	64.037 ($KL_1M_{4,5}$ +)	63.870		with $KL_2N_{2,3}$	64.195 (KL_1O_{1-3} + $KL_1N_{6,7}$)	64.495 ($KL_3M_{4,5}$ + $KL_1O_{4,5}$ + KL_3N_{1-7})				
65.48							68.05			
	73.42	72.85								
73.30 (4)	73.87 (4)									
73.93										
75.46	76.04									

4.08 (40)				0.50 (15)	0.84 (15)					
0.38 (10)	0.10 (5)			0.08 (3)						
1.7(3)				0.55 (20)		0.3 (2)		0.55 (20)		

Table XI (cont.)

Intensities (cont.)

Ref.	Line / Z	KL_1M_1	KL_1M_2	KL_1M_3	KL_2M_1	KL_1M_4	KL_1M_5	KL_2M_2	KL_2M_3	$KL_2M_{4,5}$	KL_3M_1
89)	49	0.42	0.62		0.20	0.865 (+KL_3M_1)				with $KL_3M_{2,3}$	with $KL_2M_{2,3}$
30)	55	0.97(15)			0.1(1)				0.62 (10)		
100)	69	0.39 (4)	0.30 (3)	0.29 (5)	0.24 (5)	0.05 (2)	0.09 (3)		0.44 (4)	0.16 (7)	0.21 (4)
104)	71	0.41	0.29	0.44					0.56		0.26
103)	71	0.40 (5)	0.31 (5)	0.52(10)		0.45(4)				0.09 (3)	0.14 (3)
105)	$74^{\dagger}$	0.43 (14)	0.36 (14)	0.43(14)					0.22 (7)		0.36 (14)
106)	75	0.44 (8)	0.30 (6)	0.62(12)				very weak	0.38 (8)	0.32(6) (KL_2M_4 only)	0.32(6)
108)	78	0.35 (5)	0.32 (4)	0.47(2)		with KL_2M_3		with KL_1M_3+ KL_2M_1	0.35 (5) (+$KL_1M_{4,5}$)	0.11(5)	0.11(5)
116)	81	0.52	0.42	0.57	0.09				0.40 (5)	0.13 (3)	0.62
144)	83		0.6	0.6					0.5		
34)	83	$0.4^{\dagger\dagger}$	0.5		0.8				0.7		
120)	$83^{\dagger}$	0.80^{Δ}	0.16		0.27						
120)	$84^{\dagger}$	0.63^{Δ}			1.40				1.90		

Note: The published probable errors are often below the published value to conserve space.

Symbols:

$^{\times}$ All intensities normalized to KL_1M_1 = 1.

$^{+}$ Energies corrected +0.04 keV for energies < 22.5 keV and +0.05 keV for energies > 22.5 keV (see ref. 30)).

† Intensities renormalized to KL_1L_1 = 1.

†† Probable error: ±(10 - 20%).

$^{\Delta}$ Probable error: ±(30 - 40%).

§ Probable error: ±(0.01 - 0.10%).

KL_3M_2	KL_3M_3	KL_1N_1	$KL_1N_{2,3}$	$KL_3M_{4,5}$	KL_2N_1	$KL_2N_{2,3}$	KL_3N_1	$KL_3N_{2,3}$	KL_3O	Other
1.28 (+$KL_2M_{4,5}$)		0.15 (KL_1N)		0.20	0.15 (KL_2N)		0.25 (KL_3N)			
0.93(15)			0.19(5) (+$KL_1M_{4,5}$)			0.13 (5)		0.27 (5)		
0.44 (5)	0.50 (6)	0.12 (3)	0.12 (5)	0.30(9) (+KL_1O $KL_1N_{4,5}$)		0.20 (6)				
0.56	0.60			0.28 (KL_3M_4 only)			0.26			0.24 (KL_3N_4)
0.42 (5)	0.55(5)		0.05 (2)	0.21(3) (+$KL_1N_{4,5}$+KL_1O)		0.12(2)	0.21(3)		0.04(2) (+$KL_3N_{4,5}$)	0.03(2) (KL_2O + $KL_2N_{4,5}$)
0.50 (22)	0.64(22)				0.36(14)		0.14(7)			
0.59 (12)	0.60 (12)			0.32(6)			0.16(8)			
0.44 (9)	0.40 (12) (+$KL_1N_{4,5}$)	0.10(5)		with $KL_2N_{2,3}$	0.10(5) (+KL_1O_{1-3} $KL_1N_{6,7}$)	0.25(7) (+$KL_3M_{4,5}$ $KL_1O_{4,5}$ KL_3N_{1-7})				0.18(6) (KL_3N_{1-7})
	0.42			0.2			0.29(3)			
	0.6	0.8								
0.7	1.0									
0.16										

$\Delta E^Z(L_p M_q)$ in the expression for the $KL_p M_q$ Auger electron energy

$$E^Z(KL_p M_q) = E^Z(K) - E^Z(L_p) - E^Z(M_q) - \Delta E^Z(L_p M_q).$$

Here $E^Z(K)$, $E^Z(L_p)$, and $E^Z(M_q)$ are the K-, L_p-, and M_q-atomic electron binding energies for the atom of atomic number Z. Mladjenovič and Slätis obtained $\Delta Z = 0.59$, 0.58, and 0.61 for the M_1-, M_2-, and M_3-subshells, respectively. For $Z = 74$, Gallagher et al.[105] found $\Delta Z \cong 1.00$ for all the M_{1-3} subshells; Newton[106] gives $\Delta Z = 0.75(20)$ for the KLX-Auger lines for all L-vacancies.

4.2.2. Relative Intensities of the KLM-Auger Lines

Pincherle[57] carried out calculations on the transition probabilities for some of the KLM-Auger components. These were performed assuming j–j coupling and using unscreened hydrogenic wave functions. The calculations do not include all the possible transitions, however, and are therefore incomplete.

The relative intensities of the KLM-spectral lines were computed for five different Z-values by Asaad and Burhop[16] for intermediate coupling between the two inner shell vacancies of the atom in the final atomic states (see Table V).

The amount of experimental data on the KLM-Auger spectra thus far available is rather meager. A comparison of Kaye's measured intensities[102] with the values for $Z = 70$ interpolated from the theoretical results gives reasonable agreement except for the KL_1M_1 and KL_1M_2 lines. His measured spectrum was analyzed assuming pure j–j coupling. Figure 11 shows the KLM-, KLN-, and KLO-Auger spectra for $Z = 70$ as measured by Kaye.[102] Gizon[100] has recently carried out similar measurements for $Z = 69$.

In several measurements of the KLM-Auger spectra a semicircular spectrograph in combination with photographic detection was employed; the relative intensities of the observed lines were difficult to estimate and are subjected to errors of $\sim 20\%$. The work was performed by Johnson and Foster[37] and Geoffrion and Nadeau[30] for $Z = 47$, Gallagher et al.[105] for $Z = 74$, Newton[106] for $Z = 75$, Mladjenovič and Slätis[34] for $Z = 83$, and the same authors[120] for $Z = 83$, 84.

4.3. Other K-Series Groups

Experimental results on the energies and intensities of the line components making up the KLN-, KMM-, and so on, Auger groups are few. The close proximity of the lines in each of these groups, as well as the overlapping between the groups, makes it difficult to carry out detailed spectrum investigations. The component line energies, however, can be satisfactorily approximated by the application of the method of Bergström and Hill[29] using $\Delta Z = 1$.

Measurement of the relative intensities of some of the KLN,O groups and their components have been carried out and are included in Table XI. Measurements on the KMM-spectrum include the determinations of Johnson and Foster[37] for $Z = 47$, Krisciokaitis and Haynes[89] for $Z = 49$, Geoffrion and Nadeau[30] for $Z = 55$, Gallagher et al.[105] for $Z = 74$, and Sujkowski[116] for $Z = 81$.

Determination of the KLX and KXY group intensities (where the X and Y refer to shells lying outside the L-shell) relative to that of the KLL-Auger group have been of more interest. The $KLL{:}KLX{:}KXY$ group intensity ratio values can be used in determining Q-values in electron capture decay of nuclei[139] and to aid in estimating atomic level widths[140] (see Chapter 6). Figure 12 shows the experimentally determined ratios KLM/KLL or KLX/KLL, and KXY/KLL as a function of Z, and Table XII gives their numerical values and literature sources.

The work of Pincherle[57] gives calculated transition probabilities for the KLL- and some of the KLM,N, KMM,N-, and KNN- Auger group components. A few rather strong transitions are not included in the calculations, and values derived for the KLX/KLL and KXY/KLL Auger group intensity ratios from the included transition probabilities are only approximate. It is found that these intensity ratios give values that are larger by a factor of 2 to 3 than the experimental values for $Z \gtrsim 35$.

The nonrelativistic calculations by Geoffrion and Nadeau (see Ref. 141) for the most intense KLL, KLM, KLN, KLO, KMM, KMN, KMO, and KNN transitions, using hydrogenlike wave functions, agree well with Pincherle's calculations for the KLL, KLM, and KLN transitions, but discrepancies arise for the others. The KLX/KLL and KXY/KLL intensity ratio values are also much larger than experimental values for all Z values.

Rubenstein[63] calculated the KLX/KLL and KXY/KLL intensity ratios in his work on the Auger effect. His values, included in Figure 12, lie below the general trend of values and, in the case of KLX/KLL they are found close to the values calculated by Asaad and Burhop[16] for KLM/KLL; these last authors used the KLM-transition probabilities of Rubenstein in their calculations, and agreement is expected.

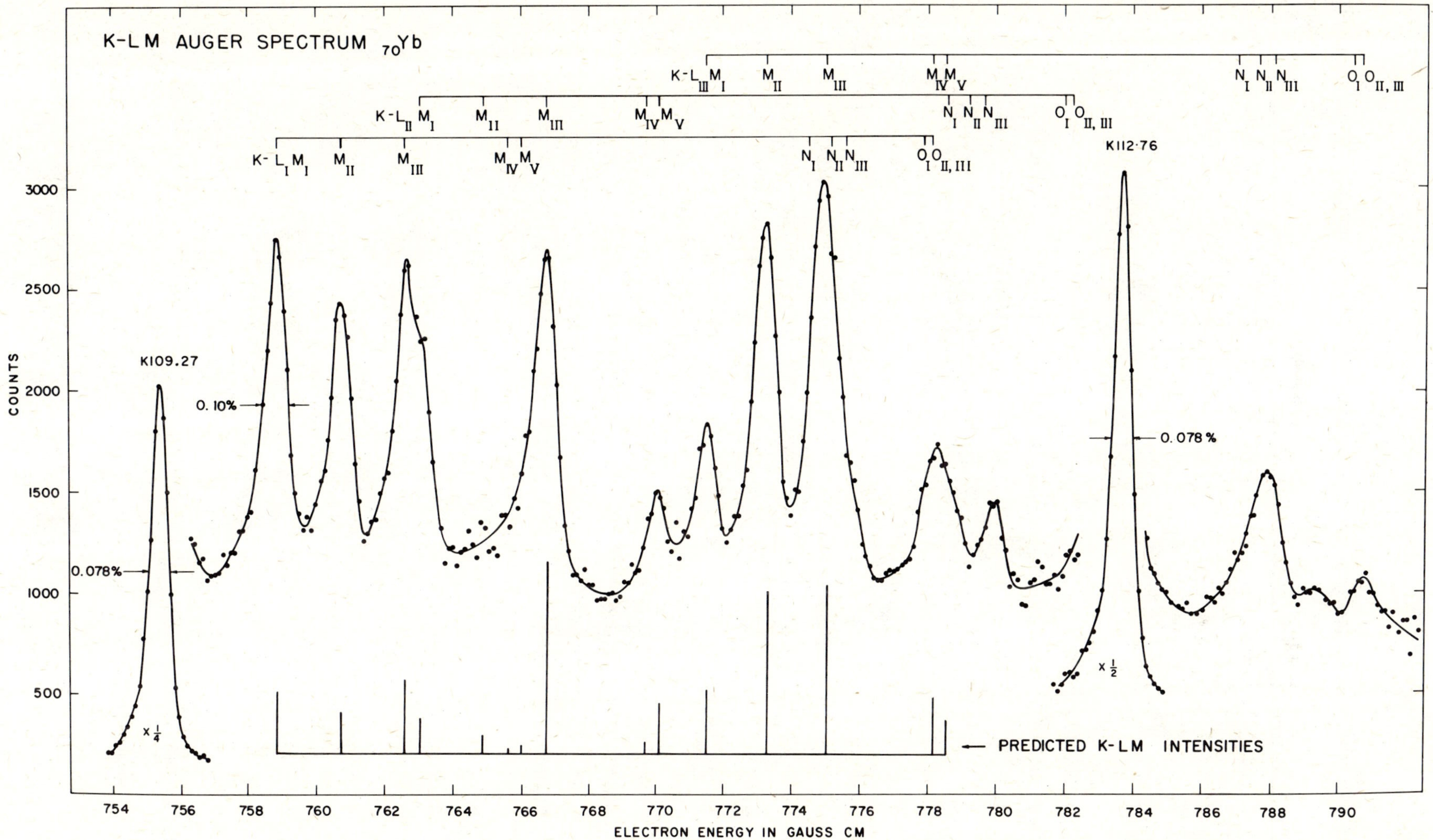

Figure 11. The *KLM*, *N*, *O* Auger spectra in ytterbium ($Z = 70$) as measured by Kaye[102]. The differences in widths of the Auger and the superimposed *K*-conversion lines are obvious. The *KLM*-intensities predicted by Asaad and Burhop[16] (see Table V) are inserted for comparison.

Table XII.

Summary of experimentally determined values of the K-Auger group intensity ratios.

Ref.	Z	KLX/KLL	KXY/KLL	Ref.	Z	KLX/KLL	KXY/KLL
141)	17	0.156(10)	0.009(1)	154)	59	0.55(3)	0.13(2)
79)	26	0.20(1)	--	154)	61	0.55(10)	--
55)	26	0.19(1)	--	93)	62	0.47[+]	--
145)	28	0.42	--	94)	63	0.41[+]	0.054
77)	29	0.303(10)	0.0166(10)	95)	63	0.60(3)	0.086(15)
146)	29	0.292(10)	0.016(2)	96)	65	0.48(7)	--
38)	29	0.30(2)	0.017(4)	96)	65	0.37(6)[+]	0.11(4)
26)	32	0.31[+]	--	154)	65	0.46(5)	--
91)	33	0.24(4)[+]	--	154)	65	0.43	--
147)	33	0.33(12)[+]	0.013(5)[++]	155)	66	0.45(10)	--
80)	35	0.33(2)	0.026(3)	154)	66	0.59(10)	--
121)	38	0.28	--	98)	67	0.56(5)	0.070(5)
128)	43	0.48(6)	0.050(10)	154)	67	0.55(6)	0.09(3)
149)	43	$0.35^{+(11)}_{-(4)}$	0.03(1)	154)	68	0.59(6)	0.24(7)
150)	47	0.49	0.054	156)	70	0.49(4)	--
37)	47	0.71(14)	0.06(3)	121)	70	0.64(5)	--
150)	48	0.45	0.056	157)	71	0.466(65)	0.097(24)
148)	49	0.42(10)	0.05(2)	103)	71	0.51(4)	--
151)	49	0.417(16)	0.076(8)	105)	74	0.45(12)	0.10(3)
89)	49	0.431(3)	0.0513(7)	106)	75	0.57	--
90)	52	0.44(2)	--	108)	78	0.548(44)	--
147)	53	0.51(9)	--	158)	78	0.82(3)	0.25(12)
152)	54	0.51(4)[+]	--	107)	78	0.403[+]	0.035
152)	55	0.48(5)[+]	--	113)	80	0.496(15)	0.094(3)
30)	55	0.41(3)	0.08(1)	151)	80	0.71(15)	--
24)	55	0.48(3)	--	159)	81	0.53	0.063
153)	55	0.48	--	116)	81	0.61(5)	0.089(10)
91)	56	0.42(20)	0.081(12)	34)	83	0.64(15)	--
151)	56	0.40(4)	0.08(2)	121)	84	0.61(8)	--
154)	57	0.44(10)	--	121)	84	0.55(3)	--
154)	57	0.32	--	123)	93	0.58(3)	0.085(8)

[+]KLM/KLL only [++]KMM/KLL only

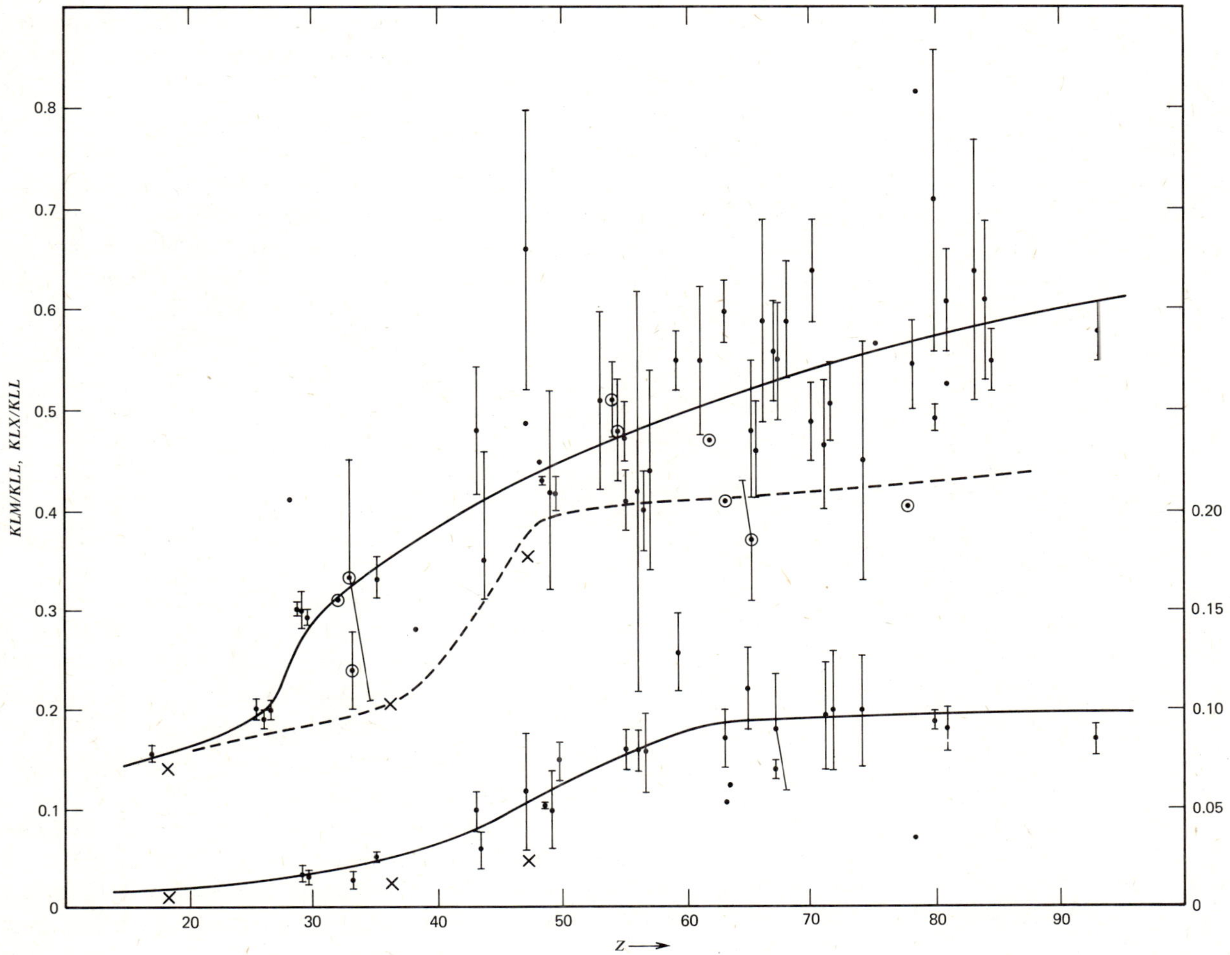

Figure 12. Experimental *KLX/KLL* and *KXY/KLL* Auger intensity ratio values compared with theoretical estimates. The heavy lines are the loci of approximate mean values. Circled dots, *KLM/KLL* intensity ratios; solid dots, *KLX/KLL* and *KXY/KLL* ratios; crosses, theoretical estimates of *KLX/KLL* and *KXY/KLL* ratios by Rubenstein;[63] dashed curve, *KLM/KLL* theoretical estimate by Asaad and Burhop.[16]

5. OTHER AUGER SERIES AND COSTER–KRONIG TRANSITIONS

Past investigations of the *L*-, *M*-, and so on, Auger, and Coster–Kronig spectra have been of considerably less interest than studies of the *K*-Auger series. This was primarily because of experimental difficulties associated with the investigation of low energy electron spectra and because of the complexities of the spectra of these series.

For electron sources in the solid state it is necessary to employ special and more exacting source preparation techniques to ensure a thin, uniform layer of activity or converter material upon a substrate. The collection of the activity in the form of an ionized gas, as is done in preparing the activities thorium B (^{212}Pb) and radium B (^{214}Pb), makes it possible to obtain very thin, yet intense, activity deposits. Because quality sources of these activities are rather easy to prepare, there has been extensive investigation of the low energy conversion and Auger spectra of these isotopes and their decay products.

There have been numerous investigations of the very low energy Auger and Coster–Kronig electron spectra from the rare gases (argon, krypton, and xenon). Initial ionization is caused by electron or ion bombardment or photoionization. Electron energy losses and spectral line broadening due to straggling become negligible, since the target material is in the gaseous state at low pressure.

As compared to the *K*-Auger series, the *L*- and *M*-series are far more complex, not only because there is more than one atomic subshell in which a single initial vacancy can be created, but also because the atom may become multi-ionized (if other Auger transitions occur before an L- or M-Auger transition takes place).

The occurrence of Coster–Kronig transitions gives rise to two unique characteristics of L-Auger spectra. First, through transitions of the Coster–Kronig type the initial ratio of L-subshell vacancies is altered before these vacancies are finally filled. This will result in (*a*) a marked decreased in the intensities of the L_1MM, L_1MN, and so on, transitions, (*b*) an intensity increase in the L_3MM, L_3MN, and so on, transitions, and (*c*) a possible intensity increase or decrease in the L_2MM, and so on, transitions. In fact, even if the initial L-shell vacancies are produced primarily in the L_1-subshell, as by the internal conversion of a magnetic dipole nuclear transition of energy less than the K-electron binding energy, a high L_1L_3X- Coster–Kronig transition probability will often cause the L-Auger spectrum to reveal L_3-Auger transitions of much greater intensity than the L_1-Auger transitions.

Second, the L-Auger electron energy for a given transition taking place in an atom that is in a state of double ionization (or triple, in the event of successive KLL- or KLX-, and LLX-transitions) will be of an an energy that differs from that for the same transition in a singly ionized atom.

Although L-Auger investigations are numerous and energy determinations of the Auger transition lines have been made with varying degrees of accuracy, precise intensity measurements are relatively few. Again, this is primarily owing to the earlier widespread use of photographic electron detection and to the associated spectrum line intensity determination difficulties. Through the employment of other electron detection methods, however, more recent investigations have enabled workers to make quite accurate and detailed spectrum analyses.

In the investigation of the multicomponent L- and M-Auger spectra, it is advantageous to employ a spectrometer of high resolving power. However, the ultimate limit of the detail in which a spectrum can be investigated is dependent on the instrumental resolving power, the complexity of the spectrum, and the natural widths of the spectral components. Therefore, highest spectrometer resolutions are not always most suitable at very low electron energies.

Semicircular spectrometers were often used in earlier investigations of low energy electron spectra at high resolution. The detection of the low energy electrons was usually made possible by the employment of photographic emulsion plates with or without pre- or postacceleration (see Chapter 2, Sections 2.1. and 6.5).

When the detector is a Geiger–Müller counter, which is usually the case when double-focusing spectrometers are used for electron spectrum analysis, the counter window is often too thick to allow effective detection of low energy electrons. This is sometimes remedied by the use of pre- or postacceleration techniques to increase the electron energy to a value that ensures detector window penetration. Acceleration techniques may also be advantageously applied when emulsions are used for electron detection, since blackening per electron in the emulsion (i.e., the electron track length) increases with energy.

With time, new window techniques have made it possible to produce sufficiently thin counter windows to allow effectively 100% transmission of low energy electrons down to $\sim$5 keV. Below this value it is still necessary to take into consideration window absorption effects.

The recently introduced windowless continuous-channel electron multiplier is now fairly widely used. Although electrons can be detected from 10 keV down to 250 eV with greater than 50% efficiency, the detection efficiency of the device must be calibrated before relative intensity measurements are performed (see Chapter 2, Section 6.4).

Theoretical treatments of L-Auger transition probabilities have been presented on a limited number of occasions, and these have been confined to a few individual atomic cases. It is only in these cases that comparison between theory and experiment can be made.

The following subsections present a review of theoretical and experimental work carried out on the L-Auger effects (Sections 5.1.1, 5.1.2, and 5.3) and M-Auger effects (Sections 5.1.3 and 5.3); comparison between the theoretical and experimental is made whenever possible.

Finally, a survey is presented of the small amount of theoretical and experimental work done on Coster–Kronig transitions is included (Sections 5.2 and 5.2.1), together with considerations of the Z-value ranges where the occurrence of the various Coster–Kronig transitions is allowed.

Descriptions of the experimental investigations that have been carried out on L- and M-Auger and Coster–Kronig spectra (Section 5.3) contain information on the type of instrumentation and, in some cases, the source preparation techniques.

The reader is also referred to the reviews by Sujkowski and Slätis,[142] Listengarten,[10] and Dionisio[143,11,12] for earlier treatments of the subjects of L- and M-Coster–Kronig and Auger effects.

5.1. Theoretical Transition Energies and Probabilities of *L*- and *M*-Auger Transitions

5.1.1. *L-Auger Transition Energies*

Auger transitions for the *L*-series arise when a vacancy in one of the *L*-subshells is filled by an electron from an outer shell and the difference between the initial and final state energies is carried off by the simultaneous emission of a second outer shell electron. The necessary initial *L*-shell vacancy can be created by methods cited in Section 2. Vacancies may also be caused by the occurrence of (*a*) $K\alpha$ X-ray or (*b*) *KLL*- or *KLX*-Auger transitions, whereby a *K*-shell vacancy beomes shifted to the *L*-shell. In case *a* the original single atomic vacancy in the *K*-shell is transferred to the L_3- or L_2-subshell with the simultaneous emission of a $K\alpha_1$ or α_2 X-ray. An L_2XY- or L_3XY- Auger electron may then be emitted with the filling of the created *L*-subshell vacancy. In certain *Z*-value regions a Coster–Kronig transition may cause the L_2-vacancy to shift to the L_3-shell. (This phenomenon is discussed later.) In case *b* there may be the simultaneous formation of two vacancies in the *L*-shell (as the result of a *KLL*-Auger transition) or one hole in the *L*-shell and a second in the *M*, *N*, and so on, shells (resulting from *KLM*, *KLN*, and so on Auger transitions). The atom is thus doubly ionized, and therefore the *L*-Auger electron energy for a given transition taking place in the presence of a "spectator" vacancy[160] differs slightly from the value that would be obtained if the atom were singly ionized.

For elements lying in certain defined *Z*-value regions of the periodic table, an initially singly ionized atom with the vacancy in the L_1- or L_2-subshell may become doubly ionized through the occurrence of an $L_1L_{2,3}X$ or L_2L_3X Coster–Kronig transition (see Section 5.2). Following such a transition, the initial *L*-subshell vacancy will have shifted to another *L*-subshell, and a second vacancy will have been created in an atomic orbital external to the *L*-shell.

In the *LMM*-Auger spectrum, which is of greatest intensity and primary interest among the *L*-series groups, 45 transitions are energetically possible, assuming extreme *j–j* coupling between the two final *M*-shell vacancies. This coupling approximation is never valid, however.

The energies of *LMM*-Auger electrons range from about 18 keV for fermium ($Z = 100$) to below 1 keV for elements lighter than zinc ($Z = 30$). The calculation of the approximate *L*-Auger transition energies for an atom that is initially singly ionized can be made in a manner similar to the method used for calculating *KLL*-Auger electron energies introduced by Bergström and Hill.[29] In analogy to equation 6 in Section 4.2.1, we obtain for the *LMM* series

$$\Delta Z = \frac{\Delta E^Z(M_p M_q)}{E^{Z+1}(M_q) - E^Z(M_q)}, \tag{7}$$

where the denominator is the difference in the M_q electron binding energy for the atoms of atomic number Z and $Z+1$. For purposes of spectral line identification, it is usually assumed that $\Delta Z = 1$ (e.g., Refs. 28, 142), and the validity of this assumption has been discussed by Burhop.[6] However, for the heavier elements, values of $\Delta Z \neq 1$ have been obtained (e.g., Refs. 77, 161).

Krisciokaitis and Haynes[89] found that, for the *L*-Auger spectrum of ^{113}In, a choice of $\Delta Z = 0.55$ and 0.67 provided good energy estimates for transitions of the types $L_{1,2,3}M_{1,2,3}M_{1,2,3}$ and $L_{1,2,3}M_{4,5}M_{4,5}$, respectively. These values are similar in magnitude to those obtained for *KLL*-spectra for high *Z*-values (see Section 4.1.1). The measured *L*-Auger line positions may actually be center of gravity estimates for multiplets, since for this *Z*-value the coupling is intermediate between *j–j* and *L–S*.

In their analysis of the *L*-Auger spectrum of ^{210}Bi, Haynes et al.[160] used a graphical analysis of the energy spectrum. This analysis, to be described, shows that ΔZ, for a given set of transitions $L_iM_pM_q$, increases with increasing *p* and *q*.

We may also express the $L_iM_pM_q$ Auger transition energies as follows:

$$E(L_iM_pM_q(^{2S+1}L_J)) = E(L_i) - E(M_pM_q(^{2S+1}L_J)), \tag{8}$$

where near *L–S* or intermediate coupling is assumed. The value $E(L_i)$ is the energy of the atom singly ionized in the L_i-subshell, and $E(M_pM_q(^{2S+1}L_J))$ is the energy of the final doubly ionized (in the valance shell) atomic state, which can sometimes be calculated from the tables by Moore[162] (see Refs. 51 and 163).

Large-scale theoretical calculations on the *L*-Auger electron transition energies are not yet available. There are, however, indications that such will soon appear.

The preceding paragraphs are valid only if the atom is singly ionized initially in the L_i-subshell. However, when the L_i-vacancy is produced by a *KLL*-Auger or an *L*-Coster–Kronig transition or both, and the final L_i-Auger transition takes place in an atom that is initially doubly of triply ionized, the energies of such transitions may be more clearly analyzed in an alternative manner, as follows.

The effects of multiple ionization on *L*-Auger

spectra have been extensively analyzed and treated recently by Haynes et al.[160] in connection with their investigation of the L-Auger spectrum of ^{210}Bi following the beta-decay of radium D (^{210}Pb). Initial L-shell vacancies, caused by internal conversion of the 46.5-keV $M1$-transition in ^{210}Bi, are formed primarily in the L_1-subshell—the conversions that take place in the L_2- and L_3-subshells are approximately 0.1 and 0.01 of the L_1-value, respectively. Therefore, the observed $L_3\,MM$-Auger lines of dominating intensities result mostly from the filling of L_3-vacancies created by $L_1\,L_3\,M_{4,5}$ Coster–Kronig transitions.

The energies of these $L_3\,MM$-transitions, which occur in the presence of the additional $M_{4,5}$-vacancy, were found to be shifted to lower energies with respect to the energies expected from the energy analysis of the $L_{1,2}\,MM$ spectral components. Such a transition taking place in the presence of the $M_{4,5}$ spectator vacancy was designated by the symbol $L_3\,MM(M_{4,5})$ and is called an Auger-vacancy-satellite line by Haynes et al.[160] This Auger-vacancy-satellite transition may also be written $L_3\,M_{4,5}$–$MMM_{4,5}$.

In obtaining an energy analysis of Auger-vacancy- and X-ray satellites, one is often concerned with the states of multi-ionized atoms. Using j–j coupling notation, the energy of an atom $E_{pqr\cdots}$ with vacancies $p, q, r, \ldots,$ one can write

$$E_{pqr\cdots} = E_p + E_q + E_r + \cdots + \Delta E_{pqr\cdots}. \tag{9}$$

If the coupling is not purely j–j, the values $E_{pqr\cdots}$ and $\Delta E_{pqr\cdots}$ are dependent on the coupling between $p, q, r, \ldots.$ In X-ray and Auger-vacancy-satellite spectra, the components due to different coupling are not resolved, and $E_{pqr\cdots}$ and $\Delta E_{pqr\cdots}$ are used to indicate the average over all different coupling substates of the state $pqr\cdots$.

In the first approximation we can write,

$$\Delta E_{pqr} = \Delta E_{pq} + \Delta E_{qr} + \Delta E_{rp} + \cdots. \tag{10}$$

In this notation the energies of ordinary X-rays and ordinary Auger electrons are

$$\text{X-ray} \qquad E_{i\to p} = E_i - E_p = h\nu, \tag{11}$$

$$\text{Auger electron} \qquad E_{i\to pq} = E_i - E_p - E_q - \Delta E_{pq}. \tag{12}$$

The energies of the same transitions with a "spectator" vacancy, v, present are as follows:

$$\text{satellite X-ray} \qquad E_{i,v\to p,v} = E_i - E_p + \Delta E_{i,v} - \Delta E_{p,v} = h\nu' \tag{13}$$

vacancy satellite Auger electron

$$E_{i,v\to pq,v} = E_i - E_p - E_q + \Delta E_{i,v} - \Delta E_{pq} - \Delta E_{p,v} - \Delta E_{q,v}. \tag{14}$$

The satellite displacements in the two cases are

$$\Delta E_x = E_{i,v\to p,v} - E_{i\to p} = \Delta E_{i,v} - \Delta E_{p,v} \tag{15}$$

and

$$\Delta E_a = E_{i,v\to pq,v} - E_{i-pq} = \Delta E_{i,v} - \Delta E_{q,v} = \Delta E_x - \Delta E_{q,v}. \tag{16}$$

A plot was made of $(\Delta E_{pq})_{\text{exper}}/[E(L_i M_5 M_5) - E(L_i M_1 M_1)]$ against $[E(L_i M_p M_q) - E(L_i M_1 M_1)]/[E(L_i M_5 M_5) - E(L_i M_1 M_1)]$, as predicted for $\Delta Z = 1$, for the experimental results of Haynes et al. for ^{210}Bi, for the values $i = 1, 2$. The pattern of points could be approximated by a straight line, determined by a least-squares analysis, showing a decrease in $(\Delta E_{pq})_{\text{exper}}$ with decreasing energy. The graph was used to make new $L_1\,MM$ and $L_2\,MM$ assignments and new position predictions, thus eliminating a number of ambiguities. The same quantities for $i = 1$ obtained from the data of Sujkowski and Slätis[142] for thorium C (^{212}Bi) were plotted in the same way. A similar least-squares analysis gave a line parallel to that calculated for ^{210}Bi for $i = 1$, but shifted slightly toward larger $(\Delta E_{pq})_{\text{exper}}$ values. This displacement was attributed to a difference in calibration between the two investigations. The disagreement in the absolute value of ΔE_{pq} thereby arising makes it more suitable to use the value of ΔE_{pq} relative to $\Delta E_{4,5}$, the spectral line that is the strongest and most accurately measured in all the L_i-spectra. One obtains, then, the quantity

$$\Delta E_{\text{rel}} = \Delta E_{pq} - \Delta E_{4,5} = -E_{i\to pq} - E_p - E_q + E_{i\to 4,5} + E_4 + E_5. \tag{17}$$

Values of ΔE_{rel} from the $L_{1,2,3}\,MM$- Auger energy for ^{212}Bi and from that of the $L_1\,MM$- Auger energy data for ^{210}Bi were calculated and the average was determined for each final $M_p M_q$ state. The values for the $L_2\,MM$- and $L_3\,MM$- Auger energy data for ^{210}Bi were then calculated in the same way and compared with the first average values. (The L_2-data for ^{210}Bi was treated separately, since more than 50% of the L_2-vacancies are produced by $L_1 L_2\,N$- Coster–Kronig transitions. Also, essentially all L_3-vacancies are produced by $L_1 L_3\,M_{4,5}$ Coster–Kronig transitions from initial L_1-vacancies.) It was found that the average energy deviation for the L_2-subshell was negligible; however, for the $L_3\,MM$- Auger lines, which are actually $L_3\,MM(M_{4,5})$ lines, the deviation amounted to -36 ± 5 eV. It could further be shown that this deviation agrees with energy displacement values from X-ray satellite data.

There is also the possibility (about 50%) that the $M_{4,5}$-vacancy becomes filled before the L_3-vacancy. If this is through an Auger transition (which is most likely in this case— ~95%, Ref. 36), there will arise two spectator vacancies. These may in turn result in further Auger or Coster–Kronig transitions, increasing the number of atomic electron vacancies before the L_3-vacancy is finally filled. Auger electron displacements caused by two spectator vacancies for bismuth vary, according to Haynes et al., from 100 eV (N_1N_1), 50 eV ($N_{4,5}N_{4,5}$), 47 eV ($N_{4,5}N_{6,7}$), and 44 eV ($N_{6,7}N_{6,7}$), the last three having the greatest populations, down to nearly zero for pairs of valence electrons. The observed broadening of the $L_3MM(M_{4,5})$ peaks reported by Haynes et al. is thus explained, for the L_3MM- Auger peaks will not suffer a mean displacement much larger than 40 eV as a result of the $M_{4,5}$-vacancy becoming filled before the L_3-vacancy.

Auger-vacancy satellites may also be produced from a double L-shell vacancy state resulting from an initial KLL-Auger transition. These satellites are expected to be difficult to observe in the elements of high Z-value because of low K-Auger yields. It would be more advantageous to investigate the L-Auger spectra of elements of lower Z-value, and larger K-Auger yields, in searching for such satellites.

From equation 16 we have, in the first approximation,

$$\Delta E_a = E_{i,v \to pq,v} - E_{i \to pq} = \Delta E_{i,v} - \Delta E_{p,v} - \Delta E_{q,v} \tag{18}$$

for the satellite energy displacement of the $L_iM_pM_q$-Auger line owing to the presence of a spectator vacancy v. Here both i and v designate L-levels, whereas p and q stand for higher levels. Values for $\Delta E_{i,v}$ can be obtained from KLL-Auger energies (Table II, or VIIIa), from K and L binding energies, X-ray satellite data, and from the use of equation 15. The Other terms may be obtained from

$$\Delta E_{p,v} = E^{Z+1}(L_p) - E^Z(L_p), \tag{19}$$

$$\Delta E_{q,v} = E^{Z+1}(L_q) - E^Z(L_q). \tag{20}$$

The values of ΔE_a for the LL–LMM (i.e., $L_iMM(L_j)$) satellites are always positive, since coupling between two L-levels is much greater than that between an L-level and an M-level, and the vacancy satellite will be displaced toward higher energies from the main line. In an investigation of the L-Auger spectrum of ^{113}In, Krisciokaitis and Haynes[89] looked for evidence of such satellites (see Section 5.3.2).

5.1.2. *L-Auger Transition Probabilities*

As is the case for the KLL-Auger spectral components, the theoretical intensities of L-Auger spectral components are expected to vary markedly with Z-value and to depend on the kind of coupling. Analyses of L-Auger spectra have often been made assuming j–j coupling, which is justified for the heavy elements. Intermediate coupling should be used for lighter elements instead of the extreme j–j coupling. The complexity of the L-spectrum makes the application of intermediate coupling a formidable task, and even now the calculations that have been carried out have assumed only extreme coupling. Among these, extensive theoretical calculations have been performed for only a limited number of elements.

Rubenstein[63] carried out extensive calculations on the LMM-, LMN-, and LNN-transition probabilities for argon ($Z = 18$). krypton ($Z = 36$), and silver ($Z = 47$) assuming extreme L–S coupling.

Some time later Asaad[164] calculated the L-Auger transition probabilities for the same elements using the transition amplitudes computed by Rubenstein.

Supposing we have an initial vacancy in L_1, L_2, L_3, or $L_{2,3}$ (p-vacancy) shell; then the distribution of final vacancies does not depend on the coupling scheme, since the sum to all possible vacancies of the same subshell (or subshells) is always the same. But the coupling scheme for initial vacancies (L-shell and Auger electron) depends of course on the vacancy (L_2-, L_3-vacancy for j–j coupling, $L_{2,3}$-vacancy for L–S coupling). So, just by writing j–j coupling, one assumes j–j coupling for the initial and final states. The coupling scheme is fixed for the initial vacancy (j–j coupling for an L_2- or L_3-vacancy or L–S coupling for an $L_{2,3}$-vacancy; for an L_1-vacancy it is the same for both coupling schemes). Then the coupling scheme can be changed from L–S to intermediate to j–j, provided one stays within the final configuration; the total transition probability is always the same. Even with j–j or L–S coupling for both the initial and final states, the total transition rate to all final vacancies of the same subshell (or subshells) is the same for an initial L_2- or L_3-vacancy (j–j coupling) or for an initial p-vacancy (L–S coupling). Now also we change the coupling scheme for the initial vacancy, and even then the total transition rate is independent of the coupling scheme.

Recent relativistic calculations by Talukdar and Chattarji[165] of the $L_1L_2M_{4,5}$ Coster–Kronig transition probabilities have been presented for the Z-values from 32 to 41.

5.1.3. *M-Auger Transition Energies and Probabilities*

The energies of the M-Auger transitions vary from 1 to 5 keV for uranium ($Z = 92$) to 70 to 420 eV and below for zirconium ($Z = 40$) and lighter elements. Except for a few recent articles, reports of the detection and investigation of such transitions have been made in connection with the study of L-Auger spectra of the elements of interest to the investigator.

For high Z-values, the general M-Auger energy region overlaps with that including the L-Coster–Kronig spectrum, making the analysis of either quite difficult. For very low Z-values, on the other hand, there is a partial overlapping of the M-Auger spectrum with the M-Coster–Kronig lines.

Energies of the M-Auger transition lines can be estimated by using an equation analogous to that used for K-Auger energies, equation 1. For the heavier elements, however, the MNN- MNO-, and MOO-Auger spectra together are composed of a multitude of overlapping transition lines, which results in a composite structure showing only a few intensity maxima (see Figure 13 and also Section 5.2.1). Figure

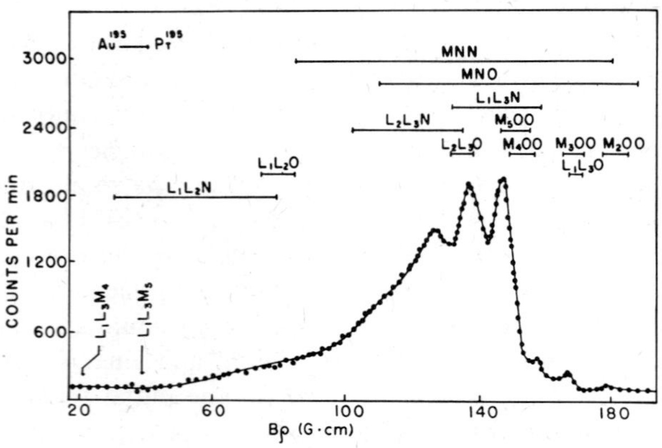

Figure 13. The L-Coster–Kronig and M-Auger electron spectra in platinum ($Z = 78$) measured by Toburen and Albridge.[108] Note the absence of the $L_1L_3M_{4,5}$ components.

13 reveals that there is also an overlapping of these lines with $L_1L_3\,N,O$- and $L_2L_3\,N,O$-Coster–Kronig transition lines.

For M-Auger investigations of the heavier rare gas elements in the gaseous state, where electron straggling effects can be made negligible, it is possible to investigate M-Auger spectra in great detail (see, e.g., Ref. 163). The energies of certain transitions from an initial single M-vacancy to the final doubly ionized state can be calculated in the same manner employed for L-Auger transitions (equation 8) using known M-subshell binding energies together with the energy of the final doubly ionized atomic state as calculated from the tables of Moore;[162] Moore's values pertain only to final vacancies in the valence shell. On the other hand, the energy of the final doubly ionized atomic state, NN, NO, and so on, can be calculated from the measured MNN-, MNO-, and so on, transition energy and from the binding energy, in a neutral atom, of the M-subshell electron in question.

The energies of electron lines resulting from transitions of the type M_iN_j–$N_pN_qN_j$, in atoms that are initially doubly ionized, can be calculated from the differences between the energies of the atom in the initial state and in the final state. The energy of the initial atomic state can be calculated from the measured energy of an Auger or Coster–Kronig transition having this state as the final state (e.g., of an $L_kM_iN_j$-Auger or $M_kM_iN_j$-Coster–Kronig transition) and the initial state electron binding energy in the neutral atom (i.e., L_k or M_k binding energy, respectively). The energies of many of the final triply ionized states can be found in Moore's tables.[162]

To date few calculations have been made on M-Auger transition probabilities. Rubenstein[63] studied krypton and silver, assuming L–S coupling for the initial and final states, and Mehlhorn[163] used Rubenstein's transition amplitudes to investigate initial M_2- and M_3-vacancies for silver, assuming j–j and L–S coupling for the initial and final states, respectively (Ref. 163 provides more information).

The recent calculations by Asaad and Mehlhorn[166,167] of the L_2MM and L_3MM transition rates in argon, using j–j coupling for the initial vacancies and L–S for the final ones, made it possible to derive the M_2NN and M_3NN transition rates in krypton, since they correspond so closely to the former transition in argon.[126] The work was performed using Rubenstein's amplitudes for $Z = 47$. These values are collected in Table XVI, Section 5.2.1. Also included are the $M_1M_{2,3}M_{4,5}$ and $M_1M_{4,5}M_{4,5}$ Coster–Kronig transition probabilities.

5.2. Coster–Kronig Transitions

The Auger transitions first described by Coster and Kronig[7] can occur in certain Z-value regions of the periodic table. An initial vacancy in the $L_{1,2}$ (or $M_{1,2,3}$) subshell may become filled by an electron from an outer L- (or M-) subshell, the energy difference freeing an electron from a shell lying beyond the L- (or M-) shell. The freed energy may be taken by any M, N, and so on, atomic electron whose binding energy is less than the initial and final L- (M)-subshell vacancy binding energy difference.

An atom that is initially ionized in the L_1- or L_2-subshell may become doubly ionized if a Coster–

Kronig transition occurs. The energy of an X-ray emitted in filling the final L-subshell vacancy will be slightly different from the energy emitted if the atom were singly ionized. Coster and Kronig[7] postulated this to be the origin of $L\alpha$ and $L\beta$ X-ray satellites. Richtmeyer and Ramberg[168] calculated the separations and relative intensities of these satellites for gold ($Z = 79$) on the basis of this hypothesis. The agreement of their calculations with experimental results confirmed Coster and Kronig's theory (see also Ref. 169). Pincherle[170] carried out similar but more extensive research of the $L\alpha$ and $L\beta$ X-ray satellite lines on the assumption that they originate in doubly ionized atoms.

Cooper[171,172] presented a study of the possible Coster–Kronig transitions and the elements in which they occur. It was shown that the satellite intensity for the $L\alpha$ X-ray definitely increased as a function of atomic number for $Z > 73$ and $Z > 78$; the phenomenon was attributed to $L_1L_3M_5$- and $L_1L_3M_4$- transitions, which were allowed in the first and second region, respectively. It was also pointed out that increases occur in certain X-ray line widths for the other Z-value regions where Coster–Kronig transitions are allowed.

The analogous Auger-vacancy-satellite phenomenon is treated in Section 5.1.1.

5.2.1. *L- and M-Coster–Kronig Transition Energies*

The energy of an ejected $L_iL_pX_q$- Coster–Kronig electron in an element of atomic number Z can be approximated by

$$E^Z(L_iL_pX_q) = E^Z(L_i) - E^Z(L_p) - E^{Z+1}(X_q), \quad (21)$$

where $E^Z(L_i)$ is the binding energy of the L_i electron in an atom of atomic number Z; $E^Z(L_p)$ is the L_p electron binding energy; $E^{Z+1}(X_q)$ is the binding energy of an M_q, N_q, and so on, electron in an atom of atomic number $Z+1$. The choice of the X_q-subshell level binding energy for the $Z+1$ atom is made to account for the change in screening due to the initial vacancy in the L_i-subshell. The various Coster–Kronig transitions are energetically possible only for certain Z-value regions. These regions can be approximately determined by applying equation 21.

An equation similar to equation 21 can be given for the approximate $M_iM_pM_q$- Coster–Kronig transition energies.

The following relationship for Coster–Kronig electron transition energies may also be useful:

$$E[L_iL_pX_q(^{2S+1}L_J)] = E(L_i) - E[L_pX_q(^{2S+1}L_J)], \quad (22)$$

$$E[M_iM_pX_q(^{2S+1}L_J)] = E(L_i) - E[M_pX_q(^{2S+1}L_J)], \quad (23)$$

when the symbols are the same as for equation 8. Again, many of the values for $E(L_pX_q)$, if $X_q \equiv L$, and $E(M_pX_q)$, if $X_q \equiv M$, and if these L's and M's are in the valence shell, can be obtained from the tables by Moore[162] (see Refs. 163 and 173). Equation 22 or equation 23 can be used for determining any one of the three parameters when the other two are known.

The regions of atomic number Z, where the most important LLX- and MMX-transitions are allowed, as determined by equation 21, are collected in Table XIII.

The relation between the Coster–Kronig phenomenon and the L-Auger effect following nuclear internal conversion transitions was first pointed out by Kinsey.[174] He calculated that an initial L_1-subshell vacancy is filled through an $L_1L_3M_{4,5}$-Coster–Kronig transition with a probability of about 50% for $Z \gtrsim 74$. The calculations were based on an analysis of experimentally determined atomic level widths in terms of a theoretical radiation width, a Coster–Kronig transition width, and a constant width representing the remaining nonradiative transitions.

In a later investigation Kinsey[175] established from measurements on bismuth and thallium that Coster–Kronig transitions of the type L_1L_2N occur for about 12% of the initial L_1-ionization.

The weakness of L_1MM- and L_1MN-Auger transitions, as observed in L-Auger spectrum investigations, is attributed to large L_1L_3M, N-transition probabilities. From this large Coster–Kronig transition probability, one would expect very strong spectral electron lines lying at energies less than 1 keV.

These low energy Coster–Kronig transition lines have been sought in a number of low energy Auger spectrum investigations (see also, Sections 5.1.1 and 5.1.2). Albridge and Hollander,[28] for example, in their study of the decay of ^{233}Pa to ^{233}U, assigned all eight of the spectral lines lying between 0.9 and 3.8 keV to Coster–Kronig transitions of types LLM, N, O. Toburen and Albridge[108] found intense lines in the region of 0.9 to 2.0 keV in the decay of ^{195}Au to ^{195}Pt (see also, Section 5.3.2). These lines were interpreted as superimposed M-Auger and L_2L_3N-, L_2L_3O-, L_1L_3N-, and L_1L_3O- Coster–Kronig transition lines (see Figure 13). No evidence of Coster–Kronig transitions of the types $L_1L_3M_{4,5}$ was found. However, considering the data in Table XIII, one would expect these transition lines to be quite intense.

Table XIII.

The regions of Z value where the most important L and M series Coster-Kronig transitions are allowed.

Transition	Z	Transition	Z
$L_1L_2M_1$	<30	$M_1M_2N_{6,7}$	$\lesssim$93
$L_1L_2M_2$	<33	$M_1M_3N_1$	$\lesssim$53
$L_1L_2M_3$	<34	$M_1M_3N_{2,3}$	<56
$L_1L_2M_4$	<42	$M_1M_3N_{4-7}$	all Z with N_{4-7} electrons
$L_1L_2M_5$	<42	$M_1M_{4,5}N_{1-7}$	all Z with N_{1-7} electrons
$L_1L_2N_1$	<70	$M_2M_3N_1$	$\lesssim$31
$L_1L_2N_2$	<77	$M_2M_3N_{2,3}$	$\lesssim$35
$L_1L_2N_3$	<82	$M_2M_3N_{4,5}$	$\lesssim$53
$L_1L_2N_4$	<93	$M_2M_3N_{6,7}$	all Z with $N_{6,7}$ electrons
$L_1L_2N_5$	$\lesssim$95	$M_2M_4N_1$	<86
$L_1L_3M_1$	<32	$M_2M_4N_{2-7}$	all Z with N_{2-7} electrons
$L_1L_3M_2$	<36	$M_2M_5N_{1-7}$	all Z with N_{1-7} electrons
$L_1L_3M_3$	<37 & >90	$M_3M_4N_1$	<56
$L_1L_3M_4$	<50 & >76	$M_3M_4N_2$	<66
$L_1L_3M_5$	<51 & >73	$M_3M_4N_3$	<72
$L_1L_3N_{1-7}$	all Z with N_{1-7} electrons	$M_3M_4N_4$	<83
$L_2L_3M_3$	>97	$M_3M_4N_5$	$\lesssim$85
$L_2L_3M_4$	$\lesssim$31 & >92	$M_3M_4N_{6,7}$	all Z with $N_{6,7}$ electrons
$L_2L_3M_5$	$\lesssim$31 & >90	$M_3M_5N_1$	<58
$L_2L_3N_{1-7}$	all Z with N_{1-7} electrons	$M_3M_5N_2$	<71
	--------	$M_3M_5N_3$	<77
$M_1M_{4,5}M_{4,5}$	$\lesssim$45	$M_3M_5N_4$	<89
$M_{2,3}M_{4,5}M_{4,5}$	$\lesssim$38	$M_3M_5N_5$	<91
$M_1M_2N_1$	<48	$M_3M_5N_{6,7}$	all Z with $N_{6,7}$ electrons
$M_1M_2N_{2,3}$	<53	$M_4M_5N_{4,5}$	$\lesssim$47
$M_1M_2N_{4,5}$	$\lesssim$73	$M_4M_5N_{6,7}$	<78

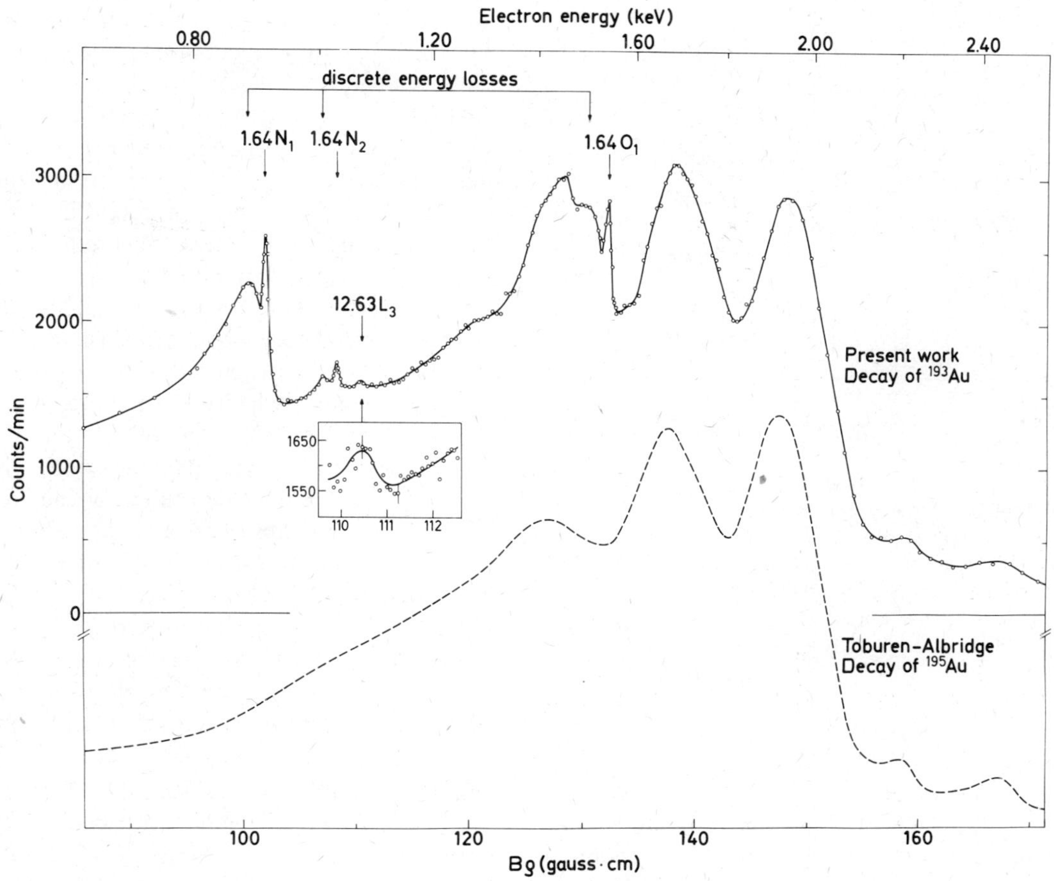

Figure 14. The overlapping 1.64-keV N and O and 12.63-keV L_3 conversion lines in the decay of ^{193}Au to ^{193}Pt with the L-Coster–Kronig and M-Auger electron spectrum in platinum as measured by Johannsson et al[176]. The spectrum for the same energy region obtained by Toburen and Albridge[108] (see Figure 13) is inserted for comparison.

A later study by Johansson et al.[176] of the MXY-Auger energy region for the decay of ^{193}Au to ^{193}Pt established the presence of N and O conversion lines from a 1.64-keV transition. The MXY-Auger spectrum shape as observed by Toburen and Albridge[108] was used to analyze the coinciding MXY and 1.64-O_1 spectral lines (see Figure 14). No attempt was made to investigate the $L_1L_3M_{4,5}$ Coster–Kronig energy region. (See also ref. 176a)

Mehlhorn[177,178] has investigated the L_1-Coster–Kronig and Auger electron spectra for argon ($Z = 18$) using electron bombardment for initial ionization. The spectra for the L_1MM, $L_1L_{2,3}M_1$, and $L_1L_{2,3}M_{2,3}$ transition groups were measured and analyzed with respect to the intensities and energies of the components. Again, analogous to the case of KLL-Auger transitions in elements of low Z-values, the coupling between the final MM-vacancy states following L_1MM-Auger transitions is not far from extreme Russel–Saunders coupling. The coupling for the final $L_{2,3}M$-vacancy states following $L_1L_{2,3}M$-Coster–Kronig transitions is really intermediate. Therefore Mehlhorn examined both the influence of mixed coupling and the effect of configuration mixing on the energies and intensities of the transitions.

Figure 15*a* shows the measured $L_1L_{2,3}M_{2,3}$-Coster–Kronig electron spectrum. The vertical lines indicate the energies and intensities of the component lines as calculated under the assumption of intermediate coupling in the final doubly ionized atomic state ($L_{2,3}M_{2,3}$). Configuration interaction was included only in the L_1MM calculations corresponding to the KLL-calculations. The necessary Coulomb interaction and spin-orbit matrix elements were calculated with Hartree–Fock–Slater type wave functions. The thin solid line is obtained after summing the line components, assuming them to be Lorentzian in shape and having a half-width of 2.93 eV. (This value follows from Rubenstein's calculated transition probability of the L_1-vacancy.) The discrepancy between the two

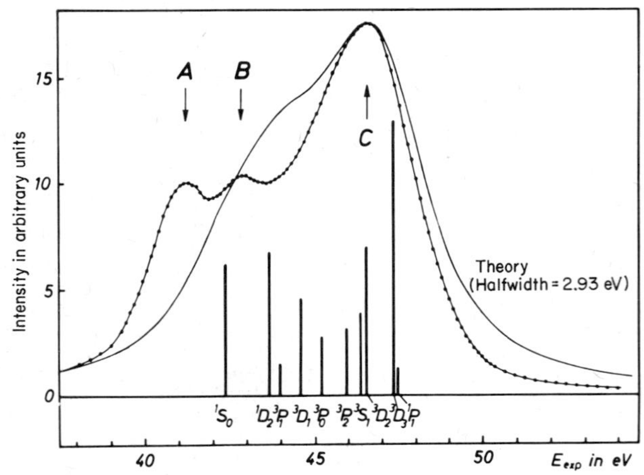

Figure 15a. The $L_1L_{2,3}M_{2,3}$ Coster–Kronig spectrum for argon measured by Mehlhorn[178] (the curve through points) corrected for dispersion and scattering effects. The theoretical spectral energies and intensities of the components are inserted. The solid line is obtained by summing the spectral components (assuming them to have a Lorentzian distribution of 2.93 eV half-width) and the sum spectrum is folded with the spectrometer function.

curves is primarily attributed to failure to use the correct natural line width and the correct Coulomb and spin-orbit matrix elements.

Figure 15*b* illustrates the agreement between the experimental spectrum and the sum of the components whose energies and intensities were obtained after considering the effects of mixed coupling, by varying the Coulomb matrix elements, redetermining the spin-orbit interaction, varying the Coster–Kronig line

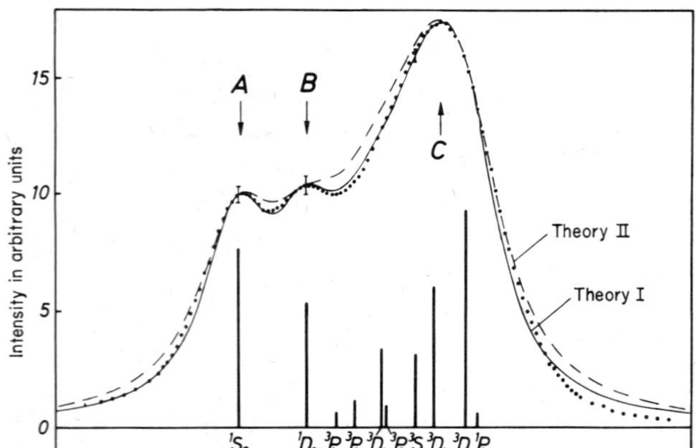

Figure 15b. The poor agreement obtained in Figure 15*a* between the theoretical and experimental profiles was improved by altering certain of the matrix elements and the ^3S, ^1D, and and ^3D component intensity values used in the fit ("Theory II"). The natural width of the Coster–Kronig lines was decreased to 2.0 eV. The "theory II" curve was obtained using a natural line width of 2.4 eV. (See text and Ref 178.) The experimental intensity and energy values thus obtained are entered in Table XIV.

width, and damping slightly the intensities of Rubenstein's L–S coupling calculations. The maxima A and B were shown to be due only to the $^1S_0$ and $^1D_2$ lines, respectively, whereas the maximum C is primarily due to the $^3D_3$ line. The experimental energies and relative intensities of the $L_1L_{2,3}M_{2,3}$ lines, as in Figure 15*b*, and the $L_1L_{2,3}M_1$ lines are given in Table XIV. Also given are the relative theoretical intensities, determined from calculations including intermediate coupling (vertical lines in Figure 15*a*). The transition amplitudes were taken from Rubenstein.[63] The relative intensities are normalized to that of the $^3D_3$ level set equal to unity, which is a practical choice, since this state cannot mix with any other state of the $L_{2,3}$ $M_{2,3}$ configuration and its transition probability must remain constant independent of the coupling.

Table XV compares the relative and absolute probabilities for transitions, both theoretical[63] and experimental, of the L_1-shell in argon.

The $M_{2,3}M_{4,5}N_{1-3}$ Coster–Kronig transitions for krypton have been investigated by Mehlhorn.[163] Initial vacancies were caused by electron impact, and the resulting electron spectrum was analyzed with an electrostatic spectrometer. The $M_{2,3}M_{4,5}N_{2,3}$ components were partially overlapped on the lower energy side by weaker Auger vacancy-satellite transitions, but they were observed to be of appreciable strength.

Spohr et al.[179] used a similar experimental arrangement to study the $M_{2,3}M_{4,5}N_{2,3}$ transition components. Their intensity values agreed well with those calculated by Mehlhorn,[163] assuming *j*–*j* coupling for the initial vacancy and L–S coupling for the final configuration and using Rubenstein's[63] transition amplitudes for $Z = 47$. The experimental results from the two investigations are compared in Table XIX. The $M_{2,3}$ $M_{4,5}$ $N_{2,3}$ Coster–Kronig[163] and $M_{4,5}$ $N_{2,3}$ $N_{2,3}$ Auger[130] lines appear in Figure 16. The difference in widths is attributed partially to the larger M_2 and M_3 natural level widths as compared to the widths of the other levels involved. Also, broadening is expected, since these two lines ($M_2M_{4,5}N_{2,3}$ and $M_3M_{4,5}N_{2,3}$) are actually composed of many lines given by intermediate coupling of the two final vacancies $M_{4,5}N_{2,3}$.

Furthermore, lines due to $M_1M_{4,5}M_{4,5}$ transitions are expected to coincide in position with the $M_{2,3}$ $M_{4,5}N_{2,3}$ Coster–Kronig transition lines and the Auger-vacancy satellites, which lie to slightly lower energies; since their intensities were low, these lines were not observed.[163]

There are few theoretical treatments of the L-Coster–Kronig effect in the literature. As early as

Table XIV.

Relative intensities of the $L_1L_{2,3}M$ Coster-Kronig transitions in argon reported by Mehlhorn[178]. The theoretical values have been calculated[178] using the transition amplitudes by Rubenstein[63].

Transition	Experimental energies (eV)	Relative intensities (Intermediate coupling)	
		Experiment	Theory
$L_1L_{2,3}M_1(^1P_1)$	28.7(2)	0.62(6)	1.84
$(^3P_0)$	--	--	0.007
$(^3P_1)$	30.5(2)	0.47(5)	2.05
$(^3P_2)$	--	--	0.035
$L_1L_{2,3}M_{2,3}(^1S_0)$	41.14(15)	0.82	0.475
$(^3S_1)$	45.91(15)	0.34	0.297
$(^1P_1)$	47.56(15)	0.067	0.103
$(^3P_0)$	44.28(15)	0.13	0.212
$(^3P_1)$	43.77(15)	0.077	0.117
$(^3P_2)$	45.10(15)	0.12	0.242
$(^1D_2)$	43.00(15)	0.57	0.521
$(^3D_1)$	45.00(15)	0.37	0.353
$(^3D_2)$	46.41(15)	0.65	0.538
$(^3D_3)$	47.26(15)	1.0	1.0

Table XV.

Relative and absolute probabilities for transitions of the L_1 shell in argon as given in ref.[178].

Transition	Experimental		Theory[63]	
	relative	absolute$^+$	relative	absolute$^+$
L_1MM	1.0	3.5(4)	1.0	3.08
$L_1L_{2,3}M_1$	3.8(6)	13.5(2.4)	16.6	51.12
$L_1L_{2,3}M_{2,3}$	14.4(1.7)	50.8(6.8)	16.2	49.85
L_1M			<0.003	<0.01
Total		67.8(7.3)		104.05
$\Gamma(L_1)$ in eV^{++}		1.84(0.20)		2.93

$^+$In atomic units $\times 10^{-3}$.

$^{++}$1 eV corresponds to 15.1×10^{14} sec^{-1}.

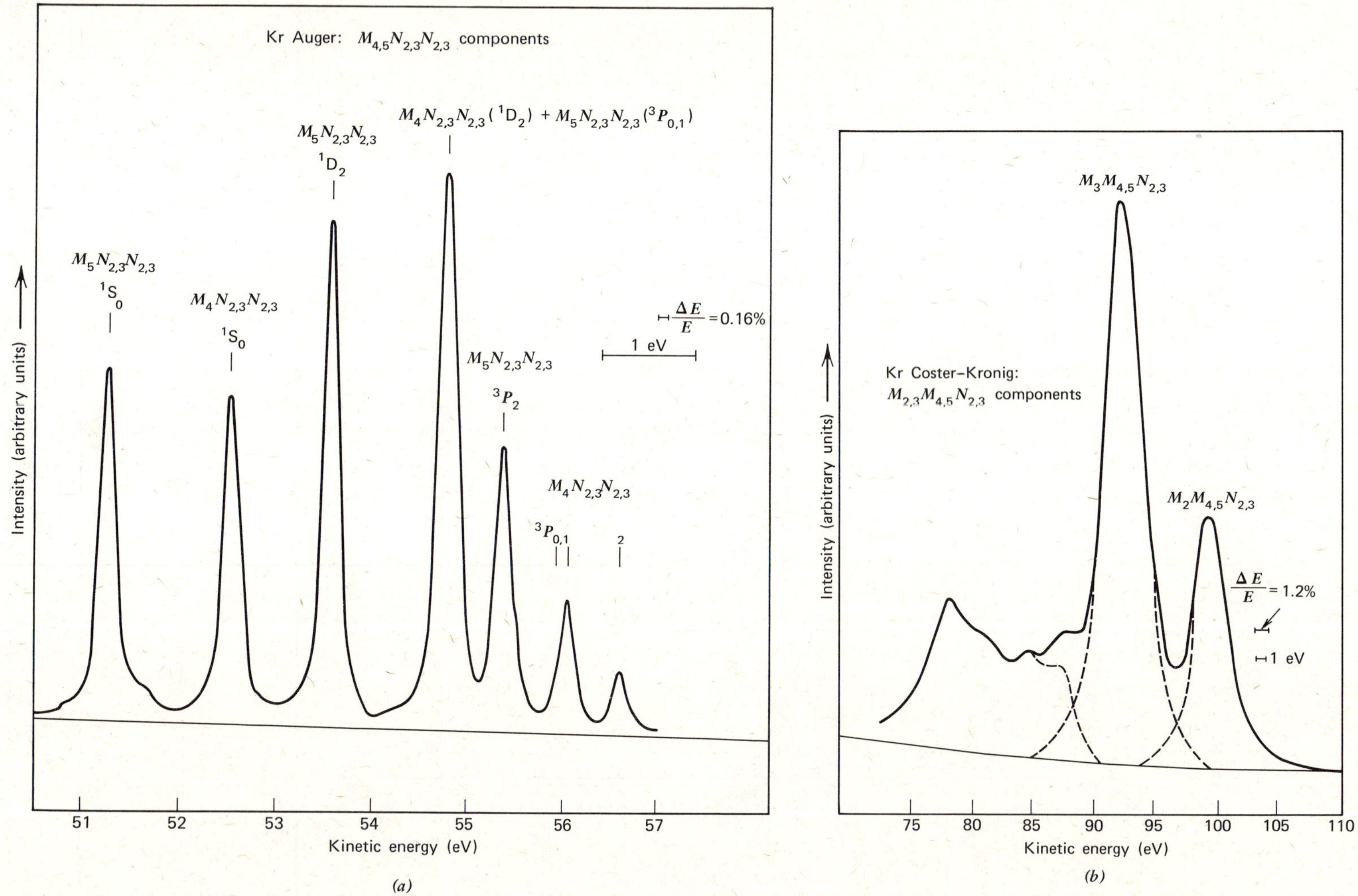

Figure 16. A comparison of (*a*) the $M_{4,5}N_{2,3}N_{2,3}$ Auger Components for krypton [130] with (*b*) the krypton $M_{2,3}M_{4,5}N_{2,3}$ Coster-Kronig components.[163] The obvious difference in energy widths is due mainly to the factors: (1) different natural widths of the $M_{2,3}$ and $N_{2,3}$ atomic levels and (2) the multiplicity of final vacancy states. See also Chapter VI. (These spectra were kindly provided by W. Mehlhorn.)

1935 Pincherle[57] calculated *LLM* transition rates using unscreened hydrogenic wave functions. This was followed by the work of Ramberg and Richtmeyer,[62] who based their calculations of *LLM* transition rates for gold on a Thomas–Fermi field.

Rubenstein[63] presented the L_1-Coster–Kronig transition probabilities for both argon and krypton, those for the M_1-subshell for krypton and silver, and those for the $M_{2,3}$-subshells for silver. These were calculated by using nonrelativistic self-consistent-field functions and assuming L–S coupling. The obtained values for the M_1- and $M_{2,3}$-Auger and Coster–Kronig transition rates obtained from his calculations are given in Table XVI. The same table lists the M_2NN- and M_3NN-Auger transition probabilities for silver calculated by Mehlhorn,[126] assuming L–S coupling in the initial state, j–j coupling in the final atomic configuration and employing Asaad's formulas[67] and Rubenstein's[63] transition amplitudes for $Z = 47$.

In 1963 Asaad[164] presented his calculation of L_1-Coster–Kronig and *LMM*-, *LMN*-, and *LNN*-Auger transition probabilities for argon, krypton, and silver. Extreme j–j coupling was assumed for the initial and final vacancies.

In 1963 Callan[180,181] published calculations on the $L_1L_{2,3}M_{4,5}$ Coster–Kronig transitions based on a method similar to that used in his earlier calculations on *KLL*-Auger transitions.[59] Screened hydrogenic wave functions were employed, L–S coupling was assumed, and no relativistic effects were included. The calculations covered the elements from scandium ($Z = 21$) to tin ($Z = 50$) and a number of other elements from tungsten ($Z = 74$) to thorium ($Z = 90$). The main objective of the work was to investigate the sensitivity of the Coster–Kronig transition rates to variation in energy levels and screening parameters.

In 1965 Asaad[182] published calculations of the $L_1L_{2,3}M_{4,5}$ transition rates using Callan's $L_1L_{2,3}M_{4,5}$ transition amplitudes[180,181] introducing intermediate coupling for the $L_{2,3}M_{4,5}$ final vacancies. The calculations were made for all elements from $Z = 29$ to $Z = 50$ and also for seven elements in the range from $Z = 76$ to $Z = 90$. Unsurprisingly for the elements with which all transitions to all the final states of the doubly ionized atom are possible, the total transition rates were close to those calculated by Callan's method;[180] the agreement between the total transition rates based on j–j and intermediate coupling was also good, again as could be expected. Since all transitions are not energetically possible, however, the total transition sum for these elements based on the intermediate coupling theory can be up to about 15% less than that based on the extreme j–j coupling theory.

5.3. L- and M-Auger Investigations

Many of the early L-Auger series investigations were carried out using radioactive sources of thorium B (^{212}Pb), radium B (^{214}Pb), and radium D (^{210}Pb) and their decay products. These sources were popular because of the ease with which uniform, thin radioactive deposits could be obtained by collecting upon a source backing the recoil atoms after alpha-decay of the parent nuclei.

Since so much work has been done in investigating the L- and M-Auger spectra resulting from the decay of these isotopes, Section 5.3.1 consists entirely of a review of this work. Section 5.3.2. is a review of the research on the L-, M-, and N-Auger spectra for other elements. For a further treatment and review of early work on the L-Auger spectra for elements having $Z = 77$ to 84, see the work of Dionisio.[183,11,12]

For simplicity, the following abbreviations are used in referring to the experimental procedures and instrumentation used in the investigations.

Relating to the electron source
- RS = Atomic vacancy produced by electron capture and/or internal conversion in a radioactive source,
- X = atomic vacancy produced by X-ray bombardment,
- El = atomic vacancy produced by electron bombardment.

Relating to the employed spectrometer
- SCS = semicircular spectrometer,
- ICDF = iron-core double focusing spectrometer,
- IFDF = iron-free double focusing spectrometer,
- MLS = magnetic lens spectrometer,
- ES = electrostatic spectrometer,
- Poa = postacceleration techniques applied,
- Pra = preacceleration techniques applied.

Electron detection method used:
- Ph = photographic emulsion plates,
- C = electrical counting, G–M counter employed,
- Co = electrical counting, other counter employed.

5.3.1. *The L- and M-Auger Groups Arising from the Decay of Thorium B (^{212}Pb), Radium B (^{214}Pb), and Radium D (^{210}Pb)*

The first study of L-Auger electron spectra was reported in 1925 by Black,[184] who compared the L-Auger spectrum from the radium B (^{214}Pb) decay products with that from the thorium B (^{212}Pb) decay products (SCS; Ph). Ellis[144] also worked during the

Table XVI.

M_1- and $M_{2,3}$-Auger and M_1-Coster-Kronig transition rates for krypton and silver as calculated by Rubenstein[63] assuming L-S coupling. (atomic unit $\times 10^{-3}$ = 4.1×10^{13} sec^{-1})

Initial M_1 vacancy: Final configuration	Term	Transition rate: Krypton	Transition rate: Silver	Initial $M_{2,3}$ vacancy: Final configuration	Term	Transition rate: Silver
$M_{2,3}M_{4,5}$	P	1.99	--	$M_{4,5}N_1$	D	16.3
"	F	--	--	$M_{4,5}N_{2,3}$	P	6.61
$M_{2,3}N_{2,3}$	S	26.8	--	"	D	0.00
"	D	0.4	--	"	F	0.00
$M_{4,5}M_{4,5}$	S	0.32	3.2	N_1N_1	S	0.06
"	D	0.04	11.4	$N_1N_{2,3}$	P	0.87
"	G	17.7	0.0	$N_{2,3}N_{2,3}$	S	0.25
$M_{4,5}N_1$	D	43.7	31.0	"	P	0.87
$M_{4,5}N_{2,3}$	P	2.64	--	"	D	0.75
"	F	5.26	--			
N_1N_1	S	2.09	0.54			
$N_1N_{2,3}$	P	11.6	--			
$N_{2,3}N_{2,3}$	S	1.00	--			
"	D	2.62	--			

M_2NN and M_3NN Auger rates for krypton assuming j-j coupling for the initial vacancy (M_2 or M_3) and L-S coupling for the final vacancies[126]. (atomic units $\times 10^{-3}$)

Final configuration	Term	Initial M_2 vacancy: Transition rate krypton	Initial M_3 vacancy: Transition rate krypton
N_1N_1	$(^1S_0)$	0.061	0.061
$N_1N_{2,3}$	$(^1P_1)$	0.500	0.500
"	$(^3P_0)$	0.111	0.007
"	$(^3P_1)$	0.229	0.073
"	$(^3P_2)$	0.034	0.295
$N_{2,3}N_{2,3}$	$(^1S_0)$	0.247	0.247
"	$(^1D_2)$	0.744	0.744
"	$(^3P_0)$	0.193	0.048
"	$(^3P_1)$	0.433	0.219
"	$(^3P_2)$	0.241	0.602

early 1930s on the same isotopes (SCS; Ph), drawing results and conclusions similar to those of Black.

In 1938 Flammersfeld[185] studied the *L*-Auger and internal conversion electron spectra from the thorium B decay products (SCS; C). The resolving power was not high enough to render more than four lines in the *L*-Auger energy region distinguishable. Some time later, in 1950, Butt[186] reinvestigated this spectrum (MLS; Poa; C). He ascribed five of the eight detected lines to the *L*-Auger effect in bismuth ($Z = 83$) and the other three to that in thallium ($Z = 81$).

When Kobayashi[187] investigated the radioactive decay of radium D, using a spherical condenser type electrostatic spectrometer set for a resolution of 1.3%, he found 12 lines which he interpreted as belonging to the *L*-Auger spectrum. He also observed two uninterpreted bands at 11.7 to 12.6 and 12.6 to 13.5 keV. It was suggested that a band at 15 to 16 keV and a weak band at 4 to 5 keV had resulted from internal conversion of gamma-transitions.

The conversion electron spectrum in radium D was also investigated by Bashilov et al.[188] using a nonuniform field magnetic spectrometer and a "ketone" (C); the *L*-Auger spectrum was also measured. There were 16 lines found, but not interpreted, in the *L*-Auger electron energy region. The energy determinations of the lines, which are not very accurate at these low energies, do not agree well with the values obtained in other investigations.

In 1956 a number of investigations were reported of the *L*-Auger spectrum from thorium B and its decay products.

Moussa and Bellicard[189] (IFDF; C), wrote of 20 L_2- and L_3-Auger lines originating in either thallium or bismuth. They attributed the near absence of L_1MM-Auger lines to a large L_1-Coster–Kronig transition probability.

Rosenblum et al.[27] investigated the *L*-Auger spectrum from the decay products of thorium B as well as that from ^{237}Np ($Z = 93$) (SCS; Pra; Ph). The thorium B spectrum yielded 29 lines and two unresolved bands lying between the energies of 6.20 and 14.76 keV. A band was also observed to lie between 1 and 3 keV, conceivably arising from *M*-Auger transitions. A similar low energy band, as well as a number of *L*-Auger lines, were seen in the electron spectrum of ^{237}Np and were compared with results of three earlier investigations. There was no interpretation of the lines observed by the authors.

Also in 1956, Burde and Cohen[190] investigated the *L*-Auger spectrum of the products of thorium B decay, employing a novel technique: A thin-lens iron-free spectrometer with a resolving power of about 1% (C) was employed with a counter window sufficiently thin to transmit a good percentage of the *L*-Auger electrons without the use of an acceleration arrangement. Burde and Cohen were able to separate the ^{208}Tl and ^{212}Bi *L*-Auger electrons by making a recording of the spectrum of those in coincidence with alpha-particles originating in the decay of ^{212}Bi. Hence they could confine their measure to the ^{208}Tl *L*-Auger electron spectrum. It was then possible to obtain the ^{212}Bi *L*-Auger spectrum by subtracting that for ^{208}Tl from the composite spectrum. The spectra were analysed and interpreted.

Nuclear recoil effects on the *L*-Auger electrons were demonstrated by measuring the ^{208}Tl coincidence *L*-Auger spectrum for various relative alpha-particle and Auger electron acceptance directions.

It was pointed out by the authors that the *L*-Auger spectrum for radium D and its decay products as measured by both Bashilov et al.[188] and Kobayashi[187] resembled in shape their own spectrum for ^{208}Tl, but not for ^{212}Bi. This is reasonable, since the original *L*-subshell vacancy distribution in ^{210}Bi (the decay product of radium D) results from the internal conversion of a 46.52-keV *M*1 gamma-transition in that isotope. The initial vacancies in the ^{208}Tl *L*-subshells are very similar in distribution to those in ^{210}Bi, for they are produced almost exclusively by internal conversion of a 39.85-keV *M*1 transition.

In 1957 Geoffrion and Nadeau[30] measured the low energy electron spectrum of the thorium B decay products (SCS; C). Using a resolution of 0.3%, they reported positive identification of 37 *LMM*- and *LMN*-Auger lines in the region from 5 to 13 keV. Here the two strongest lines were assigned to the $L_3M_4M_5$ and $L_2M_4M_5$ transitions in bismuth. These assignments were also made by Moussa and Bellicard,[189] and the same peaks were assigned to the $L_3M_4M_4$ and $L_3M_4N_{4,5}$ transitions by Burde and Cohen.[190]

Bellicard[77] (IFDF; C) investigated the *L*-Auger spectrum of thorium B and its decay products. Uniform sources were made by evaporation of the activity onto aluminized formvar films (thickness 10 μg/cm^2). Of the 26 recorded lines of energies between 5.9 and 12.5 keV, 21 were interpreted. Bellicard also estimated the number of *L*-subshell holes in ^{212}Bi ($Z = 83$) and ^{208}Tl ($Z = 81$), after internal conversion and with $K\alpha$ X-ray and Coster–Kronig transitions considered.

Using a resolution of 0.7%, Risch[161] (MLS; C) studied the *L*- and *M*-Auger electron spectra of thorium C (^{212}Bi) and thorium C″ (^{208}Tl). A thin zapon plus formvar window gave an electron detection cut-off at ~1 keV. The estimated Auger line energies for $Z = 83$ using the Bergström–Hill method and setting $\Delta Z = 1$ were found to be too high as compared with experimental values. A value of $\Delta Z = 0.8(0.2)$ gave better agreement. For $Z = 81$, however, a mean value of $\Delta Z = 1.1$ was obtained. The difference in these ΔZ-values indicates that, compared to the Bergström–Hill estimates for $\Delta Z = 1$, the Auger lines in thallium ($Z = 81$) were 30 to 40 eV lower in energy that the $Z = 83$ lines. Two possible origins were assumed to be at last partially responsible for these energy shifts.

1. Simultaneous final L_3 and $M_{4,5}$ vacancies remain following $L_1L_3M_{4,5}$ Coster–Kronig transitions. This should modify the L_3MM-transition energies. This effect is expected to be stronger in thallium than in bismuth because of the high conversion coefficient of the 39.85 keV transition for the L_1-subshell, as compared to those for the L_2- and L_3-subshells. Coster–Kronig transitions would then be of primary importance in the formation of vacancies in the L_3-subshell. This plausible explanation is also given by Haynes et al. (Section 5.1.1).
2. Nuclear recoil of ^{208}Tl after alpha-decay of ^{212}Bi gives a Doppler broadening of the Auger electron lines from the former isotope. In addition, electrons originating in an atom that has recoiled into the backing material will suffer energy losses, introducing additional broadening of spectral lines.

These two phenomena and their influence on the L-Auger spectrum were also discussed in 1958 by Sujkowski and Slätis.[142] Using a resolution better than 0.1%, they noticed a similar energy discrepancy, as well as an excessive broadening of the ^{208}Tl Auger lines. They gave similar explanations of the cause, discussing the second point more fully. According to Sujkowski and Slätis, measuring the energy of an electron emitted from an atom that is recoiling away from the source surface will produce a momentum corresponding to the vector sum of the two velocities. This will yield a Doppler energy shift of about 112 eV for a 10-keV electron. An atom recoiling into an aluminum backing will come to rest in $\sim 2 \times 10^{-13}$ sec[191] and will be embedded in the backing at the moment of decay of the 39.85-keV excited nuclear state, whose half-life has most recently been measured to be $6.46(79) \times 10^{-12}$ sec.[192] The L-Auger electrons emitted after this transition will then have to travel a small distance through the backing, thereby suffering energy losses. It is estimated[142] that the maximum loss is of the order of 50 eV (see Chapter 5).

In the investigation by Sujkowski and Slätis[142] (SCS; Ph) a total of 54 lines were observed and interpreted, the ones due to electrons originating in ^{208}Tl being broadened as stated previously. The energies were very accurately determined using the well-established energy of the thorium $(B + C + C'')$ A-line (the 39.85 L_1 line in ^{208}Tl) for calibration. The line intensities were estimated in terms of very strong, strong, and so on, with no correction being made for the variation with electron energy of the detection efficiency of the photographic emulsion.

The L-Auger spectrum is very complex, since it is a composite of three separate spectra (i.e., L_1-, L_2-, and L_3-spectral series components) plus possible Auger-vacancy-satellite lines.

It would be interesting to cause an initial vacancy in the L_3-subshell but not in the L_1- and L_2-subshells in order to investigate the process of de-excitation of an atom singly ionized in the L_3-subshell. It would then be possible to investigate the L_3-Auger spectrum group free from disturbing electrons from the L_1- and L_2-Auger spectra and from possible multiple ionization effects due to Coster–Kronig transitions. This refinement has in effect been achieved by Risch,[193] who studied the L_3-Auger spectrum of bismuth ($Z = 83$) by irradiating the metal with the prefiltered continuous X-ray spectrum from an Y_2O_3 anode. A rubidium nitrate X-ray filter was used to strongly decrease the intensity of the X-radiation of energies greater than a value lying between the L_2- and L_3-subshell binding energies for bismuth. The number of L_2-ionizations was about ten times less than that of L_3-ionizations. The L_3-Auger spectrum would be due almost exclusively to transitions occurring in singly ionized atoms. Risch[193] used an instrument that had been used earlier (Ref. 161) with the detector window transmission cutoff at an electron energy of 1.4 keV. No intensity corrections were made, but for calibration purposes Risch took the energy values for the $L_3M_2M_3$ and $L_3M_4M_5$ lines (from Ref. 142), as well as the energy of the bismuth $M_3(YK\alpha_1)$ photoelectron line.

It was reported that of the LMN-lines the unresolved $L_3M_5N_{4,5}$ component was most intense. It was also determined that the intensity ratio between the L_3MN and L_3MM Auger electron groups was 0.46(2), and that the intensity ratio $I(L_3M_5M_{4,5})/I(L_3MM + L_3MN) = 0.30(1)$.

The preceding experimental procedure may be taken one step further. One may irradiate the target with monoenergetic $K\alpha_1$ X-rays, obtained by using a crystal monochromator, from an element for which this X-ray energy lies between the L_2- and L_3-electron binding energies of the element to be investigated. Hence one may investigate the L_3-Auger series of dysprosium ($Z = 66$) using copper $K\alpha_1$ X-rays. The construction of a semicircular spectrograph with a monoenergetic X-ray irradiation arrangement has been reported.[194]

In 1965 Slätis[195] reported the construction of a pre- and postacceleration arrangement to be used in conjunction with his semicircular spectrograph. The apparatus allowed all the spectral electrons to strike the photographic emulsion plate with the same final energy and about the same incidence angle; the blackening of the plate, then, was a function of the electron density alone.

In 1967 Haynes et al.[160] reported their investigation of the *L*-Auger spectrum of ^{210}Bi following the decay of radium D (^{210}Pb). The activity's half-life (22 years), the high spectrometer resolution (0.18%), and the very thin source used in the study made it necessary to sacrifice statistical accuracy in order to obtain a detailed measurement of the spectrum.

The isotope ^{210}Pb (radium D) decays by beta-decay with 80% probability to the first excited, 46.50-keV state in ^{210}Bi (radium E), which decays in turn by an *M*1-transition to the ground state. This 46.50-keV *M*1 transition is highly converted in the L_1-subshell, the internal conversion in the L_2- and L_3-subshells being smaller by factors of approximately 10 and 100, respectively. Although the initial number of vacancies in the L_3-subshell is negligible compared to that in the L_1-subshell, the L_3-Auger spectral lines are more intense than those of the L_1- and L_2-Auger spectra. This strong intensity of the L_3-Auger lines is primarily attributed to the shift of initial L_1-holes to the L_3-subshells caused by the occurrence of L_1L_3M-Coster–Kronig transitions, which leave the atom in a doubly ionized state. Effectively, all L_3-Auger transitions filling the L_3-subshell hole will take place in the presence of a second atomic vacancy, which will cause a decrease in the emitted L_3-Auger electron energy. By considering this displacement to be analogous to the displacement of X-ray satellites from the main diagram lines, the energy decrease for the L_3MM-spectrum is estimated at approximately 35 eV. It was shown in the *L*-Auger spectrum measurements that virtually the whole L_3MM-spectrum is displaced by an average of about this amount toward lower energies. Indication of spectral line shifts were also found in the L_3NN spectrum. In Table XVII are given the *LMM*, as well as *LMN*, *O*- and *LNN*, *O*-Auger and the Auger-vacancy satellite line energies originating in ^{210}Bi. Energy calibration was obtained by measuring the internal conversion lines of the 46.50-keV transition in ^{210}Bi (see also, Section 5.1.1).

It is interesting to compare the *L*-Auger spectra originating in ^{210}Bi following the beta-decay of ^{210}Pb (radium D) with that originating in ^{212}Bi following the beta-decay of ^{212}Pb (thorium B).

In the first case, as mentioned previously, primary vacancies in the *L*-subshell arise through internal conversion of the *M*1 46.50-keV transition in ^{210}Bi. In the second case, initial vacancies in the *K*- and *L*-shells are created through the internal conversion of the higher energy 115-, 239-, and 300-keV M_1 transitions. Here, then, primary vacancies are created mainly in the *K*-shell, whereupon they shift almost entirely to the L_2- and L_3-subshells through $K\alpha_{1,2}$ X-ray transitions. The initial vacancy distribution among the K, L_1, L_2, L_3 shells and subshells can be shown to be 32.6, 5.7, 0.67, and 0.04%, as expressed in percentage of ^{212}Pb decays.[196] In the case of ^{210}Pb, the corresponding quantities are 0, 44.5, 4.7, and 0.41%.[196] Auger-satellite-vacancy lines should then be strongly visible in the ^{210}Bi *L*-Auger spectrum, whereas they will be weak in the *L*-Auger spectrum of ^{212}Bi.

The *M*-Auger spectra for these elements lie in the energy region between some hundreds of electron volts and some 3 to 4 keV. The investigation of these spectra has seldom been reported, but the reader is referred to the account of the decay of thorium B by Juillard.[197] Here some 40 low energy spectral components were found and their probable identities considered.

5.3.2. *L*- and *M*-Auger Groups for Other Elements

The investigation of the *L*-Auger spectrum of ^{203}Tl, following the electron capture decay of ^{203}Pb, has been reported by Sujkowsky and Melin.[198] The activity was collected after mass separation onto an aluminum foil ($\sim$1.5 mg/cm^2 thickness) (ICDF; C). The counter window transition was about 85%, and the cut-off energy was $\sim$1.6 keV.

Considerable analytical efforts were made in terms of L_1-, L_2-, and L_3-Auger group components, but discrepancies remained rather large. Any influence of double ionization of an atom after a Coster–Kronig transition on the following L_2- or L_3-Auger transition energies was assumed to be negligible. It was found that the L_1-Auger lines were relatively weak and that their assignments were least certain. Measurement was made of the energies of a total of 29 *L*-Auger lines or line groups as well as of the relative intensities of many of the strongest of these.

Aside from the work cited in Section 5.3.1, little was done on the investigation of *L*-Auger spectra of elements other than bismuth and thallium until 1954. In that year Brodie[199] used the same apparatus (MLS; Poa; C) that had been used earlier[186, 200]

Table XVII.

The analysis of the L-Auger and Auger-satellite line spectrum of ^{210}Bi following the beta decay of RaD(^{210}Pb) as reported by Haynes et al.[160]. The transitions in parenthesis are considered to be of negligible intensity.

Energy	Identification	Intensity from analysis	Inten. est.	Energy	Identification	Intensity from analysis	Inten. est.
6132(30)	$L_3M_1M_3(M_{4,5})$	15.0(4.0)		9861(15)	$L_2M_3M_5$	2.0(0.8)	1
6431(20)	$L_3M_2M_3(M_{4,5})$	24.4(4.0)			$L_3M_5N_1(M_{4,5})$		1
6740(30)	$L_3M_1M_5(M_{4,5})$	3(2)		9958(12)	$L_1M_3M_3$	4.8(1.2)	
6948(20)	$L_3M_3M_3(M_{4,5})$	30.4(5.0)		10018(12)	$L_1M_2M_5$		weak
7007(20)	$L_3M_2M_5(M_{4,5})$	18(5)			$L_3M_5N_2(M_{4,5})$	8.8(1.6)	8.8
7433(10)	$L_3M_3M_4(M_{4,5})$	38(4)			$L_3M_4N_3(M_{4,5})$		
7543(10)	$L_3M_3M_5(M_{4,5})$	40.8(4.0)		10078(25)	$L_3M_3N_{6,7}(M_{4,5})$	4.2(2.0)	
7896(40)	$L_3M_4M_4(M_{4,5})$	4.4(3.0)		10121(25)	$L_3M_5N_3(M_{4,5})$	7.8(2.0)	
8011(5)	$L_3M_4M_5(M_{4,5})$	100(8)		10196(20)	$L_2M_4M_4$	8.4(2.0)	
8135(8)	$L_3M_5M_5(M_{4,5})$	56(6)		10239(20)		14.6(3.0)	
8620(20)	$L_1M_1M_2$	7.0(1.2)			$(L_3M_4N_4(M_{4,5}))$		
8669(25)	$L_3M_1N_3(M_{4,5})$	5.8(1.2)		10298(25)	$L_3M_4N_5(M_{4,5})$	17(3)	
8743(20)	$L_2M_2M_3$	3.8(1.0)		10345(5)	$L_2M_4M_5$		25.2
9000(40)	$L_2M_1M_5$	1(1)			$L_3M_5N_4(M_{4,5})$	30.2(3.0)	5
	$(L_3M_1N_3(M_{4,5}))$				$L_3M_5N_5(M_{4,5})$		
9125(15)	$L_1M_1M_3$	12.0(1.2)		10580(15)	$L_3M_4N_{6,7}(M_{4,5})$	5.0(1.4)	
9182(35)	$L_1M_2M_4$	11.6(2.4)		10636(10)	$L_3M_5N_{6,7}(M_{4,5})$	4.6(1.0)	
	$(L_3M_2N_4(M_{4,5}))$			10779(35)	$L_3M_4O_{4,5}(M_{4,5})$	1.4(1.4)	
9262(20)	$L_1M_3M_3$	7.8(1.2)			$L_3M_5O_{4,5}(M_{4,5})$		
	$(L_3M_2N_5(M_{4,5}))$			10917(25)	$L_1M_4M_4$	0.9(1.0)	
9359(15)	$L_2M_2M_5$	9.4(1.8)	6.3	11020(5)	$L_1M_4M_5$	29(3)	
	$L_3M_3N_2(M_{4,5})$		3.1	11148(20)	$L_1M_5M_5$	3(1)	~all
9472(25)	$L_1M_2M_3$	7.6(1.6)			$L_2M_2N_2$		small fract.
9558(30)	$L_3M_3N_3(M_{4,5})$	8(2)		11277(20)	$L_2M_2N_3$	1.2(0.8)	
	$(L_3M_2N_{6,7}(M_{4,5}))$			11459(25)	$L_2M_2N_4$	1.0(0.6)	
9631(20)	$L_1M_1M_4$	19.4(3.0)		11547(10)	$L_2M_2N_5$	3.6(1.0)	
9719(30)	$L_1M_1M_5$	14(3)			$(L_2M_3N_1)$		
	$(L_3M_2O_{4,5}(M_{4,5}))$			11640(20)	$L_1M_1N_3$	2.0(0.8)	
	$L_2M_3M_4$		10.3	11701(15)	$L_2M_3N_2$	3.6(1.0)	1.2
9750(10)	$L_3M_4N_1(M_{4,5})$	20.6(2.4)			$L_1M_2N_1$		2.4
	$L_3M_3N_4(M_{4,5})$		10.3	11882(8)	$L_1M_1N_4$	5.2(0.8)	
	$L_3M_3N_5(M_{4,5})$						

Table XVII (cont.)

12191(10)	$L_3N_2N_5(M_{4,5})$ $L_2M_4N_2$	3.2(0.8)	1.6 1.6
12239(12)	$L_1M_2N_5$ $(L_1M_3N_1)$ $L_2M_5N_2$	5(1)	1.7 3.3
12309(10)	$L_2M_4N_3$ $(L_3N_3N_4(M_{4,5}))$ $(L_3N_3N_5(M_{4,5}))$	3.4(0.8)	
12387(20)	$L_2M_5N_3$	0.5(0.5)	
12507(10)	$L_2M_4N_4$ $L_3N_4N_4(M_{4,5})$ $L_3N_4N_5(M_{4,5})$	6(1)	4 2
12576(20)	$L_3N_5N_5(M_{4,5})$ $(L_2M_4N_5)$ $(L_3N_3N_{6,7}(M_{4,5}))$	1.8(0.6)	
12620(15)	$L_2M_5N_4$ $(L_3N_2O_{4,5}(M_{4,5}))$	2.0(0.6)	
12720(10)	$L_1M_4N_1$ $(L_1M_3N_4)$ $(L_1M_3N_5)$	2.4(0.6)	
12809(10)	$L_2M_4N_{6,7}$ $L_1M_5N_1$ $(L_3N_4N_{6,7}(M_{4,5}))$ $(L_3N_5N_{6,7}(M_{4,5}))$	3.4(0.8)	1.7 1.7

12867(15)	$L_1M_4N_2$	2.0(0.6)	
	$(L_3N_4O_{4,5}(M_{4,5}))$ $L_3N_5O_{4,5}(M_{4,5})$ $L_2M_5N_{6,7}$ $L_1M_5N_2$ $L_1M_4N_3$	possible small distr'd intensity	
13039(20)	$L_1M_3N_{6,7}$	1.2(0.6)	
13197(20)	$L_1M_4N_5$ $(L_1M_4N_4)$	3.4(0.8)	
13261(20)	$L_1M_5N_4$ $L_3N_{6,7}O_{4,5}(M_{4,5})$	3.2(0.6)	1.6 1.6
13400(20)	$L_1M_5N_5$ $L_3O_{4,5}O_{4,5}(M_{4,5})$	0.4(0.2)	0.2 0.2
13560(20)	$L_1M_4N_{6,7}$	0.6(0.3)	
13632(20)	$L_1M_5N_{6,7}$	1.0(0.4)	
13705(20)	$L_1M_4O_{4,5}$	0.2(0.2)	
13810(15)	$L_1M_5O_{4,5}$	1.2(0.4)	
14459(20)	$L_2N_2N_5$	0.6(0.3)	
14588(20)	$L_2N_3N_5$ $L_1N_1N_2$	1.6(0.8)	0.8 0.8
14718(20)	$L_1N_1N_3$ $(L_3N_2N_{6,7})$	0.6(0.4)	

to study the conversion lines of the 28.8- and 40.7-keV transitions in ^{233}U, which were superimposed upon the *L*-Auger spectra. The presence of the conversion lines of a 17.2-keV transition was also reported.

The spectrum was measured again later by two research groups. Albouy and Valadares[201] reported their measurements of the *L*-, *M*-, and *N*-Auger spectra in ^{233}U (SCS; Pra; Ph). The presence of the 28.43-, 40.23, and 57.9-keV (but not ~17-keV) gamma-transition conversion lines were reported. The authors' results were compared with those of earlier measurements. Regarding the measured low energy electron spectrum, they reported a very intense, narrow band at 0.2 to 0.3-keV. This was deemed attributable to three causes: Coster–Kronig transitions, the Auger effect in the *O*-atomic shell, and secondary electrons from electon bombardment of the source surface. It was considered likely that a band at 0.5 to 3 keV and weak lines at 3.6 to 4.0 keV belonged to the *M*-and *N*-Auger spectra. The identification of the six strongest *L*-Auger components was made using the results obtained by Sujkowski and Slätis[142] as a guide.

Albridge and Hollander[28] (SCS; Pra; Ph) carried out an extensive investigation of the Auger electron spectrum of ^{233}U. A ^{233}Pa source was prepared by dipping a 0.25-mm diameter platinum wire into a solution having very high specific activity. There was but little line broadening noticed, even for electrons of initial energies below 1 keV. In calculating the Auger line energies, ΔZ was assumed equal to 1. The several lines found at energies from 0.94 to 3.78 keV were assigned to Coster–Kronig transitions. In uranium, L_1L_2M and L_2L_3M (except for M_5) transitions are energy forbidden; $L_1 L_3 M_{1,2}$ transitions are also forbidden, but not $L_1 L_3 M_{3-5}$.

The L-Auger spectrum for ^{199}Hg ($Z = 80$) has also been measured following the beta-decay of ^{199}Au (by Haynes and Achor[202] and Nall et al.[113]) and following the electron capture decay of ^{199}Tl (by Jung and Svedberg[115]).

Haynes and Achor[202] used a lens spectrometer set for rather low resolving power to study the beta-spectrum and conversion lines in the disintegration of ^{199}Au. The measured L-Auger spectrum was analysed into five main components.

Nall et al.[113] (IFDF; C) used a window of 10 μg/cm^2 collodium film on a Lektromesh screen for their measurements. The activity in the chloride form was evaporated onto a formvar (10 μg/cm^2) plus aluminum (10 μg/cm^2) backing. With a spectral line width of $\sim 1\%$ at 8 keV, they determined the relative intensities of some 39 spectral lines and line groups.

Jung and Svedberg[115] (ICDF; C) worked with a resolution setting of 0.4% and a G–M counter window of 50 μg/cm^2 VYNS (a commercial plastic film). The activity, ^{199}Tl, was collected directly after mass separation upon a 2 mg/cm^2 aluminum foil. The penetration of the ions into the backing caused a marked spectral line broadening at low electron energies, and the measured L-Auger line width became $\sim 1.5\%$. The L-Auger line intensities were not corrected for electron absorption in the counter window.

Using two instruments similar to each other (SCS; Pra; Ph), Dionisio investigated the electron spectra arising in the decay of ^{191}Au to ^{191}Pt and ^{193}Hg to ^{193}Au. In both cases, described in Refs. 109 and 111, respectively, several Coster–Kronig and M-Auger lines were observed, in addition to a number of L-Auger lines and low energy conversion lines.

In 1963 Päschke[203] studied the L-Auger spectrum for gold ($Z = 79$), using electron impact to cause the initial L-subshell ionization. The bombarding electrons had a 55-keV energy. For electron spectrum analysis a stabilized variable field spectrometer with a resolution of 0.5% was employed, together with a G–M counter fitted with a thin window giving negligible absorption of 6-keV electrons.

The arrangement was first tested using a copper target and the K-Auger spectral groups were measured. The target metal was evaporated onto thin zapon foil of thicknesses between 2 and 30 μg/cm^2. It was found that the target material thickness should be kept between 2.5 and 6 μg/cm^2 in order to avoid both source charging and excessive electron energy loss effects.

An 8-μg/cm^2 gold target was then used for the study of the L-Auger spectrum from that element. The effectively negligible energy losses made it possible to analyze some multicomponent lines (e.g., the $L_3 M_4 M_4 - L_3 M_4 M_5$ doublet). A total of 18 lines and groups were interpreted.

Toburen and Albridge[108] have reported the investigation of the K-, L-, and M-Auger and L-Coster–Kronig spectra of platinum following the electron capture decay of ^{195}Au. The energies and intensities of a total of 55 Auger peaks were measured in the energy interval from 5 to 12 keV. A thin G–M counter window of zapon (~ 35 μg/cm^2) supported on a 90% transmitting copper mesh was used together with a postacceleration arrangement giving $\sim 100\%$ window transmission for energies as low as 5 keV. Electrons of 3 and 0 keV initial energies were detected with efficiencies of 92 and 62%, respectively. A reproduction of the measured L-Coster–Kronig and M- Auger spectra of platinum appears in Figure 13. It is interesting to compare this figure with that on page 1541 of Ref. 13, which is the MXY-Auger spectrum in the decay of platinum isotopes measured by Dionisio using a thin formvar window (cut off at ~ 1 keV) and no pre- or postacceleration.

Stockendal[117] (ICDF; C) registered separately the L-Auger spectra from ^{205}Pb and ^{206}Pb following the electron capture decay of the ^{205}Bi and ^{206}Bi, respectively. The sources were made by collecting the bismuth isotopes onto a thin aluminum foil after electromagnetic separation at an ion energy of 40 keV. The measured spectra for the two activities were compared to establish the presence of a 26.22-keV ($M2$) transition in ^{205}Pb. The superposition of the transition's L-conversion lines upon the Auger spectrum was clearly visible. Similar situations, in which spectrum analysis is complicated by the superposition of Auger and conversion lines, have been reported in other articles. For example, Ryde et al.,[204] in a study of the decay of ^{159}Tb, found the K- and L_1-conversion lines of the 58.0-keV transition in ^{159}Dy among the L- and K-Auger lines, respectively.

Monnand and Moussa[95] suggested that a comparison of their measured L-Auger spectrum for ^{153}Eu ($Z = 68$) showed the probable presence of the M_3-conversion line of a 5.8-keV ($M2$) transition in that isotope. The question could be decided by comparing the spectrum with the L-Auger spectra for other isotopes of europium, but the suggestion was not followed up. The authors presented the measured energies and probable identities of some 13 electron spectral lines.

Muir and Boehm[205] proved very nicely the presence of a 6.25(30) keV ($E1$) transition in ^{181}Ta by

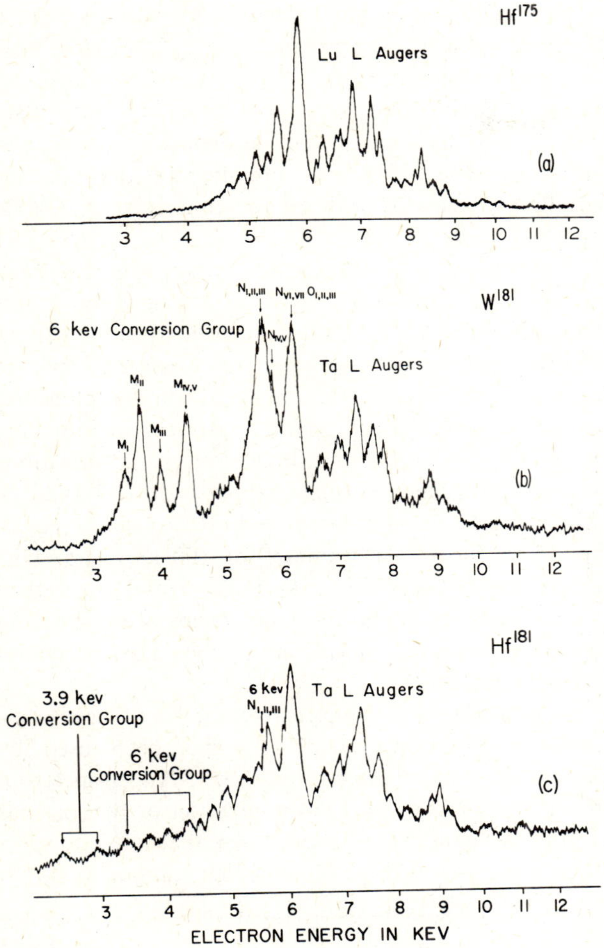

Figure 17. The *L*-Auger spectrum in ^{181}Ta($Z = 73$), obtained by Muir and Boehm[205], following (*b*) the electron capture decay of ^{181}W, showing superimposed *M*- and *N*-conversion lines of the 6-keV transition in ^{181}Ta, and (*c*) following the beta-decay of ^{181}Hf, showing several internal conversion components of the 3.9-keV and one of the 6-keV transition in ^{181}Ta. The *L*-Auger spectrum in ^{175}Lu, (*a*), which is free from superimposed conversion lines, following electron capture decay of ^{175}Hf appears for purposes of comparison.

comparing the low energy electron spectra of that isotope as fed by the beta-decay of ^{181}Hf and by the electron capture decay of ^{181}W. An iron-free semicircular spectrometer (resolution ~0.8 %) (C) was used in the study. Evidence of conversion lines of a 3.9-keV transition in ^{181}Ta following the beta-decay of ^{181}Hf was also seen. Figures 17 *a*, *b*, and *c* show the low energy electron spectra following these two decays together with that of ^{175}Lu following the electron capture decay of ^{175}Hf. A comparison of the spectra establishes the presence of the 6-keV transition. Using the same instrument, Henriksen[206] has also investigated the neighboring *L*-Auger and low energy conversion line spectrum in ^{169}Tm.

The *L*-Auger spectrum in ^{175}Lu following the electron capture decay of ^{175}Hf was examined by Gizon et al.[207] This spectrum, which is free of conversion lines, is used to identify conversion lines in this energy region in the radioactive decay of other isotopes of the same or close-lying elements.[208] The same procedure is now being employed to investigate low energy conversion line spectra from isotopes in the $Z = 92$ region.[208]

Figure 14 shows the *MXY*-Auger energy region of the electron spectrum arising in the decay of ^{193}Au to ^{193}Pt as measured by Johansson et al.[176,209] (IFDF; Co). Guided by the measurements of Toburen and Albridge,[108] the superimposed 1.63-O_I conversion and *M*-Auger lines were analyzed (previously discussed; see also, Figure 13), for the resolving powers of the instruments used in both investigations were similar. The low energy electrons were detected with a continuous-channel electron multiplier; the detection efficiency of the detector, which varies with electron energy, was not known. The conversion lines of the 12.65-keV transition in the same decay were seen to overlap with the *L*-Auger spectrum (see also, Ref. 107).

The last-named conversion lines were also observed earlier by Wapstra et al.[210] in an investigation of the decay of the gold isotopes$^{193,194,195,196(\text{and } 196m),199}$Au. In that investigation, although the spectrum was measured with a resolution of 0.1 to 0.4 % and down to an energy of ~1 keV, the presence of the 1.63-keV transition in ^{193}Au was not observed. The primary objective was, however, to investigate the decay of 12-isomeric state in ^{196}Au. Here again, many conversion lines arising in the decay of this isomer overlapped with the *L*-Auger spectrum.

Sébille and Widemann[211] (SCS; Pra; Ph) studied the electron spectrum produced in the decay of ^{187}W to ^{187}Re between the energies of 0 and 23 keV. Besides the low energy Coster–Kronig (0–0.13 keV) and the *M*-Auger spectrum (having intensity maxima at 1.10, 1.30, 1.39, and 1.68 keV) in the form of bands, the presence of the *K*-72-keV conversion line at 0.30 keV was also observed. This conversion line was proven to exist by comparing the 0 to 1.0 keV region of this spectrum with that from the decay of ^{182}Ta to ^{182}W. In addition, the energies of some 20 *LMM*-, *LMN*-, and *LNN*-Auger lines were measured; no lines in the energy region between 11.56 and 23 keV were observed.

Valentin[104] (SCS; Pra; Ph) investigated the *K*- and *L*-Auger electron spectra in ^{172}Lu and ^{173}Lu, noting six *LMM*- and *LMN*-Auger lines and four Coster–Kronig transition lines at the energies of 1.180, 1.379, 1.554, and 1.582 keV.

The *L*-Auger spectrum following the beta-decay of

^{137}Cs has been investigated twice,[212,213] the more complete treatment being given by Marguin and Moussa[91] (IFDF; Poa; C). The resolution was set at 0.31%. A postacceleration voltage of 1500 V was employed for electrons of energies below ~5 keV, and a thin window (aluminum-covered formvar on Lektromesh, 50% transmission, cut-off 1.6 keV) G–M counter was used for electron detection. Six peaks were observed between the electron energies of 3.0 and 4.7 keV. Using the same instrument, the authors[156] also measured the *L*-Auger spectrum for ^{170}Yb ($Z = 70$). The measured spectrum, corrected for detector window absorption, showed three main components that were identified by their energies as resulting from the $L_3M_3M_{4,5}$, $L_3M_4M_{4,5}$, and $L_2M_4M_{4,5}$ Auger transitions.

Erman and Sujkowski,[43] in their investigation of the decay of ^{133}Xe, registered the *L*-Auger spectrum from ^{133}Cs (ICDF; Poa; C). The mass-separated ^{133}Xe (50-kV acceleration) was collected on an aluminum backing. The spectrum was not analyzed nor were line intensities measured because of electron absorption uncertainties in the detector window.

The study of the *L*-Auger spectrum of ^{113}In, following the electron capture decay of ^{113}Sn, has been reported by Krisciokaitis and Haynes[89] (IFDF; Poa; C). Here *single* *L*-shell vacancies can be created by either (*a*) electron capture, (*b*) internal conversion, or (*c*) $K\alpha$ X-ray transitions following the initial production of a *K*-shell vacancy. *Double* ionization of the *L*-shell can occur through *KLL*-Auger transitions following initial *K*-shell vacancies. Evidence of vacancy satellites of the type L_3L_2-$M_4M_5L_2$ and L_2L_3-$M_4M_5L_3$ was sought in the *LMM*-spectrum of ^{113}In. The predicted intensities of these satellites were sufficient to permit their detection. Only a small indication of the presence of type L_3L_2-$M_4M_5L_2$ was found, but type L_2L_3-$M_4M_5L_3$ was quite conclusively identified. It is interesting to note that the intensities of these vacancy satellite lines—L_3-$M_4M_5(L_2)$ and L_2-$M_4M_5(L_3)$, using the alternative representation—relative to their respective Auger peaks (L_3-M_4M_5 and L_2-M_4M_5) indicate that the L_2-level tends to fill first when there are simultaneously L_2- and L_3-vacancies. However, from Auger and Coster–Kronig yields it can be shown that a solitary L_3-vacancy has a slightly higher Auger transition probability than a single L_2-vacancy. The presence of a L_2L_3-*MMM* triple electron transition was sought in the appropriate energy region, but none was found[213a] (see also, Section 9.3). Evidence also existed for the possible, and in some cases probable, presence of the vacancy satellites associated with all the other strong *LMM*-spectral lines.

The *LMM*-Auger spectrum for krypton has been reported by Krause[214] (ES; Co). The krypton atoms were photoionized in the gaseous state by molybdenum X-rays, and the electron spectrum analyzed using an energy resolution of 2%. After correcting for intensities of superimposed lines from other groups, the following relative intensities of the three prominent transitions were found: $2p3d3d/2p3p3d/2p3p3p = 100/42(3)/20(2)$, which are to be compared with Rubenstein's corresponding theoretical values[63] of 100/13.4/24.5. Discrepancies between the experimental and theoretical values were attributed to inadequacies of the electron wave functions used in calculating the Auger transition probabilities.

A rotationally symmetrical electrostatic spectrometer (energy resolution 2.75%) has been used by Mehlhorn[85] to investigate the *LMM*-Auger spectrum from argon ($Z = 18$). Photoionization of the gaseous argon was caused by tungsten X-radiation. The spectrum consisted of three obvious line components:

	Line 1	Line 2	Line 3
Energy (eV)	202.5(5)	189.7(6)	176.9(6)
Auger transition	$L_{2,3}$-$M_{2,3}M_{2,3}$	$L_{2,3}$-$M_1M_{2,3}$	$L_{2,3}$-M_1M_1
Relative intensity	1.0	0.25(2)	0.15(2)

By knowing that the excitation energy of a singly ionized argon atom is 27.5 eV,[215] it was possible to calculate the energy of the $L_{2,3}M_{2,3}M_{2,3}$ spectral component $= E_{L_{2,3}} - E_{M_{2,3}} - (E^+_{M_{2,3}})_{M_{2,3}} = 202.6(1.1)$ eV, in agreement with the experimental value. The binding energy of an $M_{2,3}$-electron (denoted by the outer subscript) in an atom that is singly ionized in the $M_{2,3}$-subshell (inner subscript) is represented by $(E^+_{M_{23}})_{M_{2,3}}$. It was also found that $(E^+_{M_1})_{M_1}$ equals $(E^+_{M_{2,3}})_{M_1}$ and that $(E^+_{M_1})_{M_{2,3}}$ is equal to $(E^+_{M_{2,3}})_{M_{2,3}}$; that is, the binding energies of the *M*-shell electrons have fixed values for a single *M*-shell hole, independent of its subshell location.

Later Mehlhorn,[177,178] employing electron bombardment for creation of initial vacancies, extended his investigation of argon to the L_1-Auger and associated L_1M-*MMM*-Auger-satellite transitions, as well as of the L_1-Coster–Kronig electron spectra (see also,

Table XVIIIa.

Experimental energies and relative intensities of the L_1MM Auger transitions in argon[178] compared with theoretical values[63].

Transition	Experiment		Theory
	Energy in eV	Relative intensity	Relative intensity
$L_1M_1M_1(^1S_0)$	255.9(5)	0.6(2)	0.42
$L_1M_1M_{2,3}(^1P_1)$	265.3(5)	0.8(1)	0.33
$(^3P_0)$, $(^3P_1)$, $(^3P_2)$	268.9(5)	1.0	1.0
$L_1M_{2,3}M_{2,3}(^1S_0)$	279.0(5)	0.08(1)	0.083
$(^1D_2)$	281.4(5)		0.008
$(^3P_0)$, $(^3P_2)$	283.0(5)	0.00(1)	0.0

Section 5.2.1). The superposition of the L_1MM- and L_1M-MMM-spectra made it difficult to obtain reliable values for the relative intensities of the various L_1MM components. However, the $L_1M_{2,3}M_{2,3}$ group intensity was observed to be very much lower than that of the $L_1M_1M_{2,3}$ group, in agreement with theory.[63] The theoretical intensity of the $L_1M_1M_1$ line seemed to be lower than the experimental value, but the superimposed $L_1M_{2,3}$-$MMM_{2,3}$ spectrum made this questionable.

Table XVIIIa lists the experimental intensities and energies of the L_1MM-Auger transitions in argon as measured by Mehlhorn.[177,178] Also included for comparison are the theoretical intensity values calculated by Rubenstein.[63] The comparison between the experimental absolute and relative total intensities of the L_1MM-Auger and the $L_1L_{2,3}M_1$ and $L_1L_{2,3}M_{2,3}$ Coster–Kronig transitions in argon with their relative and absolute theoretical values[63] appear in Table XV, Section 5.2.1. It is seen that the agreement between both sets of absolute and relative values is within experimental error limits except for the $L_1L_{2,3}M_1$-Coster–Kronig transition; the intensity of this group is a factor of about 4 smaller than the theoretical value.

Table XVIIIb gives the experimental relative intensity and energy values for the argon $L_{2,3}MM$-spectrum as measured by Stalherm.[130] Again, 4-keV electron bombardment was used for initial ionization, and a resolution of 0.17% ($\Delta E/E$) revealed Auger-vacancy-satellite lines of appreciable intensity in some of the spectral regions. Spectral line groups were analyzed into their Auger and Auger-vacancy-satellite line components by using the isolated $L_3M_{2,3}M_{2,3}$ ($^1S_0$) line shape as a standard. Figure 18a shows the result of the unfolding procedure in the $L_{2,3}M_{2,3}M_{2,3}$ spectral line region. As can be seen, there is a measurable splitting of the ^3P-triplet components. The $L_{2,3}M_1M_1$ and $L_{2,3}M_1M_{2,3}$ spectral regions exhibited rather strong superimposed Auger-vacancy-satellite lines. Figure 18b shows the L_2-$M_1M_1(^1S_0)$ region, where several secondary spectral components are present. According to Carlson and Krause,[216] some of these may be caused by an $L_{2,3}M_1M_1$ transition where another M-shell electron is simultaneously excited to a higher bound orbital (see Section 9). The intensities of the Auger line components falling in these groups and given in Table XVIIIb are obtained by the unfolding procedure. Absolute energies for the components can be obtained by using the energies of the main components of this spectrum, as explained previously.

Also using the same electrostatic spectrometer set for a resolution of 0.7% and employing electron impact to create the initial atomic subshell vacancies, Mehlhorn[163] investigated the M-Auger (M_{2-5} subshells) and Coster–Kronig (M_{1-3} subshells) transitions in krypton ($Z = 36$). A gaseous target was again used, which made it possible to investigate Auger lines of energies down to 25 eV. The absolute energies and relative intensities of a large number of transitions were measured: 22 lines of the $M_{4,5}$-spectrum, 14 of the $M_{2,3}$-spectrum, and 2 of the M_1-spectrum. In addition

Table XVIIIb.

Relative energies and intensities of the L_2MM and L_3MM Auger spectrum for argon[130]. In more recent calculations including configuration interaction the relative $L_2M_1M_1$ and $L_3M_1M_1$ transition intensities become increased[166)167)].

Transition	Relative energy (eV) Experiment	Relative energy (eV) Optical data[162)]	Relative intensity Experiment	Relative intensity Theory[63)]
$L_2M_1M_1(^1S_0)$	2.14(1)	--	0.49(1)	0.50
$L_2M_1M_{2,3}(^1P_1)$	11.50(1)	11.50	0.58(2)	4.39
$(^3P_0)$	15.04(3)	15.06	0.31(1)	1.43
$(^3P_1)$	15.10(3)	15.12	0.60(2)	2.86
$(^3P_2)$	--	15.25	<0.07	0.04
$L_2M_{2,3}M_{2,3}(^1S_0)$	25.22(2)	25.23	0.85(4)	2.45
$(^1D_2)$	27.62(6)	27.62	3.29(7)	12.69
$(^3P_0)$	29.14(6)	29.16	0.59(2)	2.37
$(^3P_1)$	29.20(6)	29.22	1.37(3)	5.33
$(^3P_2)$	29.37(6)	29.35	0.76(2)	2.96
$L_3M_1M_1(^1S_0)$	0.0	--	1.00	1.00
$L_3M_1M_{2,3}(^1P_1)$	9.37(1)	9.37	1.11(2)	8.78
$(^3P_0)$	--	12.93	<0.02	0.016
$(^3P_1)$	12.98(3)	12.99	0.26(1)	1.47
$(^3P_2)$	13.12(3)	13.12	1.28(3)	7.17
$L_3M_{2,3}M_{2,3}(^1S_0)$	23.08(2)	23.10	1.68(3)	4.90
$(^1D_2)$	25.47(3)	25.49	6.4(1)	25.39
$(^3P_0)$	27.02(5)	27.03	0.27(3)	1.19
$(^3P_1)$	27.08(5)	27.09	1.48(5)	5.33
$(^3P_2)$	27.21(5)	27.22	4.26(8)	14.81

to the *M*-Auger lines, Mehlhorn saw a number of Auger-vacancy-satellite lines due to transitions taking place in krypton atoms, which were initially doubly ionized (see Section 4.1.3). The measured relative line intensities for the $M_{2,3}$-spectrum were compared with Rubenstein's[63] theoretical values for *M*-Auger transitions for silver ($Z = 47$), given in Table XVI.

Spohr et al.[179] reinvestigated the *M*-Auger spectrum in krypton. They employed a spherical electrostatic spectrometer and created the initial atomic vacancy in the *M*-shell by electron impact. The spectrum was investigated at a resolution of ~0.12%. Spectrometer calibration was accomplished by using the energy values of the $M_5N_1N_{2,3}$ and $M_5N_{2,3}N_{2,3}$ transitions as determined from optical data. The measured intensities of the spectral lines were not corrected for the energy dependent detection efficiency of the continuous-channel electron multiplier that was employed.[217a] Figure 16 shows the $M_{4,5}N_{2,3}N_{2,3}$ Auger line region compared to that of the $M_{2,5}M_{4,5}N_{2,3}$ Coster–Kronig line region as obtained by Stalherm[130] and Mehlhom[163]. The energy scales are not identical in the two cases, but the difference in spectral line widths is obvious, being related in part to the difference in natural widths of the $M_{2,3}$ and $N_{2,3}$ atomic subshells and in part to the multicomponent final $M_{4,5}N_{2,3}$ vacancies due to intermediate coupling.

The $M_{4,5}NN$ spectrum of krypton was measured

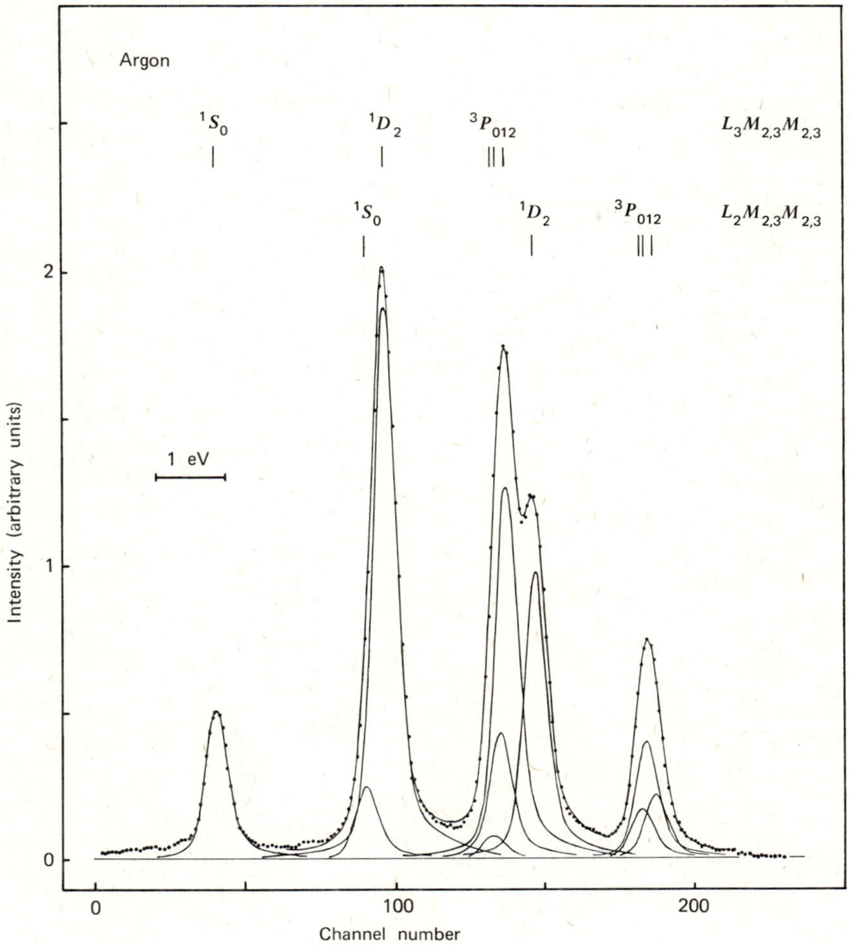

(*a*)

Figure 18a. The $L_{2,3}M_{2,3}M_{2,3}$ Auger spectral energy region for argon as measured by Stalherm.[130] After corrections for scattering and dispersion, the measured line shapes are separated into the several components.

by Stalherm[130] using the same instrument and atomic ionization technique as used by Mehlhorn[163] but with the energy resolution setting of about 0.16%. Figure 19*a* shows the obtained spectrum, which includes a number of unidentified lines. Figure 19*b* contains the graphical analysis of the complex structure of the $M_{4,5}N_1N_1$ line group.

The measured energies and intensities of the $M_{4,5}NN$-Auger and $M_{2,3}M_{4,5}N_{2,3}$ Coster–Kronig lines, as presented in Refs. 179, 130, and 163, are presented in Table XIX. Also included are the line energies as calculated from optical data and presented in Refs. 179 and 130.

The Auger spectra involving the two outermost shells in the four lightest noble gases have been investigated: the *KLL* in neon, *LMM* in argon, the *MNN* in krypton, and the *NOO* in xenon.[217] (Included in Ref. 217a.) The shells of neon and argon have energies considerably higher than those of krypton and xenon. It is interesting and informative to compare the $M_{4,5}NN$-spectrum for krypton and $N_{4,5}OO$-spectrum for xenon to see the effect of differences in coupling.

6. AUTOIONIZATION

The resonant excitation of atomic autoionizing levels has been observed in experiments of many types. There are several methods of investigating these levels, and improvements made in recent years have allowed researchers to obtain very exact data.

An autoionizing atomic state is one in which the atomic excitation energy is in excess of the first ionization limit of the atom, but the atom is not in an ionized state. This excitation state can arise either through multiple electron excitation or through the transfer of an inner orbital electron to a free peripheral orbital. Autoionization,[219] then, is the nonradiative deexcitation of such an autoionizing atomic state.

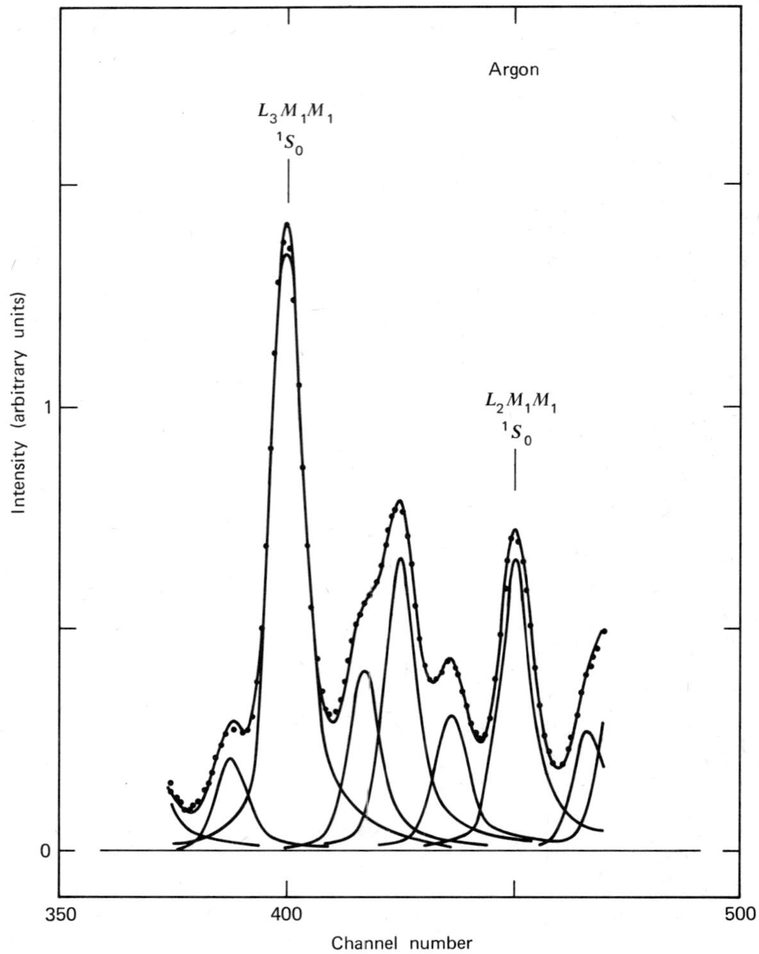

(b)

Figure 18b. The $L_{2,3}M_1M_1$ Auger region for argon. The analysis was done through separating the composite structure, using a standard spectral line shape as a pattern. The several extra lines originate from *LM–MMM* type Auger-vacancy-satellite transitions and/or *L–MMM* double Auger transitions (see Section 9).

The transition is Auger-like and leaves the atom in a final state of single ionization.

The term autoionization has been used on occasion to refer to the atomic electron shake-off effect associated with alpha and beta nuclear decay. The effect to be treated here, under this heading, however, is closely related to the Auger effect.

For an atomic electron to be excited from the atomic core to a lightly bound outer atomic orbital, the atom under investigation must be in the gaseous state or in an isolated environment.

Experimental studies of the characteristics and properties of autoionizing levels have been performed almost exclusively with respect to their excitation. In particular, the study of excitation through electron scattering and photon absorption experimental techniques provides much detailed information. The articles by Burke serve to review the work done in the field up to 1965[220] and through 1967[221] (see also Ref. 222).

However, the study of the electrons ejected through the autoionization decay of these excited states is also expected to provide important information on their properties. Only since about 1963 have electron spectrometry experimental methods been employed to this end. New advances in electron spectrometric techniques are making it possible to investigate the electron decay spectra of autoionizing levels with an accuracy that is comparable to that obtained using optical spectroscopy methods.

6.1. The Nature of Atomic Autoionizing Levels

The existence of atomic excitation levels lying at energies between the first and second ionization limits for the atom under investigation was established only a short time[223,224] after the discovery of the Auger

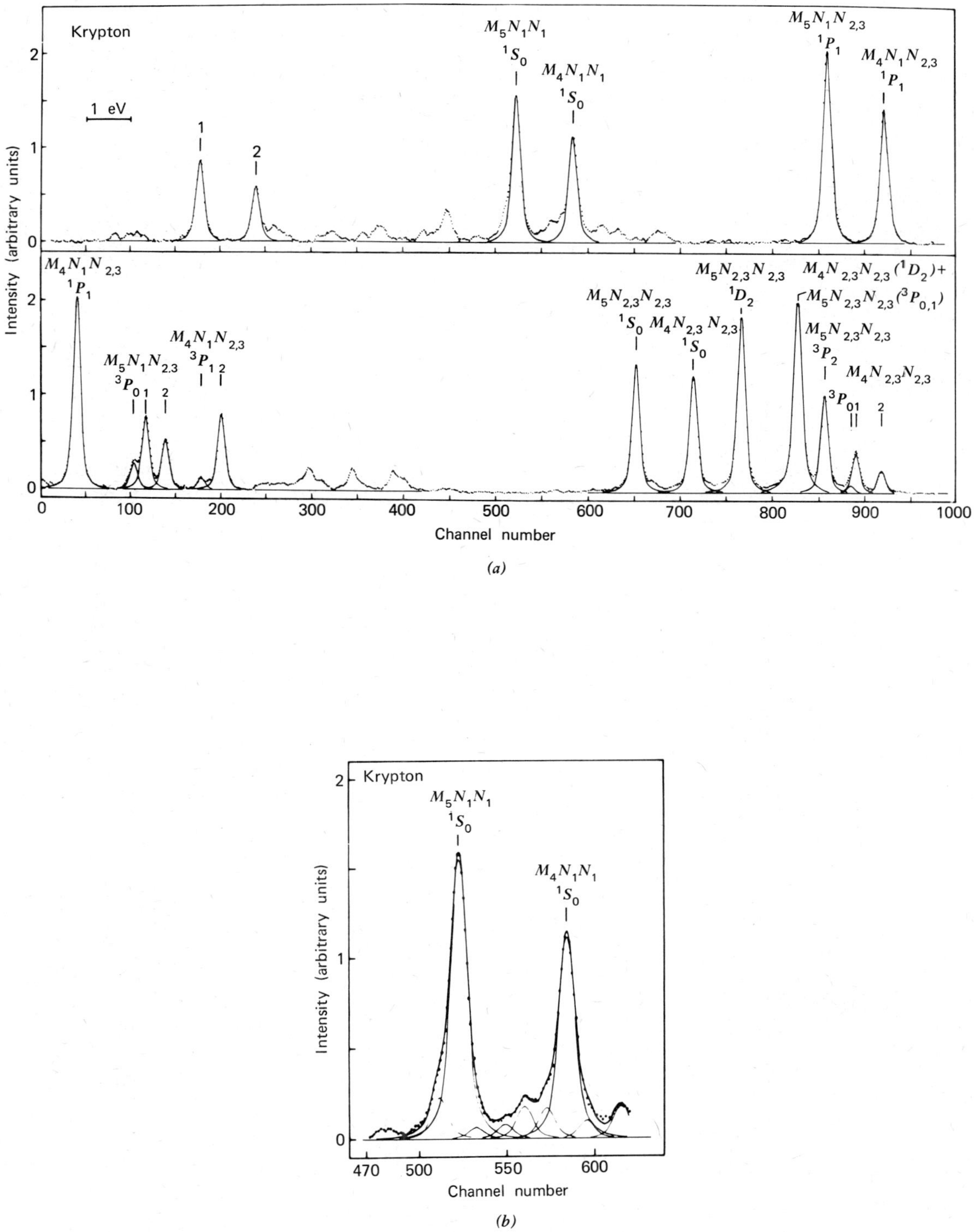

Figure 19. (*a*) The $M_{4,5}N_{1,2,3}N_{1,2,3}$ Auger electron spectrum for krypton as measured by Stalherm.[130] The analysis of the $M_{4,5}N_1N_1$ region is shown in greater detail in *b*. (See also Figure 16.)

Table XIX.

Low energy $M_{4,5}$-Auger and $M_3M_{4,5}N_{2,3}$ and $M_2M_{4,5}N_{2,3}$ Coster-Kronig lines of krypton as determined by Mehlhorn[163], by Spohr et al.[179], and by Stalherm[130]. In the second case the energy dependence of the measured intensities is disregarded and the errors in these are estimated to be ∿20%. The energy values of certain of the transitions calculated by Stalherm using optical spectroscopic data from Codling and Madden[218] and Moore[162] are entered for comparison.

	Energies in eV				Relative intensities		
	Experimental			Calc.[218]			
Transition	ref.[179]	ref.[163]	ref.[130]	ref.[162]	ref.[179]	ref.[163]	ref.[130]
	24.06(3)	24.06(3)	--	--	0.7	1.70(5)	--
	25.30(3)	25.27(3)	--	--	0.5	0.63(3)	--
	25.73(3)	25.53(6)	--	--	0.2	0.85(2)	--
	26.02(6)	--	--	--	0.1	--	--
	26.96(6)	26.87(10)	--	--	0.1	0.19(2)	--
	27.31(6)	--	--	--	0.0	--	--
	27.64(6)	--	--	--	0.1	--	--
	27.99(3)	27.96(7)	--	--	0.2	0.44(3)	--
	28.93(6)	--	--	--	0.1	--	--
	29.45(3)	29.37(7)	--	--	0.3	0.80(5)	--
	30.67(6)	--	--	--	0.1	--	--
$M_5N_1N_1(^1S_0)$	30.95(2)	30.91(4)	30.95(1)	--	1.6	2.83(5)	1.0
	31.72(6)	--	--	--	0.2	--	--
	31.95(6)	32.0(4)	--	--	0.1	2.36(10)	--
$M_4N_1N_1(^1S_0)$	32.18(2)	32.13(4)	32.18(1)	--	1.2	1.63(4)	0.71(3)
	32.80(3)	--	--	--	0.2	--	--
	33.15(6)	--	--	--	0.1	--	--
	34.04(6)	34.05(12)	--	--	0.2	0.17(2)	--
	--	35.27(12)	--	--	--	--	--
$M_5N_1N_{2,3}(^1P_1)$	(set at 37.67)	37.70(2)	37.67(1)	37.67	2.4	3.32(3)	1.26(4)
$M_4N_1N_{2,3}(^1P_1)$	38.91(2)	38.92(2)	38.91(1)	38.89	1.8	2.31(3)	0.87(3)
$M_5N_1N_{2,3}(^3P_0)$	40.15(3)		40.20(1)	40.19	0.2		0.12(1)
$(^3P_1)$	40.44(2)	40.56(12)	40.46(2)	40.46	0.7	1.78(2)	0.33(2)
$(^3P_2)$	40.86(3)		40.90(2)	40.89	0.5		0.22(1)
$M_4N_1N_{2,3}(^3P_0)$	41.31(6)		--	41.41	0.1		<0.01
$(^3P_1)$	41.67(6)	42.08(12)	41.68(2)	41.68	0.1	0.98(2)	0.065(3)
$(^3P_2)$	42.12(2)		42.12(2)	42.11	0.7		0.34(2)
	42.90(6)	--	--	--	0.1	--	--

Table XIX. (cont.)

	43.14(6)	--	--	--	0.1	--	--
	43.48(6)	--	--	--	0.1	--	--
	43.79(6)	--	--	--	0.1	--	--
	44.01(3)	44.09(12)	--	--	0.2		--
	44.30(6)	--	--	--	0.1		--
	44.61(6)	--	--	--	0.0	1.62(3)	--
	44.96(3)	44.94(12)	--	--	0.2		--
	45.21(6)	--	--	--	0.1		--
	45.86(3)	45.91(12)	--	--	0.2		--
	46.10(6)	--	--	--	0.1	--	--
$M_5N_{2,3}N_{2,3}(^1S_0)$	(set at 51.15)	51.15(2)	51.15(1)	51.16	1.6	1.69(2)	0.55(2)
$M_4N_{2,3}N_{2,3}(^1S_0)$	52.38(2)	52.36(3)	52.39(1)	52.38	1.4	1.47(2)	0.50(2)
$M_5N_{2,3}N_{2,3}(^1D_2)$	53.44(2)	53.42(3)	53.43(1)	53.44	2.2	2.38(3)	0.74(3)
$M_4N_{2,3}N_{2,3}(^1D_2)$				54.66			
$M_5N_{2,3}N_{2,3}(^3P_0)$	54.68(2)		54.66(3)	54.60	2.6		1.11(6)
$(^3P_1)$				54.70			
$(^3P_2)$	55.26(2)	54.74(10)	55.23(2)	55.26	1.2	4.63(4)	0.41(2)
$M_4N_{2,3}N_{2,3}(^3P_0)$	55.93(2)		55.81(3)	55.82	0.5		0.032(2)
$(^3P_1)$			55.92(2)	55.91			0.16(1)
$(^3P_2)$	56.51(3)		56.48(2)	56.48	0.2		0.096(4)
$M_3M_{4,5}N_{2,3}$	92.6(4)	92.3(2)	--	91.9$^+$	16	5.2(2)$^{++}$	--
$M_2M_{4,5}N_{2,3}$	99.9(4)	99.6(2)	--	100.8$^+$	6	2.0(2)$^{++}$	

$^+$These values were presented in ref.[179)].

$^{++}$These intensities are relative to each other and not to those listed above.

effect. Lines observed in the ultraviolet emission spectrum for helium were interpreted as arising from transitions from initial doubly excited atomic states, which overlap energetically with the first ionization continuum, to normal bound atomic states. Theoretical interpretations of the positions and widths of these lines correlated poorly with experimental results.

The observation of resonant states in helium in inelastic electron scattering exeriments, due to double electron excitation, was first observed in about 1935 by Whiddington and Priestley.[225,226]

In 1935 Beutler[227] noted anomalous photon absorption excitation in argon, krypton, and xenon, to discrete excited levels lying between the $^2P_{1/2}$ and $^2P_{3/2}$ singly ionized atomic states. These anomalous absorption lines were pointed out by Fano[228] (see also the work of Rice[229]) to be very asymmetric because of configuration mixing in forming the stationary state of double electron excitation in the atom.

The general theoretical treatment of the nature of autoionizing levels has been performed by a number of workers.[230–235] The most extensive application of this theory, however, has been made in attempting to interpret and to understand the double electron excitation of helium—the lightest atom in which such excitation can arise.

A neutral atom that becomes excited through the transfer of a deep lying atomic electron to an outer excited level, to an energy that is in excess of the ionization potential of the atom, may release energy through either (*a*) an X-ray transition, (*b*) an Auger-type transition, or (*c*) an autoionization process. In cases (*a*) and (*b*) the atomic core vacancy becomes filled through an X-ray or Auger transition, and the excited electron is undisturbed in its excited level except for a small readjustment owing to the core's further ionization. This phenomenon differs from the usual Auger or X-radiation process in that the transition takes place in a neutral, excited atom.

In case (*c*), however, a transition takes place whereby the initial core vacancy becomes filled by a second atomic electron, say, and the electron in the bound outer orbital becomes expelled. In cases *b* and *c* the atom is left in a final singly ionized excited state.

Considering now the case of helium, atomic excitation to an energy in excess of the first ionization potential will arise only through the simultaneous excitation of both atomic electrons. The $2s^2(^1S)$ state is the lowest excitation configuration of helium that can autoionize [to the $1sEs(^1S)$ state, where E represents the total energy of the system above that of the ground state;[232] the superscript *o* would mean that the term is "odd," i.e., that the configuration contains an odd number of *p*, *f*, *h*, etc., electrons; a superscript *e* would denote an even term].

A simple case of configuration interaction is the interaction of a single discrete state with a single continuum of states with which it overlaps; however, the properties of the interaction are unique and intrinsically important. Neglecting first configuration interaction, we can signify the discrete and continuous states by φ and $\chi_{E'}$ respectively. The exact problem can be solved by choosing these two functions so that the energy matrix of the system satisfies the following conditions:[236,230]

$$\langle\varphi|H|\varphi\rangle = E_\varphi, \tag{24a}$$

$$\langle\chi_{E'}|H|\varphi\rangle = V_{E'}, \tag{24b}$$

$$\langle\chi_{E''}|H|\chi_{E'}\rangle = E'\,\delta(E''-E'). \tag{24c}$$

The δ-function in equation 24c implies previous diagonalizing in the zero approximation of the submatrix within the more limited subset of states $\chi_{E'}$. The continuous range of values of E' includes the discrete energy level E_φ.

Upon including configuration interaction, the exact eigenfunction ψ_E may be written as a linear combination of φ and χ_E:

$$\psi_E = a\varphi + \int b_{E'}\chi_{E'}\,dE', \tag{25}$$

where a and $b_{E'}$ are both functions of E and can be determined in connection with the diagonalization of the energy matrix.

A study of the energy dependence of the eigenvector ψ_E and, more important, the resulting variation of the probability of excitation of the stationary state ψ_E, reveals a rapid change in transition probability between the initial state ψ_0 and the state ψ_E as E passes through a resonance. The associated matrix element $\langle\psi_E|T|\psi_0\rangle$, where T is a suitable transition operator, is composed of terms that interfere with opposite phases on the two sides of the resonance.

The ratio of the transition probability to the final state ψ_E to the transition probability of the unperturbed continuum becomes

$$\frac{|\langle\psi_E|T|\psi_0\rangle|^2}{|\langle\chi_E|T|\psi_0\rangle|^2} = \frac{(q+\varepsilon)^2}{1+\varepsilon^2}. \tag{26}$$

The quantities ε and q are defined as follows:

$$\varepsilon = \varepsilon_0 - \varepsilon_1 = \frac{E - E_\varphi}{\frac{1}{2}\Gamma_E} - \frac{1}{\frac{1}{2}\Gamma_E}\mathbf{P}\int\frac{\Gamma_{E'}/2\pi}{E - E'}\,dE' \tag{27a}$$

$$= \frac{(E - E_r)}{\frac{1}{2}\Gamma_E}, \tag{27b}$$

where **P** indicates that for $E' = E$ the "principle part of" the function is taken and the second term in equation 27a represents the line's shift in units of $\Gamma_E = 2\pi|V_E|^2$ which is the line width function (i.e., the width of the autoionized state φ); $\frac{1}{2}\Gamma_E$ is the full width at half maximum of the discrete state at energy E_r.

$$q = Q + q_1 = \frac{\langle\varphi|T|\psi_0\rangle}{\pi V_E^*\langle\chi_E|T|\psi_0\rangle} + \frac{\mathbf{P}\int V_{E'}\langle\chi_{E'}|T|\psi_0\rangle|(E-E')\,dE'}{\pi V_E^*\langle\chi_E|T|\psi_0\rangle}. \tag{28}$$

The function of equation 26 is shown in Figure 20 as a function of ε for several values of q. The maximum height of the curve for each value is normalized to unity. The curve for $q = 0$ is the inverse of that for $|q| = \infty$, whereas the curve for $|q| = 1$ is the well-known symmetrical resonance curve.

From the structural form of the transition operator written in terms of a reduced Green's function, Davis[237] has expressed the shift and width of an isolated resonance in terms of the level shift operator. From this operator a dispersion relation is obtained relating the shift and width of an autoionization resonance structure. To the lowest order of the shift operator, the expression reduces to that of Fano.[230]

Because of the interference of an autoionizing state

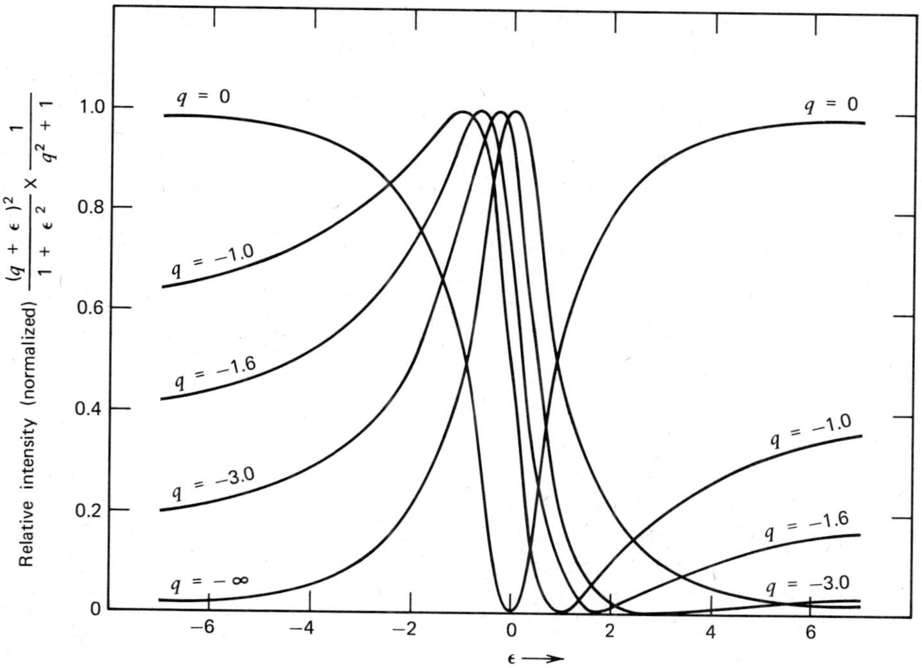

Figure 20. The function $(q+\varepsilon)^2/1+\varepsilon^2$ as a function of ε and q normalized to unit maximum intensity (See equation 26).

with the photoionization continua, the absorption cross section has the form

$$\sigma = \sigma_1 \frac{(q+\varepsilon)^2}{1+\varepsilon^2} + \sigma_2 . \qquad (29)$$

Here the background cross section σ_1 is that associated with the fraction of available continua that interacts with the discrete state; the remaining background cross section σ_2, is associated with the fraction of the noninteracting continua. When ε is large (far from the resonance), then $\sigma = \sigma_1 + \sigma_2$. The correlation index ρ is defined by $\sigma_1/(\sigma_1 + \sigma_2) = \rho^2$.

It was shown by Comes and Sälzer[233] that the integration of the excess transition probability is

$$\int I(E)\,dE = \int [|\langle\psi_E|T|\psi_0\rangle|^2 - |\langle\chi_E|T|\psi_0\rangle|^2]\,dE$$

$$= |\langle\varphi|T|\psi_0\rangle|^2 = |\langle\chi_E|T|\psi_0\rangle|^2 \frac{\pi}{2} Q^2\Gamma_E, \qquad (30)$$

assuming q, $|V_E|^2$, and the second term in equation 27a to be energy independent. The quantity Q is defined in equation 28. The final expression in equation 30 differs from that in Refs. 230 and 232 because an error in these two works has been corrected. More complete discussions of transition probabilities and a treatment of the special situations featuring configuration interaction of a discrete state with states of two or more continua or a set of discrete states that experience configuration interaction with a single continuum are contained in Refs. 233, 230, and 232.

The excitation of an inner core electron of an atom to a very loosely bound outer atomic orbital may be followed by an Auger-type transition without involving the excited electron (case b, previously defined). This transition probability is then not dependent on the low probability of interaction of the outer excited electron with electrons of the core.

In treating this problem, the continuum states χ_{jE} may be represented in terms of the set of one-electron states used to represent the discrete states φ_n (see Ref. 232 for details). The continuum states can be classified into two groups: one group of states, j_2, consists of one electron ionized and a second still in the excited orbital φ_n (case b); the other group, j_1, consists of states of a singly ionized atom with no electron in the excited orbital φ_n (case c). Defining a symbol N_n^2 in terms of the mean separation of energy levels adjacent to E_n [$N_n^2 = \frac{1}{2}(E_{n+1} - E_{n-1})$ and $E_n = E_{\lim} - I_H/(n-a)^2$, the Rydberg formula] and defining the reduced wave function $\bar{\varphi} = \varphi_n/N_n$, one may factor $\langle\varphi_n|H|\chi_{j_1E}\rangle$ in the form $N_n\langle\bar{\varphi}|H|\chi_{j_1E}\rangle$. The matrix elements $\langle\varphi_n|H|\chi_{j_2E}\rangle$ cannot be factored, however, since the outer electron occupies the space over which the integration must be carried out. Therefore, this last expression does not become smaller with increasing n but is independent of n, since one outer orbital of φ_n and χ_{j_2E} each overlap exactly. For the total width of the level having the initial vacancy, we obtain the sum total of the partial widths due to the two autoionizing processes:

$$\Gamma_n = 2\pi \sum_{j_1} |\langle \varphi_n | H | \chi_{j_1 E} \rangle|^2 + 2\pi \sum_{j_2} |\langle \varphi_n | H | \chi_{j_2 E} \rangle|^2$$
$$= N_n{}^2 \overline{\Gamma} + \Gamma_2 . \qquad (31)$$

The lines eventually merge into each other, unless Γ_2 vanishes, since the ratio of line width to line separation $\Gamma_n / N_n{}^2 = \overline{\Gamma}_1 + \Gamma_2 / N_n{}^2$ increases indefinitely with increasing n and decreasing $N_n{}^2$.

6.2. Classification of Autoionizing Levels

Autoionizing levels excited in the photoabsorption spectrum for helium by Madden and Codding[238] were classified by Cooper et al.[231] in a quite useful manner. The observed series of autoionizing levels, lying in the 60 to 65 eV energy region and converging to the $n = 2$ level of He^+, were identified as $^1P^o$ levels. However, two separate helium series $2snp$ and $2pns$ might be expected to converge to this limit, since the $n = 2$ level of He^+ is degenerate (in a nonrelativistic approximation), as are the $n = 3$, 4, etc., levels. As a result of this degeneracy, the interaction between the electrons of the doubly excited helium atom is so large that a close coupling interaction rather than a perturbation procedure must be used in treating the interaction.

If the electron-electron interaction is strong enough to remove the near degeneracy of the $2snp$ and $2pns$ states, then in the application of perturbation theory the independent-electron wave functions $u(2pns)$ may be replaced by the pair of functions[231]

$$\psi(2n\pm) = \frac{1}{\sqrt{2}} [u(2snp) \pm u(2pns)], \qquad (32)$$

where the + (−) quantum number corresponds to radial motions of the two electrons in (out of) step with each other. The following additional properties also characterize the functions 32: the ± designation does not describe a basic symmetry of the compound state wave function, but it is a suitable means of classifying these states. The validity of this classification has been demonstrated by Altick and Moore.[239] For each pair the + level is presumed to lie above the minus level. The optical transitions from the $1s^2$ state to the minus level and radiationless transitions from both the plus and the minus levels to the $1sEp$ continuum are all quasi-forbidden. Hence transitions of the type $1s^2 \rightarrow sp2n+$ are those observed in Ref. 238 for helium. Transitions designated as belonging to the type $1s^2 \rightarrow sp2n-$ have also been reported weak but not of negligible intensity (see Ref. 240).

The autoionizing $^{1,3}P^o$ states in helium are classified into the series: plus, minus, and $2pnd$.[241,242] The last of these has not yet been found experimentally.

Lipsky and Russek,[243] using the truncated diagonalization method, carried out very complete calculations of some of the properties of the lowest autoionizing states of helium with symmetries $^1S^e$, $^1P^o$, or $^3P^o$. For the ^1S states calculations were made for configurations up to g-g and for the ^1P states up to f-g. The calculated transition energies for these three autoionization states and experimental and theoretical values from other sources were compared.

The $^{1,3}S^e$ and $^{1,3}P^o$ autoionizing states lying below the $n = 2$ threshold of H^- and of helium were investigated by O'Malley and Geltman,[244] using a variational approach, and the $^1S^e$ autoionizing states were treated by Holøien and Midtdal.[245–248] In the former work[244] the classification of autoionizing states does not include explicitly the ± sign of the Cooper et al.[231] classification.

J. F. Perkins[249] has proved that the orthogonalization method of Holøien and Midtdal[246] is equivalent to the Q-operator procedure developed by O'Malley and Geltman.[244]

Calculations have also been carried out on the energy values of the $^{1,3}D^e$ autoionizing states in helium below the $n = 2$ level of He^+.[250] Cooper et al.[251] have shown that there are three series converging on this $n = 2$ limit for each of the $^1D^e$ and $^3D^e$ states. Analogous to the $^{1,3}P^o$ states, the $^{1,3}D^e$ states are classified into plus, minus, and $2pnf$ states. Here the $2pnp$ and $2snd$ pairs mix strongly to form the ± series.

Altick and Moore[252] also performed calculations on the same states, comparing their estimated line widths with those obtained by Cooper et al.[251]

For comparisons of calculated energies of the $^{1,3}S^e$ and $^{1,3}P^o$ levels for helium, see Ref. 243; the energies of the $^{1,3}D^e$ levels are compared in Ref. 251.

Calculations on the autoionizing levels between the $n = 2$ and $n = 3$ levels of He^+ have been carried out by Ormonde et al.[253] Determination was made of the $^1S^e$, $^1P^o$, and $^1D^e$ partial cross-section contributions to the total electron scattering cross section for He^+ using the truncated diagonalization method of Lipsky and Russek.[243] Here three series are found for the $^1S^e$ case in analogy to the case for levels below the $n = 2$, He^+ limit: The $3sns$ and $3pnp$ series interact to give $3n\pm = 3sns \pm 3pnp$; the $3n-$ states then interact with the $3dmd$ states to form states designated by $3nmd\pm = 3dmd \pm 3n-$. Level energies and widths were calculated for the ^1S levels.

Excited autoionizing levels in the other rare gases[218] can be classified in a manner similar to that used for the helium case. Atomic autoionizing level excitation of the types

$$ms^2mp^6(^1S) \rightarrow msmp^6(^2S)np(^{1,3}P^o), \qquad (33a)$$

$$ms^2mp^6(^1S) \rightarrow msmp^6(^2S)ns(^{1,3}S^e), \qquad (33b)$$

with $m = 2, 3, 4, 5$ for neon, argon, krypton, and xenon, respectively, and with $n > m$, may be caused by ultraviolet absorption and electron and ion bombardment techniques, to be described. Here the configuration of the $msmp^6$ core excitation state ^2S is followed in the notation by that of the state of the externally bound electron and finally by the configuration of the total excited atomic state $^{1,3}P^o$ and $^{1,3}S^e$.

Other states are also capable of becoming excited, namely,

$$ms^2mp^6(^1S) \rightarrow msmp^6(^2S)nd(^1D^o), \qquad (33c)$$

for argon, krypton, and xenon with $m = 3$, 4, and 5, respectively. For each value of m, n takes the values m, $m + 1$, and so on. The series limits for equations 33 are then, in X-ray terminology, the L_1-edge for neon, M_1 for argon, and so on.

Excitations of the types

$$\left.\begin{aligned} ms^2mp^6(^1S) &\rightarrow ms^2mp^5(^2P_{3/2,\,1/2})nd, \\ ms^2mp^6(^1S) &\rightarrow ms^2mp^5(^2P_{3/2,\,1/2})ns \end{aligned}\right\} \qquad (34)$$

have been investigated using photon absorption techniques. Figure 21 shows the photoabsorption spectrum for krypton between the Kr^+ $(P_{3/2})$ and $(P_{1/2})$ ionization energy limits as measured by Comes at al.[234] The Fano characteristics of the line shapes are obvious. The d states are seen to be much broader than the s states.

The discrete excited levels represented, for example, in equation 33a may autoionize only into the continuum states

$$\left.\begin{aligned} &ms^2mp^5(^2P)Ed(^{1,3}P_1^o),\ (^3D_1^o) \quad \text{and} \\ &ms^2mp^5(^2P)Es(^{1,3}P_1^o). \end{aligned}\right\} \qquad (35)$$

Among these only transitions to the $^{1,3}P^o$ states would be allowed by the selection rules $\Delta L = \Delta S = \Delta J = 0$ and no parity change (see Section 6.3.3). The final ion configuration ms^2mp^5 can be in one of the doublet states $^2P_{3/2,1/2}$, for each interacts separately with the continuum electron Es or Ed to form the possible total configurations denoted.

Autoionizing levels of the type

$$(m-1)d^{10}ms^2mp^6 \rightarrow (m-1)d^9ms^2mp^6(^2D_{5/2,\,3/2})np \qquad (36)$$

can be excited by photon absorption in krypton ($m = 4$) and xenon ($m = 5$). Here $n > m$, and the series limits $n \rightarrow \infty$ become the $M_{4,5}$ and $N_{4,5}$ edges for krypton and xenon, respectively. These levels can autoionize to the atomic states

$$(m-1)d^{10}ms^2mp^5(^2P_{3/2,\,1/2})Es,\ Ed. \qquad (37)$$

In ion impact excitation of autoionizing levels, multielectron excitation states may also be produced. The large number of such possible states gives rise to an autoionization spectrum of great complexity (see Section 6.3.3).

Double electron excitation states in atoms result in most cases from L–S coupling of the unpaired electrons. In such cases L–S coupling notation is used to specify the various configurations, as in equations 33.

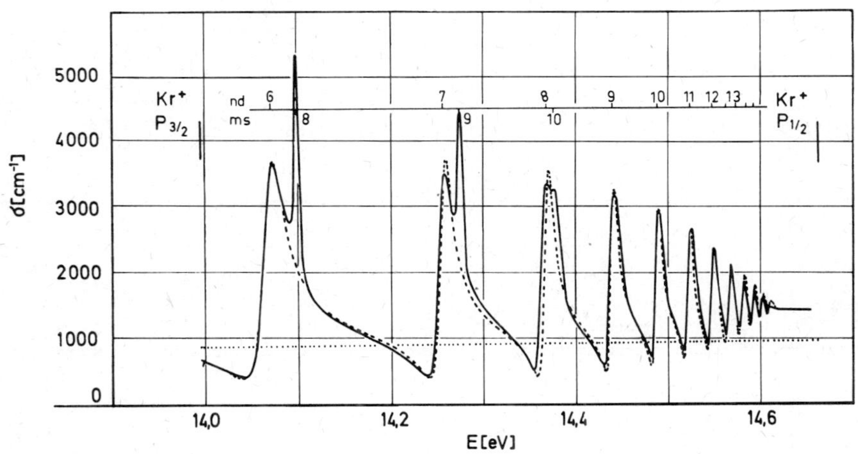

Figure 21. The comparison of the measured (solid curve) and the calculated (dashed curve) absorption coefficients for krypton between the $^2P_{3/2}$ and $^2P_{1/2}$ ionization limits, as measured by Comes et al.[234] The superposition of two series due to absorption excitation of a $4p$-electron to an ms or nd excited state is evident. Note the larger difference in spectral line widths for the two series.

Table XX.

Autoionizing state excitation (experimental)

1a) Photo-absorption investigations.

Absorbant	Autoionizing state	Other ref.
He	260) 238) { 2s2p($^1P^o$) (E=60.123(15) eV; q=-2.80(25); Γ=38(4) meV) (sp,23+)($^1P^o$) (E=63.651(7) eV; q=-2.0(1.0); Γ=8(4) meV) (sp,2n+)($^1P^o$) (n>3) (q positive) (sp,3n+)($^1P^o$) (n>2) (q positive) (sp,4n+)($^1P^o$) (n>3) (q positive) } 2s2p($^1P^o$) (E=60.0(1) eV; q = -1.87)[265][230]	267)
H_2, D_2, HD	ref.[269][270][271]	
Be	2p3p(^1P) (E=11.91 eV; q=∿+5.0; Γ=0.0594 meV)[256] 2p4d(^1P) (E=12.49 eV; q=∿-3.1; Γ=0.193 meV)[256]	257)
N_2	(in atomic N) 2s2p^3($^5S^o$)np($^4P^e$)[273]	
O_2	(in atomic O) 2p^3($^2P^o$)ns($^3P^o$); 2p^3($^2P^o$)nd($^3P^o$),($^3D^o$); and others[272]	
Ne	254) { 2s2p^{6}3p($^1P^o_1$) (E=45.546(8) eV; q=-1.6(2); Γ=13(2) meV; ρ^2=0.70(7)) 2s2p^{6}4p($^1P^o_1$) (E=47.121(5) eV; q=-1.6(3); Γ=45(15) meV; ρ^2=0.70(7)) 2s2p^{6}5p($^1P^o_1$) (E=47.692(5) eV; q=-1.6(5); Γ=2(1) meV; ρ^2=0.70(14)) 2p^{4}3s3p($^1P^o_1$) (E=44.979(8) eV; q=-2.0(3); Γ=10(3) meV; ρ^2=0.17(5));etc. } 2s2p^{6}np($^1P^o_1$) (q negative)[260]	
Ar	3s3p^{6}np($^1P^o_1$) ($\lvert q \rvert$∿large)[238]; 3p^{5}nd (n>8)[234]	261)
Kr	4s4p^{6}np($^1P^o_1$); 3d^{9}4s^{2}4p^{6}np($^1P^o_1$); 4s4p^{5}5snp($^1P^o_1$); 3d^{9}4s^{2}4p^{5}nl n'l'[274] 4p^{5}ns (n>5); 4p^{5}nd (n>7)[234]	261) 218) 262)
Xe	5s5p^{6}np($^1P^o_1$); 4d^{9}5s^{2}5p^{6}np($^1P^o_1$); 4d^{9}5s^{2}5p^{5}nl n'l'[274] 5p^{5}ns; 5p^{5}nd[234]	261) 218) 262)
Ca	3dnp($^{1,3}P^o_1$) (n>4) (τ=3 - 26×10^{-14} sec.)[275]; 3dnp,nf[277]	276)
Sr	279) 263) { 4dnp,nf, others. q ranges from +0,1 5p6s($^3P^o_1$)) to +27.5 (4d8p($^3P^o_1$)) and -13.7 (4d($^2D_{5/2}$)4f[1/2]$^o_1$(?)); τ=8 to 300×10^{-15} sec. }	
Ba	5dnp (n>7); 5dnf[279]	
Cd	4d^{9}5s^2($^2D_{5/2}$)np($^1P^o$) (n=5,6); 4d^{9}5s^2($^2D_{3/2}$)mp($^3P^o$) (m=5,6,7)[280]	281)
Ga[282], Tl[283][284], In[285], Zn[287], Pb[288];		

Table XX. (cont.)

1b) Inelastic electron scattering investigations.

Target (gas or vapor)	Autoionizing state	Other ref.
He	$2s^2(^1S^e)$; $2s2p(^3P^o)$; $2p^2(^1D^e)$; $2s2p(^1P^o)$; etc.[286] $2s^2(^1S^e)$; $2p^2(^1D^e)$ [289] (see also ref.[290])	268) 265) 291) 292) 293) 297) 296) 298)
$He^+(1s)$	(sp,3n+) (^1S, ^1P, ^1D) [299]	
Ne	$2s2p^6ns(^1S)$, $2s2p^6np(^1P)$ (n>2)[286,289,290]	
Ar	$3s3p^6np(^1P^o)$ (n>3); $3s3p^63d(^1D^e)$[289,290]	286) 303)
Kr, Xe	ref.[286,290]	
K, Rb, Cs	K: $3p^54s^2$ Rb: $4p^55s^2$; $4p^55s4d$; etc. Cs: $5p^56s^2$; $5p^56s5d$; etc. } 304)	
Hg, Ag	ref.[309]	
CH_4	E = 19.4(2) eV[305,306]	307)
Ethane n-Propane	ref.[306]	

In certain cases, such as for $2s^22p^6(^1S) \rightarrow 2s^22p^5(^2P)nd, ns$, in neon, *L–S* couplings does not hold, and it is advantageous to use the *jl* notation employed by Moore in her tables of atomic energy levels.[162] In this case, then, the series that converge on the limit are [254]

$$2s^22p^5(^2P^o_{3/2})ns[1\tfrac{1}{2}]^o,\ nd[\tfrac{1}{2}]^o,\ \text{and}\ nd[1\tfrac{1}{2}]^o.$$

(Note that in Ref. 254 the first [*jl*] notation is in error.)

The series converging on the upper limit are

$$2s^22p^5(^2P^o_{1/2})ns'[\tfrac{1}{2}]^o\ \text{and}\ nd'[1\tfrac{1}{2}]^o.$$

Autoionizing levels in beryllium are of interest because of the similarity between this element's ground state valence electron structure and the electron structure of helium. Beryllium has the configuration $1s^22s^2(^1S)$; the first excited state of the singly ionized atom $2p$ lies about 4 eV above its ground state $2s$, whereas in the helium ion the $2s$ and $2p$ levels are degenerate.

Moores[255] has performed a theoretical treatment of autoionizing and bound states of neutral beryllium. The positions and widths of six series of $2pnl$ autoionizing states of neutral beryllium were calculated by solving close-coupling equations for the $Be^+ + e^-$ system.

Altick[256] has investigated the photoionization cross section for beryllium as a function of energy near threshold. Resonant absorption structure was seen for excitation of the $2pnd(^1P)$ ($n = 3, 4$) autoionizing levels (see Table XX).

6.3. Excitation, Decay, and Study of Autoionizing Atomic Levels

The study of the excitation and the decay of autoionizing levels has been confined to atoms in gaseous or isolated state. The rare gases, being inert and monatomic, have been most frequently investigated. However, mass-separated ions or atoms of several heavier elements have also been examined (see Ref. 258).

The methods for the study of autoionizing atomic levels can be divided into three main groups:

1. Through the excitation of autoionizing levels
   a. by photoabsorption (see Section 6.3.1)
   b. by inelastic electron scattering (see Section 6.3.2)

In case a the absorption of electromagnetic radiation by the element of interest is measured as a function of wavelength (or energy). In case b electron resonant energy losses and scattering amplitudes are measured. Recently Chock and Fano[311] analyzed theoretical and experimental data on two-electron transitions in helium associated with electron, proton, and ion bombardment excitation.

In addition to these means of exciting autoionizing atomic states, there are those of proton and atomic or ion collision processes. When the excitation of autoionizing levels is caused by these methods, the electron emission spectra arising from the decay of these levels form the basis for their study. Therefore these means of excitation are included in item 2.c.

2. Through the de-excitation of autoionizing levels (see Section 6.3.3)
   c. by the measurement of autoionizing electron spectra
   d. by the measurement of end ion products
3. Through the measurement of elastic electron scattering cross sections (see Section 6.3.4)

Another phenomenon that is closely related to the creation of autoionizing levels is that of the creation of a short-lived resonant negative ionic state in electron scattering experiments. Evidence of the existence of such a state will appear as a resonance structure in the elastic scattering cross section. Such negative ion states may have configurations resembling those of autoionizing states. For a recent treatment of electron scattering in gases, with a very useful reference list, see the review article by Kuyatt.[259]

6.3.1. *Photoabsorption*

Very detailed information about the excitation of autoionizing atomic states can be obtained by investigating the photoabsorption coefficient in the vicinity of resonance energies. The excitation process may be represented by

$$h\nu + A \rightarrow A^* \rightarrow e^- + A^+,$$

where a photon excites an atom to an autoionizing state, which then decays by electron emission to a singly ionized atomic state.

In photoabsorption, excitation selection rules allow only transitions having $\Delta L = 1$. In helium, then, excitation to only the levels $nsmp(^1P^o)$ $(n \geq 2, m \geq 2)$ are allowed from the $1s^2(^1S^e)$ ground state.

Madden and Codling[238,260] have carried out extensive measurements on two-electron excitation states in helium. They observed absorption resonances in the photoionization continuum of helium which indicate the excitation of autoionizing levels. The continuous synchrotron radiation from a 180-MeV electron synchrotron was used as the light source, and the radiation intensity spectrum was analyzed with an optical spectrograph. Fano-type absorption curve structures, equation 26, were clearly observed. The narrow, weak resonance excitation of the states $(sp, 2n-)$ were seen, together with the much stronger $(sp, 2n+)$ resonance absorption structures. A number of weaker resonances were also seen which belonged to series having the $n = 3$ and $n = 4$ states of He^+ as limits.

Fitting the Fano shape to the measured absorption line shapes, the q and Γ parameters were obtained for several of the resonance states (see Table XX). Thereafter, the measured energies and transition probabilities and theoretical values for these were compared by other authors.

Fano and Cooper[232] have attempted to analyse the various contributions to the value and sign of q for some of these transitions. They also considered the relative widths of the $(sp, 2n\pm)$ and $3s3p$ states. It was shown by a crude estimate that $q \approx +1.7$ for the $3s3p$ state.

Comes et al.[234] have investigated the autoionizing excitation states of energies lying between the $^2P_{3/2,1/2}$ ionization limits in argon, krypton, and xenon. Photoabsorption cross-section techniques were used. The spectral components consisted of two superimposed series, one quite broad and the other narrow. Figure 21 reproduces the absorption spectrum for krypton where absorption resulting from the excitations to the two final state series $4s^24p^5ns$ $(n > 7)$ and $4s^24p^5nd$ $(n > 5)$. The Fano shapes of the broader lines are obvious.

Absorption line shapes for neon, to final $2s2p^6np$-$(^1P_1^o)$ states, were observed to be similar to those for helium, to final $2snp(^1P_1^o)$ states, whereas those for argon, to final $3s3p^6np(^1P_1^o)$ states, were of different character.[238] These last had the appearance of emission lines (i.e., $|q| \gg 1$). See also the work of Samson[261] for studies of $ns \rightarrow (n+m)p$ electron excitation in argon $(n = 3,\ m > 3)$, krypton $(n = 4,\ m > 4)$, and xenon $(n = 5, m > 5)$.

The photoexcitation of autoionizing levels in neon has recently been investigated in great detail by

Codling et al.[254] The absorption spectrum was measured using synchrotron radiation of energies from 20 to 150 eV and using photographic plates for detection. Structure appeared in three distinct energy regions.

The lowest energy series are those involving the excitation of a $2p$ electron to an ns or nd bound level and lie at about 20 eV. Since L–S coupling does not hold for these states of neon, jl notation is used to designate the final configurations.[162]

In the region 44 to 49 eV, single-electron excitations of the type $2s^22p^6(^1S_0) \rightarrow 2s2p^6(^2S)np(^1P_1^o)$ were found; and the appearance of a single $^1P_1^o$ series, unless the $^3P_1^o$ series cannot be resolved from the $^1P_1^o$ series, implies that the L–S coupling is dominant. The parameters of the three strongest resonances belonging to this series were determined (see Table XX).

Two series of two-electron resonance absorption lines were found. One of these, lying between 44 and 66 eV in energy was due to transitions of the type $2s^22p^6 \rightarrow 2s^22p^4nln'l'$, where n, l, n', and l' are the quantum numbers of the two excited electrons. Several such excitation series were noted, and were identified by applying an L–S coupling analysis. The L–S coupling assumption for two-electron excitation in neon was supported by several spectral characteristics. Nearly all the observed resonances were identified on this basis. The parameters of one of these resonances, to the final $2s^22p^43s3p(^1P_1^o)$ configuration, were determined (see Table XX). In addition, several anomalies were observed in the spectra. For example, the $2s^22p^4(^1D)3s(^2D)4p(^1P_1^o)$ member of the $2s^22p^4(^1D)3snp$ series was broader than the first $(n = 3)$ member; subsequent members were also broader than the width of the first member had led researchers to expect. It was suggested that the effect resulted from an additional broadening of the $n \geq 4$ terms owing to the interaction of these resonances with the continuum $2s^22p^4(^3P)3sEp$, which is not available to the $2s^22p^4(^1D)3s3p(^1P_1^o)$ level.

Finally, excitations of the type $2s^22p^6 \rightarrow 2s2p^5nln'l'$ (73–89 eV) were also studied. Again L–S coupling was assumed. Only two weak resonances, to final $2s2p^5(^{1,3}P)3p^2(^1P_1^o)$ configurations, were identified.

Inner p-shell excitation has also been observed in xenon. Several terms of the series $4p^64d^{10}5s^25p^6(^1S_0) \rightarrow 4p^54d^{10}5s^25p^6ns(^1P_1^o)$ have been reported.[262]

Two-electron excitation has also been found in krypton and xenon[262] and in neon.[254]

Photoabsorption measurements have been performed for several of the heavier metals (see Table XX), usually while in the vapor state. In one of the most recent of these studies Garton et al.[263] worked with strontium ($Z = 38$), investigating and identifying a large number of photoabsorption transitions and determining values of q, Γ, and so on, for most of them. The variety of absorption line shapes is of interest, since q-values range from near zero to rather large values, both positive and negative (see Table XX). Absorption line shape analysis gave lifetime values ranging from about 8×10^{-15} to 3×10^{-13} sec with the majority of the values lying in the (1 to 3) $\times 10^{-13}$ sec range. The treatment of strontium is the most thorough to date.

Photoelectron spectra from oxygen using helium resonance radiation (21.2 eV) shows a π_g2p spectral group between 12 and 13 eV and a π_u2p group between 16 and 18 eV. However, when neon resonance radiation is used (16.9 eV), there appears an additional envelope of spectral lines extending downward from 15 eV and overlapping with the π_g2p group.[264] These lines were attributed to the decay of autoionization states excited in the latter case but not in the former.

6.3.2. *Inelastic Electron Scattering*

Atomic excitation resulting from a collision process with an electron of a defined initial energy will be reflected in the energy loss suffered by the electron. The excitation process of interest is then

$$e^- + A \rightarrow A^* + e^-.$$

The process of atomic autoionizing level excitation by electron bombardment is not strictly subject to optical selection rules. Simpson and Mielczarek[308] have shown, however, and as might be expected, that single electron excitation of helium by 50-eV electron bombardment at low scattering angles favors dipole transitions; in addition, excitation of triplet states becomes more probably with increasing scattering angle. Therefore, for this energy and small scattering angles, the Born approximation is still valid, whereas for large scattering angles the occurrence of optically forbidden transitions is more probable (see also, Ref. 310).

Simpson et al. [286] investigated the resonant inelastic scattering of electrons in the excitation of autoionizing levels in helium, neon, argon, krypton, and xenon. Electrons scattered in the forward direction were measured for incident energies of 90 to 100 eV. Optically allowed transitions were most prominent, although optically forbidden excitation to triplet states were also noted.

In a more recent article, Simpson et al.[289] investigated the excitation of optically forbidden states in helium, neon, and argon. They used a small scattering angle ($\Delta\theta \approx 10^{-2}$ rad about the forward direction) and initial energies from 60 to 400 eV. As expected, optically forbidden transition probabilities to ^1D and ^1S states with respect to those for allowed transitions (to ^1P states) increased with decreasing impact energy. For both helium and neon the ratio of forbidden to allowed transition probabilities increased a great deal toward lower energies. For argon, however, both positive and negative peaks were seen in the spectrum, the intensities and characteristics of some of these varying strongly with energy.

Silverman and Lassettre,[265] who investigated the energy loss spectrum of 500-eV electrons bombarding helium as a function of scattering angle, found that the single electron $1s \rightarrow 2s$ forbidden transition increased with scattering angle to an intensity about equal to the $1s \rightarrow 2p$ allowed transition at the scattering angle of 15.3°. The double-electron excitation of $(sp, 23+)(^1P^o)$ at 63.5 eV was also observed. A second observed peak lying at 60.0 eV could be primarily attributed to the excitation of the $2s2p(^1P^o)$ state of energy 60.1 eV; but it could also result to an appreciable extent, at 500-eV bombardment energy, from the forbidden excitation of the $2p^2(^1D)$ state of energy 60.0 eV.[311] This conclusion is based on the relative intensity measurements for the two transitions as measured by Simpson et al.[289]

Boersch et al.[293] have measured (Figure 22) the electron energy loss spectrum for helium from about 20 to 65 eV. The initial bombardment electrons were of 25-keV energy and the electron energy loss spectrum (scattering angle $\leq 10^{-4}$ rad) was analyzed with a modified electron microscope (see Chapter 2, Section 5.3.1). The weak intensity energy loss due to the forbidden transition $1s \rightarrow 2s$ is also visible. The Fano shape of the two-electron excitation of the $2s2p(^1P)$ level is plainly observable.

Figure 23 gives the relative positions of the various singlet and triplet S and P single-electron excitation states in helium. The various two-electron autoionizing state series at higher energies are also shown, and the relative positions of He^- states appear to the right.

Ehrhardt et al.[312] and Andrick et al.[313] investigated the characteristics of inelastic electron scattering excitation of final $^{1,3}S$ and $^{1,3}P$ states in helium via resonance excitation of autoionizing states in He^- in the energy range of 22.4 to 22.9 eV. (Refs. 312 and 313 provide a detailed treatment of the topic.) Using a double electrostatic electron energy analyzer, independent energy selection is made both before and after the scattering process.

The scattering resonances investigated were those due to the single-electron excitation of $n' = 2$ states in helium via the creation and decay of short-lived

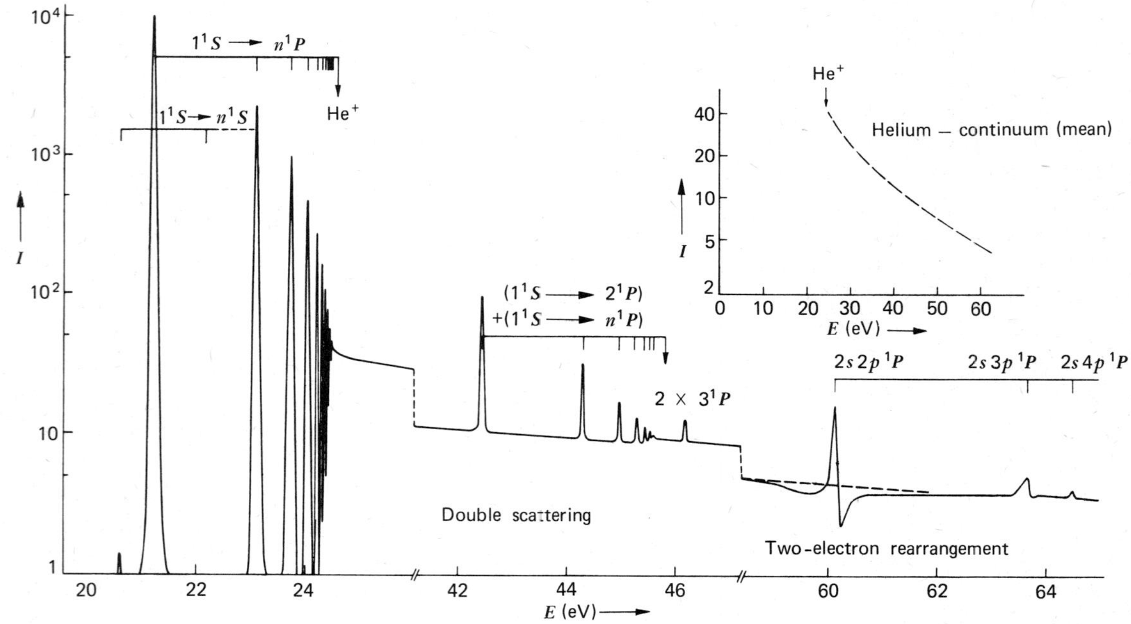

Figure 22. The energy loss spectrum (25-keV electrons, scattering angle $\leq 10^{-4}$ rad) as measured by Boersch et al.[293] The allowed transition series $1s \rightarrow np$ is strongest; energy losses due to excitation in two individual atoms are also evident (double scattering). The intensity ratio for transitions to the $1s2p$ and $1s3p(^1P_1)$ states are not in agreement with theory, nor are the shape and the intensity of the underlying continuum,[294] apparently because of a nonlinear transmission–energy-loss function for the instrument employed in these preliminary measurements. In Geiger's[295] work, however, no such discrepancy is present. Note the first term in the optically forbidden transitions $1s \rightarrow ns$ and, at higher energies, the energy loss excitation of the $2snp(^1P)$ autoionizing levels.

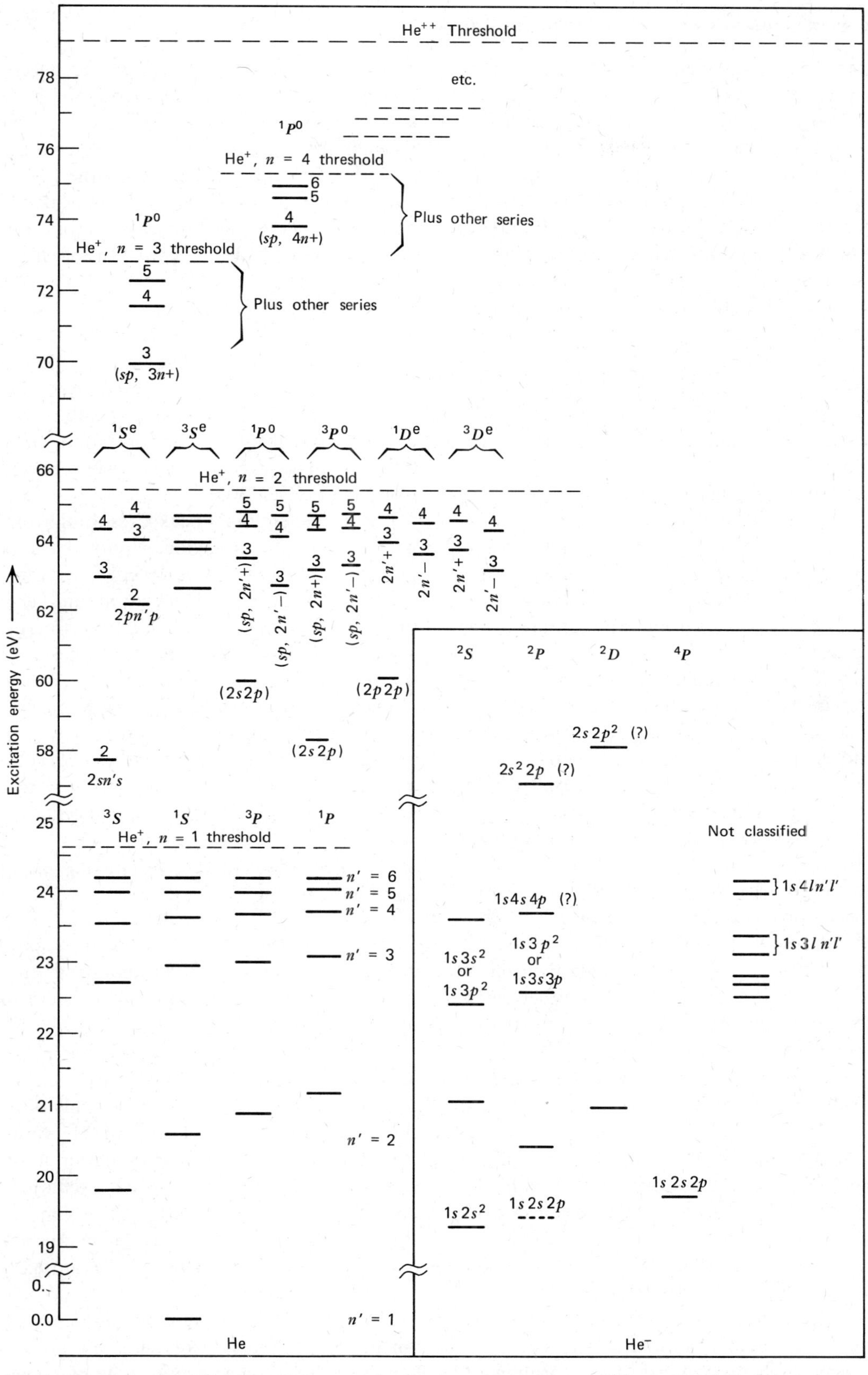

Figure 23. The energy level schemes of He and He^-. In the He case, single-electron excitation levels lie below the first ionization limit at about 24.6 eV. Levels lying above that limit are related to two-electron excitation and can decay through autoionization. In the He^- case, all the levels can decay, of course, through autoionization of the negative ion state. The production of the ^2S, ^2P, and ^2D states is observed in electron scattering measurements; the ^4P state is metastable against decay via autoionization (see text).

excited states in He^- (see Figure 23). By selecting the difference between impact and scattered electron energies, $He^- \rightarrow He^* + e^-$, autoionization transitions to a given final state can be selected; the resonance energy profile for the creation of this final state can be investigated by altering the impact electron energy, while keeping the above-mentioned energy difference constant. This variable bombardment energy measurement was carried out for scattering angles between 0 and 145°. The characteristics of two of the He^- levels, among five all lying in close proximity to one another, were determined.

The measured resonance shapes associated with the various He^- states were found to vary drastically with scattering angle, sometimes even disappearing for certain values.

In an experiment slightly resembling the one just described, Daly and Powell[299] carried out measurements on resonances seen in the electron bombardment of He^+ in the $2s$ excited state. These were attributed to the creation of autoionizing states of the type $(sp, 3n+)(^1P)$ $(n = 3, 4, 5)$ through the capture of the bombarding electrons.

Among several of the resonance scattering phenomena are those in which an electron is absorbed by a neutral atom, which becomes excited to an autoionizing level, and then decays to an excited neutral atomic state. Since the final atomic state is an excited one, the autoionization electron does not have the same energy as the impact electron and the scattering process is called inelastic.

Extensive data on the electron impact excitation of helium has been presented by Moussa et al.[300]

Experimental results of the study of a metastable N^- state has led to the theoretical investigation of this metastable ion[301] (see Ref. 301 for other references). It is found that the $N^-(^1D)$ state of nitrogen lies 0.844 eV below the first excited $N(^2D)$; decay to the $N(^4S)$, with the emission of an electron, is forbidden, and the lifetime of the parent state will be long.

Calculations of cross sections for electron impact excitation of $2s2p^6\ 3s, p, d$ autoionizing states in highly ionized iron Fe^{+16}, which has a neon electron configuration, have been carried out by Bely and Bely.[302] Calculations were also made for excitation cross sections for $2p \rightarrow 3\,s, p$, and d transitions.

In processes of elastic resonance scattering (see Section 6.3.4) the bombarding electron combines for a short time with the neutral atom, forming an excited negative ionic state where an atomic orbital electron is also excited. The excited state then decays to the original ground state of the neutral atom through an autoionization process, the emitted electron having the same energy as it did before the collision. The survey article by Ehrhardt[314] covers electron-ion collision processes and provides a reference list.

Little work has been done on the investigation of autoionizing levels in more complicated molecules. Measurements of photoionization cross sections have, however, indicated the existence of such states in methane, ethane, *n*-heptane.[305,306]

The measurement of the angular dependent inelastic scattering of 70-eV electrons on methane enabled Ehrhardt and Linder[305] to demonstate the existence of an autoionizing level lying at an excitation energy of 19.5(2) eV in the neutral molecule. This level then decays through autoionization to the stable CH_4^+ ion. The level was previously reported by other workers (see Ref. 305).

Electron energy loss peaks, resulting from the bombardment of alkali halide vapors, have been interpreted by Geiger and Pfeiffer[315] as the results of combined molecular dissociation and excitation of the freed alkali and/or halide atoms. Energy considerations revealed that certain of the final atomic excitation states were autoionizing levels.

The resonance scattering of low energy electrons on carbon monoxide and molecular hydrogen was investigated by Ehrhardt[316] with regard to the angular scattering distributions. The results were compared with those obtained earlier for molecular nitrogen.

6.3.3. Electron Emission in Autoionizing Level De-Excitation

In about 1964 the study of electron spectra arising in the decay of excited autoionizing atomic levels became an important source of information on the properties of these states.

The study of the initial autoionizing level excitation, however, can be carried out by using methods outlined in Section 6.3, 1.a and 1.b For the study of the electron emission de-excitation of these levels, however, electron, proton, and ion impact excitation must be added to this list of methods of initial excitation. Electron impact excitation is essentially the same as inelastic electron scattering, although the bombarding energy is usually much in excess of the atomic excitation energy of interest.

Regarding the electron emission decay of an autoionizing level, the rate of decay of an autoionizing level is related to the matrix element of the perturbation $\mathscr{H}$, coupling the discrete and continuum states; this perturbation may be owing to either electric or magnetic interactions between the atomic electrons. By first-order perturbation theory, the transition rate to a continuum state χ_c of energy E_c from a discrete state φ_d of energy E_d is

$$P^{(1)} = \left(\frac{2\pi}{\hbar}\right) |\langle \chi_c | \mathscr{H} | \varphi_d \rangle|^2 \, \delta(E_d - E_c).$$

For Coulomb interaction, the transition rate ranges from about 10^{13} to 10^{15} sec^{-1}, which is the usual range for allowed transitions.

However, if the first-order matrix element vanishes, as when the Coulomb operator relating a given discrete state to a continuum vanishes, the transition rate must be estimated by second-order perturbation theory:

$$P^{(2)} = \frac{2\pi}{\hbar} \left| \sum_i \frac{\langle \chi_c | \mathscr{H} | \varphi_i \rangle \langle \varphi_i | \mathscr{H} | \varphi_d \rangle}{E_i - E_d} \right|^2 \delta(E_d - E_c),$$

where the summation is over the intermediate states φ_i. Such a state is said to be "metastable" against decay via autoionization.

Let L, S, J, and m_J symbolize the quantum numbers for the discrete state and L', S', J', and $m_{J'}$ those for the continuum state. The parity is odd or even when the sum of the orbital quantum numbers over all electrons is odd or even, respectively. Setting $\Delta L = L' - L$, $\Delta S = S' - S$, and so on, assuming L–S coupling, and assuming all interaction distances to be of the order of the Bohr radius, the selection rules for a perturbation $\mathscr{H}$ become:

Returning to the light elements, certain excited states, such as the excited $1s2s2p(^4P_{5/2})$ state in lithium and in He^- are metastable and decay only through spin–spin interaction[312]. On the other hand, the $(^4P_{3/2,1/2})$ states may decay with a higher decay rate through spin–spin, spin–orbit, and spin–other-orbit interactions; they may also be indirectly coupled through these interactions to the continuum because of mixing with short-lived doublets (^2P) of the same configuration. Therefore much shorter lifetimes are expected for the $^4P_{3/2,1/2}$ states. Since the decay rates differ so greatly, a small amount of mixing of doublet and quartet states will strongly affect the lifetime of the latter.

These selection rules relate to transitions between initial excited states and final ion-plus-ejected-electron continuum states. However, since it is not possible to know the angular momentum of the emitted electron, Rudd[323,324] has suggested that the selection rules be restated in a manner such that the initial excited state is related to the possible final ionic states. For L–S coupling these rules are:

1. If neither is an S state, then the autoionizing transition is not forbidden.
2. If one of the states is an S state, then if the parities of the configurations are the same (different), the other state must be an $S, D, G, \ldots$ $(P, F, H, \ldots)$ state.
3. Multiplicities, $2S + 1$, of the states must differ by one unit, since the spin of the emitted electron is $\frac{1}{2}$.

Interaction	ΔL	ΔS	ΔJ	Parity Change	Relative Transition Rate	Range in τ(sec)
Coulomb	0	0	0	No	1	10^{-13}–10^{-15}
Spin–orbit, Spin–other-orbit	0, ±1	0, ±1	0	No	α^4	10^{-4}–10^{-6}
Spin–spin	0, ±1, ±2	0, ±1, ±2	0	No	α^4	10^{-4}–10^{-6}
Hyperfine	0, ±1, ±2	0, ±1	0, ±1	No	$\alpha^4(m_o/M)^2$	10^2–10^0

In the table, α is the Sommerfeld fine-structure constant, m_o is the electron mass, and M is the proton mass.

It was seen earlier in this chapter that excitation states for low Z-values can be suitably described using the L–S coupling scheme. Any autoionization metastability required by this coupling scheme is then satisfied. However, for high Z-values strict L–S coupling breaks down because of the large spin–orbit interaction. This breakdown makes it possible, for example, to excite ^4F and ^4P levels through optical excitation.[317,318,320]

In addition to the first three rules, the following can be stated about excitation of neutral states:

4. In cases of pure L–S coupling the multiplicity can change at most by a number $2N$, where N is the number of electrons in the projectile.
5. In atoms not subject to pure L–S coupling, or in atoms arising from the dissociation of molecules, there is no restriction on the change of spin, except that total spin must be conserved.

The application of external electric or magnetic fields affects the decay rate of autoionizing levels.

In the former case a metastable autoionizing level may mix with a shorter lived level of the same multipolarity and opposite parity (Stark effect). An electric field may, then, couple a metastable quartet state to an intermediate quartet state that has opposite parity and is coupled to the continuum. Feldman and Novick[318] have shown that, for the alkali metals, the "Stark quenching rate" of a metastable level is proportional to the square of the field strength and becomes greater with increasing atomic number. The quenching ionization of excited atoms near a metallic surface has recently been investigated by Chaplik.[325]

Upon application of an external magnetic field, different fine structure levels of a given state may become coupled to one another (Zeeman effect). This coupling may greatly reduce the longer of the levels if states of different lifetimes are involved.

Although the preceding discussion concerned autoionizing excitation states in single atoms or ions, there is a much different situation in the case of excited autoionizing molecular systems.

Bombardment of molecules by photons, electrons, or ions may cause excitation of vibrational, rotational, and/or excited-electron ionized states of the target molecule (also dissociation, if the excitation energy is high enough). Photoionization of a molecule A can produce both vibrational and rotational states in the singly ionized ion. Electron impact may cause vibration-rotational excitation, with or without simultaneous ionization, through a scattering collision during which the electron is at no time captured by the molecule. However, resonance electron scattering occurring at defined energies creates a negative metastable molecular ionic state, with or without vibration-rotational excitation A^{-*}, which decays to a neutral molecular state, ground or excited, through the autoionization process. In ion-molecular collisions the situation is made much more complex by the temporary creation of a compound molecule with associated strong molecular orbital perturbation, vibration, rotational, and electron excitation and electron ionization processes.

It has been demonstrated that autoionization processes are in direct competition with radiative modes of molecular de-excitation. Observed evidence for de-excitation of vibrational states of H_2^* through autoionization[326] is discussed later. Dissociation-autoionization competition has also been found.[327,328] Seemingly, the multifarious roles that autoionization plays counter to other modes of the de-excitation of excited molecular states have only begun to be appreciated.

For an example of the above-mentioned roles played by autoionization, consider a diatomic molecule in which one of the electrons is highly excited to a Rydberg state. If the quantum number n of this state is great enough so that the orbital radius for the electron is large compared to the ion core dimensions, then the molecular system is in many regards similar to a hydrogen atom. However, besides a Coulomb field potential, the ionic core generates potentials of higher multipolarity in contrast to the hydrogen atomic situation. If the core is in a state of vibrational and/or rotational excitation, and if this excitation energy is in excess of the binding energy of the Rydberg electron, the excited system can decay by autoionization.

This autoionizing process is very similar to nuclear internal conversion, wherein the energy freed through the occurrence of a nuclear transition is given to a bound atomic electron, which is thereby expelled from the atom (see Chapter 4).

Exploiting this similarity, Russek et al.[329] have calculated the autoionizing lifetimes for Rydberg states of molecular hydrogen and the deuterium hydride molecule.

The electrostatic potential experienced by the excited electron as supplied by the ionic core is an extended charge distribution; therefore it will be a monopole Coulomb potential varying with distance from the ion core center as r^{-1}; a quadrupole potential varying as r^{-3}; for heteronuclear cases, a dipole potential falling off as r^{-2}; and, since the core is polarizable, a polarization potential that diminishes as r^{-4}. Among these, only the monopole electron-ion core interaction was included in the calculations, since it is by far the most important in determining the motion of the Rydberg electron.

From the calculations, certain quasi-selection rules were obtained:

1. All transitions except those for which the vibrational quantum number v decreases by 1 are unlikely; transitions where v decreases by 2 are 100 times longer.
2. In general, a change of angular momentum of the ion core does not affect the transition rate.
3. Transitions except those having $\Delta l = 0$ are unlikely; the 0–0 transition is always expressly forbidden.

Measurements on the photoionization cross sections of hydrogen, deuterium, and deuterium hydride, determined by Diebler et al.[269] (see also, Refs. 330–333), showed structure in the form of peaks, which was attributed to excitation of the above Rydberg type of autoionization states. These were unquestionably $l = 1$ Rydberg states of the electron due to the nature of the photoabsorption process. Russek et al.[329] calculated that these Rydberg states had mean lifetimes of $\sim 10^{-10}$ sec in H_2^*, whereas Rydberg $l = 0$ states in HD and H_2 had mean lifetimes of the order of 10^{-8} and 10^{-7} sec, respectively.

The de-excitation mode just described for a vibrationally and rotationally excited molecule in an electronic excitation state is only one of several. Taking the hydrogen molecule as an example, electron impact can cause ionization through one of a number of possible processes, most of them involving at one state an autoionizing transition.

1. Vibration-rotational and electronic excitation take place to a level R, lying at an energy above the ionization limit:

$$H_2(^1\Sigma_g{}^+, v = 0, K) + e^- \rightarrow H_2^*(R, v', K') + e^-,$$

where v and K are the vibrational and rotational quantum numbers. This excited state may decay in a short time (with respect to the radiative lifetime) to the electronic ground state, with or without vibration-rotational excitation:

$$H_2^*(R, v', K') \rightarrow H_2{}^+(^2\Sigma_g, v'', K'') + e^-.$$

This type of autoionizing transition includes the special case treated by Russek et al.[329] presented previously. Evidence of this process has been found for H_2 and N_2 by Chupka and Berkowitz.[334,335]

2. A short-lived excited negative ionic state may be created, with or without vibration-rotational excitation:

$$H_2(^1\Sigma_g{}^+, v = 0, K) + e^- \rightarrow H_2{}^{-*}(R', v''', K''').$$

This state may decay through autoionization to the original molecular ground state

$$H_2{}^{-*}(R', v''', K''') \rightarrow H_2(^1\Sigma_g{}^+, v = 0, K) + e^-,$$

which is the pair of processes involved in elastic resonance electron scattering.

A decay transition to an excited electronic state in the two cases given will necessitate the original excitation energy to be in excess of the energy of that state.

Other possibilities of molecular ionization are:

3. The molecule becomes ionized but remains in the ground electronic state:

$$H_2(^1\Sigma_g{}^+,\ v = 0,\ K) + e^- \rightarrow H_2{}^+(^2\Sigma_g{}^+,\ v',\ K') + 2e^-.$$

4. An excited negative molecular state is created, as in 2, which is followed by double-electron emission to the molecular positive ion

$$H_2{}^- \rightarrow H_2{}^+ + 2e^-.$$

There is no proof of the existence of this process.

In the theoretical study of inelastic atomic collisions, we must consider the variation of the natures and energies of the possible temporary atom-ion or atom-atom molecular states during impact. The work by Lichten[336–338] and by Russek [339] treats the topic and provides further references (see also Ref. 340).

In distinction to electron spectrum analysis, measurements of autoionizing level properties can also be performed by mass-separation analysis of the state of ionization of bombarded atom or ion after it autoionizes. Information about the excitation cross section and energy can be obtained by varying the impact energy. Also, lifetimes of the excited states are sometimes determined by time delay measurements between the excitation onset and detection of the subsequent ions.

Table XXI summarizes investigations of ionizating levels as carried out either by autoionization electron spectrum analysis, denoted by c, or by subsequent atom or ion detection techniques denoted by d. (See also the recent thesis of van der Wiel.[340a])

A new phenomenon of induced molecular dissociation in an external electrostatic field has recently been reported by Comes and Wenning.[341a] The effect has been shown to occur in molecular hydrogen and molecular deuterium, and it is made possible through configuration mixing induced by the electrostatic field. By virtue of this mixing, treated more specifically in another article by the same authors[341b], a new dissociation path becomes accessible to the excited molecular state. It is also shown that predissociation of excited molecules can be caused by collision through a similar effect due to molecular fields (van der Waal's fields).

6.3.3.1. EXPERIMENTAL INVESTIGATIONS. Berry[341] observed autoionization electron emission following He^+–He, Ne, Ne^+–He, Ne, Ne–Ne, Ar^+–Ar, Xe–Xe, Xe^+–Xe, Kr–Kr, Kr^+–Kr, ion-atom and atom-atom collisions. Ion energies were between 0.3 and 3.0 keV, and a cylindrical condenser electron energy analyzer giving instrumental linewidth of about 3 eV at 25-eV electron energy was used. With this resolution, however, it was difficult to identify the autoionizing state decay.

In 1964 Rudd[342] reported the observation of some of the autoionizing levels in helium through the measurement of the energy spectrum of electrons emitted after bombardment with protons (75 keV) and hydrogen molecular ions, $H_2{}^+$, (75 keV), at an energy resolution of about 3%. Deceleration of the electrons before analysis provided resolutions of 1 to 1.5%. The

Table XXI.

2) Decay of autoionizing states (experimental).

Target	Bombardment particle (energy)	Decaying autoionizing level	Other ref.
He	H^+ c) (25-250 keV)	$2s^2(^1S)$; $2s2p(^1P)$[342,353]	341) 369)
	H_2^+ c) (25-250 keV)	$2s^2(^1S)$; $2s2p(^3P)$; $2p^2(^1S,^1D)$; $2s2p(^1P)$[342,353] $2sns(^1S)$ (n>1); $2snp(^3P)$ (n>1)[348]	
	e^- c) (4 keV)	349) { $2s^2(^1S)$; $(sp,2n+)(^1P)$ (2<n<6); $3s3p(^1P)$; $(sp,34+)(^1P)$; $2s2p(^1P)$ (E=60.00(15) eV, q= { −1.60(10) / −1.90(10) }) }	
	He^+ c) (1-4 keV)	$2s2p(^1P)$ (ion field: $\tau=2.36(17)\times10^{-15}$ sec. and $2.70(69)\times10^{-15}$ sec.) 343)	
	Ne^+ (1-4 keV)	(in Ne) ref. 343) { $2s2p^6 3s(^1S)$ (E=21.8 eV; $\tau=3.5(1.2)\times10^{-15}$ sec.) $2s2p^6 4s(^1S)$; $2s2p^6 5p(^1P)$ }	
N^{5+} O^{6+}	(see text) d)	362) { $1s2s(^1S)$ $\tau=(1\text{-}8)\times10^{-7}$ sec. (Be^{2+}, B^{3+}, C^{4+} also " " " $\tau=(0.5\text{-}4)\times10^{-7}$ sec. investigated) }	
Li	e^- c)d)	318) 320) { $1s2s2p(^4P)$ (E=57.3(3) eV; $\tau=5.1(1.0)\times10^{-6}$ sec.)	304)
Na		$2p^5 3s3p(^4D)$ (E=31.8(3) eV)	322)
K		$3p^5 4s3d(^4F)$ (E=19.9(3) eV; $\tau=90(20)\times10^{-6}$ sec.)	319)
Rb		$4p^5 5s4d(^4F)$ (E=15.8(3) eV; $\tau=75(20)\times10^{-6}$ sec.)	
Cs		{ $5p^5 6s7s(^4P)$ " "$5d(^4F)$ " "$4f(^4G)$ } (E=12.6(3) eV; $\tau=40(15)\times10^{-6}$ sec.) }	
C^{+3} N^{+4} O^{+5}	(see text) d)	$1s2s2p(^4P_{5/2})$ $\tau=11.3(2)\times10^{-8}$ sec.[362] $1s2s2p(^4P_{5/2})$ $\tau=5.5(2.0)\times10^{-8}$ sec.[360]; $7.0(3)\times10^{-8}$ sec.[362] $1s2s2p(^4P_{5/2})$ $\tau=4(2)\times10^{-8}$ sec.[360]; $4.4(8)\times10^{-8}$ sec.[362]	B^{+2} 362) Be^+ "
He, Ne, Ar, Kr, Xe	e^- c) (3-5 keV)	Ar[351]: { $3s3p^6 nd(^1D)$ (n>2) (q>1) $3s3p^6 np(^1P)$ (n>3) (q<1) }	370)
Ne	H^+, He^+, Ne^+ c) (150-250 keV)	{ Ne^+: $2s2p^5(^{1,3}P)ns,np$; Ne: $2s2p^6(^2S)ns,np$; $2p^4(^3P)3s,3pnp$; $2p^4(^1S)3s(^2S)nd$ } 53) 367)	371) 372)
Ne, Ar	H_2^+ c) (25-250 keV)	Ne[353] { $2s2p^6 ns(^1S)$ (n>2); $2s2p^6 np(^1P)$ (n>2); $2s^2 2p^4 3s^2(^3P)$ }	
Mg, Ca, Sr, Ba	e^- d) (0.5-2 keV)	ref.[373,375]	
Ar	H^+, Ar^+ c) (100-300 keV)	Lines for ∿5 to 20 eV and ∿120 to 220 eV[355,359]	372) 374)
Cu, Au	Ar^+, Ne^+ c) (60,80,90 keV)	ref.[366]	

Table XXI. (cont.)

O_2	H^+, He^+ c) (100 keV)	(in O)[324] $2p^3(^2P)ns''(^{1,3}P), nd''(^{1,3}P, ^{1,3}D, ^{1,3}F)$; $2p^3(^2D)ns'(^3D), np'(^3P, ^3F), nd'(^3S, ^3D, ^3G)$	264) 266)
H_2	$h\nu$ c) (He res.)	ref.[334]	
H_2, N_2	"	ref.[335]	

c): autoionization electron spectrum measurement techniques.
d): atom or ion detection techniques.

electrons were detected in a direction of 160° from the ion beam direction.

There was found a striking difference between the electron spectra following bombardment by the two projectiles: the spectrum for H_2^+ bombardment showed that the $2s2p(^3P)$ state had been excited, whereas excitation of this state had not taken place in the case of H^+ bombardment. The difference was attributed to a transition from the ground state to the triplet state—a change in spin of one unit—which can occur only if an orbital electron can be interchanged with one from the beam projectile; no such change can occur when the bombarding particle is a proton (see also Section 6.3.3).

Furthermore, the relative probability of excitation of the nearly coinciding $2p^2(^1D)$ and $2s2p(^1P)$ states appears to alter strongly for the two methods of excitation, and excitation of the $2p^2(^1S)$ was also caused for H_2^+ bombardment. The peak-to-background ratios for four of the most prominent transitions in the H_2^+ spectrum were given as a function of ion beam energy.

In 1970 Crooks and Rudd[342a] reported experimental evidence for charge transfer to continuum states in H^+–He bombardment at 100 to 300 keV. A prominent electron peak was observed in the forward direction with an ejection velocity equal to the velocity of the incident proton.

The electron energy spectra following He^+–He and Ne^+–He collisions have been investigated by Barker and Berry.[343] A maximum ion energy of 4.3 keV was used. Electrons were accepted at 90° to the beam direction, and a 90° electrostatic energy analyzer with an energy resolution of 1.33% was employed.

In the former case, He^+–He, the collision times for the ion energies used—and therefore the duration of existence of the possible intermediate states of the He_2^+ molecular ion—will be on the order of the lifetime of these states if they are autoionizing. An energy analysis of the possible He_2^+ molecular states,[343,336] shows that the energies of these states vary drastically as a function of the He^+–He nuclear separation (see also Ref. 337). Hence if the lifetimes of the autoionizing states are comparable to the collision times, the emitted electrons for a given transition may be continuously distributed over an extended energy range. The He^+–He case is in opposition to that in which collision times are much less than the lifetimes of the autoionizing levels, as in Rudd's experiments (previously cited), where electron peaks appear at the energies expected for the isolated atom.

In view of the preceding considerations, the position and shape of the autoionizing peak for the $2s2p$ state in helium will vary with the energy of the He^+ ion beam. An analysis of the line shift with ion energy gave a mean lifetime value of $2.70(69) \times 10^{-15}$ sec. An analysis of the variation of the half-width of the autoionization line with ion energy gave a value of $2.36(17) \times 10^{-15}$ sec. These values are to be compared with the experimental value by Madden and Codling[260,238] of $1.72(18) \times 10^{-14}$ (see Table XX). Burke, McVicar, and Smith[240,344,345] have calculated the mean lifetime of this state to be 1.6×10^{-14} sec. The difference between the mean life obtained here and those obtained by other measurements may result from shortcomings in the assumptions made about the nature of the ion field at the autoionizing atom and the effects on the lifetime of the state. Newton et al.[346,347] have shown that, in the case of the rare gas ions in autoionizing excited states (especially Ar^{++} ions), the decay probability for the state is greatly enhanced when the ion is near a metal surface. A comparable effect of the ion field on the autoionization probability of the $2s2p$ state in helium may also exist.

Enhanced ionization of the ion He^{+*} when the electron was in a very highly excited bound level was also observed in the vicinity of a metallic surface.

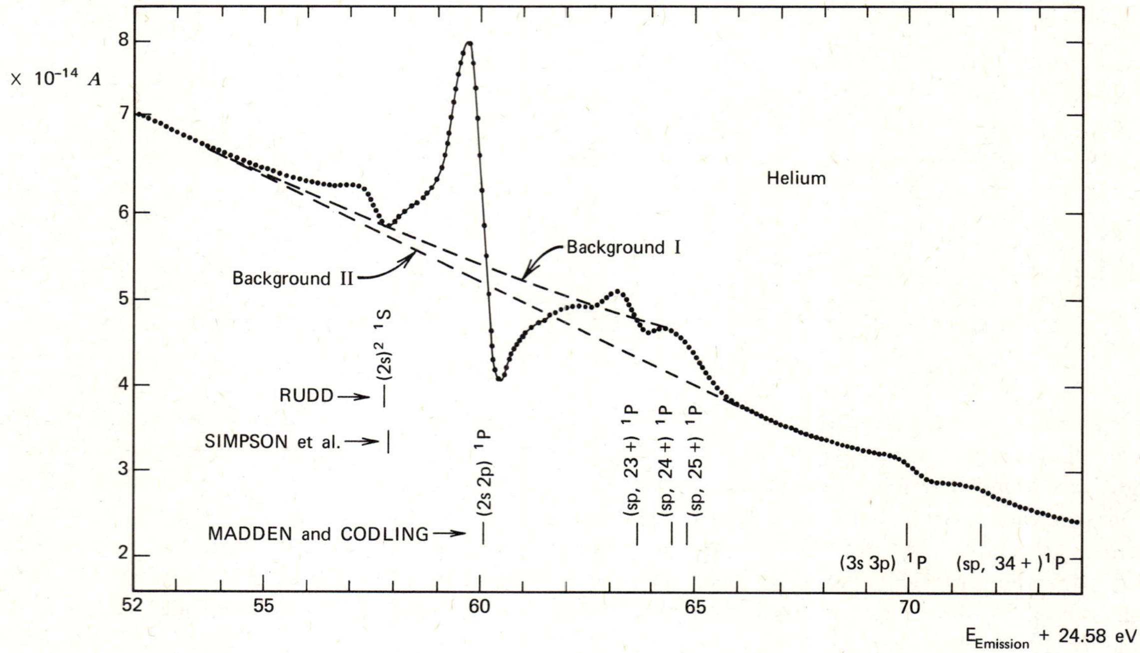

Figure 24. The autoionization electron spectrum for helium as measured by Mehlhorn.[349] The positions of the spectral components are compared with measurements by other methods (see text). The Fano shape of the $2s2p(^1P)$ state decay electron line is obvious.

However, autoionization is manifestly impossible in a one-electron atomic system.[347]

In the Ne^+–He impact measurements[343] a number of autoionization peaks were observed for neon. Several of these were identified by the comparison of their energies with results of measurements by other workers. The lifetimes of two of the autoionizing levels were determined as previously described, but only one level was identified (see Table XXI).

Additional results on the autoionizing levels in helium by H_2^+ impact were reported soon afterward by Rudd[348]: the $2sns(^1S)$ and $2snp(^3P)$, each with $n = 2$ through 5. Agreement with theoretical results was good.

Mehlhorn[349] has reported the investigation of the autoionization spectrum of helium, where excitation of the autoionizing levels was caused by electron impact (4-keV bombarding energy). The electron spectrum was analyzed using a cylindrical electrostatic spectrometer. In the obtained spectrum (Figure 24), the $2s2p(^1P) \rightarrow 1sEp(^1P)$ transition was the most intense. Also inserted are the excitation energy values for the various states as measured by Madden and Codling[238,260] (photoabsorption), Simpson et al.[286] (inelastic electron scattering), and Rudd[342] (proton, ion bombardment). Here for the first time the asymmetric nature of autoionization electron emission lines was demonstrated. The experimental values for q and Γ are presented in Table XXI; the two values for q are derived assuming two different background intensities. There is also evidence of the decay of other levels, as seen in Figure 24. The measured energies of these transitions agree well with values taken from other measurements. It can be seen that autoionization occurs the most strongly where the two electrons are in orbitals close to each other and lie near the nucleus (see also Ref. 350).

The investigation of autoionizing levels in helium has been repeated using a spherical condenser electrostatic spectrometer at a much higher resolution setting.[351] Impact of 4-keV electrons caused the initial excitation and the energy region between about 59 and 67 eV in Figure 24 was studied. The spectral electrons were analyzed in a direction of 90° to the incident electron beam. Although the structure in the 63 to 65 eV region was more clearly resolved, it remained rather complex.

An extensive investigation of the metastable autoionizing levels in the alkali metals has been performed by Feldman and Novick.[318,320] These levels (listed in Table XXI) are situated between the first and second ionization limits for the respective atoms. The excitation of the metastable states was caused by electron bombardment, and the energies, excitation cross sections, and natural lifetimes of the states were measured. The lifetimes of the states all lay in the 10^{-5} sec region and corresponded to spin-spin interaction decay.

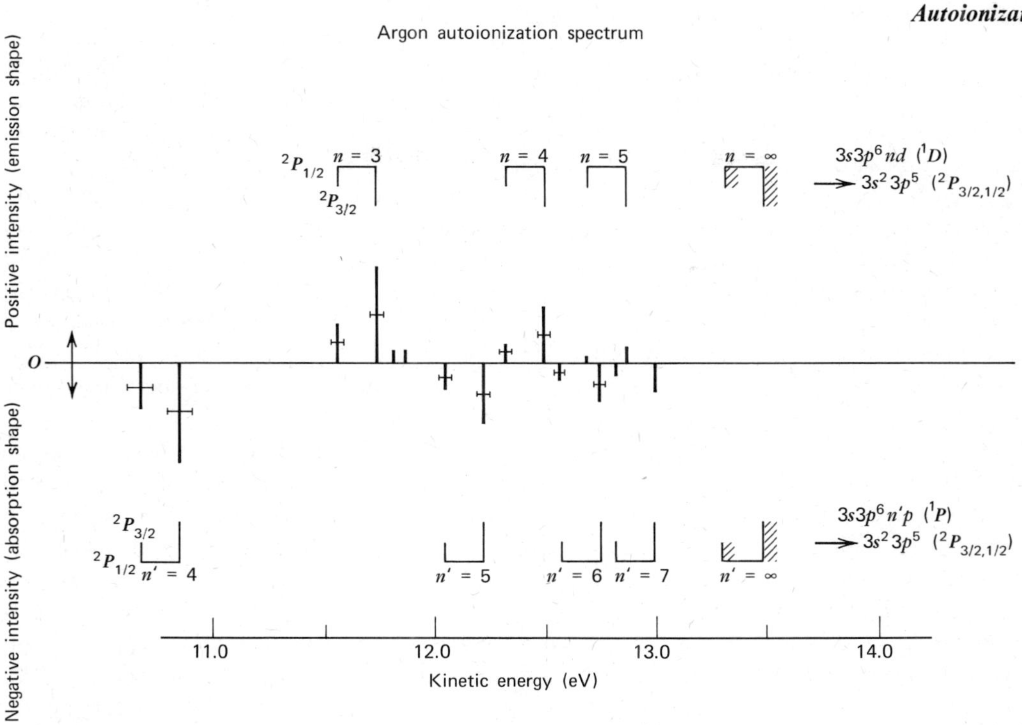

Figure 25. A schematic representation of the main components of the electron spectrum from the decay of autoionizing levels in argon. Note that one of the two series of components is "positive" in intensity ($|q|$ large) while the other series of "negative" intensity (i.e. an absorption profile in essence; $|q|$ small). Also shown as cross-bars are the approximate halfwidths of the stronger components. (See text.) (Data from Ref. 351.)

Subsequently, the energies and lifetimes of the metastable autoionizing $1s2s2p(^4P_J)$ states of ^{6}Li and ^{7}Li were investigated rigorously by Feldman et al.[322] The Zeeman quenching curve versus applied magnetic field showed anticrossings between sublevels of same magnetic quantum number but different values of J. These anticrossings occurred at different field values for the two isotopes, and the differences in the quenching were clearly related to the respective spin interactions of ^{6}Li and ^{7}Li. More information is contained in the authors' compilation and comparison of experimental and theoretical data on $1s2s2p(^4P)$ lifetimes for He^-, Li, Be^+, and other $Z^{+(Z-3)}$ elements up to O^{+5}.

The measured metastable state lifetime of lithium (see Table XXI) agrees well with theoretical calculations of its value by Pietenpol,[354] 5×10^{-6} sec, and by Manson[356] (see Ref. 318), 5.88×10^{-6} sec.

This metastable state in lithium, the $1s2s2p(^4P_{5/2})$ state, is analogous to the $1s2s2p(^4P_{5/2})$ state in He^-. The excitation of the latter is found in evidence in electron resonance scattering measurements on helium. For comparison, the lifetime of the metastable state in He^- has been measured to be $18.2(2.7) \times 10^{-6}$ sec by Nicolas et al.,[357] or three to four times longer than that of the lithium metastable state. Recent calculations of the lifetime of this He^- level by Laughlin and Stewart[358] give 5.5×10^{-4} sec, in strong disagreement with the previously given experimental value.

In addition, the analogous $1s2s2p(^4P_{5/2})$ states of N^{+4} and O^{+5} have been investigated by Dmitriev et al.[360] Beams of nitrogen and oxygen ions were passed through thick celluloid films to obtain states of high ionization. After the ion species had been separated, the one of interest was allowed to collide with gas atoms and the charge spectrum was then analyzed. By varying the collision (charge exchange) chamber-to-analyzer distance, the lifetimes could be determined. The lifetimes of these states were measured and were found to be shorter than that for lithium—$5.5(2.0) \times 10^{-8}$ and $4(2) \times 10^{-8}$ sec, respectively. These values compare well with those calculated by Manson[356] of 5.8×10^{-8} and 3.1×10^{-8} sec for N^{+4} and O^{+5}, respectively.

Dmitriev et al.[362] have employed the same method to investigate the lifetimes of the helium and lithium-like ions of beryllium, boron, carbon, nitrogen and oxygen; the values for the N^{+4} and O^{+5} are more accurately determined here than in Ref. 360. The helium-like ions were assumed to be in $1s2s(^1S_0)$ and $1s2s(^3S_1)$ states, the former state's lifetime being approximately determined. The C^{3+}, N^{4+}, and O^{5+} ions had lifetimes corresponding to that expected for the $1s2s2p(^4P_{5/2})$ level. The Be^+ ion showed a shorter lifetime, attributed to $1s2s2p(^4P_{3/2,1/2})$ configurations, and the B^{2+} ions had lifetimes indicating the mixing of two states having two different lifetimes.

A theoretical investigation of the electron impact excitation cross section of only the 2*s* electron in N^{+4} has been presented by Burke et al.[363]

The relative population of the $3s3p4s(^4P^o_{1/2,3/2})$ short-lived and $3s3p4s(^4P^o_{5/2})$ long-lived autoionizing levels excited in aluminum by electric discharge has been studied by Kiselevskii and Grankova.[361]

Bely[365] has calculated the energies and reduced energy-dependent ionization cross sections (i.e., normalized to the hydrogen ionization probability) for sodium and sodium-like ions. Only the decay of the autoionizing states $2s^22p^53snl$ $(n > 2)$ and $2s2p^63snl$ were considered. It was seen that the effect of autoionization compared to usual ionization for these ions increased strongly for increasing Z-values, that for sodium being very small.

The electron decay of autoionizing levels excited in neon by H^+ (150-keV), He^+ (150-keV), and Ne^+ (200-keV) bombardment has been investigated by Edwards and Rudd.[367,53]

Removal of an electron from the neon (^1S) ground state by H^+ bombardment will create a doublet autoionizing state in singly ionized Ne^+, whereas removal by He^+ or Ne^+ bombardment may give rise to a quartet state through electron exchange. On this basis, doublet autoionizing states in singly ionized atoms will arise in all three spectra, and quartets will appear only through He^+ and Ne^+ bombardment.

The first three members of each of the $2s2p^5(^3P)np$-$(^2D, ^4P) \rightarrow 2p^4(^3P)$ emission series, were found for the three projectiles; the quartet series was not seen in H^+ bombardment excitation. The three members were situated below the series limit energy of about 25.5 eV. Superimposed upon these series were the $2s2p^5(^1P)$-$np(^2?)$ and $2s2p^5(^1P)ns(^2P) \rightarrow 2s^22p^4(^1D)$ series.

Strong lines due to transitions of the types $2s2p^6(^2S)$-$np(^1P)$ and $2s2p^6(^2S)ns(^1S) \rightarrow 2p^5(^2P)$ were seen in the H^+ bombardment spectrum, whereas in the He^+ spectrum the transitions $2s2p^6(^2S)ns(^3S)$ and $2s2p^6$-$(^2S)np(^3P) \rightarrow 2p^5(^2P)$ appeared to have dominating strength.

Lines due to transitions of energies between about 28.5 and 31.5 eV, from initial $2s^22p^4(^3P)3p(^2S)np(^?P)$ (in He^+ and Ne^+ bombardment) and $2s^22p^4(^3P)3p$-$(^4D)np(^3?)$ (Ne^+ bombardment) states, were also visible in the spectra. The Ne^+ bombardment spectrum also revealed $2s^22p^4(^1S)3s(^2S)nd(^?D) \rightarrow 2p^5(^2P)$ lines in the 32.5 to 34 eV energy range.

The spectra could be satisfactorily analyzed for almost all autoionizing levels on the basis of L–S coupling, indicating that divergences from coupling in this limit are very small.

It was noted that Ne^+ impact formed two electron excitations more strongly than did He^+ or H^+. Inner shell excitations increased in probability with decreasing mass of the bombarding particle.

The Uppsala group[351] has investigated the autoionizing electron spectra for neon, argon, krypton, and xenon. The uncertainty of the absolute energy was about 1 eV, and no energy calibration was performed.

The neon spectrum showed doublet structure corresponding to transitions to final $2s^22p^5(^2P_{3/2,1/2})$ doublet states, and all the spectral components were of positive intensity.

The spectrum for argon, however, was found to consist of spectral components of both positive and negative intensity (see Figure 25 for a schematic representation of the spectrum). The doublet structure obviously resulted because the $3s^23p^5(^2P_{3/2,1/2})$ states were involved in the transitions. If calibration is effected by assuming that the most intense positive couplet is due to the autoionizing decay of the $3s3p^63d(^1D)$ state, the other positive couplets correspond in energy to transitions of the types $3s3p^6nd(^1D)$ $(n > 3) \rightarrow 3s^23p^5$-$(^2P_{3/2,1/2})$. With this energy calibration, the negative intensity lines correspond in energy to transitions of the type $3s3p^6np(^1P)$ $(n > 3) \rightarrow 3s^23p^5(^2P_{3/2,1/2})$ according to data in Ref. 238, where absorption excitation to those states appeared to resemble an emission series. There is in fact a similarity between Figure 25 and the electron energy loss spectrum for argon in Ref. 289, where certain of the series are of positive intensity and others are of negative intensity. However, there was no indication[351] of the decay of the $3s3p^64s$-(^1S) state, which is excited by low energy electron bombardment. Note that there is evidence of additional structure near the first positive couplet; too, the positive peaks in Figure 25 are narrower than the negative peaks.

Also seen in the argon spectrum were fairly strong unidentified emission lines of comparable strengths, lying at about 6.5 eV and 8.0 eV. The former is the same as that seen by Rudd and Lang[353] in their electron emission spectrum for both H^+ and $H_2{}^+$ bombardment. However, they saw no line of similar intensity at 8 eV. Ogurtsov et al.[352] have interpreted the line lying at 6.25 eV as being due to a $3s3p^5$-$(^3P)4s(^4P)4p(^3P) \rightarrow 3p^4(^3P)4p(^2P) + e^-$ transition in the doubly excited atom. Other transitions in the same region are also due to transition from initially doubly excited atoms to final singly excited ion states.

Rudd and Lang[353] measured the electron emission energy region for neon from 19 to 28 eV for $H_2{}^+$ bombardment excitation. Two emission series were identified.

Rudd et al. investigated the electron energy spectra

from H^+–Ar[355] and Ar^+–Ar[355,359a] collisions, both projectiles having energies between 100 and 300 keV. Both spectra showed two sets of line structure: one at about 120 to 220 eV, the other below 20 eV.

At about 206 eV the H^+–Ar (200-keV H^+) electron spectra showed a number of $L_{2,3}M_{2,3}M_{2,3}$ Auger transition lines resulting from internal shell ionization during impact.

The study of $L_{2,3}MM$-Auger electrons emitted in H^+–Ar and He^+–Ar bombardment experiments was followed up later by Volz and Rudd.[359a] The reader is also referred to Ref. 359b for similar studies in Ne^+–Ne and Ar^+–Ar collision experiments.

The electron spectra for Ar^+–Ar revealed a large number of peaks lying in the energy range of 5 to 20 eV. The spectrum was actually composed of two identical sets of peaks, one originating in the excited target atoms and the other, Doppler shifted with respect to the first set, originating in excited projectile atoms. The strong 6.5 eV line was also visible in the spectrum.

In the spectrum at 100 to 200 eV some of the lines showed the same Doppler shift relationship to the rest of the lines, but when the bombarding energy was varied, one set of lines shifted and the other remained stationary. The multicomponent nature of this spectra did not support the purely statistical model of electron ejection (see Refs. 355, 364, 365). The inadequacy of this model has also been pointed out by Snoek et al.[366] However, the presence of the Auger peaks at about 200 eV and the low energy autoionization peaks confirm the Fano–Lichten model,[368] in which the main energy dissipation mechanism for ion-atom collisions is via the formation of an excited molecular ion, which decays by autoionization and/or Auger processes.

An interesting and pertinent summary article on electron spectra in light atoms, primarily in association with ion impact work, has been presented by Rudd.[323] Also, the work of Ogurtsov et al.[352] reports the measurement of the autoionization electron spectrum between 5 and 550 eV as obtained from H^+–Ar, Ne^+–Ar, Ar^+–Ne, and Ar^+–Ar collision experiments.

6.3.4. Elastic Electron Scattering

Negative ionic states of atoms or molecules can be created in the form of compound states through the capture of an impact electron. After a short time, the negative ion state may autoionize to the original pre-impact ground state, whereupon the electron is emitted with an energy identical to the original impact energy.

For electron bombardment at energies below the inelastic scattering threshold, elastic scattering can be described in terms of (*a*) phase shifts associated with impinging electrons of angular momentum $l\hbar$ and (*b*) interference of different partial waves. The angular-dependent scattering cross section can be calculated and compared with experimental determinations of the same quantities.

Ramsauer and Kollath[376] have investigated the angular and impact energy dependence of scattered low energy electrons passing through atomic and molecular gases. Postscattering energy selection was not employed, and no distinction between elastic and inelastic scattering could be made. For the lower energies, however, the measurements of these workers are in fact of elastic electron scattering in the gases investigated. Hoeper et al.[377] performed theoretical calculations of the angular scattering distributions and total scattering cross sections for helium, neon, and argon and found very good agreement with the experimental results of Ramsauer and Kollath for energies from 1.0 eV up to 0.6 eV below the inelastic scattering threshhold. The computed data on the energy dependence of phase shifts for helium and neon agreed well with the phase shifts determined from experimentally determined resonance parameters by Andrick and Ehrhardt.[378]

Resonance scattering, on the other hand, results from the formation of a definite state of the short-lived negative ion. When the negative ion decays to the original ground state of the target atom, the resonance scattering is elastic; it is inelastic if the decay is to an excited state, of either electronic or vibration-rotational excitation, in the scattering atom or molecule (see Section 6.3.3).

The profile of an elastic resonance will have the Fano shape (equation 26). The quantity q for a given resonance can vary strongly as a function of scattering angle. For example, the $|q|$ value for the elastic resonance scattering profile for helium at 19.3 eV varies from >1 for $\sim 0°$ scattering[379,381] (see Figure 26*a*) to unity at $\sim 10°$, and to ~ 0 at 90°[378] (Figure 26*b*; see also Ref. 381), and the scattering amplitude for a given resonance changes drastically with scattering angle. In this connection, Andrick and Ehrhardt[378] found that the elastic resonance doublet at ~ 16 eV in neon, exciting probably the $2s^2 2p^5 3s^2 (^2P_{3/2,1/2})$ states in Ne^-, gives a positive electron resonance profile at 8°, and this profile decreases in amplitude with increasing scattering angle to zero at 90°. Kuyatt et al.,[379] however, noted a negative intensity for this scattering profile for 0° scattering angle. Again, for the ~ 11-eV resonance doublet in argon, involving probably the $3s^2 3p^5 4s^2 (^2P_{3/2,1/2})$ autoionizing state in Ar^-,[378,379] we see that $q \sim -1$ for 0°, $|q| > 1$ for 10°, $q \sim +1$ at 60°, and that the scattered electron intensity decreases rapidly all the while to near zero at 75°.[378]

It is usually difficult to identify the configuration of

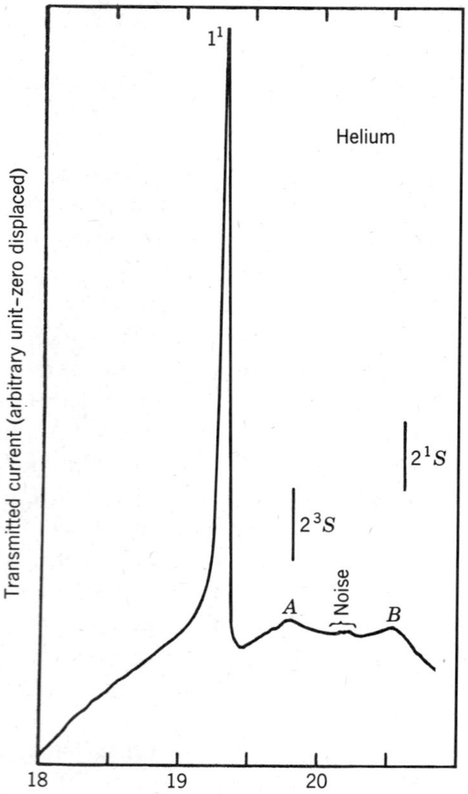

(*a*)

Figure 26a. The energy dependent electron transmission current through helium for near 0° scattering measured by Kuyatt et al.[379] The line has a Fano shape with $|q| > 1$. The structures *A* and *B* arise from single-electron excitation in helium (see Figure 23).

the excited autoionizing level involved in electron resonance scattering. A systematic procedure for the identification of the most likely configuration candidates for a given negative ion state has been presented by Fano and Cooper.[382]

In general, however, a given He^- level (e.g., He^- $1s2sns(^2S)$) must lie below the limit helium core configuration—the $1s2s(^1S)$ level in this case. An estimate of the energy of an He^- level[379] with respect to the helium ground state level can be roughly estimated by first knowing the energy of the *doubly excited* state associated with the He^- level. The two excited electrons in helium are then moving in a $Z \cong 2$ field. By taking the energy difference between the doubly excited helium state and the He^{++} limit and dividing it by $Z^2 = 4$, the energy of the He^- state relative to the He^+ ionization limit is obtained. The $1s2s^2(^2S)$ state in He^- will have the energy

$$[\text{He}^+ \text{ lim } (24.6 \text{ eV})] - \tfrac{1}{4}\{\text{He}^{++} \text{ lim } (79 \text{ eV}) - E[2s^2(^1S)(57.8 \text{ eV})]\} = 19.35 \text{ eV}$$

which is very close to the experimental value. In a theoretical treatment of this level by Kwok and Mandl,[383] the results are in favorable agreement with experiment.

Resonance scattering measurements on atomic hydrogen[384,385,387] reveal an elastic scattering resonance lying at about 9.70 eV. This state probably has a $2s^2(^3P)$ configuration, calculated[267] to lie at 9.78 eV. At 0° scattering, the resonance is a positive intensity structure, whereas for larger angles it becomes of negative intensity. The $2s^2(^1S)$ state is calculated[248] to lie at 9.556 eV and measured[287] to lie at 9.56 eV.

Experiments also reveal elastic scattering states for the other rare gases. Besides the Ne^- and Ar^- resonance states mentioned, resonance structure for krypton and xenon[381,379] and for mercury[379] have been seen. Many of these resonance profiles resemble Fano shapes with $|q| \approx 1$; the energies of the resonances in these cases and in those listed in Table XXII were taken as the mean between the intensity maximum and minimum positions.

Position and widths of autoionizing levels in neutral magnesium and calcium have been calculated by Burke and Moores.[380] They applied the close-coupling approximation to the problem of electron scattering by Mg^+ and Ca^+ ions. The results in the latter case were compared with the experiment.

The three lowest lying resonances in mercury have been discussed by Fano and Cooper,[382] who showed that the excitation of the $6s6p^2(^4P_{1/2,3/2,5/2})$ configurations was responsible (see Table XXII).

Elastic electron scattering resonances have also been seen for molecules. An increase in electron transmission (0° scattering) has been found at 11.48(5) eV for molecular nitrogen.[389] The structure showed evidence of excitation of vibrational levels in the final N_2 molecule.

Kuyatt et al.[390] have measured resonance transmission (0° scattering angle) with vibrational structure for H_2, HD, and D_2 at energies between 11 and 13 eV. There was indication of two resonance structure series in both H_2^- and HD^-, whereas only one was observed for D_2^-. The case of H_2 is one of the excitation possibilities cited by McGowan et al.[326] (see also Section 6.3.3), and it decays by autoionization to the neutral molecular ground state.

Scattering of electrons from gaseous molecules usually leaves the molecule in a state of vibrational excitation, with the scattered electron suffering an energy loss. However, Heideman et al.[391] demonstrated that this is not the case in elastic resonance scattering at 0° in N_2 for energies of 1.8 to 3.5 eV. The transition cross section over this region gives evidence of the

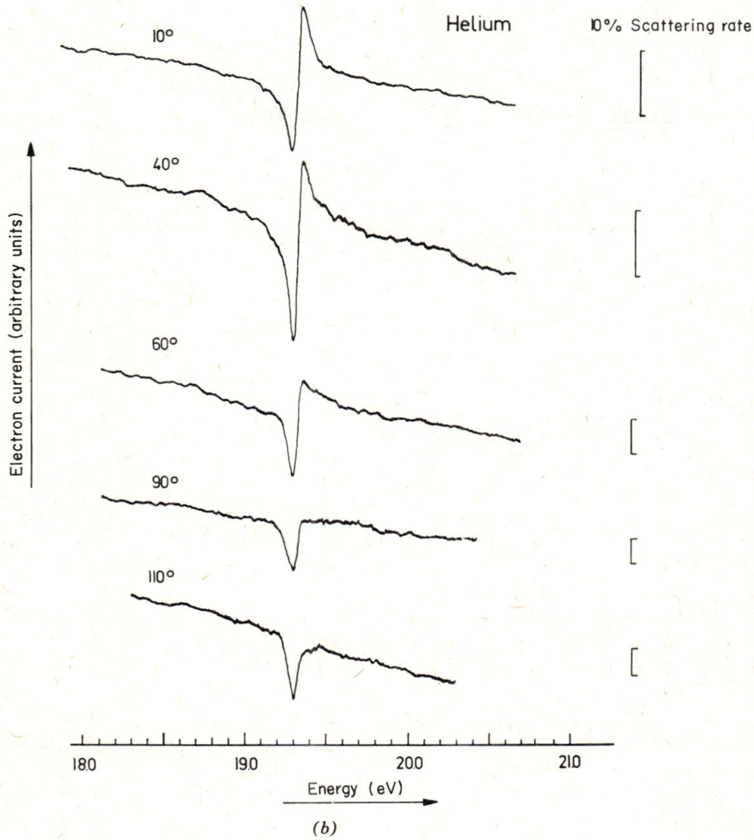

Figure 26b. The angular dependence of elastic resonance scattering of electrons by helium measured by Andrick and Ehrhardt.[378] The Fano shape varies strongly with scattering angle.

excitation of consecutive vibrational states. Since no energy loss is suffered by the scattered electron, excited N_2^- vibrational states appear to decay directly to the N_2 ground state.

Andrick and Ehrhardt[378] and Ehrhardt and Willmann[392] investigated the angular dependence of low energy elastic scattering by N_2 via vibrational states in N_2^-, also in the 1.8 to 3.5 eV energy region. They saw that the elastic scattering amplitude decreased with increasing scattering angle more rapidly for the higher vibrational states than for the lower. Similar measurements were also made for the gaseous molecule nitric oxide.

7. THE ANGULAR DISTRIBUTION OF AUGER ELECTRONS

An angular dependence of Auger electron emission is expected. The possibility has been referred to a number of times in the literature, but not until recently has it been investigated, either theoretically or experimentally.

Mehlhorn[397] has considered the polarization of X-radiation emitted in filling an inner atomic shell vacancy. He showed that, in the ionization of an atomic electron ($n, l > 0; j > \frac{1}{2}$) by electron or proton impact, there will be an alignment of the ionized atom with respect to the ionizing particle beam direction. This alignment will then give rise to a nonisotropic angular distribution and polarization of following emitted X-radiation (e.g., L_3-radiation). Therefore Auger electrons should also be emitted in a nonisotropic angular distribution.

The preliminary results of a recent investigation by Rudd[398] of the angular distribution of *L*-Auger electrons from argon show that most of the lines are nearly isotropically distributed in angle. Any deviations seem to be small.

Cleff and Mehlhorn[399] are also in the process of making Auger angular distribution measurements. See ref. 399a for recent results from this investigation.

8. FLUORESCENCE YIELDS

The filling of an initial atomic vacancy may result in the radiative emission of an X-ray in direct competition with allowed nonradiative transitions. However, since the experimental investigation of fluorescence yields is of greater interest to workers in the

Table XXII.

3) Elastic electron scattering (experimental)

Target	Scatt. angle	Scattering resonances (negative ion states)	Other ref.
H	70°–160°	$2s^2(^1S)$ (9.56(1) eV; Γ=40 meV)[387]	393)
	0° 70°,90°,94°	$2s^2(^3P)$ { (9.70(15) eV)[384] (9.73(12) eV[385]; Γ=10 meV[387]) }	394) 395) 314) 388)
H_2	5°–110°	(in H_2^-): $^2\Sigma_u^+$ [316]	
He	0°	[379] { $1s2s^2(^2S)$ (E=19.31(3) eV) $1s2s2p(^2P)$ (E≈19.45 eV) (others) }	394) 385) 386) 314)
	72°	$1s2s^2(^2S)$ (E=19.30(5) eV)[381]	
	10°–110°	$1s2s^2(^2S)$[378]	
Ne	0°	$2s^22p^53s^2(^2P_{3/2,1/2})$ (E=16.04(2) eV, 16.135(20) eV, resp.)[379]	314)
	8°–90°	" " " " " 378)	
Ar	0°	$3s^23p^54s^2(^2P_{3/2,1/2})$ (E≈11.09(5) eV, 11.25(5) eV, resp.)[379]	314)
	10°–75°	" " " " " 378)	
Kr	0°	$4s^24p^55s^2(^2P_{3/2,5/2})$ E≈9.465(10) eV, 10.10(1) eV, resp.)[379]	
Xe	0°	$5s^25p^56s^2(^2P_{3/2,1/2})$ (E≈7.755(11) eV, 9.02(4) eV, resp.)[379] (others)	
Hg	0°	$5d^{10}6s6p^2(^4P_{1/2,3/2,5/2})$ (E≈4.07(10) eV, 4.30(10) eV, and 4.89(10) eV resp.)[379] (others)	
H_2,HD,D_2	0°	(E = 11–13 eV)[390]	392) 314)
CO_2	∼0°	CO_2^-: $^2\Pi$ (E > 1.5 eV)[396]	392) 314)
N_2	10°–110°	N_2^-: $^2\Pi_g$ (v=1 to 8) (E=1.89 to 3.58 eV)[392]	314) 396)
CO	5°–110°	CO^-: $^2\Pi$ (E > 1 eV)[316]	
NO	10°–110°	ref.[392]	396)
O_2	∼0°	ref.[396]	392)

field of X-ray physics than to those interested in electron spectrometry, an extensive treatment of the subject is outside the scope of this book. The objective of the limited treatment presented here is to give an understanding of the relation of fluorescence yields to the Auger and Coster–Kronig effects and the part it plays in atomic de-excitation. Any results of investigations in the field of X-ray physics are referred to in summarized form, without the details of experimental procedures.

For more complete and specific treatments of fluorescence yields, the reader is referred to the articles by Listengarten[10] and Fink et al.[400]

8.1. The *K*-Shell Fluorescence Yields

Wentzel[56] showed that the total *K*-Auger transition probability W_{aK}, is almost independent of the atomic number, especially for intermediate and large *Z*-values where the inner atomic shells are complete. Later, nonrelativistic treatments confirmed this result, although it was found that relativistic effects greatly increased this probability (see Section 4.1.2).

The X-radiation emission probability ω_K, called the *K*-fluorescence yield, is defined by the ratio

$$\omega_K = 1 - a_K = \frac{W_{XK}}{W_{XK} + W_{aK}}, \tag{38}$$

where W_{aK} is the *K*-Auger yield and W_{XK} is the *K*-series X-radiation probability. Since W_{XK} is proportional to Z^4, because X-radiation is electric dipole, we may write

$$\omega_K = \frac{Z^4}{Z^4 + b} \quad \text{or} \quad \left(\frac{\omega_K}{1 - \omega_K}\right)^{1/4} = BZ, \tag{39}$$

where b and B are constants. It was pointed out later (e.g., Boyles et al.[151]) that this expression did not fit experimental data well, and that screening and relativistic effects should be taken into consideration. This was done in 1955 by Burhop,[36] who obtained the semiempirical expression

$$\left(\frac{\omega_K}{1 - \omega_K}\right)^{1/4} = B\left(Z + \frac{A}{B}\right) + CZ^3 = A + BZ + CZ^3. \tag{40}$$

The constants A, B, and C are to be determined by a least-square fit of equation 40 to experimentally determined values for ω_K. The screening term A/B is of course negative. For low *Z*-values, $5 \leq Z \leq 18$, the variation of ω_K with Z is best predicted[400,401] by equation 39. For $Z < 5$, ω_K takes values well below those predicted by theory.[400,402]

Values of the constants A, B, and C have been determined by a number of authors since 1955, adjusting to available experimental determinations of the *K*-fluorescence yield. In 1960 Listengarten[10] compared the available experimental ω_K values measured to that time with values obtained from equation 40, using the empirical values for A, B, and C calculated by Burhop,[36] Laberrique-Frolow and Radvanyi,[148,403] and Wapstra et al.[404] These three sets of constants are found in Table XXIII.

Rubenstein and Snyder[64] carried out theoretical calculations of ω_K for argon ($Z = 18$), krypton ($Z = 36$), and silver ($Z = 47$); the latter two agreed well with the measurements made later by Heintze,[405] although agreement with earlier measurements was poor (see Ref. 405).

Listengarten's[66] relativistic treatment of the Auger effect in terbium ($Z = 65$), thallium ($Z = 81$), and uranium ($Z = 92$) included the calculation of the fluorescence yields for these elements (see Section 4.1.2). Expressing the *K* X-ray and *K*-Auger transition probabilities in terms of the radiative and total widths of the *K*-atomic level, $\omega_K = \Gamma_{X_K}/\Gamma_{T_K} = \Gamma_{X_K}/(\Gamma_{X_K} + \Gamma_{A_K})$, Listengarten used his theoretical values for a partial *K*-level width $\Gamma_{A_K}(KLL)$. The radiative width is Γ_{X_K} and Γ_{T_K} is the total width of the *K*-level. The quantity $\Gamma_{A_K}(KLL + KLX + KLY)$ was obtained by use of the experimental values of $[I(KLX) + I(KXY)]/[I(KLL)]$ (see Section 4.3). The value thus calculated for W_K for $Z = 65$, 81, and 92 compared better with those tabulated by Wapstra et al.[404] than with those of Laberrique-Frolow and Radvanji.[148,403] The values of W_K for elements lying between these atomic numbers were determined by interpolation between the three. It is found that experimental values obtained for heavy elements agree well with the *K*-fluorescence yields obtained from the relativistic calculations by Listengarten.[66]

Callan[406,140] calculated the *K*-level widths of atoms having atomic numbers in the range $16 \leq Z \leq 83$. The empirical equation $\Gamma_R = \sum_n A_n Z^4$ was used to compute the radiative widths, where A_n are empirical coefficients

Table XXIII.

Values of the constants in equation (41) calculated by various authors for n = 1/4 and n = 1/3.

Authors	Constants $A \times 10^2$	$B \times 10^2$	$C \times 10^6$	n
Burhop[36]	-4.4	3.46	-1.35	1/4
Laberrique-Frolow and Radvanyi[148,403]	-2.17	3.318	-1.14	1/4
Wapstra et al.[404]	-6.4	3.40	-1.03	1/4
Bailey and Swedlund[407]	-10.19	3.377	1.178	1/3

for the $1s$-np dipole transitions. For partially filled atomic shells, the full shell coefficient A_n was multiplied by the fraction to which the shell is filled. This approximation accounts adequately for the fractional contribution to the widths due to an unfilled shell. The total Auger widths were determined from KLL-widths, which were corrected for KXY/KLL ratios and for relativistic effects among the heavy elements.

The results of Callan[140] agree well with those of Listengarten[66] in the high Z-value region; they lay slightly higher than those of Wapstra et al.[404] and Burhop[36] in the intermediate Z-value region. Although the K-fluorescence yields are relatively insensitive to the detailed partial and total widths, there is enough sensitivity to distinguish between the empirical versions based on a single aZ^4 radiative width[404,36] and Callan's $\sum_n A_n Z^4$ widths.[140]

Fink et al.[400] have collected the experimentally determined K-fluorescence yield values in table form. Bailey and Swedlund,[407] in a report of their measurements of the K X-ray fluorescence yields of nine elements from $Z = 13$ to $Z = 27$, and $\bar{\omega}_L$ for $Z = 39$ and $Z = 47$, have found (including their values complemented by those listed by Fink et al.), using an equation of the form

$$\left[\frac{\omega_K}{1 - \omega_K}\right]^n = A + BZ + CZ^3 \tag{41}$$

and taking $n = \frac{1}{4}$ and $\frac{1}{3}$ the best fit to experimental data to be given by $n = \frac{1}{3}$ and the constants A, B, C given in the last row of Table XXIII. The standard deviation from the experimental values was 2.4%, and the largest deviation was ~5.3% at $Z = 30$. The screening factor is then $A/B \cong -3.0$. In addition it was found that using $n = \frac{1}{4}$ gave not only a poorer fit but a positive screening term, as well.

The least-squares fit of Callan's[140] calculated ω_K-values (equation 41) also gives a positive screening term when $n = \frac{1}{4}$ is assumed. However, there is indication in Callan's work that the Auger transition rate is proportional to Z rather than being Z-independent. This contradicts Burhop's assumptions leading to the derivation of equation 40; therefore, considering equations 38 and 39, the value of $n = \frac{1}{3}$ becomes more suitable.

In order to obtain a set of ω_K-values of highest accuracy, Bambynek et al.[408] have reviewed all published measured values and chosen from these a total of 33 values on the basis of accuracies and reliabilities of the experimental procedures employed. From these they obtained values for A, B, and C; the exponent giving the best fit was $\frac{1}{4}$. From these it was possible to calculate ω_K for all Z-values and also their associated error limits. See also the more recent work on fluorescence yields by McGuire.[82a]

8.2. The L-Shell Fluorescence Yields

The de-excitation of an initial vacancy in the L-shell is much more complicated than that of an initial K-shell vacancy because there are three L-subshells, each corresponding to a unique quantum state. Similar to the situation for the K-shell, an initial vacancy in one of the L-subshells can be filled by an L X-ray or L-Auger transition. For certain regions of the periodic table, the situation is further complicated by the possibility that a Coster–Kronig transition will occur. Here the initial L-subshell vacancy is filled by an electron from an outer L-subshell, the released energy being taken by the outer shell electron that is expelled from the atom. Hence an initial L_1-subshell vacancy may move to the L_2- or L_3-subshell before it is filled by an electron from the outer shell. The abrupt variation in the mean L-fluorescence yield in certain Z-value regions was first directly attributed by Lay[409] to the occurrence of Coster–Kronig transitions.

Another complication is that initial vacancies in the L-shell are not confined to one subshell alone. That is, there is usually a specific ratio for probabilities of initial ionization among the three subshells, and the L-fluorescence yield measured is an average value, $\bar{\omega}_L$. The mean L-fluorescence yield is therefore dependent on the process by which the initial vacancies are formed, since the fluorescence yields for the three subshells ω_{L_1}, ω_{L_2}, and ω_{L_3}, differ, generally, in magnitude.

If the primary vacancy distribution in the L-subshells is $N_1 : N_2 : N_3$ where $N_1 + N_2 + N_3 = 1$, then the final vacancy distribution V_1, V_2, and V_3, which results from the primary distribution because Coster–Kronig transitions have occurred, will be

$$\begin{aligned} V_1 &= N_1, \\ V_2 &= N_2 + f_{12} N_1, \\ V_3 &= N_3 + f_{23} N_2 + (f_{13} + f_{12} f_{23}) N_1, \end{aligned} \tag{42}$$

where f_{12}, f_{13}, and f_{23} are the Coster–Kronig transition probabilities that shift subshell vacancies from the L_1 to the L_2, L_1 to the L_3, and L_2 to the L_3 subshells, respectively. Thus the mean L-fluorescence yield can be expressed as

$$\bar{\omega}_L = V_1 \omega_1 + V_2 \omega_2 + V_3 \omega_3 . \tag{43}$$

If all the Coster–Kronig transition probabilities are equal to zero, $V_1 + V_2 + V_3$ is equal to unity; otherwise the sum is larger than unity.

On the other hand, if the mean L-fluorescence yield is to be expressed in terms of the primary vacancy distribution, we must write

$$\bar{\omega}_L = N_1 v_1 + N_2 v_2 + N_3 v_3, \tag{44}$$

where v_i ($i = 1, 2, 3$) is the fraction of all L X-rays resulting from one initial vacancy in the L_i-subshell, regardless of the subshells in which they may occur. The quantity ω_i is then the fraction of L_i X-ray transitions that result from an initial vacancy in the L_i-subshell. We can thus write the following relations between the v_i's and the ω_i's:

$$\begin{aligned} v_1 &= \omega_1 + f_{12}\omega_2 + (f_{13} + f_{12}f_{23})\omega_3, \\ v_2 &= \omega_2 + f_{23}\omega_3, \\ v_3 &= \omega_3. \end{aligned} \tag{45}$$

Equations 45 show that if the Coster–Kronig transition probabilities f_{12}, f_{13}, and f_{23} are equal to zero or negligible, equations 42 or 45 are greatly simplified. In order to determine the three L-subshell fluorescence yields, separate measurements of $\bar{\omega}_L$ would have to be made for three different initial vacancy ratios $N_1 : N_2 : N_3$. When Coster–Kronig yields are included, the procedure for determining ω_1, ω_2, and ω_3 becomes more complicated. It is possible, however, to adjust the experimental conditions to reduce this complexity (see Section 8.4.)

The Auger yield of the L_i-subshell a_i is defined as the probability that an L_i-subshell vacancy is filled through a nonradiative transition in which the inital vacancy is transferred to an outer atomic shell (M, N, etc. shell). One may write the relationship between the fluorescence, Auger, and Coster–Kronig yields as follows:

$$\begin{aligned} \omega_3 + a_3 &= 1, \\ \omega_2 + a_2 + f_{23} &= 1, \\ \omega_1 + a_1 + f_{12} + f_{13} &= 1. \end{aligned} \tag{46}$$

Also

$$\bar{a}_L + \bar{\omega}_L = 1, \tag{47}$$

where $\bar{a}_L$ is the mean L-subshell Auger yield, and

$$\bar{a}_L = a_1 V_1 + a_2 V_2 + a_3 V_3. \tag{48}$$

The total width of an L_i-subshell energy level Γ_{T_i} in which there is an initial vacancy is the sum of all the partial decay modes by which the vacancy may be filled. It can be seen that

$$\omega_i = \frac{\Gamma_{X_i}}{\Gamma_{T_i}}, \qquad \alpha_i = \frac{\Gamma_{a_i}}{\Gamma_{T_i}}, \qquad f_{ik} = \frac{\Gamma_{ik}}{\Gamma_{T_i}}, \tag{49}$$

where Γ_{a_i} and Γ_{ik} are the Auger and fluorescence yield widths, respectively (see also Chapter 6).

In addition, two important experimentally determinable partial L-shell fluorescence yields should be mentioned. One is the L-shell fluorescence yields following K X-ray emission ω_{KL}. The quantity is defined as $\omega_{KL} = N_2 v_2 + N_3 v_3$, where N_2 and N_3 are the intensitites of the $K\alpha_2$ and $K\alpha_1$ X-rays, respectively, relative to the total K X-ray intensity. Since $N_2 : N_3$ is nearly equal to 1:2, and assuming f_{23} to be small, we may write

$$\omega_{KL} \approx \tfrac{1}{3}\omega_2 + \tfrac{2}{3}\omega_3. \tag{50}$$

This average quantity is often measured in experiments when the $K\alpha_1$ and $K\alpha_2$ X-rays can not be resolved in the X-ray detector.

Recent measurements[410,411] indicate that the value of f_{23} remains close to 0.2 throughout the $Z = 50$ to 90 region. This is in opposition to Listengarten's[10] original plot, showing a falling off to zero of this quantity for $90 \gtrsim Z \gtrsim 60$. Listengarten's original plot of such parameters (Figure 27) is altered accordingly.

A second often determined quantity is the partial L-shell fluorescence yield following L-electron capture ω_{LL}. For allowed and first forbidden L-electron capture nuclear transition, $N_1 \approx 1$, N_2 and $N_3 \ll 1$, and we can define

$$\omega_{LL} \approx v_1 = \omega_1 + f_{12}\omega_2 + (f_{13} + f_{12}f_{23})\omega_3. \tag{51}$$

Listengarten[10] presented curves showing the fluorescence and Coster–Kronig yields for the three L-subshells for atomic numbers from $Z = 47$ to $Z = 98$. The curves were constructed in a manner similar to that of Kinsey[174,175] by choosing radiation, Auger, and Coster–Kronig widths for each Z-value in this region, normalized to experimental data for $Z = 47$, 54, 55, 56, and 82. The Auger widths Γ_a were assumed to increase linearly from $Z = 47$ to 82 and more rapidly from $Z = 82$ upward. The radiation widths by Massey and Burhop,[61] Γ_X were increased by 40, 20, and 10% for $Z = 51$, 79, and 92, respectively. Finally, Listengarten took advantage of the Coster–Kronig

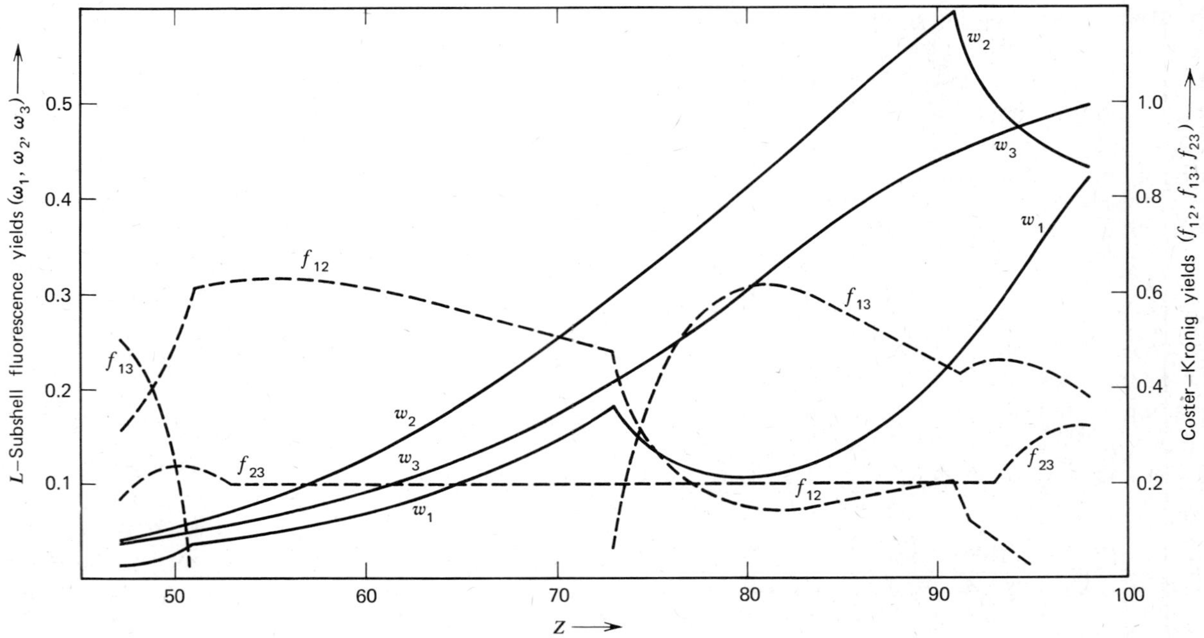

Figure 27. Fluorescence yields ω_i and Coster–Kronig yields f_{ik} for the L-subshells as a function of Z (according to Listengarten[10]). The value $f_{23} \approx 0.2$ for $50 \gtrsim Z \gtrsim 92$ has been indicated by recent X-ray work.[410]

width Γ_{ik}, which changes sharply at $Z = 51$, 73, 91, and 92, as shown in Figure 27. The curve for ω_3 was considered most accurate (error of $< 10\%$ for $Z = 73$–92; ~ 15–20% for the other Z-values). The curves for ω_1, ω_2, and f_{13} have an uncertainty of about 20%, and the curves for f_{23} and f_{12} may contain errors greater than 40%.

A comparison of experimental data with these curves was performed by Fink et al.,[400] who also published extensive tables of experimental values of L-shell fluorescence and Coster–Kronig yields obtained prior to May 10, 1966. For measurements on $Z = 71$ to 83, 90, and 92 regarding L_2- and L_3-fluorescence yields, see Ref. 411.

8.3. The *M*-Shell Fluorescence Yields

Only a limited amount of investigatory work has been done on the fluorescence yields of the M-shell, and it is restricted to the elements from $Z = 76$ to 92.[412] For a survey see Ref. 400.

8.4. Primary Vacancy Creation

As discussed in Section 2, primary atomic vacancies may be created in many ways, and some are more advantageous than others when the object is to measure fluorescence yields. The initial vacancies are usually created by bombarding the atom with photons, electrons, or heavy charged particles of energies sufficient to cause atomic ionization. The method chosen for causing ionization is not crucial in measuring K-shell fluorescence yields. This, however, is not the case in producing L-shell ionization, since the various methods give rise to different initial L-subshell vacancy distributions. In each case this ratio must be known in order to be able to determine the individual L-subshell fluorescence yields.

It is possible to take advantage of selective ionization of only one or two of the subshells—for example, using photoionization by a monoenergetic X-ray beam. In certain situations internal conversion after the alpha- or beta-decay of a radioactive nucleus gives rise to a known ratio of L-subshell vacancy production probabilities.

These possible methods for producing an initial atomic vacancy, described in the next three sections, may be listed as follows: photoelectric ionization, ionization by electron and heavy charged particle impact, and ionization by internal conversion or electron capture.

8.4.1. Photoelectric Ionization

In most of the early work on fluorescence yield determinations, X-radiation was used to cause the initial atomic vacancy by photoionization. Measuring the attenuation of an X-ray beam by an irradiated sample and the intensity of the simultaneously produced fluorescence K X-rays, would permit an approximate determination of K-shell fluorescence yields.

Actually, the measurement of the X-ray beam attenuation is not always reliable, since absorption can result from a number of other interaction mechanisms.

For higher atomic shells we must also know the ratio of initial vacancies caused among the subshells. If the X-ray energies used to ionize the L-shell are 2 to 5 times the mean L-shell electron binding energy, the approximate primary vacancy ratio among the L-subshells is $N_1:N_2:N_3 = 1:2:3$.[413]

The X-ray intensity below a certain energy value can be reduced greatly if an absorber is used. If the energy limit for absorption is chosen to lie between the binding energies of the L_2- and L_3-subshell electrons of the element under investigation, $N_1 = N_2 \cong 0$, and ionization will be restricted to the L_3-subshell and higher levels.[193] One can also select a monoenergetic X-ray line from an X-ray continuum by employing a crystal monochromator so that ionization is confined to the L_3-subshell and atomic levels of lower binding energies (see Section 5.3.1).

For a further treatment of photoionization, see Chapter 7.

8.4.2. Ionization by Electron and Heavy Charged Particle Impact

Ionization of atomic shells can be accomplished by electron bombardment; however, it is difficult to measure the rate of vacancy production directly. This amount must be calculated by employing the theory of ionizing collisions (see Bethe[414,415]).

For electron bombardment energies much higher than the atomic electron binding energy, the probability of electron ionization in the L-subshells is proportional to the subshell electron populations or $N_1:N_2:N_3 = 1:1:2$. For lower energies, Bethe has derived an expression for the relative ionization in each of the L-subshells. More information on this problem is given by the several authors in Ref. 416.

Multiple (almost exclusively double) initial atomic ionization may occur when electron bombardment is employed for fluorescence yield determinations. Auger spectrum investigations clearly exhibit the influence of initial double ionization, which is not always the case in X-ray investigations because of the lower energy resolution characteristic of these studies.

Bombardment using heavy charged particles (protons, alpha-particles) has not been used extensively for initial atomic vacancy creation in fluorescence yield determinations, although X-ray production cross sections have been measured with this ionization process. The energy range of bombarding particles has been from 0.1 to 4.3 MeV. See Ref. 400 for further discussion.

8.4.3. Ionization by Internal Conversion or Electron Capture

Atomic ionization resulting from internal conversion or electron capture is of value in determining fluorescence yields. The primary vacancy distribution is often known with good accuracy from the type of nuclear transition involved[417,139] (see also Chapter 4).

In the case of electron capture decay to excited levels in the daughter nucleus, if the decay energy is not high enough to allow K-capture, vacancies in only the L-shells and higher shells will be created. If the transition is allowed and first forbidden, most of the L-shell primary vacancies will be created in the L_1-subshell. From the coincidence rate between the gamma-rays emitted in the decay of the excited nuclear levels in the daughter nucleus and the L X-rays, we can determine a partial L-shell fluorescence yield ω_{LL}, as defined in Section 8.2. When K-capture is allowed, coincidence between K and L X-rays can be used to determine the partial L-shell fluorescence yield ω_{KL}.

In a like manner for internal conversion processes, K–L X-ray coincidence techniques can be used to determine ω_{KL}.

The multipolarity of a nuclear transition, as well as its energy, will determine the ratio of primary vacancies produced in the L-subshells. Since these primary L-subshell vacancy ratios for the various nuclear transition multipolarities differ greatly, it may be difficult to use internal conversion for fluorescence yield determinations. However, when a nuclear decay scheme is well known, it is sometimes possible to measure the fluorescence yields of certain of the atomic subshells. As an example, in the case of the alpha-decay of ^{242}Cm to ^{238}Pu, all gamma-transitions in the daughter nucleus are electric quadripole ($E2$), and L-conversion is primarily in the L_2- and L_3-subshells. Initial K-conversion will also lead to the L-shell vacancy production, mainly in the L_2- and L_3-subshells. Exploiting this situation, Salgueira et al.[418] ascertained the fluorescence and other yields of the L_2-subshell in plutonium by (a) measuring (α)–(L X-ray) coincidences and the L_2 and L_3 X-ray intensity ratios and (b) assuming a likely value for ω_3.

8.5. *K*-Shell Vacancy Shifts to Other Major Shells

An initial K-shell can become filled through the occurrence of X-ray or Auger transitions. The number of L-shell vacancies arising from the filling of a K-shell vacancy can be written

$$n_{KL} = k\omega_K + a_K[2I_a(KLL) + I_a(KLX)], \qquad (52)$$

where k is the ratio of the $K\alpha$ X-ray intensity to the total K-series X-ray intensity, a_K is the Auger yield $(1-\omega_K)$, and $I_a(KLL)$ and $I_a(KLX)$ are the indicated K-Auger intensities relative to the total K-Auger intensity. Values for n_{KL} and n_{KL_i} $(i = 1, 2, 3)$ as a function of atomic number have been published by Robinson and Fink,[417,139] Listengarten,[10] and Wapstra et al.[404]

Vacancy shifts from the K-shell to the M-shell and from the L-shell to the M-shell are denoted by n_{KM} and n_{LM}, respectively. Data on L-Auger yields and other pertinent quantities are sparse, and therefore accurate values for these vacancy shifts are not yet available.

The determination of n_{KL} varies in complexity with that of the nuclear decay modes leading to, and the decay scheme of the nucleus of, the atom in question. For the simple case of beta-decay, followed by a single gamma-ray, Ramaswamy[419] has shown that

$$n_{KL} = \frac{(N_{KL}/N_K)(L/K)}{\omega_K \varepsilon_K - (N_{KL}/N_L)}, \tag{53}$$

where ε_K is the K X-ray detection efficiency of the scintillation counter (including solid angle), N_{KL} is the number of K X-ray–L X-ray coincidences, N_K and N_L are the number K and L X-rays, respectively, and L/K is the L X-ray to K X-ray intensity ratio.

9. SECONDARY EFFECTS ASSOCIATED WITH THE AUGER PROCESS

In the gaseous state, initial ionization of an atom may be caused by photo- or electron impact ionization processes, as well as by ion bombardment (not considered here). The ionization processes that may take place in the atomic shells, where the binding energies are lower than the initial photon or electron kinetic energy, may be classified as *primary* ionization processes. An additional mode of ionization to be included in this group is the Auger process. *Secondary* ionization, then, is that which gives rise to the ejection of additional electrons during photoionization, electron bombardment, or Auger processes. These secondary processes can be grouped under the following headings:

1. Electron shake-off accompanying sudden changes in effective charge.
2. A direct collision with a bound electron, experienced by a photoelectron or Auger electron as it emerges from the atom.
3. The electron correlation process, such as the double Auger process, which is inadequately explained by the usual theoretical treatments of the phenomena 1 and 2. (See also the interesting article by Krause et al.[419a] on the "Manifestation of Atomic Dynamics through the Auger Effect.")

Multiple ionization (in particular double) associated with photoionization is a secondary effect coming mainly under points 1 and 2. That associated with electron impact behaves as a secondary process having a probability in agreement with electron shake-off calculations, as compared to the primary processes of single ionization.[420]

Nuclear beta, positron, and alpha decay, as well as internal conversion, can also cause atomic ionization through secondary ionization by process 1 or 2.

Final atomic ionization states are produced primarily through a cascade of Auger transitions following initial vacancy production in an inner shell. Final ionization states arising through primary processes can be predicted from knowledge of fluorescence yield values and the transition probabilities for the various competing Auger processes. Discrepancies between these predicted ionization state production probabilities and experimentally determined values are caused mainly by secondary processes.

Final atomic ionization states following initial photoionization, electron capture, beta-decay, and so on, are covered in the extensive work on the topic performed by Carlson, Krause and others: (He, Ne, Ar),[421,422] (Ne),[423] (Kr),[424] (Xe, Hg).[425] For the production of ionic states following electron impact ionization, see the work of Ziesel and coworkers: (Ne, Na, Mg),[426] (Kr, Rb, Sr),[375] (Ba),[373] (Zn),[427] and the results of Schram et al. (He, Ne, Ar, Kr, Xe),[420] and El-Sherbini (He, Ne, Ar),[428] and references to earlier work presented in these articles. See also the recent article by Carlson et al.[428a]

Kishinevsky and Parilis[429] have considered from a theoretical viewpoint the occurrence of Auger ionization when collisions arise between multicharged ions and neutral atoms. Three possible channels were considered: direct ionization, ionization after preliminary single charge exchange, and ionization after double charge exchange. The relative contributions of these processes to total Auger ionization were studied, as well as their dependence on ion charge and velocity. Highly ionized states of atoms recoiling, after alpha-decay, from a solid substrate have been investigated in an interesting manner by Perrin, de Wieclawik, and others (see Chapter 5, in which the subject is treated in detail). The work of Kishinevsky and Parilis[429] suggest an area of investigation; namely, the Auger electron emission spectrum expected to arise from collisions of these highly charged recoil ions and neutral gaseous atoms.

However, of main importance to the present chapter are the investigations by Carlson and Krause of Auger transitions involving the ejection of two electrons. It was found that for neon the readjustment of an initial K-vacancy results in the ejection of two electrons 7.5% of the time.[430] Double Auger processes occurred 9(1) and 31(4)% of the time for initial vacancies in the L-shell of argon[431] and the M-shell of krypton,[432] respectively. In these processes the secondary electrons were ejected from the outermost atomic shell and may be at least partially attributed to electron shake-off. However, the experimentally determined probabilities for double Auger processes are about an order of magnitude larger than the 0.5% predicted by theory.

Carlson and Krause[216] investigated the electron emission spectrum from argon in the 80 to 226 eV kinetic energy region. Initial L_2 and L_3, but not L_1, subshell vacancies were created by carbon $K\alpha$ X-ray band irradiation, and Auger-vacancy-satellite line production was avoided because the X-ray energies were generally too low to cause both L- and M-shell electron ejection simultaneously.

Figure 28 shows the $L_{2,3}MM$ spectrum obtained for argon. Since the instrumental resolution was quite low, details of the spectral components are not discernible. The $2p3p3p$ line region, for example, appears in detail in the high resolution measurement by Stalherm (Figure 18*a*). In Figure 28 we see a definite contribution to the spectrum due to $2p$-$3p3p3p(^4S)$ double Auger transitions (the hatched part). The region lying immediately above the energy limit for this type of transition and below the lowest Auger component $[2p\text{-}3s3s(^1S)]$ reveals an appreciable electron intensity. This is probably owing to transitions of the type $2p$-$3p3p3p^*$, where the asterisk indicates that the third electron remains in the atom in a higher orbital. The $2p$-$3s3s(^1S)$, or $L_{2,3}M_1M_1(^1S_0)$ region, as measured at high resolution by Stalherm, is shown in Figure 18*b*. At least some of the extra spectral components visible lying near or between the $L_{2,3}M_1M_1(^1S_0)$ components in Figure 18*b* may arise from $2p$-$3p3p3p^*$ type transitions.

The experimentally determined ratio of double to single Auger processes in argon was measured to be 0.12(4), in agreement with an earlier experimental determination[431] of 0.10(1). The double Auger process results in the ejection of two electrons from the outermost atomic shell and bears a resemblance to electron shake-off. However, transition probability calculations

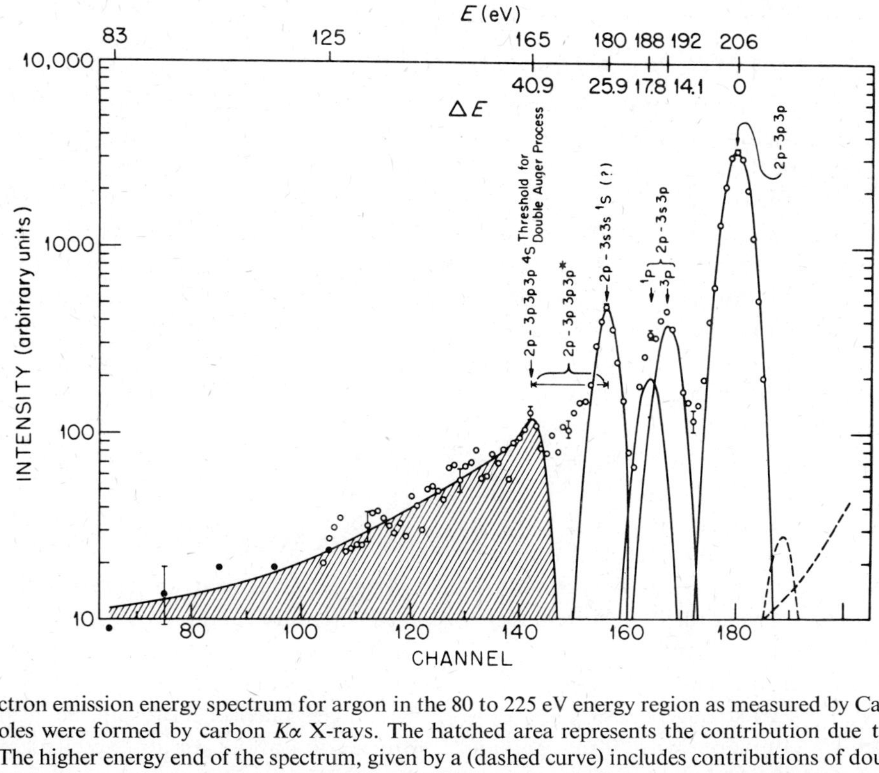

Figure 28. The electron emission energy spectrum for argon in the 80 to 225 eV energy region as measured by Carlson and Krause.[216] Initial L_2 and L_3 holes were formed by carbon $K\alpha$ X-rays. The hatched area represents the contribution due to the double Auger process $2p$-$3p3p3p$. The higher energy end of the spectrum, given by a (dashed curve) includes contributions of double-electron ejection arising from photoionization in the M-shell of argon. Between 180 and 165 eV there may be a series of lines corresponding to the transitions $2p$-$3p3p3p^*$, where one of the $3p$-electrons is removed from the atom and another is placed into a higher orbital.

for the process, as based on single-electron wave functions, lie far below the experimental values. This discrepancy may exist because correlation effects among electrons removed from the same shell have been overlooked.

Of course the mean number of electrons ejected either by primary or secondary processes following the production of an inner shell vacancy in a heavy atom depends on the number of electrons in the close environment of the atom or molecule. If an initial ionization takes place in an inner atomic shell of an atom in a molecular compound, the compound will decompose into a number of ions and radicals as the net charge of the molecule as a whole increases through a cascade of Auger transitions. Adding up the total final positive charge, we can estimate the mean number of electrons emitted. More electrons will be ejected from a compound including a given heavy atom in which the initial ionization occurs than will be ejected by that same atom when it is isolated at the instant it suffers an identical initial ionization. Carlson and White[433] found that X-radiation of vaporized $Pb(CH_3)_4$, producing vacancies primarily in the *L*- and *M*-shells of lead, caused the ejection of an average of 18.7 electrons. This figure is much in excess of the average of 9 electrons emitted from vaporized mercury (having $Z = 80$ compared to $Z = 82$ for lead) when irradiated by X-radiation from the same source.

9.1. Simultaneous X-ray and Auger Electron Emission

Åberg and Utriainen[434] have recently observed a broad X-ray emission structure with several maxima on the lower energy side of the $K\alpha_{1,2}$ line in elemental magnesium, aluminum, silicon, and sulfur. The intensity of the highest peak was about 0.05% of the $K\alpha$ line intensity. It was found that the structure increased in intensity to a limiting energy value, where it fell off sharply; just before the fall-off two strong peaks appeared (unresolved for aluminum).

It was found that the several maxima could be related by the relationship $h\nu + E_{\text{kin}}(L) = E(KLL)$, where $h\nu$ is the emitted photon energy, $E_{\text{kin}}(L)$ is the kinetic energy of the *L*-electron, and $E(KLL)$ is the Auger energy. As $E_{\text{kin}}(L)$ varies from 0 to $E(KLL)$, a continuous band of photon energies is given, which is initiated by an edge at $h\nu = E(KLL)$. This initiation edge corresponds to the formation of an excitation; that is, the *L*-electron is left in a final excitation state in the solid (see Chapter 8). This edge, actually a doublet, gives rise to the two strong peaks.

The phenomenon was compared with X-ray Raman bands reported by Suzuki,[435] pointing up a similarity between the two.

9.2. Cross-Auger Transitions

X-ray emission spectra for metal compounds have exhibited certain satellite lines that may be interpreted as a cross- or hybrid-transition between cation and anion. Weich[436] has noticed structure in his silicon, $K\beta$- and $L_{2,3}$-emission spectra from SiC; the energy was that to be expected for a C $2s \rightarrow$ Si $2p$ cross-transition. Ramqvist et al.[437] have observed similar X-ray structure in VC, TiN, and TiC, and they attributed the phenomenon to cross transitions involving the metallic $1s$ or $2p$ shell and the carbon or nitrogen $2s$ shell, the former transition being very weak because it is forbidden. The $K\beta'$ satellite investigated in compounds of magnesium, aluminum, and silicon by Linkoaha et al.[438] are suggested by them to be of the same origin.

Åberg et al.[438a] have decided that from energy considerations the Na^+ $K\beta_1$ X-ray emission line for NaF, NaCl, and NaBr is to be identified with a cross transition between cation and anion. (See this work for reference to earlier work on cross transitions.)

Now, the nonradiative transition associated with this cross transition is also to be expected: for example, a $K(\text{Ti})$–$L_1(N)L_1(N)$ hybrid Auger transition in TiN.

9.3. Questionable Origin of Electron Lines at ~450 eV Emitted in Ar^+–Ar Collisions

Ogurtsov[352] has found multicomponent electron emission lines over an energy region of some 50 eV at about 450-eV[439] energy arising in Ar^+–Ar collisions. These lines were seen during an investigation of autoionization state excitation and decay. The energy region corresponds to that expected due to improbable *LL-MMM* Auger transitions where two initial *L*-shell vacancies are simultaneously filled by two *M*-shell electrons, with the released energy being carried off by a third *M*-shell electron. The validity of this interpretation, which is highly speculative from all points of view, is undergoing further test. (See also Section 5.3.2 regarding the possibility of similar transitions in ^{113}In.)

10. NUCLEAR, ATOMIC, AND OTHER APPLICATIONS OF THE AUGER EFFECT

The experimental determination of a number of parameters associated with nuclear decay requires a knowledge of basic quantities relating to the Auger effect. There are also a number of situations in which the Auger phenomenon can be utilized for gaining information about an atomic effect or a condition of practical interest. These are summarized in the remainder of this chapter.

10.1. *K*-Capture Probabilities

The *K*-capture probability can be determined in certain nuclear decay schemes by employing coincidence measurements between *K*-Auger electrons and *K*- and *L*-conversion electrons.[440]

Measurements of *KLL*-Auger and *K*-conversion electron intensity have also been used[97] to ascertain the *K*-capture to total-electron capture probability ratio in the decay of ^{160}Er.

10.2. Determination of Relative *L*- to *K*-Capture Probabilities

The relative probability of *L*-capture to *K*-capture P_{LC}/P_{KC}, is related to the *K* and *L* X-ray intensity ratio as follows:

$$\frac{I(L\ \text{X-rays})}{I(K\ \text{X-rays})} = \frac{P_{LC}}{P_{KC}} + n_{KL}\frac{\omega_L}{\omega_K}, \tag{54}$$

where n_{KL} is defined in Section 8.5 (see also Refs. 441 and 442). It is also possible to determine P_{KC}/P_{total} or P_{LC}/P_{total} with coincidence techniques (see Ref. 139).

10.3. Relative Probabilities of *K*-capture and Positon Emission

It is easily seen that the ratio of *K*-capture to positon emission is

$$\frac{P_{KC}}{P_{\beta^+}} = \frac{I(K\ \text{X-rays})}{\omega_K I(\beta^+)} = \frac{I(K\text{-Augers})}{a_K I(\beta^+)}. \tag{55}$$

See also Ref. 443.

10.4. Electron Energy Calibration

Auger electron lines appear in most internal conversion spectra. The energies of these lines for a number of elements have been measured with a high accuracy (up to 10^{-3}–10^{-4}), and they can be conveniently used as energy calibration lines.[41, 119] However, for the lower atomic mass elements, $Z \precsim 25$, the energies of Auger electrons are much too dependent on the chemical state of the atom (see Section 4.1.3.3).

10.5. *K*-Conversion Coefficient Determination

When the internal conversion of only one gamma-transition is the origin of all *K*-atomic vacancies, the *K*-conversion coefficient α_K can be calculated from the intensities of the *K* X-rays and the gamma-rays detected in a scintillation crystal of detection efficiencies ε_γ and ε_{KX} by using the relationship:[444]

$$\alpha_K = \frac{I(K\ X\text{-rays})}{I(\gamma\text{-rays})} \times \frac{\varepsilon_\gamma}{\varepsilon_{KX}\,\omega_K}. \tag{56}$$

In more complicated cases with several γ-rays and/or electron capture feeding, coincidence techniques may be very effective.[445]

10.6. Determination of a Low Energy *K*-Conversion Line Intensity

Internal conversion in the *K*-shell can in some cases produce a conversion line of an energy so low that it is difficult to measure the line intensity accurately. In such a case it is possible to estimate the unknown *K*-conversion line intensity by measuring the *K*-Auger electron intensity and the other *K*-conversion line intensities and by knowing ω_K. The ratio of this intensity value to the intensity of the *L*-shell conversion lines *K*/*L* will give information about the multipolarity of the nuclear transitions.

10.7. Nuclear Level Lifetime Determinations Using Coincidence Techniques

In simple decay schemes the measurement of coincidences between "prompt" *K*-Auger electrons (following electron capture in the parent nucleus) and conversion electrons arising in the decay of the excited level in the daughter nucleus can be used to determine the lifetime of that level. This has been done, for example, by Marelius et al.[446] in the measurement of the lifetime of the 166-keV level in ^{139}La.

10.8. Determination of Electron Shake-off Probabilities During Alpha-Decay

The intensity of the Auger spectrum in ^{206}Pb following the alpha-decay of ^{210}Po will give an estimate of the probability of production of atomic vacancies caused by electron shake-off. The decay leads essentially 100% of the time to the ground state in ^{206}Pb ($\sim 1.2 \times 10^{-3}$% to the 803-keV, 2+ level). Measurement of the *K*-Auger spectrum intensity, relative to the alpha-decay rate, would give the electron shake-off probability for the *K*-shell electrons, but these electron lines are relatively weak because of the high *K*-fluorescence yield value. On the other hand, the *L*-Auger line intensities may be measured more advantageously, since ω_L is lower than ω_K. In an experiment of this kind Juillard and Moussa[447] (see also Ref. 197) obtained the ratio of (1.3–2.5) × 10^{-3} *L*-Auger electrons : alpha-disintegration. Some of the *L*-shell vacancies arise, of course, as a result of X-ray or Auger transitions filling an original *K*-shell vacancy created by shake-off.

10.9. Determination of the Probability of Electron Shake-off Resulting from *K*-shell Photoionization in Neon

An initial *K*-vacancy in neon ($Z = 10$) will be filled about 99% of the time by the *KLL*-Auger process leaving the atom in a state of double ionization. Therefore, neon ions of charge larger than +2 will have arisen because of secondary processes of ionization. Measurement of relative abundances of neon ions resulting from X-irradiation of neon has given evidence for a considerable contribution of secondary ionization processes following photoionization in the *K*-shell.[69] (The process of *K*-photoionization and the *KLL*-Auger process are considered primary ionization events; any additional electrons ejected during either of these events are here called secondary electrons. See Section 9.)

From data on secondary ionization accompanying the *KLL*-Auger process[430] and from data on the change in the neon ion charge spectrum as a function of photoionizing X-ray energy, Carlson and Krause[423] were able to determine the electron shake-off probability accompanying *K*-shell photoionization. At higher photon energies the relative abundances of ions of charges greater than +2 began to decrease, indicating a reduction in the probability of shake-off associated with *K*-photoionization. Specifically, it was found that when a photoelectron from the *K*-shell leaves with a velocity greater than 1.5×10^9 cm/sec, the probability for a single electron shake-off is 16%. This value agrees well with that calculated on the sudden approximation basis. Multiple electron ejection calculations showed agreement with the observed abundances of the more highly charged ions.

10.10. Determination of Electron Impact Ionization Cross Sections of Inner Atomic Levels

Electron impact ionization cross sections for the *K*-atomic shell can be determined by measuring the emission intensity of *K*-fluorescence radiation or of *K*-Auger electrons following *K*-shell ionization. For elements of low *Z*-value the *K*-fluorescence yield ω_K is very small, and its experimental value usually has an uncertainty of more than 50%. The value of the *K*-Auger yield is, on the other hand, $a_K = 1 - \omega_K$, and has then an uncertainty of only a couple percent. Measuring the ionization cross sections of the inner shells by the Auger process has advantages over the former method when no accurate fluorescence yields are available.

Glupe and Mehlhorn[448] took advantage of the relative certainty of the *K*-Auger yield in their determination of the ionization cross section for the *K*-shell of carbon (in CH_4), nitrogen, oxygen, and neon. Gaseous targets were used and the *K*-Auger electron intensities were determined by employing an electrostatic spectrometer set for an energy resolution of 1.2%. The ionization cross section was measured for bombarding electron energies from 1.5 times the *K*-shell binding energy to 4800 eV. The extension of these investigations to *K*-shell of carbon to neon, the *L*-shell of argon, and the *M*-shell of krypton is in progress.[126]

10.11. Intensity Standard for the Investigation of Chemical Effects on Internal Conversion

The intensities of Auger transition lines involving the deeper lying atomic shells of heavy atoms will be negligibly sensitive to the chemical state of the atom. On the other hand, internal conversion probabilities of nuclear transitions for atomic electrons lying in the outermost orbitals are sensitive to the chemical state of the atom. In the heavier elements, *K*-Auger electron line intensities may be used as an intensity standard for measuring chemical effects on internal conversion.

Bocquet et al.[449] have investigated the chemical effects on outer shell (*N* and *O*) internal conversion of the 23.87-keV transition in ^{119}Sn. The energy positions of these conversion lines fall in the KL_1M_{1-3} Auger energy region. However, it was possible to analyze meaningfully the superposition of the lines and to determine the $N : KL_1M_{2,3}$ and $N : O$ intensity ratios for various chemical states of tin. The former ratio was found to remain essentially constant, but the latter varied greatly.

10.12. Study of the Interaction of Noble Gas Ions with Solid Surfaces

Electrons may be released from a solid surface in an Auger type of neutralization of ionization produced by noble gas ion bombardment (sputtering). The spectrum of Auger electrons emitted per ion per electron volt of electron energy varies greatly as a function of surface purity conditions. The results are very sensitive to the concentrations of impurities absorbed in the surface, and they are used to evaluate procedures for cleaning surfaces. Investigations for certain metals, silicon, germanium, or superconducting metals, have been reported.[450–457] Information can also be obtained on the temperature dependent energy distribution of the bound electrons in the solid.[454]

10.13. Auger Processes Involving Valence Band Electrons

Auger transitions between an initial inner shell vacancy (*K*, *L*, etc.) and two electrons in the valence band of a semiconductor or in the conduction band of a metal, will cause discrete maxima in the secondary electron energy spectrum that results from electron bombardment. For the semiconductors, assuming a valence bandwidth of *W*, the final Auger electron energy will range between the limiting values for both final vacancies in the top of the band and for both in the bottom of the band (these giving the maximum and minimum Auger energies, respectively). Hence the contribution to the Auger electron energy width due to the bandwidth will be $2W$. Also, for any given allowed final energy there will be a continuum of final state pairs contributing to the transition probability. For the conductor, W will refer to the width of the filled portion of the conduction band.

Lander[458] has investigated this type of Auger transition and its contribution to the secondary electron spectra. The metals investigated were beryllium, aluminum, nickel, copper, barium, and platinum; the semiconductors were carbon and oxides of beryllium, aluminum, nickel, copper, and barium. (See also Refs. 459–464. For a recent and extended treatment of the many-body character of the electron gas in simple metals, see Ref. 464a).

10.14. Band-to-Band Auger Recombination in Semiconductors

The Auger effect plays an important role in the filling of a created vacancy in the inner band of semiconductors. Beatty and Landsberg[464] have shown that this type of de-excitation occurs across the low energy gap in InSb. In the case of germanium, the band structure is more complicated and the band-to-band energy gap is larger. The band-to-band Auger recombination in semiconducting germanium is strongly temperature dependent. Here it was found that the Auger recombination process was not direct, as for InSb, but indirect. That is, an Auger recombination process transfers an initial hole to a split-off band, where it may reach thermal equilibrium with the lattice. This hole may also de-excite through direct radiative recombination with electrons of the conduction band.[465,466] It has also been shown through calculations that the Auger process is not negligible in high band gap materials (e.g., GaAs) when it is heavily doped.[467] (See also Ref. 464a.)

10.15. Intraband Auger Transitions

Quantum mechanical calculations show that the Auger mechanism may play an important role in recombination when an electron in the upper energy portion of a band fills a lower lying hole in the same band and the freed energy is given to a second electron in the same band, lifting it to a higher band. The estimated cross section for this process indicates that, for low temperatures and high impurity concentrations, the Auger process may compete strongly with radiative or other types of recombination.[468]

Intraband Auger transition probabilities may also play a role in the creation of the low energy tail on L X-ray spectra in semiconductors.[436] Experiments have indicated that the intensity and extent of this tail diminish as either the valence-conduction band gap increases or the valence bandwidth decreases.

10.16. Auger Recombination of Excitons Bound to Neutral Donors

At low temperatures, the measured lifetimes of excitons bound to neutral donors in GaP and silicon are shorter than predicted from absorption measurements. The phenomenon is explained by a believed predominance of the Auger recombination process over the radiative. In this process, one bound electron recombines with the bound hole with the simultaneous excitation of a second bound electron to the conduction band.[469]

10.17. Two-Electron Luminescence Decay of Exciton-Neutral Donor and Other Complexes

Dean et al.[470] have described a radiative exciton decay process by which excitons bound to neutral donors (sulfur, selenium, and tellurium) in GaP decay leaving the donor in an excited state. It is also possible for a second donor electron to become ejected into the conduction band.

Borgardus and Hebb[471] have investigated the temperature dependence of the exciton-neutral donor radiative recombination in GaAs and found an apparent influence by nonradiative Auger recombination.

10.18. Determination of Surface Structure and Thin Layer Thicknesses

By predetermining Auger electron spectra emitted from clean metallic surfaces, it is possible to determine the thickness of a uniform deposition of a second metal deposited on that surface.[472] This is made possible by measuring the Auger electron spectrum intensity and comparing it with an intensity-thickness calibration function. The energy region of interest is from 0 to 1000 eV, and only the surface of the sample is studied.

By combining this method with low energy electron diffraction investigation techniques, Palmberg and Rhodin[472] have shown that both detection of surface impurities and determination of surface structure can be effected.

The Auger method of analysis has been used by Harris[473,474] to study surface migration and segregation and processes of diffusion.

10.19. Autoionization Involving $^2P_{1/2} \rightarrow {}^2P_{3/2}$ Transitions in Excited Rare Gases

A peripheral electron in a rare gas atom may become excited and form an $ns^2np^5(^2P_{1/2})m(s, d)$ ($n =$ 2, 3, etc., for neon, argon, etc., and $m > n$) excited atomic state, with the binding energy of the $m(s, d)$ electron less than the $^2P_{1/2,3/2}$ spin-orbit splitting energy for the singly ionized ns^2np^5 state. Under such circumstances, an autoionization transition can take place to the final, lower energy $ns^2np^5(^2P_{3/2})$ ionized state with the emission of an electron. Peresse et al.[475] proved that this process is important in the production of Xe^+ ions in the ionization threshold region. Evidence of the autoionization phenomenon was observable for this element because of the large size of the $^2P_{1/2,3/2}$ energy splitting (ca. 1.30 eV). Autoionization of this type was not seen earlier for krypton and argon because of limitations by the smaller splittings.[476]

10.20. Auger Resonance Interference Analog in Raman Scattering for One- and Two-Phonon States

Resonance interference between one- and two-phonon states in crystals, observed through Raman scattering measurements, may be formulated in full by using the Fano[230–232] treatment and description of one- and two-electron excitation states in the outer shells of atoms. This analogy has been treated very nicely by Rousseau and Porto[477] and applied to analyze their experimental Raman spectrum for $BaTiO_3$. The strength of two-phonon processes is greatest in scattering with certain tensor components, but both the one- and two-phonon processes are strongly dependent on the phonon propagation directions. This results in a large range of values of q for a given resonance, depending on just the scattering and the phonon propagation direction. Indeed, it was found that for the resonance in the 175 cm^{-1} region, q ranged between -0.9 and infinity for different scattering geometries. Both q and Γ were evaluated for the several geometries.

10.21. Determination of Inner Shell Binding Energies from Auger Energies

It follows from earlier sections that an Auger electron energy can in general be represented as

$$E[X_i Y_j Z_k(^{2S+1}L_J)] = E(X_i) - E[Y_j Z_k(^{2S+1}L_J)],$$

where X_i is the initial vacancy and $Y_j Z_k$ is the final double vacancy state having the quantum numbers J, L, and S. When Y_j and Z_k are the outermost electrons of a gaseous atom, such as of a noble gas, then the last value in the equation can be obtained from Moore's tables.[162] This value, plus the measured Auger energy, will then give the binding energy of the X_i-shell atomic electron. This has been done for the K-shell in neon, L_1-, L_2-, L_3-shells in argon and the M_1-, M_2, M_3-shells in krypton (see Chapter 7, Section 3.4).

Recently the K-Auger spectrum for vaporized sodium was measured.[126] The K-binding energy for sodium in the gaseous state may be obtained in a similar manner using the measured KL_3M_1 Auger energy and appropriate double ionization values from Moore's tables.

REFERENCES

1. S. Rosseland, *Z. Phys.* **14** (1923) 172.
2. *Compt. Rend.* **177** (1923) 169; Auger learned, after the appearance of this article, of Rosseland's work through a letter to him from Niels Bohr. (Private communication from P. Auger, 1971.)

3a. P. Auger, *Compt. Rend.* **180** (1925) 65.

3b. P. Auger, *Ann. Phys.* (*Paris*) **6** (1926) 183.

4. H. R. Robinson, *Proc. Roy. Soc.* (*London*) **A104** (1923) 455.
5. R. L. Mayo and H. R. Robinson, *Proc. Roy. Soc.* (*London*) **A173** (1939) 192.
6. E. H. S. Burhop, *The Auger Effect and Other Radiationless Transitions*, Cambridge University Press, 1952.
7. D. Coster and R. De L. Kronig, *Physica* **2** (1935) 13.
8. B. G. Pettersson, J. E. Thun, and T. R. Gerholm, *Nucl. Phys.* **24** (1961) 223.
9. I. Bergström, in *Beta- and Gamma-Ray Spectroscopy*, K. Siegbahn (Ed.), North Holland Publishing Company, Amsterdam, 1955.
10. M. A. Listengarten, *Izv. Akad. Nauk SSSR*, *Ser. Fiz.* **24** (1960) 1041 [*Bull. Acad. Sci. USSR*, *Phys. Ser.* **24** (1960) 1050].
11. J. S. Dionisio, *Ann. Phys.* (*Paris*) **8** (1963) 747.
12. J. S. Dionisio, *Ann. Phys.* (*Paris*) **9** (1964) 29.
13. I. Bergström and C. Nordling, "The Auger Effect," in *Alpha-*, *Beta-*, *and Gamma-Ray Spectroscopy*, K. Siegbahn (Ed.), North Holland Publishing Company, Amsterdam, 1965, p. 1523.
14. A set of lectures given by W. Mehlhorn, September, 1969, at the Department of Physics, University of Nebraska, Lincoln. Copies are available through Prof. M. E. Rudd, at that address.
15. E. Sokolowski, C. Nordling, and K. Siegbahn, *Ark. Fys.* **12** (1957) 301.
16. W. N. Asaad and E. H. S. Burhop, *Proc. Phys. Soc.* (*London*) **71** (1958) 369.

17. G. T. Ewan, J. S. Geiger, R. L. Graham, and D. R. MacKenzie, *Can. J. Phys.* **37** (1959) 174.

18. R. L. Graham and J. S. Merritt, *Can. J. Phys.* **39** (1961) 1058.

19. G. T. Ewan and J. S. Merritt, *Can. J. Phys.* **38** (1960) 324.

20. R. L. Graham, I. Bergström, and F. Brown, *Nucl. Phys.* **39** (1962) 107.

21. G. T. Ewan and R. L. Graham, "Internal Conversion Processes at Very High Resolution," in *Alpha-, Beta-, and Gamma-Ray Spectroscopy*, K. Siegbahn (Ed.), North Holland Publishing Company, Amsterdam, 1965, p. 983.

22. The energies for this spectrum were kindly provided through private communication from Dr. R. L. Graham, 1967.

23. G. T. Ewan, R. L. Graham, and L. Grodzins, *Can. J. Phys.* **38** (1960) 163.

24. R. L. Graham, F. Brown, G. T. Ewan, and J. Uhler, *Can. J. Phys.* **39** (1961) 1086.

25. O. Hörnfeldt, A. Fahlman, and C. Nordling, *Ark. Fys.* **23** (1962) 155.

26. M. Ference, Jr., *Phys. Rev.* **51** (1937) 720.

27. S. Rosenblum, J. S. Dionisio, and M. Valadares, *J. Phys. Radium* **17** (1956) 112.

28. R. G. Albridge and J. M. Hollander, *Nucl. Phys.* **27** (1961) 554.

29. I. Bergström and R. D. Hill, *Ark. Fys.* **8** (1954) 21.

30. C. Geoffrion and G. Nadeau, *Can. J. Phys.* **35** (1957) 1284.

31. E. U. Condon and G. H. Shortley, *The Theory of Atomic Spectra*, Cambridge University Press, 1953.

32. S. Fraga and C. Fisk, Tech. Rept. TC–6801, Dept. of Chemistry, Univ. of Alberta.

33. C. Fisk and S. Fraga, *Can. J. Phys.* **46** (1968) 1140, 2228; *Can. J. Phys.* **47** (1969) 637.

34. M. Mladjenovič and H. Slätis, *Ark. Fys.* **9** (1954) 41.

35. D. Hartree and W. Hartree, *Proc. Roy. Soc.* (*London*) **A149** (1935) 210.

36. E. H. S. Burhop, *J. Phys. Radium* **16** (1955) 625.

37. F. A. Johnson and J. S. Foster, *Can. J. Phys.* **31** (1953) 469.

38. J. B. Bellicard, A. Moussa, and S. K. Haynes, *Nucl. Phys.* **3** (1957) 307.

39. W. N. Asaad, *Proc. Roy. Soc.* (*London*) **A249** (1959) 555.

40. A. Fahlman and C. Nordling, *Ark. Fys.* **26** (1964) 248.

41. O. Hörnfeldt, *Ark. Fys.* **23** (1962) 235.

42. K. Siegbahn, C. Nordling, A. Fahlman, R. Nordberg, K. Hamrin, J. Hedman, G. Johansson, T. Bergmark, S.-E. Karlsson, I. Lindgren, and B. Lindberg, *Nova Acta Reg. Soc. Sci. Upsal. Ser. IV*, **20** Uppsala (1967); Tech. Rept. AFML-TR-68-189, Wright-Patterson Air Force Base, Ohio, October, 1968.

42a. H. Hillig, Diplomarbeit, Westfälische Wilhelm-Universität, Institut für Kernphysik, Münster, 1970.

42b. N. Weichert, to be published; private communication from W. Mehlhorn, 1970.

43. P. Erman and Z. Sujkowski, *Ark. Fys.* **20** (1961) 209.

44. C. W. Nestor, T. C. Tucker, T. A. Carlson, L. D. Roberts, F. B. Malik, and C. Froese, Oak Ridge Nat. Lab. Rept.–4027 (UC–32), December, 1966.

45. W. N. Asaad, *Z. Phys.* **203** (1967) 362.

46. U. Gelius and L.-A. Nordqvist, UUIP–495, University of Uppsala, Internal Report, August, 1966.

47. P. Erman, I. Bergström, Y. Y. Chu, and G. T. Emery, *Nucl. Phys.* **62** (1965) 401.

48. W. N. Asaad, *Nucl. Phys.* **66** (1965) 494.

49. A. Fahlman, R. Nordberg, C. Nordling, and K. Siegbahn, *Z. Phys.* **192** (1966) 476.

50. H. Körber and W. Mehlhorn, *Phys. Letters* **13** (1964) 129.

51. H. Körber and W. Mehlhorn, *Z. Phys.* **191** (1966) 217.

52. W. Mehlhorn, D. Stalherm, and H. Verbeek, *Z. Naturforsch.* **23a** (1968) 287.

53. A. K. Edwards and M. E. Rudd, *Phys. Rev.* **170** (1968) 140.

54. R. G. Albridge, K. Hamrin, G. Johansson, and A. Fahlman, *Z. Phys.* **209** (1968) 419.

55. W. Mehlhorn and R. G. Albridge, *Z. Phys.* **175** (1963) 506.

56. G. Wentzel, *Z. Phys.* **43** (1927) 524.

57. L. Pincherle, *Nuovo Cim.* **12** (1935) 81, 162.

58. E. H. S. Burhop, *Proc. Roy. Soc.* (*London*) **A148** (1935) 272.

59. E. J. Callan, *Phys. Rev.* **124** (1961) 793.

60. E. J. Callan, P. Nikolai, and W. L. McDavid, *Atomic Collision Processes*, M. R. C. McDowell (Ed.), North Holland Publishing Company, Amsterdam, 1964, p. 348.

61. H. S. W. Massey and E. H. S. Burhop, *Proc. Roy. Soc.* (*London*) **A153** (1936) 661.

62. E. G. Ramberg and F. K. Richtmyer, *Phys. Rev.* **51** (1937) 913.

63. R. A. Rubenstein, Ph.D. thesis, University of Illinois (1955); I am grateful to Prof. J. N. Snyder for kindly sending me a copy of Dr. Rubenstein's thesis.

64. R. A. Rubenstein and J. N. Snyder, *Phys. Rev.* **97** (1955) 1653.
65. M. A. Listengarten, *Izv. Akad. Nauk SSSR, Ser. Fiz.* **25** (1961) 793 [*Transl. Bull. Acad. Sci. USSR* **25** (1961) 803].
66. M. A. Listengarten, *Izv. Akad. Nauk SSSR, Ser. Fiz.* **26** (1962) 182 [*Transl. Bull. Acad. Sci. USSR* **26** (1962) 182].
67. W. N. Asaad, *Nucl. Phys.* **44** (1963) 399.
68. W. N. Asaad, *Nucl. Phys.* **44** (1963) 415.
69. M. O. Krause, M. L. Vestal, W. H. Johnston, and T. A. Carlson, *Phys. Rev.* **133** (1964) A 385.
70. D. Chattarji and B. Talukdar, *Phys. Rev.* **174** (1968) 44.
70a. Private communication from C. Briançon, 1970.
71. E. Sokolowski and C. Nordling, *Ark. Fys.* **14** (1959) 557.
72. G. D. Archard, private communication, 1964, in Ref. 48; 1963, in Ref. 182.
73. S. Hagström, C. Nordling, and K. Siegbahn, in *Alpha-, Beta-, and Gamma-Ray Spectroscopy*, K. Siegbahn (Ed.), North Holland Publishing Company, Amsterdam, 1965, p. 845.
74. A. Fahlman, K. Hamrin, R. Nordberg, C. Nordling, and K. Siegbahn, *Phys. Rev. Letters* **14** (1965) 127; R. Nordberg, K. Hamrin, A. Fahlman, C. Nordling, and K. Siegbahn, *Z. Phys.* **192** (1966) 462.
75. W. Mehlhorn and W. N. Asaad, *Z. Phys.* **191** (1966) 231.
76. B. Cleff and W. Mehlhorn, Rept. No. K-133 at the meeting of the German Physical Society in Bad Pyrmont, 1967; *Z. Phys.* **219** (1969) 311.
77. J. B. Bellicard, *Ann. Phys.* (*Paris*) **2** (1957) 419.
78. J. B. Bellicard and A. Moussa, *Compt. Rend.* **241** (1955) 1202.
79. J. B. Bellicard and A. Moussa, *J. Phys. Radium* **A18** (1957) 115.
80. P. Erman, E. C. O. Bonacalza, and J. Rossi, *Ark. Fys.* **26** (1963) 35.
81. H. Slätis, *Ark. Phys.* **37** (1968) 25.
82. H. Slätis and M. Rockberger, *Ark. Fys.* **40** (1969) 49.
82a. E. J. McGuire, *Phys. Rev.* **185** (1969) 1; *Phys. Rev.* **A2** (1970) 273; Sandia Corporation Research Reports SC-RR-70-429, August 1970; SC-RR-70-50, February 1970; SC-RR-69-137, August 1969.
83. M. O. Krause, T. A. Carlson, and R. D. Dismukes, *Phys. Rev.* **170** (1968) 37.
84. Private communication from T. A. Carlson, 1969.
84a. M. O. Krause, F. A. Stevie, L. J. Lewis, T. A. Carlson, and W. E. Moddeman, *Phys. Letters*, **31A** (1970) 81.
85. W. Mehlhorn, *Z. Phys.* **160** (1960) 247.
86. Y. Y. Lui and R. G. Albridge, *Nucl. Phys.* **A92** (1967) 139.
87. M. S. Freedman, F. T. Porter, and F. Wagner, *Phys. Rev.* **151** (1966) 886.
88. R. E. Johnston, J. H. Douglas, and R. G. Albridge. *Nucl. Phys.* **A91** (1967) 505.
89. R. J. Krisciokaitis and S. K. Haynes, *Nucl. Phys.* **A104** (1967) 466.
90. R. W. Casey and R. G. Albridge, *Z. Phys.* **219** (1969) 216.
91. P. Marguin and M. Moussa, *J. Phys. Radium* **21** (1960) 149.
92. J. S. Geiger, private communication in Ref. 20.
93. B. V. Bobykin and N. I. Novik, *Izv. Akad. Nauk. SSSR* **21** (1957) 1556 [*Bull. Acad. Sci. USSR Phys. Ser.* **21** (1957) 1546].
94. T. Suter and P. Reyes-Suter, *Ark. Fys.* **20** (1961) 393.
95. E. Monnand and A. Moussa, *Nucl. Phys.* **25** (1961) 292.
96. L. Persson, H. Ryde, and K. Oelsner-Ryde, *Ark. Fys.* **21** (1962) 193.
97. M. P. Avotina, E. P. Grigor'ev, B. S. Dzhelepov, and A. V. Zolotavin, *Izv. Akad. Nauk SSSR, Ser. Fiz.* **29** (1965) 1098 [*Bull. Acad. Sci. USSR, Phys. Ser.* **29** (1965) 1100.]
98. H. Ryde, L. Persson, and K. Oelsner-Ryde, *Ark. Fys.* **23** (1963) 171.
99. J. M. Cork, M. K. Brice, D. W. Martin, L. C. Schmid, and R. G. Helmer, *Phys. Rev.* **101** (1956) 1042.
100. J. Gizon, *Nucl. Phys.* **A108** (1968) 425.
101. M. I. Babenkov and B. V. Bobykin, *Izv. Akad. Nauk SSSR, Ser. Fiz.* **32** (1968) 1998.
102. G. Kaye, *Nucl. Phys.* **68** (1965) 529.
103. J. Gizon and J. Boutet, *Nucl. Phys.* **A103** (1967) 9.
104. J. Valentin, *Compt. Rend.* **254** (1962) 858.
105. C. J. Gallagher, Jr., D. Strominger, and J. P. Unik, *Phys. Rev.* **110** (1958) 725.
106. J. O. Newton, *Phys. Rev.* **117** (1960) 1529.
107. G. T. Ewan, *Can. J. Phys.* **35** (1957) 672.
108. L. M. Toburen and R. G. Albridge, *Nucl. Phys.* **A90** (1967) 529.
109. J. S. Dionisio, *Compt. Rend.* **254** (1962) 257.

109a. J. L. Wolfson and A. P. Baerg, *Can. J. Phys.* **42** (1964) 1781.
110. J. W. Mihelich, *Phys. Rev.* **88** (1952) 415.
111. J. S. Dionisio, *Compt. Rend.* **254** (1962) 1972.
112. J. J. H. Park and P. Christmas, *Can. J. Phys.* **45** (1967) 2621.
113. J. C. Nall, Q. L. Baird, and S. K. Haynes, *Phys. Rev.* **118** (1960) 1278.
114. C. J. Herrlander, R. Stockendal, and R. K. Gupta, *Ark. Fys.* **17** (1960) 315.
115. B. Jung and J. Svedberg, *Nucl. Phys.* **20** (1960) 630.
116. Z. Sujkowski, *Ark. Fys.* **20** (1961) 243.
117. R. Stockendal, *Ark. Fys.* **17** (1960) 553.
118. M. Mladjenovič and H. Slätis, *J. Phys. Radium* **16** (1955) 638.
119. A. I. Zhernovoi, E. M. Krisiuk, G. D. Latyshev, A. S. Remennyi, A. G. Sergeev, and V. I. Fadeev, *Zh. Eksp. Teor. Fiz.* **32** (1957) 682 [*Russ. J. Exp. Theor. Phys.* **5** (1957) 563].
120. M. Mladjenovič and H. Slätis, *Ark. Fys.* **8** (1954) 65.
121. P. R. Gray, *Phys. Rev.* **101** (1956) 1306.
122. C. Briançon, *Compt. Rend.* **267** (1968) 348; thesis, University of Paris (Orsay), 1970.
123. D. Hoffman and B. J. Dropesky, *Phys. Rev.* **109** (1958) 1282.
124. J. O. Rasmussen, F. L. Canavan, and J. M. Hollander, *Phys. Rev.* **107** (1957) 141.
125. J. M. Hollander, W. G. Smith, and J. W. Mihelich, *Phys. Rev.* **102** (1956) 740.
126. Private communication from W. Mehlhorn, 1968–1969.
127. O. Hörnfeldt and A. Fahlman, *Ark. Fys.* **22** (1962) 412.
128. P. Erman, B. Sigfriedsson, T. A. Carlson, and K. Fransson, *Nucl. Phys.* **A112** (1968) 117.
129. C. Nordling, *Ark. Fys.* **15** (1959) 397.
130. D. Stalherm, Diplomarbeit, Westfälische Wilhelms-Universität, Institut für Kernphysik, Münster, 1967. (I thank Dr. W. Mehlhorn for kindly sending a copy of this thesis.)
131. Private communication from E. M. Rudd, 1968.
132. A. Fahlman, K. Hamrin, G. Axelson, C. Nordling, and K. Siegbahn, *Z. Phys.* **192** (1966) 484.
133. A. Fahlman, K. Hamrin, R. Nordberg, C. Nordling, and K. Siegbahn, *Phys. Letters* **20** (1966) 159.
134. A. Fahlman, K. Hamrin, J. Hedman, R. Nordberg, C. Nordling, and K. Siegbahn, *Nature* **210** (1966) 4.
135. C. A. Coulson, and F. A. Gianturco, *J. Phys.* [*Proc. Phys. Soc.* (*London*)] **1B** (1968) 605; F. A. Gianturco, *J. Phys.* [*Proc. Phys. Soc.* (*London*)] **1B** (1968) 614.
136. C. Nordling, E. Sokolowski, and K. Siebgahn, *Ark. Fys.* **13** (1958) 483.
137. Private communication from R. Spohr, 1968.
138. D. Stalherm, B. Cleff, H. Hillig, and W. Mehlhorn, *Z. Naturforsch* **24a** (1969) 1728.
138a. T. A. Carlson, W. E. Moddeman, B. P. Pullen, and M. O. Krause, *Chem. Phys. Letters* **5** (1970) 390.
138b. W. E. Moddeman, thesis, Univ. of Tennesee, Oak Ridge National Laboratory report ORNL-TM-3012, September 1970.
139. B. L. Robinson and R. W. Fink, *Rev. Mod. Phys.* **32** (1960) 117.
140. E. J. Callan, "On *K*-level Widths," in *Proc. Conf. Role of Atomic Electrons in Nuclear Transformations*, Warsaw, 1963, III, p. 419, (1965).
141. P. Erman, J. Rossi, E. C. O. Bonacalza, and J. Miskel, *Ark. Fys.* **26** (1964) 135.
142. Z. Sujkowski and H. Slätis, *Ark. Fys.* **14** (1958) 101.
143. J. S. Dionisio, in *Proc. Conf. Role of Atomic Electrons in Nuclear Transformations*, Warsaw, 1963, p. 440 (1965).
144. C. D. Ellis, *Proc. Roy. Soc.* (*London*) **139** (1933) 336.
145. E. Plassman and F. Scott, *Phys. Rev.* **84** (1951) 156.
146. J. F. Perkins and S. K. Haynes, *Phys. Rev.* **92** (1953) 687.
147. E. Grigor'ev, A. Zolotavin, V. Klement'ev, and R. Sintsyn, *Izv. Akad. Nauk SSSR, Ser. Fiz.* **23** (1959) 159; [*Bull. Acad. Sci. USSR, Phys. Ser.* **23** (1959) 153].
148. L. Laberrique-Frolow and P. Radvanyi, *J. Phys. Radium* **17** (1956) 944.
149. J. Ravier, P. Marguin, and A. Moussa, *J. Phys. Radium* **22** (1961) 249.
150. O. Huber, F. Humbel, H. Schneider, and A. de Shalit, *Helv. Phys. Acta* **25** (1952) 3.
151. C. D. Broyles, D. A. Thomas, and S. K. Haynes, *Phys. Rev.* **89** (1953) 715.
152. I. Bergström, *Ark. Fys.* **5** (1952) 191.
153. I. Bergström and S. Thulin, *Phys. Rev.* **79** (1950) 538.
154. B. B. D'yakov and I. M. Rogachev, *Izv. Akad. Nauk SSSR, Ser. Fiz.* **26** (1962) 191 [*Bull. Acad. Sci. USSR. Phys. Ser.* **26** (1962) 191].

155. E. P. Grigor'ev, *Izv. Akad. Nauk SSSR, Ser. Fiz.* **23** (1959) 159 *Bull. Acad. Sci. USSR, Phys. Ser.* **23** (1959) 153].

156. P. Marguin and A. Moussa, *J. Phys. Radium* **21** (1960) 17.

157. A. O. Burford, J. F. Perkins, and S. K. Haynes, *Phys. Rev.* **99** (1956) 3.

158. R. Steffen, O. Huber, and F. Humbel, *Helv. Phys. Acta* **22** (1949) 167.

159. T. O. Passell, M. C. Michel, and I Bergström, *Phys. Rev.* **95** (1954) 999.

160. S. K. Haynes, M. Velinsky, and L. J. Velinsky, *Nucl. Phys.* **A90** (1967) 573.

161. K. Risch, *Z. Phys.* **150** (1958) 87.

162. C. E. Moore, "Atomic Energy Levels," Vol. II, (*U.S.*) *Circ. 467 Nat. Bur. Std.* (1952).

163. W. Mehlhorn, *Z. Phys.* **187** (1965) 21.

164. W. N. Asaad, *Proc. Conf. Role of Atomic Electrons in Nuclear Transformations*, Warsaw, 1963, p. 371 (1965).

165. B. Talukdar and D. Chattarji, *Phys. Rev.* **174** (1968) 44.

166. W. N. Asaad and W. Mehlhorn, *Z. Phys.* **217** (1968) 304.

167. W. Mehlhorn and D. Stalherm, *Z. Phys.* **217** (1968) 294.

168. F. K. Richtmyer and E. G. Ramberg, *Phys. Rev.* **51** (1937) 925.

169. *Handbuck der Physik*, Vol. 30, S. Flügge (Ed.), Springer-Verlag, Berlin, 1957, p. 199.

170. L. Pincherle, *Phys. Rev.* **61** (1942) 225.

171. J. N. Cooper, *Phys. Rev.* **61** (1942) 235.

172. J. N. Cooper, *Phys. Rev.* **65** (1944) 155.

173. W. Mehlhorn, *Phys. Letters.* **15** (1965) 46.

174. B. B. Kinsey, *Can. J. Res.* **26** (1948) 404.

175. B. B. Kinsey, *Can. J. Res.* **26** (1948) 421.

176. A. Johansson, G. Malmsten, A. Marelius, B. Nyman, H. Pettersson, and B. Svahn, *Phys. Letters.* **26B** (1967) 83. See also Ref. 209.

176a. M. J. Zender, W. Pou, and R. G. Albridge, *Z. Phys.* **218** (1969) 245.

177. W. Mehlhorn, Rept. No. K-132 at the meeting of the German Physical Society in Bad Pyrmont, 1967.

178. W. Mehlhorn, *Z. Phys.* **208** (1968) 1.

179. R. Spohr, private communication, 1967; R. Spohr, T. Bergmark, N. Hagnusson, and K. Siegbahn, University of Uppsala, Physics Institute Report UUIP-560, October 1967.

180. E. J. Callan, *Rev. Mod. Phys.* **35** (1963) 524.

181. E. J. Callan, *Proc. Conf. Role of Atomic Electrons in Nuclear Transformations*, Warsaw, 1963, p. 398, (1965).

182. W. N. Asaad, *Nucl. Phys.* **63** (1965) 337.

183. J. S. Dionisio, *J. Rech. Cent. Nat. Rech. Sci.* **48** (1959) 231.

184. D. H. Black, *Proc. Cambridge Phil. Soc.* **22** (1925) 838.

185. A. Flammersfeld, *Z. Phys.* **114** (1939) 227.

186. D. K. Butt, *Proc. Phys. Soc.* (*London*) **A63** (1950) 986.

187. Y. Kobayashi, *J. Phys. Soc. Japan* **8** (1953) 440.

188. A. A. Bashilov, B. S. Dzhelepov, and L. S. Chervinskaya, *Izv. Akad. Nauk SSSR, Ser. Fiz.* **17** (1953) 428.

189. A. Moussa and J. B. Bellicard, *Compt. Rend.* **242** (1956) 1156.

190. J. Burde and S. G. Cohen, *Phys. Rev.* **104** (1956) 1085.

191. J. Burde and S. G. Cohen, *Phys. Rev.* **104** (1956) 1093.

192. K. D. Sevier, *Nucl. Phys.* **61** (1965) 601.

193. K. Risch, *Z. Phys.* **159** (1960) 89.

194. A. Fahlman and K. Siegbahn, *Ark. Fys.* **32** (1966) 111; private communication from A. Fahlman, 1965.

195. H. Slätis, *Ark. Fys.* **29** (1965) 485.

196. C. M. Lederer, J. M. Hollander, and I. Perlman, *Table of Isotopes*, 6th ed., John Wiley & Sons, Inc., New York, 1967.

197. A. Juillard, thesis, University of Grenoble, 1961.

198. Z. Sujkowski and O. Melin, *Ark. Fys.* **20** (1961) 193.

199. W. D. Brodie, *Proc. Phys. Soc.* (*London*) **A67** (1954) 397.

200. W. D. Brodie, *Proc. Phys. Soc.* (*London*) **A67** (1954) 265.

201. G. Albouy and M. Valadarès, *Compt. Rend.* **250** (1960) 2877.

202. S. K. Haynes and W. T. Achor, *J. Phys. Radium* **16** (1955) 635.

203. R. Päschke, *Z. Phys.* **176** (1963) 143.

204. H. Ryde, L. Persson, and K. Oelsner-Ryde, *Ark. Phys.* **23** (1962) 195.

205. A. H. Muir and F. Boehm, *Phys. Rev.* **122** (1961) 1564.

206. H. E. Henrikson, California Institute of Technology, Spec. Tech. Rept. No. 24 to the U.S. At. Energ. Comm. (October, 1956), unpublished.

207. J. Gizon, A. Gizon, and J. Valentin, *Nucl. Phys.* **A120** (1968) 321.

208. Private communication from A. Gizon and J. Gizon, 1969.

209. E. Svahn, A. Johannsson, B. Nyman, G. Malmsten, and H. Pettersson, *Z. Phys.* **210** (1968) 466.

210. A. H. Wapstra, P. F. A. Goudsmit, J. F. W. Jansen, J. Konijn, K. E. G. Löbner, G. J. Nijgh, and S. A. de Wit, *Nucl. Phys.* **A93** (1967) 527.

211. C. Sébille and F. Widemann, *Compt. Rend.* **258** (1964) 5636.

212. A. Burford and S. K. Haynes, *Bull. Amer. Phys. Soc.* **3** (1958) 208.

213. P. Marguin, *J. Phys. Radium* **21** (1960) 469.

213a. Private communication from S. K. Haynes, 1969.

214. M. O. Krause, *Phys. Letters* **19** (1965) 14.

215. Landolt-Börnstein, *Zahlenwerte und Funktionen*, Springer-Verlag, Berlin, 1950, Vol. I/2, p. 171.

216. T. A. Carlson and M. O. Krause, *Phys. Rev. Letters* **17** (1966) 1079.

217. T. Bergmark and R. Spohr, 1967 (private communication).

217a. *ESCA, Applied to Free Molecules*, K. Siegbahn, C. Nordling, G. Johannsson, J. Hedman, P. F. Hedén, K. Harmin, U. Gelius, T. Bergmark, L. O. Werme, R. Manne, Y. Baer, North-Holland Publishing Company, Amsterdam, 1969.

218. K. Codling and R. P. Madden, *Phys. Rev. Letters* **12** (1964) 106.

219. A. G. Shenstone, *Phys. Rev.* **38** (1931) 873.

220. P. G. Burke, *Advan. Phys.* **14** (1965) 521.

221. P. G. Burke, *Advan. At. Mol. Phys.* **4** (1968) 173.

222. *Autoionization*, A. Temkin (Ed.), Mons Book Company, Baltimore, Md. 1966.

223. K. T. Compton and J. C. Boyce, *J. Franklin Inst.* **205** (1928) 497.

224. P. G. Kruger, *Phys. Rev.* **36** (1930) 855.

225. R. Whiddington and H. Priestley, *Proc. Roy. Soc. (London)* **A145** (1934) 462.

226. R. Whiddington and H. Priestley, *Proc. Leeds Phil. Soc.* **3** (1935) 81.

227. H. Beutler, *Z. Phys.* **93** (1935) 177.

228. U. Fano, *Nuovo Cim.* **12** (1935) 154.

229. O. K. Rice, *J. Chem. Phys.* **1** (1933) 375.

230. U. Fano, *Phys. Rev.* **124** (1961) 1866.

231. J. W. Cooper, U. Fano, and F. Prats, *Phys. Rev. Letters* **10** (1963) 518.

232. U. Fano and J. W. Cooper, *Phys. Rev.* **137** (1965) A1364.

233. F. J. Comes and H. G. Sälzer, *Phys. Rev.* **152** (1966) 29.

234. F. J. Comes, H. G. Sälzer, and G. Schumpe, *Z. Naturforsch.* **23a** (1968) 137.

235. J. Macek, *J. Phys.* [*Proc. Phys. Soc. (London)*] **1B** (1968) 831.

236. F. Prats and U. Fano, in *Atomic Collision Processes*, M. R. C. McDowell (Ed.), North-Holland Publishing Company, Amsterdam, 1964, p. 600.

237. J. Davis, *Phys. Letters* **24A** (1967) 388.

238. R. P. Madden and K. Codling, *Phys. Rev. Letters* **10** (1963) 516.

239. P. L. Altick and E. N. Moore, *Phys. Rev. Letters* **15** (1965) 100.

240. P. G. Burke, D. D. McVicar, and K. Smith, *Phys. Rev. Letters* **11** (1963) 559.

241. M. Gailitis and R. Damburg, *Zh. Eksp. i. Teoret. Fiz.* **44** (1963) 1644 [*Russ. J. Exp. Theor. Phys.* **17** (1963) 1107].

242. M. J. Seaton, *Proc. Phys. Soc. (London)* **A77** (1961) 174.

243. L. Lipsky and A. Russek, *Phys. Rev.* **142** (1966) 59.

244. T. F. O'Malley and S. Geltman, *Phys. Rev.* **137** (1965) A1344.

245. E. Holøien, *Proc. Phys. Soc. (London)* **88** (1966) 538.

246. E. Holøien and J. Midtdal, *J. Chem. Phys.* **45** (1966) 2209.

247. E. Holøien and J. Midtdal, *J. Chem. Phys.* **45** (1966) 3897.

248. E. Holøien and J. Midtdal, *J. Chem. Phys.* **46** (1967) 4170.

249. Private communication from E. Holøien, 1967.

250. P. G. Burke, D. D. McVicar, and K. Smith, *Phys. Letters* **12** (1964) 215.

251. J. W. Cooper, S. Ormonde, C. H. Humphrey, and P. G. Burke, *Proc. Phys. Soc. (London)* **91** (1967) 285.

252. P. L. Altick and E. N. Moore, *Proc. Phys. Soc.* (*London*) **92** (1967) 853.

253. S. Ormonde, W. Whitaker, and L. Lipsky, *Phys. Rev. Letters* **19** (1967) 1161.

254. K. Codling, R. P. Madden, and D. L. Ederer, *Phys. Rev.* **155** (1967) 26.

255. D. L. Moores, *Proc. Phys. Soc.* (*London*) **91** (1967) 83.

256. P. L. Altick, *Phys. Rev.* **169** (1968) 21.

257. N. Swanson and K. Codling, *J. Opt. Soc. Amer.* **58** (1968) 1192.

258. J. D. Morrison, *Chem. Phys.* **40** (1964) 2488.

259. C. E. Kuyatt, *Methods of Experimental Physics*, Vol. **7A**, *Atomic Interactions*, L. Marton (Ed.), Academic Press, New York, 1968. (The author is indebted to Dr. C. E. Kuyatt for sending a copy of this article prior to publication.)

260. R. P. Madden and K. Codling, *Astrophys. J.* **141** (1965) 364.

261. J. A. R. Samson, *Phys. Rev.* **132** (1963) 2122.

262. K. Codling and R. P. Madden, *Appl. Opt.* **4** (1965) 1431.

263. W. R. S. Garton, G. L. Glasdalen, W. H. Parkinson, and E. M. Reeves, *J. Phys.* [*Proc. Phys. Soc.* (*London*)] **1B** (1968) 114.

264. W. C. Price, in *Molecular Spectroscopy*, P. Hepple (Ed.), The Institute of Petroleum, Adlard & Son, Ltd., Dorking, 1968, p. 221.

265. S. M. Silverman and E. N. Lassettre, *J. Chem. Phys.* **40** (1964) 1265.

266. J. Peresse and F. Tuffin, *Compt. Rend.* **265** (1967) 1314.

267. P. G. Burke and K. Smith, in *Atomic Collision Processes*, M. R. C. McDowell (Ed.), North-Holland Publishing Company, Amsterdam, 1964.

268. C. J. Joachain and M. H. Mittleman, *Phys. Rev.* **140** (1965) A432; *Phys. Rev.* **151** (1966) 7.

269. V. H. Diebler, R. M. Reese, and M. Krauss, *J. Chem. Phys.* **42** (1965) 2045; P. H. Doolittle and R. L. Schoen, *Phys. Rev. Letters* **14** (1965) 348.

270. F. J. Comes, *Phys. Letters.* **24A** (1967) 465.

271. F. J. Comes and U. Wenning, *Z. Naturforsch.* **24a** (1969) 587, 1227.

272. F. J. Comes, F. Speier, and A. Elzer, *Z. Naturforsch.* **23a** (1968) 125.

273. F. J. Comes and A. Elzer, *Phys. Letters.* **25A** (1967) 334; F. J. Comes and A. Elzer, *Z. Naturforsch.* **23a** (1968) 133.

274. R. P. Madden and K. Codling, *J. Opt. Soc. Amer.* **54** (1964) 268.

275. R. W. Ditchburn and R. D. Hudson, *Proc. Roy. Soc.* (*London*) **A256** (1960) 53.

276. G. H. Newsom, *Proc. Phys. Soc.* (*London*) **87** (1966) 975; G. H. Newsom and B. W. Shore, *J. Phys.* [*Proc. Phys. Soc.* (*London*)] **1B** (1968) 742.

277. W. R. S. Garton and K. Codling, *Proc. Phys. Soc.* (*London*) **86** (1965) 1067.

278. W. R. S. Garton and K. Codling, *Proc. Phys. Soc.* (*London*) **75** (1960) 87.

279. W. R. S. Garton and K. Codling, *J. Phys.* [*Proc. Phys. Soc.* (*London*)] **1B** (1968) 106.

280. J. Berkowitz and C. Lifshitz, *J. Phys.* [*Proc. Phys. Soc.* (*London*)] **1B** (1968) 438.

281. G. V. Marr and J. M. Austin, *Proc. Roy. Soc.* (*London*) **A310** (1969) 137.

282. W. R. S. Garton, *Proc. Phys. Soc.* (*London*) **65** (1952) 268.

283. G. V. Marr, *Proc. Roy Soc.* (*London*) **A224** (1954) 83; G. V. Marr and R. Heppinstall, *Proc. Phys. Soc.* (*London*) **87** (1966) 293.

284. H. Beutler and W. Demeter, *Z. Phys.* **91** (1934) 143, 202, 218.

285. G. V. Marr and R. Heppinstall, *Proc. Phys. Soc.* (*London*) **87** (1966) 547.

286. J. A. Simpson, S. R. Mielczarek, and J. W. Cooper, *J. Opt. Soc. Am.* **54** (1964) 269.

287. G. V. Marr and J. M. Austin, *J. Phys.* [*Proc. Phys. Soc.* (*London*)] **2B** (1969) 107.

288. R. Heppinstall and G. V. Marr, *Proc. Roy. Soc.* (*London*) **A310** (1969) 35.

289. J. A. Simpson, G. E. Chamberlain, and S. R. Mielczarek, *Phys. Rev.* **139** (1965) A1039.

290. F. M. J. Pichanick and J. A. Simpson, *Phys. Rev.* **168** (1968) 64.

291. H. R. M. Moussa and F. J. de Heer, *Physica* **36** (1967) 646.

292. P. G. Burke, J. W. Cooper, and S. Ormonde, *Phys. Rev. Letters* **17** (1966) 345.

293. H. Boersch, J. Geiger, and B. Schröder, *Abh. Deut. Akad. Wiss. Berlin*, no. 1 (1967), Akademie-Verlag, Berlin.

294. Private communication from L. Vriens, 1969.

295. J. Geiger, *Z. Phys.* **175** (1963) 530.

296. P. M. Becker and J. S. Dahler, *Phys. Rev. Letters* **10** (1963) 491; P. M. Becker and J. S. Dahler, *Phys. Rev.* **136** (1964) A73.

297. S. M. Silverman and E. N. Lassettre, see Ref. 230.

298. G. J. Schulz and J. W. Philbrick, *Phys. Rev. Letters* **13** (1964) 477.

299. N. R. Daly and R. E. Powell, *Phys. Rev. Letters* **19** (1967) 1165.

300. H. R. M. Moussa, F. J. de Heer, and J. Schutten, *Physica* **40** (1969) 517.

301. H. F. Schaefer and F. E. Harris, *Phys. Rev. Letters* **21** (1968) 1561.

302. O. Bely and F. Bely, *Solar Phys.* **2** (1967) 285.

303. G. C. Hale and J. B. Hasted, *Nature* **217** (1968) 945.

304. G. Brink, *Phys. Rev.* **127** (1962) 1204.

305. H. Ehrhardt and F. Linder, *Z. Naturforsch.* **22a** (1967) 11.

306. H. Ehrhardt, F. Linder, and T. Tekaat, *Advan. Mass Spectrom.* **4** (1967) 705.

307. H.-U. Chun, *Z. Naturforsch.* **23a** (1968) 1415.

308. J. A. Simpson and S. R. Mielczarek, *J. Chem. Phys.* **39** (1963) 1606.

309. H. Boersch, J. Geiger, H. Hellwig, and H. Michel, *Z. Phys.* **169** (1962) 252.

310. L. Vriens, J. A. Simpson, and S. R. Mielczarek, *Phys. Rev.* **165** (1968) 7.

311. D. P. Chock and U. Fano, Informal Rept., Dept. of Physics, University of Chicago, Ill., 1966.

312. H. Ehrhardt, L. Langhaus, and F. Linder, *Z. Phys.* **214** (1968) 179.

313. D. Andrick, H. Ehrhardt, and M. Eyb, *Z. Phys.* **214** (1968) 388.

314. H. Ehrhardt, "A Survey of Phenomena in Ionized Gases," Int. At. Energ. Agency Rept., Vienna (1968) 61.

315. J. Geiger and H.-C. Pfeiffer, *Z. Phys.* **208** (1968) 105.

316. H. Ehrhardt, L. Langhaus, F. Linder, and H. S. Taylor, *Phys. Rev.* **173** (1968) 222.

317. H. Beutler and K. Guggenheimer, *Z. Phys.* **87** (1933) 176; *Z. Phys.*, **88** (1933) 25; H. Beutler, *Z. Phys.* **91** (1934) 131.

318. P. Feldman and R. Novick, *Phys. Rev.* **160** (1967) 143. (See also Ref. 320.)

319. G. Sprott and R. Novick, *Phys. Rev. Letters* **21** (1968) 336.

320. P. Feldman and R. Novick, *Phys. Rev.* **168** (1968) 270.

321. G. N. Estberg and R. W. LaBahn, *Phys. Letters* **28A** (1968) 420.

322. P. Feldman, M. Levitt, and R. Novick, *Phys. Rev. Letters* **21** (1968) 331.

323. M. E. Rudd, "Electron Spectra from Light Atoms," invited paper, presented to the American Physical Society, June 17, 1968.

324. M. E. Rudd and K. Smith, *Phys. Rev.* **169** (1968) 79.

325. A. V. Chaplik, *Zh. Eksp. i Teor. Fiz.* **54** (1968) 332 [*Russ. J. Exp. Theor. Phys.* **27** (1968) 178.]

326. J. W. McGowan, M. A. Fineman, E. M. Clarke, and H. P. Hanson, *Phys. Rev.* **167** (1968) 52.

327. W. P. Jesse and R. L. Platzman, *Nature* **195** (1962) 720.

328. R. A. Harris, *J. Chem. Phys.* **39** (1963) 978.

329. A. Russek, M. R. Patterson, and R. L. Becker, *Phys. Rev.* **167** (1968) 17.

330. G. R. Cook and P. H. Metzger, *J. Opt. Soc. Amer.* **54** (1964) 968.

331. H. Beutler and H. O. Jünger, *Z. Phys.* **100** (1936) 80.

332. J. T. Dowell and T. E. Sharp, *J. Chem. Phys.* **47** (1967) 5068.

333. F. J. Comes and H. O. Wellern, *Z. Naturforsch.* **23a** (1968) 881; F. Comes and U. Wenning, *Z. Phys. Chemie Neue Folge* **70** (1970) 139; *Z. Naturforsch.* **25a** (1970) 237; F. J. Comes and K. W. Weber, *Z. Naturforsch.* **24a** (1969) 1941.

334. W. A. Chupka and J. Berkowitz, *J. Chem. Phys.* **51** (1969) 2341.

335. J. Berkowitz and W. A. Chupka, *J. Chem. Phys.* **51** (1969) 4244.

336. W. Lichten, *Phys. Rev.* **131** (1963) 229.

337. W. Lichten, *Phys. Rev.* **164** (1967) 133.

338. W. Lichten, "Resonant Charge Exchange," in *Advances in Chemical Physics*, Vol. 13, I. Prigogine (Ed.), John Wiley & Sons, Inc., New York, 1967, p. 41.

339. A. Russek, *Phys. Rev.* **132** (1963) 246.

340. P. G. Burke, *J. Phys.* [*Proc. Phys. Soc.* (*London*)] **1B** (1968) 586.

340a. M. J. van der Wiel, thesis, F.O.M., Amsterdam, Holland (1971).

341. H. W. Berry, *Phys. Rev.* **127** (1962) 1634.

341a. F. J. Comes and U. Wenning, *Z. Naturforsch.* **25a** (1970) 406.

341b. F. J. Comes and U. Wenning, *Chem. Phys. Letters* **5** (1970) 195

342. M. E. Rudd, *Phys. Rev. Letters* **13** (1964) 503.

342a. G. B. Crooks and M. E. Rudd, *Phys. Rev. Letters* **25** (1970) 1599.

343. R. B. Barker and H. W. Berry, *Phys. Rev.* **151** (1966) 14.

344. P. G. Burke, D. D. McVicar, and K. Smith, *Proc. Phys. Soc.* (*London*) **84** (1964) 749.

345. P. G. Burke, D. D. McVicar, and K. Smith, *Proc. Phys. Soc.* (*London*) **83** (1964) 397.

346. A. S. Newton, A. F. Sciamanna, and R. Clampitt, *J. Chem. Phys.* **47** (1967) 4843.

347. A. S. Newton, A. F. Sciamanna, and R. Clampitt, *J. Chem. Phys.* **46** (1967) 1779.

348. M. E. Rudd, *Phys. Rev. Letters* **15** (1965) 580.

349. W. Mehlhorn, *Phys. Letters* **21** (1966) 155.

350. P. G. Burke and D. D. McVicar, *Proc. Phys. Soc.* (*London*) **86** (1965) 989.

351. T. Bergmark, R. Spohr, N. Magnusson, L. O. Werme, C. Nordling, and K. Siegbahn, Internal Rept.-589, Inst. of Phys., Uppsala, Sweden, 1969.

352. Private communication from G. N. Ogurtsov, 1969. (See also Ref. 439.)

353. M. E. Rudd and D. V. Lang, *4th. Int. Conf. Physics of Electron and Atomic Collisions*, University of Laval, Quebec, 1965, p. 153.

354. J. L. Pietenpol, *Phys. Rev. Letters* **7** (1961) 64.

355. M. E. Rudd, T. Jorgensen, Jr., and D. J. Volz, *Phys. Rev.* **151** (1966) 28.

356. S. T. Manson, *Phys. Letters* **23** (1966) 315.

357. D. J. Nicolas, C. W. Trowbridge, and W. D. Allen, *Phys. Rev.* **167** (1968) 38.

358. C. Laughlin and A. L. Stewart, *Proc. Phys. Soc.* (*London*) **88** (1968) 151.

359. M. E. Rudd, T. Jorgensen, Jr., and D. J. Volz, *Phys. Rev. Letters* **16** (1966) 929.

359a. D. J. Volz and M. E. Rudd, *Phys. Rev.* **A2** (1970) 1395.

359b. R. K. Cacak and T. Jorgensen, Jr., *Phys. Rev.* **A2** (1970) 1322; R. K. Cacak, Q. C. Kessel, and M. E. Rudd, *Phys. Rev.* **A2** (1970) 1327.

360. I. S. Dmitriev, D. I. Vinogradova, V. S. Nikolaev, and B. M. Popov, *Zh. Eksp. i. Teor. Fiz. Pis'ma v Radoktslyu SSSR* **3**(1966) 35 [*J. Exp. Theor. Phys. Letters USSR* **3** (1966) 20].

361. L. I. Kiselevskii and D. A. Grankova, *Opt. Spectrosc.* **17** (1964) 179.

362. I. S. Dmitriev, V. S. Nikolaev, and Ya. A. Teplova, *Phys. Letters* **26A** (1968) 122.

363. P. G. Burke, J. H. Tait, and B. A. Lewis, *Proc. Phys. Soc.* (*London*) **87** (1966) 209.

364. E. Everhart and Q. C. Kessel, *Phys. Rev.* **146** (1966) 27.

365. O. Bely, *Proc. Phys. Soc.* (*London*) **88** (1966) 587; O. Bely, *J. Phys.* [*Proc. Phys. Soc.* (*London*)] **1B** (1968) 23.

366. C. Snoek, R. Geballe, W. F. v. d. Weg, P. K. Rol, and D. J. Bierman, *Physica* **31** (1965) 1553.

367. A. K. Edwards, thesis, University of Nebraska, Lincoln, 1967, (a copy of which was kindly sent by Prof. M. E. Rudd).

368. U. Fano and W. Lichten, *Phys. Rev. Letters* **14** (1965) 627.

369. E. W. Thomas and G. D. Bent, *Phys. Rev.* **164** (1967) 143, 151.

370. R. Fox, W. Hickman, and T. Kjeldaas, *Phys. Rev.* **89** (1953) 555.

371. Q. C. Kessel, M. P. McCaughey, and E. Everhart, *Phys. Rev. Letters* **16** (1966) 1189.

372. D. Jaeck, F. J. De Heer, and A. Salop, *Physica* **36** (1967) 606.

373. J.-P. Ziesel and R. Abouaf, *J. Chem. Phys.* **64** (1967) 702.

374. Q. C. Kessel and E. Everhart, *Phys. Rev.* **146** (1966) 16.

375. J.-P. Ziesel and R. Abouaf, *J. Chim. Phys.* **64** (1967) 695.

376. C. Ramsauer and R. Kollath, *Ann. Phys.* **12** (1932) 529, 837 (see for references to earlier articles by the same authors).

377. P. S. Hoeper, W. Franzen, and R. Gupta, *Phys. Rev.* **168** (1968) 50.

378. D. Andrick and H. Ehrhardt, *Z. Phys.* **192** (1966) 99.

379. C. E. Kuyatt, J. A. Simpson, and S. R. Mielczarek, *Phys. Rev.* **138** (1965) A385.

380. P. G. Burke and D. L. Moores, *J. Phys.* [*Proc. Phys. Soc.* (*London*)] **1B** (1968) 575.

381. G. J. Schulz, *Phys. Rev.* **136** (1964) A650.

382. U. Fano and J. W. Cooper, *Phys. Rev.* **138** (1965) A400.

383. K. L. Kwok and F. Mandl, *Proc. Phys. Soc.* (*London*) **86** (1965) 501.

384. G. J. Schulz, *Phys. Rev. Letters* **13** (1964) 583.

385. H. Kleinpoppen and V. Raible, *Phys. Letters* **18** (1965) 24.

386. G. E. Chamberlain, *Phys. Rev. Letters* **14** (1965) 581.

387. J. W. McGowan, *Phys. Rev. Letters* **17** (1966) 1207.

388. R. Y. Damburg and R. K. Peterkop, *Zh. Eksp. i. Teor. Fiz.* **44** (1962) 244 [*J. Exp. Theor. Phys.* **17** (1963) 167].

389. H. Ehrhardt and K. Willmann, *Z. Phys.* **203** (1967) 1.

390. C. E. Kuyatt, J. A. Simpson, and S. R. Mielczarek, *J. Chem. Phys.* **44** (1966) 437.

391. H. G. H. Heideman, C. E. Kuyatt, and G. E. Chamberlain, *J. Chem. Phys.* **44** (1966) 355.

392. H. Ehrhardt and K. Willmann, *Z. Phys.* **204** (1967) 462.

393. A. Temkin and R. Pohle, *Phys. Rev. Letters* **10** (1963) 22, 268.

394. G. J. Schultz, *Phys. Rev. Letters* **10** (1963) 104.

395. J. Macek and P. G. Burke, *Proc. Phys. Soc.* (*London*) **92** (1967) 351; J. Macek, *Proc. Phys. Soc.* (*London*) **92** (1967) 365.

396. M. J. W. Boness and J. B. Hasted, *Phys. Letters* **21** (1966) 526.

397. W. Mehlhorn, *Phys. Letters.* **26A** (1968) 166.

398. See Ref. 131.

399. Private communication from B. Cleff, 1969.

399a. W. Mehlhorn, presented at the "*VIIth International Conference on the Physics of Electronic and Atomic Collisions*," Amsterdam, July 29–30, 1971.

400. R. W. Fink, R. C. Jopson, H. Mark, and C. D. Swift, *Rev. Mod. Phys.* **38** (1966) 513.

401. G. D. Archard, *Proc. 2nd Symp. X-Ray Microscopy and Microanalysis*, Elsevier Publishing Company Ltd., Amsterdam, 1960, p. 331.

402. A. J. Campbell, *Proc. Roy. Soc. (London)* **A274** (1963) 319.

403. J. Laberrique-Frolow and P. Radvanyi, *Compt. Rend.* **242** (1953) 901.

404. A. H. Wapstra, G. J. Nijgh, and R. van Lieshout, *Nuclear Spectroscopy Tables*, North Holland Publishing Company, Amsterdam, 1959.

405. J. Heintze, *Z. Phys.* **143** (1955) 153.

406. E. J. Callan, *Bull. Amer. Phys. Soc.* **7** (1962) N416.

407. L. E. Bailey and J. B. Swedlund, *Phys. Rev.* **158** (1967) 6.

408. W. Bambynek, B. Crasemann, R. W. Fink, H. U. Freund, H. Mark, R. E. Price, P. V. Rao, and C. D. Swift, to be published in *Rev. Mod. Phys.* (1972); private communication from R. W. Fink (1969).

409. H. Lay, *Z. Phys.* **91** (1934) 533.

410. Private communication from P. V. Rao, 1969.

411. R. E. Price, H. Mark, and C. D. Swift, *Phys. Rev.* **176** (1968) 3; H. U. Freund and R. W. Fink, *Phys. Rev.* **178** (1969) 1952; P. V. Rao, R. E. Wood, J. M. Palms, and R. W. Fink, *Phys. Rev.* **178** (1969) 1997.

412. R. C. Jopson, H. Mark, C. D. Swift, and M. A. Williamson, *Phys. Rev.* **137** (1965) A1353.

413. M. A. S. Ross, A. J. Cochran, J. Hughes, and N. Feather, *Proc. Phys. Soc. (London)* **A68** (1955) 612.

414. H. Bethe, *Ann. Phys.* **5** (1930) 325.

415. H. Bethe, *Handbuch der Physik*, J. Springer-Verlag, Berlin, 1933, Vol. 24, p. 273.

416. C. Victor, *Ann. Phys. (Paris)* **6** (1961) 183; H. W. Darwin, *Z. Phys.* **164** (1961) 513; M. Gryziński, *Phys. Rev.* **138** (1965) A336; R. Peterkop and V. Veldre, *Advances in Atomic and Molecular Physics*, Vol. 2, Academic Press, New York, 1966, p. 263.

417. B. L. Robinson and R. W. Fink, *Rev. Mod. Phys.* **27** (1955) 424.

418. L. Salgueiro, J. G. Ferreira, J. J. H. Park, and M. A. S. Ross, *Proc. Phys. Soc. (London)* **77** (1961) 657.

419. M. K. Ramaswamy, *Nuovo Cim.* **49B** (1967) 194.

419a. M. O. Krause, T. A. Carlson, and W. E. Moddeman, presented at the conference on "Les Processas Electroniques simple et Multiple du Domaine X-et UV," Paris, September, 1970.

420. B. L. Schram, A. J. H. Boerboom, and J. Kistemaker, *Physica* **32** (1966) 185; B. L. Schram, *Physica* **32** (1966) 197.

421. T. A. Carlson, *Phys. Rev.* **156** (1967) 142.

422. See Ref. 83 and related article 5 cited therein.

423. T. A. Carlson and M. O. Krause, *Phys. Rev.* **140** (1965) A1057.

424. M. O. Krause and T. A. Carlson, *Phys. Rev.* **158** (1967) 18.

425. T. A. Carlson, W. E. Hunt, and M. O. Krause, *Phys. Rev.* **151** (1966) 41.

426. J.-P. Ziesel, *J. Chim. Phys.* **62** (1965) 328.

427. J.-P. Ziesel, *J. Chim. Phys.* **64** (1967) 706.

428. T. M. El-Sherbini, FOM-Inst. voor Atoom- en Molecuulfysica, Amsterdam, (February, 1969).

428a. T. A. Carlson, W. E. Moddeman and M. O. Krause, *Phys. Rev.* **A1** (1970) 1406.

429. L. M. Kishinevskiĭ and E. S. Parilis, *Zh. Eksp. i. Teor. Fiz.* **55** (1969) 1932 [*J. Exptl. Theor. Phys. USSR*, **28** (1919)1020.]

430. T. A. Carlson and M. O. Krause, *Phys. Rev. Letters* **14** (1965) 390.

431. T. A. Carlson and M. O. Krause, *Bull. Amer. Phys. Soc.* **10** (1965) 455.

432. M. O. Krause and T. A. Carlson, *Phys. Rev.* **149** (1966) 52.

433. T. A. Carlson and R. M. White, Oak Ridge Nat. Lab. Rept.-3994, (June, 1966); *J. Chem. Phys* **44** (1966) 4510.

434. T. Åberg and J. Utriainen, Physics Lab., Technical University of Helsinki Rept. No. 5/1969; *Phys. Rev. Lett.* **22** (1969)1346.

435. T. Suzuki, *J. Phys. Soc. Japan*, **22** (1962) 1139.

436. G. Wiech, *Z. Phys.* **216** (1968) 472.

437. L. Ramqvist, B. Ekstig, E. Källne, E. Noreland, and R. Manne, *J. Phys. Chem. Sol.* **30** (1969) 1849; B. Ekstig, Thesis, Physics Dept., University of Uppsala, 1970.

438. M. Linkoaho, T. Åberg, G. Graeffe, and J. Utriainen, Physics Lab., Technical University of Helsinki Rept. No. 3/1969.

438a. T. Åberg, G. Graeffe, J. Utriainen, and M. Linkoaho, *J. Phys. C. (Solid St. Phys.)* **3** (1970) 1112.

439. G. N. Ogurtsov, I. P. Flaks, and S. V. Avakyan, *6th Int. Conf. Physics of Electronic and Atomic Collisions*, Boston, 1969, p. 274.

440. H. Leutz and K. Ziegler, *Z. Phys.* **166** (1962) 582.

441. R. W. P. Drever and A. Moljk, *Phil. Mag.* **2** (1957) 427.

442. M. Langevin, *J. Phys. Radium* **16** (1955) 516.

443. R. Bouchez, *Physica* **18** (1952) 1171.

444. I. Bergström, K. Nybö, S. Thulin, A. H. Wapstra, and B. Åström, *Ark. Fys.* **7** (1953) 239.

445. P. Sparrman, T. Sundström, and A. Marelius, *Nucl. Instr. Meth.* **41** (1966) 138.

446. A. Marelius, P. Sparrman, and S.-E. Hägglund, *Nucl. Phys.* **A95** (1967) 632.

447. A. Juillard and A. Moussa, *J. Phys. Radium* **22** (1961) 677.

448. G. Glupe and W. Mehlhorn, *Phys. Letters* **25A** (1967) 274.

449. J. P. Bocquet, Y. Y. Chu, O. C. Kistner, and M. L. Perlman, *Phys. Rev. Letters* **17** (1966) 809.

450. H. D. Hagstrum, *J. Phys. Chem. Solids* **14** (1960) 33.

451. H. D. Hagstrum, *J. Appl. Phys.* **32** (1961) 1015.

452. H. D. Hagstrum, Y. Takeishi, and D. D. Pretzer, *Phys. Rev.* **139** (1965) A526.

453. Y. Takeishi and H. D. Hagstrum, *Phys. Rev.* **137** (1965) A641.

454. R. A. Craig, *Nuovo Cim.* **27** (1963) 257.

455. D. W. Vance, *Phys. Rev.* **164** (1967) 372; *Phys. Rev.* **169** (1968) 252, 263.

456. F. M. Propst and E. Lüscher, *Phys. Rev.* **132** (1963) 1037.

457. F. M. Propst and E. Lüscher, *Rev. Sci. Inst.* **34** (1963) 574.

458. J. J. Lander, *Phys. Rev.* **91** (1953) 1382.

459. P. J. Landsberg, D. A. Evans, and C. Rhys-Roberts, *Proc. Phys. Soc. (London)* **83** (1964) 325.

460. P. T. Landsberg, *Proc. Phys. Soc. (London)* **A62** (1949) 806.

461. C. G. Morgan, F. L. Jones, and D. K. Davies, *Proc. Phys. Soc. (London)* **86** (1965) 403.

462. A. R. Beattie and P. T. Landsberg, *Proc. Roy. Soc. (London)* **A249** (1959) 16.

463. J. Pirenne and P. Longe, *Physica* **30** (1964) 277.

464. A. R. Beattie and P. T. Landsberg, *Proc. Roy. Soc. (London)* **A249** (1959) 16.

464a. P. Longe and A. J. Glick, *Phys. Rev.* **177** (1969) 526; B. Bergersen, F. Brouers, and P. Longe, Report 1C/70/79, International Center for Theoretical Physics, Miramare, Trieste, Italy; I am indebted to Dr. P. Longe for discussions of these problems.

465. R. Conradt, *Z. Phys.* **209** (1968) 445.

466. R. Conradt and W. Wardelich, *Phys. Rev. Letters* **20** (1968) 8.

467. L. R. Weisberg, *J. Appl. Phys.* **39** (1968) 6096.

468. E. I. Tolpygo, K. B. Tolpygo, and M. K. Sheinkman, *Fiz. Tver. Tela* **7** (1965) 1790 [*Sov. Phys.-Solid State* **7** (1965) 1442].

469. D. F. Nelson, J. D. Cuthbert, P. J. Dean, and D. G. Thomas, *Phys. Rev. Letters* **17** (1966) 1262.

470. P. J. Dean, J. D. Cuthbert, D. G. Thomas, and R. T. Lynch, *Phys. Rev. Letters* **18** (1967) 122.

471. E. H. Borgardus and H. B. Bebb, *Phys. Rev.* **176** (1968) 993.

472. P. W. Palmberg and T. N. Rhodin, *J. Appl. Phys.* **39** (1968) 325.

473. L. A. Harris, *J. Appl. Phys.* **39** (1968) 1419.

474. L. A. Harris, *J. Appl. Phys.* **39** (1968) 1428.

475. J. Peresse, F. Tuffin, and G. Le Coz, *Phys. Letters* **28A** (1969) 356.

476. J. Peresse, F. Tuffin, and G. Sinou, *Phys. Letters* **25A** (1967) 773.

477. D. L. Rousseau and S. P. S. Porto, *Phys. Rev. Letters* **20** (1968) 1354.

Chapter Four

Internal Conversion

1. INTRODUCTION

The phenomenon of de-excitation of an atomic nucleus is a problem of central interest in the field of nuclear physics. The possible modes of de-excitation are directly dependent on the nature of the nuclear excitation state, the nuclear level scheme, and the relative energies of those levels. Higher excitation states, and in some cases metastable nuclear states, can decay by particle emission by the nucleus whereby a new nuclide becomes formed as the daughter product (e.g., beta- and alpha-decay). Also, transitions may occur between states in the same nuclide, the excitation energy difference between these states being carried away either by radiation emission or by the nonradiative process of internal conversion. It is these last two processes, in competition with each other, which are of interest in this chapter.

The study and investigation of the de-excitation of a nuclear state by either gamma-radiation or internal conversion provide valuable information regarding the nuclear properties of the states involved. More specifically, the electromagnetic radiative mode is one that is almost exclusively dependent on the properties of the radiating nucleus. On the other hand the internal conversion mode, wherein the pertinent nuclear state energy difference is transferred to a bound orbital electron, will depend in addition on the atomic electron density in the vicinity of the nucleus. These two modes of nuclear de-excitation are in direct competition with each other.

In the past, internal conversion has been treated primarily from the point of view of nuclear physics and within the framework of nuclear theory. The phenomenon is, however, one of a nuclear-atomic nature, and its investigation may be expected to provide important information to the field of atomic physics. The accurate electron spectrometric investigation of internal conversion electron spectra may, for example, provide information on: atomic level widths (Chapter 6, Section 3), atomic electron binding energies (Chapter 7, Section 3.3), atomic electron densities in the vicinity of the nucleus (Section 5), the chemical and physical states of the atom at the moment of the internal conversion event (Section 5.2), screening of the nuclear charge by atomic electrons (Section 3.1). In addition, internal conversion electron lines of previously accurately determined nuclear transition energies can often be used as energy calibration standards. Such nuclear transition energy standards are collected in tabular form in Appendix G. The alternative momentum ratio calibration method, however, does not require accurate knowledge of gamma-energies (see Chapter 7, Section 3.2.3).

In the following sections of this chapter is presented a treatment of the internal conversion phenomenon, with emphasis on the ways in which atomic properties influence that phenomenon. The chapter is divided into the following sections and topics: the definition and calculation of internal conversion coefficients, Section 2; conversion line form characteristics and analysis, Section 3; applications of conversion electron spectrum analysis in atomic physics research, Section 4; and finally, the effects of changes in electron density near the nucleus, Section 5.

2. GENERAL CONSIDERATIONS

Considering processes of lowest nonvanishing orders of perturbation theory (i.e., neglecting all higher order effects), electromagnetic transitions between two nuclear states are primarily of two complementary categories:

1. Gamma-ray emission of an energy E_γ and angular momentum L—An initial excited nuclear state having an angular momentum J_i may decay to a final state of lower energy and angular momentum J_f by emitting a gamma-ray of such multipole order L that $|J_i - J_f| \leq L \leq J_i + J_f$ as necessitated by conservation of angular momentum.

Each multipole order has two possible classes of

radiation which are of opposite parity: the electric 2^L-pole (EL) and the magnetic 2^L-pole (ML). The "multipolarity" of a transition specifies both its class and multipole order. In addition to the necessary conservation of angular momentum is that of parity π. The EL-radiation has as parity change $\pi_i \pi_f = (-1)^L$, and the ML-radiation has $\pi_i \pi_f = (-1)^{L-1}$. Here $\pi_i \pi_f = +1$ denotes no parity change, whereas the negative value denotes such a change. Additionally, radiation of multipole order O is prohibited; two-quantum $O \rightarrow O$ transitions may take place, but this is extremely unlikely.

The probability for a transition between two nuclear states via gamma-emission is dependent on the energy and multipolarity of that transition as well as on the nature of the wave functions of the initial and final nuclear states. The probability for a gamma-transition decreases rapidly with increasing multipolarity, regardless of the nuclear wave functions involved, subject to the selection rules. Therefore a transition will take place preferentially via a gamma-transition of the lowest multipolarity allowed by the selection rules; that is, $L = |J_i - J_f|$. A competitive mixture of the two lowest multipolarities, of opposite class due to parity considerations, may also arise when these transition probabilities are similar.

Gamma-transition probability estimates can be obtained from calculations based on one of several nuclear models. These models make it possible to describe nuclear particle excitations or collective excitation states (e.g., rotational or vibrational excitations). These alter in relative applicability for various regions of the periodic table. The radiative decay lifetime for the several multipolarities alter for gamma energies between 10 and 100 keV from $\sim 10^{-8}$ to $\sim 10^{-12}$ sec ($E1$ and $M1$) to $\sim 10^{8}$ to $\sim 10^{-2}$ sec ($E3$ and $M3$) and greater for higher orders.

2. Internal conversion of an atomic orbital electron to a continuum state of positive energy $E = E_\gamma - E_B - \phi_w$—the final state has an angular momentum j_f, which is the resultant of L and the total angular momentum j_i of the electron in the initial bound state. Here E_B is the binding energy of the converted electron (here defined as positive) and ϕ_w is the work function of the material of which the conversion electron energy analysis instrument is constructed (see Chapter 7, Section 3.2.1). For internal conversion, $|j_i - j_f| \leq L \leq |j_i + j_f|$, where lowercase j's are used in referring to atomic electron angular momenta.

The internal conversion process is in direct competition with gamma-ray emission. The ratio of the sum total internal conversion electron intensity of a nuclear transition in the various allowed atomic electron shells to the corresponding gamma-emission intensity is the definition of the transition's total conversion coefficient:

$$\alpha_{\text{tot}} = \frac{N_{\text{c.e.}}}{N_\gamma}.$$

The conversion electron intensity $N_{\text{c.e.}}$ per gamma-ray can be broken down into the contributions from the various atomic shells and subshells in which conversion can take place:

$$\alpha_{\text{tot}} = \alpha_K + \alpha_{L_1} + \alpha_{L_2} + \cdots.$$

Of course, internal conversion can involve only those electrons whose binding energies are less than $E - \phi_w$, where ϕ_w is the work function of the conversion electron source. Certain nuclear transitions, such as the 30-eV transition in ^{235m}U, are of such low energies that internal conversion can take place only in the outermost atomic levels (see Section 5). It is also possible that near threshold ($E_\gamma \simeq E_B$), the excited electron is transferred from the bound state to, in the case of a metallic radioactive sample, a bound vacancy near the Fermi level.

Nuclear $J_i = 0 \rightarrow J_f = 0$ transitions may occur through internal conversion. However, since there is no multipole field associated with either the initial or final nuclear state, the energy transfer rate is related to the overlap of the atomic electron and nuclear wave functions, i.e., the probability that the electron finds itself within the nuclear volume.

3. THEORETICAL CALCULATIONS OF INTERNAL CONVERSION COEFFICIENTS

Several penetrating discussions and treatments of the subject of internal conversion have been presented.[1–6] These treatments usually consider the phenomenon from the nuclear point of view. Here, however, emphasis is placed on the role played by atomic electrons in the process, the nucleus being essentially the source of transition energy and the origin of the multipole field which causes the internal conversion transitions. (From this point on, the abbreviation ICC is used to stand for internal conversion coefficient.)

Viewing the problem from this direction and following the notation of Rose,[3,7,8] an angular momentum eigenstate of the electron may be written as

$$\Psi = \begin{pmatrix} -f_\kappa \Omega_{-\kappa}(r) \\ g_\kappa \Omega_\kappa(r) \end{pmatrix}, \tag{1}$$

where κ is the nonvanishing quantum number defined as

$$\kappa = \pm(j + \tfrac{1}{2}) \tag{2}$$

for

$$j = l \pm \tfrac{1}{2}. \tag{3}$$

The symbols g_κ and f_κ, at times referred to as the large and small components of the electron wave function, represent the radial function for the continuum state electron; $g_{\kappa'}$ and $f_{\kappa'}$ those of the bound electron wave function. The expressions for electric and magnetic 2^L-pole ICC are, respectively,*

$$\alpha(EL) = \frac{\alpha\omega}{16\pi}\frac{L}{(L+1)(2L+1)}\sum_\kappa C_{\kappa\kappa'} \times \left|\frac{\kappa'-\kappa}{L}(R_3+R_4)+(R_4-R_3)+R_1+R_2\right|^2 \quad (4)$$

$$\beta(ML) = \frac{\alpha\omega}{16\pi}\frac{1}{L(L+1)(2L+1)}\sum_\kappa B_{\kappa\kappa'}|R_5+R_6|^2, \quad (5)$$

where ω is the nuclear transition energy and the κ-values in equation 4 are the negatives of those in equation 5. The radial integrals, which involve the initial and final electron state wave functions, and the multipole expansion G_L, are

$$\begin{aligned} R_1 &= \int_0^\infty g_\kappa g_{\kappa'} G_L r^2\,dr, \\ R_2 &= \int_0^\infty f_\kappa f_{\kappa'} G_L r^2\,dr, \\ R_3 &= i\int_0^\infty f_\kappa g_{\kappa'} G_{L-1} r^2\,dr, \\ R_4 &= i\int_0^\infty g_\kappa f_{\kappa'} G_{L-1} r^2\,dr, \\ R_5 &= \int_0^\infty f_\kappa g_{\kappa'} G_L r^2\,dr, \\ R_6 &= \int_0^\infty g_\kappa f_{\kappa'} G_L r^2\,dr, \end{aligned} \quad (6)$$

where

$$G_L(\omega r) = (2\pi)^{3/2} i^L \frac{1}{(\omega r)^{1/2}} H^{(1)}_{L+1/2}(\omega r). \quad (7)$$

Since in equation 7 $H^{(1)}_{L+1/2}(\omega r)$ is the spherical Hankel function of the first kind,

$$G_L(\omega r) = 8\pi \sum_{k=0}^{L} \frac{(-1)^k (L+k)!}{k!(L-k)!}\frac{e^{i\omega r}}{(2i\omega r)^{k+1}}. \quad (8)$$

The first ICC calculations, made for the K-shell by Rose et al.,[9] neglected both effects of the finite size of the nucleus—the nucleus was regarded as a point charge—and effects of screening of the nuclear charge.

* Relativistic units are used, where $\hbar = c = m_o = 1$ and $e^2 = \alpha \cong 1/137$; m_o is the electron rest mass.

The consequences of including the first effect in the calculations are twofold. First, the atomic electron wave function is altered and is no longer singular at the origin; this is a purely electrostatic effect. Second, the expression for the IC and gamma-transition probabilities contain non-equivalent terms, which make them dependent in differing manners on details of the nuclear model. Knowledge of nuclear currents or wave functions is particularly necessary. This is called a dynamic effect of the nuclear size. The first effect is usually much greater than the second, except for a very limited number of strongly forbidden transitions.

In the calculation of K-shell ICC, effects of screening are small. For the other shells, however, screening effects become of much greater importance, and corrections for these are imperative. The manner in which such corrections may be made are several, but the most common were by applying the Thomas–Fermi–Dirac statistical model,[3,10,11] the Hartree–Fock or Hartree–Fock–Slater self-consistent-field,[12] or the Hartree[13] potentials, all in the nonrelativistic approximation. Relativistic effects cannot readily be included in a statistical theory of the atom, for a self-consistent-field calculation must be used. Furthermore, of the methods just named, the H-F or H-F-S potentials are preferable over T-F-D statistical model, but the most accurate results are given by a completely self-consistent relativistic H-F treatment.[14,15]

In all ICC calculations, use is made of the Dirac relativistic equation, which is solved using one of the above-mentioned potentials outside of the nucleus.

Alternatively, ICC calculations can be made neglecting screening and correcting for the omission by the application of a correction factor. This procedure is considered further in Section 3.1.

Table I surveys the ICC calculations published since the appearance of the first treatment of the effect for the K-shell by Rose et al.[9] in 1951. Most of the earlier calculations cover conversion electron final energies down to approximately 25 keV. The more recent reports (e.g., Refs. 6, 15, 16) extend these energies down to ~1 keV; threshold values have been investigated by O'Connell and Carroll.[17–19]

The test of the accuracy of theoretical ICC values lies basically in their comparison with experimental values. In recent years such comparisons have been carried out mostly on pure $E2$ transitions, which take place between rotational levels in nuclei in the deformed region of the periodic table.[20–23] This choice is made to ensure purity of transition multipolarity. High degrees of instrumental and experimental precision are required in order to measure intensity ratios of $K/L/M$, L_1/L_2, M_1/M_5, and so on to an accuracy of ~5% or conversion coefficient values to ~10%.

Table I.

ICC calculations since 1951

Year	Ref.	Authors	Atomic shells	Screening	Multi-polarity	Nuclear size	Comments
1951	9)	Rose et al.	K	no	E(1-5) M(1-5)	Point nucleus	See for references to earlier work.
1956 1958	10)	Sliv & Band	K, L_{1-3}	T-F-D	E(1-5) M(1-5)	Finite size	Assumed surface distr. of nuclear currents.
1958	3)	Rose	K, L_{1-3}, M_{1-5}	T-F-D (K, L) no (M)	E(1-5) M(1-5)	Fin. nu.(K,$L_{1,2}$) Pt. nu.(L_3,M)	Dynamic eff. of finite nuclear size neglected (no penetration model).
1963	18)	O'Connell	K	no	E(1-5) M(1-5)	Point nucleus	Threshold value of ICC (eff. of finite size and screening considered).
1965	17)	O'Connell & Carroll	K, L_{1-3}, M_{1-5}	no	E(1-5) M(1-5)	Point nucleus	Threshold value of ICC.
1965	60)	Borisoglebskii	M_{1-3}	H (non-rel.; Z=80) T-F-D (Z=81)	E(1,2) M(1,2)	Finite size	Z=80, 81.
1966	61)	Bhalla	K,L_{1-3}	H-F-S (SCF) non-rel.	M(1-4)	Finite size	Two gamma energies for Z=39.
1966	14)	Bhalla	K, L_{1-3}, M_{1-5}	H-F-S (SCF) relativistic	E2	Finite size	Six transitions in rare earth region.
1966	12)	Bhalla	K, M_{1-5}	H-F-S (SCF) Non-rel.	M4	Finite size	Z=52; gamma energy = 81.78 keV; investigated screening effect.
1966	62)	Bhalla	N_1/N_2 N_1/N_3	" "	E2, M1	Finite size	Calculations for 7.85 keV transition in ^{239}Pu.
1967	26)	Pauli & Alder	K, L, M	T-F-D	E1	Finite size	Investigated experimental anomalies.
1967	19)	O'Connell & Carroll	K, L, M	no	E(1-5) M(1-5)	Point nucleus	Threshold values of ICC.
1967	15)	Hager & Seltzer	K, L, M	H-F-S (SCF) relativistic	E(1-4) M(1-4)	Finite size	From 1 keV electron kinetic energy.

Table I (cont.)

1967	5)	Pauli	L_{1-3}	T-F-D	E(1,2) M(1,2)	Point and finite nucleus	Investigated nuclear size effects; Z=61-95, 5 values.
1967	11)	Pauli	K, L, M	T-F-D	E(1-4) M(1-4)	Finite size	Z=60-96 (K, L); Z=72-96 (M).
1967	13)	Borisoglebskii	$M_{4,5}$, $N_{1,2}$, O_1	T-F-D (Z=81) Hartree (Z=80) non-rel.	E(1,2) M(1,2)	Finite size	Z=80, 81.
1968	16)	Dragoun et al.	N_{1-5}	H-F-S (SCF) non-rel.	E(1-4) M(1-4)	Finite size	Z=60 to 100, from 1 keV electron kinetic energy.
1969	6)	Pauli	K, L, M,N	Screening is input program quantity.	E(0-6) M(1-6)	Finite size	Program described.
1969	63)	Dragounová et al.	All sub-shells for shells K through Q.	no	All multi-polarities for L≠0.	Point and finite size.	Program described, based on formulae derived by O'Connell and Carroll[17)19)].
1969	33)	Dragoun et al.	$N_{6,7}$, O_{1-7}, P_{1-5}, Q_1	no; and approx. screening corr.	E(1-4) M(1-4)	Point nucleus and correction for size.	Screening based on equation (9).

Abbreviations: T-F-D: Thomas-Fermi-Dirac model; SCF: Self-consistent field calculation; H: Hartree model; non-rel.: non-relativistic calculations; H-F-S: Hartree-Fock-Slater model.

Such comparisons have been made for several multipolarities in the work of Listengarten,[2] Dragoun and Jahn,[24,25] Pauli and Alder,[26] Sharff-Goldhaber and McKeown,[27] Hornshøj et al.,[28] and Bogdanovič et al.[29,30]

For a review of experimental methods for determining ICC, see the article by Subba Rao.[31]

3.1. Screening Functions

Each electron in a multielectron atom is associated with an electrostatic potential, since the electrons are in the field of the nucleus. The magnitude of this potential will vary with distance from the nucleus as

$$V(r) = \frac{\alpha Z}{r}\,\phi(r),$$

where $\alpha Z/r$ is the magnitude of the electrostatic potential at a distance r from a nucleus of charge Z; $\phi(r)$ is the screening function, which alters the hydrogenic expression to account for the decrease in potential with increasing r owing to screening of the nuclear charge by atomic electrons. The magnitude of the electrostatic potential may also be expressed as

$$V(r) = \frac{\alpha[Z - \sigma(r)]}{r},$$

where the potential at r is diminished by that portion of the electron density distribution lying at radii less than r. The nuclear charge is thereby screened by the amount $\sigma(r)$, the screening factor. For a given electron shell, the binding energy of an electron of that shell may be expressed as the sum of the potential and kinetic energies, $E = PE + KE = -\alpha e(Z - \bar{\sigma})/2R$, where $\bar{\sigma}$ and R are the screening factor and orbital radius for the electron respectively. Screening factors are also calculable for ICC for the various multipolarities, as well as for a variety of atomic coupling and hyperfine interactions (see Chapter 7, Section 4.1).

The first K-shell ICC calculations by Rose et al.[9] omitted screening effects. However, this omission for that shell is not of great consequence, since the screening factor is quite a bit less than unity. For the outer shells, however, the importance of screening increases.

Rose's tables of 1958[3] included K- and L-ICC, with screening included, and M-ICC without screening. In order to compare the experimentally measured M-ICC values for a given transition in an isotope of atomic number Z with the corresponding computed values, one may compare with the ICC for the element of atomic number $Z - \sigma$, where σ is the screening of the nuclear electrostatic field by the atomic electron cloud.

In general, corrections of ICC for screening can be made by applying the expression

$$\alpha = \alpha_0 \left(\frac{\Psi}{\Psi_0}\right)^2, \tag{9}$$

where α is the corrected ICC; α_0 is that calculated without screening; Ψ and Ψ_0 are the screened and unscreened orbital electron wave functions in the nuclear vicinity. This procedure was employed with surprising success by Dragoun et al.[32] in determining the ICC for N-, O-, and P-shell electrons and for 14 transitions of different energies and multipolarities. Because calculated ICC values for these outer shells agreed so well with the experimental values (to within a factor of 2 and often to less than 50%) extensive calculations were carried out. The ICC values for the $N_{6,7}$, O_{1-7}, P_{1-5}, and Q_1 atomic subshells calculated according to equation 9 have been presented,[33] where $E(1–4)$ and $M(1–4)$ multipolarities and 10 to 500 keV gamma-transition energies were considered. Also included were tabulations for finite nuclear size corrections and of screening correction factors.

In 1964 Chu and Perlman[34] treated this problem in another manner and obtained a semiempirical screening correction for the $M4$-81.78 and 88.46-keV transitions, in the M_1, $M_{2,3}$, and $M_{4,5}$ subshells, of ^{121m}Te and ^{123m}Te ($Z = 52$), respectively. By comparing the theoretical coefficients where screening is neglected with the corresponding experimental values, it was possible to determine σ_{M_i}, where $Z_{\text{eff}\,M_i} = Z - \sigma_{M_i}$, for the $3s$, $3p$, and $3d$ electrons. These screening values are much smaller than screening values used in evaluating wave functions for the computation of electron binding energies in moderately heavy atoms[35] (see Ref. 34).

The difference between screening factors used in calculating atomic electron binding energies and those associated with the calculations of ICC have been shown to be real and to be expected.

O'Connell and Carroll[36] considered screening effects from a different point of view, employing a semiempirical method where two different types of Z_{eff} were considered: Z_b (the effective atomic number seen by bound electrons) and Z_c (that seen by continuum state electrons following conversion). The effective charge Z_b was taken as the effective charge associated with the binding energy of the atomic electron in question. It was noted that the Z_b's were generally smaller than Z_{eff} obtained by Chu and Perlman.

The quantity Z_c was set equal to Z_b since, although $Z_c \to 0$ as the conversion electron goes to infinity, the greatest contribution to the transition matrix element occurs at distances on the order of the original atomic orbital radius distance. Hence $Z_b = Z - \sigma_{BE}$ in these calculations; the consequences of this choice have been treated by several authors.

Dragoun and Jahn[24] compared the large and small components of the 3*s* electron wave function in tellurium as obtained by different computation procedures. Figures 1*a* and *b* show rg_{-1} and rf_{-1} respectively as functions of radial distance from the nucleus. The curves for Rose's calculation ($\sigma = 0$) are the least expanded, whereas those for O'Connell and Carroll's ($\sigma = 26.4$) are greatly expanded toward larger *r*. The other wave functions, using Hartree–Fock and semi-empirical potentials, lie between these two limiting cases, but closer to the former. It was pointed out by the authors that the screening functions associated with

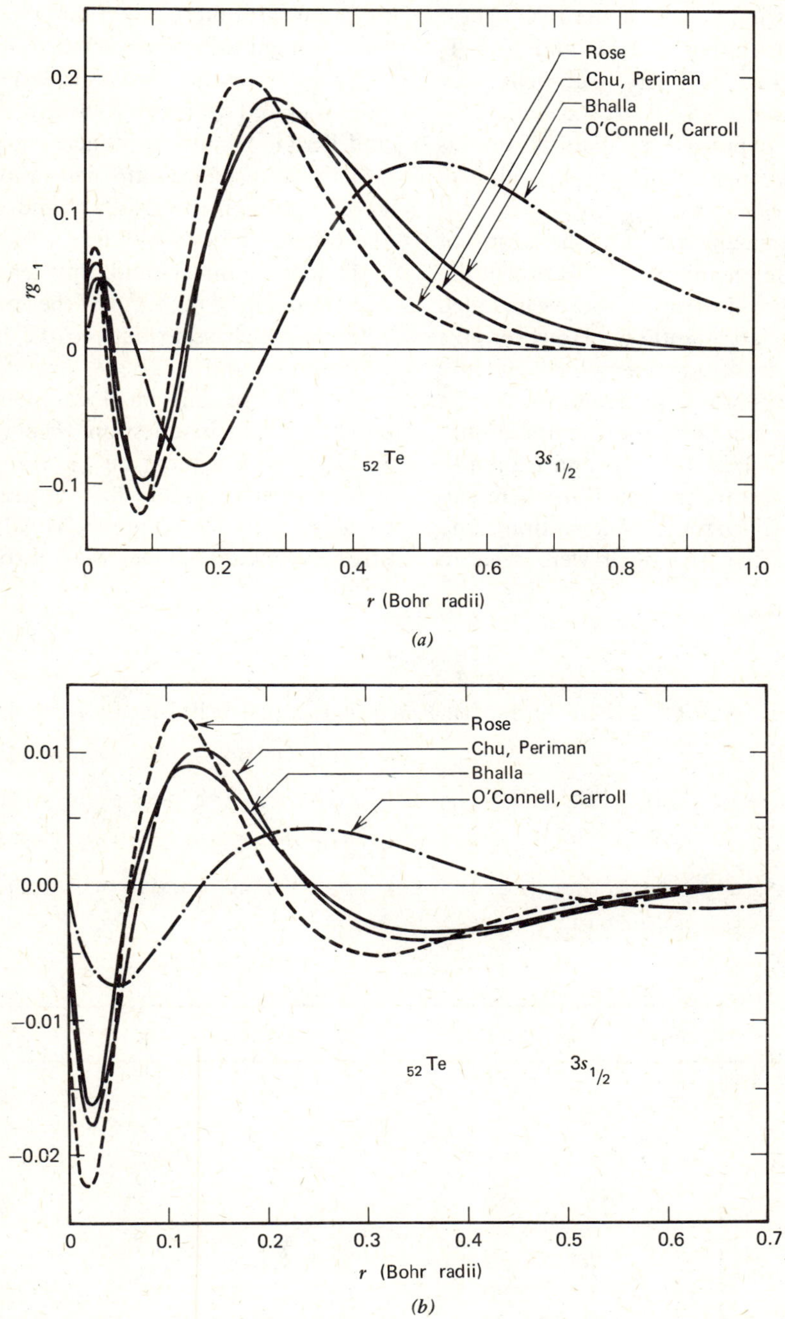

Figure 1. (*a*) Product of the electron coordinate *r* and the large component g_{-1} of the 3*s*-electron wave function in tellurium: full curve, Bhalla's[12] calculation with a Hartree–Fock potential; dashed curves, wave functions in pure Coulomb fields according to Rose[3] ($\sigma = 0$), Chu and Perlman[34] ($\sigma = 7.0$), and O'Connell and Carroll[36] ($\sigma = 26.4$). (*b*) Product of the electron coordinate *r* and the small component f_{-1} of the 3*s* electron wave function in tellurium. (From Ref. 24.)

various properties of the wave functions (e.g., $\sigma(BE)$, $\sigma(\bar{r})$, $\sigma(r^2)$) differ, as is shown by the calculations by Garstang and Mayers.[37] Therefore, we might expect a σ-value associated with ICC which differs strongly from those associated with the above-mentioned properties (see Chapter 7, Section 4.1). In addition, it was shown that the screening models of Hartree–Fock, Hartree–Fock–Slater, and Thomas–Fermi–Dirac give slightly differing values of the ICC. These differences for the nonrelativistic H–S and T–F–D models can be as large as 3.5% ($L_{2,3}$-subshells) and 6% ($M_{2,3}$-subshells) for $Z = 78$.[23]

This view was also expressed by Pauli[23a] in his treatment of the significance of atomic screening in L-ICC. There he also calculated σ_{M_1}, σ_{M_2}, and σ_{M_3} for two different gamma-energies for the atom of interest to Chu and Perlman and for an $M4$ transition. The screening factors agreed well with those calculated semiempirically by the latter authors. However, the existence of a gamma-transition energy dependence of these screening factors was also demonstrated. In addition, the screening numbers for the three subshells M_{1-3} for the $M1$, $M2$, $M4$, $E1$, and $E2$ multipole orders and for two transition energies were calculated. The screening values for the $E1$ multipole and the $M_{1,3}$ subshells and the M_{tot} shell were seen to differ strongly from the average, while the other values agree well.

For illustration purposes, the screening factors for the M_{1-5} subshells in $Z = 65$ and for a nuclear transition energy of $k = 0.10\ m_o c^2$ are listed in Table II for the multipolarities $E(1–5)$ and $M(1–5)$. These values were obtained by comparing Hager and Seltzer's[15] theoretical values (with screening) with those by Rose[3] (without screening). Slight differences due to finite size effects might also be expected.

Table II reveals that σ is larger for lower multipolarities and decreases to near the corresponding Chu and Perlman values with increasing L. In addition, for $M4$ the agreement with these authors' values is very good, although the Z-values and gamma-energies for the two sets of values differ.

To try to understand this tendency qualitatively, one should note the form of the components R_i, equations 6, which contribute to the ICC values through equations 7 and 8. The R_i are weighted integrations over the large and small components of the atomic wave function in question. The weighting factor G_L or G_{L-1} is dependent on the transition multipolarity according to equation 8. It is seen that integrations included in the R_i expressions will be weighted more strongly toward smaller radii through the factors G_L

Table II.

Screening factors σ(EL) and σ(ML) for multipolarities L=1 to 4 in the M-subshells of Tb (Z = 65). The transition energy is assumed to be 51 keV. Calculations with screening[15] are compared with those without[3], and the screening factor is taken as the Z-value difference between where the non-screened value agrees with the tabulated screened value.

Multipolarity	Subshell M_1	M_2	M_3	M_4	M_5
E1	26	11	13	25	14.5
E2	9	7	8.5	10.2	14
E3	5	7	5.5	19	9.5
E4	6	7	7	8	9.5
M1	4.5	6.5	9.5	11	15.5
M2	6	6.5	6	8.5	11.5
M3	4.6	8	5.5	9	8.5
M4	7	9	6	9	7.5

and G_{L-1}. The nature of the wave functions at small radii will contribute more strongly with increasing multipolarities. The multipole interaction with the orbital electron will be associated with the smaller screening values at smaller radii. Therefore we might expect the tendency revealed in Table II toward decreasing screening values associated with internal conversion of increasing multipolarity.

The reader is also referred to Chapter 7, Section 4.1, for a comparison and discussion of screening factors associated with several properties of atomic electron wave functions and interactions.

4. INTERNAL CONVERSION ELECTRON LINE ANALYSIS

The analysis of IC electron spectral lines consists essentially of determining the energy (or momentum) position and intensity of the line.

An IC electron line shape is dependent on a number of factors: the width, which results from the natural atomic level width; the nuclear level width (negligible for low energy transitions); the spectrometer window function; and inelastic electron scattering in the radioactive source. These factors and their influence on the IC line shapes and positions, as well as the relative reliability of various techniques for determining this latter, are treated in Chapter 6, Section 3 and Chapter 7, Sections 3.2.2. and 3.2.3. See also Chapter 8, Section 7.

Energy calibration of a spectrum is usually accomplished by the simultaneous measurement of standard electron lines of accurately known energy. These may be Auger lines (always present in IC electron spectra) or IC lines of gamma-transitions listed in Appendix G. (The manner of obtaining these kinetic energies is explained in that appendix text.) Alternatively, if no calibration standard is available, the momentum ratio method (Chapter 7, Section 3.2.3) may be used.

The determination of IC line intensities is a problem of a different nature. Often the subshell components overlap to such an extent that accurate analysis is possible only if the individual line shapes and separations are known.[32, 38–40] The former may be determined as described in Chapter 6 and the latter may be obtained from differences in binding energies of the atomic levels involved (Appendix F).

If the IC lines are close to one another and the natural atomic level widths differ negligibly, then the log plot of a line shape will need only to be shifted to account for energy and intensity differences. Figure 2 illustrates this procedure. The ^{170}Yb 84.26-keV transition L-lines are shown plotted both linearly and logarithmically;[40a] the line profiles in the log plot are identical to each other. The logarithmic plots of IC lines that lie far apart will be dissimilar because of (a) different proportions of instrumental to natural width contributions to the total measured line width, and (b) different degrees of contributions of inelastic scattering to the total line shape. These are treated in Chapter 6, Section 3. The extent and nature of other contributions, such as thermal vibration in the source, are found in Chapter 6, Section 4.

5. EFFECTS OF ELECTRON DENSITY NEAR THE NUCLEUS

The variation of the finite values of atomic wave function amplitudes in the vicinity of the nucleus gives rise to two effects: alteration of essentially nuclear properties and processes, such as the isomer shift, and alteration of nucleus-atomic interactions and processes (i.e., mainly the IC process). In addition, atomic shell binding energies can be grossly changed through placing a free atom in some high degree of ionization, which may in turn effect the internal conversion process.

The study of the effect of alteration of essentially nuclear properties and processes is best made through the use of the Mössbauer effect. Usually we are interested in transitions of low energy[41] (e.g., 14.4 keV in ^{57}Fe, 6.2 keV in ^{181}Ta, described in Ref. 42, 24 keV in ^{119}Sn, etc.), but even much higher energy levels have been used: 58 keV in ^{159}Tb, 74.5 keV in ^{161}Dy, 79.5 keV in ^{158}Gd have been studied by Atzmony et al.,[43] Khurgin et al.,[44] and Fink,[45] respectively. This field has been covered by Mössbauer[46] and by De Benedetti et al.,[47] whose works also list reviews of related research.

5.1. Influence on the Isomer Shift

The calculation of an atomic system's total energy is usually carried out by assuming that the nucleus is a point charge Ze. Now, if the nucleus is allowed to expand to its true finite size in the presence of the negative charge density due to the presence of the atomic electron cloud, the total energy of the atom will decrease. It can be shown[47] that the energy of a gamma-transition is influenced by the finite radii of the initial and final states. The final nuclear state of a Mössbauer transition is always the ground state, and for constant-electron charge density at the nucleus may be expressed as

$$\Delta E_\gamma = E_{em} - E_{ab} = \tfrac{2}{5}\,\pi Z e^2 \left|\psi(0)\right|^2_{em} [R_{em}^2 - R_{ab}{}^2] \quad (10)$$

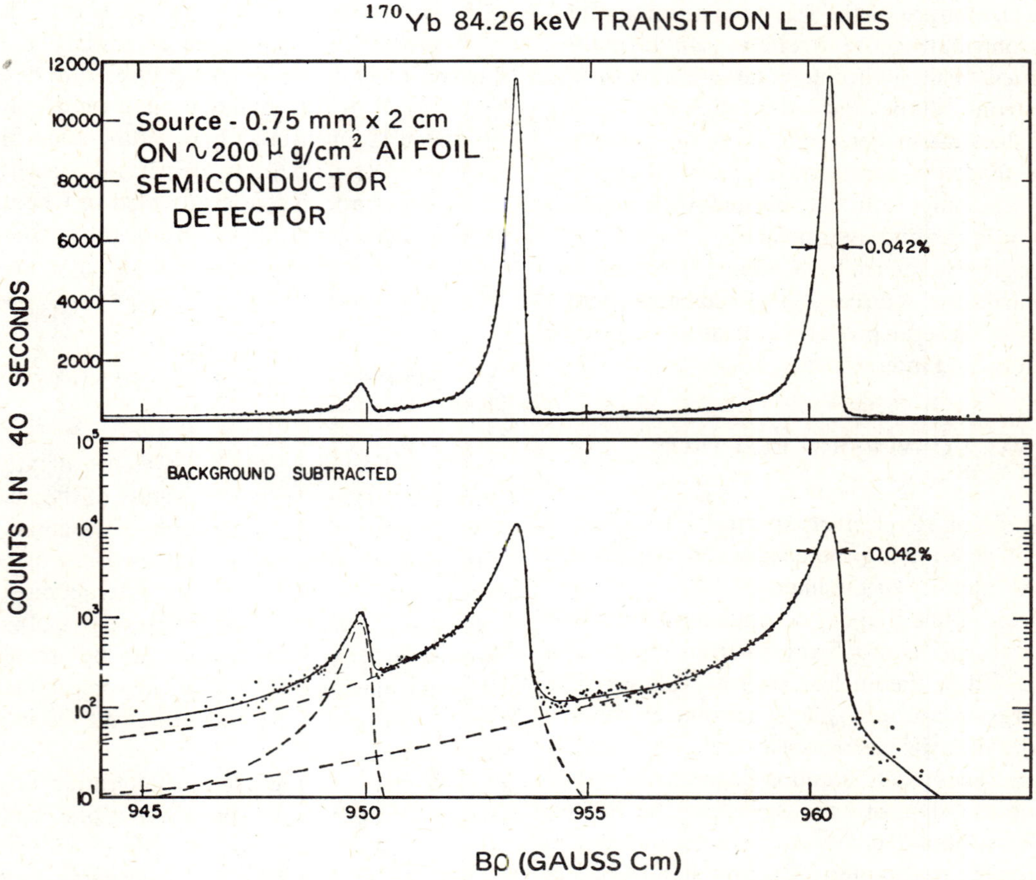

Figure 2. Spectrum of the *L*-conversion lines of the 84.26-keV *E*2 transition in ^{170}Yb reported by Gelletly et al.[40a] This set of data was obtained using a cooled Si(Li) detecter in combination with the $\pi\sqrt{2}$ ($\rho_0 = 100$ cm) iron-free spectrometer at Chalk River. A possible systematic effect on intensity ratios due to detector response was being sought, but none was found.

where $|\psi(0)|^2_{em}$ is the electron charge density near the emitter nucleus and R_{em} and R_{ab} are the emitter and absorber nuclear radii. In the Mössbauer effect one actually measures the difference between ΔE_γ for the absorber and for the emitter. This is called the isomer shift and can be expressed

$$\delta = \frac{4\pi}{5} Ze^2 R(R_{em} - R_{ab})\left\{\left|\psi(0)\right|^2_{ab} - \left|\psi(0)\right|^2_{em}\right\}, \quad (11)$$

where δ is the experimentally measured isomer shift, and $R_{em} - R_{ab}$ is the change in nuclear radius associated with the isomeric transition from the excited to the ground state; $|\psi(0)|^2_{ab}$ and $|\psi(0)|^2_{em}$ are the electron densities at the nucleus for the absorber and emitter, and $R \cong R_{em} \cong R_{ab}$. For further considerations, see Ref. 48.

It can be seen from equation 10 that interpretation of the isomer shift requires knowledge of the changes of both the nuclear radius and the electron density at the nucleus. Therefore, if the electron density change can be determined by an independent method, the isomer shift can be interpreted solely in terms of the nuclear radius change.

The IC process, which is dependent on the interaction of the nuclear multipole field with the atomic electron density about the nucleus, is expected to give information about this density. Information on the variation of the atomic electron density with chemical and physical property changes can also be obtained. Photoelectron and Auger electron spectral measurements may provide this necessary information to some degree, as well.

5.2. Influence on Internal Conversion

This section can be divided into two subsections treating the alteration of IC probabilities (i.e., line intensities) and the alteration of IC electron energy values and the influence on IC probabilities.

5.2.1. Internal Conversion Probabilities

Internal conversion of a nuclear transition may take place in atomic levels of binding energies less than the transition energy, aside from a possible work-function correction. As the transition energy lessens, two changes occur in the decay process. First, the gamma-transition probability decreases rapidly, and internal conversion becomes increasingly the dominant decay process. Second, the IC process is confined more and more to the outer atomic shells of sufficiently low binding energies to permit participation in IC. Since the outermost atomic levels are most strongly influenced by the chemical state of the material, the low energy IC process itself is also affected thus.

Nuclear transitions of very low energy are therefore necessary in order to observe changes in decay rates, or half-lives, for a nuclear level. Notable examples are the 30-eV transition in ^{235m}U (see Refs. 49 and 50), and the low energy transitions in ^{90}Nb, ^{193}Pt, ^{99}Tc, and ^{125}Te (relatively high energy transition), studied by Cooper,[51] Marelius,[52] Bainbridge et al.,[53] and Malliaris and Bainbridge respectively.[54] To this list should also be added the early measurements on ^{7}Be made by Leininger et al.[55] In this research half-lives were measured as a function of the chemical or physical state of the elemental atoms. Table III provides a summary of the results.

Each of the references listed in Table III contains a discussion of the specific causes for the lifetime changes. Quantitatively, the chemical binding—involving the valence electron—alters the value of $|\psi(0)|^2$ for that outer electron, and the nuclear decay rate changes accordingly.

The electrons involved in chemical bonding usually contribute very little to the total IC process. For dipole transitions, however, major contributions to the IC probability are made by *s*-electrons. Outer shell conversion probabilities will then be functions of these *s*-electron densities near the nucleus. Perlman and Emery[56] have used calculations by J. Mann[57] to show that, on electron density distributions in various tin ions, outer shell electron densities near the nucleus may undergo significant alterations with the ionization (or chemical) state of the atom. The calculations revealed, for example, that the removal of one of the two 5*s*-valence electrons results in a change in 5*s*-density near the nucleus from 2.44 to 1.40. The remaining 5*s*-electron tends to move in to slightly smaller radii. Effects of atomic ionization on wave functions having ($n < 5$) are almost negligible.

Therefore, for dipole transitions, *s*-electrons contribute strongly to the total conversion probability. Outer shell conversion line intensities are functions of *s*-electron densities near the nucleus; they are therefore directly related to the isomer shifts observed in the Mössbauer effect.

Perlman and Emery[56] have also supplied insight into the probable reasons for several of the decay rate changes due to chemical effects.

A case in point, studied by Cooper,[51] is the large magnitude (3.6%) of the chemical effect on ^{90m}Nb decay, and it was suggested that this unusually high percentage may have been caused by an ordinarily unimportant factor: the transition energy 2.38(36) keV is in the *L*-subshell binding energy region—L_1, 2.698; L_2, 2.465; and L_3, 2.371 keV. Binding energies depend on chemical form; indeed, changes as great as 5 eV sometimes occur for oxidized forms, compared with reduced forms. Therefore, it may be that the IC in one of the subshells is strongly altered or even prohibited by changes in the chemical form. It should be remembered, however, that the "binding energy change" due to change in chemical state is primarily owing to electrostatic interaction between the electron and the ionic lattice environment after escape from the atom. The change in eigenvalue for the electron energy in the atom is quite small compared with this (see Chapter 5, Sections 3 and 3.2 and Chapter 7, Sections 4 and 4.1). However, the binding energy, defined as that energy required to remove the electron from the atom to infinity, may become larger than the gamma transition energy, and this change will prohibit IC in that subshell.

The article by Perlman and Emery[56] contains a valuable discussion of similar considerations of other chemical effects on ICC.

The techniques used to measure the relative changes in decay rates were usually carried out using differential methods.[50,53,58] Here two identical electron detectors (multipliers or proportional counters) measure the radiation from the two sources of approximately the same intensity. The output currents of the detectors are balanced at time zero, and the drift from this balance is recorded as a function of time. A slight difference in the two lifetimes (due, say, to a chemical state difference) can be evaluated by making the measurement, interchanging the sources, repeating the measurement, and analyzing the obtained data.

The 2.38(36)-keV transition in ^{90m}Nb is the only one listed that does not lead directly to the ground state. Here then it is possible to measure the gamma intensity depopulating the lower level in that transition.[51] The isomeric state has a measured half-life of 24 sec. A change in ^{90m}Nb half-life, by a change of chemical state, will show a measurable restabilization of the measured gamma intensity.

The influences of the physical effects of pressure and temperature on nuclear decay rates are provided in Table III. The reader is also referred to the report by

Table III.

Chemical and physical alterations of total nuclear decay rates.

Isotope (trans. energy in keV)	Ref.	Chemical effects — Results (measured change)
^{7}Be	55)	$\frac{\lambda(Be) - \lambda(BeF_2)}{\lambda(Be)} = 7.41(47)\times10^{-4}$
^{90m}Nb (2.38(36))	51)	$\frac{\lambda(Nb) - \lambda(\text{fluoride complex})}{\lambda(Nb)} = 3.6(4)\times10^{-2}$
^{99m}Tc (2.15(3))	53)63)64)	$\frac{\lambda(Tc) - \lambda(KTcO_4)}{\lambda(KTcO_4)} = -2.4(1)\times10^{-3}$
" "	" "	$\frac{\lambda(KTcO_4) - \lambda(Tc_2S_7)}{\lambda(Tc_2S_7)} = 2.7(1)\times10^{-3}$
^{125m}Te (109.4)	54)	$\frac{\lambda(Te) - \lambda(Ag_2Te)}{\lambda(Te)} = 2.59(18)\times10^{-4}$
" "	"	$\frac{\lambda(Te) - \lambda TeO_2)}{\lambda(TeC_2)} = 0.36(17)\times10^{-4}$
^{193m}Pt (1.642(2))	52)	$\frac{\lambda(Pt) - \lambda(PtCl_x)}{\lambda(PtCl_x)} = 4(2)\times10^{-2}$ $(x \geq 1)$
^{235m}U (0.030(3))	50)49)	$\frac{\lambda(U\ in\ Pt) - \lambda(U\ in\ C)}{\lambda(U\ in\ C)} = 3.18(50)\times10^{-3}$
" "	" "	$\frac{\lambda(U\ in\ Pt) - \lambda(U\ in\ Si)}{\lambda(U\ in\ Si)} = 2.21(36)\times10^{-3}$
		Physical effects
		Pressure
^{7}Be	65)	$\frac{\lambda(10^5\ atm) - \lambda(1\ atm)}{\lambda(1\ atm)} = 0.2\times10^{-2}$
^{90m}Nb (2.38(36))	51)	" " $= 6.3(7)\times10^{-3}$
^{99m}Tc (2.15(36))	66)	" " $= 2.3(5)\times10^{-4}$
^{131}Ba	65)	" " $= 6.6\times10^{-4}$
		Temperature
^{99m}Tc (2.15(36))	67)51)	$\frac{\lambda(4.2°K,\ Normal) - \lambda(Room\ temp.)}{\lambda(Room\ temp.)} = 1.3(4)\times10^{-2}$
" "	" "	$\frac{\lambda(4.2°K,\ Supercond.) - \lambda(Room\ temp.)}{\lambda(Room\ temp.)} = 6.4(4)\times10^{-2}$
^{90m}Nb (2.38(36))	68)	$\frac{\lambda(4.2°K,\ Supercond.) - \lambda(4.2°K,\ Normal)}{\lambda(4.2°K,\ Normal)} = -1.95(55)\times10^{-3}$

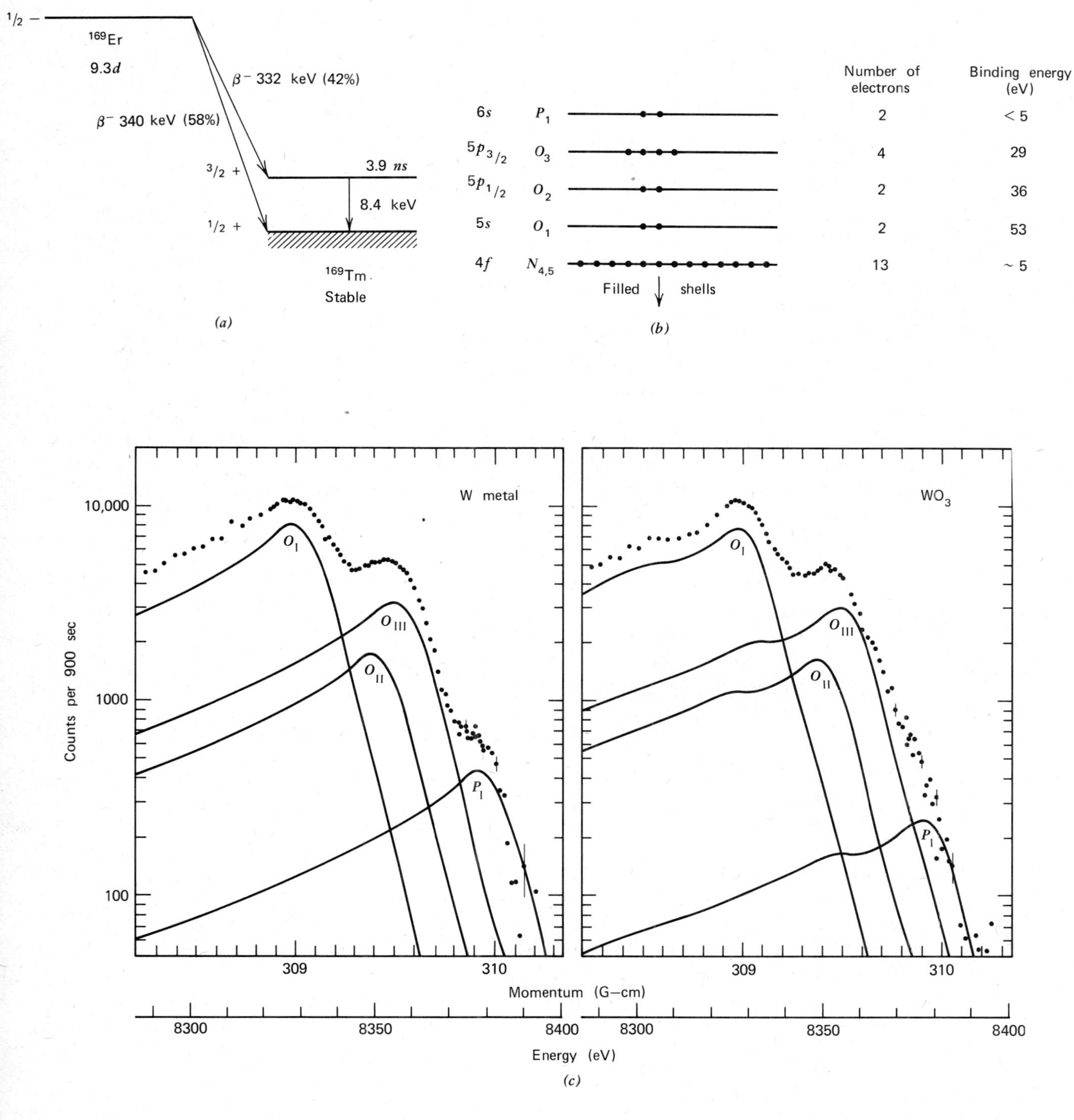

Figure 3. (*a*) Decay scheme of ^{169}Er. (*b*) Electron configuration of the outer shells in a free thulium atom. (*c*) Analysis of the *O* and *P* conversion lines of the 8.4-keV transition in ^{169}Tm. In the left-hand figure the activity is imbedded in tungsten metal, and in the other it is placed in WO_3. The shape of the master curve was obtained in both cases from a simultaneous run of the associated *N*-spectrum. The momentum and energy scales should be multiplied with 1.0012 and 1.0024, respectively. (From Ref. 69.)

Bergamini et al.,[59] who investigated the physical state effect (liquid versus solid) on the decay of ^{131}I and daughters.

We would expect the degree of ionization of an atom to effect the decay rate of a nuclear transition, largely because of the removal of electrons available for participation in the IC decay process. de Wieclawik[68a] has carried out a measurement of the half-life of the 59.5-keV level in ^{237}Np following the alpha-decay of ^{241}Am. He compared the decay half-lives the +1 charged and +15 charged recoil atoms. No difference in half-life was measured. (See Chapter 5, Section 3.3 for a discussion.)

5.2.2. *Conversion Line Measurements*

If the nuclear transition is of sufficiently high energy to allow direct measurement of the IC electron spectrum, the relative probability of conversion in the various shells may be determined directly as a function of chemical or physical state.

The 8.4-keV transition in ^{169}Tm may be used as an example[69]: Figure 3*a* shows the points of interest to the problem in the decay of ^{169}Er to ^{169}Tm, and Figure 3*b* gives the population of the outer atomic levels in this element. (Note the order with respect to binding energy.)

As can be seen in Figure 3*c*, the atoms of ^{169}Tm in metallic tungsten and in an oxide WO_3 show different IC electron spectra with regard to the P_1 (i.e., 6*s*) IC peak. The relative intensities of the conversion lines from the 3*s*, 4*s*, and 5*s* levels showed very good agreement with the relativistic Hartree–Fock–Slater wave function calculations of the relative densities of these electrons in the vicinity of the nucleus. The agreement for the 6*s* level was somewhat poorer. Similar measurements have been carried out by Bocquet et al.[70,71] on ^{119}Sn.

It has been shown that IC electron spectrum measurements for the outer atomic shells are sensitive to the degree of penetration of these electrons into the center of the atom. Since this penetration is dependent on the chemical and/or physical state of the atom, the high resolution investigation of IC spectra for the purpose of learning about the environmental conditions and chemical state of the emitting atom appears to be a potential research method.

REFERENCES

1. M. E. Rose, in *Beta- and Gamma-Ray Spectroscopy*, K. Siegbahn (Ed.), North-Holland Publishing Company, Amsterdam, 1955.
2. M. A. Listengarten, *Bull. Acad. Sci., USSR, Phys. Ser.* **22** (1958) 755.
3. M. E. Rose, *Internal Conversion Coefficients*, North-Holland Publishing Company, Amsterdam, 1958.
4. E. L. Church and J. Weneser, *Ann. Rev. Nucl. Sci.* **10** (1960) 193.
5. H.-C. Pauli, *Helv. Phys. Acta* **40** (1967) 713.
6. H.-C. Pauli, "A Computer Program for Internal Conversion Coefficients and Particle Parameters," Niels Bohr Inst. Rept., Copenhagen, 1968.
7. M. E. Rose, *Relativistic Electron Theory*, John Wiley & Sons, Inc., New York, 1960.
8. M. E. Rose, *Multipole Fields*, John Wiley & Sons, Inc., New York, 1955.
9. M. E. Rose, G. H. Goertzel, B. I. Spinrad, J. Harr, and P. Strong, *Phys. Rev.* **83** (1951) 79.
10. L. A. Sliv and I. M. Band, in *Alpha-, Beta-, and Gamma-Ray Spectroscopy*, K. Siegbahn (Ed.), North Holland Publishing Company, Amsterdam, 1965, Vol. 2, p. 1639.
11. H.-C. Pauli, Purdue Univ. Rept. COO-1420-137, Lafayette, Indiana, 1967.
12. C. P. Bhalla, in *Internal Conversion Processes*, J. H. Hamilton (Ed.), Academic Press, New York, 1966, p. 373.
13. L. A. Borisoglebskiĭ, Jadernia fiz., SSSR, **5** (1967) 379 [*Sov. J. Nucl. Phys.* **5** (1967) 266].
14. C. P. Bhalla, *Phys. Rev.* **157** (1967) 1136.
15. R. S. Hager and E. C. Seltzer, "Internal Conversion Tables, I: *K*-, *L*-, *M*-Shell Conversion Coefficients," California Institute of Technology, Rept. CALT-63-60 (1967), *Nucl. Data* **A4** (1968) 1.
16. O. Dragoun, H.-C. Pauli, and F. Schmutzler, Max-Planck-Institut für Kernphysik, Heidelberg, Rept. (MPIH-1968) V-14 (even *Z* values); *Nucl. Data* **A6** No. 3 (1969) (all *Z* 60–100).
17. R. F. O'Connell and C. O. Carroll, *Phys. Rev.* **138** (1965) B1042.
18. R. F. O'Connell, *Nucl. Phys.* **45** (1963) 142.
19. R. F. O'Connell and C. O. Carroll, *Nucl. Data* **A3** (1967) 287.
20. S.-E. Karlsson, I. Andersson, Ö. Nilsson, G. Malmsten, C. Nordling, and K. Siegbahn, *Nucl. Phys.* **89** (1961) 513.
21. Ö. Nilsson, S. Högberg, S.-E. Karlsson, and G. M. El-Sayad, *Nucl. Phys.* **A100**(1967) 351.
22. I. Andersson, S.-E. Karlsson, G. Malmsten, Ö. Nilsson, and E. Aisenberg, *Ark. Fys.* **30** (1965) 488; J. H. Hamilton, *Phys. Letters* **20** (1966) 32; M. Arnoux and A. Gizon, *Compt. Rend.* **264** (1967) 1518.
23. O. Dragoun and H.-C. Pauli, Max-Planck-Institut für Kernphysik, Heidelberg, Rept. V-10 (1967).

23a. H.-C. Pauli, *Nucl. Phys.* **A109** (1968) 94.
24. O. Dragoun and P. Jahn, *Nucl. Phys.* **A101** (1967) 305.
25. O. Dragoun and P. Jahn, *Phys. Letters* **24B** (1967) 146.
26. H.-C. Pauli and K. Alder, *Z. Phys.* **203** (1967) 255.
27. G. Scharff-Goldhaber, and M. McKeown, *Phys. Rev.* **158** (1967) 1105.
28. P. Hornshøj, B. I. Deutch, and A. Miranda, *Nucl. Phys.* **95** (1967) 65.
29. H. Bogdanovič, M. Mladjenovič, and R. Stepič, *Z. Phys.* **216** (1968) 267.
30. H. Bogdanovič, M. Mladjenovič, and R. Stepič, *Nucl. Phys.* **A106** (1968) 209.
31. B. N. Subba Rao, *Nucl. Instr. Meth.* **45** (1966) 22.
32. O. Dragoun, P. Jahn, H. Allers, M. Vobecký, and H. Daniel, Max-Planck-Institut für Kernphysik, Heidelberg, Rept. V-11 (1968); *Phys. Letters* **28B** (1968) 251.
33. O. Dragoun, Z. Plajner, and F. Schmutzler, Max-Planck-Institut für Kernphysik, Heidelberg, Report V-5 (1969).; O. Dragoun "Int. Conf. Radioactivity in Nuclear Spectroscopy," Vanderbilt Univ., Nashville, Tenn. (August 11-15, 1969).
34. Y. Y. Chu and M. L. Perlman, *Phys. Rev.* **135** (1964) B319.
35. A. C. Douglas, D. R. Hartree, and W. A. Runciman, *Proc. Cambridge Phys. Soc.* **51** (1955) 486.
36. R. F. O'Connell and C. O. Carroll, in *Internal Conversion Processes*, J. H. Hamilton (Ed.), Academic Press, New York, 1966, p. 333; *Nuovo Cim.* **38** (1965) 1431.
37. R. H. Garstang and D. F. Mayers, *Proc. Cambridge Phil. Soc.* **62** (1966) 777.
38. O. Dragoun and P. Jahn, Max-Planck-Institut für Kernphysik Rept. V-4 (1967); O. Dragoun, C. Ribordy, and O. Huber, *Nucl. Phys.* **A124** (1969) 337.
39. P. Erman, G. T. Emery, and M. L. Perlman, *Phys. Rev.* **147** (1966) 858.
40. H. J. Hennecke, J. C. Manthuruthil, O. Bergman, and C. R. Cothern, *Phys. Rev.* **159** (1967) 955.
40a. W. Gelletly, J. S. Geiger, and R. L. Graham, *Phys. Rev.* **157** (1967) 1043.
41. S. Ofer and I. Nowik, *Nucl. Phys.* **A93** (1967) 689.
42. C. Sauer, E. Matthias, and R. L. Mössbauer, *Phys. Rev. Letters* **21** (1968) 961.
43. U. Atzmony, E. R. Bauminger, and S. Ofer, *Nucl. Phys.* **89** (1966) 433.
44. B. Khurgin, S. Ofer, and M. Rakavy, *Nucl. Phys.* **A110** (1968) 577.
45. J. Fink, *Z. Phys.* **207** (1967) 225.
46. R. L. Mössbauer, in *Alpha-, Beta-, and Gamma-Ray Spectroscopy*, K. Siegbahn (Ed.), North-Holland Publishing Company, Amsterdam, 1965.
47. S. De Benedetti, F. de S. Barros, and G. R. Hoy, *Ann. Rev. Nucl. Sci.* **16** (1966) 31.
48. V. I. Goldanskii, *Mössbauer Effect and Applications to Chemistry*, Van Nostrand Company, Princeton, N. J., 1966; R. A. Uhler and R. A. Sorensen, *Nucl. Phys.* **86** (1966) 1.
49. M. N. De Mevergnies, *Phys. Letters* **26B** (1968) 615.
50. H. Mazaki and S. Shimizu, *Phys. Letters* **17** (1966) 275; *Phys. Rev.* **148** (1966) 1161.
51. J. A. Cooper, thesis, Lawrence Radiation Laboratory, Berkeley, Calif., UCRL Rept. 16910.
52. A. Marelius, *Ark. Fys.* **37** (1968) 427.
53. K. T. Bainbridge, M. Goldhaber, and E. Wilson, *Phys. Rev.* **90** (1953) 430.
54. A. C. Malliaris and K. T. Bainbridge, *Phys. Rev.* **149** (1966) 958.
55. R. F. Leininger, E. Segré, and C. Wiegand, *Phys. Rev.* **76** (1949) 897; *Phys. Rev.* **81** (1951) 280.
56. M. L. Perlman and G. T. Emery, "Int. Conf. Radioactivity in Nuclear Spectroscopy," Vanderbilt Univ., Nashville, Tenn. (August 11–15, 1969).
57. J. B. Mann, Los Alamos Lab. Rept. LA-3690 (1967); LA-3691 (1968).
58. E. Rutherford, *Wein. Beu.* **120** (1911) 303 (see Ref. 53).
59. P. G. Bergamini, G. Palmas, F. Piantelli, and M. Rigoto, *Phys. Rev. Letters* **18** (1967) 468.
60. L. A. Borisoglebskiĭ, *Zh. Eksp. i Teor. Fiz. SSSR* **48** (1965) 1347 [*Russ. J. Exp. Theor. Phys.* **21** (1965) 900].
61. C. P. Bhalla, *Z. Phys.* **196** (1966) 26.
62. C. P. Bhalla, in *Processes of Bases of Spin-Parity Assignment*, N. B. Grove (Ed.) Academic Press, New York, 1966, pg. 104.
63. N. Dragounová, O. Dragoun, and G. Heuser, private communication from O. Dragoun (1969) (to be published).
63a. B. S. Dzhelepov and O. E. Kraft, *Vestnik Leningrad Univ., Ser. Mat. Fiz. Khim.* **10** (1955) 97.
64. J. C. Slater, *Phys. Rev.* **84** (1951) 1261.
65. W. B. Gogarty, S. S. Kistler, and E. B. Christiansen, Off. Nav. Res. Tech. Rept. VII, Contr. No. 1288(62), Proj. Nr-0520357, June, 1963 (see Ref. 53).
66. K. T. Bainbridge (1952) (see Ref. 53).

67. D. H. Byers and R. Stump, *Phys. Rev.* **112** (1958) 77.
68. A. Olin and K. T. Bainbridge, *Phys. Rev.* **179** (1968) 450.
68a. W. de Wieclawik, private communication, 1969.
69. T. A. Carlson, P. Erman, and K. Fransson, *Nucl. Phys.* **A111** (1968) 371.
70. J. P. Bocquet, Y. Y. Chu, O. C. Kistner, M. L. Perlman, and G. T. Emery, *Phys. Rev. Letters* **17** (1966) 809.
71. J. P. Bocquet, Y. Y. Chu, G. T. Emery, and M. L. Perlman, *Phys. Rev.* **167** (1968) 1117.

Chapter Five

Consequences and Applications of Atomic Recoil after Alpha-Decay

1. INTRODUCTION

Atomic recoil after nuclear reactions or nuclear decay can be exploited to study and investigate a number of nuclear phenomena. The heavier the bombardment or decay particle, the greater the associated atomic recoil velocity, and the stronger the recoil effect. Atomic recoil effects in solids following gamma-decay are important to resonance absorption by nuclei and in relation to the Mössbauer effect. For the gaseous state, atomic recoil following gamma-decay is, however, still of minor interest.

Post-beta-decay[1] or internal conversion atomic recoil effects are measurable for higher decay energies. The maximum and average recoil energies of ^{88}Rb following the beta-decay of gaseous ^{88}Kr, for example, have been measured to be 51.5(2) and 29(1) eV, respectively, the former in agreement with the maximum beta-energy 2.6 MeV.[2] The atomic recoil velocity then has the maximum value of close to 4×10^6 cm/sec or about 1 order of magnitude smaller than recoil velocities caused by alpha-decay.

In the beta-decay of the lighter ^{6}He to ^{6}Li, the recoil energies of the light atoms have energies up to 1425 eV.[3] The article by Snell[4] reviews work done up to 1962 on the subject of recoil and other consequences of radioactive decay processes other than alpha.

Coincidence techniques can also be applied in the measurement of nuclear level lifetimes following alpha-decay. One can approach this problem from two different points of view. Alpha-particle–recoil-atom coincidences can be used to gate conversion electrons whose energy spectrum is measured. In this method either an electrostatic potential is placed about the source or a stopping foil is used, to obtain a recoil path length versus electron kinetic energy shift dependence; these techniques are treated further in Section 2.2. In the alternative method, the time-dependent alpha-particle–conversion-electron coincidence measurement is carried out. Here the recoil atom is not of interest. This procedure has recently been refined by Korolev et al.,[5] who employed microwave techniques together with alpha-electron coincidence using electrostatic electron selection; it is thereby possible to measure half-lives down to $\sim 5 \times 10^{-11}$ sec by using microwave time shutters.

Studies of a variety of atomic and nuclear effects and properties have taken advantage of the atomic recoil associated with alpha-decay. Among these, the most developed application is the determination of the lifetimes of low energy nuclear states. This is done by measuring the distance traveled by an atom in an excited nuclear state, recoiling after alpha-decay before the excited state decays through a given decay transition. The parent atoms are usually deposited on a foil surface, by vacuum deposition or electrostatic collection, in the purest form and thinnest layer possible (Figure 1*a*). The atomic recoil velocity distribution is strongly dependent on the source thickness, and although in many cases difficulties of this origin prevail, under suitable conditions the effects can be taken into account in evaluating the experimental data.

Consider an atom recoiling from the substrate out into the vacuum. The initial charge of the alpha-decay recoil atom will be close to +1, as is shown in Sections 2.2.1 and 3.1. The nuclear state populated through the alpha-decay transition may decay by gamma-emission or by internal conversion. In the former case, relative to the laboratory coordinates, the gamma-ray will suffer an energy change of a magnitude and sign dependent on its emission direction with respect to the recoil direction. This Doppler frequency shift is very small and is not considered further here.

In the IC case, measurement of the conversion electron will reveal an energy lying within a small limit from the conversion electron energy for an atom

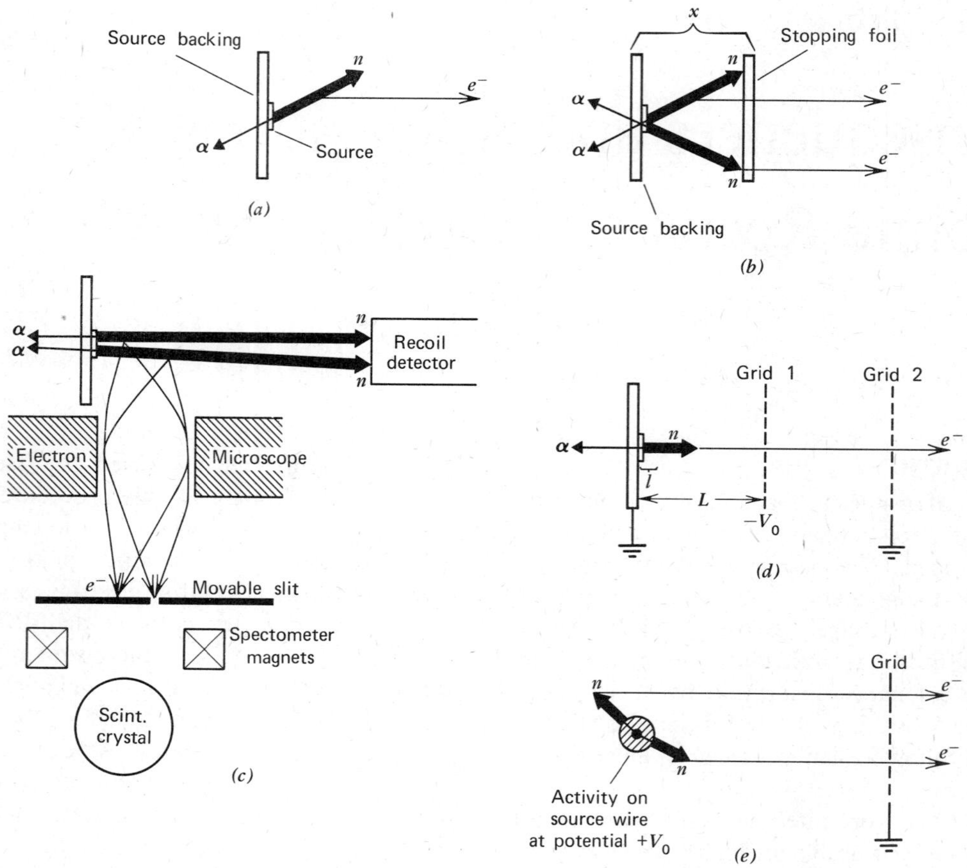

Figure 1. The various source geometries and experimental arrangements used in the alpha-recoil measurement of nuclear lifetimes (not drawn to scale).

at rest. The magnitude and sign of this difference will depend on the relative recoil-conversion electron directions and the source geometry. The total conversion line will then consist of electron intensity contributions from electrons emitted from atoms recoiling out of the source (which give rise to a Doppler shift in energy) and electrons emitted from atoms which have recoiled into the source backing and come to rest before emitting the conversion electron. The shape of this conversion line can be altered by placing a thin foil close enough before the source to stop a large fraction of the atoms before they decay. The measurement of this line shape alteration as a function of stopper distance from the source will give information about the lifetime of the state. This technique is treated in Section 2.2.1.

The shape of the conversion line contribution due to the atoms recoiling out of the source can also be altered by having an electric field directly before the source. In this way the final measured electron energy will be related to the distance the recoil atom travels before the decay takes place. The spectral line shape changes as a function of applied potential; these changes can be analyzed to obtain the nuclear state lifetime (Section 2.2.3).

Coincidence measurements between alpha-particles, which have traveled through the source backing, and conversion electrons, accepted in a direction parallel to that of the recoil atom, provide a more straightforward means of measuring the nuclear state lifetime.

The techniques just given have yielded measurements of lifetimes down to about 10^{-10} sec. The use of a cylindrical geometry, in combination with an electrostatic field, provides field gradients high enough to allow the measurement of lifetimes as short as 10^{-12} sec (see Section 2.2.3.2). These measurement techniques were reviewed in 1966 by Novakov.[6]

Measurement of the atomic recoil distance after Coulomb excitation by backscattered heavy ions has been exploited by Ashery et al.[7,8] for half-life determinations. The decay of the state of interest is indicated when a gamma-ray of given energy, or an X-ray following IC nuclear decay, emitted nearly perpendicular to the recoil direction is noted. Radiation intensity measurements are made using a gamma-detector, accepting rays from only a small portion of the recoil path, as a function of detector position along that path. Recoil velocities are on the order of 3×10^8 cm/sec.

Attenuation of recoil atoms as a function of the time lapse after nuclear excitation can also be accomplished using a gaseous medium. Blaugrund et al.[9] determined the lifetimes of three levels in ^{22}Na, following bombardment of ^{19}F by alpha-particles [(α, n) reaction], by measuring gamma-ray energy shifts as a function of stopper gas pressure. The gamma-rays were detected with a Ge(Li) detector. Excited nuclear level half-lives of the order of 10^{-11} to 10^{-12} sec were measured (see Ref. 9 for further references to similar work carried out using solids for recoil attenuation).

It is also possible to use a high resolution electron spectrometer and either an electrostatic field or a stopper foil to investigate Doppler energy shifts of conversion electrons following heavy ion bombardment of a thin target layer deposited onto a foil. The atomic recoil velocities involved in heavy ion reactions are much higher than those following alpha-decay, and the associated momentum shifts are correspondingly greater. The prospect of applying the electrostatic method to half-life determinations following ion reactions has been considered by Novakov et al.[10]

Following internal conversion in the recoiling atom, the created atomic vacancy will cascade, primarily through Auger transitions, to the outer atomic orbital ions (see Chapter 3, Section 9) leaving the atom in a state of high ionization. Acceleration of these final atomic ions in an electric field and the application of techniques of time-of-flight of the multi-ionized recoil atoms will allow one to determine the relative abundance of atomic ionization states. Experimental work on this topic is treated in Section 3.1.

If, after the occurrence of an initial IC nuclear transition that results in a multi-ionizing cascade, a second IC transition takes place, the energy distribution of these conversion electrons will be still more complicated than that for the first IC transition. The atomic binding energy of the internally converted electron will increase with increasing degree of atomic ionization at the instant of transition. This will displace the Doppler-shifted electrons to lower energies, the magnitude depending on the degree of ionization. The conversion line shapes for the near-neutral and highly ionized atomic states show strong structural differences (Section 3.2).

The related question of whether the lifetime of a nuclear transition, which is highly internally converted, is altered to any extent as the atom becomes more and more ionized has been investigated by the use of time-of-flight and coincidence measurement techniques (Section 3.3).

2. POST-ALPHA-DECAY ATOMIC RECOIL

Atomic recoil energies after nuclear alpha-decay lie in a range between approximately 80 and 200 keV. These energies correspond to recoil velocities from 10^7 to 10^8 cm/sec, depending on the alpha-decay energy and the mass of the recoil atom.

The alpha-decaying activity is usually deposited in a thin layer on a solid conducting backing, and the recoiling atoms travel either into the backing, where they are quickly brought to rest, or out into the vacuum, where they travel unhindered.

2.1. Atomic Recoil into Solid Media

The range and stopping time of an atom that recoils into a source substrate can be determined from measurements of stopping powers of the materials of interest. For a short account of work that has been done on the subject, the reader is referred to the article by Marx,[11] who studied heavier atoms travelling through materials of a wide range of atomic mass. (See also Refs. 12–14.) Energy losses and charge states of lighter boron, carbon, and sulfur ions traveling through thin metallic foils of various atomic masses have been investigated by Bethge et al.;[15] for a measurement of the range of sodium ions in aluminum, see Ref. 16.

The velocity of an atom recoiling after alpha-decay can be calculated from momentum considerations

$$V_n = \frac{1}{M}(m'v) = \frac{1}{M}\sqrt{2m'E},$$

where M is the mass of the recoil atom, $m' = m/\sqrt{1-\beta^2}$ ($\beta = v/c$), and m and E are the alpha-particle rest mass and the atomic recoil energy, respectively. Since $v/c \gtrsim 7\%$, relativistic effects on the alpha-particle mass are small; relativistic effects on the recoil atomic mass are negligible.

For high recoil velocities—$V_n \gtrsim v_0 = e^2/\hbar = 2.2 \times 10^8$ cm/sec, the first orbital electron velocity in a hydrogen atom—the recoil atom traversing the stopping medium is slowed mainly through electronic collisions (i.e., energy losses due to the ionization of atomic electrons in the stopping material). For $V_n \lesssim v_0$, most of the energy is transferred elastically to the atom of the stopping material; ionization still occurs, however. The ionization process for very low atomic velocities is so small that an "ionization defect" effect is produced toward the end of the particle range (see, e.g., Ref. 17). In the present treatment we are dealing with recoil velocities lying in the $V_n < v_0$ region.

Energy losses due to elastic collisions occur generally in large portions, compared to electronic stopping, and the original recoiling atoms also experience considerable deviations in flight direction. Therefore, range straggling is to be expected, especially if the stopping material atomic mass is greater than that of the projectile.

Firsov[18] found the following expression most suitable for his numerical results for the biatomic repulsive potential:

$$V_n(r) = \frac{Z_1 Z_2 e^2}{r}\,\Phi\left(\frac{r}{a}\right), \qquad a = \frac{0.8823 a_0}{(Z_1^{1/2} + Z_2^{1/2})^{3/2}}, \tag{1}$$

where $Z_1 e$ and $Z_2 e$ are the nuclear charges of the colliding atoms, Φ is the Fermi function, and a_0 is the first Bohr radius (5.29×10^{-9} cm). This potential was used by Biersack[19] to treat elastic atomic collisions and the related energy loss; the collision integral was simplified by introducing the dimensionless variable ε,

$$\varepsilon = \frac{aM_2 E}{Z_1 Z_2 e^2 (M_1 + M_2)}, \tag{2}$$

where M_1 and M_2 are the mass numbers of the recoil atom and the stopping material atom respectively.

Biersack[19] obtained for the stopping power S the expression

$$S = -\frac{dE}{ds} = \frac{4\pi a N M_1 Z_1 Z_2 e^2}{M_1 + M_2}\,\frac{\ln \varepsilon}{2\varepsilon(1 - \varepsilon^c)}, \tag{3}$$

where N is the number of stopping atoms per cubic centimeter and $c = -1.49$ (i.e., $\sim -\frac{3}{2}$). For high projectile energies ($\varepsilon > 0.3$) the electronic stopping power becomes important and an extra term must be added to equation 3.

The total range, or path length, of a recoil atom is obtained from the integration of the inverse stopping power. A general range energy curve is obtained for all possible combinations of recoil atoms and stopping materials (assumed to be monatomic) by replacing s, together with the remaining constants in equation 3, by a variable

$$\rho = \frac{4\pi a^2 N M_1 M_2 s}{(M_1 + M_2)^2}, \tag{4}$$

so that we obtain from equation 3

$$-\frac{d\varepsilon}{d\rho} = \frac{\ln \varepsilon}{2\varepsilon(1 - \varepsilon^c)}. \tag{5}$$

From the integration of the inverse of this expression between ε_{max} and zero, one can determine ρ as a function of ε_{max} (see Ref. 19). Integration of the inverse of equation 5 multiplied by the recoil velocity, in suitable units, will give the mean time necessary for the recoil atom to come to rest.

Such calculations of mean stopping time have not yet been carried out, but we can form an idea of the order of magnitude of this quantity. Marx[11] measured the range, in a number of materials, of ^{208}Pb recoil atoms after the 8.78-MeV alpha-decay of ^{212}Po. The recoiling lead atoms had an energy of 169 keV, which corresponds to a recoil velocity of about 4×10^7 cm/sec. For the metals investigated as stopping materials, the element was evaporated onto a thin formvar foil which was traversed first by the recoil atoms. As a result of straggling in the formvar, the energy of the atom was reduced to about 125 keV or about 3.4×10^7 cm/sec. Extrapolating the mean recoil energy versus traversed foil thickness, we obtain for gold about 7.8×10^{-7} cm for the "mean" range value. Assuming that the recoil atom loses energy linearly with penetration thickness, we calculate the stopping time to be $\approx 5 \times 10^{-14}$ sec. This value is an overestimation in consideration of equation 5, which shows that the deceleration of the recoil atom becomes stronger with decreasing energy. From a similar calculation, we find that ^{208}Pb recoil atoms of an original energy of 169 keV have a mean range of $\sim 5 \times 10^{-6}$ cm in a carbon foil (no formvar support foil) and come to rest after $\approx 2 \times 10^{-13}$ sec. Again this is an overestimate.

In the following sections of this chapter, it is assumed that the atoms recoiling after alpha-decay into the activity substrate have come to rest at the time that they undergo nuclear decay, the half-life region of interest being $> 10^{-13}$ sec.

Recoil ranges in solids are sometimes estimated by using the simple power-law expression $x \propto v^n$, where x is the mean range. Wolfson and Dixon[20] used this expression for treating recoil atomic straggling in the activity layer (see Section 2.2.3.1). They set $n = 2$ for the atomic velocities associated with alpha-decay on the strength of work by Lindhard et al.[21] Here, then, $x \propto E$, the recoil energy.

2.2. Alpha-Decay Atomic Recoil Effects on Internal Conversion Lines

It was established more than a decade ago that conversion electrons emitted by an atom recoiling after alpha-decay are measurably different in energy from those emitted by a stationary atom. Figure 1*a* shows schematically the experimental arrangement for measuring electrons from an alpha-decay radioactive source. The alpha-particle emission and atomic recoil are in the opposite directions. If an internal conversion electron is emitted by the recoil atom, with momentum p_0 and energy E_0 with respect to the recoiling atom, the momentum of the electron as measured by the spectrometer will be, in the laboratory frame of reference,

$$p = p_0 + \Delta p = p_0 \left[1 + \frac{V_n}{v_e} \cos\theta\right], \tag{6}$$

where θ is the angle between the electron emission and atomic recoil directions, and v_e and V_n are the velocity of the conversion electron, referred to the atom, and the atomic recoil velocity, respectively. It is assumed here that the source is thin enough to make straggling of the recoil atom negligible.

Atoms recoiling into the source backing will come to rest before internal conversion takes place, and emitted conversion electrons, which straggle as they emerge from within the source backing, will thus have a momentum only very slightly less than p_0.

Figure 2 shows the L_1- and L_2-conversion lines of the 39.85-keV transition in ^{208}Tl as measured by Ewan.[22] The L_2-line clearly reveals the effect of atomic recoil Doppler broadening on the high energy side. The effect on the L_1-line is obscured by the superimposed *K*-115.1 keV conversion line from ^{212}Bi. (See also Section 2.2.3.2). However, the L_2-shoulder disappears when an absorber thick enough to stop the recoil atoms is placed on the source, which proves that this shoulder is caused by an atomic recoil effect. At the same time, only part of the L_1-line shoulder disappears since a *K*-conversion line underlies it. The shoulder of the L_1-39.85 keV conversion line was measured earlier, using a high instrumental resolution,[23] but it was attributed entirely to the *K*-115.1 keV line in ^{212}Bi (see also Ref. 24 and work listed there for earlier treatments of conversion line shape changes due to atomic recoil).

The unusual structure of electron conversion lines due to atomic recoil following nuclear alpha-decay can be disturbing in IC electron energy analysis. This characteristic has been reduced or removed by depositing a thin layer of a metallic conductor over the alpha-decaying activity. Effectively all recoil atoms will be brought to rest before the occurrence of lower energy nuclear transitions.[25]

Not long after it had been established that atomic recoil causes the conversion line shape alteration, attempts were made to exploit the effect in order to measure the lifetime of low energy nuclear transitions occurring after alpha-decay. Since the first successful attempts in this direction, by Burde and Cohen[26,27] using recoil stopping foils (see Section 2.2.1), more and more sophisticated methods have been introduced, and the experimental results have been increasingly accurate and reliable. These methods are reviewed and discussed in the following sections.

2.2.1. *Stopping Foil Methods*

The Doppler shift of conversion electrons emitted from recoiling atoms can be made to disappear if the atoms are stopped before they decay by internal conversion. This may be achieved by placing a thin foil—thick enough to stop recoiling atoms but thin

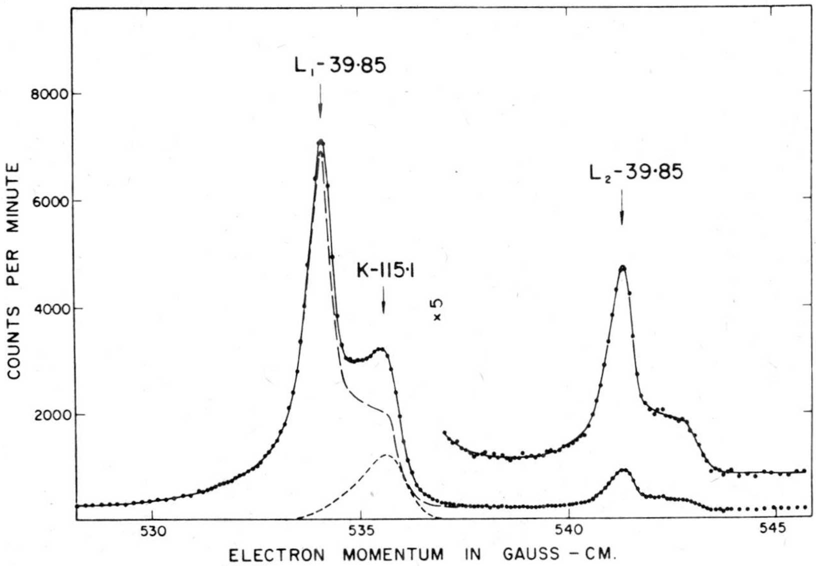

Figure 2. The L_1 and L_2 conversion lines of the 39.85-keV transition in ^{208}Tl as measured by Ewan.[22] A flat source backing was used and the extension of the line profiles to higher energies due to atomic recoil effects is plainly visible. A comparison of the L_1- and L_2-line shapes reveals the presence of the *K*-line of the 115.1-keV transition in ^{212}Bi.

enough to transmit conversion electrons without appreciable energy loss—in contact with the radioactive deposit. As this foil is then moved to increasing distances from the source, the recoil shoulder will appear more and more strongly.

Burde and Cohen[26,27] used this method to measure short nuclear half-lives. A schematic drawing of their experimental arrangement appears in Figure 1*b*. A thin stopper foil is placed at a certain distance in front of the source. The resolution of the lens spectrometer employed by the investigators ($\sim 1\%$) was not high enough to show the change in the conversion line shape in detail. Therefore, the conversion line centroid position was measured as a function of the source-stopping foil distance.

The source backing was of very thin material, to allow the alpha-particles to pass through and be detected by a scintillation crystal. Coincidence measurements between conversion electrons and alpha-particles made it possible to select electrons from atoms recoiling almost perpendicularly out from the source.

Taking n to be the fraction of nuclei that decay before stopping and $1 - n$ to be the fraction that decay after stopping, the conversion line will be the superposition of the two electron line components at two different energies. If the alpha-particle detection solid angle is small, then the intensity $(1 - n)$ will be exponentially dependent on the source-stopper distance x:

$$(1 - n) \propto \exp\left(\frac{-x}{V_n T_{1/2}}\right), \tag{7}$$

where $T_{1/2}$ is the half-life of the nuclear state.

The measurement of the picosecond half-life of the 39.85-keV transition in ^{208}Tl was carried out[27] by using the method just described. The L_1-conversion electrons were detected in coincidence with the 6-MeV alpha-particles. The measurements were tedious, since the distance traveled by the recoiling ^{208}Tl atom is about 0.3 μm/10^{-12} sec. The value of $1.0(5) \times 10^{-12}$ sec was obtained for the half-life.

Straggling of the recoil atoms in the source was considered to be the largest source of error in the measurement. This was primarily a result of the activity collection technique—recoil collection from ^{228}Th(RdTh). The details of the collection technique were not given,[27,28] but the activity was probably collected in the conventional way on a negatively charged source backing after the alpha-decay of ^{228}Th(RdTh). This activity would consist primarily of ^{224}Ra(ThX). This isotope may acquire multiple positive charge as a result of Auger transitions in the atom following the internal conversion of the 84.4-keV transition, or it may become singly or doubly charged after alpha-decay if no internal conversion transition follows. Secondarily, activity would also be collected later in the decay chain as ^{216}Po, or ^{212}Pb, after the alpha-decay of ^{220}Rn(Tn) (a rare gas).

It was shown later by Siekman and de Waard[29] that Burde and Cohen's half-life value determination should be corrected for the effect of penetration of the source atoms into the backing when they are collected using the electrostatic source preparation technique. Siekman and de Waard found that the final thorium B (^{212}Pb) source was "thick" partly as a result of the diffusion of radon gas into the source foil and partly because of atomic recoil associated with the alpha-decay chain $^{220}\text{Rn} \xrightarrow{\alpha} {}^{216}\text{Po} \xrightarrow{\alpha} {}^{212}\text{Pb}$. Estimating that the 39.85-keV level lifetime should be a factor of about 4 larger than the value obtained by Burde and Cohen, they proceeded to remeasure this quantity,[30] taking special care during the source preparation to avoid the penetration of the activity into the backing. Recoil activity from a RdTh source was collected on a clean platinum electrode and dissolved in a dilute acid solution. A droplet of the activity solution was then put on a backing foil and dried. It is not clear whether all effects of atomic recoil were removed in this way, but certainly one of the recoil steps in coming to the final thorium B was deleted. They remeasured the half-life, using essentially the same technique as Burde and Cohen, and obtained $T_{1/2} = 2.6(1.0) \times 10^{-12}$ sec.

The point brought up by the last authors about recoil effects on the source thickness properties is well taken. The activity of ^{212}Pb, which decays to ^{212}Bi by beta-decay, the parent of ^{208}Tl, is obtained from the original

$$
\begin{aligned}
{}^{228}_{90}\text{Th(RdTh, 1.91 y)} &\xrightarrow[5.4\text{ MeV}]{\alpha} {}^{224}_{88}\text{Ra(ThX, 3.64 d)} \\
&\xrightarrow[5.7\text{ MeV}]{\alpha} {}^{220}_{86}\text{Rn(Tn, 55 sec)} \\
&\xrightarrow[6.3\text{ MeV}]{\alpha} {}^{216}_{84}\text{Po(ThA, 0.155 sec)} \\
&\xrightarrow[6.78\text{ MeV}]{\alpha} {}^{212}_{82}\text{Pb(ThB, 10.64 h).}
\end{aligned}
$$

When a negative collection potential is used to collect the recoil atoms, the collection of recoil ^{224}Ra is expected to be strong. Further alpha-decay of these collected atoms, nuclei will result in additional driving of some of the atoms into the source backing. When the rare gas ^{220}Rn is reached in the decay chain, most of the atoms will not escape because they are not close enough to the source surface.

However, ^{224}Ra atoms in which internal conversion

did not occur, will have charges of mostly +1 and +2 (see Section 3), and those escaping electrostatic collection will decay to ^{220}Rn (rare gas), possibly to be collected later in the decay chain. In any case, "source thickening" due to atomic recoil seems to be unavoidable in this source preparation procedure.

Studies have been made of the charge of a recoiling atom in the absence of IC processes. Contrary to what might be expected, 0, +1, +2, charges of the recoil atoms are most representative, and there are virtually no negative charges.

The charge distribution of recoils from ^{226}Th alpha-decay has been investigated by Gunter et al.[31] They collected the ^{226}Th activity, following alpha-decay of ^{230}U, on a negatively charged platinum plate in an air atmosphere. Using a 180° double-focusing alpha-particle spectrograph, they measured the recoil energies and charges and the alpha energies. The recoil charge distribution for alpha-decay of ^{226}Th to the ground state of ^{222}Ra, either directly or via the first excited nuclear state (and subsequently decays by gamma-emission), showed a maximum of 50% for +1 and about half that intensity for both +2 and 0. The intensities of the −1 and −2 charged recoils were <0.03%. Charges of from +3 to +6 following alpha-decay directly to the ground state were of intensities between 1 and 10%. The +2 charge was probably favored to some extent because it has a closed electronic configuration (that of the radon atom).

Szucs and Delfosse[32] found in their study of ^{216}Po recoils after alpha-decay of gaseous ^{220}Rn that the charge distribution had a maximum at zero; the relative intensities for charges −1, +1, +2 were each a factor of about 30 less in intensity.

From the results just cited, then (see also Table I) as we proceed from radium to polonium as alpha-decay products, the mean recoil charge changes from +1 to ~0, and the relative abundance of negative recoil ions goes from ~0 to ~3%. This shift in distribution can be attributed to the stability of the closed shell configuration—that of Ra^{+2}, Rn^0, and Po^{-2}.

Bergman[33] investigated the properties of thorium B sources prepared by electrostatic collection in air using a positive collection potential. He found that the decay rate of the activity collected at 1000 V had increased over one week from an apparent half-life of 11.1 hr to one of 32 hr. A source made using a potential of 150 V showed a constant half-life of 10.6 hr (within statistical errors) over the same period. In the latter case, then, primarily ^{212}Pb is collected after the decay of gaseous ^{220}Rn; the short half-life of ^{216}Po probably made it impossible to collect this isotope directly in any appreciable quantity.

Therefore, electrostatic collection of the activity according to Ref. 33 will give ^{212}Pb sources of high purity for low collection potentials. If the source is used immediately after collection at 1000 V, ~100% of the ^{212}Pb decaying there will have been freshly collected.

If alpha-decay occurs and no IC processes follow, charges of 0 to +2 for the recoil ions are obtained. Collisions with gas atoms in the recoil chamber may result in charge transfer and/or impact ionization, resulting in a lower ionized recoil atom and ionized atmospheric gas molecules or atoms. When internal conversion occurs after alpha-decay, Auger, Coster–Kronig, and X-ray cascading to the outermost shells will result in highly ionized recoil atoms. However, these atoms will doubtless pick up electrons through collisions with the atoms of the gaseous atmosphere. All final recoil atoms should therefore be of much lower state of ionization than if they were to recoil into a high vacuum.

This source preparation technique was used for the cylindrical electrostatic 39.85-keV level half-life determination for ^{208}Tl described in Section 2.2.3.2. The

Table I.

Ref.	Decay	Z (daughter)	Relative abundances of final low charges −2	−1	0	+1	+2
46)	$^{241}Am \rightarrow ^{237}Np$	93	--	--	~1	~4	~1
31)	$^{226}Th \rightarrow ^{222}Ra$	88	≤0.0012	≤0.0012	~1	~2	~1
41)	$^{223}Ra \rightarrow ^{219}Rn$	86	--	--	~9	~8	~1
32)	$^{220}Rn \rightarrow ^{216}Po$	84	≤0.05	~1	~30	~1	~1
42) 43)	$^{212}Po \rightarrow ^{208}Pb$	82	--	--	~1	~2.2	~0.8
" "	$^{212}Bi \rightarrow ^{208}Tl$	81	--	~0	~1	~1	~0

obtained half-life value was appreciably larger than that from the previous measurements, since the ^{212}Pb had been freshly collected.

The half-life of the 84.4-keV level in ^{224}Ra, following the decay of ^{228}Th, was also measured by Siekman and de Waard.[30] Here they used a ^{228}Th source where nuclear recoil source thickening effects were absent. Using their same method, they obtained the value of $T_{1/2} = 7.5(1.5) \times 10^{-10}$ sec, in agreement with other measurements.

In all the measurements cited, instrumental momentum resolutions of about 1% were employed. An instrument of higher resolving power will show the conversion line shape in detail (see Figure 2). With such an instrumental resolution, line shape changes, especially those of the shoulder, could be readily detected by placing a stopping foil in front of the source. The shoulder shape change as a function of source-stopper foil distance can be calculated for various half-life values and fitted to experimental electron line shapes (see Ref. 34).

2.2.2. *Electron Optical Method*

Severiens and Richter[35] have employed an electron microscope lens for the electron optical enlargement of the distance traveled by the recoil atom perpendicularly away from the source. The electrons emitted from the recoiling atoms are focused by the microscope, magnifying the source-decay position distance about 25 times, onto a plain where a movable slit allows origin position selection of the electrons. The electrons that pass the slit enter a simple beta-spectrometer and are counted by a plastic scintillator.

Another new feature of the apparatus is that the conversion electron is counted only if it is in delayed coincidence with the de-excited nucleus, which is detected by a suitable counter. If the recoil detector is placed in front of the source at a distance of about 4 cm, the detected recoil atoms leave the source in a roughly perpendicular direction. One can then select only those recoils of a velocity uniquely determined by the source-counter distance and the coincidence circuit delay.

With this set-up it is possible to measure half-lives down to 2×10^{-11} sec, and additional alteration could shift this limit to the 10^{-12} sec range.

The 84.4-keV level in ^{224}Ra was measured using the Severiens–Richter apparatus. The source was prepared by dropping the activity in solution (cleaned of daughter products in a cation exchange column) on a flat, polished gold source backing. The composite L_2- and L_3-conversion line group was accepted and counted by the spectrometer. A recoil velocity of 2.40×10^7 cm/sec was chosen, which is less than the maximum value stated in Table II, and the half-life of the state was measured to be $6.0(1.6) \times 10^{-10}$ sec, in agreement with other measurements.

2.2.3. *High Resolution Electrostatic Field Methods*

The method outlined in Section 2.2.1 of using atomic recoil for nuclear half-life determinations has two main drawbacks which reduce the reliability and accuracy of measurement results. First, the low resolving power of the spectrometers employed made it impossible to investigate the conversion line shape changes caused by the recoil effect. The change in shape of the high energy recoil shoulder, rather than the line centroid shift, would be expected to lead to a more accurate lifetime determination.

Second, the use of thin foils to stop the recoiling atoms introduces two uncertainties into the experiment. The influence of straggling on transmitted conversion electrons will cause an undesirable spreading out of the electron energy spectrum, although this effect will be constant for the series of measurements employing a given stopper foil. Furthermore, the source-foil distance is subject to uncertainty, especially when the quantity $V_n T_{1/2}$ is of the order of 1 μm; here spacer dimensions become subject to rigid accuracy requirements.

Source-decay site distances can be measured accurately by establishing an electric field over the region where the nuclear decay of the recoil atom takes place. A conversion electron will then become shifted in energy by an amount related to the electric potential at the decay site. By giving to the potential gradient of this field the correct orientation, the high energy shoulder of the conversion line will become spread out toward higher energies. From the analysis of this shoulder shape alteration one can determine the lifetime of the state involved. The following subsections give treatments of these electrostatic methods.

2.2.3.1. PLANAR GEOMETRY. The application of an electric field in the recoil measurement of nuclear lifetimes was first presented in 1964 by Novakov et al.[10] There they reported the method together with its application in the measurement of the first excited 42.8 keV state in ^{240}Pu.

The arrangement used is shown schematically in Figure 1*d*. It can be seen there that the maximum Doppler shift is observed when both the alpha and electron emission directions are perpendicular to the source plane. It is also seen that a conversion electron emitted from an atom that has recoiled a distance *l* from the source plane will be decelerated as it travels to the

Table II.
Summary of excited nuclear state halflives measured by atomic recoil after alpha-decay.

Ref	Parent nucleus	Daugh. nucleus	α-energy feeding state (MeV) & (state energy (keV))	Nuclear trans. energy (keV) & multi-polarity	Conv. line meas.	Maximum recoil velocity (∿10^7 cm/sec)	Meth; Instr. ----- Geometry	Measured halflife (sec.)
27)	^{212}Bi	^{208}Tl	6.05 (39.85)	39.85 M1(+E2)	L_1	3.27	Coinc; LS Planar	$1.0(5)\times10^{-12}$
30)	"	"	"	"	"	"	"	$2.6(1.0)\times10^{-12}$
37)	"	"	"	"	L_2	"	ESF; SCS Cylindrical	$6.49(79)\times10^{-12\dagger}$
30)	^{228}Th	^{224}Ra	5.34 (84.4)	84.4 E2	L_2+L_3	2.87	Coinc; LS Planar	$7.5(1.5)\times10^{-10}$
35)	"	"	"	"	"	"	Coinc; EIM Planar	$6.0(1.6)\times10^{-10}$
20)	^{241}Am	^{237}Np	5.44 (103.0)	59.57 M1(+E2)	L_3	2.74	ESF; IFDF Planar	$8(4)\times10^{-11}$
10)	^{244}Cm	^{240}Pu	5.76 (42.8)	42.88 E2	L_2, L_3	2.78	ESF(coinc); IFDF Planar	$1.6(2)\times10^{-10}$

LS = Lens spectrometer
SCS = Semi-circular spectrometer
EIM = Electron microscope
IFDF = Iron-free double-focusing spectrometer
ESF = Electrostatic field

†This value is estimated to be several procent too large (see text).

first grid and then accelerated toward grid 2. The net gain in electron kinetic energy will be equal to the potential energy it has at the site of the decay.

Consider first the case, where the recoiling excited nucleus travels in the direction of the spectrometer baffle ($\theta = 0°$) with a velocity V_n. The distance traveled in the time t, the time lapse between alpha particle emission and conversion electron emission, will be $l = V_n t$. When the electron reaches grid 1 it will have an energy

$$E_i - \left(1 - \frac{l}{L}\right) V_0 e,$$

where E_i is the initial electron energy, e is the electronic charge, V_0 is the voltage applied to grid 1, and L is the distance from the source to grid 1. While traveling from grid 1 to grid 2, the electron will receive an increase in energy of the amount eV_0. The final total energy of the conversion electron after passing grid 2 will be the sum of the above-mentioned two quantities, or

$$E_f = E_i + \frac{l}{L} V_0 e,$$

where $V_0 e$ is expressed in electron volts. Setting $E_f - E_i = \varepsilon'$, we obtain

$$l = \frac{\varepsilon'}{(eV_0/L)}, \tag{8}$$

and the time t required for the recoil atom to travel the distance l is

$$t = \frac{\varepsilon'}{(eV_0/L)V_n}. \tag{9a}$$

After one half-life $T_{1/2}$ the distance recoiled will be $\varepsilon'_{1/2}$, and

$$T_{1/2} = \frac{\varepsilon'_{1/2}}{(eV_0/L)V_n}. \qquad (9b)$$

For atoms recoiling perpendicular to the source plane the exponential time relationship of recoil decays will necessitate the energy distribution to be exponential as well. Therefore the slope of the semilog plot of the high energy shoulder extension brought about by the introduction of an electric field will be proportional to the half-life of the nuclear level.

The acceleration of the recoil atom by the electric field is very small. A typical recoil atom of charge +2, for example, will increase in energy by about 400 eV in a field of 40000 V/cm if it decays after a time of 4×10^{-10} sec. This will increase its velocity ~0.2%, a negligible amount.

The 42.8-keV level in ^{240}Pu, following the 5.759-MeV alpha-decay of ^{244}Cm, was the level of interest to the investigators. The half-life of this level had been measured previously by other methods to be about 10^{-10} sec. Assuming negligible recoil straggling, a time lapse of this length will allow the daughter plutonium atoms to travel 2.87×10^{-3} cm (28.7 μm) before decaying. In order to obtain measurable extensions of the conversion line recoil shoulder, potentials of up to $V_0 = -12$ kV were applied over a distance of $L = 2.6$ to 5 mm. One half-life of 10^{-10} sec would then give a maximum energy shift of $\varepsilon'_{1/2,\,\max} \approx 130$ eV.

Alpha-particle–conversion-electron coincidence measurements were made to aid the determination of the half-life of the first excited 42.8-keV level in ^{240}Pu. The instrumental momentum resolution was ≈0.1%. The alpha-particles were detected after they had traveled through the thin source backing, and nearly perpendicular to it, opposite in direction to that of the spectrometer acceptance baffle. Figure 3 shows the semilog plot of the 42.8-L_3 conversion line measured with coincidences for three different potentials (including zero). The finite sizes of the alpha-counter and source and a slight misalignment of the alpha-detector made it necessary to include a small $\cos\theta$ correction to V_n. The correction was estimated to be ≈ 5%, and the corrected recoil velocity becomes $\overline{V}_n \cong 0.95\ V_{n,\,(\text{ideal})} \cong 2.54 \times 10^7$ cm/sec.

Considering now the measurement of the single conversion electron line shape, we will expect the slope of the semilog plot of the high energy shoulder extension for a given applied potential to be steeper than that obtained under identical conditions for the coincidence measurements. This is because in the former case conversion electrons from atoms recoiling isotropically with θ contribute to the measured line shape.

In the ideal case where there is no recoil atom straggling and where the instrumental resolution is very high, the momentum distribution of the high energy shoulder of the conversion line will take the form of a rectangular Doppler shoulder. This shape can be obtained by proper integration of equation 6. Since Δp in equation 6 is small compared to p_0, the electron intensity from p_0 to $p_0 + \Delta p$, and correspondingly from E_0 to $E_0 + \Delta E$, may be considered constant. Expressing the electron energy above E_0 in units of shoulder width ΔE, $\varepsilon = \varepsilon'/\Delta E = (E - E_0)/\Delta E$. The Doppler energy shift χ of electrons having a forward velocity component $V = V_n \cos\theta$ is $\chi = \cos\theta$.

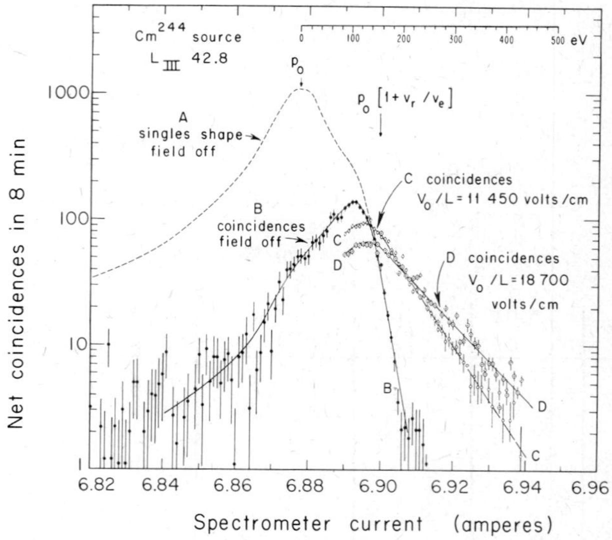

Figure 3. A semilog plot of the 42.8-L_3 line shapes after ^{244}Cm alpha-decay as observed with the alpha–conversion electron coincidence method of Novakov et al.[10] With the application of the electric field, the distribution of electrons in the Doppler shoulder is altered and takes on an exponential high energy tail.

When an electric field V_0/L is applied, the intensity contribution $I(\varepsilon)$ from an element $d\chi$ at χ becomes an exponential starting at $\varepsilon = \chi$, and has the half-value

$$\varepsilon_{1/2} = \varepsilon_{1/2,\,\max}\,\chi, \qquad (10)$$

where

$$\varepsilon_{1/2,\,\max} = T_{1/2}\,V_n \frac{eV_0}{L}(\Delta E)^{-1}. \qquad (11)$$

A spectrometer set to record electrons in the interval $\delta\chi$ at ε will show the intensity contribution from one element $d\chi$ (for $\chi < \varepsilon$) given by

$$I(\varepsilon)\,\delta\varepsilon = \delta\varepsilon I_0(\chi)e^{-\lambda(\varepsilon-\chi)}, \qquad (12)$$

where

$$\lambda = \frac{\ln 2}{\varepsilon_{1/2}} \quad \text{and} \quad I(\chi) = \lambda\,d\chi.$$

The total contribution of all elements is then

$$I(\varepsilon)\,\delta\varepsilon = \delta\varepsilon \int_0^k \frac{0.639}{\varepsilon_{1/2,\,\max}} \times \frac{\exp\left[-0.693(\varepsilon-\chi)/\varepsilon_{1/2,\,\max}\,\chi\right]}{\chi}\,d\chi$$

$$= \delta\varepsilon\,\frac{0.693}{\varepsilon_{1/2,\,\max}}\exp\frac{0.693}{\varepsilon_{1/2,\,\max}}\left[-Ei\left(\frac{-0.693\varepsilon}{\varepsilon_{1/2,\,\max}}\,\frac{1}{k}\right)\right], \tag{13}$$

where

$$Ei(-x) = \int_{-\infty}^{-x} \frac{e^t}{t}\,dt,$$

and the integration limits are $k = 1$ for $\varepsilon \geq 1$, and $k = \varepsilon$ for $\varepsilon < 1$.

Figure 4 shows the results of the evaluation of equation 13 for various values of $\varepsilon_{1/2,\,\max}$, and their asymptotic half-slopes for $\varepsilon \to \infty$. The inset in Figure 4 sets forth the relationship between the ratio of the half-slope $\varepsilon_{1/2}$ at ε to the asymptotic value $\varepsilon_{1/2,\,\max}$ for several values of ε. It should be noted that this asymptotic half-slope value is identical to that given by equation 9b and setting $\varepsilon'_{1/2} = (E_f - E_i)\varepsilon_{1/2}$.

Using Figure 4, a measured half-slope can be translated into its asymptotic value, $\varepsilon_{1/2,\,\max}$, which together with equation 9b will yield the half-life.

Novakov et al.[10] applied the procedure described previously to evaluate the half-life of the 42.8-keV state in ^{240}Pu. The average of the half-life values obtained in this way was about 10% higher than the value from the coincidence method. Considering uncertainties in the magnitude of the applied voltage and the source-grid distance, the final half-life value of $1.6(2) \times 10^{-10}$ sec was obtained.

Effects of straggling of the recoil atoms on the measured half-life value were not considered in the work, but evidence of the presence of such straggling was found.

Wolfson and Dixon[20] have measured the half-life of the 103-keV level in ^{237}Np using the same geometry as used by Novakov et al.[10] However, in this case the parent nucleus, ^{241}Am, has a 460-yr half-life and decays with a 12.7% feeding of the 103-keV level of interest in ^{237}Np (to be compared with 17.6 yr and 23.3%, respectively, in the $^{244}Cm \xrightarrow{\alpha} {}^{240}Pu$ case). Therefore, for the same source dimensions and quantity of activity the situation is worsened with respect to intensity by a factor of 48. Source-thickness effects will limit the extent to which a source intensity may be increased, and an increase in source dimensions will decrease the resolution. Hence Wolfson and Dixon had to work with a weak source.

A treatment of the change in shape of the recoil shoulder of a conversion line for a thin source gave a result identical to that obtained by Novakov et al.[10] although it was expressed in another form. When the

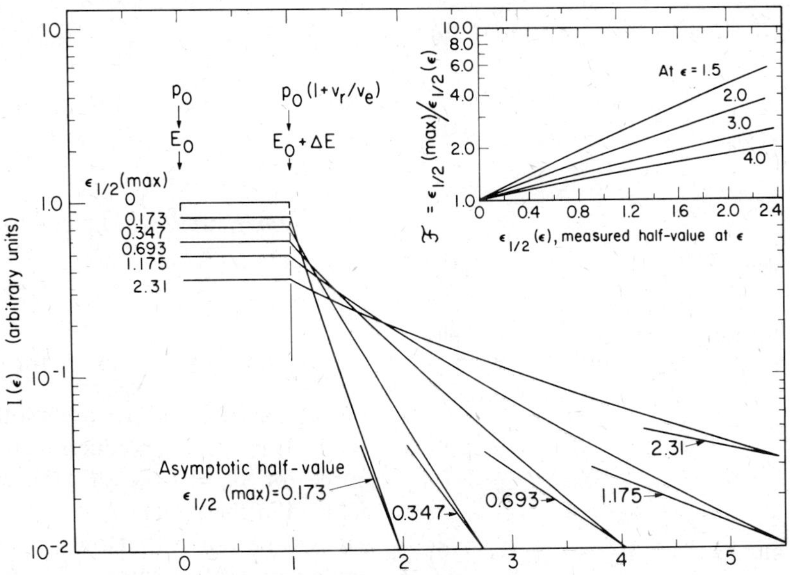

Figure 4. Theoretical shapes of the Doppler-shift electron distribution for various values of the parameter $\varepsilon_{1/2}(\max)$, which is the limiting slope at $\varepsilon \to \infty$. The apparent slope $\varepsilon_{1/2}$ at finite values of ε is shorter than $\varepsilon_{1/2}(\max)$ and is not constant. The value of $\varepsilon_{1/2}(\max)$ for an experimental curve may be deduced by determining the apparent half-value $\varepsilon_{1/2}$ at different values of ε and multiplying by correction factors such as those plotted in the upper right-hand inset. (From Ref. 10.)

counting rate was too low to allow a significant determination of the asymptotic slope, the line intensity fraction that is shifted from the shoulder to higher energies was calculated as a function of level half-life (i.e., the decrease in intensity of the shoulder in Figure 4 as a function of half-life for a given applied voltage).

At the other extreme, infinitely thick sources were considered, that is, sources of thicknesses exceeding the atomic recoil range. Here it is necessary to determine the velocity distribution function, $g(\theta, V_n)$, for the recoil atoms emerging from the source. When this is done, the analysis can be carried out in a manner analogous to that for the thin sources. Here, then

$$g(\theta, V_n)\, d\theta\, dV_n = \frac{\left(\frac{2}{R}\right) \sin\theta \cos\theta \, d\theta \, dV_n}{dV_n/dx}, \quad (14)$$

where the factor $\sin\theta$ expresses the initial isotropy of the recoiling atoms, $\cos\theta$ expresses the effective recoil path length in the source for an angle θ, and (dV_n/dx) is the velocity decrease per unit traveled path length in the source material; $(2/R)$, where R is the recoil range in the source material, is a normalization factor.

Slowing of the recoil atoms in the source owing to straggling was accounted for by using the $x \propto V_n^n$ relationship. In general, then, we may write

$$g(\theta, V_n) = 2nV_n^{n-1}V_n^{-n} \sin\theta \cos\theta. \quad (15)$$

Transforming the expression from the coordinates (θ, V_n) to (η, ζ) using $\zeta = V_n \cos\theta$ and $\eta = V_n \sin\theta$, and integrating over all values of η, we obtain for $n \neq 2$

$$f(\zeta) = \left(\frac{2n}{n-2}\right) V_n^{-n}\zeta(V_n^{n-2} - \zeta^{n-2}) \quad (16)$$

and for $n = 2$,

$$f(\zeta) = 4V_n^{-2}\,\zeta \ln \frac{V_n}{\zeta}, \quad (17)$$

The intensity distribution, with no electric field applied, in units of relative energy shift ε for $n \neq 2$ is

$$F_{\text{off}}(\varepsilon) = \frac{2n}{n-2}\varepsilon(1 - \varepsilon^{n-2}) \qquad \text{for} \quad 0 < \varepsilon < 1 \quad (18)$$

and for $n = 2$,

$$F_{\text{off}}(\varepsilon) = -4\varepsilon \ln \varepsilon \qquad \text{for} \quad 0 < \varepsilon < 1. \quad (19)$$

Note that the atomic recoil slowing-down time in the source is negligible compared to the nuclear half-life of interest. Applying the electrostatic field, the relative energy shift is given by

$$\varepsilon = \frac{\zeta}{V_n} \frac{1 + eV_0 t}{Lp_0}, \quad (20)$$

where

$$\varepsilon(V_0 = 0) = \frac{\zeta}{V_n} = \cos\theta = \frac{\text{electron energy Doppler shift}}{\text{maximum energy shift}}$$

and $dE/dp = v_e$, where $E = m_o c^2$, the total energy of the electron.

Introducing the parameter

$$y = \frac{\lambda L p_0}{eV_0} = \frac{\lambda L\, \Delta E}{eV_0 V_n} \quad (21)$$

and combining equations 20 and 21 with equation 16, we obtain for $n \neq 2$

$$F(\varepsilon) = \left(\frac{2n}{n-2}\right) ye^y[\varepsilon Ei_2(y) - \varepsilon^{n-1}Ei_n(y)] \qquad \text{for} \quad 0 < \varepsilon < 1 \quad (22)$$

and

$$F(\varepsilon) = \left(\frac{2n}{n-2}\right) ye^y[Ei_2(y\varepsilon) - Ei_n(y\varepsilon)] \qquad \text{for} \quad 1 < \varepsilon < \infty \quad (23)$$

where

$$Ei_n(x) = \int_1^\infty e^{-xu}u^{-n}\, du.$$

For $n = 2$, we define the new function

$$L_2(x) = \int_1^\infty e^{-xu}u^{-2} \ln u\, du,$$

which can be expressed in an expansion based on Ei_n functions.

We have

$$F(\varepsilon) = 4ye^y[\varepsilon L_2(y) - (\varepsilon \ln \varepsilon)Ei_2(y)] \qquad \text{for } 0 < \varepsilon < 1 \quad (24)$$

and

$$F(\varepsilon) = 4ye^y L_2(y\varepsilon) \qquad \text{for} \quad 1 < \varepsilon < \infty. \quad (25)$$

Equations 24 and 25 are plotted in Figure 5 for $y = 5$. The recoil shoulder distribution for a thick source for $V_0 = 0$ (i.e., $F_{\text{off}}(\varepsilon)$, equation 19) is also included in the figure.

It will be noted that the asymptotic slopes on the semilog plot of equations 24 and 25, for increasing $y\varepsilon$, will provide a measure of y and therefore of $T_{1/2}$. Clearly the treatment represents an extension of the method by Novakov et al.[10] to include effects of atomic recoil straggling in a thick source.

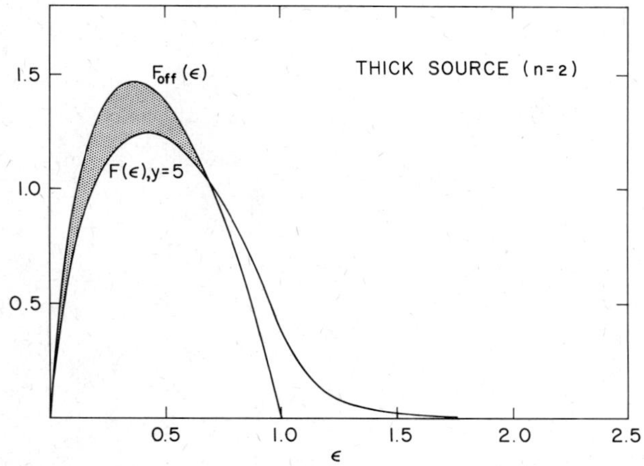

Figure 5. The theoretical Doppler-shifted distributions for conversion electrons from a thick source for $n = 2$. Application of voltage to grid 1 "pulls" counts from the shaded area and deposits them in an approximately exponential distribution at values of $\varepsilon > 1$. The case $y = 5$ is chosen as an illustration. (From Ref. 20.)

The calculation of the shaded area in Figure 5 is easily carried out; For all values of n, this area is

$$G(y) = (\varepsilon_c)^2[1 - ye^y Ei_2(y)] = 2(\varepsilon_c)^2 \varepsilon^y Ei_3(y), \quad (26)$$

where

$$(\varepsilon_c)^{n-2} = \frac{2Ei_3(y)}{nEi_{n+1}(y)} \quad \text{for} \quad n \neq 2 \quad (27)$$

and

$$\varepsilon_c = \exp\left[\frac{-yL_2(y)}{2Ei_3(y)}\right] \quad \text{for} \quad n = 2. \quad (28)$$

In Figure 6 is plotted $G(y)$ for $n = 1, 2, 3$, to show the effect of different values of the exponent n. Also included for comparison is the function for a thin source.

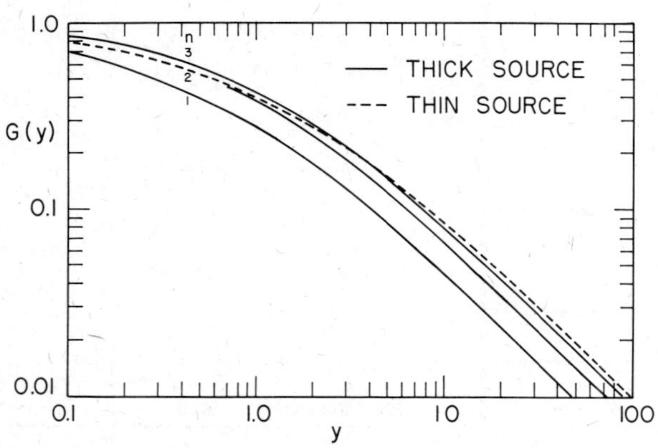

Figure 6. Plot of $G(y)$ as a function of y for both thick and thin sources. (From Ref. 20.)

Measurement of the 43.4-L_3 conversion line in ^{237}Np (the 43.4 keV transition depopulates the 103 keV level) was made using the geometry in Figure 1*d* with the instrumental momentum resolution of $\approx 0.05\%$, $V_0 = -10.75$ kV, and $L = 3.1$ mm. Broadening due to the natural width of the L_3 atomic level was accounted for. The ratio of the shaded area pulled out of the line profile to the Doppler area was determined (see Figure 7). For $n = 2$ in Figure 6, the obtained value of $G(y) = 0.31$ gives the value $y = 1.4$. From this the value of $T_{1/2} = 7.7 \times 10^{-11}$ sec was obtained for the 103 keV level; an uncertainty of 4×10^{-11} sec was estimated.

The 103-keV level in ^{237}Np is fed additionally by a highly internally converted 55.6-keV transition from the 158.6-keV level. It can be determined from the decay scheme for ^{237}Np that about 8% of the feeding of the 103-keV level takes place via alpha-decay to the 158.6-keV level.

The measured half-life of the 103-keV level, using the recoil method just described, will be slightly larger than the true half-life of that level because of the contribution from the 158.6-keV level half-life. A hypothetical half-life of the 158.6-keV level of ∞ sec will increase the apparent 103-keV level half-life by about 25%. But from an estimate, the 158.6-keV level half-life has a value of about 8×10^{-11} sec; therefore, the apparent 103 keV level half-life should be decreased by about 8% to obtain the true value.

This last point shows the importance of knowing in detail the manner in which a level of interest is both populated and depopulated in the decay scheme. Alpha-decay feeding of higher energy levels that decay to the level of interest may result in an apparent level half-life much in excess of the true half-life.

2.2.3.2. CYLINDRICAL GEOMETRY. Deposition of an alpha-decaying activity on a small-diameter metallic wire, rather than a flat foil, will lead to changes in conversion electron energy distributions. In this case, conversion electrons from atoms recoiling from the back side of the wire, as viewed from the spectrometer acceptance baffle, may be detected if the nuclear half-life involved is long enough to allow the atom to travel beyond the limb of the source wire before decaying. This effect is, of course, also dependent on the wire diameter and the recoil velocity. On the other hand, conversion electrons emitted after beta-decay or electron capture will not cause atomic recoil velocities of the magnitude of those caused by alpha-decay. Therefore, for a given total activity, the intensity of IC decay of levels fed by alpha-transitions will vary with source-wire diameter, whereas the levels fed by

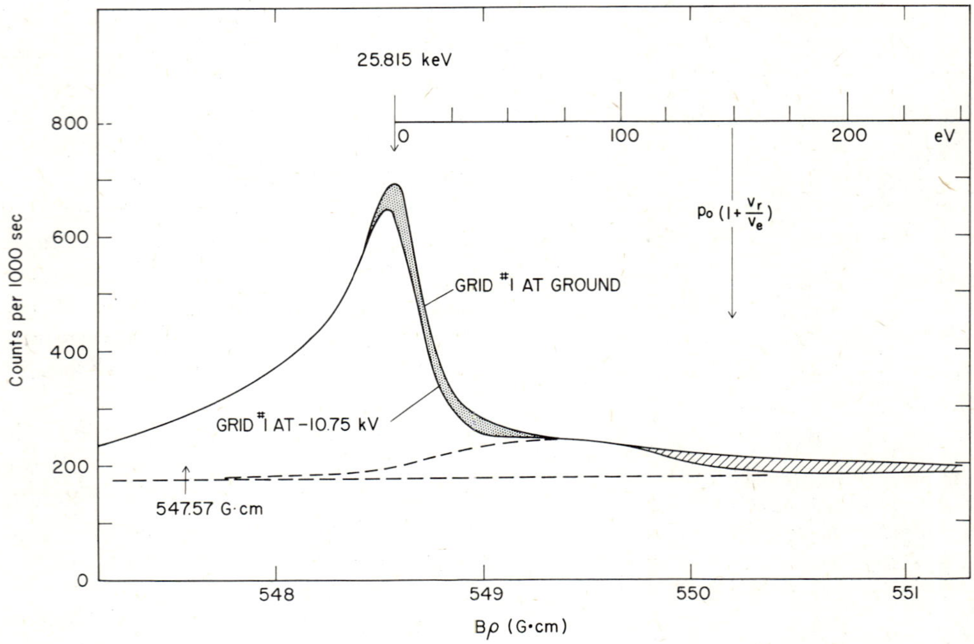

Figure 7. The 43.4-L_3 conversion line in ^{237}Np as measured by Wolfson and Dixon[20] for electric potentials of zero and -10.75 kV applied to grid 1. The broken curve is the low energy portion of the Doppler-shifted distribution with grid 1 grounded. The shaded area is due to counts "pulled out" of the Doppler-shifted distribution by acceleration. The hatched area should be equal to the shaded area. In the region of the peaks of the curves, the intensity differences are larger than the intensity of the Doppler-shifted distribution at the same momentum values. This discrepancy may be due in part to an error in determining the experimental peak positions of the lines and in part to the procedure by which the Doppler-shifted continuum is obtained.

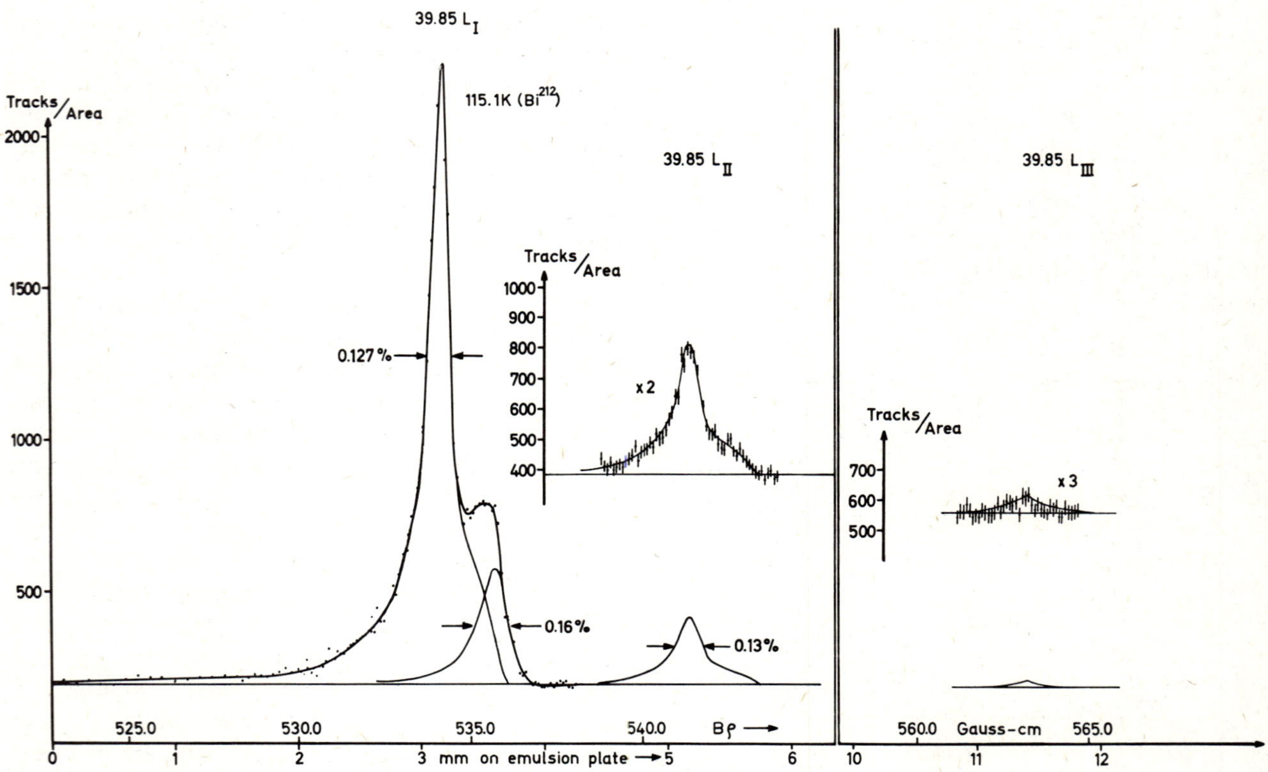

Figure 8. The 39.85-*L*-conversion line region ^{208}Tl, obtained using a wire of 25-μm diameter as source backing (cf. spectrum in Figure 2, obtained using a flat source). (From Ref. 36.)

beta-decay or electron capture will depend only slightly, if at all, on the source-wire diameter. (This is discussed later.)

Figure 8 shows the L-conversion line energy region[36] of the 39.85-keV transition in ^{208}Tl. A cylindrical source was used, as in Figure 1*e*. The activity, which was collected electrostatically using a positive collection potential, was very thin. It is interesting to compare the 39.85-L_1 line shape obtained in this measurement with that using a planar source (Figure 2). With the planar source, the shoulder is nearly flat, whereas with the cylindrical source it is strongly sloped. This is mainly because of recoil straggling and the difference between the source geometries. In the planar case recoils causing the greatest electron energy shift recoil perpendicularly from the foil toward the baffle slit. In the cylindrical case, baffle-directed recoils may come from near the limb of the wire, as seen from the baffle, leaving the source obliquely. Also, for the cylindrical geometry, atoms recoiling from the limb of the wire, but perpendicularly to the wire surface, will experience little energy loss, but their decay will not contribute to the conversion line shoulder. Since this will result in an increased net straggling in the latter geometry over the former for the baffle direction recoils, conversion electrons from these recoil atoms will be detected closer to the momentum p_0 of the line peak.

Considering the source geometry in Figure 1*e*, the application of a potential V_0 to the wire will give a potential $V(r)$ at a point at a distance r from the wire axis though close to the wire surface, expressed as

$$V(r) = V_0 \frac{\ln (r'/r)}{\ln (r'/a)} \cong V_0 \frac{\ln (2b/r)}{\ln (2b/a)}, \qquad (29)$$

where r' is the distance of the point in question from the wire axis image mirrored in the grid plane, a is the source-wire radius, and b is the wire axis-to-grid distance. This geometry makes it possible to obtain a large potential gradient near the wire surface. The application of +500 V to the source, assuming $b = 0.8$ mm and $a = 4\ \mu$m, will produce a radial potential gradient at 1 μm from the wire surface of $\sim 2 \times 10^5$ V/cm. The difference between the potentials at this point and at the wire surface will be about 19 V. It will take a recoil atom $\sim 3 \times 10^{-12}$ sec to travel this distance.

Applying a potential to the source will cause the entire conversion line to shift by the corresponding energy. Nevertheless, changes in the conversion line shape as a whole or in parts can be inspected by superimposing the two measured line profiles for two potentials. This is done in Figure 9 for the 39.85-L_2 conversion line.[37] It can be seen also that introducing an

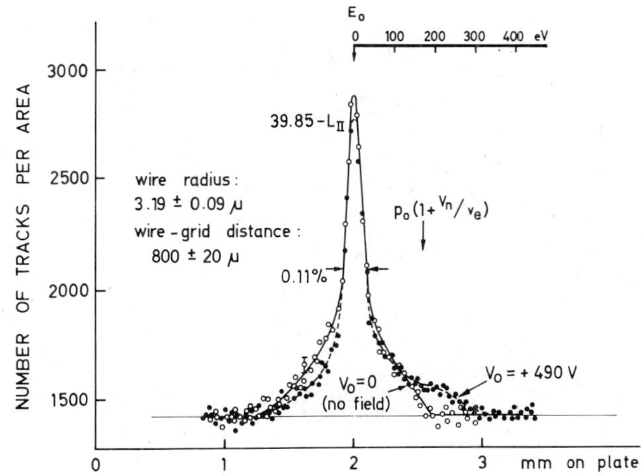

Figure 9. The shapes of the 39.85-L_2 conversion line ^{208}Tl, with a grounded source and as a result of applying a positive potential. The wire radius is 3.19(9) μm and the wire-grid distance is 800(20) μm. (From Ref. 37.)

electrical potential at the decay site causes a shift to higher energies of atoms recoiling with a velocity component away from the baffle direction. A semicircular spectrograph set for a momentum resolution of $\sim 0.03\%$ was used in the investigation. Nuclear research emulsions were used for electron detection and the spectrum was determined using the electron track counting method (see Chapter 2, Section 6.5.2).

The analysis of the experimental data was carried out as follows. It is possible to calculate the energy distribution of the emitted electrons, for a given nuclear lifetime and recoil velocity and assuming the absence of straggling, after the electrons have passed the grounded grid as in Figure 1*e*. We begin by considering the atoms that recoil in a plane that is perpendicular to the axis of the source wire and contains the point of origin of the recoil atom. First note the intensity contribution to the resulting extension of the high energy shoulder of the conversion line due to atoms recoiling with various values of the angle α for a given value of θ, the former being measured in a plane that is perpendicular to the source-wire's axis (see Figure 10). Obviously α ranges between $\alpha = \theta + \pi/2$ and $\alpha = \theta - \pi/2$; the intensity distribution for negative values of θ is equal to that for positive θ, and therefore the calculations need be carried out only for positive θ. Letting V_n be the recoil velocity of the atom, the electron being emitted from an atom originating from the wire surface at $\alpha = \alpha$ and decaying at the position (R, θ) will obviously have a momentum in the direction of the entrance baffle as given by equation 6. The energy increase of a conversion electron due to recoil is

$$\Delta E_n = \xi \cos \theta, \qquad (30)$$

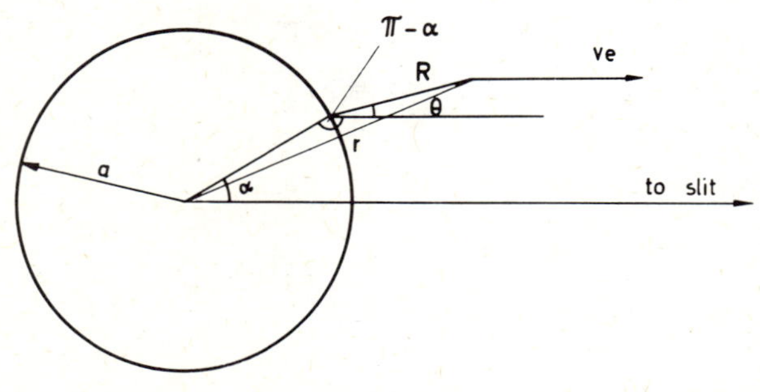

Figure 10. A cross-sectional view of the source wire and the relations between the quantities in the calculation of the change in conversion line shape.

where ξ is the maximum value of the Doppler energy shift. Equation 29 can then be used to find an additional increase in the energy of this conversion electron due to acceleration in the electrostatic field:

$$e\,\Delta V = eV_0\left(1 - \frac{\ln 2b/r}{\ln 2b/a}\right)\text{ eV}, \tag{31}$$

where

$$r = [a^2 + R^2 + 2aR\cos(\alpha - \theta)]^{1/2}. \tag{32}$$

If the excited nuclear state has the lifetime τ, which is related to the half-life by $T_{1/2} = \tau \ln 2$, then the electron emission probability per unit distance r traveled, at a point described by the coordinates α, R, and θ, will be*

$$\frac{dN}{dr} = \frac{r\,dN/dR}{R + a\cos(\alpha - \theta)} = \frac{N_0 r}{\tau V_n}\frac{\exp(-t/\tau)}{R + a\cos(\alpha - \theta)}$$
$$= \frac{N_0 r}{\tau V_n}\frac{\exp(-R/\tau V_n)}{R + a\cos(\alpha - \theta)}. \tag{33}$$

In order to take into account the atoms that recoil at an angle β to the plane described previously, we must realize that expression 31 remains unaltered, whereas expression 30 and the quantity R in equation 32 and in the denominator of equation 33 are multiplied by $\cos\beta$:

$$\Delta E_n = \xi\cos\theta\cos\beta, \tag{34}$$

$$r = [a^2 + R^2\cos^2\beta + 2aR\cos\beta\cos(\alpha - \theta)]^{1/2}, \tag{35}$$

$$\frac{dN}{dr} = \frac{N_0 r}{\tau V_n}\frac{\exp(-R/\tau V_n)}{R\cos\beta + a\cos(\alpha - \theta)}. \tag{36}$$

* This equation and equation 36 differ from those given in Ref. 37 because of an omission in that work. The lifetime determination is not affected by this omission, however, since the determination is related to the slope of the distribution and not to its absolute value.

First, the decay probability as a function of distance r from the wire center was calculated for alpha-decay sites included in $-\pi/2 \le \alpha \le \pi/2$ with the recoil direction defined by $\theta = \beta = 0$. The abscissa is also expressed in energy units using equation 31. Second, on the same plot are included the contributions for θ taking values between 0 and $\pi/2$. Each contribution is shifted along the energy scale to lower energies by the amount $E'' = (\xi - \Delta E_n)$, equation 30, and the decay curve for that contribution is calculated for the abscissa region $r \ge a$. Correction of the intensity by multiplying with an adjustment function is made in order to adjust the E''-scale to the E-scale values. Next, this step is repeated for values of $0 < \beta \gtrsim \pi/2$, consistently adjusting the intensity distributions when changing to the E-scale. Here, equations 34 to 36 are used; the last differ from the corresponding equation in Ref. 37 because of an error in that work (see footnote, this page).

If the experimental shoulder extension lying above $E_0 + \xi$ is expressed not in energy units but in units of r using equation 31, its semilog plot may be fitted by a shape obtained as described previously for a suitable value of τ (see Figure 11).

The calculations were also carried out for thick sources where mean recoil velocities were diminished by as much as 30%. The calculated curves were adjusted to the experimental shoulder extensions, the nuclear half-life was determined, and all final values agreed. The mean value of the nuclear half-life is included in Table II.

The correction of the error just mentioned is expected to decrease the final half-life value obtained in Ref. 37 by several percentage points. In addition, the overall accuracy of these calculations was diminished because they were carried out by hand.

The effect mentioned earlier of atomic recoil on electron line intensities measured using a wire source strengthens as the wire radius is reduced to $V_n T_{1/2}$ and less. For the 39.85-keV level half-life, this value amounts to about 2 μm. However, a small effect is expected even for radii as large as 12 μm. The intensity ratio of the 115.1-K line to the 39.85-L_1 line will thus depend on the source-wire dimension and the half-life of the 39.85-keV level (see Figure 8). Calculations have shown that, for the half-life value determined in Ref. 37, the measured I(115.1-K) to I(115.1-K + 39.85-L_1) intensity ratio obtained using a wire source of 25-μm diameter should be increased from the measured value of 15.7(1.9) to 17.4(2.3)%.[38] The change is not as large as the quoted error limits, and the corrected value is still in good agreement with other experimental values.

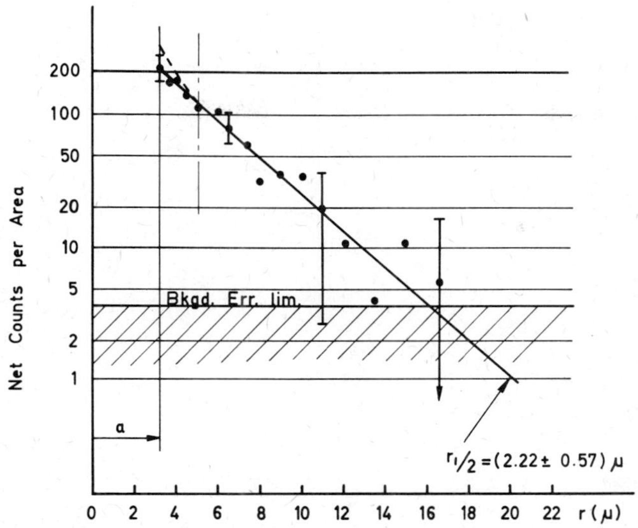

Figure 11. The line extension in Figure 9 above the momentum value $p_0(1 + V_n/V_e)$ as analyzed according to the procedure described in the text. The theoretical curve is dashed. Only that part of the experimental curve lying above $r = 5$ μm was used in the analysis, because of its approximate linearity. (From Ref. 37.)

3. CHARGE STATES OF RECOIL ATOMS AFTER ALPHA-DECAY

The charge spectra of recoil atoms after the alpha-decay of ^{228}Th and its decay products, as well as ^{226}Th and its decay products, are treated in Section 2.2.1. There emphasis is placed on the recoil atom charge spectrum resulting when there occur no subsequent transitions in the daughter nucleus through internal conversion. These charges lie almost exclusively in the 0 to +2 range. This can be seen in Table I, where the relative abundance of these low charges of recoil atoms after alpha-decay are collected. These charge distributions result from electron shake-off at the moment of the alpha-decay.

However, if a nuclear transition were to take place through internal conversion, the created vacancy would cascade to the outer shells via X-ray and, primarily, Auger and Coster–Kronig transitions. When several IC transitions take place in cascade, the final atomic ionic state for heavy atoms can reach charges beyond +30 (Section 3.1).

The cascade of an inner shell vacancy to the outer shell has been investigated earlier, the initial inner shell ionization being produced by internal conversion, electron capture, or photoionization. Snell and Pleasanton[39] were able to explain final charge states of gaseous atoms initially ionized in an inner shell, in terms of successive Auger, Coster–Kronig, and X-ray transitions. Later, similar work was performed by Carlson, Krause, and others.[40] From such measurements it is possible to construct a scheme that predicts for any atomic number the final mean ionic charge due to a rearrangement cascade occurring following an initial vacancy in any of the K to M_5 levels.[40] The mean values of final ionic charge and of the charge distribution for higher atomic numbers have been shown by de Wieclawik to be in good agreement with the results of his work.[41]

We expect nuclear transitions taking place through internal conversion in an atom already in a highly ionized state to exhibit effects of this charging. As the ionizing state increases, decreased shielding of the nuclear charge will cause the binding energies of the inner atomic shell electrons to increase. As a result, internal conversion line shapes will become more complex through the addition of satellite structure (Section 3.2).

A change in low energy nuclear transition half-life, where internal conversion probability is high, might also be detectable if the ionization state of the atom is great enough (Section 3.3).

3.1. Charge States of Recoil Atoms Following Alpha-Decay

The charge spectra of recoil ^{208}Tl and ^{208}Pb atoms following the beta-decay of ^{212}Pb ($^{212}\text{Pb} \xrightarrow{\beta^-} {}^{212}\text{Bi} \xrightarrow{\alpha} {}^{208}\text{Tl}$ and $^{212}\text{Pb} \xrightarrow{\beta^-} {}^{212}\text{Bi} \xrightarrow{\beta^-} {}^{212}\text{Po} \xrightarrow{\alpha} {}^{208}\text{Pb}$, respectively) have been studied by Perrin and de Wieclawik.[41–43] They employed a time-of-flight measurement arrangement, illustrated schematically in Figure 12. Here delayed coincidence between alpha-particles (detected with the solid state dectector) and the recoil atoms (detected with the multiplier) were analyzed with a time-to-pulse height converter. (The entire layout in Figure 12 is not used for these charge spectra measurements.) Separation of the recoil

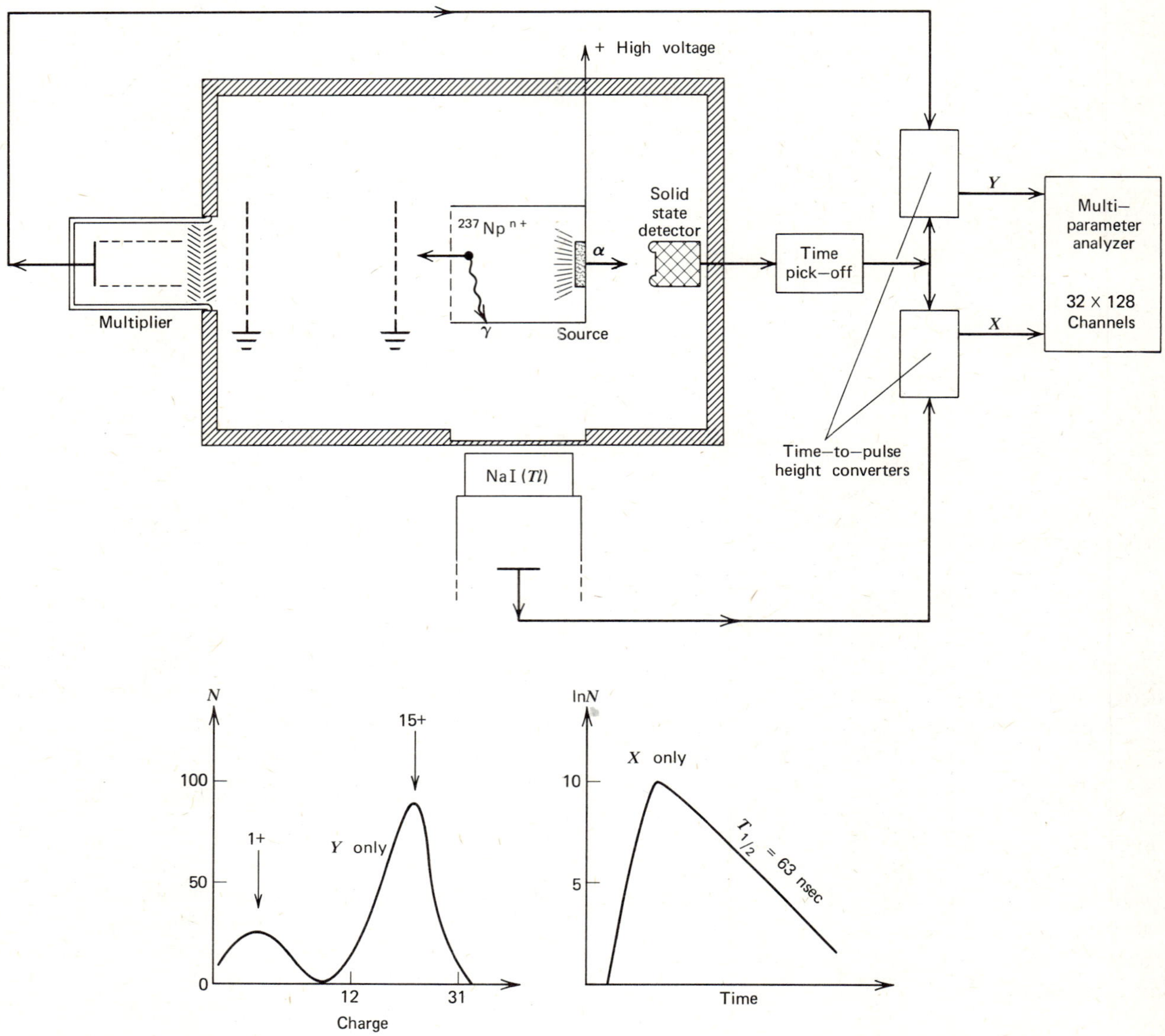

Figure 12. The experimental arrangement used to measure recoil atomic charge spectra (only solid state detector and multiplier combination) and the half-life of the 59.5-keV level in ^{237}Np (using entire set-up).[54] In the latter case the information sent into the multiparameter analyzer through the two inlets is shown at the bottom of the figure.

thallium and lead atoms, independent of charge, is made possible by their different recoil velocities, and, hence, times of flight.

The atomic charge states were separated by accelerating the charged recoil atoms through a potential difference of 25 to 30 kV. The ^{208}Pb spectrum showed an abundance ratio of about 1 : 2.2 : 0.8 for recoil atomic charges of 0, +1, +2, respectively. The ^{208}Tl showed $\sim 0 : \sim 1 : \sim 1 : \sim 0$ for charges of -1, 0, $+1$, and $+2$. These results contradict those of Cano and Dressel[44] for ^{206}Pb.

The mean charge for ^{208}Tl was 12 ± 1, the error indicating a charge state calibration uncertainty. These highly charged states arise primarily as a result of vacancy cascades to outer shells following internal conversion of the 39.85-keV transition.

Similar experiments were carried out by de Wieclawik and Perrin[45] for the alpha-decay of ^{227}Th to ^{223}Ra, using the same apparatus and applying a high voltage for charge separation. Coincidence measurements between final multicharged atoms and alpha-particles feeding the various excited levels in ^{223}Ra were made. Thus it was possible to investigate the mean charge state of the recoil atoms following a single IC transition or two or three such transitions in cascade. The transition energies were low: 29.9, 31.6,

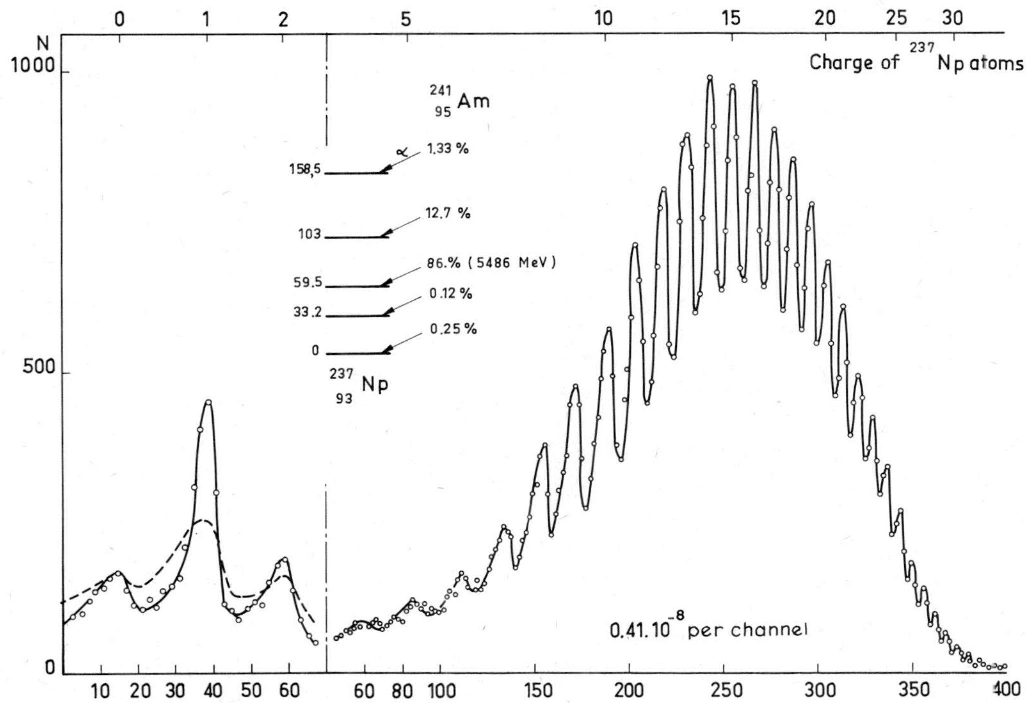

Figure 13. Time of flight spectrum for multicharged $^{237}Np^{n+}$ recoil atoms: source activity, 1.2 μCurie/cm²; flight path, 106 cm; acceleration potential, 12 kV.[46] The dashed distribution was obtained using a 20 times stronger (hence, thicker) source.

61.5, and 113.1 keV. The IC probability was therefore high. For one internal conversion, the recoil charge distribution is centered at +12, as in the case of ^{208}Tl. For two (coincidences with alpha-decay feeding of the second excited level) and three (alpha-decay feeding of a higher level) IC transitions in cascade, the charge distributions lie centered at +20 and +25, respectively.

The recoil charge spectrum for ^{237}Np following the alpha-decay of ^{241}Am and obtained by de Wieclawik[46] with use of 12 kV for recoil charge separation is shown in Figure 13. Clearly, the charge state resolution is very high, indicating the thinness of the source. The maximum charge lies at +33. As mentioned previously the highest charge states result from nuclear IC transition cascades with subsequent repeated Auger, Coster–Kronig, (and X-ray) transition cascades. Charge distribution centers for alpha-decay to the 59.9- and 103-keV levels were measured and analyzed in terms of charge distribution contributions from one and two IC transitions, in the first case, and one, two, and three such transitions in the latter. The recoil charge distributions for the three contributions (resulting from single, double, and triple internal conversions in cascade) were taken from the measurements of these in the investigation of ^{223}Ra recoil atoms described earlier.[45] However, the three recoil charge distributions for ^{237}Np were found to center at +14, +21, and +26 (i.e., at values higher than for the ^{223}Ra).

The study of the recoil atomic charge spectrum in coincidence with the various preceding alpha-decays will give information about the number of successive IC processes leading to the ground nuclear state. By such measurements and by studying the charge spectrum in coincidence with each of the three alpha transitions from ^{223}Ra to ^{219}Rn, de Wieclawik[47] was able to deduce several properties of the level scheme of the daughter isotope.

3.2. Complex Structure of Internal Conversion Lines due to Atomic Ionization

During the study of the IC electron spectrum following the alpha-decay of ^{227}Th to ^{223}Ra, it was found that the conversion lines of the 29.90-keV transition had an anomalous form.[48,49] Figures 14 and 15*a* indicate that the L_2 and L_3-conversion lines of the 29.90-keV transition have extra structure on the low energy side, as compared to the L_2- and L_3-conversion lines of the 31.61-keV transition.[50] That the satellite structure *S* was associated with the conversion line, and not with the presence of other IC lines, was supported by the equality between the ratios of: (*a*) the Doppler plus satellite contributions *to* the intensity portion due to atoms recoiled into the source backing

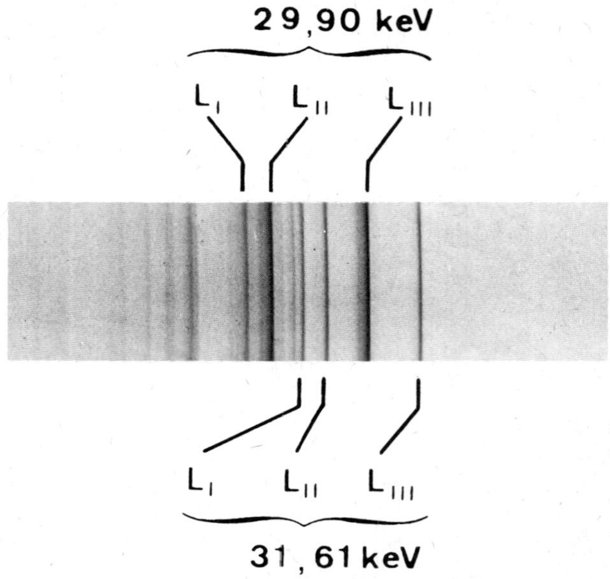

Figure 14. The ^{237}Np 29.90-*L* and 31.61-*L* conversion line energy region measured using a semicircular spectrograph.[41,54] Comparison of the intensity distributions of the two sets of lines reveals the complexity of the structure of the former due to atomic multi-ionization states of the recoil atoms.

(for a 29.90-*L* line) and (*b*) the Doppler contribution *to* the backing stopped recoil portion (for a 31.61-*L* line).

The effect is interpreted as being of the same origin as X-ray and Auger-vacancy-satellite lines. In an atom recoiling after alpha-decay, an initial IC transition will lead, as shown in Section 3.1, to a state of high atomic ionization. Subsequent internal conversion, in, say, the *L*-shell of the highly ionized atom, will result in the emission of an electron of higher atomic binding energy than would have been emitted if the atom had been neutral.

Figure 15*a* shows the profiles of the L_2- and L_3-conversion lines of the 29.90-keV transition in ^{223}Ra compared with the 31.61-L_2 and L_3 lines for the same isotope. There are also low intensity L_2- and L_3-conversion lines of a 29.6-keV transition on the low energy tails of the L_2- and L_3-conversion lines of the 29.90-keV transition; the presence of these must be considered when a more detailed analysis of the line shape is desired.

The foregoing interpretation of the origin of the conversion line satellite structure *S* was checked further by applying a positive electrostatic potential to the source and grounding the spectrograph entrance slit. Then IC electrons emitted from a recoil atom immediately following alpha-decay will be detected at a greater energy above the conversion electron peak arising from backing-stopped recoils than if the potential were removed. When the transition is second in a series of two, a conversion electron for the same transition is of lower original kinetic energy owing to the higher state of ionization of the recoiling atom, but it will be shifted by an amount of energy even larger than in the former case because a longer time has lapsed between the alpha-decay and the conversion electron's ejection (i.e., effectively two nuclear lifetimes). When the transition of interest is the third in a cascade series, the original conversion transition energy is even lower, owing to the higher mean state of atomic ionization, but the electron will be accelerated correspondingly more as it is emitted still further away from the source. Therefore, the *R* and *S* recoil components of the conversion line in Figure 15*a* will tend to merge as the applied voltage is increased—which indeed they do, proving their origin to be that stated previously.[50]

The experimental results just cited, and their interpretation, were subsequently tested by Gelletly and Geiger.[51] Using the Chalk River iron-free instrument, they measured and compared the shapes of the M_3- and L_3-conversion lines for both the 29.9- and 31.6-

keV transitions in ^{223}Ra. The 29.9-M_3 profile agreed well with Figure 14, but the 31.6-M_3 line showed a narrower width relative to the 29.9-M_3 line, the absence of structure on the low energy side, and an increase in intensity of the high energy shoulder. There were similar but less pronounced tendencies in the L_3-lines.

These workers also carried out $\alpha - e^-$ coincidence measurements detecting the alpha-particle and the conversion electron emitted in opposite directions. Comparison of both the 29,9-M_3 and 31.6-M_3 lines showed strong structure lying below the main peak in the former line; this structure was absent in the latter. The same measurement was made on the 29.9-L_3 and 31.6-L_3 lines with similar results, as shown in Figure 15*b*, where a broad component lies a mean value of 216 eV below the 29.9-L_3 peak. This component arises from internal conversion in the L_3-subshell of ions of an average charge $\sim +12$. This indicates that, for a radium atom charged to approximately $+12$, the

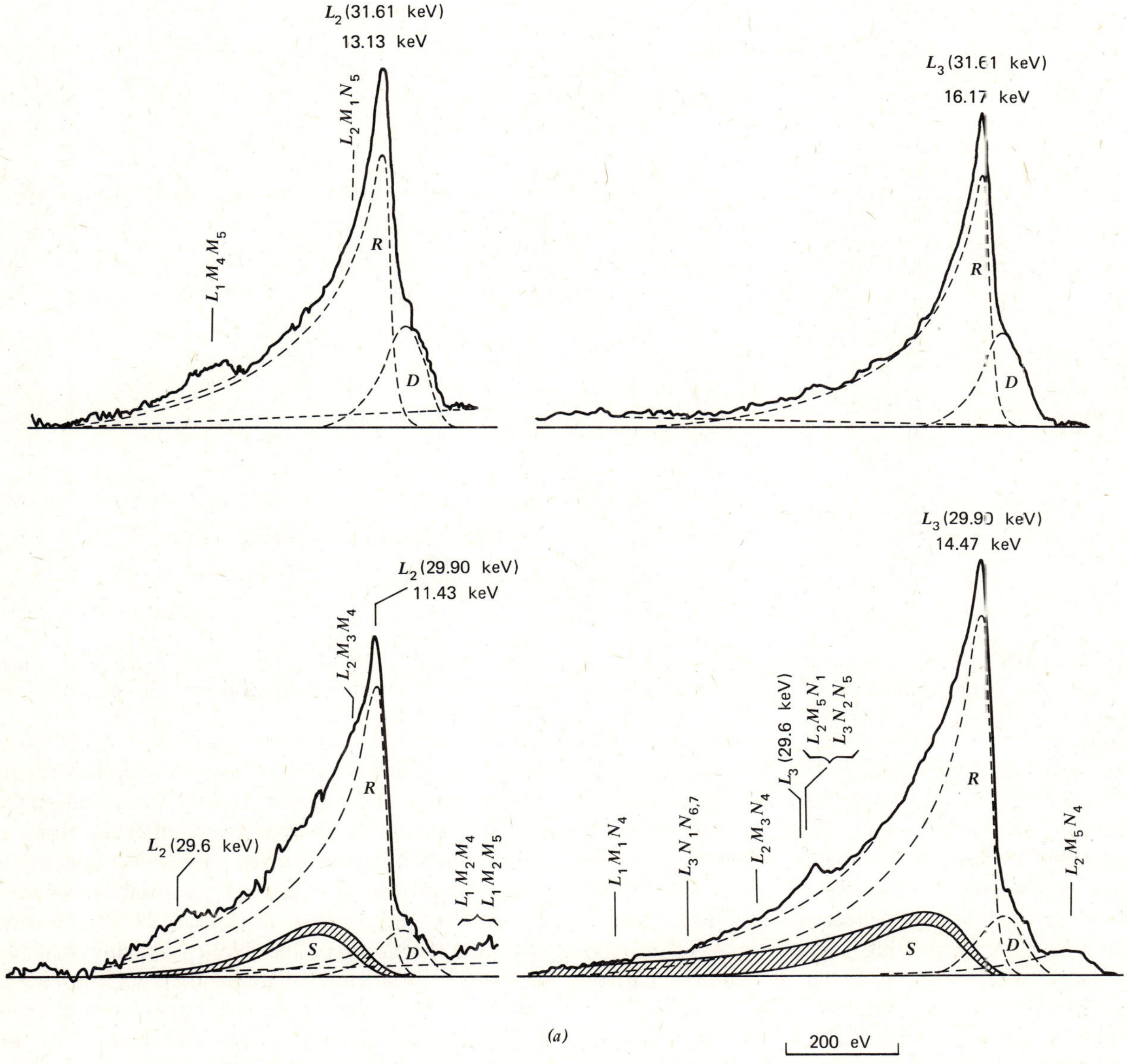

Figure 15*a*. Comparison of the different composite L_2- and L_3-29.90 and L_2- and L_3-31.61 keV conversion lines following the alpha-decay of ^{227}Th to ^{223}Ra. The former show a satellite band component *S* due to internal conversion that occurs in the highly ionized recoil atom; this structure is not present in the latter two profiles. The spectral components *R* are those from atoms that had recoiled into the backing and come to rest; the *D*-components are the Doppler-shifted electrons originating in recoiling atoms. Several weak *L*-Auger lines and the weak L_2- and L_3-conversion lines of the 29.6-keV transition in the decay add to the spectral structure. (Due to Briançon.[52])

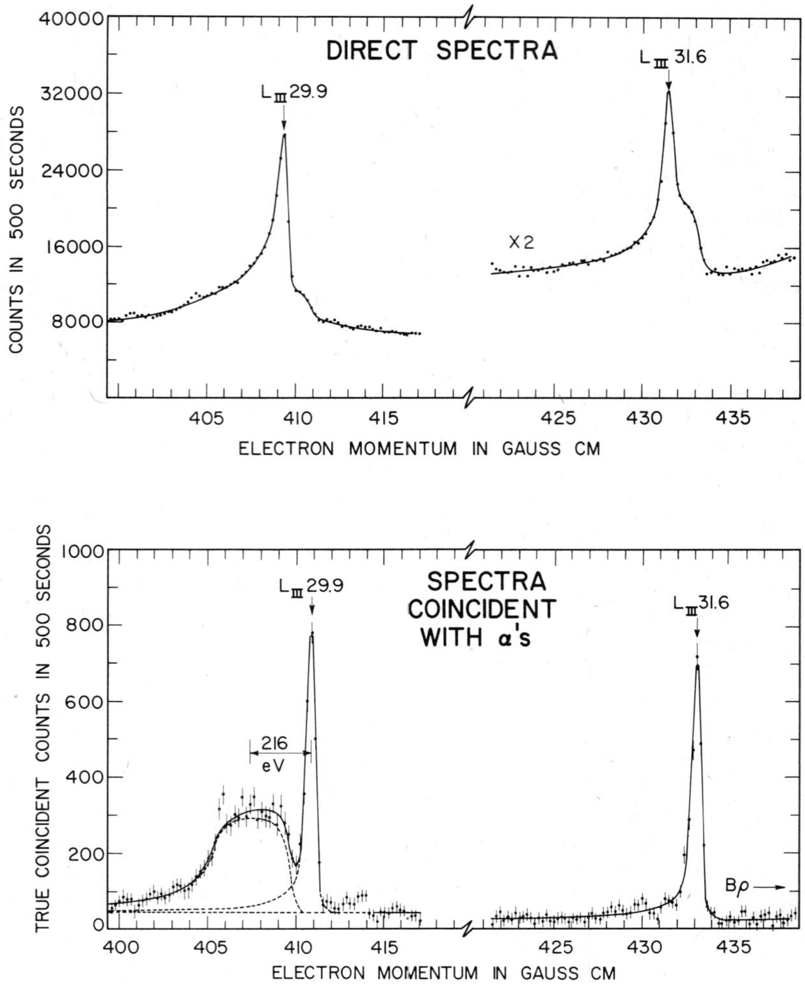

(b)

Figure 15*b*. The L_3-29.9 keV and L_3-31.6 keV conversion lines in ^{223}Ra following alpha-decay of ^{227}Th:[51] upper graph, singles spectrum; lower graph, spectrum in coincidence with alphas from the decay of ^{227}Th. Note the strong, broad structure to the lower side of the L_3-29.9 keV line (see text).

energy required to remove the L_3-electron from the atom is 216(25) eV greater than for the neutral atom. According to Gelletly and Geiger,[51] the binding energy increase for the L_2-subshell is 239(32) eV and for the M_3-subshell it is 240(38).

The investigation of the effect of multiple ionization on atomic electron binding energies in recoil atoms was more recently investigated in an alternative manner by Briançon and Walen,[52,53] who pointed out that several weak $L_i MM$ and $L_i MN$ ($i = 1, 2, 3$) Auger lines will be positioned about the conversion lines of interest. These lines are shown in Figure 15*a*. In addition, and even more important, the L_2- and L_3-conversion lines of a 29.6-keV transition in ^{223}Ra will lie just some 300 eV below the 29.9-L_2 and L_3-peaks.

Briançon and Walen have investigated the satellite structure (*S* in Figure 15*a*) using a modified experimental system allowing one to determine if indeed this structure is due to electron emission from atoms in recoil. Such an investigation may be conducted as already described in the work of Briançon et al.[50] The effect of the presence of the weak 29.6-L_2 line does not alter the proof of the 29.9-L_2 satellite structure presence.[52] Briançon has found that, for the 29.9-L_2 and L_3 conversion lines, the component maxima from atoms of a charge $\sim +12$ (see Section 3.1) are situated some 160(30) eV toward lower energies from the main component. This is in strong disagreement with the results of Gelletly and Geiger[51] cited previously. Although the discrepancy between the two sets of results may have been caused by a tendency to neutralization of the recoiling ions due to collision with residual gas atoms in the vacuum chamber, this explanation was considered insufficient in view of the quality of vacuum and the mean life of the 29.9-keV level (10^{-10} sec): the recoil path length in one mean life would be only a few tens of microns. Instead, it was decided that this binding energy change discrepancy had occurred because a fast plastic scintillator with mediocre resolution had been used for alpha-detection. Therefore, the alpha-groups feeding the 79.8-keV level, which decays as $79.8\text{ keV} \xrightarrow{(29.6\text{ keV})} 59.2\text{ keV} \xrightarrow{(20.3\text{ keV})} 29.9\text{ keV} \rightarrow$ ground, and the 61.5-keV level, decaying as $61.5\text{ keV} \xrightarrow{(31.61\text{ keV})} 29.9\text{ keV} \rightarrow$ ground, will be detected simultaneously. The broad structure in the coincidence spectrum for the 29.9-L_2 line in Figure 15*b* may then include a contribution from the 29.6-keV transition, shifting the broad component to lower energies. The experimental value of 160(30) eV (for the change in L_2- or L_3-level energies quoted earlier) in recoil ions of +12 charge is in good correlation with the value of ~140 eV for an ionic charge of +10 obtained using a relativistic Hartree–Fock–Slater calculation.[52]

Finally, an experimental arrangement used by Briançon and Walen[52,53] to investigate the energy spectrum of conversion electrons from only the recoil atoms in flight merits consideration; this was devised to study the structure of conversion lines for transitions in ^{237}Np following the alpha-decay of ^{241}Am. Here the activity-covered surfaces were placed about the vacant "source position region" in a semicircular spectrograph. Recoil atoms would then be allowed to travel into the region, and electrons emitted by them would be accepted into the spectrograph through a baffle arrangement; a collimation arrangement ensured that the recoil directions were nearly at right angles to the electron acceptance direction, and Doppler broadening of the electron lines was thereby reduced. A thin grounded grid was placed across the outer edges of the collimator and only 0 or +1 charged recoil ions were allowed to enter the "source position region," and a 14-kV potential was placed between the activated surface and the grid. The results of the measurement of the L_2-59.5 keV conversion line in ^{237}Np appear in Figure 16, and the significance of each of the various sets of points is stated in the legend. The calculated low energy tail for 0-kV potential was that to be obtained for a monoenergetic line for the set geometry of the instrument. The low intensity of the experimental tail for 14-kV applied voltage shows that the highly charged ions are efficiently removed. The total experimentally obtained shape of the 59.5-keV line follows a single internal conversion event 34.5% of the time and a second in a cascade of two 5% of the time. Note also that for the 33-keV transition, for example, the respective percentages are 2.7 and 18.4%; the conversion lines of this transition consist mainly of first satellite components. This is clearly seen in the photographic exposures of the spectrum.[52]

Figure 16 shows that the satellite peak S_1 associated with a single preceding internal conversion event is centered at about 250-eV below the direct peak *D*. This larger percentage shift in energy, compared to that for the radium case, occurs mainly because the mean ionic charge at the instant of internal conversion is ~ +14, or two units greater than it was in the case of radium.

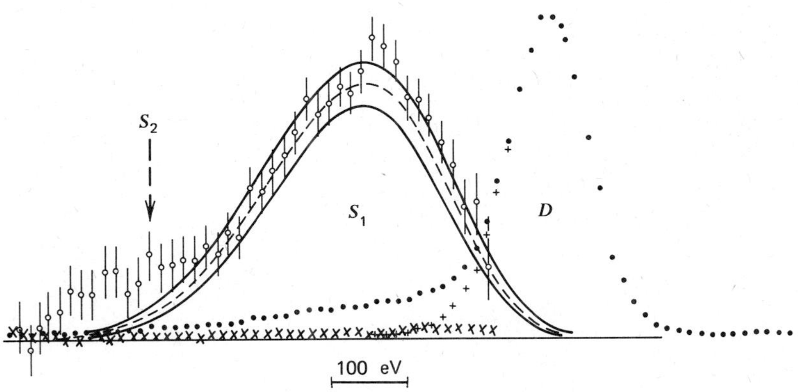

Figure 16. The structure of the L_2-59.5 keV conversion electron line from recoiling ^{237}Np after ^{241}Am alpha-decay.[52, 53] Here *D* is the component due to first-time internal conversion in the recoiling atoms; S_1 is that due to a second (in a cascade of two) conversion; the dashed arrow shows the calculated position of the S_2 satellite maximum (third internal conversion in a cascade of three). Solid circles, experimental points for 0-kV applied voltage; pluses, calculated low energy tail of the *D*-component for 0 kV; ×'s, residue of experimental points for 14 kV before subtraction of the calculated tail of the *D*-component for 14 kV; open circles, difference between the three previous curves times a factor 10; dashes, theoretical curve with uncertainty limits (solid curves).

3.3. Effect of High Ionization State on Low Energy Nuclear Transitions

The half-life of an excited nuclear state is dependent on both the gamma-transition probability and the IC probability. Low energy transitions are highly converted and the total decay probability of the nuclear state is related to the number of atomic electrons of low enough binding energy to take part in the transition. For an extremely low transition energy, where all but the outermost atomic electrons have binding energies in excess of the transition energy in question, changes in the chemical state of the atom can measurably affect the internal conversion and hence nuclear transition rates (see Chapter 4, Section 5.2).

The number of electrons on hand to participate in internal conversion may also be reduced by placing the atom in a state of high ionization. For low energy transitions, say less than 1 keV, the transition rate will be expected to differ appreciably between the neutral atom and an atom in the charge state of +15.

de Wiecklawik[54] has measured the half-life of the 59.5-keV level in ^{237}Np following the alpha-decay of ^{241}Am using the total apparatus shown in Figure 12. Applying a +20-kV potential to the source for recoil charge separation, measurement could be made of coincidences between alpha-particles and recoil atoms, the alpha-particles leading to the 103-keV level being those detected. A sodium iodide (Tl) scintillation crystal was used to detect 59.5-keV gamma-rays, indicating decay from that level to the ground state.

Two sets of coincidences were measured: between the 5.443-MeV alpha-particles and recoil atoms of charge +1 and of charge +15 following the transition (gamma or IC, respectively) from 103-keV level to the 59.5-keV level. These coincidences were then used to gate the time-to-pulse height converter signals for the time delay measurement between the alpha-particle and 59.5-keV gamma-ray emissions. This time delay is essentially that due to the 63×10^{-9} sec half-life of that level, since the contribution from the 103-keV level half-life ($\sim 0.08 \times 10^{-9}$ sec) will be negligible.

The half-lives obtained from the two sets of measurements for the 59.54-keV level for the atomic ionization states +1 and +15 agreed within experimental error limits, about 10%.

REFERENCES

1. D. L. Baulich and J. F. Duncan, *Proc. Phys. Soc.* (*London*) **268** (1962) 181.
2. J. C. Jacobsen and O. Kofoed-Hansen, *Phys. Rev.* **73** (1948) 675.
3. C. H. Johnson, F. Pleasanton, and T. A. Carlson, *Phys. Rev.* **132** (1963) 1149.
4. A. H. Snell, in *Alpha-, Beta-, and Gamma-Ray Spectroscopy*, K. Siegbahn (Ed.), North-Holland Publishing Company 1965, p. 1545.
5. G. A. Korolev, A. A. Vorob'yov, Yu. K. Zalite, A. K. Lebedev, B. V. Grigor'ev, V. A. Smirnov, and L. M. Vasil'eva, A. F. Ioffe Institute Rept. 151, Leningrad, 1968.
6. T. Novakov, *Nucl. Instr. Meth.* **43** (1966) 189.
7. D. Ashery, N. Assaf, G. Goldring, A. Sprinzak, and Y. Wolfson, *Nucl. Phys.* **77** (1966) 650.
8. D. Ashery, N. Assaf, G. Goldring, A. Sprinzak, and Y. Wolfson, *Nucl. Phys.* **A101** (1967) 51.
9. A. E. Blaugrund, A. Fisher, and A. Schwarzschild, *Nucl. Phys.* **A107** (1968) 411.
10. T. Novakov, J. M. Hollander, and R. L. Graham, *Nucl. Instr. Meth.* **26** (1964) 189; *Bull. Amer. Phys. Soc.* **9** (1964) 9.
11. D. Marx, *Z. Phys.* **195** (1966) 26.
12. L. Winsberg and J. M. Alexander, *Phys. Rev.* **121** (1961) 518.
13. I. Bergström, J. A. Davies, B. Domeij, and J. Uhler, *Ark. Fys.* **24** (1963) 99, 389, 413.
14. L. Morbitzer and A. Scharmann, *Z. Phys.* **185** (1965) 488.
15. K. Bethge, P. Sandner, and H. Schmidt, *Z. Naturforsch.* **21a** (1966) 1052.
16. J. A. Davies and G. A. Sims, *Can. J. Chem.* **39** (1961) 601.
17. K. D. Sevier, *Nucl. Instr. Meth.* **14** (1961) 318.
18. O. B. Firsov, *Zh. Eksp. i. Teor. Phys. SSSR*, **32** (1957) 1464, **33** (1957) 696, **34** (1958) 447. [*Russ. J. Exp. Theor. Phys.*, **5** (1957) 1192; **6** (1958) 534; **7** (1958) 308].
19. J. P. Biersack, *Z. Phys.* **211** (1968) 495.
20. J. L. Wolfson and W. R. Dixon, *Nucl. Phys.* **64** (1965) 433.
21. J. Lindhard and M. Scharff, *Phys. Rev.* **124** (1961) 128; J. Lindhard, M. Scharff, and H. E. Schiøtt, *Mat. Fys. Medd. Dan. Vidensk. Selsk.* **33**, No. 14 (1963).
22. G. T. Ewan, A.E.C. of Canada Progr. Rept. PR-P-56, Chalk River National Laboratories (October–December, 1962); *Can. J. Phys.* **41** (1963) 2202; private communication, 1964.
23. E. M. Krisiuk, G. D. Latyshev, and A. G. Sergeev, *Izv. Akad. Nauk SSSR, Ser. Fiz.* **20** (1956) 367 [*Bull. Acad. Sci. USSR, Phys. Ser.* **20** (1956) 335].

24. A. V. Gnedich, L. N. Kriukova, V. V. Murav'eva, V. S. Shpinel', and V. I. Shumshurov, *Izv. Akad. Nauk SSSR, Ser. Fiz.* **22** (1958) 867. [*Bull. Acad. Sci. USSR, Phys. Ser.* **22** (1958) 861].

25. J. L. Wolfson and J. J. H. Park, *Can. J. Phys.* **42** (1964) 1387.

26. J. Burde and S. G. Cohen, *Phys. Rev.* **101** (1956) 495.

27. J. Burde and S. G. Cohen, *Phys. Rev.* **104** (1956) 1093.

28. J. Burde and S. G. Cohen, *Phys. Rev.* **104** (1956) 1085.

29. J. G. Siekman and H. de Waard, *Phys. Rev.* **107** (1957) 1731.

30. J. G. Siekman and H. de Waard, *Nucl. Phys.* **8** (1958) 402.

31. K. Gunter, F. Asaro, and A. C. Helmholz, *Phys. Rev. Letters* **16** (1966) 362.

32. S. Szucs and J. M. Delfosse, *Phys. Rev. Letters* **15** (1965) 163.

33. C. Bergman, *Nucl. Instr. Meth.* **52** (1967) 122.

34. G. T. Ewan and R. L. Graham, in *Alpha-, Beta-, and Gamma-Ray Spectroscopy*, K. Siegbahn (Ed.), North-Holland Publishing Company, 1965, p. 951.

35. J. C. Severiens and S. W. Richter, *Nucl. Phys.* **13** (1959) 239.

36. K. D. Sevier, *Ark. Fys.* **25** (1963) 87.

37. K. D. Sevier, *Nucl. Phys.* **61** (1965) 601.

38. K. D. Sevier, *Ark. Fys.* **28** (1964) 397.

39. A. H. Snell and F. Pleasanton, *Phys. Rev.* **100** (1955) 1396; *Phys. Rev.* **102** (1956) 1419; *Phys. Rev.* **111** (1958) 1338.

40. M. O. Krause, M. L. Vestal, W. H. Johnston, and T. A. Carlson, *Phys. Rev.* **133** (1964) A385; M. O. Krause and T. A. Carlson, *Phys. Rev.* **149** (1966) 52; T. A. Carlson, W. E. Hunt, and M. O. Krause, *Phys. Rev.* **151** (1966) 41; M. O. Krause and T. A. Carlson, *Phys. Rev.* **158** (1967) 18.

41. W. de Wieclawik, thesis, University of Paris, Orsay, 1969.

42. W. de Wieclawik and N. Perrin, *Compt. Rend.* **262** (1966) 511.

43. N. Perrin and W. de Wieclawik, *Izv. Akad. Nauk SSSR, Ser. Fiz.* **31** (1967) 80. [*Bull. Acad. Sci. USSR, Phys. Ser.* **31** (1967) 86.]

44. G. L. Cano and R. W. Dressel, *Phys. Rev.* **139** (1965) A1883.

45. W. de Wieclawik and N. Perrin, *J. Phys. (Paris)* **29** (1968) C1-104.

46. W. de Wieclawik, *Compt. Rend.* **266** (1968) 577.

47. W. de Wieclawik, *Compt. Rend.* **268** (1969) 1268.

48. C. Briançon, M. Valadares, and R. J. Walen, *J. Phys. (Paris)* **29** (1968) C1-103.

49. M. Valadares, R. J. Walen, and C. Briançon, *Compt. Rend.* **263** (1966) 313.

50. C. Briançon, M. Valadares, and R. Walen, private communication from M. Valadares, 1967; *Compt. Rend.* **265** (1967) 1496.

51. W. Gelletly and J. S. Geiger, *18th Ann. Nat. Conf. Nuclear Spectroscopy*, Riga, 1968; Private communication from J. S. Geiger, 1969.

52. C. Briançon, thesis, University of Paris, Orsay, 1969.

53. C. Briançon and R. J. Walen, *Int. Conf. Radioactivity in Nuclear Spectroscopy*, Nashville, 1969.

54. W. de Wieclawik, private communiction, 1969.

Chapter Six

Natural Widths of Atomic Levels

1. INTRODUCTION

The natural widths of atomic levels have been a subject of investigation since the early days of X-ray spectroscopy. Individual X-ray spectral lines can be analyzed in terms not only of their energies and intensities, but also of their intrinsic energy shapes or distributions. Once corrections have been made for instrumental contributions to the experimental line shape, the analysis of the shape of a spectral line profile can be accomplished in terms of profile form, width, and degree of asymmetry. The first two parameters are most strongly related to the probabilities of X-ray transitions and possible competing Auger or Coster–Kronig transitions. There are also contributions to all three parameters due to effects of possible multiple atomic ionization and the temperature and chemical state of the sample under investigation. In addition, effects of spin coupling between unpaired atomic electrons and with the nucleus may also contribute to the width.

As applied to an energy or other distribution, "width" can be defined to cover a variety of parameters. In this chapter, "width" or "natural width" refers to the full width of a distribution as measured at half-maximum intensity (the FWHM), unless otherwise stated.

The approximate natural width of an atomic or nuclear level is related to the mean life of the state by the Heisenberg uncertainty principle of $\tau\Gamma \cong \hbar$, where τ is the mean lifetime in question and Γ is the corresponding width in energy units. Since most low energy nuclear states have mean lives of $\tau > 10^{-13}$ sec, we expect the natural widths to be < 0.007 eV and the energy distributions of gamma-rays to be very sharp.[1] For atomic states, however, the natural widths are expected to be much larger ($\sim$70–0.07 eV), since an atomic vacancy is filled in a time interval of the order of 10^{-17} to 10^{-14} sec.

In the past, knowledge of X-ray widths has provided the only means of estimating atomic vacancy lifetimes. Recently, however, Carlson and White[2] have determined that the half-life of an initial vacancy in iodine caused by photoionization by 8-keV X-rays is 1.7×10^{-15} sec, with photoionization in the L-shell $\sim 90\%$ and $\sim 10\%$ in the M-shell; initial ionization in the other shells is negligible. This result was obtained by measuring the recoil energy spectrum of H^+ formed following the interaction of the X-rays with hydrogen iodide gas molecules.

When corrections have been made for instrumental broadening, the measured spectral line widths in X-ray emission result from the combined effects of the widths of the atomic states associated with the transition. The way in which these widths combine to give the measured widths of the X-ray energy distribution is dictated by the shape of the energy distributions of these atomic states.

In X-ray absorption spectra, a decrease in the intensity of the X-radiation passing through the absorber will be caused by the excitation of bound atomic electrons to unoccupied bound or virtual atomic levels. The multiplicity of the possible final levels gives rise to an absorption spectrum which is often of complex structure; the precise interpretation and analysis may be a difficult task.

If the energy of the incident X-ray is increased to the extent that the electron is removed from the initial bound atomic state to the positive energy continuum, the electron energy may be investigated with the aid of a suitable electron spectrometer. Here then the electron line width will be due to the combined effects of the width of the X-radiation, the natural width of the level from which the electron was ejected, and broadening effects owing to the resolving power of the spectrometer and resulting from electron energy losses (elastic or inelastic scattering; see also Chapter 8, section 5) in the converter. These last may be estimated in some cases or reduced to a minimum with careful preparation of the converter and/or proper choice of the irradiating X-ray energy. Irradiation of a converter with gamma-rays will also produce photo-electron spectra. The characteristics of the photo-electron lines so produced will be much the same as those produced with X-radiation, except for the much reduced broadening contribution due to the energy width of the converted radiation.

The width of internal conversion electron energy distributions will be caused by the combined effects of the gamma-transition probability (width), which is negligible, the width of the initial bound atomic state of the electron, the instrumental window curve, and broadening effects of electron scattering in the source material.

Auger transition line widths arise from the width contributions of the three atomic levels participating in the transition. Width value estimates obtained by assuming that the Auger electron line width is equal to the sum of these agree well with measured Auger line width values after correcting for instrumental and scattering contributions.[3–6] Furthermore, measured Auger line widths can, in some cases give accurate information about the widths of the levels participating in the transition.[7]

In determining natural level widths employing any of the above-named procedures, corrections of the experimental spectral line shape—be it an X-ray emission or absorption line or a conversion or photoelectron line—must first be made in order to obtain the true spectral energy or momentum distribution. These changes in the spectral energy distribution are caused primarily by scattering in the electron or X-ray source and by broadening effects due to the finite resolving power of the spectrum analyzing instrument; hence they arise after the electron or quantum has been emitted from the atom and are independent of the emission or transition process. Therefore, we expect corrections to be greater and more important for electron spectra than for X-ray spectra.

The primary contributions to an atomic level width, as deduced from X-ray or electron spectrum measurements, are due to X-ray and Auger-type transition probabilities. Secondarily there are atomic broadening effects arising from perturbing interactions between the electron cloud of the emitting atom and its environment. In addition, we may expect a small Doppler broadening owing to the effect of thermal heat motion.

The factors that contribute to the widths of atomic levels as obtained from X-ray and conversion and photoelectron spectrum measurements may be enumerated as follows:

1. X-ray, Auger, and Coster–Kronig transition probabilities (Section 4.1).
2. Chemical state and lattice effects (Section 4.2).
3. Effects of multiple ionization (Section 4.3).
4. Effects due to coupling between atomic electrons and the nucleus (Section 4.4).
5. Effects of thermal motion (Section 4.5).

Earlier review articles also treat this topic.[8–10]

2. X-RAY EMISSION AND ABSORPTION SPECTRA

The shapes and natural widths of X-ray-emission lines can be calculated following the procedures of Weisskopf and Wigner[11] and Hoyt[12] (see also Ref. 13).

The shape of an X-ray-emission line can be calculated in terms of the classical model by treating the radiating atom as a damped harmonic oscillator with a time-dependent displacement $A(t)$ given by

$$A(t) = A_0 e^{-2\pi\Gamma t} \cos(2\pi\nu_0 t - \varphi), \qquad (1)$$

where A_0 is the initial amplitude of the oscillator, ν_0 its undamped frequency, $2\pi\Gamma$ the damping factor, and φ a phase constant. The rate at which energy is lost by the oscillator, (i.e., the damping factor) is a function of the oscillation frequency

$$2\pi\Gamma = \frac{8\pi^2}{3} \frac{e^2\nu_0{}^2}{m_o c^3}. \qquad (2)$$

When the oscillation is not damped too heavily (i.e., $\Gamma \ll \nu_0$) the Fourier analysis of the function in equation 1 gives for the intensity $|a_\nu|^2$ of emitted radiation of frequency

$$J(\nu)\, d\nu = |a_\nu|^2\, d\nu = \frac{\Gamma}{2\pi} \frac{d\nu}{(\nu_0 - \nu)^2 + (\Gamma/2)^2}. \qquad (3)$$

The mean life τ of an atom in its excited state, which in the classical representation is the oscillator at its initial amplitude, is equal to the reciprocal of the damping factor

$$\tau = \frac{1}{2\pi\Gamma}. \qquad (4)$$

The treatments of the problem by Weisskopf and Wigner and by Hoyt were based on the Dirac quantum theory of the radiation field and led to a formula similar to equation 3 for the natural shape of the emission line. The quantity Γ, however, took on a new meaning. Denoting the initial state by A and the final state by B, the energy distribution of the radiation emitted in the transition will, according to Weisskopf and Wigner, take the form

$$J_{AB}(\nu)\,d\nu = \frac{\Gamma_A + \Gamma_B}{2\pi} \frac{d\nu}{(\nu_{AB} - \nu)^2 + [(\Gamma_A + \Gamma_B)/2]^2} \quad (5)$$

where

$$\nu_{AB} = \nu_A - \nu_B = \frac{E_A - E_B}{h} \quad (6)$$

and

$$\Gamma_A = \frac{1}{2\pi\tau_A} = \sum_{E_C \le E_A}{}' (C)\gamma_{AC}$$

$$\Gamma_B = \frac{1}{2\pi\tau_B} = \sum_{E_D \le E_B}{}' (D)\gamma_{BD}. \quad (7)$$

The quantity ν_{AB} is the frequency value at the center of gravity of the emission line and is equal to the energy difference between the atomic states E_A and E_B divided by Planck's constant h. The widths Γ_A and Γ_B are also expressed in units of frequency. The quantity $2\pi\gamma_{AC}$ is the probability per unit time that a spontaneous transition will take place from a state A to any other state C of lower or equal energy; this includes also Auger and Coster–Kronig radiationless transitions.

The frequency distribution for a transition A → B can be represented by

$$J_{AB}(\nu)\,d\nu = \int_{\nu'=0}^{\infty} J_A(\nu')J_B(\nu' - \nu)\,d\nu'\,d\nu. \quad (8)$$

Thus the frequency distributions for the initial and final states A and B have the form,

$$J_j(\nu)\,d\nu = \frac{\Gamma_j}{2\pi} \frac{d\nu}{(\nu_j - \nu)^2 + (\Gamma_j/2)^2}, \quad (9)$$

where (j = A, B). Inserting these equations into equation 8 and performing the integration, we obtain equation 5. The transition line width is therefore equal to the sum of the widths of the initial and final states.

The frequency distribution in equation 5 or 9 is commonly called a dispersion, Hoyt, or Lorentzian distribution. When two such curves are "folded" one into the other, a procedure represented by equation 8, the resulting distribution is also Lorentzian, with a width equal to the sums of the widths of the original two curves, or

$$\Gamma_{AB} = \Gamma_A + \Gamma_B. \quad (10)$$

In contrast, the Gaussian distribution has the general shape

$$G(\nu) \propto \exp\,[-(\nu_j - \nu)^2 \ln 2/(\Gamma/2)^2].) \quad (11)$$

The folding of two Gaussian curves having widths Γ_a and Γ_b results in a Gaussian curve having a width Γ_{ab}, where

$$\Gamma_{ab}^2 = \Gamma_a^2 + \Gamma_b^2. \quad (12)$$

Analysis of X-ray emission line shapes necessitates correction for the instrumental width contributions. The emission line shape can be determined from the measured line shape if the instrumental line intensity distribution is known. This last is that shape measured with the instrument if the natural width of the energy distribution of the X-ray-emission line is vanishingly small. The instrumental line shape seldom has a Lorentzian distribution. Therefore the measured width is neither the sum of the instrumental width and the natural line width of the line nor the root of the sum of their squares. Richtmyer et al.,[14] for example, used the empirical relation $\Gamma_T = \Gamma_O - \frac{1}{2}\Gamma_I$ for their work on the widths of the L-series X-ray lines of gold. This was a compromise between the two shape components contributing to the line shape: the Gaussian (instrumental width) and the Lorentzian (natural width). Here Γ_T is the true width, Γ_O is the observed width, and Γ_I is the instrumental width. See Ref. 15 for more references to additional articles treating the analysis of experimental line shapes.

The analysis of absorption spectra can also give information about the natural widths of spectral lines. Here again, we can apply the theory of Weisskopf and Wigner to predict the shape of an absorption limit.

In the absorption process, a bound electron is transferred from a given inner atomic shell to an unoccupied bound orbital or a continuum state subject to the selection rules of the excitation process. The nature of the possible final bound states to which the electron is allowed to transfer (e.g., conduction band, bound orbitals) depends on the electron configuration of the element or compound making up the sample, as well as its physical state (i.e., gas, liquid, solid). The excitation of an initially bound electron may also take place to the continuum.

In the investigation of X-ray absorption spectra, we may interpret the absorption spectrum according to a one-electron model. Here an inner electron is excited either to a bound orbital beyond the closed electron core or to a continuum state. An oscillator strength f_n, which is related to the transition probability of that state,[16] is assigned to the resonance absorption transition of a given core electron to the nth bound orbital. Each resonance absorption transition, lying below the continuum edge, has associated with it an oscillator strength. It has been shown[17] that the oscillator density for discrete states below the absorption edge and that for the continuum above the edge match at the edge.

If an exponential decay of excited atomic states is assumed, the shape of a given atomic level resonance line arising from the excitation of that level can be obtained from equation 3, substituting $E = h\nu$

$$J(E) = \frac{\text{constant}}{1 + 4[(E_0 - E)/\Gamma]^2}, \quad (13)$$

where the distribution is centered at $E = E_0$. The width Γ is the total contribution of the widths of the states involved in the transition. However, since we are assuming that the width of the initial state is much larger than that of the final state, the latter may be neglected. The contribution of all resonance transitions to the absorption coefficient can be written

$$\mu_R(E) = \sum_{n=n'}^{\infty} A_n J(E). \quad (14)$$

This may be rewritten in the form

$$\mu_R(E) = \frac{C_1}{1 + 4[(E - E_{n'})/\Gamma]^2} + \frac{C_2}{1 + 4[(E - E_{n'+1})/\Gamma]^2} + \sum_{n=n'+2}^{\infty} A_n J(E), \quad (15)$$

where the first two terms are separated out in preparation for the analysis to follow. It is assumed here that the resonance transitions to energy states n' and $n' + 1$ cause spectral lines that are at least partially separable from the total resonance spectrum.

For transitions to the continuum state, the contribution to the absorption coefficient, μ_c, near the ionization edge for the initial state of interest, may be written

$$\mu_c(E) = \int_0^{\infty} \frac{B(E')\,dE'}{1 + 4[(E - E')/\Gamma]^2} \quad (16)$$

where $B(E')$ is a continuous function of E' and proportional to the oscillator density for the photoelectric effect.

As mentioned previously, the oscillator densities for resonance transitions and for transitions to the continuum state match at the absorption edge.[18] Therefore it is reasonable to approximate the sum of resonant transitions to states having $n > n' + 1$ by extending the integration equation 16 by an amount ε to the lower side of the edge. The quantity ε is to be determined empirically.

The total absorption coefficient is then

$$\mu_{\text{total}}(E) = \mu_R + \mu_c + \mu_{OS}$$

$$= \frac{C_1}{1 + 4[(E - E_{n'})/\Gamma]^2} + \frac{C_2}{1 + 4[(E - E_{n'+1})/\Gamma]^2} + \int_{-\varepsilon}^{\infty} \frac{B(E')\,dE'}{1 + 4[(E - E')/\Gamma]^2} + \sigma, \quad (17)$$

where $\sigma = \mu_{OS}$ is the absorption coefficient due to ionization in shells lying outside the atomic level of interest. For $B(E')$ in the vicinity of the edge, we may choose[16] empirically the function $\alpha/(1 + E'/\eta)$, which approaches zero for increasing energy. Clearly, the function chosen for $B(E')$ diminishes with increasing E and becomes negative for $E' < -\eta$. However, η is large compared to the energy region of interest, and $B(E')$ varies slowly there. Richtmyer et al.,[14] in their calculation of X-ray absorption spectral distributions, set the function $B(E')$ equal to a constant. Both the final expressions obtained in these investigations include the function

$$\tfrac{1}{2} + \frac{1}{\pi}\tan^{-1}\frac{2(E - \varepsilon)}{\Gamma}, \quad (18)$$

which by itself quite accurately describes the characteristic shape of the absorption edge.

The K-absorption spectrum for gaseous argon, for example, shows a resolvable line representing a $1s \rightarrow 4p$ resonant transition.[16] (See Figure 1*a* and also

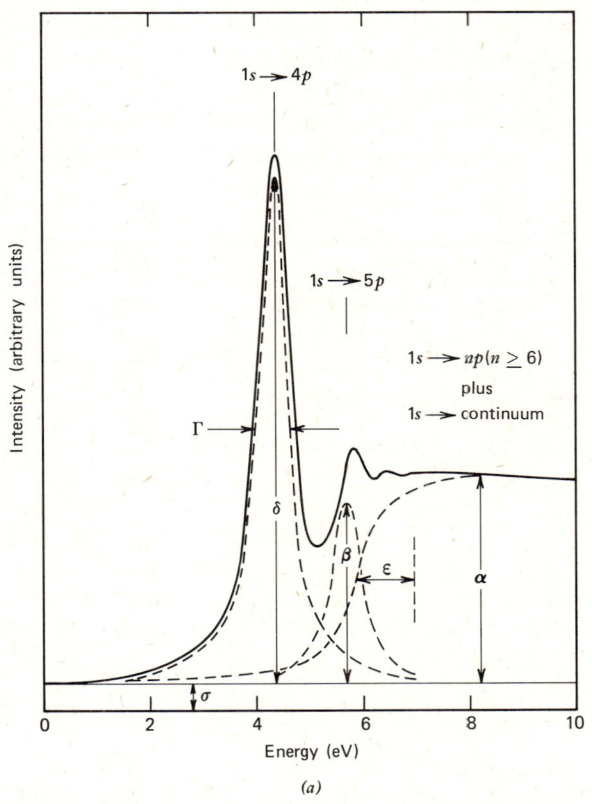

Figure 1*a*. The K X-ray absorption spectrum of argon (measured by Parratt[20] and by Watanabe[16]) after correcting for the window function. The spectrum is analyzed into the two strongest resonance absorption lines ($1s \rightarrow 4p$ and $1s \rightarrow 5p$) and the component due to $1s \rightarrow np$ ($n \geq 6$) resonant transitions and to $1s \rightarrow$ continuum transitions.

Ref.[19,20].) The shape of this peak was shown to be Lorentzian. The second peak is due to the $1s \rightarrow 5p$ resonant transition, whereas the absorption continuum lying to yet higher energies is due to $1s \rightarrow np$ ($n \geq 6$) and $1s \rightarrow$ continuum transitions. The underlying background arises from photoionization in the *L*- and *M*-shells. The first two terms in equation 17, then, represent the absorption peaks in this spectrum, and the *K*-level width may be evaluated directly from the more intense of these.

In some cases no resonance transition structure is discernible near the absorption edges for atomic levels under investigation. Examples of this appear in Figure 1*b*, which displays the *L*-absorption spectra for xenon near the L_1-, L_2-, and L_3-edges, as reported by Watanabe.[18] The L_2- and L_3-limits show intensity maxima at the edges; but for the L_1-limit, the ratio of the intensity of the maximum of the absorption structure near the limit to the intensity of the higher energy continuum is close to unity. Here then the absorption spectrum near the limit can be fitted by an arctangent curve and the level width may thereby be obtained.[14,18] For the L_1-edge in Figure 1*b*, the value $\Gamma_{L_1} = 1.1(3)$ eV is found. For an extensive treatment of X-ray emission and absorption in gases and solids, see the work by Barinski and Nepedow.[21a]

In the foregoing treatment it is assumed that the X-ray de-excitation follows an exponential decay. It has been shown, however, that if the decay of an excited state is fast, or if the initial state is a statistical state of an ensemble of systems (atom and field oscillators), the distribution of the X-ray may not be Lorentzian.[21,22] There has been no experimental indication yet that X-ray intensity distributions are other than Lorentzian in shape, except for $K\alpha_{1,2}$ line asymmetries and certain unique $K\beta_{1,3}$ line shapes (see Sections 4.1 and 4.2).

When investigating electron line shapes where the natural line width is quite small ($\gtrsim 0.05$ eV), Doppler broadening due to the thermal motion of the emitting atom may be important enough to merit consideration. The intensity distribution of the electron line may then be unfolded, in the ideal case, into the Gaussian (Doppler) and Lorentzian shape components. For an extensive treatment of this problem, see the work of Zemansky[23] and of Farach and Teitelbaum,[24] as well as Section 4.5 of this chapter.

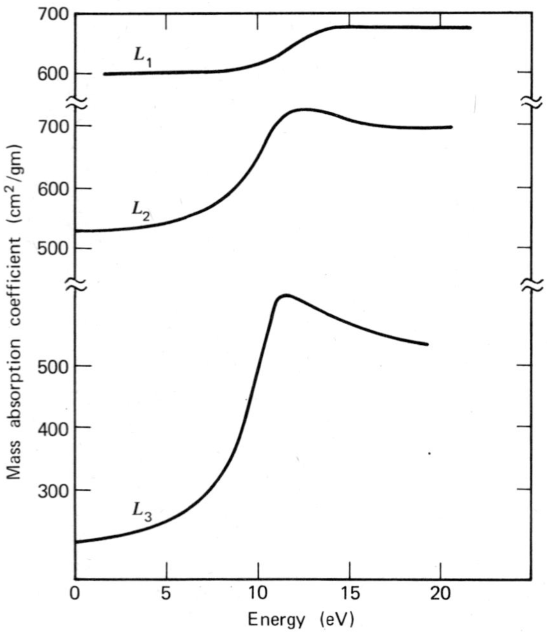

Figure 1*b*. The xenon *L*-absorption spectra near the L_1, L_2, and L_3-edges. (Measured by Watanabe.[18])

3. PHOTOELECTRON, INTERNAL CONVERSION, AND AUGER ELECTRON LINES

When the energy of the radiation quanta striking a target is raised sufficiently so that the atomic electron of interest is ejected into the positive energy continuum, investigation of the electron energy distribution can be made in terms of the distribution of the excited electron's original atomic state, the energy distribution of the irradiating quanta, and electron straggling effects. Assuming that the converting radiation is monoenergetic and that electron straggling effects are absent, the energy distribution of the ejected electrons from a given atomic shell *C* will be given by equation 9 with $j = C$.

The X-radiation spectrum usually used in external conversion experiments consists of a number of characteristic X-ray transition lines superimposed upon an X-ray continuum. The photoelectron spectrum resulting from irradiating a target with such a radiation spectrum will then be composed of a multitude of photoelectron lines lying on a continuous electron background. The idealized photoelectron peak due to external conversion in a given atomic shell *C* —assuming that energy losses owing to straggling are negligible and using energy as the variable parameter—will have a shape given by

$$J_{\mathrm{AB;C}}(E)\,dE = \int_{E'=0}^{\infty} J_{\mathrm{AB}}(E')J_{\mathrm{C}}(E'-E)\,dE'\,dE, \quad (19)$$

where $J_{\mathrm{AB}}(E')$ is the known distribution of the impinging X-ray, and $J_{\mathrm{C}}(E)$ is the energy distribution for the atomic level C. If these two distributions are

Lorentzian in form, then the width of the photoelectron intensity distribution, now expressed in units of energy, will be

$$\Gamma_{AB;\,C} = \Gamma_{AB} + \Gamma_C = \Gamma_A + \Gamma_B + \Gamma_C, \tag{20}$$

and its shape will then be given by

$$J_{AB;C}(E)\,dE = \frac{\Gamma_A + \Gamma_B + \Gamma_C}{2\pi} \times \frac{dE}{(E_{AB;C} - E)^2 + [(\Gamma_A + \Gamma_B + \Gamma_C)/2]^2}, \tag{21}$$

where $E_{AB;C} = E_A - E_B - E_C$.

As pointed out earlier, the natural widths of gamma-ray energy distributions are usually very small. Thus external conversion electron spectra from an atomic level C employing gamma-rays of suitable energy from a gamma-ray source, would give photoelectron energy distribution as in equation 9 with $j = C$, (assuming the absence of any electron energy loss due to straggling). However, difficulties in producing a sufficiently strong gamma-ray source, together with the disadvantageously low photoelectron cross-sectional values involved, usually make this investigation procedure unfeasible.

On the other hand, certain radioactive isotopes may be employed directly as low energy conversion electron sources, and these can be used to advantage in studying atomic level widths. Such investigations were begun in the early 1930s on the conversion line spectrum of the decay of thorium B (see Refs. 25–28). If straggling effects are negligible, the energy distribution of the internal conversion electron ejected from the atomic energy level C will be

$$J_C(E)\,dE = \frac{\Gamma_C}{2\pi}\frac{dE}{(E_C - E_\gamma - E)^2 + (\Gamma_C/2)^2}, \tag{22}$$

where Γ_C, E_C, and E_γ are the natural atomic level width, the binding energy of the C-shell electron, and the gamma-ray energy, respectively.

The instrumental window curve (described later) causes an alteration of the "true" electron line shape when an electron spectrometer is used to obtain the energy distribution of an internal conversion or photoelectron line.

Since the instrumental window curve will probably not be Lorentzian in shape, and assuming that the electron line shape to be measured is Lorentzian, the widths of the observed line shape will be neither the sum (equation 10) nor the root of the sum of the squares (equation 12) of the widths of the two distributions. Mladjenovič[29] found that the expression

$$\Gamma_L^{1.5} = \Gamma_{sp}^{1.5} + \Gamma_e^{1.5} \tag{23}$$

was satisfied by his data. This expression is a special case of the general empirical formula[30]

$$\Gamma_L^n = \Gamma_{sp}^n + \Gamma_e^n, \tag{24}$$

where Γ_L is the experimentally observed line width, Γ_{sp} is the width of the spectrometer window curve, Γ_e is the natural width of the atomic level involved, and n takes a value lying between 1 and 2 (see equations 10 and 12 and accompanying text). Dijkstra and de Vries[31] and Nordling et al.[30] found n about equal to unity. Ewan and Graham[32] have found that $n = 1.2$ is a satisfactory value for the Chalk River spectrometer at momentum resolution settings of 0.05 to 0.12%. Gizon[33] also found $n = 1.2$ for the iron-free instrument at Grenoble, whereas Jahn et al. (e.g., in Ref. 34) find $n = 1.1$ for their $\pi\sqrt{13/2}$ iron-free spectrometer.

Letting now the line intensity be expressed as a function of the more convenient parameter, momentum p, we can measure an electron line having a momentum distribution $f(p)$ with a magnetic spectrometer having a window curve $S(p/p_0)$. The window curve $S(p/p_0)$ is derived from the experimental shape $S(p)$ of a monoenergetic electron line of momentum p_0 as measured by the instrument for a given adjustment, source and counter geometry. Initial accurate knowledge of $S(p/p_0)$ will make it possible to determine $f(p)$ by unfolding the measured electron line intensity distribution $F(p)$. However, $S(p/p_0)$ may vary greatly between experiments because of differences in source properties, instrumental geometry, and magnetic field shape. For this reason, instead of determining $f(p)$ it is more suitable to compare the relative shapes of the K, L, M, and so on, conversion lines. An L-line, for example, may be shifted with proper scaling, to a K-line position and folded with Lorentzian curves of various widths until the best fit with the K-line is achieved (see below).

Expressing the Lorentzian function (equation 22), the assumed form of the interval conversion line, in a slightly altered form, we have

$$J_0(E) \propto \frac{1}{(E_0 - E/(\Gamma_0/2))^2 + 1}, \tag{25}$$

where the center of the distribution lies at E_0. The momentum distribution of the line $f(p)$ can be derived from this in terms of electron momentum p, and has the form

$$f(p) \propto \frac{1}{\left[\dfrac{(E_0/p_0)p_0 - (E/p)p}{\Gamma_0/2}\right]^2 + 1}. \tag{26}$$

For narrow distributions $f(p)$, we can set $E/p \cong E_0/p_0$ so that, to a good degree of accuracy,

$$f(p) \propto \frac{1}{1 + [(p_0 - p)/\varepsilon_0]^2} \tag{27}$$

where $\varepsilon_0 = (p_0/E_0)(\Gamma_0/2)$, and Γ_0 is expressed in units of energy. Folding this function with the window curve, the momentum distribution of the line will be

$$F_0(p) \propto \int_0^{+\infty} \frac{1}{1 + \{[p_0 - (p - q)]/\varepsilon_0\}^2} S\left(\frac{q}{p}\right) dq. \tag{28}$$

This line distribution can now be shifted and scaled in shape to the momentum value of the j-line position by replacing p by p/a and p_0 by p_j.

In practice, however, the entire electron line is shifted and scaled in a single step: the line at p_0 is shifted to the p_j momentum position, and its entire distribution is scaled so that its width is diminished by a factor $a = p_j/p_0$. At the same time, of course, the included natural width of the shifted line is reduced. If an L-line is shifted and scaled some distance to lower energies, the natural width component may become negligibly small (see Figure 2 and accompanying text).

After the line is shifted and scaled, it is folded with Lorentzian functions of various widths until the resulting shape compares well with that of the line positioned at p_j.

Thus we see that the shifted and scaled electron line will have a narrower distribution than it would have if just that conversion line were to lie at the p_j position. This is because the natural width component was also diminished in the scaling, and the width of the Lorentzian function that is finally folded with it is expected to be larger than the difference in widths of the two atomic levels being compared.

Usually source thickness has an important influence on experimentally determined line shapes. Since such influences on the experimental line shape can vary strongly with electron energy, the most satisfactory comparisons of line shape are obtained when the conversion lines being compared are similar in energy.

For other treatments of the folding procedure, see Refs. 29, 31, 32, 35, and 36. A Fourier analysis method may also be used for correcting for instrumental broadening.[37]

Figure 2 gives the analysis of the M_3-4.825 keV conversion line in ^{151}Sm. The L_2-25.70 keV conversion line was scaled in momentum to the position of the M_3-4.821 keV line.[38] The best fit is given by folding the L_2-line with a Lorentzian curve having $\Gamma = 7.7$ eV, although the fit to the experimental data is not perfect.

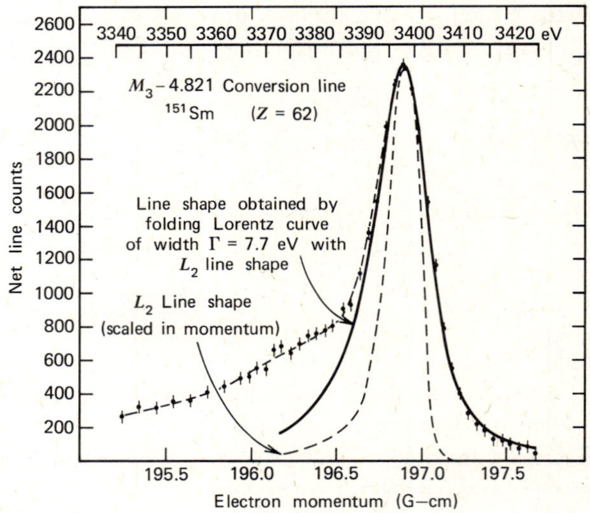

Figure 2. The net counts on the M_3-4.821 keV conversion line in ^{151}Sm as presented by Geiger et al.[38] The narrow dashed curve is the L_2-25.70 keV line shape, scaled to the M_3-line momentum value. The solid curve was obtained by folding a Lorentzian curve of $\Gamma = 7.7$ eV with the scaled L_2-line shape. This width value, which gives the best fit to the M_3-line, closely approximates the M_3 level width (see text).

Failure to match the M-line low energy tail occurs because the L_2-line shape has been taken from a higher energy, where straggling effects are much less prominent. The instrumental momentum resolution used, 0.12%, corresponds at the M_3-line position to an energy resolution of ≈ 8 eV, which was sufficient to allow a partial separation of the low energy tail resulting from the "zero-energy-loss" line peak. At the L_2-25.70 conversion line position (≈ 18.4 keV), the instrumental resolution will correspond to an energy resolution of ≈ 44 eV, and the "zero-energy-loss" peak and electrons suffering inelastic scattering will not be resolved from each other.

The Γ-value of the Lorentzian curve giving the best fit to the M_3-line will approximate the M_3-level width more closely than it will the M_3,L_2-level width difference. This is because the L_2-25.70 keV line width is composed of a ≈ 44-eV instrumental width and a ≈ 2-eV L_2-natural width. An L_2-line at the energy of the M_3-line (≈ 3.4 keV) would have an instrumental width contribution of ≈ 8 eV plus the ≈ 2 eV L_2-natural width. The line shape obtained by scaling the 25.70-L_2 line in a momentum consists then of an instrumental contribution of ≈ 8 eV plus the original natural width scaled down to ≈ 0.3 eV. Therefore, the width of the scaled line will closely approximate the instrumental line width.

Narrow low energy conversion lines, which exhibit

structure on the low energy side because of characteristic energy losses and straggling, shift slightly in energy when folded with a Lorentzian distribution in comparing the resulting composite profile with a K-conversion line shape. This shift was noticed by Kaye[36] (see Figure 3) when the L_3-19.384 keV conversion line (lying at ≈ 10.4 keV) of ^{171}Yb was scaled in momentum to the position of the K-72.365 keV conversion line (lying at ≈ 11.0 keV). After the L_3-19.384 keV line shape had been folded with a Lorentzian distribution of width $\Gamma = 27$ eV, it fit the K-72.365 keV conversion line shape very well, as shown in Figure 3. This line shape fit was expected, since similar effects of source thickness are included in the shapes of the two electron lines. The slight mismatch on the low energy side of the K-line is due to a slight overestimation of the source thickness contribution to the K-line shape. The solid line is the folded curve giving the best fit, and the broken line is the L_3-line scaled in momentum to the K-line position. There is a change in peak position of the unfolded L_3-line when folded with the Lorentzian line shape to obtain the solid curve. This shift in peak position corresponds here to a line energy shift of 4 eV and so must be accounted for in precision electron line energy measurements.

Mehlhorn et al.[39] have determined the K-level width for neon ($Z = 10$) and the $L_2 \approx L_3$ level widths for argon ($Z = 18$) from shape measurements and analyses of the $KL_{2,3}L_{2,3}(^1D_2)$ and $L_3M_{2,3}M_{2,3}$-$(^1D_2)$ Auger transition lines, respectively. The window curve of the spectrometer was determined by comparing the measured and theoretical shapes of elastically scattered electrons, 2000-eV primary energy, on the target gas. The elements were in the gaseous state and very accurate measurements were possible. The values are entered in Figures 4 and 5 and Tables I and II. Most of the Γ_{L_2} and Γ_{L_3} values in Figure 5 were obtained by subtracting Callan's[42] Γ_K values (see Figure 4 and Table I) measured from $K\alpha_1$ and $K\alpha_2$ X-ray width values. The values in Figure 5 are collected in Table II.

4. THE ATOMIC LEVEL WIDTHS

The preceding sections present the theoretical shapes of ideal X-ray and photoelectron and conversion electron energy distributions before they are spectrometrically analyzed—ideal, in that they were assumed to be Lorentzian in shape. The widths of these Lorentzian distributions were shown to be equal to the sum of the widths of the levels taking part in the transitition. In photoelectron spectra lines, the additional energy width of the X-radiation is included in the width of a measured photoelectron line.

The natural width of an atomic level is related by the uncertainty principle to the lifetime of a vacancy created in that level (see Section 1). An initial atomic vacancy becomes filled through the occurrence of an X-ray, Auger, or Coster–Kronig transition (see Chapter 3). In special cases, however, the transition

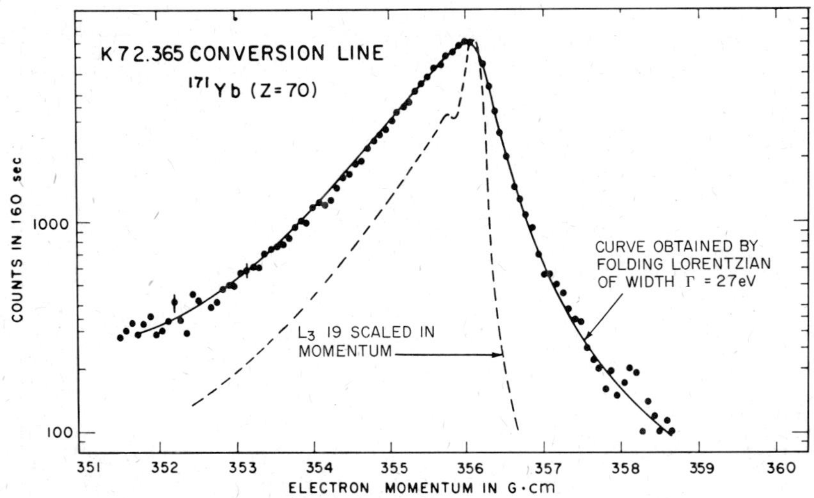

Figure 3. The net counts on the K-72.365 keV conversion line as shown on a semilog plot by Kaye.[36] The broken line is the L_3-19 keV conversion line scaled to the K-line momentum. The position of the L_3-line shifts upon folding. The difference in position between this L_3-peak and the folded curve is a measure of the shift in peak position caused by the energy loss component of the K-level (see text). The scaled L_3-19 line shape folded with a Lorentzian curve having $\Gamma = 27$ eV gives the best fit to the K-72.365 line shape, except for a slight discrepancy on the low energy side (see text).

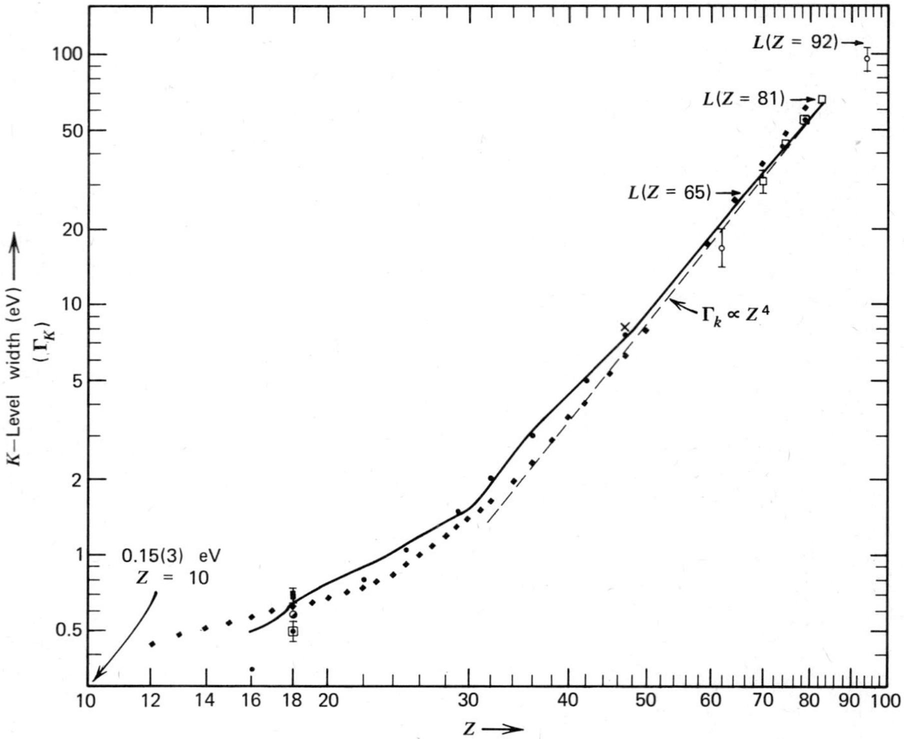

Figure 4. Theoretical and experimental widths of the *K*-atomic levels as taken from the literature. The values plotted in the figure are collected in Table I and discussed in the text. Solid curve, empirical Callan data;[42] L, theoretical Listengarten values;[41] solid squares, X-ray absorption data of Parratt;[20] small solid circles, X-ray emission data of Parratt;[20] crosses, X-ray emission data of Parratt;[43] small solid diamonds, X-ray emission data of Blokhin;[10] plusses, X-ray absorption data of Watanabe;[16] large open circles, X-ray absorption data of Parratt;[44] open squares, data of Gokhale;[45] small open circles, electron conversion line width data.

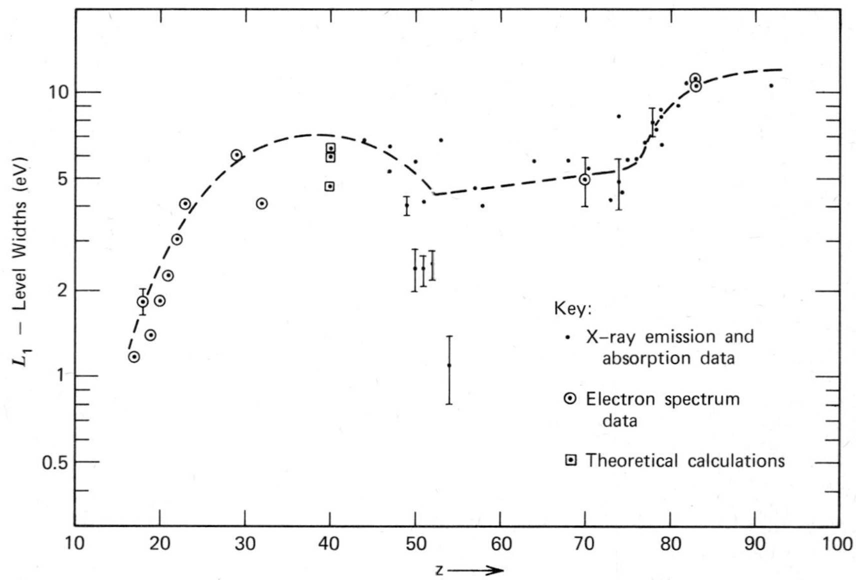

Figure 5*a*. Experimental natural line widths of the L_I-subshells. Solid circles, X-ray emission and absorption data; ⊙, electron spectrum data; ⊡, theoretical calculations.

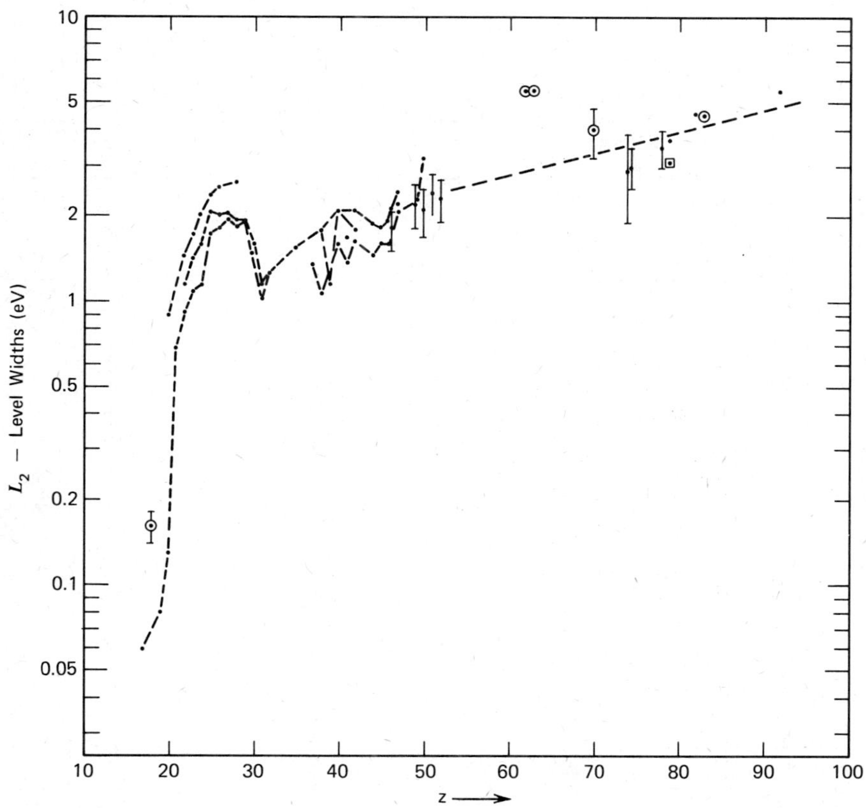

Figure 5*b*. Experimental natural line widths of the L_2-sybshells. Symbols same as in Figure 5*a*.

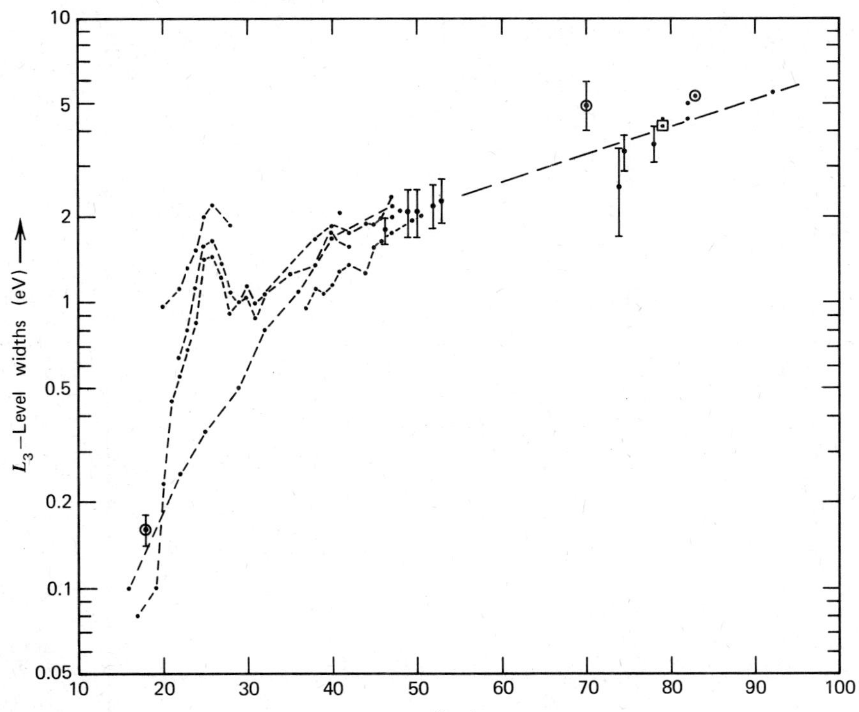

Figure 5*c*. Experimental natural line widths of the L_3-subshell. Symbols same as in Figure 5*a*.

Table I.

K level widths (eV)

Theoretical			Experimental			
			X-ray spectra			Electron spectra
Z	Callan ref.[42]	Listengarten ref.[41]	Blokhin ref.[10]	Parratt ref.[20]	Others	
10	-		-			0.15(3)[39]
12	-		0.44			
13	-		0.48			
14	-		0.51			
15	-		0.54			
16	0.51		0.57	0.35		
17	0.55		0.60			
18	0.65		0.63	0.50(5)	0.58[44] 0.68(3)[16]	
19	0.71		0.65			
20	0.79		0.68			
21	0.84		0.71			
22	0.89		0.74	0.8		
23	0.95		0.79			
24	1.03		0.84			
25	1.10		0.92	1.05		
26	1.19		1.00			
27	1.27		1.09			
28	1.35		1.19			
29	1.45		1.30	1.5		
30	1.53		1.40			
31	1.75		1.53			
32	1.96		1.67	2.0		
33	2.19		-			
34	2.46		1.98			
35	2.82		-			
36	3.18		2.37	3.0		
37	3.44		-			
38	3.72		2.82			
39	4.02		-			
40	4.34		3.55			
41	4.68		-			
42	5.05		4.10	5.0		
43	5.44		-			
44	5.87		-			
45	6.32		5.38			
46	6.81		-			
47	7.32		6.28	7.5	8.0[43]	
48	7.88		-			
50	-		7.98			
54	12.41		-			
55	-		12.0			
60	-		17.7			
62	-		-			17(3)[28]
65	-	27.1	26			
70			36			32.0(6.4)[36]
74	40.39		-		42.5[45]	
75	-		48			

Table I. (cont.)

79	52.03		61	54	54 [14)45)]	
80	54.61		-			
81	-	65.7	-			
83	63.11		-		67(5) [47)] 65.3 [45)]	
92	-	112.8	-			
94	-		-			96(11)†

† $\Gamma_K - \Gamma_{L_2}$ from ref. [9)35)] and Γ_{L_2} = 7(1) eV, estimated from figure 5b.

probabilities for these can be altered appreciably for a given atom by changing its physical environment or chemical state.

Other effects, not expected to alter appreciably the lifetime of an initial atomic vacancy, will give rise to an additional broadening of an electron or X-ray emission line. These effects are due to thermal motion of the atoms in the source and to possible spin coupling with unpaired atomic electrons and with the nucleus.

The asymmetry of K X-ray lines in the region $Z \cong 23$ to $Z \cong 32$, a Z-value region that includes the elements having ferromagnetic properties, must be considered separately.

4.1. X-Ray, Auger and Coster–Kronig Transition Probabilities

A quasi-hydrogenic behavior is exhibited for atoms over the entire Z-value range. This approximation is reasonable for the inner atomic levels, and it increases in validity with increasing atomic number. For hydrogenic atoms, X-ray transition rates are proportional to $C_n E^{2\Delta l+1}(1/Z^2)$, where E is the transition energy (proportional to Z^2) and C_n is a constant related to the 1s-np radiative dipole transition rate (the change in orbital momentum $\Delta l = 1$). Therefore the X-ray transition rates are proportional to Z^4.

The deviations from the pure hydrogenic cases are particularly strong for low atomic numbers and for the outer atomic shells, where effects of screening are important. Toward high Z-values, the hydrogenic behavior assumption increases in validity for the inner atomic shells. However, although the Z^4-dependence is approached toward higher Z-values, this is not an asymptotic limit, since strongly increasing relativistic effects cause the dipole transition rate to decrease slightly.

The relative contribution to the K-level width due to Auger transition probabilities increases toward the lower Z-values. Since Auger transition probabilities are only weakly dependent on atomic number, their increasing importance for $Z < 40$ causes a stronger reduction in slope of the variation of Γ_K with Z in this region (see also Ref. 32).

Three theoretical calculations have been performed on K-atomic level widths. Pincherle[40] reported calculations of the K-level widths for neon, magnesium, argon, nickel, and krypton. A quantum mechanical treatment was employed and effects of screening were neglected. The obtained values were about twice as large as the corresponding values derived in later theoretical and experimental determinations.

Listengarten[41] used the results of his relativistic calculations on the KLL-Auger effect for the atomic numbers $Z = 65$, 81, and 92 to determine the total K-level widths for these elements. The results of the calculations are included in Table I and Figure 4.

Callan[42] has presented calculations of K-level widths for $16 \leq Z \leq 83$. The relative widths were computed empirically using the equation $\Gamma_R = \sum_n A_n Z^4$, where A_n are the empirical coefficients for the $1s \rightarrow np$ transitions, suitably modified for the normal filling of levels. Total K-Auger widths were derived from theoretical KLL-widths corrected for both KXY/KLL Auger transition intensity ratios and relativistic effects for the heavy atoms. The widths thus obtained agreed within 5% with those reported by Parratt[20] for all but the lightest atoms, where the agreement was to within 0.15 eV. These values are included in Table I and Figure 4. Meisel and Nefedow,[48] using a $(Z - \beta)^4$ relationship, have obtained a good fit between calculated K-and L_3-widths, from X-ray data. The K-level width values as determined by X-ray emission and absorption measurements are presented in Table I and Figure 4.

Parratt's[20] $K\alpha_{1,2}$ X-ray distributions for $23 \gtrsim Z \gtrsim 31$, corrected for asymmetry before the widths

Table II.

L_1, L_2, L_3 level widths (eV)

Z	L_1	L_2	L_3
16		-0.1[64]	-0.07[64], 0.1[20]
17	1.18[57]++	0.06[64]	0.08[64]
18	1.84(20)[7]†	0.16(2)[7][39]†	0.16(2)[7][39]†
19	1.40[57]++	0.08[64]	0.10[64]
20	1.83[57]++	0.13[64], 0.90[56]	0.23[64], 0.97[56]
21	2.25[57]++	0.68[64]	0.45[64]
22	3.05[57]++	0.92[64], 1.16[10], 1.44[56]	0.55[64], 0.64[10], 1.11[56], 0.25[20]
23	4.08[57]++	1.09[64], 1.42[10], 1.72[56]	0.68[64], 0.79[10], 1.31[56]
24		1.15[64], 1.59[10], 2.02[56]	0.85[64], 1.12[10], 1.52[56]
25		1.72[64], 2.04[10], 2.33[56]	1.42[64], 1.54[10], 1.99[56], 0.35[20]
26		1.80[64], 2.00[10], 2.51[56]	1.45[64], 1.65[10], 2.18[56]
27		1.93[64], 2.03[10]	1.21[64], 1.36[10]
28		1.93[64], 1.84[10], 2.61[56]	0.92[64], 1.07[10], 1.88[56]
29	6.01[54]++	1.91[64], 1.91[10]	1.01[64], 1.01[10], 0.5[20]
30		1.56[64], 1.50[10]	1.13[64], 1.04[10]
31		1.14[64], 1.02[10]	0.98[64], 0.87[10]
32	4.06[55]++	1.24[64], 1.27[10]	1.06[64], 1.06[10], 0.8[20]
35		1.56[64]	1.26[64]
36			1.1[16]
37		1.35[45]	0.95[45]
38		1.79[64], 1.78[10], 1.07[45]	1.36[64], 1.68[10], 1.11[45]
39		1.14[64]	1.08[45]
40	6.4, 6.0, 4.7[58]+	2.03[64], 2.05[10], 1.59[45]	1.69[64], 1.85[10], 1.15[45]
41		1.69[10], 1.34[45]	2.09[10], 1.29[45]
42		1.78[64], 2.08[10], 1.61[45]	1.57[64], 1.76[10], 1.37[45], 1.7[20]
44	6.8[50]††	1.84[10], 1.45[45]	1.94[10], 1.28[45]
45		1.82[10], 1.59[45]	1.92[10], 1.54[45]
46		1.89[10], 1.58[45], 1.8(3)[53]	1.99[10], 1.63[45], 1.8(2)[53]
47	5.3[43], 6.5[50]††	2.42[10], 2.06[45], 2.2[43]	2.32[10], 1.77[45], 2.0[43], 2.2[20]
48		2.15[45]	2.10[38]
49	4.0(3)[53]	2.2(4)[53], 2.25[45]	2.1(4)[53], 1.95[45]
50	2.4(4)[53], 5.8[50]††	2.1(4)[53], 3.17[45]	2.1(4)[53], 2.05[45]
51	2.4(3)[53], 4.1[50]††	2.4(4)[53]	2.2(4)[53]
52	2.5(3)[53]	2.3(4)[53]	2.3(4)[53]
53	6.9[50]††		

Table II. (cont.)

54	1.1(3)[20]		
57	4.6[17]††		
58	4.0[50]††		
62		5.5[38]†	
63		5.5[65]†	
64	5.8[51]††		
68	5.8[51]††		
70	5(1)[36]†, 5.4[52]††	4.0(8)[36]†	5(1)[36]†
73	4.2[52]††		
74	4.9(1.0)[60], 8.4[52]††	2.9(1.0)[59], 3.0(5)[60]	2.6(9)[59], 3.4(5)[60]
	4.5[52]††		
75	5.8[52]††		
76	5.9[52]††		
77	6.7[52]††		
78	8.0(1.0)[60], 7.4[52]††	3.5(5)[60]	3.6(5)[60]
79	8.7[62,14], 6.5[62,40]+	3.7[14], 3.1[62,40]+	4.4[14], 4.4[20], 4.2[62,40]+
	8.3[52]††		
81	9.0[52]††		
82	11.0[61]	4.6[61]	4.5[60], 5.1[61]
83	10.6[29]†, 11.11[28]†	4.5[29]†	5.4[29]†
92	10.8[46]	5.5[63]	5.6[63]

† From electron spectrum measurements.

†† $L\beta_3$ and $L\beta_4$ X-ray width values with Γ_{M_3} and Γ_{M_2} from figure 6 subtracted.

\+ Theoretical calculation.

++ From photoelectron spectral line width differences and Γ_{L_3} values (see text).

were determined, are included. (See Section 4.2). Also included in Figure 4 and Table I are the few values obtained by conversion electron line width measurements. The *K*-shell photoelectron line widths, uncorrected for ionizing X-radiation and other width contributions, for $Z = 5$ to 17 (not $Z = 10$) have been reported by Hagström and Karlsson.[49]

Coster–Kronig, as well as Auger and radiative, transition probabilities contribute to the L_1- and L_2-atomic level widths. For this reason Γ_{L_1} is significantly larger than Γ_{L_2} and Γ_{L_3}, and Γ_{L_2} is slightly larger than Γ_{L_3}, for Z-value regions where $L_1L_{2,3}X$ and L_2L_3X Coster–Kronig transitions, respectively, are allowed.

Determinations of L_1-level widths are limited in number. From X-ray data, the values may be obtained from the widths of low energy L_1 X-ray series.[50,51] Table II and Figure 5*a* include the average L_1-level width values, for several Z-values from 44 to 81, determined here from measured $L\beta_3$ and $L\beta_4$ X-ray widths and M_2- and M_3-level widths taken from Figure 6. Also included are the L_1-level width values arrived at in the same manner from *L* X-ray widths measured by Cooper[52] for $70 \lesssim Z \lesssim 81$, and which

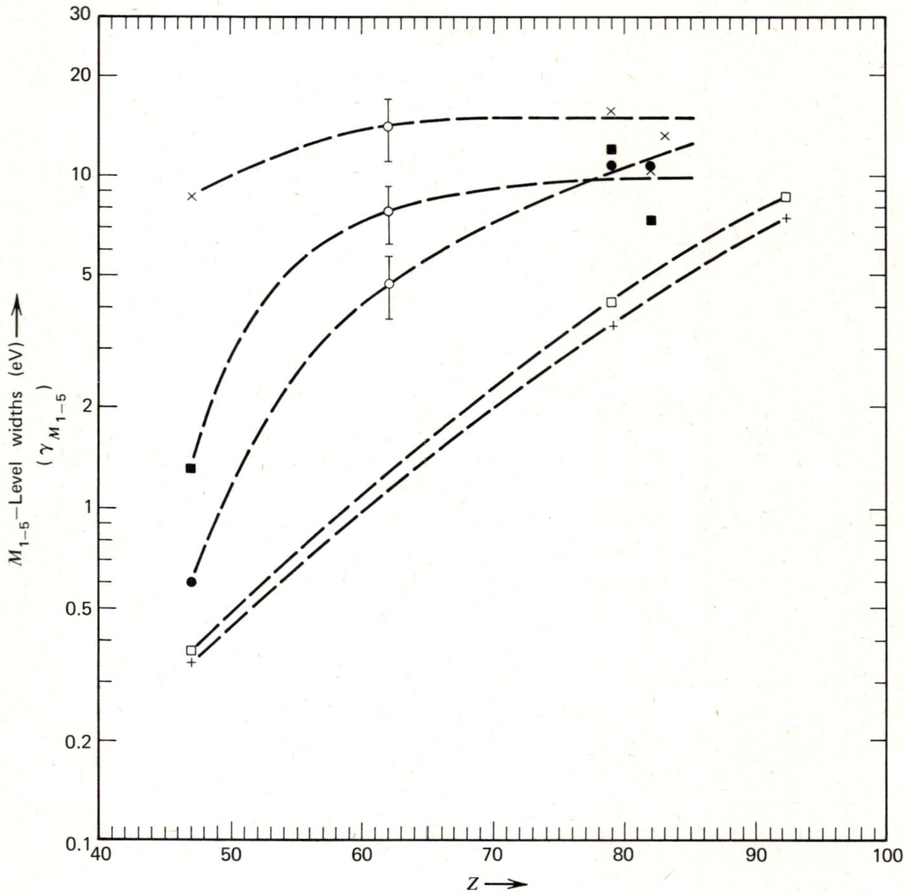

Figure 6. Experimental natural widths of the M-atomic subshells. The plotted values are also given in Table III. The open circles represent electron spectrometric determinations from conversion line data. Level widths as follows: ×, M_1; ●, M_2; ■, M_3; □, M_4; +, M_5.

show a strong increase in Γ_{L_1} for the $Z \gtrsim 75$ region. Noreland's[53] treatment of X-ray absorption and emission measurements indicate a similar strong L_1-level width increase for $Z \gtrsim 50$. (See Table II and Figure 5*a*.)

X-Ray absorption and internal conversion electron line measurements may provide width values for any of the L (also M, etc.) subshells. The former method is limited in accuracy by the complexity of X-ray-absorption spectra and associated difficulties in interpretation. The latter method, of course, provides level width values only for those subshells in which the nuclear transition is sufficiently strongly converted.

It is interesting to consider the difference between Parratt's[20] Γ_{L_3} values (from X-ray emission line distributions corrected for asymmetry) with the values obtained when the correction is not applied.[42,10] The difference is about 1.0 eV at $Z = 25$ and falls off rapidly toward both $Z = 21$ and $Z = 31$ (see Figure 5*c*).

Table II and Figures 5*a*, *b*, *c* include the L-subshell level widths either as published in the literature or as obtained from subtracting the K-level widths, as calculated by Callan[42] (Table I and Figure 4), from measured $K\alpha_{1,2}$ X-ray width data. Determinations of L-level widths from electron spectroscopic measurements are also included: from internal conversion line widths,[36] $Z = 70$; from $\Gamma_K - \Gamma_{L_1, L_2}$ conversion line width differences and Callan's[38] Γ_K values, $Z = 83$; from Auger line widths,[7] $Z = 18$.

The widths of the L_2- and L_3-atomic levels are approximately the same size. The two values differ by less than 0.4 eV for the lighter elements except for the vicinity of $Z = 28$ where the difference becomes as great as 1 eV. The increase of Γ_{L_2} for $Z \gtrsim 31$ may be related to $L_2 L_3 M_{4,5}$ Coster–Kronig transitions that are allowed here (see Section 5.2.1, Chapter 3).

For larger atomic numbers the L_2 and L_3 level width differences are quite small. From Table II, however, we see that $\Gamma_{L_2} < \Gamma_{L_3}$ by an average of about 0.4 eV for $Z > 70$.

The values for Γ_{L_1} are consistently larger than the corresponding Γ_{L_2} and Γ_{L_3} values. It is possible to establish the general variation of the L_1-level width with Z-value from experimental data for $Z \lesssim 45$, but there is still little known about Γ_{L_1} for the region $10 < Z \gtrsim 45$. However, measurements[54] of the photoelectron emission from the L-subshells of copper $(Z = 29)$ indicate that Γ_{L_1}–$\Gamma_{L_3} \approx 5$ eV. From the spectrum from a similar investigation[55] for germanium $(Z = 32)$ we obtain Γ_{L_1}–$\Gamma_{L_3} \approx 3$ eV. Combining these values with those of Callan[42] for Γ_{L_3} yields the Γ_{L_1} values listed in Table II.

A more recent investigation of the photoelectron spectra resulting from conversion of the aluminum $K\alpha_{1,2}$ X-rays in the $L_{1,2,3}$ subshells of chlorine $(Z = 17)$ and of the elements from potassium $(Z = 19)$ through vanadium $(Z = 23)$ has given additional information on the L-level widths in these elements.[57] Although the widths of the measured electron lines have not been corrected for contributions from the resolving power of the spectrometer ($R_{mom} = 0.02$–0.04%) and the energy distribution of the X-rays (width about 1 eV), the difference between the measured level width values for a given element should give a good approximation of the natural width difference. The $\Gamma_{L1} - \Gamma_{L2}$ and $\Gamma_{L1} - \Gamma_{I3}$ spectral line width differences, respectively, for the elements measured were: chlorine—1.1, 1.1; potassium—1.2, 1.3; calcium—1.5, 1.6; scandium—1.5, 1.8; titanium—2.1, 2.5; vanadium—2.9, 3.4 eV. Taking the latter value in each case and adding the L_3-width values obtained from X-ray widths and the empirical K-widths as listed by Callan,[42] we obtain the Γ_{L_1} values of Table II and Figure 5*a*. These elements were in either an unambiguous oxide or chloride form (of the most common oxidation state for each) when the photoelectron measurements were carried out.

In his investigation of the sensitivity of $L_1L_{2,3}M_{4,5}$ Coster–Kronig transition rates to variations in level energies and screening parameters, Callan[58] estimated the natural width of the L_1-level in zirconium $(Z = 40)$. This was done by multiplying the results of three theoretical determinations of the LLM-transition rate (for the Watson-Freeman, Mayers, and Löwdin-Appel bases) by a factor 1.2 to account for radiation, LLN, and other transition contributions to the width. These three values are entered in Figure 5*a* and Table II.

The increases in Γ_{L_1} in Figure 5*a* for $Z \gtrsim 51$ and $Z \lesssim 74$ can be associated with $L_1L_3M_{4,5}$ transitions that are energetically allowed in these regions (see Chapter 3, Section 5.2.1).

Experimentally determined M-level widths are few. These data, obtained from X-ray emission and absorption results and internal conversion line analysis, are included in Table III and Figure 6. It is interesting to note that $\Gamma_{M_i} > \Gamma_{L_i}$ $(i = 1, 2, 3)$ for $Z \lesssim 55$.

The widths of the outer atomic levels are more strongly affected by atomic environmental perturbations than are the inner level widths. This is because of the closer proximity of the levels to, and their greater interaction with, valence electrons of the atom. The widths of the L-shell and the M-shell for higher atomic numbers are expected to be less affected by the chemical and physical state of the atom, since they are much more strongly bound than are the valence electrons.

The $K\beta_{1,3}$ X-ray emission spectrum of manganese, involving the $M_{2,3}$-states, which lie just inside the partially completed $M_{4,5}$- and N_1-shells, was measured

Table III.
M_{1-5} level widths (eV)

Z	M_1	M_2	M_3	M_4	M_5	Ref.
47	8.6	0.6	1.3	0.37	0.34	43)
62	14(3)+	4.7(1.0)+	7.7(1.5)+	--	--	38)
79	15.5	10.7	12.1	4.2	3.6	14)
82	10.5	10.8	7.3	--	--	61)
83	13.11+	--	--	--	--	28)
92	--	--	--	8.8	7.5	46)

+ From conversion line measurements.

by Schnopper[19] (see Ref. 20). After he had corrected for the instrumental resolving power and the contribution to the shape by the K-state energy distribution (assumed to be Lorentzian), he found that the remaining density of M-states was more satisfactorily resolved into M_2- and M_3-components that were Gaussian rather than Lorentzian in shape. Also, underlying the M_2, M_3 composite structure was found a low broad secondary component extending to lower energies.

The complex low energy secondary structure $K\beta'$ accompanying the $K\beta_{1,3}$ emission lines was first ascribed to the outer electron orbitals having been left in excited states as a result of the electron bombardment creation of the initial K-shell hole.[20] This interpretation was later tested by Schnopper[66] by measuring the $K\beta_{1,3}$ plus secondary structure for ^{55}Mn produced by K-capture in ^{55}Fe. Here a K-shell electron is captured by the ^{55}Fe nucleus, and the outer regions of the resulting excited ^{55}Mn atom are left unperturbed during the vacancy creation, which is not the case for electron bombardment K ionization. No difference was found in the shapes of the $K\beta_{1,3}$ plus $K\beta'$ spectrum for the two modes of K-vacancy creation. This shows that the $K\beta'$ secondary structure is to be identified only with the mode of decay of the K-state and the final electron configuration, not with the initial state configuration (see Ref. 66).

Källne and Ekstig[67] have recently tested the possibility of magnetic dipole interaction as the cause of the $K\beta'$ structure. They measured the $K\beta'$ structure for iron in FeGe, which changes in atomic spin from 1 to 2 in passing through a critical temperature near 7°C. The $K\beta_{1,3}$ plus $K\beta'$ spectrum was measured at temperatures well above and below this critical temperature and the results were compared. The absence of a change in the structure showed that the $K\beta'$ satellite structure was not of this origin. A further interpretation of the structure's origin necessitates more investigation and work. (See also Refs. 68, 21a)

Widths of the N- and O-levels have been determined on a number of occasions from X-ray measurements.[14,43,60–62,69] The N_{1-3} level width values vary from a few electron volts for silver[43] to about 10 eV for gold,[62] whereas $\Gamma_{N_{4,5}}$ values for these two elements are about 2 and 6 eV, respectively. The values of Γ_{O_1} and $\Gamma_{O_{4,5}}$ for lead are reported to be $\sim$10 and $\sim$3 eV, respectively.[61]

However, there has been recent indication that the $N_{2,3}$-level widths vary strongly in a certain Z-value region. Ekstig and Källne[67] have measured the $L\gamma_{2,3}$ (i.e., $L_1 \rightarrow N_{2,3}$) X-ray transitions for several elements from $Z = 40$ to 50, and found that the widths of the X-ray lines increased strongly with atomic number. The widths, uncorrected for an instrumental broadening of less than an electron volt, were: $Z = 41$, 6.5; $Z = 42$, 7.3; $Z = 46$, 14.2; $Z = 47$, 12.4; $Z = 48$, 19; $Z = 49$, 21.5; $Z = 50$, 20.3 eV. On consulting the L_1-level widths for this region (Figure 5*a*), we see that for $Z \sim 40$ the $L\gamma_{2,3}$-widths can be primarily ascribed to the width of the L_1-atomic level, assuming these X-ray widths to be basically due to the sum of the widths of the two levels in the transition, i.e., $\Gamma_{L\gamma_{2,3}} \sim \Gamma_{L_1} + \Gamma_{N_{2,3}}$. However, for $Z \gtrsim 46$ the $N_{2,3}$-widths increase strongly. The reason for the sudden change at this atomic number, where the 4d-shell becomes closed, is unclear. It would also be interesting to investigate the analogous $M\gamma_{2,3}$ ($M_1 \rightarrow O_{2,3}$) X-ray line widths in the $Z \approx 78$ region, where the 5d-shell closes.

4.2. Chemical and Lattice Effects on Level Widths

Measurements have shown that the $K\alpha_{1,2}$ X-rays for low Z-values, particularly for the region from $Z \cong 23$ to $Z \cong 31$, are asymmetric. The X-ray intensity distribution can be analyzed into a main symmetric Lorentzian component and a secondary component lying to lower energies from the center of the sym-

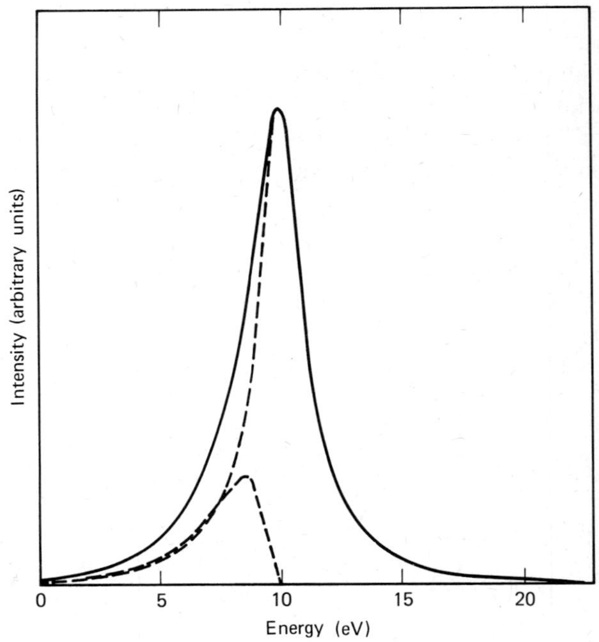

Figure 7. An asymmetric $K\alpha_1$ X-ray line profile, after the instrumental broadening effects have been corrected for, as presented by Parratt[20]. The shape of the asymmetric part was then estimated by using the high energy side of the line as a guide.

metric portion. (See Figure 7.) The general effect of the presence of this asymmetric part on the $K\alpha_{1,2}$ X-ray line shapes for this Z-value region can be seen in Figures 5*b* and 5*c*. Parratt,[20] in his determination of the *K*- and L_3-level widths, corrected for this asymmetry and obtained a set of L_3-width values smaller than those derived from uncorrected data. Parratt's values increase smoothly over the entire range of Z (see Table II and Figure 5); the method for correcting for the asymmetry is illustrated in Figure 7.

Parratt[20] has suggested that the most probable explanation for the presence of the asymmetric component is that the low energy component to the X-ray line is due to the creation of excitation of states in the valence electron configuration. The same explanation is also proposed for the weak component underlying the $K\beta_{1,3}$ X-ray emission of manganese. It is interesting to compare this theory with the one regarding the excitation lines in Auger spectra (Chapter 3, Sections 4.1.3–4.1.3.2).

As is shown in Section 4.1, excitation states are not the cause of the $K\beta'$ X-ray spectral component, and it is highly unlikely that such states will be created. There is, however, another possible explanation for the structure: namely, exchange polarization effects. For elements having a partially filled *d*-shell, these effects will cause a splitting of inner shells by electrostatic interaction with the *d*-shell electrons (e.g., splitting of the L_3-level in manganese owing to interaction of the $2p3d$ electrons[70]). Nefedov[71] has carried these considerations further and has made calculations of natural $K\alpha$ X-ray line widths for compounds of elements in the $Z \sim 23$ atomic number region, and his figures agreed very well with experimental values. The exchange splitting of the *K*-level was estimated to be negligible. The degree of asymmetry is proportional to the number of unpaired 3*d*-electrons, and the exchange interaction was related to the variation of $L_{2,3}$ doublet splitting with chemical states, and to the different signs of the $K\alpha_1$ and $K\alpha_2$ energy shifts for certain compounds. (See also Ref. 72.)

Recently measurements of manganese $K\alpha_1$ in manganese fluoride and $K\alpha_{1,2}$ in chromium salts have shown complex shapes aside from the asymmetry and even splittings. These are treated further in Chapter 7, Section 4.2.1. See also Ref. 21a.

According to the foregoing explanation, then, the widths of $K\alpha$ X-ray lines, which exhibit asymmetric shapes, will be expected to depend on the nature and characteristics of the valence electron configuration. This is in turn related to the chemical state of the element under investigation, as well as to the physical state of the investigated sample.

The variation of the $K\alpha_{1,2}$ X-ray line asymmetry for a given element has been investigated as a function of its chemical state.[43,56,73–76] Snyder[77] showed that the $K\alpha_{1,2}$ lines emitted by cobalt, nickel, and copper were broader for the element when in a nonzero oxidation state than when in the elemental state. The same comparison for bromine showed no differences. Sometimes, however, comparable increases of line asymmetry for nickel and cobalt upon change from the pure metal to a compound could be accompanied by increases in $K\alpha_{1,2}$ line widths in the ratio of as much as nearly 3 : 1.

Auger electron line widths have also been shown to be dependent on the chemical state of the element of interest. Nordling et al.[30] found that the width of the $KL_{2,3}L_{2,3}$ line for pure copper *decreased* by 23% upon oxidation of the metal; this change was accompanied by a 25% reduction of *K*-Auger intensity.

Bearden and Snyder[60] investigated the widths of *L* X-ray transitions involving the valence band of tungsten for the pure metal and for the oxide (WO_3). No change was observed.

Parratt[78] has studied the effects of the chemical and physical states of an emitter on $K\alpha$ X-ray lines by noting the effect of alloying the metal of interest with a second metal. Several alloys of nickel, iron, manganese, chromium, and titanium were used. It was found that the width of the *K* X-ray lines of element A either increased or decreased upon alloying with the element B, depending upon the nature of B. A target of Mn–Ni alloy showed a maximum $K\alpha_1$ line width for manganese when the elements were in a 1 : 1 concentration ratio; the width decreased from this value for both increasing and decreasing nickel concentration. The maximum width of the manganese $K\alpha_1$ X-ray for the alloy was 23% greater than for pure manganese. The width of $K\alpha_1$ was generally more susceptible than that of $K\alpha_2$ to changes in atomic environment. Although the peak intensities changed, the ratio of total intensities remained constant.

Reference may be made to the work of Meisel and co-workers,[79] who have done extensive investigations of the $K\alpha$ and $K\beta$ X-ray transitions in elements between $Z = 21$ and $Z = 28$. The effects of the chemical state and the alloying of these metals on the X-ray line energies, asymmetries, and, in some cases, the X-ray line widths were studied.

4.3. Effects of Multiple Ionization

Satellite lines observed in X-ray spectroscopy studies often arise because a given X-ray transition occurs in an atom that is in an initial state of double or higher ionization. Initial multi-ionized atomic states can be caused by electron bombardment of the sample (also by photoionization) or through Auger or Coster–Kronig transitions that take place in an initially singly ionized atom (see Chapter 3, Sections 5, 5.1).

As a result of the associated change in shielding of the nuclear charge and altered atomic electron binding energy, the absence of the second atomic electron at the instant of occurrence of the X-ray transition will cause a slight X-ray energy shift. The size of this shift will depend on both the nature of the X-ray transition and the atomic shells in which the initial vacancies are situated. The superposition of several X-ray lines, slightly separated from one another, may form a complex line structure or a line broadening, depending on the size of the shifts. (See also Refs. 80, 81, 29.)

Analogously, Auger-vacancy-satellite lines arise when an Auger transition takes place in an atom that is in an initial state of double or greater ionization (see Chapter 3, Sections 4.1.3.1, 5.1).

4.4. Effects Due to Coupling Between Atomic Electrons and the Nucleus

Breit[82] has considered the possibility of a broadening of X-ray lines due to a coupling interaction between the nucleus and the unpaired electron in an atomic level that is initially singly ionized (see also Refs. 83, 69). If a single initial ionization has occurred in the K-shell, there should be a spin-spin (since $l = 0$) interaction between the nucleus and the remaining K-shell electron. The coupling between a nonzero spin I and the K-electron spin $S = J = \frac{1}{2}$ will give $F = I \pm \frac{1}{2}$. The energy difference is given as[82]

$$\Delta E = 1.04\left(\frac{Z}{100}\right)^3 (I + \tfrac{1}{2})\left(\frac{\mu}{I}\right)(2\rho^2 - \rho)^{-1}\ \text{eV}. \quad (29)$$

where μ is the nuclear magnetic moment expressed in nuclear magnetons. The final term is a relativistic correction factor where $\rho = (1 - \alpha^2 Z^2)^{1/2}$, $\alpha = 2\pi e^2/hc$ being the fine structure constant. Assuming $I = 9/2$ and $\mu = 9/2$ proton units, the K terms, for the following Z-values, are split as follows:[82]

	Uranium	Bismuth	Tungsten	Molybdenum
Z	92	83	74	42
ΔE (eV)	22.5	12.6	7.31	0.896

The corresponding splitting of the L_2-shell of uranium, neglecting screening, was 1/5.64 of the K-level splitting, or about 4 eV. This fraction approaches $\frac{1}{8}$ for lighter elements.

Gokhale[83] studied the broadening effects on K X-ray emission lines due to this hyperfine interaction, confining his calculations to elements in the intermediate Z-value region. Here it was calculated that, for the ground state of niobium ($Z = 41$, $\mu = 6.16$ nm, and $I = 9/2$), $\Delta E = 0.563$ eV. It was also suggested that, if the measurement of hyperfine structures in the K X-ray series is feasible, it should be possible to determine magnetic moments of nuclei of known spin. (See also Refs. 84, 85.)

4.5. Effects of Thermal Motion

Doppler broadening of photoemission lines due to the thermal motion of atoms or molecules under investigation is important in optical spectroscopy. For X-ray or electron lines of much higher energies, the effect is usually negligible. For hydrogen gas at 300°K, the broadening effect on a photoelectron line of 5-eV energy will be ≈20 meV[86], which is obtained using the formula

$$\Delta E = 3.338\left(\frac{m_o EkT}{M}\right)^{1/2} = 7.23 \times 10^{-4}\left(\frac{ET}{M}\right)^{1/2}\ \text{eV},$$

where m_o and E are the electron mass and kinetic energy in electron volts, T is the temperature in degrees Kelvin, and the mass M is in atomic mass units. This value varies with the inverse square root of M of the emitting atom or molecule. The Doppler broadening contribution to a spectral line shape will be Gaussian in form.[87] Hence X-rays emitted from gaseous atoms will have intensity profiles displaying both Gaussian and Lorentzian contributions,[88] if the energy distributions of the atomic levels involved are Lorentzian. (For the effect of gas pressure on electron energy distributions, see Ref. 89.)

Electron energy changes of small magnitude, suffered during passage through a solid material, have been investigated by Geiger.[90] Effectively monoenergetic electrons (25 keV) were passed through thin layers of lithium fluoride at room temperature (300°K). An interaction of the transversing electron with the thermal lattice vibrations of the solid caused the occurrence of both positive and negative energy changes. That is, both energy gains and energy losses were observed, and the losses were the more intense. The most probable energy loss and gain were both of the same magnitude and varied from 0.042 to 0.050 eV, which is the energy of the lattice vibrational quantum, for layer thicknesses from 240 to 700 Å. The energy loss peak was several times as intense as the energy gain peak. But with increasing temperature, the two intensities tended toward equaling each other, although their energies did not change appreciably. The intensities of the loss and gain peaks, however, were very small compared to the traversing electron beam intensity. (See also Chapter 8, Section 5.)

One would also expect that photoelectrons ejected from solid materials would experience, with small probability, energy gains and losses of the magnitude of the thermal lattice vibration energy. Such effects have not yet been investigated.

REFERENCES

1. J. S. Levin and D. J. Hughes, *Phys. Rev.* **101** (1956) 1328.
2. T. A. Carlson and R. M. White, *J. Chem. Phys.* **44** (1966) 4510.
3. R. J. Krisciokaitis and S. K. Haynes, *Nucl. Phys.* **A104** (1967) 466.
4. J. L. Wolfson and A. P. Baerg, *Can. J. Phys.* **42** (1964) 1781.
5. P. Erman, I. Bergström, Y. Y. Chu, and G. T. Emery, *Nucl. Phys.* **62** (1965) 401.
6. B. Cleff and W. Mehlhorn, Rept. No. K-133 at the meeting of the German Physical Society, Bad Pyrmont, 1967.
7. W. Mehlhorn, Rept. No. K-132, at the meeting of the German Physical Society, Bad Pyrmont, 1967; *Z. Phys.* **208** (1968) 1.
8. A. H. Compton and S. K. Allison, *X-Rays in Theory and Experiment*, 2nd ed., D. Van Nostrand Company, New York, 1935.
9. G. T. Ewan and R. L. Graham, in *Alpha-, Beta-, and Gamma-Ray Spectroscopy*, K. Siegbahn (Ed.), North-Holland Publishing Company, Amsterdam, 1965, p. 951.
10. M. A. Blokhin, *The Physics of X-Rays*, 2nd ed., Moscow, 1957; A.E.C. Tech. Rept.-4502.
11. V. Weisskopf and E. Wigner, *Z. Phys.* **63** (1930) 54.
12. A. Hoyt, *Phys. Rev.* **36** (1930) 860; *Phys. Rev.* **40** (1932) 477.
13. J. A. Prins, *Physica* **2** (1935) 231.
14. F. K. Richtmyer, S. W. Barnes, and E. G. Ramberg, *Phys. Rev.* **46** (1934) 843; see also L. G. Parratt *Phys. Rev.* **46** (1934) 749.
15. L. G. Parratt, C. F. Hampstead, and E. L. Jossem, *Phys. Rev.* **105** (1957) 1228.
16. T. Watanabe, *Phys. Rev.* **139** (1965) 1747.
17. Y. Sugiura, *J. Phys. Radium* **8** (1927) 113.
18. T. Watanabe, *Phys. Rev.* **137** (1965) A1380.
19. H. W. Schnopper, M.S. thesis, Cornell University, 1958, cited in Ref. 20.
20. L. G. Parratt, *Rev. Mod. Phys.* **31** (1959) 616; see also Ref. 42.
21. R. G. Winter, *Phys. Rev.* **123** (1961) 1503.

21a. R. L. Barinski and W. I. Nefedow, "Röntgenspektroskopische Bestimmung der Atomladungen in Molekülen," (Translated from Russian) Akademische Verlagsgesellschaft, Leipzig, 1969. I thank Professor Meisel for kindly providing me with a copy of this work.

22. M. Mizushima, D. Robert, and L. Galatry, *J. Phys. Radium* **26** (1965) 194.
23. M. W. Zemansky, *Phys. Rev.* **36** (1930) 219.
24. H. A. Farach and H. Teitelbaum, *Can. J. Phys.* **45** (1967) 2913.
25. G. Lindström, *Ark. Fys.* **4** (1951) 1.
26. H. Slätis and G. Lindström, *Phys. Rev.* **88** (1952) 1429.
27. H. Slätis, *Ark. Fys.* **6** (1953) 415.
28. K. Siegbahn and K. Edvarson, *Nucl. Phys.* **1** (1956) 137.
29. M. Mladjenovič, *Ark. Fys.* **8** (1954) 27.
30. C. Nordling, E. Sokolowski, and K. Siegbahn, *Ark. Fys.* **13** (1958) 483.
31. J. H. Dijkstra and C. de Vries, *Nucl. Phys.* **23** (1961) 524.
32. See Ref. 9, p. 968.
33. J. Gizon, *Nucl. Phys.* **A108** (1968) 425.
34. P. Jahn, W. Todt, and H. Daniel, *Phys. Letters* **20** (1966) 395.
35. R. L. Graham, G. T. Ewan, and J. S. Geiger, *Nucl. Instr. Meth.* **9** (1960) 245.
36. G. Kaye, *Nucl. Phys.* **68** (1965) 529.

37. J. S. Rollett and L. A. Higgs, *Proc. Phys. Soc. (London)* **79** (1962) 87.
38. J. S. Geiger, R. L. Graham, and J. S. Merritt, *Nucl. Phys.* **48** (1963) 97; *Conf. Role of Atomic Electrons in Nuclear Transformations*, Warsaw, 1963, Vol III p. 484.
39. W. Mehlhorn, D. Stalherm, and H. Verbeek, *Z. Naturforsch.* **23a** (1968) 287.
40. L. Pincherle, *Nuovo Cim.* **12** (1935) 81, 162.
41. M. A. Listengarten, *Izv. Akad. Nauk SSSR, Ser. Fiz.* **26** (1962) 182. [*Bull. Acad. Sci. USSR, Phys. Ser.* **26** (1962) 182]; *Izv. Akad. Nauk SSSR, Ser. Fiz.* **25** (1961) 792 [*Bull. Acad. Sci. USSR, Phys. Ser.* **25** (1961) 803].
42. E. J. Callan, *Bull. Amer. Phys. Soc.* **7** (1962) N416; *Conf. Role of Atomic Electrons in Nuclear Transformations*, Warsaw, 1963, Vol III, p. 419, (1965); Aeronautical Res. Lab. Rept., Wright-Patterson A.F.B., Dayton, Ohio, 1963.
43. L. G. Parratt, *Phys. Rev.* **54** (1938) 99.
44. L. G. Parratt, *Phys. Rev.* **56** (1939) 295.
45. B. G. Gokhale, thesis, Paris, 1952, (results cited in Ref. 46); see also Ref. 42.
46. W. Lasker, *Ann. Phys (Paris)* **3** (1958) 258.
47. G. D. Latyshev, A. G. Sergeyev, E. M. Krisyuk, L. A. Ostretsov, Yu. S. Egorov, and N. M. Shirshov, *Izv. Akad. SSSR, Nauk Ser. Fiz.* **20** (1956) 345 [*Bull. Acad. Sci. USSR, Phys. Ser.* **20** (1956) 324].
48. A. Meisel and W. Nefedow, *Ann. Phys. (Leipzig)* **9** (1961) 48.
49. S. Hagström and S.-E. Karlsson, *Ark. Fys.* **26** (1964) 451.
50. D. Coster and A. Bril, *Physica* **9** (1942) 84.
51. A. Bril, *Physica* **13** (1947) 481.
52. J. N. Cooper, *Phys. Rev.* **61** (1942) 234.
53. E. Noreland, *Ark. Fys.* **26** (1964) 341.
54. K. D. Sevier, A. Fahlman, and K. Siegbahn, *Ark. Fys.* **29** (1965) 367.
55. R. Carlson, A. Fahlman, and K. Siegbahn, *Ark. Fys.* **32** (1966) 103.
56. L. G. Parratt, *Phys. Rev.* **44** (1933) 695.
57. Private communication from K. Hamrin, 1968; see also K. Siegbahn, C. Nordling, A. Fahlman, R. Nordberg, K. Hamrin, J. Hedman, G. Johansson, T. Bergmark, S.-E. Karlsson, I. Lindgren, and B. Lindberg, *Nova Acta Reg. Soc. Sci. Upsal. Ser. IV*, **20** (1967); Tech. Rept. AFML-TR-68-189, Wright-Patterson A.F.B., Dayton, Ohio (October, 1968).
58. E. J. Callan, *Rev. Mod. Phys.* **35** (1963) 524.
59. F. K. Richtmyer and S. W. Barnes, *Phys. Rev.* **46** (1934) 352.
60. J. A. Bearden and T. M. Snyder, *Phys. Rev.* **59** (1941) 162.
61. R. E. Shrader, *Phys. Rev.* **49** (1936) 644.
62. E. G. Ramberg and F. K. Richtmyer, *Phys. Rev.* **51** (1937) 913.
63. J. H. Williams, *Phys. Rev.* **45** (1934) 71.
64. See Ref. 42 (used X-Ray data from Ref. 74) and L. G. Parratt, *Rev. Sci. Instr.* **6** (1935) 387.
65. Private communication from R. L. Graham in Ref. 38.
66. H. W. Schnopper, *Phys. Rev.* **154** (1967) 118; H. W. Schnopper and L. G. Parratt, *Röntgenspektren und Chemische Bindung*, International Symposium (September 1965) Leipzig, p. 314.
67. B. Ekstig and E. Källne, private communication, 1969; B. Ekstig, thesis, Institute of Physics, University of Uppsala, 1970.
68. B. Ekstig, E. Källne, E. Noreland and R. Manne, University of Uppsala, Institute of Physics Report UUIP-702, April, 1970.
69. J. H. Williams, *Phys. Rev.* **37** (1931) 1431.
70. L. K. Izraileva, *Izv. Akad. Nauk SSSR, Ser. Fiz.* **25** (1961) 954 [*Bull. Acad. Sci. USSR, Phys. Ser.* **25** (1961) 965].
71. V. I. Nefedov, *Izv. Akad. Nauk SSSR, Ser. Fiz.* **28** (1964) 816 [*Bull. Acad. Sci. USSR, Phys. Ser.* **28** (1964) 724].
72. C. S. Fadley and D. A. Shirley, *Phys. Rev. Letters* **21** (1968) 980.
73. S. K. Allison, *Phys. Rev.* **44** (1933) 63.
74. L. G. Parratt, *Phys. Rev.* **50** (1936) 1.
75. J. A. Bearden and C. H. Shaw, *Phys. Rev.* **48** (1935) 18.
76. C. Mande, A. S. Nigavekar, and P. Chivate, *Indian J. Phys.* **41** (1967) 897.
77. T. M. Snyder, *Phys. Rev.* **59** (1941) 168.
78. L. G. Parratt, *Phys. Rev.* **45** (1934) 364.
79. A. Meisel, M. Köstler, and A. Merkel, *J. Prakt. Chem.* **34** (1966) 112; A. Meisel and D. Köstler, *Z. Phys. Chem.* **231** (1966) 99; A. Meisel, *Röntgenspektren und Chemische Bindung*, International Symposium, (September, 1965), Leipzig, p. 212; A. Meisel, *Phys. Status Solidus* **10** (1963) 365.
80. F. K. Richtmyer, *Rev. Mod. Phys.* **9** (1937) 391.
81. L. Pincherle, *Phys. Rev.* **61** (1942) 225.

82. G. Breit, *Phys. Rev.* **35** (1930) 1447.
83. B. G. Gokhale, *Ann. Phys. (Paris)* **7** (1952) 852.
84. J. H. van der Tuuk, *Z. Phys.* **44** (1927) 737.
85. R. E. Watson and A. J. Freeman, *Hyperfine Interactions*, A. J. Freeman and R. B. Frankel (Eds.), Academic Press, 1967, p. 53.
86. H. A. Bethe, *Rev. Mod. Phys.* **9** (1937) 69; see p. 140 of this reference.
87. J. I. Gersten and H. M. Foley, *J. Opt. Soc. Amer.* **58** (1968) 933.
88. S. N. Dobryakov and Ya. S. Lebedev, *Dokl. Akad. Nauk SSSR*, **182** (1968) 68 [*Sov. Phys.—Doklady* **13** (1969) 873.]
89. C. E. Kuyatt, J. A. Simpson, and S. R. Mielczarek, *Phys. Rev.* **138** (1965) A385.
90. J. Geiger, *Physikertagung*, Munich, 1966, p. 108; H. Boersch, J. Geiger, and W. Stickel, *Phys. Rev. Letters* **17** (1966) 379; *Z. Phys.* **212** (1968) 130.

Chapter Seven

Atomic Electron Binding Energies

1. INTRODUCTION

Because of the intimate relation between X-ray energies and the electron binding energies of the atomic levels involved in the X-ray type of transition, the experimental investigation and determination of atomic electron binding energies made great strides following the establishment and development of the field of X-ray physics. The electron binding energy values obtained from X-radiation absorption and emission measurements accumulated rapidly and were soon collected into the first binding energy tables.

Simultaneously, in the theoretical treatments of single and multielectron atomic systems great advances were based on the work of Bohr and others, including the introduction, development, and application of the quantum theory (Section 2). Increasing sophistication in atomic theory and theoretical computational methods and the recent advent of modern time-saving computers are making it possible to calculate electron binding energies with ever-increasing accuracy (Section 2).

Increased accuracy in electron binding energy determinations due to improvements of X-radiation investigation techniques and of various electron spectrometric investigational methods frequently give absolute values with accuracies to less than 1 eV; relative energy measurements can be made to ~0.1 eV or better, depending on the method. With such accuracy in binding energy measurements, it has become possible to investigate second-order variations in these owing to chemical effects on nuclear charge screening (Section 4) and lattice and other influences (Sections 2, 4.1.1, and 4.2). The relation between the electron binding energies of an atom and its chemical state has been developed into a chemical analysis method. This has been applied with special success using techniques of photoelectron spectrometry (Section 4.1.2).

The method of determination of binding energies from various electron spectrometry procedures is emphasized in this chapter; treatments of X-radiation techniques are found in other works.[1-7] The binding energy determination techniques of primary concern here appear in Section 3.2, together with a discussion of the nature and importance of the Fermi level position, effects of electron straggling in the source, and absolute electron energy determinations.

Appendix F presents a table of atomic electron binding energy values. The table is basically the same as that reported by Bearden and Burr except for the addition of several more recent photoelectron spectrometry measurement values and values obtained by the other electron spectrometry methods discussed in Section 3.

2. THEORETICAL CALCULATION OF ATOMIC ELECTRON BINDING ENERGIES

The theory of atoms containing one or two electrons has been treated thoroughly during the past few decades through the application of quantum mechanics. Single-electron atoms (i.e., hydrogen-like) have been studied using both the nonrelativistic theory of Schrödinger and the relativistic theory of Dirac. The validity of the results of these calculations can be tested by comparing them with corresponding experimentally determined quantities for hydrogen and hydrogen-like atoms. Conversely, the high accuracy of experimental level energies for such atoms allows us also to test and judge the validity of theoretical correction terms that account for quantum electrodynamic and other higher order effects.

The theoretical treatment of the hydrogen-like atom is useful to investigate the trends and characteristics of atomic energy levels. Calculated results can be extended to fit experimental values for multielectron atoms by introducing electron-electron interaction terms and nuclear charge shielding corrections.

This chapter is primarily concerned with the experimental determination of atomic electron binding energies. The only aspects of the theory presented are those which aid in understanding the general characteristics of these quantities as determined experimentally. For complete treatments of the quantum mechanical theory of atomic structures, the reader is referred, for instance, to the works of Dirac,[8] Hartree,[9] Slater,[10] and Lindgren,[11,12] as well as to the review article by Bethe and Salpeter.[13]

The nonrelativistic Schrödinger equation for an electron in the central field of a nucleus of charge Ze and of infinite mass is

$$\nabla u + \frac{2m_o}{\hbar^2}\left(E + \frac{Ze^2}{r}\right)u = 0,$$

where u is the single electron wave function and r is the electron-nucleus radial distance. After separation into spherical polar coordinates $u = R(r) \times Y(\vartheta, \varphi)$, the equation can be solved exactly. Solution of the radial portion of the equation will give the Balmer formula for the energy levels of a hydrogen-like atom

$$E = -\frac{Z^2}{n^2}\,\mathrm{Ry}, \tag{1}$$

where n is the principal quantum number and Ry designates the Rydberg unit of energy. The level energies do not depend on the additional quantum numbers l (angular momentum quantum number) or m (magnetic quantum number). The l-independence is special for the hydrogen atom case where the Coulomb potential is due only to the nucleus-electron interaction. Hence, for a given value of n, the levels of different l are degenerate. The m-independence exists because no spacial orientation is preferred; this would not be true if the atom were situated in an external magnetic field.

The energy difference between two atomic states, and therefore the emission radiation energy of an atomic transition from quantum state n to n', will be

$$h\nu = E(n) - E(n') = Z^2\left(\frac{1}{n'^2} - \frac{1}{n^2}\right)\mathrm{Ry}, \tag{2}$$

where ν is the frequency of the emission radiation.

In 1913 Moseley[14] showed, in his study of K X-radiation from elements for $20 \leq Z \leq 30$, that the frequency of each X-ray line is approximately proportional to the square of the atomic number of the emitting element. According to this relationship, known as the Moseley law,

$$\sqrt{\frac{\nu}{R}} \propto Z, \tag{3}$$

in agreement with the quantum mechanical expression of equation 2. (In the original work $\sqrt{\nu}$ was plotted against Z. The use of the dimensionless quantity $\sqrt{\nu/R}$, where R is the Rydberg constant, was introduced by Sommerfeld.) It was later shown that Moseley's law also holds for the L- and M-radiation series.

Actually, for the elements investigated by Moseley, the quantity Z^2 should be replaced by $(Z - \sigma)^2$, where σ is the screening of the nuclear charge by the atomic electron cloud. The screening correction for the K series is close to unity, whereas for the L-series it becomes about 7.4. In a like manner, for a given major shell of principal quantum number n, screening of the nuclear charge is different for subshells of unlike l-quantum number (e.g., the L_1–L_2 binding energy difference is primarily of this origin, but additional small variations can be caused by nuclear finite size effects.[15]) It was first shown by Hertz (see Ref. 16), that

$$\Delta\sqrt{\frac{\nu}{R}} \cong \frac{\Delta\sigma_1}{n}, \tag{4}$$

where $\Delta\sigma_1$ is a constant and represents the difference in screening between two levels of same n and j but different l-quantum numbers. This explains why certain curves in a Moseley diagram are nearly parallel. Equation 4 expresses the *screening doublet law*.

Hence in the multielectron atom degeneracy is removed for levels of like n (and j) and different l as a result of differences in screening.

The Dirac theory of the atom takes into account two additional considerations that cause an actual hydrogen atom to undergo the fine structure splitting that remains unexplained by the Schrödinger theoretical treatment. These are the electron spin and the effects of relativity.

The Dirac differential equation for an electron in a Coulomb field is $Hu = Eu$, where

$$H = V(r) - \beta E_0 - c\boldsymbol{\alpha} \cdot \mathbf{p}. \tag{5}$$

$E_o = m_o c^2$ is the electron rest mass energy, $\mathbf{p} = -i\hbar$ grad is the electron momentum operator, and $V(r) = -e^2Z/r$, the potential energy of an electron at a distance r from the nucleus. The Cartesian components $(\alpha_1, \alpha_2, \alpha_3)$ of the vector operator $\boldsymbol{\alpha}$ together with $\beta \equiv \alpha_4$ are the Dirac operators α_i. The exact solution of equation 5 can be calculated. The expression for the binding energy (B.E.) of an electron in a hydrogen-like atom will then be

$$\mathrm{B.E.} = m_o c^2\left\{\left[1 + \left(\frac{\alpha Z}{n - k + \sqrt{k^2 - \alpha^2 Z^2}}\right)^2\right]^{-1/2} - 1\right\}, \tag{6}$$

where n is again the principal quantum number, α is

Sommerfeld's fine structure constant, and $k = j + \frac{1}{2}$ is a quantum number that can take the values $(1, 2, \ldots, n)$. For each value of k or j the orbital quantum number takes the two possible values $l = j + \frac{1}{2}$ and $l = j - \frac{1}{2}$ (except for $k = n$, when $l = j - \frac{1}{2} = n - 1$ only). Equation 6 was derived some time earlier, using the older Bohr theory, by Sommerfeld.[17]

Equation 6 may be expanded in powers of the quantity $(Z\alpha)^2$, which is usually small compared to unity. Considering only the first two terms of the expansion,

$$\text{B.E.} \cong -\frac{Z^2}{n^2}\left[1 + \frac{(\alpha Z)^2}{n}\left(\frac{1}{k} - \frac{3}{4n}\right)\right] \text{Ry}. \qquad (7)$$

The first term in equation 7 is identical to the right-hand side of equation 1. Including the screening factor σ, letting binding energy be positive, and neglecting the second term in equation 7, the useful expression

$$\sqrt{\text{B.E.}} - \frac{Z}{n} - \frac{\sigma}{n} = \sqrt{\text{B.E.}} - aZ - b = M_1 \qquad (8)$$

is derived. Here a and b are constants. If the level energies vary as Z^2, the plot of expression 8 versus Z, a modified Moseley diagram,[18,19] will give a straight line. The constants a and b are chosen to give the line a convenient position in the diagram.[19]

For the inner levels of heavier elements, however, the modified Moseley diagram has a parabolic shape, demonstrating the importance of relativistic effects (see Figure 1, top). Hagström[19] remedied this by including the second term in equation 7. He was able to obtain a linear instead of parabolic trend of values when he plotted

$$((\text{B.E.})^{1/2} - AZ - B)^{1/2} - CZ - D = M_2 \qquad (9)$$

as a function of Z (see Figure 1, bottom), where A, B, C, and D are constants to be determined from the experimental data. The general horizontal trend for K-level binding energies breaks between $Z = 71$ and 72, and begins to slope downward for higher Z-values. It is at this atomic number that the rare earth series ends with the filling of the $4f$ shell, and the $5d$ shell begins to fill. From a plot of this type we can estimate with good accuracy the binding energy value of a level not yet measured (e.g., that for $Z = 61$ in the figure).

A plot of equation 9 for the K-shell and in the region above $Z = 71$ shows a series of short-range monotonic variations of differing slopes.[20] These changes in slope for a given level are owing to the change in screening constant with Z value.

Relativistic effects, rather than screening effects, are mainly responsible for the spin-doublet splitting. Returning to equation 6 the energies of levels having the

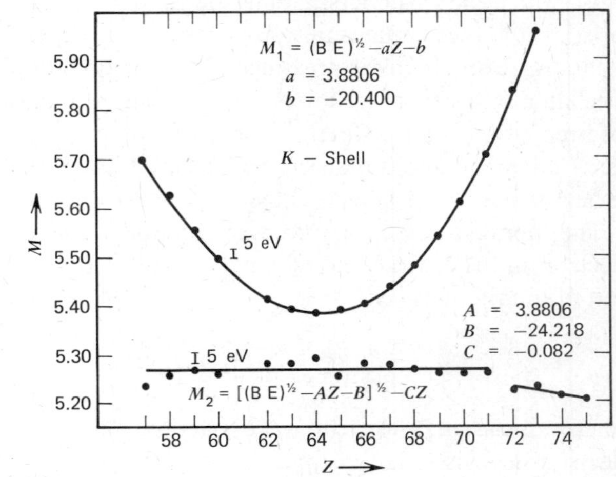

Figure 1. Modified Moseley diagrams of K-shell binding energies in the rare earth region of the periodic table using photoelectron data. Upper curve obtained with equation 8; lower lines represent the same experimental binding energy values fitted by equation 9. (From Ref. 19.)

same n- and l-values and different j-values can be determined. The energy difference between these as a function of atomic number will give spin-doublet energy splittings that can be compared with experimental values. As an example, for the L_2–L_3 splitting equation 6 gives

$$\Delta(\text{B.E.}) = \frac{m_0 c^2}{2}\left[\sqrt{4 - \alpha^2 Z^2} - (2 + 2\sqrt{1 - \alpha^2 Z^2})^{1/2}\right]. \qquad (10)$$

By replacing Z with $(Z - d)$ and fitting the values Δ(B.E.) in equation 10 to experimental data, we can determine the screening factor d for a given n, l-spin couplet. Bergvall and Hagström[21] obtained $d = 3.58$ for L_2–L_3 splitting in the rare earth region. Fahlman and Hagström[20] extended the analysis to the $Z = 73$ to 83 region, using the same value for d, and obtained good agreement with experimental values derived from both photoelectron and $K\alpha_{1,2}$ X-ray emission measurements. For the same Z-value region, Fahlman and Hagström also compared the M_2–M_3 and M_4–M_5 binding energy differences from photoelectron measurement data with values calculated using equation 6. They found that the constant d took the values 8.27 and 13.59, respectively. (These values do not, however, agree with corresponding Slater and other screening constant values—see Section 4.1.) The initial comparison proved to be extremely good, the two sets of values usually agreeing to within 1 eV. Spin-doublet splitting values for elements not yet measured and lying in Z-value regions for which d is known, can be calculated quite accurately with the aid of equation 6.

In recent years much progress has been made on the theoretical calculation of energy levels in multielectron atoms and ions. The complexity of such analyses has been made less prohibitive by the use of approximations and the employment of electronic computers to reduce the time necessary to perform these calculations.

Calculations on multielectron atoms can be carried out using the Schrödinger or the Dirac equation, provided screening of the nuclear charge by inner shell electrons is taken into account. This is usually done by employing a screening approximation. Earliest efforts along these lines resulted in the Thomas–Fermi, Thomas–Fermi–Dirac, and other statistical models as used, for example, by Shalitin[22,23] and by Yonei.[24] However, later self-consistent Hartree–Fock calculations treat the problem of screening more accurately and, along with several modified procedures, are now used extensively for atomic calculations.

The most common approximation is that in which each atomic electron is considered to move independently in the field of the nucleus and in the field of the other electrons. Here the field of the average electron distribution derived from the individual electron wave functions is the same as the field used in evaluating this same wave function. The atomic field determined in this way is thus termed self-consistent.[9]

The simplest self-consistent-field (SCF) method is the Hartree method, wherein the wave function of an atom of N electrons is given by the product of the individual single-electron wave functions. The approximate wave function so obtained, however, is not in conformity with Pauli's exclusion principle.

This point was later taken into account by expressing the approximate wave function of the atom as a determinant of one-electron wave functions. The expansion of the determinant consists of terms representing the permutation of all electrons among all one-electron wave functions, each term having the coefficient ± 1. This approach leads to the Hartree–Fock method.

In the Hartree–Fock method as simplified by Slater[25] (called the Hartree–Fock–Slater method), Slater showed that the Hartree–Fock equation can be regarded as an ordinary Schrödinger equation for the motion of an electron if the exchange term of the potential is replaced with its average value for all occupied orbitals. He proposed further that this average be approximated by the exchange potential of an electron gas.

The central potential for an atom given by the Slater correction is not ultimately the best. One of the modifications of this correction, by Lindgren,[11,12] minimizes the total energy of the atom. Wave functions calculated by using the Schrödinger equation and this modified potential correction are quite close to those obtained by the exact Hartree–Fock method.

Other modifications of the Slater exchange potential have been suggested,[26–29] but although improvements are realized, these improvements are inferior to that achieved with the Lindgren optimized potential.

The Hartree, Hartree–Fock, and Hartree–Fock–Slater methods can also be used in relativistic self-consistent-field calculations. Here the starting point is the Dirac equation 5, instead of the Schrödinger equation. The relativistic calculations of bound state atomic wave functions have been carried out by Liberman et al.,[30] Nestor et al.,[31] Tolliver et al.,[32] Schonfelder,[33] Gelius and Nordqvist,[34] and Smith and Johnson.[35] See also the calculations for $Z = 2$ to 88 by Froese-Fischer[36] and for $Z = 114$, 126, 140 by Tucker et al.[37] In most of these articles, comparison is made between results obtained with both relativistic and nonrelativistic calculations and using several different methods. Calculations have also been performed for ions with inner shell vacancies (cf. Bagus[38] and Refs. 31, 39). For calculations on molecular electron charge distributions and chemical bonding see Ref. 53.

Two points must now be considered. First, theoretical binding energies calculated according to the above-named methods are approximations that seldom agree with experiment. This infrequent agreement is partially due to the second point, namely, the calculations are carried out for the isolated atom or ion.

Returning to the first point, if the orbitals of atomic electrons other than that of interest are assumed to remain fixed during the ionization process by which the binding energy is measured, the total energy of the final atom with a single vacancy in the given level will differ from that of the original neutral atom by the interaction between the atomic electron cloud and the electron filling the level about to be vacated. This change in total energy is, by Koopmans's theorem,[40] equal to the eigenvalue of that electron in the atom. Relativistic Hartree–Fock values of binding energies for all shells for $Z = 2$ to 103 have been calculated using this method by Carlson and Pullen.[41] (The relativistic values were obtained by multiplying the nonrelativistic Hartree–Fock solutions by a relativistic correction factor.) Here the orbitals of the remaining atomic electrons are considered to be "frozen" during the ionization. These calculations have recently become superseded by eigenvalues (binding energies), radial expectation values and potentials for free atoms from $Z = 2$ to 126 as calculated from relativistic Hartree–Fock–Slater atomic wave functions.[41a]

If the ionization process is one that occurs rapidly

as compared with the relaxation time of the atom, the ionization energy will indeed be the eigenvalue of the atomic electron. However, when the relaxation time is short relative to the ionization process duration, the energy of the ionized atomic state must be calculated separately; the binding energy will then be the difference between these two (atomic and ionic) total energies.

The binding energy of an atomic electron in a given level, determined by taking the difference between the total energies of the neutral atom and of the atom with a vacancy in that level, is, as expected, in better agreement with experimental values when inner orbitals (i.e., having higher binding energies) are involved. Both relativistic and nonrelativistic calculations of electron binding energies using this method have been reported.[29–31,34,42] A comparison of the level values obtained by the various authors is presented in Ref. 39. However, for the outer atomic electrons the situation is more closely described by the Koopmans theorem.

The foregoing statements can be illustrated by comparing experimental level energy values with the self-consistent-field calculations of Bagus[38] on argon. The binding energy values for the 1*s*, 2*s*, 2*p*, 3*s*, and 3*p* levels were calculated by two methods and the results compared as follows: (*a*) the difference in the SCF calculation of total energies for the neutral atom and the ion, and (*b*) the calculation of the binding energy of the electron in question in the neutral atom. It was seen that the outer level energy calculated by method *b* agreed better with experimental values than those calculated by method *a*; the opposite was true for the inner level energies. According to Mulliken,[43] this is because the removal of an inner level electron makes the SCF orbitals considerably different from those for the neutral atom. Therefore again, the "frozen orbital" approximation of Koopmans[40] is poor for inner levels. The binding energy values calculated by Bagus for argon are 1*s*, 3195.2; 2*s*, 324.8; 2*p*, 248.9; 3*s*, 33.2; 3*p*, 16.1 eV. The last value was calculated by method *b*, the others by method *a*. These values agree fairly well with the experimental values for argon gas in Appendix F.

A detailed comparison of the relative validity of the "frozen" and "relaxed" orbital approximations for calculating binding energies has not yet been reported. Limited treatments are given in Sections 4.2.1. and 4.2.2 of this chapter.

Auger electron energies can be calculated from the difference between total atomic energies for the related one and two inner shell vacancy states in much the same way used with the binding energy values previously named. Such calculations have been carried out and presented in Refs. 31 and 34. A comparison of theoretical and experimental *KLL*-Auger energy values for neon ($Z = 10$) is given in Table III of Chapter 3.

Consider now the second point—that theoretical calculations are performed for the isolated atom or ion. The energy difference between the neutral and ionic atomic states is the work required to remove the given electron from the atom to infinity, where it has zero final kinetic energy. These calculations of atomic electron binding energies correspond best to atoms in the gaseous or vapor states.

For atoms in the solid state, the energy of an atomic level is that energy required to remove an electron in that level to the lowest vacant level in the atom (e.g., to the Fermi level in a metal). In this case, however, the binding energy of the electron in its initial atomic level depends on its physical environment as well as the chemical state of the atom (Sections 4.1.1 and 4.1.2).

The influence of the physical environment is also important, for the comparison between theoretical and experimental binding energies of electrons for the rare gas elements and for neighboring elements give some indication of the significance of these solid state effects.

Coulthard[42] has calculated the binding energies of the 2*s*-electrons in neon and magnesium. The value of 49.4 eV for neon is in excellent agreement with experimental determinations of this value for gaseous neon atoms. However, the calculated energy value for this level in magnesium (97.5 eV) is in severe disagreement with the photoelectron measurement value of 89.4 eV by Nordberg et al.[44] for the same metal. The disagreement is not explainable by chemical shift effects, since there is no detectable difference in L_1-shell binding energy between the metallic and oxide states.[44]

The measurement of photoelectron energies for K and L_1 (using copper $K\alpha_1$ X-radiation) and $L_{2,3}$ (using chromium $K\alpha_1$ X-radiation) levels of argon in a solid metallic environment has been reported.[39] The argon sample was made by bombarding a metallic foil with argon ions accelerated in a gaseous discharge. The binding energy values for the L_1 and $L_{2,3}$ (not resolved) shells, according to Figure III : 20 in Ref. 39, are $\approx$313 eV and $\approx$237 eV, respectively. (The binding energy scale in the figure was apparently not corrected for the spectrometer work function of about 4.5 eV; this correction was made in obtaining the foregoing values.) The *K*-binding energy for argon was about 3198 eV, which is in fair agreement with value measurements made with gaseous atoms (see Table I).

Mehlhorn and Stalherm[45,46] have determined that the L_1-binding energy in the gaseous argon atom is 326.5(5) eV, in disagreement with the previously cited value. The mean value for the $L_{2,3}$-binding energies

Table I.

Comparison of theoretical (for the isolated atom) and experimental (for both gases and solids) binding energy values to show the discrepancies and agreements between them.

Element	Level	Binding energies (eV)		
		Theoretical calculations (isol. atom)	Experimental values	
			Gaseous state	Solid state (chemical form)
Ne	L_1	49.4[42)]	48.47(10)[a)]	--
Mg	L_1	97.5[42)]	--	89.4(4)[44)] (MgO)
Ar	K	--	3202.9(3)[b)]	3198[39)] (Ar, sputtered)
"	L_1	327.0(5)[47)]	326.5(5)[45)]	313[39)] (Ar, sputtered)
"	L_2	--	250.53(25)[46)]	237[39)] (Ar, sputtered) (L_2 and L_3)
"	L_3	--	248.39(25)[46)]	
Xe	M_4	--	690.6(7)[c)]	684.0(6)[48)] (Xe, frozen)
"	M_5	--	677.6(6)[c)]	671.3(6)[48)] (Xe, frozen)
"	N_3	--	146.3(3.1)[b)] (N_2 + N_3)	140.0(6)[48)] (Xe, frozen)
"	N_4	--	69.52[d)]	63.6(6)[48)] (Xe, frozen)
"	N_5	--	67.55[d)]	61.6(6)[48)] (Xe, frozen)

a) Appendix F, ref.[9)]
b) " " " 3)
c) " " " 18)
d) " " " 28)

obtained by Mehlhorn[46] is 249.5 eV, forming another discrepancy. The synchrotron radiation absorption measurement of the L_2- and L_3-binding energy values for the gas by Sagawa et al. (see Appendix F) agree well with Mehlhorn's values.

The excellent agreement between Coulthard's[42] 2*s*-shell binding energy calculation for neon and the experimental value, on the one hand, and the disagreement between the corresponding theoretical and experimental values for the same shell in the solid metallic magnesium may be attributed primarily to solid state effects. Coulthard[47] later carried out a self-consistent-field calculation of the L_1-binding energy of argon to be compared with Mehlhorn's[45] experimental values just quoted. The theoretical evaluation gave the value 327.0(5) eV, again in excellent agreement with the experimental value (see Table I). This confirms that theoretical atomic binding energy values must be compared with measurements of the corresponding values for *isolated* atoms. Theoretical values should therefore differ from experimental values for atoms in the solid state, unless the solid state effects responsible for the binding energy changes are in some way taken into account. Similar discrepancies have been found for the $M_{4,5}$ and $N_{3,4,5}$ shells in xenon between the gaseous and frozen solid forms (see Table I).

Lotz[49] has reported calculations of the binding energies of the *K*-shell and some of the outer shells[50] of atoms and ions from $Z = 5$ to $Z = 30$. These values were obtained from experimental ionization potential values of isolated atoms and ions. Discrepancies arise between the binding energy values tabulated by Lotz for the 2*s* ($Z = 8$ and 9) and 3*s* ($Z = 17$) and experimental determinations for atoms in solid compounds (Appendix F). These values for the isolated atoms and the solid compounds (italicized) are: oxygen, 28.48, *23.7*; fluorine, 37.85, *34*; chlorine, 24.54, *17.5* eV. There is a difference of 4 to 7 eV for each of the three cases. For the rare gases neon and argon there was agreement (but the two respective M_1 binding energies for potassium agreed). Similar tendencies are found for the *K*-binding energies. Such tables of values are based on semiempirical estimates, however, and are cited here only to illustrate a qualitative disagreement.

The discrepancies between atomic electron binding energies as measured in an isolated atom and in that atom situated in a compound were considered by Fadley et al.[51,52] in their investigation of chemical effect on binding energies.

Consider the binding energy of the *i*th core level of an atom A with an assumed net charge *z* in a given solid compound X (see Figure 2). Charges are assigned to all the other atoms of the compound lattice, to establish electroneutrality. The binding energy can then be calculated in the three steps illustrated in Figure 2:

1. Removing A from the lattice to form a free ion of charge *z*.
2. Removing an *i*th-level electron from this ion.
3. Replacing the ion into the lattice.

The binding energy is then obtained from the sum of the three energies required in the cycle. The energy for 2 is the binding energy as calculated by the Hartree–Fock technique or a similar method. The sum of the energies for 1 and 3 represents the Coulomb interaction between an electron in a given level and all the other ions in the lattice. This last classical Coulomb energy is related to the Madelung constant for the solid, which depends only on the lattice structure.

The preceding statements apply primarily to chemical compounds in the solid state. The ions in these compounds have an ionicity of $\pm z$. The Madelung energy for a certain atom is the Coulomb interaction of all the electrons of the ion with the rest of the ions in the solid. This will result in a difference in the binding energies of the electrons in that *ion*, depending on whether the ion is in the lattice or removed from it and isolated. The charge on the ion itself will also cause a measured binding energy shift from the value for the neutral atom (see Section 4). Both these factors contribute to the magnesium *K*-binding energy difference between Mg (isolated) and MgO (solid), as shown in Table I.

However, in the other cases given in Table I there is no chemical binding, at least not in the usual sense. Even so, there is obviously a Coulomb interaction of the rare gas atoms with their environments, causing a discrepancy in binding energies between the gaseous and solid states. As a possible solution, one would suspect the existence of some degree of ionicity of the atoms, even though it is not due to chemical bonding.

Experimental *KLL*-Auger energies may also be compared with theoretical estimates of their values. It is seen in Table III of Chapter 3 that theoretical values

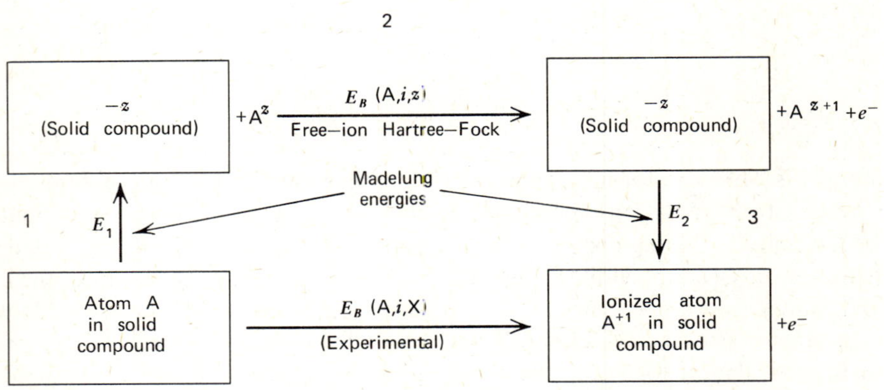

Figure 2. Energy cycle (1–3) for calculating atomic electron binding energies compared with single-energy step for experimentally determining these. (From Ref. 51.)

agree well with the experimental for gaseous neon. Both the relativistic and nonrelativistic theoretical *KLL*-Auger energy values for magnesium calculated by Gelius and Nordqvist[34] lie below the experimental values for the solid metal by 10 to 12 eV.

On the other hand, the semiempirically calculated *KLL*-Auger energy values (Table II of Chapter 3), agree well with Auger energies measured using solid samples. For gaseous neon the agreement is poor indeed, whereas the relative energy splitting of the Auger components shows accord between the experimental and semiempirical values (see Figure 2, Chapter 3).

The recent measurements by Hillig and by Weichert of the *K*-Auger spectrum of sodium in the vaporized, solid metallic, and solid oxide states have shown the *K*-shell binding energy to be different in all three cases. (See Chapter 3, Section 4.1.1, and Refs. 42a and 42b of that chapter.) See also Ref. 54 for work done on the $K\beta_{1,3}$ fluorescence spectrum in chlorine for several gaseous organic and inorganic molecules.

Alterations of atomic binding energies may also result from other causes. These causes—chemical effects, electrostatic field effects, and so on—are treated in Section 4.

3. EXPERIMENTAL DETERMINATION OF ATOMIC ELECTRON BINDING ENERGIES

Accurate values of atomic electron binding energies are necessary to experimentalists and theoreticians alike working in the fields of atomic and nuclear physics. The high precision experimental investigation of atomic and nuclear-atomic phenomena (the Auger effect, the photoelectric effect, electron shake-off, electron scattering, internal conversion, etc.) often presupposes access to tables of such values. Theoreticians dealing with these subjects need atomic electron binding energy values in order to arrive at numerical solutions of these calculations as well as to judge the merits of the various methods for calculation of the atomic binding energies themselves.

The first sets of atomic binding energy tables were based on X-ray emission and absorption measurements, and the first extensive table, by Manne Siegbahn,[55] appeared in 1931. Since then the investigation of atomic electron binding energies has improved in accuracy with advances in the field of X-ray spectroscopy. Subsequently published tables have summarized the best current sets of values.

In 1955 Slater[56] published a table of binding energies for $Z \leq 41$ derived from X-ray absorption and emission measurements. He also treated the effects of the solid state on the outer atomic level energies and widths. A more extensive table was compiled in 1957 by Sandström.[1]

Atomic level energies have, then, been determined quite accurately by using X-radiation emission and absorption data (X-ray data for the inner atomic shells and optical data for the outer[56]). In about 1958, however, the development of the photoelectron method of determining atomic level energies employing a high resolution magnetic spectrometer was begun by the research group headed by Kai Siegbahn in Uppsala. As the method was improved, it became possible to determine atomic binding energies with an accuracy approaching or even better than that obtainable with X-ray methods. An atomic energy level table including values from photoelectron measurements and, where these are not available, X-ray data, was published a few years later.[57]

In 1965 Bearden and Burr[58] presented a compilation of atomic level energies taking full advantage of all the available information from X-ray and photoelectron data. Reevaluated X-ray emission wavelength values[59] made up most of the original data from which the values were extracted. X-ray absorption data, previously used to establish the absolute scale, were substituted by photoelectron data whenever possible.

In 1967 the table of Bearden and Burr was extended and improved[59] by including new X-ray and photoelectron data and by readjusting the data to the 1965 values of atomic constants.[60] At the end of the same year the table was reproduced and extended through the addition of more recent photoelectron measurements by Siegbahn et al.[39]

In the above-mentioned tables, levels not yet determined experimentally were usually filled by interpolation on Moseley diagrams.

The most recent tables on atomic binding energies referred to here give values having quoted probable errors $\gtrsim 1$ eV for the innermost levels and $\gtrsim 3$ eV for the outermost for which values are given. Yet some of the level energies are still quite uncertain (e.g., for some of the rare gas levels and those obtained by interpolation). Surely these values will be accurately determined through future measurements.

However, several of the vacancies in the electron binding energy tables can at present be filled, and uncertain values can be improved by taking advantage of published data from still other types of measurements and determinations. Most of these measurements, it should be added, involve electron spectrometry in one way or another.

In the following subsections the various experimental sources of atomic electron binding energy values are treated. In Sections 3.1 and 3.2 the X-ray emission and absorption methods, respectively, are discussed. Section 3.3 covers the photoelectron method, whereas Section 3.4 treats the employment of Auger electron energy values; Section 3.5, the use of autoionization electron measurements; and Section 3.6, the techniques of inelastic electron scattering measurements. The manners in which binding energy values are derived from measured quantities are also included in each section.

3.1. X-Ray Emission and Absorption

Numerous treatments of the determination of atomic binding energy values from X-ray emission and absorption measurements have been published (see literature cited in Ref. 59). These articles should be consulted for extended accounts of the subjects. Both these topics are treated shortly in Chapter 6, Section 2.

Investigation of soft X-ray (ultraviolet) radiation emission and absorption is suitable for the study of electron density distributions and binding energies of the outer atomic levels and of band structure in solids. Absorption and emission spectra obtained by such measurements for solids are often quite complex and difficult to analyze to obtain meaningful outer shell binding energy values. For the monatomic rare gases, however, outer shell binding energies are accurately determined by this method.

3.1.1. X-Ray Emission

An internal shell atomic vacancy may be filled by an electron lying further out in the atom. The difference between initial and final total atomic energy, before and after the transition, is released as an X-ray. (Figure 3*a* provides a schematic description of the process.)

The atomic levels from which an electron may originate to fill the initial vacancy are restricted in accordance with selection rules for X-ray transitions. The most probable electric dipole transitions are associated with the following changes in quantum numbers: $\Delta l = \pm 1$, $\Delta j = 0$ or ± 1; $j = 0 \rightarrow j = 0$ transition is forbidden. In addition, transitions of comparatively strong intensity occur which are forbidden by the foregoing selection rules. Of primary interest among these are the electric quadrupole and magnetic dipole transitions: $\Delta l = 0$ or ± 2, $\Delta j = 0$, ± 1, or ± 2, $j = 0 \rightarrow j = 0$, $j = \frac{1}{2} \rightarrow j = \frac{1}{2}$, and $j = 0 \rightleftharpoons j = 1$ are forbidden; and $\Delta l = 0$, $\Delta j = 0$ or ± 1, $j = 0 \rightarrow j = 0$ is forbidden, respectively.

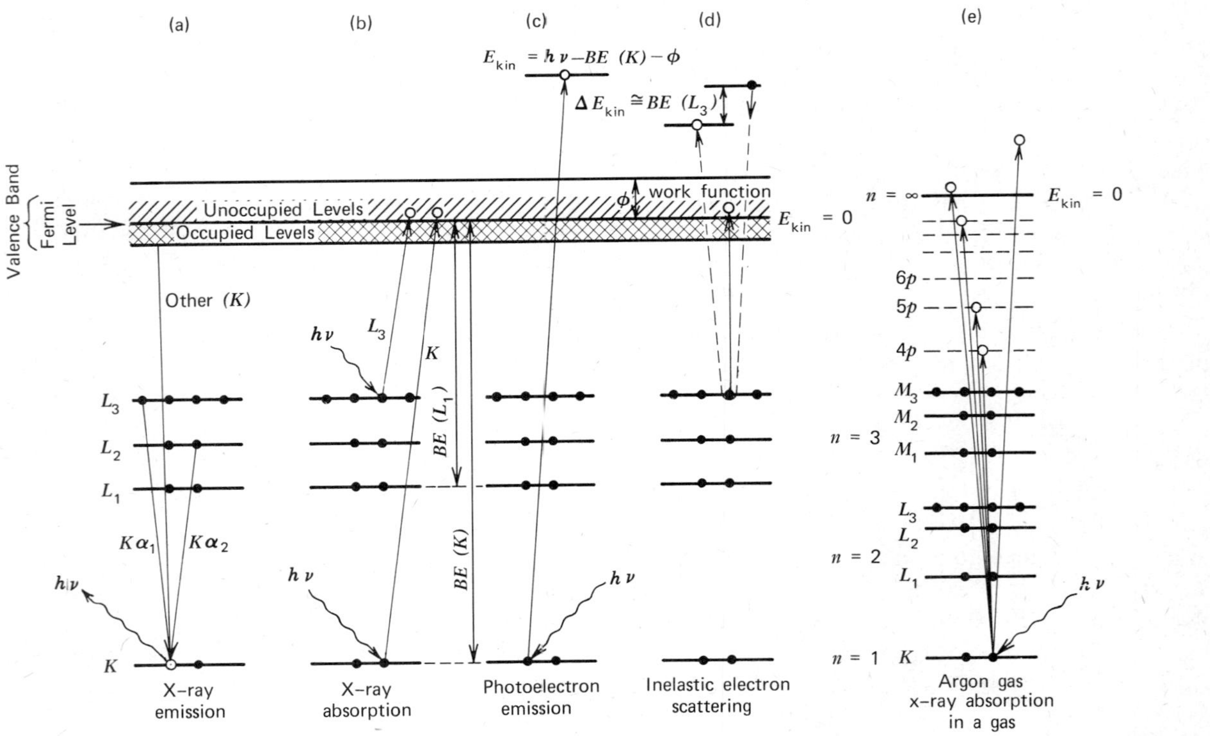

Figure 3. Schematic illustration of methods by which atomic electron binding energies can be determined. (*a*) through (*d*) pertain to atoms in solids; (*e*) illustrates X-ray absorption in an isolated atom.

The binding energy differences between two atomic levels can be determined by taking the difference between the X-ray energies involving each of the two levels and a common third level [e.g., the L_2- and L_3-binding energy difference can be determined from the difference between the $K\alpha_1$–$K\alpha_2$ (KL_3–KL_2) or $L\beta_1$–$L\alpha$ (L_2M_4–L_3M_4), etc., X-ray transitions]. Combining X-ray emission energy values with binding energy values determined uniquely by methods described in the following sections will give a set of equations relating the atomic binding energies. Solution of these equations by a least-squares procedure will give the desired level energies. For determining the absolute scale, at least one absorption measurement value (i.e., the energy required to raise the bound electron to the Fermi level) was included in the treatment. Photoelectron determinations of one or more atomic levels were included, replacing absorption measurement values where possible.

A summary of X-ray wavelengths normalized to the accurately determined tungsten $K\alpha_1$ emission line wavelength has been published by Bearden.[59] Corresponding energies were also given. This article provides a rigorous treatment of precision measurements of X-ray wavelengths.

X-Ray-emission energies are altered by the chemical state of the atom and also by the initial vacancy creation process. A treatment of some of these variations is given in Section 4.

3.1.2. X-Ray Absorption

The energy required to excite a bound atomic electron to an unoccupied level at or above the Fermi level in a solid element can be estimated through X-ray absorption techniques (Figure 3*b*). The final state of the transition electron is determined by selection rules for electric dipole transitions (i.e., $\Delta l = 1$). The accuracy of such measurements may be uncertain, owing to uncertainties concerning the properties and symmetries of unoccupied states above the Fermi level. It is, however, apparent that X-ray-absorption binding energy measurements performed with sufficient care and with modern apparatus should be comparable in accuracy with photoelectron measurement values[59] if correct interpretation of the spectra is possible.

This has been illustrated in the report of the *L*-shell absorption measurements for $Z = 46$ to $Z = 52$ performed by Noreland.[61] It was shown that *L*-subshell binding energy values for these metals obtained by absorption measurements compared exceedingly well with X-ray emission band edge measurements as well as photoelectron spectrometric measurement determinations.

The recently developed absorption spectroscopy measurements using the "synchrotron light" continuum radiation from an electron synchrotron (see, e.g., Ref. 62; for a comprehensive treatment of synchrotron radiation as a light source, see the article by Godwin[62a]) has been used to great advantage in studying atomic electron binding energies in solids and gases, as well as autoionizing atomic excitation states in gases (Section 3.5). The sample (gas or thin film) is radiated by the synchrotron radiation which is then spectrographically analyzed. The sensitivity of the technique was demonstrated by the investigation of the L_2- and L_3-subshell binding energies of aluminum in the metallic and oxide forms. The L_2- and L_3-binding energies in aluminum metal were determinated to be 73.15(2) and 72.72(2) eV, respectively.[63]

X-ray and ultraviolet absorption measurements on elements in the gaseous or vapor state can provide accurate binding energy values of even the outer levels. Here solid state effects do not interfere and the binding energy of an atomic electron is defined as that energy required to remove it from the atom to infinity with a final zero kinetic energy. Resonance structure in the absorption spectrum indicates the electron excitation transition to unfilled bound orbitals as dictated by selection rules. Figure 1*a* of Chapter 6 shows the X-ray absorption ($1s \rightarrow np$, $n \geq 4$) spectrum for argon; the absorption process appears schematically in Figure 3*e* of this chapter. Figure 1*b* of Chapter 6 gives the absorption edge regions for the *L*-subshells of xenon. In this case the allowed final electron states are so close to each other that structure cannot be detected.

Similar measurements have been carried out on the *K*-shell in neon[64] and the *M*-shell of xenon[65] (the M_4- and M_5-edge energy positions were compared with results using the photoelectron method). Anomalies in the photoabsorption spectrum to the $M_{4,5}$-levels of some of the rare earth elements are covered in Refs. 66 and 67, Ref. 68 deals with the $N_{4,5}$ levels, and Ref. 69 treats photoionization defects. The *M*-absorption study of ytterbium has been reported by Combley et al.[70]

In analyzing the X-ray-absorption spectra for compounds in the gaseous state, molecular orbital calculations may be applied.[71]

Moore's[72] tables of "atomic energy states" contain valuable information on excited levels of neutral and multiply ionized atoms in the gaseous or vapor state. The energies of two or three of the outermost levels in elements that are most commonly investigated in the monatomic gaseous state (the rare gases) as taken from these tables are included in the electron binding energy table in Appendix F.

3.2. Photoelectron Measurements

Irradiation of a solid material with photons of an energy much in excess of the atomic level energy of interest may result in the transition of an electron from that level to the positive energy continuum. The process is essentially the same as that of X-ray absorption, except that the initially bound electron is raised to an energy state usually far above any bound states in the material (see Figure 3*c*). The fact that the final state lies in the positive energy continuum makes the method seem more straightforward and therefore simpler than the X-ray absorption method. This final positive energy of the electron can be measured by using a magnetic or electrostatic spectrometer. The bound atomic level energy in a metallic conductor can then be determined if the values of the photon energy and the spectrometer material work function are known. However, obtaining atomic electron binding energy values from photoelectron energy measurements can present difficulties for semiconductor and insulator photoconverters, as discussed in Section 3.2.1. There it is shown that it is hard to define a reference level for such converters with respect to which the atomic electron binding energies should best be referred. Because of the significance of this limitation a more detailed description of the method is given.

3.2.1. Spectrometer and Converter Work Function Considerations

A solid sample of the material (assumed here to be a conductor) under investigation (the converter) is exposed to X-radiation of an accurately known energy. The photoelectrons freed by the photoelectric process will have a kinetic energy given by

$$E_{\mathrm{kin,C}} = h\nu - E_B - \phi_c = E_F - \phi_c,$$

where $h\nu$ is the photon energy and E_B is the atomic electron binding energy. The quantity ϕ_c, the work function of the converter, is that amount of energy required to raise an electron from the Fermi level edge to just outside the converter bulk, where it has zero kinetic energy. Figure 4*a* illustrates the relations among the quantities involved. The Fermi levels of the metallic spectrometer components that are in electrical contact with one another are identical. When the electron leaves the vicinity of the converter, the zero energy level position (at a work function of energy above the Fermi level) will change slightly to refer to that for the spectrometer material. During the time that the energy analysis takes place, its kinetic energy is

$$E_{\mathrm{kin,sp}} = E_F - \phi_{\mathrm{sp}}, \tag{11}$$

which is the measured value. Since the binding energy of an atomic electron in a solid converter is defined as the energy necessary to raise it to the Fermi level, its value can be obtained from the photoelectron energy by

$$E_B = h\nu - E_F = h\nu - E_{\mathrm{kin,sp}} - \phi_{\mathrm{sp}}. \tag{12}$$

The foregoing discussion includes the assumption that all the inner spectrometer surfaces are made of the same metal and therefore have the same work function value. Also, the photoelectron energy is related to ϕ_c (i.e., it is in the vicinity of the converter material) only immediately after ejection from the converter, and therefore the effect of this work function on the analyzed energy is negligible.

It is necessary to make certain that the converter material (still assumed to be a metallic conductor) is in electrical contact with the spectrometer tank metal so that the Fermi levels coincide. Poor contact may result in the introduction of a difference in converter and tank Fermi levels and an error in the measurements.

A few comments should be made about the position and nature of the Fermi level in metals and nonmetals.[73–76] The conduction band of a conductor can be represented by a large number of quantum states distributed over an energy region unique for the metal in question. All the possible states in the band are not filled by electrons, and the ratio of the occupied states to the total number of available states, as a function of state energy $F(\epsilon)$ is called the Fermi function. The Fermi function can be calculated by using quantum mechanical considerations, and the expression obtained is

$$F(\epsilon) = \frac{1}{1 + \exp{(\epsilon - \epsilon_F)/kT}}, \tag{13}$$

where ϵ_F is the Fermi level energy, k is Boltzmann's constant, and T is the temperature, in degrees Kelvin. It can be seen that this function may vary between unity and zero. For $\epsilon \ll \epsilon_F$, $F(\epsilon)$ lies close to the maximum value of 1, for $\epsilon \gg \epsilon_F$, $F(\epsilon)$ is near zero. The energy ϵ_F is defined as the inflection point in the distribution equation 13. At absolute zero temperature all states of energies $\epsilon < \epsilon_F$ are filled, and those of energies $\epsilon > \epsilon_F$ are empty. Therefore, the lowest electron vacancy in a metal is taken as being just at the Fermi level (i.e., at the energy at which half the states are filled). A bound atomic electron then will be able to leave its original atomic level if it is given enough energy to transfer to the Fermi level. From these considerations, we see that the binding energy of an atomic electron is the energy difference between the atomic and Fermi levels. And again, the work function

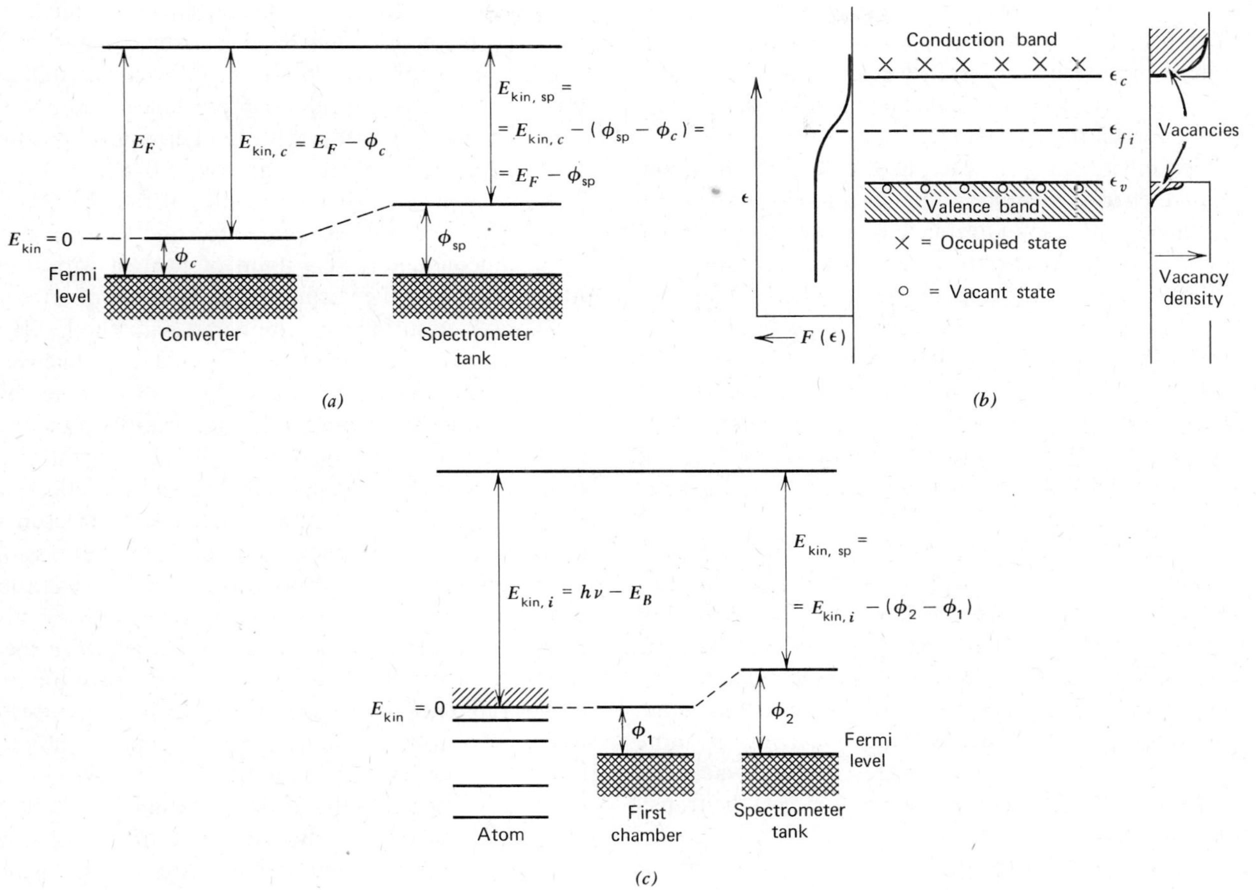

Figure 4. The relations between the various quantities involved in the photoelectron measurement of atomic electron binding energies. (*a*) The situation for a solid metallic photoconverter having a work function of ϕ_c and a spectrometer metal of work function ϕ_{sp}. (*b*) The outermost energy levels in a semiconductor: for the pure semiconductor, the intrinsic Fermi level is defined as midway between the conduction and valence bands (see text, Section 3.2.1). (*c*) The relation between the various quantities of importance in the case of a gaseous photoconverter. Here it is assumed that the first, or ionization chamber, and the spectrometer tank are constructed of metals having different work functions (Section 3.2.1).

for a metal is that energy required to remove an electron from the Fermi level to a position just outside the material bulk, its final kinetic energy being zero.

In the case of semiconductors and insulators, the situation is quite different. Here the conduction and valence bands are separated by an energy gap so large that the probability is small that an electron will be thermally excited from the valence band across the gap and into the conduction band. This probability increases with increasing temperature and with decreasing gap energy.

For pure semiconductors it is often convenient to define the Fermi level as lying at the so-called intrinsic Fermi level ϵ_{fi}. Its position is defined once more as the inflection point in the Fermi function that *spans the forbidden gap* (see Figure 4*b*). If the number of vacancies (holes) in the valence band is N_v and the number of electrons in the conduction band is N_c, then the electron concentration in the conduction band is[74]

$$n = N_c \exp\left(-\frac{\epsilon_c - \epsilon_{fi}}{kT}\right) \text{ electrons/m}^3,$$

and the vacancy or hole concentration in the valence band is

$$p = N_v \exp\left(-\frac{\epsilon_{fi} - \epsilon_v}{kT}\right) \text{ holes/m}^3.$$

At the intrinsic Fermi level we have

$$N_c \exp\left(-\frac{\epsilon_c - \epsilon_{fi}}{kT}\right) = N_v \exp\left(-\frac{\epsilon_{fi} - \epsilon_v}{kT}\right)$$

$$= \sqrt{N_c N_v} \exp\left(-\frac{\epsilon_c - \epsilon_v}{2kT}\right). \quad (14)$$

In a pure semiconductor $p = n$ and $N_c = N_v$, and ϵ_{fi} will fall at the middle of the energy gap; in the case of doped semiconductors, the hole and electron concentrations are not necessarily equal, and the intrinsic Fermi level position will then no longer lie exactly midway between the two bands.

The intrinsic Fermi level, as a definition, is a convenient energy reference point. Its relation to the binding energy of a given atomic electron or to the Fermi level for a given spectrometer material, however, is uncertain. Since the foregoing definition has the intrinsic Fermi level lying in the forbidden energy gap between the valence and conduction bands, it cannot possibly represent the lowest energy vacancy in the semiconductor. More accurately, the probabilities of electron transitions from bound atomic levels to the lowest energy states are represented on the right-hand side of Figure 4*b*.

For low temperatures or large forbidden gap, as for insulators, the lowest vacancy will lie in the conduction band and the valence band will be filled. For higher temperatures or decreasing gap, more and more holes will be created in the valence band because electrons are being transferred to the conduction band. These holes in the valence band will compete more strongly, as their number increases, with the vacancies in the conduction band, as the final level of the transition electron.

The position of the vacancy of lowest energy is strongly influenced by surface impurities. The surface band structure in a material is very sensitive, for example, to thin layers of oxide that are expected to form on the more active metals for even fairly high vacua (10^{-6} torr) and to absorption of and reaction with gases in the imperfect vacuum. Thin layers of vacuum pump oil and other impurities can also influence the characteristics of a converter surface. Care must therefore be taken to check and control the purity of surfaces of converters or, in general, any samples under investigation.

The value of the work function at the surface of a metal may vary as much as 0.4 eV in going from a metal-vacuum to a metal-air interface.[77] This change is large compared with that of ~0.03 eV estimated due to gas absorption.[77,78]

Equation 12 will allow us to use photoelectron energies to obtain the energy difference between the Fermi level of the spectrometer tank and the original bound atomic level. The position of the intrinsic Fermi level or the bottom of the conduction band is not in general known with respect to the Fermi level of a metallic holder in contact with the semiconductor. It has been shown experimentally[79] that as a metallic conductor becomes a semiconductor through oxidation, the valence band moves a smaller energy distance below then the conduction band is displaced above the conductor's Fermi level position. This implies that the intrinsic Fermi level in the semiconductor will not coincide in energy position with the Fermi level position in the original metal (see, e.g., Refs. 80–82).

Another problem is the bending of the Fermi level of the conductor, or the conduction and valence bands in a semiconductor, at a point of contact with other material or with the vacuum. This bending of levels at the converter surface, as compared with the levels in the bulk material, might be expected to influence a photoelectron spectrum depending on whether the electrons originate from the surface or bulk of the converter.[39] This matter has not yet been investigated thoroughly. Bergwall[80] has considered these effects, as well as the influences of doping on the band structure of semiconductors; his article contains further discussion.

The binding energy shifts with change in oxidation state of a metal (copper and its oxides,[83] see also Section 4.1.2) first determined by photoelectron measurement, were attributed to a shift in the zero binding energy reference position from the Fermi level position in the metal to the intrinsic Fermi level position in the semiconductor. However, there are two *K*-binding energies for the two sulfur atoms in sodium thiosulfate[84] and two or three *K*-binding energies for two or three nitrogen atoms in certain organic compounds,[85] which can be measured with respect to the same energy scale by the photoelectron spectrometric method in a single experiment. This proves that the *K*-shell binding energy is related to the oxidation state of the sulfur atom and the nitrogen atom, respectively (see Section 4.1.2). The importance of knowledge of the intrinsic Fermi level position in photoelectron determinations of binding energies and their chemical shifts is unclear and needs investigation.

Experimental data on the *K*-binding energy in nitrogen as a function of total charge q[85] (see Section 4.1.1, equation 30) indicate that equation 12 gives atomic level binding energy values that vary smoothly with the total charge of the element in question for nearly all compounds investigated. That is, the energy of an atomic level referred to the Fermi level of the spectrometer is directly related to the binding energy in question, which in turn has a strong first-order correlation with the total charge of that atom.

The situation is much different when the photoelectrons originate from atoms in the gaseous state. Figure 4*c* is an energy diagram of the case of a spectrometer divided into two chambers of different materials. The first chamber is the irradiation chamber and the second is the energy analysis chamber (i.e., the spectrometer tank). The ratio of the pressures in the two chambers usually lies between 10^{-2} and 10^{-4}, the greater pressure being where the ionization takes place. The energy required to just free the atom of an electron from the level of interest (i.e., the binding energy of that electron in the free atom) will raise the electron to the zero kinetic energy level, which coincides with the level at energy ϕ_1 above the Fermi level in the first chamber. Photoelectrons of initial energy $E_{\text{kin},i}$ will be analyzed as having a kinetic energy $E_{\text{kin},i} - (\phi_2 - \phi_1)$, where ϕ_1 and ϕ_2 are the work functions of the first chamber and the spectrometer tank material, respectively[60]. The binding energy of an atomic level in a gaseous atom irradiated with X-rays of energy $h\nu$ becomes

$$E_B = h\nu - E_{\text{kin},i} = h\nu - E_{\text{kin,sp}} - (\phi_2 - \phi_1). \quad (15)$$

If the first chamber and the spectrometer are made of the same metal, the work function is constant throughout the apparatus and $E_B = h\nu - E_{\text{kin,sp}}$, provided gas adsorption on the chamber walls is the same in the two chambers. In any case, as pointed out previously, gas adsorption effects will probably be small.

One strong advantage of gaseous targets over solid targets is that in the former case there is no converter work function to be considered. An atomic electron binding energy is measured with respect to the energy required to remove the bound electron from the atom to infinity, where its kinetic energy is zero.

3.2.2. *Photoelectron Line Position Determination*

As is shown in Chapter 6, the ideal photoelectron line shape (from a vanishingly thin source) will result from combined contributions of the Lorentzian X-ray intensity distribution, the atomic level binding energy function (also Lorentzian), and the spectrometer window function, which is usually non-Lorentzian. These contributions folded together will result in an intensity distribution whose energy is taken as the value at which its intensity is maximum. The photoelectron energy spectrum from a converter will also show evidence of electron straggling. The magnitude of this contribution and its effect on the photoelectron line shape, and therewith its peak position, will vary with photoelectron energy and converter thickness, as well as level widths and plasma loss energy values, and, of course, with the instrumental window function.

The best electron peak position must be one that is least dependent on electron source thickness effects on the line shape and one that can be determined accurately and conveniently.

In 1960 de Vries[86] considered this problem and proposed six possible line profile position definitions:

1. The center of gravity of the profile.
2. The center of the horizontal chord at half-height.
3. The intersection of the locus of midpoints of horizontal chords of the upper part of the profile with its top.
4. The intersection of the linearly extrapolated sides of the line profile.
5. The center of the high energy rise of the line profile.
6. The high energy extrapolated end point.

It can be seen that definitions 1, 2, 5, and 6 require a more exact determination of the electron background intensity than do the other definitions. The necessary measuring time may often be extensive, and instrumental drifts can result in reduced measurement precision. Definitions 1, 2, 4, and, possibly, 3 give line position values strongly dependent on electron straggling effects on the line shape. This dependence, furthermore, is different for different line energy values. Also, source thickness will influence definitions 1 and 2 more than 3 and 4, whereas, we would expect definitions 6 and probably 5 to be influenced by the natural line width.

We are therefore lead to consider only 3 and 4. Definition 4 was adopted by de Vries for his internal conversion electron line position measurements, since it was judged that definition 3 would be too sensitive to the intensities of the few measured points at the line peak. However, definition 3 is that commonly used in photoelectron line measurements. Figure 5 reveals that there is a slight difference in peak position of the N_2-subshell photoelectron line for uranium, using copper $K\alpha_1$ X-radiation, as defined by 3 and by 4; the difference here amounts to about 0.5 eV. This line position difference for the two definitions results from the presence of the line component lying toward lower energies from the peak.

An electron line shape, then, will change as a function of electron source thickness because of electron straggling. Sokolowski[87] investigated the dependence of this shift of photoelectron line position due to the presence of a close-lying electron energy loss peak. The K-photoelectron line from copper using molybdenum $K\alpha_1$ radiation (lying at a kinetic energy of $\sim$8.5 keV) was analyzed into the primary and straggling peaks. The peaks are in a ratio of 1 to 0.6, respectively. A separation of $\sim$20 eV and a width ratio of 1 : 2 of the former to the latter showed that the shift in energy of the copper K(molybdenum $K\alpha_1$) peak was $\Delta E = -0.24$ eV. When using high instrumental resolution (here $\Delta B\rho/B\rho \sim 0.5 \times 10^{-3}$) and investigating

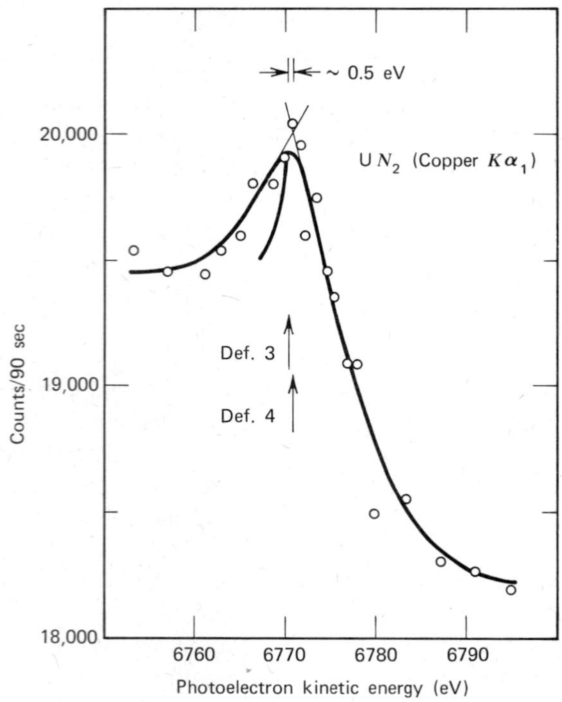

Figure 5. The photoelectron line from conversion of copper $K\alpha_1$ X-rays in the N_2-subshell of uranium. The line positions as defined by definitions 3 and 4 in the text are shown. Because of the asymmetry of the line, these positions do not coincide. (From Ref. 89.)

electron lines of narrow widths (X-ray energy distribution width ~8 eV, copper K-level width ~1.4 eV) compared to the discrete energy loss magnitudes, the loss peak intensity contribution has little effect on the photoelectron line position (see also Section 3.3).

Nordling[88] considered the problem more thoroughly later and arrived at the following expression for the energy shift of a photoelectron peak position Δx in units of the primary peak half-width:

$$\Delta x \approx -\frac{I x_0 \omega^2}{(x_0{}^2 + \omega^2)^2}. \tag{16}$$

Here it is assumed that $\Delta x \ll 1$, and I and ω are the height and the half-width of the loss peak, respectively, expressed in units of the corresponding quantities for the primary peak; x_0 is the primary peak to loss peak distance in units of primary peak half-width. This shift—Δx—is that of the primary maximum.

Internal conversion or photoelectron lines involving atomic electron levels having large natural widths may be so broad that the energy loss peak becomes integrated into the whole. Chapter 6 showed that when this occurs, electron line peak positions may be altered appreciably. (See Figure 3 and related text in Chapter 6.)

3.2.3. *Photoelectron Line Energy Determination*

An electron line position can be determined with the procedures given in Section 3.2.2. The absolute energy of an electron line, then, is taken to be the energy value of the above-defined line position, and this value can be obtained generally by using one of two different methods.

A common method of ascertaining an absolute electron line energy is to compare its spectral position with that of a standard electron line of known energy value. For an iron-free magnetic spectrometer, in the ideal case, the electron line momentum is proportional to the spectrometer magnetic field strength, which is in turn proportional to the magnetic coil current. (External fields are assumed to be effectively compensated for by Helmholtz coils placed about the instrument.) Knowledge of the calibration constant, the ratio of accepted electron momentum to the coil current, will make it possible to determine the $B\rho$-values of measured lines. This constant can be determined from the spectrometer current values for lines of accurately known momenta.

In principle, the calibration constant can also be determined by measuring directly the magnetic field value B and the radius of the central orbit ρ_0. However, determination of these quantities to one part in 10^4 and better is impractical; furthermore, their values will vary slightly between measurements because of uncertainties in positioning of the source, baffles, and counter slit. Calibration can more easily be effected by measuring the momentum values of investigated spectral lines and of standard lines of known momenta in the same measurement run. This can be done by interchanging these two electron sources.[89,90] The uncertainties of this method are reduced by the accurate interchange of the sources with respect to position, orientation, and so on. It is this method which is used most often in low energy photoelectron measurements, the standard line being either a photoelectron or an Auger line.

An alternative method can be used in photoelectron measurements if the material to be investigated is deposited in a thin or not fully continuous layer onto a backing of a previously investigated metal. Standard lines and lines of yet unknown momenta can thereby be obtained in the same spectrum and the momenta of the latter determined.

In using iron-core spectrometers, the calibration constant may vary greatly between runs because of residual magnetic fields in the spectrometer iron. The previously described calibration can be performed however (i.e., by interchanging a standard source with the investigated source during a given run).

Absolute energy determinations of electron lines made using electrostatic single- or double-focusing spectrometers can be carried out in a manner analogous to that employed when an iron-free magnetic spectrometer is used. For a solid conducting source, the energy of the transmitted electrons is proportional to the electrostatic potential applied to the spectrometer

$$E_{\text{kin}} = f \cdot U_{\text{sp}}. \tag{17}$$

Unlike the magnetic spectrometer, in which the selected electron momentum is proportional to the coil current, the proportionality factor f varies with energy setting because of the relativistic increase in electron mass. For low kinetic energies (usually $E_{\text{kin}} \gtrsim 1$ keV) the factor f alters by an insignificant amount compared to the accuracy of spectral energy measurements. For higher energies, it may be necessary to consider the change in f.

Again, an electron line of unknown energy may be determined with an electrostatic spectrometer by recording its spectral position and that of a standard line in the same run. With gaseous electron sources, charging of the gas through photoionization and possible electron straggling in the gas may cause a considerable shift in emitted electron energy. The problem is especially strong for electron impact ionization. The degree of charging will depend on the target gas pressure and on the photon intensity. The energy measured will then be decreased by an amount proportional to this charging. Both effects can be corrected for by measuring the electron line energy as a function of the target gas pressure and extrapolating to zero pressure.[45]

A second and often more convenient method of determining absolute electron line energy values using an iron-free magnetic spectrometer takes advantage of the nonlinear dependence of electron kinetic energy on its momentum. The binding energy difference between the two levels, which must be known, can be obtained from X-ray data or from atomic electron binding energy tables. Siegbahn[91] showed that, if two such electron lines have the momentum values $(B\rho)_{\text{I}}$ and $(B\rho)_{\text{II}}$ and the energy difference between the two atomic levels is ΔE, the absolute momentum values of the two lines can be found.

The relativistic equation relating the kinetic energy E of an electron to its $B\rho$-value may be written

$$E = m_o c^2 \left[\sqrt{(e/m_o c)^2 (B\rho)^2 + 1} - 1\right], \tag{18}$$

where m_o, e, and c are the electron's rest mass, the electron's charge, and the velocity of light, respectively. Two photoelectrons or internal conversion electrons expelled by conversion of the same quantity of energy will have the energy difference

$$\Delta E = E_2 - E_1 = m_o c^2 \left[\sqrt{(e/m_o c)^2 (B\rho)_2{}^2 + 1} - \sqrt{(e/m_o c)^2 (B\rho)_1{}^2 + 1}\right]. \tag{19}$$

Substituting $\alpha = (B\rho)_2/(B\rho)_1$ and solving for $(B\rho)_1{}^2$, we have

$$(B\rho)_1{}^2 = A + \sqrt{A^2 + B} \tag{20}$$

where

$$A = \frac{\alpha^2 + 1}{(\alpha^2 - 1)^2} \left(\frac{1}{ce}\right)^2 (\Delta E)^2, \tag{21}$$

$$B = \frac{4 - s^2}{(\alpha^2 - 1)^2} \frac{m_o{}^2}{e^4} (\Delta E)^2. \tag{22}$$

$$s = \frac{\Delta E}{m_o c^2}. \tag{23}$$

The quantity α is determined by the ratio of the coil currents for the two line positions. The momentum determinations increase in accuracy with increasing value of this ratio.

Carlson et al.[92] employed a generalized variation of this method to determine the energy of a gamma-ray of interest by measuring the energy values of the M_{1-5}, N_{1-5}, O_{1-3}, and P_1 conversion lines. The approximate measured $B\rho$-values for the lines were converted to energy values and the difference between pairs of these, multiplied with a factor $C \neq 1$ to account for a calibration discrepancy, were set equal to the accurate difference between corresponding tabulated binding energy values obtained from X-ray work. Solving for the value of C gave the best agreement between the two sets of values. The gamma-ray energy could then be determined with good accuracy.

3.2.4. *Low Energy Photoelectron Measurements*

Low energy photoelectron emissions, involving molecular orbitals and the band structure in solid compounds have been measured extensively by Spicer, Shay, and co-workers[93] and by Jaklevič and Juenker.[94] (See also Ref. 95.) These studies usually extend over electron emission energies from 6 to 12 eV. Since surface states are of interest here, pressures of less than 10^{-9} torr are employed to ensure against surface contamination. From the experimental data it is possible to determine the energies of many states lying between about 4 eV below to some 10 eV above the top of the valence band.

The energy distribution of photoelectrons and secondary electrons ejected from solids by low energy X-rays has been investigated by Nakhodin and Mel'nik.[96]

Low energy or ultraviolet (usually the helium resonance radiation at 21.21 eV) photoelectron spectra from gaseous atoms are of course quite similar in structure and nature to those obtained using higher energy X-rays. Spectra from molecules, however, will be further altered or complicated, since molecular[97] instead of atomic electron orbitals are present, and because of the creation of vibrational excitation states through the ionization process. Actually, the structure of the photoelectron spectrum for a given gaseous molecule will be unique for that molecule. This idea lies at base for the analysis method called molecular photoelectron spectroscopy.[98] See the recent book by Turner, et al.[98a] on this subject. Photoelectron spectrometric investigations in the electron-volt region have been employed[99,39,100] to measure vibrational levels in H_2^+. Ionization potentials,[101] binding energies associated with electron orbitals in gaseous molecules,[102] and angular distributions of photoelectrons from atoms and molecules[97,103] have also been the subjects of such research.

3.3. Internal Conversion Electron Measurements

The transition energy between two nuclear states may contribute to the expulsion of a bound electron from the atom. Emitted conversion electron energies can be analyzed by the same procedures and using the same spectrum analysis instrumentation as for the photoelectron case. Also applicable are the considerations in Sections 3.2.1, 3.2.2, and 3.2.3 concerning the instrumental work function and the electron line position and absolute energy.

Aside from electron straggling effects in the source, measured internal conversion line widths result almost entirely from the atomic level natural width and the spectrometer window contributions. This is in opposition to the photoelectron case, where the X-radiation itself can contribute strongly to the total electron line width.

In order to use conversion electron measurements to determine atomic electron binding energies, the associated nuclear transition energy must first be determined with good accuracy. This may be done by an independent method either directly through the crystal diffraction measurement of the gamma-ray energy or indirectly by the calculation of the energy from internal conversion line energy measurements involving an atomic level of previously known binding energy.

Internal conversion electron measurements can also help to ascertain a gamma-transition energy. This energy may then be used to determine the binding energies for other atomic levels in which internal conversion takes place.[104]

Consider an A, B, C sequence in nuclear levels (A of highest energy), and that both crossover A → C and cascade A → B → C transitions occur. The difference in energies for internal conversion electrons from a given atomic level for the A → C and B → C transitions will give the A → B absolute gamma-transition energy. Internal conversion lines of this gamma-transition can then be measured and the spectrum calibrated, according to Section 3.2. This procedure allowed Hollander et al.[105] to determine the binding energies of some nine atomic levels in berkelium ($Z = 97$).

The precision measurement of internal conversion spectra can be carried out effectively for high energy transitions when magnetic instruments of high resolving power are used. Atomic electron binding energy differences can be measured if the gamma-transition energy is not known accurately. Absolute conversion line momentum values can then be obtained from momentum ratio values for the conversion lines from the *K*-shell and one of the *L*-subshells as explained in Section 3.2.3.

Jahn et al.[106] measured the *L*-conversion lines of the several transitions in ^{192}Pt. The $\pi\sqrt{13/2}$ iron-free spectrometer used gave a momentum resolving power of 0.048%. Energy calibration was carried out according to the last method cited in Section 3.2.3. The $B\rho$-values for the K-, L_1-, and L_3-conversion lines of the 296- and 316-keV transitions computed in this way agreed well with one another and with other published values, but the L_2/K line momentum ratios showed strong discrepancies. The L_2-conversion line values were then used to calculate the L_2-electron binding energy in platinum, which was compared with the value listed in the table of Hagström et al.[57] The L_2-binding energy thus obtained was a mean value of 15(3) eV larger than tabulated values. This surprising result was reinvestigated by Karlsson et al.[107] using a $\pi\sqrt{2}$ iron-free instrument. The same internal conversion lines were measured with a slightly higher instrumental resolution. Here, however, no discrepancy was found in the experimentally determined binding energies compared with the tabulated values of Bearden and Burr.[55,55]

Karlsson et al.[108] reinvestigated and determined some 17 of the platinum atomic level energies using photoelectron measurement methods. Again, the *L*-subshell binding energy values agreed with tabulated values.

The energy or momentum positions of internal conversion lines are represented by their peak positions as defined in Section 3.2.2. For low energies, lines from internal conversion in narrow atomic levels can stand quite free of the characteristic energy loss structure arising from straggling in the source (see Figure 14, Chapter 3). For broader atomic levels, where the energy loss structure is integrated into the conversion line profile, correction for a resulting line shift may be called for (see Figure 3, Chapter 6).

For higher energies, $E \gtrsim 25$ keV, where contributions due to natural level width and characteristic energy losses from single plasmon excitation are small small compared to the spectrometer window, the line position is well represented by the line peak (according to Section 3.2.2) if straggling effects are small. However, straggling of conversion electrons in thick sources can result in a shift of the line position to lower energy values.

de Vries[86] demonstrated that the momentum position of the *K*-158 keV conversion line in ^{199}Hg shifted by 1.7×10^{-4} when the source thickness was increased by a factor of at least 25 from the thin source value of $<50\ \mu g/\text{cm}^2$. Figure 6 shows the *K*-conversion line profile measured for two source thicknesses. The shift in line position amounts to some 20 eV, or 2.7×10^{-4} of the kinetic energy value of the line. Note also that the shift in energy would be larger by $\sim 50\%$ if the line position definition 3 in Section 3.2.2 were used instead of definition 4.

3.4. Auger Electron Measurements

As brought forth in Sections 5.1.1 and 5.2.1 of Chapter 3, an atomic electron binding energy may be determined from Auger and Coster–Kronig transition energy measurements, provided the energy of the final two-vacancy atomic state is known. In terms of energy, then

$$E[X_i Y_j Z_k(^{2S+1}L_J)] = E(X_i) - E[Y_j Z_k(^{2S+1}L_J)], \tag{24}$$

where X_i is the initial vacancy and Y_j and Z_k are the final ones. The symbols *L*, *S*, and *J* are the quantum numbers of the final doubly ionized atomic state. The formula represents Coster–Kronig transitions if

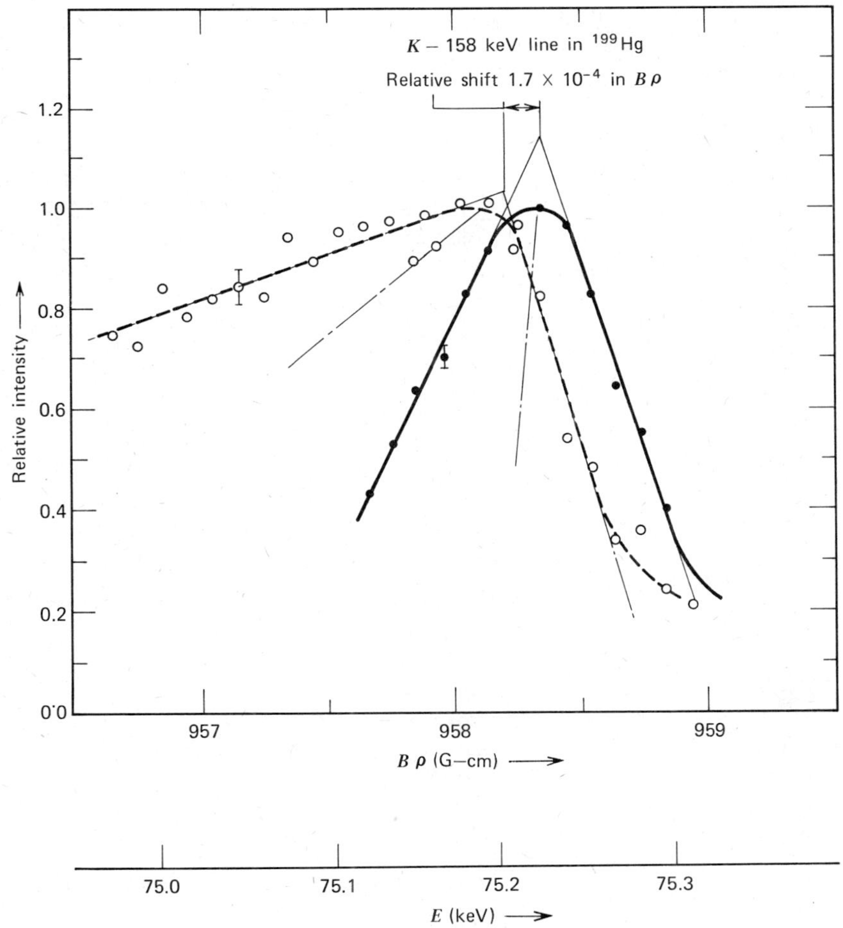

Figure 6. The *K*-conversion line of the 158-keV transition in ^{199}Hg for thin source (solid curves) and thick sources (dashed curves). The ratio of thicknesses is at least 25. See Section 3.3 for further explanation. (From Ref. 86.)

X_i and Y_j are in the same main shell and Auger transitions if they are not.

In order to determine $E(X_i)$, we must know the energy of the final state of the atom, that is,

$$E[Y_j Z_k(^{2S+1}L_j)].$$

These are not generally known for inner atomic subshell vacancies, but they are available for gaseous and vaporized atoms in the tables by Moore.[72]

Körber and Mehlhorn were able to employ this method with good success in determining the binding energy of the K-shell in neon.[109] This value was some 3 eV higher than the value tabulated at that time, although it agreed well with X-ray absorption and emission measurement values published earlier by other workers.

Later Mehlhorn determined the binding energy values of the L_1-subshell[45] and the L_2- and L_3-subshells[46] of argon and the M_1-, M_2-, and M_3-subshells[110] of krypton. These energy values were great improvements over values previously obtained.

If the final double ionized state energy is not known, binding energy differences can be obtained by measuring the energy difference of two Auger transitions having the same double vacancy final state.

3.5. Autoionizing Level Measurements

According to Section 6 of Chapter 3, autoionizing atomic excitation states are created when an atom in the gaseous or vapor state is excited into a state of energy higher than the first ionization limit. In single-electron excitation of this type, an inner shell electron of a single atom is excited to a level in a series that converges to an ionization limit. The situation is similar to that represented in Figure 1*e* (for $n < \infty$ final states). When a de-exciting autoionization transition takes place, the loosely bound electron may or may not participate. Also, this method is applicable only for atoms in the gaseous or vaporized states.

Photon absorption excitation of a given series of autoionizing levels can be utilized to obtain the original atomic binding energy of the electron involved. This was done, for example, by Codling and Madden[111] for the M_4- and M_5-levels in krypton and N_4- and N_5-levels in xenon. With excitation by synchrotron radiation, they measured the following absorption structures:

$$3d^{10}4s^24p^6(^1S_0) \rightarrow 3d^94s^24p^6(^2D_{3/2,5/2})np(^1P_1{}^o)$$

and

$$4d^{10}5s^25p^6(^1S_0) \rightarrow 4d^95s^25p^6(^2D_{3/2,5/2})np(^1P_1{}^o),$$

respectively, for the two gases. Each series is double, terminating in final $^2D_{3/5}$ and $^2D_{5/2}$ ion states. The $M_{4,5}$ and $N_{4,5}$ binding energies for the gases were found by calculating the limit excitation energy as $n \rightarrow \infty$.

3.6. Inelastic Electron Scattering Measurements

3.6.1. *Measurements in Solids*

Energy losses suffered by electrons scattered from a solid element or compound reflect, among other types of excitation, the excitation of a bound electron to a vacant level directly above the Fermi level. The energy loss of the scattered electron is then a measure of the electron binding energy difference between the initial and final electron states (Figure 1*d*). The excitation process is closely analogous to that for X-ray absorption (Figure 1*b*).

Inelastic scattering experiments were carried out during the 1940s by a number of workers;[112–114] mostly determinations of limited accuracy of the K-shell binding energies in light elements were made. Renewed interest arose in the early 1960s in connection with the investigation of plasma excitation electron energy losses in metals and their compounds.

Using the reflection technique in which electrons are scattered through 90° and measuring the energy loss spectrum with a 127° electrostatic spectrometer, Robins and Swan[115] interpreted some of the measured energy loss structure as excitation of M_1, $M_{2,3}$, or $M_{4,5}$ bound electrons to unoccupied states just above the Fermi level. They found that such structure was present in the energy loss spectra for all the elements from titanium to copper. The excitation of bound electrons in some of the $M_{4,5}$, N_1, $N_{2,3}$, $O_{4,5}$, and P_1 atomic levels of elements from gallium to lead,[116,117] from the M_1 and $M_{2,3}$ levels in potassium and calcium, and O_1, $O_{2,3}$, in barium was also observed.[118]

In all these articles the binding energy for the shell ionized was taken as the energy difference between the primary electron peak and the peak of the loss structure corresponding to the ionization. More recently, Lynch and Swan[119] measured the electron energy loss spectrum for several metals from $Z = 39$ to 47 and from $Z = 72$ to 79. For the first group, energy loss structure resulting from N_1 and $N_{2,3}$ level electron excitation was determined. For the second group, O_1 and $O_{2,3}$ electron excitation was measured. Here the inflection point of the loss structure was taken as the loss position, in analogy to the X-ray absorption edge position (see Chapter 6, Section 2).

Some time later, however, Hartley[120] showed that the intensity minimum directly preceding the loss structure is most suitable for determining the loss energy and thereby the corresponding binding energy (see Chapter 8, Section 2.4). He also presented binding energy values thus obtained for one or two subshells in a score of elements. These values are entered in Appendix F.

The electron energy loss edge due to excitation of $L_{2,3}$-level electrons from aluminum was nicely demonstrated by Swanson and Powell.[121] The binding energy value thus obtained, 72.5(5) eV, agreed well with other electron energy loss and from X-ray-absorption values. See Section 3.1.2 for the values obtained by X-ray absorption measurements.

3.6.2. *Measurements in Gases*

It was pointed out in Chapter 3, Section 6, that much can be learned about the nature of doubly excited autoionization levels in gaseous atoms through inelastic scattering of electrons in gases. Of course, energy losses due to single-electron excitation can also be investigated (e.g., see the electron energy loss spectrum for helium resulting from single and double electron excitation, Figure 22, Chapter 3). The binding energy of an outer shell electron can be determined from the energy limit of the associated Rydberg series of energy losses for the excitation of that electron. Several electron scattering measurements of excitation levels in such series for monatomic gases have been reported: hydrogen,[122,123] helium,[123–126] neon, argon, krypton, xenon,[127] and mercury.[128]

Inelastic electron scattering is especially useful in determining molecular levels of electron excitation.[129] However, for diatomic and more complicated molecules the loss spectra will also show signs of the excitation of vibrational states and, if the resolution is sufficiently high, traces of rotational structure. Investigation of the following gaseous molecules has been reported: H_2,[126,130,131] H_2O,[126] methane,[126] N_2,[123,131,132] ethane,[126,133] CO_2,[134] N_2O,[134] CO,[123,135] ethylene,[125,126] cyclohexane,[127] and benzene.[125] See also the work by Spohr et al.,[136] Boersch et al.,[137] Geiger and Wittmaack,[138] Chen,[139] Bühring,[140] Takayanagi and Geltman,[141] Crawford,[142] and Siegbahn et al.[39]

Information on ion dissociation may also be obtained from inelastic electron scattering measurements.[143]

4. ALTERATIONS OF ELECTRON BINDING ENERGY VALUES

In Section 2 we said that atomic electron binding energy values can be calculated to an accuracy dependent on the reality of the Hamiltonian operator employed (see equation 5). For lower Z-value gaseous atoms, the accuracy of correspondence between calculated binding energy values, as well as Auger energy values, and experimentally obtained values has sometimes been quite good. Agreement between theoretical values and experimental values obtained using elements in the solid state has been less satisfactory.

An atomic electron binding energy is primarily dependent on the magnitude of the Coulomb interaction with the nuclear charge. Secondary contributions arise from electron-electron interactions and effects related to the chemical state and physical environment of the atom.

Screening of the nuclear electrostatic field is experienced by an electron in, say, the M-shell, because of the atomic orbitals lying closer to the nucleus (i.e., of mainly the K- and L-shells). Classically, electrons of the M-shell, for example, are considered to lie upon an M-sphere about the nucleus, and in the first approximation these will not affect a K-electron, for example, as long as it is inside the M-shell. But as it passes the M-shell (e.g., in photoionization), it suddenly experiences a change in the *amount* of central charge—the number of M-shell electrons—and immediately the electron's potential energy decreases accordingly. Similar changes in potential energy already have been experienced upon passing the L-shell and will be experienced when passing the outer lying shells in the atom. These changes in potential energy are owing to successive increases in screening of the nuclear charge upon passing each electron shell. For an isolated atom, then, the electron will experience after its leaving the atom, an attraction due to the electrostatic interaction between itself and the remaining singly charged ion. The energy required to remove an electron in this way from its initial atomic state to infinity is defined as the binding energy.

In view of the foregoing statements, consider a singly ionized isolated atom in its lowest energy state (i.e., an outer shell electron is missing.) The amount of energy necessary to lift an inner shell electron to the atom's periphery will be essentially the same as for the neutral atom (there will be a small difference), but the energy to remove it farther to infinity will increase by a factor of two. This is because the atom is finally doubly instead of singly ionized. The difference in ionization energy between the two cases will be some 6–15 eV, depending on the atom. In Chapter 5, Section 3.2, we saw that high degrees of ionization will result in increases in binding energy of some 100 to 300 eV.

For atoms in the solid state, we expect chemical or physical alterations of the electron density in the outer shells to alter the effective ionicity of the atom in question. Binding energy alterations due to changes in chemical state are treated in Sections 4.1.1 and 4.1.2. The electron binding energy variations resulting from changes in physical state have been compared earlier (Table I and accompanying text).

However, the nuclear charge screening function increases continuously with distance from the nucleus, since the charges of the orbital electrons are distributed over appreciable ranges. Therefore, the screening value experienced by a given orbital electron depends on the "position of the electron" at the moment of ejection. It is meaningless from the quantum mechanical point of view to attempt to determine this so-called position. However, it is possible to ascertain the screening constant of an atomic shell as defined as that related to the radial distance expectation value associated with the interaction of interest in the measurement (e.g., spin-orbit or hyperfine interactions, binding energies). A rough attempt to learn the screening values related to internal conversion coefficients for the M-subshells for $Z = 65$ and 51-keV nuclear transitions for various multipolarities is presented in Chapter 4, Section 3.1. Further considerations are given in Section 4.1 of this chapter.

In addition to inner atomic shell electrons contributing to screening constants, we note a slight screening increase with increasing atomic number, for a given inner shell. This change is caused by the outer shell electrons that are added in going to heavier elements. Their wave functions, which have a small but finite probability of being near the nucleus, thus contribute to the screening of the inner shell electrons. Electrons in levels lying outside a level of interest will contribute by factors 10^{-1} to $<10^{-4}$ as strongly to the screening as will the inner electrons, depending on the level's atomic location and associated wave function. With regard to this last point, s-shell electrons will be expected to contribute most strongly, since their wave functions are finite at the origin.

We expect that chemical changes in X-ray energies, for transitions involving the innermost shells, will result from screening changes due to valence shell configuration alterations (see Section 4.1.1). Also treated in Section 4.1.1 is the X-ray-absorption technique for determining binding energies and changes in these in solids with change in chemical state; included in Table III of Section 4.1.1 is a comparison of changes in K and L_3 binding energies and $K\alpha_1$ X-ray energies for copper and aluminum and their oxides. Here the electron is lifted from its initial atomic level to the positive energy continuum or to unfilled levels at and above the Fermi level, for the respective cases. The interpretation of the spectra is somewhat difficult, since knowledge of the kinds of levels is required.

Chemical changes in observed photoelectron spectra are treated in Section 4.1.2. Such changes in Auger[144] and discrete electron energy loss spectra[145,146] have also been reported. Theoretical calculations regarding photoelectron and Auger electron energy shifts have been published as well.[39,147]

Finally, in Section 4.2, we consider other phenomena that affect binding energies or their measurements.

Appendix D presents a comparison of the natures and accuracies of several methods of chemical analysis: nuclear magnetic resonance, photoelectron spectrometry, X-ray emission and absorption methods, and others.

4.1. Atomic Screening of the Nuclear Charge

An electron in an interior atomic orbital will experience an electrostatic potential determined by a nuclear charge that is screened to some extent by the other atomic electrons. Each innermost electron decreases the effective nuclear charge as experienced by an outer electron by essentially one unit. Electrons in levels lying outside the level of interest also contribute, although slightly, to the screening, as explained in Section 4.

Slater[148] has given a rule of thumb for determining screening constant σ for a given atomic electron. That electron will then experience a Coulomb field caused by an effective central charge of $Z - \sigma$. The rules were based on the results of calculations made using simple analytic wave functions for the atomic electrons. These wave functions were nodeless, but otherwise similar to the hydrogen wave functions.

Similar calculations have been presented more recently (e.g., by Clementi and Raimondi[149] in connection with their work on self-consistent-field binding energy calculations for elements up to $Z = 36$, and by Burns[150]). Burns obtained his results for both hydrogen and single-exponential wave functions. Here the screening factor σ was obtained by directly comparing the simple analytic wave functions with Hartree–Fock wave functions that had been calculated previously. Burns compared the expectation values, or the moments $\langle r^q \rangle$. The effective nuclear charge $Z - \sigma$ can be obtained by calculating the relation between $Z - \sigma$ and $\langle r^q \rangle$ for the particular form of the wave function used. The value $Z - \sigma$ was calculated for the two wave functions, the analytical and the Hartree–Fock, for $q = -3$ to $+4$, and an average value of σ determined. A recipe for σ, similar to Slater's rules, was devised. (See also ref. 41a.)

In the work just referred to, screening factors were obtained by comparing the average values of calculated moments using hydrogenic or exponential wave functions with those obtained using Hartree–Fock wave functions. But of course, we would expect the screening factor to take on a range of values depending on which moment it is related to—this has been demonstrated in calculations by Fraga and Malli[151] using Hartree–Fock functions. These authors, together with their co-workers, have then used the same expectation values, or moments, to calculate various physical quantities: e.g., diamagnetic susceptibilities involving $\langle r^2 \rangle$, the nuclear magnetic shielding constant involving $\langle r^{-1} \rangle$, spin orbit coupling involving $\langle r^{-3} \rangle$, and so on. Such physical quantities may be said to be related to weighting functions of the form r^2, r^{-1}, r^{-3}, and so on, and therefore, qualitatively speaking, they are related to screening factors associated with the corresponding expectation values. For a given atomic or hyperfine interaction, the associated screening factor will depend on how that interaction varies with r.

From another viewpoint, the magnitude of the interaction of interest calculated for only the electron or electrons involved using hydrogenic wave functions differs from that value obtained from a Hartree–Fock calculation depending on the amount of shielding of the nuclear charge by the other atomic electrons. The size of the screening constant will depend on how the interaction varies with r.

In Table II*a* are listed several electron-electron and hyperfine interactions that vary with r in different ways and involve values of $\langle r^q \rangle$, with q taking various values.

In Table II*b* are shown screening constant values $\sigma(M_{2,3})$ for atoms in the 57 to 71 Z-value region. These values are divided into three groups: (a) and

Table IIa.

Expressions for various physical interaction quantities.

Interaction	Expression	References
Spin-orbit	$\xi_\ell = \frac{\alpha^2}{2}\left\{Z\langle r^{-3}\rangle - \sum_i \int_0^\infty \frac{1}{r^3} R_\ell^2(r) r^2 dr \int_0^r R_{\ell_i}^2(r_i) r_i^2 dr_i\right\}$	153) 154)
Magnetic hyperfine	$a_d = \left(\frac{16}{15}\right)\mu_B \mu_N (\mu_I/I)\langle r^{-3}\rangle$	154) 155)
Electric field gradient at the nucleus caused by the ith electron	$q = e\langle \sum_i (3\cos^3\theta_i - 1)\rangle\langle r_i^{-3}\rangle$	156) 157)
Diamagnetic suseptibility	$\chi_d = -\frac{N_A}{6}\alpha^2\langle r^2\rangle$	154) 158)
Nuclear magnetic screening	$S = \frac{N_A}{3}\alpha^2\langle r^{-1}\rangle$	154) 159)

Also see for nuclear quadrupole coupling constants[156)], spin-spin interactions[160) 161)], spin-other-orbit interactions[162)], octopole hyperfine interaction constants[163)], orbit-orbit interactions[164) 165)]. α is the fine structure constant, μ_B the Bohr magneton, μ_N the nuclear magneton, R_{ℓ_i} the radial wavefunction for the ith electron of angular momentum quantum number ℓ, e the electron charge, N_A Avogadro's number.

Table IIb.

$M_{2,3}$ subshell screening constants for various measurable atomic and nuclear-atomic interactions.

	Author(s)	$\sigma(M_{2,3})$	Z-value
a)	O'Connell and Carroll[166]	30-35	57-71
b)	Mande and Damle[167]	29-33	" "
c)	Slater[148]	14.75	" "
d)	Burns[150] (hydrogenic)	13.05	" "
	" (exponential)	12.12	" "
e)	Fahlman and Hagström[20]	8.3	" "
f)	Chu and Perlman[168]	7.9	53

(b); (c) and (d); (e) and (f). The second group contains the values obtained using Slater's and Burns's rules, already referred to. The values for the other two groups were obtained by adjusting theoretically calculated binding energy values (a) and (b), and from comparing relative internal conversion line intensities with theoretical predictions for the 81.78 and 88.46 keV $M4$ transitions in ^{121m}Te—(e). The value (f) was obtained by comparing the experimental $M_{2,3}$ spin-orbit splitting binding energy difference with equation 6 and solving for the screening constant. (See also Ref. 152, where screening constants were calculated for the K- and L_1-shells associated with electron binding energy $\langle r \rangle$ and $\langle r^2 \rangle$.) The screening constant values vary strongly from the one group to the next because of the nature of the interaction governing the measurable quantity involved. That the values (e) and (f) agree is merely fortuitous, since they involve entirely different types of interaction. (See also Chapter 4, Section 3.1 for a discussion of screening in internal conversion.) From Table II*b* we see that the screening correction necessary for a given physical atomic quantity depends on the kind of interaction involved and its variation with r.

Of course, with the advent of self-consistent-field calculations of atomic quantities, consideration of screening constants is not necessary. The preceding discussion, however, serves to point up the several characteristics of screening factors.

4.1.1. X-Ray Absorption and Emission Energy Dependences on the Atomic Valence State

Valence electron contributions to a change in the K X-ray absorption edge energy was first measured some 40 years ago by Lindh[169,170] and Stelling.[171] Measurements were made on phosphorus, sulfur, and chlorine in a number of chemical compounds. Research on the effect was soon taken up by other workers[172,173] and extended to X-ray emission energy dependences on chemical state for these and other elements.

It was found that for chlorine, for example, the K-shell binding energy altered by as much as 8 eV in comparing the values for the chloride (valence -1) with that for the perchlorate (valence $+7$). The observed X-ray-absorption spectra for the K-shell were complex, sometimes showing two components; but the energy value of the first K-absorption edge was clearly determinable, and its shift with valence state was unquestionable. Note, however, that the absorption edge energy will be dependent upon the initial and final state energies for the transition electron. The value of the latter will depend on the quantum nature and availability of possible final states as well as on transition selection rules. The natures and energies of these final states might be expected in some cases to change significantly with change in the chemical state of the element.

Similar results were obtained for sulfur and phosphorus. Three absorption edges were observed for the thiosulfates: in sodium thiosulfate, for example, each molecule contains two sulfur atoms, one of valence -2, the other $+6$, and the K-absorption spectrum was attributed to the superimposed contributions of absorption in these two atoms. The origin of the third edge, however, is not clear.

In the $K\alpha$ X-ray emission spectrum for sodium thiosulfate, however, two doublets lie at approximately a doublet's distance from each other. Each doublet corresponds uniquely to one of the two oxidation states of the sulfurs. It should also be noted that the $K\alpha$ energy positions for the two sulfurs in sodium thiosulfate differ slightly from the respective positions of the $K\alpha$ X-rays from a metal sulfide and a metal sulfate.[174] This shows that the difference between two atomic electron binding energies is not only dependent on the valence state of the atom, but also on the nature of the binding partner. (References 174 and 175 provide further discussion of this effect.)

In 1935 Parratt[176] carried out an interesting investigation of the $K\alpha_{1,2}$ X-ray spectrum of sulfur in a number of compounds. Initial K-vacancies in the sulfur atoms were produced by X-ray irradiation, and the resulting $K\alpha_{1,2}$ X-ray spectrum was measured with regard to line positions, line widths, and the $K\alpha_1$–$K\alpha_2$ line separation. Parratt found that some of the sulfate salts were reduced chemically during extended irradiation to sulfides and that the $K\alpha_{1,2}$ spectrum altered accordingly. The shift in the X-ray line energies and the changes in their widths were clearly visible.

In 1924 Tandberg[177] reported an investigation of the L X-ray absorption edges for iodine in various compounds. Binding energy shifts of up to +14 eV were found, with respect to the corresponding edge for the element. (See also Refs. 178, 55, 179.)

Since the 1930s a large number of articles relating to X-ray spectroscopy have treated or touched on the the dependence of atomic electron binding energies on the chemical state of the atom.[180–186] Once the effect had been established, other workers began to suggest applying it to the acquisition of information about the valence state of the atom in question. In order to do this, however, the nature of the correlation between the atomic binding energy and the valence state of the atom must be investigated.

With advances in the precision of X-ray spectroscopy measurements, it has become possible to study with increasing accuracy the much smaller variation of X-ray emission energies with valence state,[187] which will be essentially the difference in chemical shift of the binding energies of the two states involved in the transition. An estimate of the magnitude of atomic electron binding energy and X-ray energy shifts as a function of valence state can be made by calculating the screening of inner electrons by the valence electron. The probability P that a valence $n's$-electron is closer to the nucleus than an ns-electron will be[188]

$$\mathrm{P} = \int_0^\infty \left[R_{ns}{}^2(r) \int_0^r R^2_{n's}(r')\, dr' \right] dr, \tag{25}$$

where $R_{ns}(r)\, dr$ and $R_{n's}(r')\, dr'$ are the probabilities per unit time of finding the ns and $n's$ electrons in the increment dr at r and dr' at r', respectively.

If $\Delta E(Z, Z-1)$ is the energy difference in the binding energy of the inner shell in question for two elements differing by a unit nuclear charge, the change in binding energy of the electrons in that level will be[189]

$$\Delta E_{\mathrm{B.E.}} = \mathrm{P}\, \Delta E\, (Z, Z-1), \tag{26}$$

where P is given in equation 25.

If a valence electron is removed entirely from the atom, such as in the cases of a monovalent compound with 100% ionicity or the removal of the outermost electron from the isolated atom, then equation 26 will give the change in binding energy of the ns-electron.

Coulson and Zauli[187] reported in 1963 results of calculations of the sulfur $K\alpha$ X-ray doublet energy dependence on the chemical state of the atom. It was shown that the $K\alpha_{1,2}$ energies are dependent on the number of electrons in the outer (M) shell and their associated screening of the nuclear charge.

Lindgren[190] has since performed calculations on the change in binding energy of an inner level electron of an isolated sulfur atom for progressively increasing degrees of ionization. The Hartree–Fock–Slater method was used, and the energy shifts for the K- and L-shells were expressed in shift per degree of ionization. It was noted that the binding energy alteration is nearly proportional to the degree of ionization, and that the K- and L-shifts are of comparable magnitude. Also calculated were the shifts in K-shell binding energies (for elements from $Z = 10$–36) resulting from a vacancy arising in one of the various outer atomic shells. The K X-ray energy shifts were calculated in a corresponding manner, their magnitudes being a factor of 10 smaller than the K binding energy shifts. Gianturco and Coulson[191] have recently carried out similar calculations for sulfur.

More realistically, assuming that the bond is intermediate between the purely ionic and covalent cases, a compound of valence m will have m electrons, each of ionicity i, involved in the binding. We obtain, instead of equation 26,

$$\Delta E_{\mathrm{B.E.}} = im\, \mathrm{P}\, \Delta E\, (z, z-1). \tag{27}$$

In a similar manner, Lindgren's[190] energy shift calculations for sulfur can be made to fit experimental curves by relating the ionic state to the valence state, not neglecting to consider the ionicity.

Equation 25 shows that P is dependent on the overlapping of the n' (valence) and n (inner) electrons, which is in turn closely related to the nature of the valence electron wave function (i.e., on its l-quantum number in particular).

Deodhar[192] has shown by X-ray-absorption measurements that the L_3-edge for gold and platinum shifts -12 and -18 eV in going from the metal to its complexes: $NaAuCl_4$, $KAuCl_4$, $K_2Pt(CN)_6$, and K_2PtCl_6 Also, it has been shown by Fadley et al.[51,52] by Hartree–Fock calculations that in going from one oxidation state to another all the halogens' inner-core levels shift by the same amount to within $\pm 2\%$.

X-Ray-absorption measurements on the L_3-shell of lanthanum in compounds with sulfur, selenium, and tellurium, and on the $L_{2,3}$-level of chlorine in compounds with alkali metals, have been made by Russian research groups.[193]

The work of Codling and Madden,[63] referred to earlier, has demonstrated that the energy of the L_3-absorption edge for aluminum increased by 3.76(10) eV in going from the metal to the oxide, in agreement with other workers' results. Deslattes[194] has found that the *K*-absorption edge shift for aluminum showed a *K*-binding energy increase of 5.7(2) eV upon oxidation to the amorphous alumina (Al_2O_3) form. Hence the chemical shift of the *K*-edge is ~1.9 eV greater than the L_3-edge. We would then expect the energy of the $K\alpha_1$ X-ray for aluminum metal to be less than that for the oxide by about the same amount. However, Nordfors[195] has determined that the shift is only 0.27(5) eV in the expected direction (see Section 4.1.2 and Table III). Similarly, Swanson and Codling[196] used electron bombardment to study the beryllium *K*-shell binding energy in beryllium metal and beryllium oxide. (Note that for absorption data, the final level occupied by the electron may also undergo a shift due to change in chemical state. The comparisons involving X-ray absorption data are therefore to be considered of qualitative value. The comparison of $K\alpha$, X-ray emission energy and *K* and L_3 photoelectron binding energy determinations provide much more accurate energy shift data.)

A discrepancy of the same nature and of approximately this size in the case of copper and its oxide is treated in Section 4.1.2. (For a study of the effect of alloying with copper on the *K*-emission band of aluminum see Ref. 197.)

The shift of an X-radiation energy resulting from a change in chemical state of the emitting atom can be estimated by extending the foregoing treatment. From equation 27 we can obtain in a rough estimate form[198] the shift of the emission line relating to the transition filling a *K*-hole with an *L*-electron:

$$\Delta E_{K\alpha} \cong \Delta E_K - \Delta E_L$$
$$\cong im\{P_K E_K(Z, Z-1) - P_L E_L(Z, Z-1)\}. \quad (28)$$

Table III.

Comparison of X-ray emission, X-absorption, and photoelectron energies for the same shells in metals and their oxides.

	Binding energy shift (photoelectron method) (eV)		X-ray absorption (eV)		X-ray emission (eV)
	$E_B(K)$	$E_B(L_3)$	K-edge	L_3-edge	$K\alpha_1$
Al	--	--	0	0	0
Al_2O_3	--	--	5.7(3)[194)]	3.76(10)[63)]	0.27(5)[195)], 0.36(1)[a)]
			$\Delta E_K - \Delta E_{L_3} \cong 1.9$ eV		$\Delta K\alpha_1 = 0.27(5), 0.36(1)$ eV
Cu	0	0	--	--	0
CuO	4.4(5)[83)208)]	2.5(8)[83)208)]	--	--	0.11(4)[209)b)], 0.15(3)[209)b)]
	$\Delta E_K - \Delta E_{L_3} \cong 1.9$ eV				$\Delta K\alpha_1 = 0.11(4), 0.15(3)$ eV

a) K. Läuger, Thesis, Ludwig-Maximidians-Universität, München (1968).

b) Values obtained using two different methods.

The bracketed terms in equation 28 depend on the wave functions of the electrons taking part in the formation of the chemical bond and is closely related to the structure of this bond. Actually, other effects, such as the isotope shift, the isomer shift, and Doppler shifts associated with thermal motion and quantum emission recoil,[199,200] may also lead to shifts or distortions of emission X-radiation lines, thus obscuring shifts due to internal screening changes.

Petrovich et al.[189] investigated the relative importance of the above-mentioned effects relative to that of screening changes in producing *K* X-ray energy shifts. The $K\alpha_1$ lines for grey tin, and some of tin's tetravalent compounds (SnO_2, SnS_2, and $SnSe_2$) and divalent compounds (SnO, SnS, SnSe, SnTe) were investigated. The relative X-ray line shifts were plotted as a function of ionicity of the compounds in the crystalline structure form i_{cr}. The ionicity i_{cr} is related to the ionicity i of the diatomic molecular bond by the correction formula

$$(1 - i_{cr}) = (1 - i)\frac{m}{n},$$

where m is the valence and n is the number of nearest neighbors (assumed here to be equal to 6). The ionicity i of the diatomic molecular bond is[201]

$$i = 1 - e^{-0.25(\Delta\chi)^2}, \tag{29}$$

as derived from electric dipole moments of diatomic halogenic molecules, and $\Delta\chi$ is the difference between the electronegativities of the two members. The energy shifts involved lay between 0.07 and 0.2 eV. The measured shifts were approximately proportional to i_{cr}, the proportionality constant for the tetravalent series being the larger. It was concluded that the internal screening mechanism (equation 28) describes satisfactorily the chemical shifts of *K*- and *L*-levels in the heavier elements, for the influence of other effects is very small.

X-Ray shift measurements appear therefore to be a useful tool in investigating the nature of the chemical bond. This conclusion is further supported in similar investigations on tin and its oxides later reported by Gokhale et al.[202] and on fifth and sixth period elements and their compounds by Sumbaev et al.[203]

Alterations of X-ray energies, usually of $K\alpha$ radiation, with the chemical state of the emitting atom have been measured by many workers. Meisel et al.,[204–207] for example, have performed measurements on several elements in the $Z = 22$ to $Z = 28$ region both in the elemental state and in a variety of compound states. In addition to $K\alpha_1$-energy shifts, changes in the $K\alpha$-doublet energy difference and $K\alpha_1$-line widths and asymmetries were investigated. The $K\alpha_1$ X-ray shifts ranged as much as 1 eV from the metal position. Meisel and co-workers also found a splitting of the $K\alpha_1$ X-ray line itself from manganese in manganese fluoride into two components separated by ~0.5 eV[206] (see Section 4.2.1); the cause, however, was not interpreted as a screening effect.

Returning for a moment to the determination of the charge associated with an atom in a given compound, the total charge q will be the sum of the formal charge Q and the sum of the partial ionic bonds i (equation 29), shared by the atom:

$$q = Q + \sum i. \tag{30}$$

We then expect a change in an atomic electron binding energy with change in this total charge, due to altered screening of the nuclear charge. The relation between atomic binding energy and total charge, which has been investigated using photoelectron spectrometry methods, is a possible tool for chemical analysis (see Section 4.1.2).

4.1.2. Photoelectron Energy Dependences on the Atomic Valence State

The photoeffect as a process of atomic excitation can be considered an extension of the photoabsorption process for higher irradiating photon energies. The atomic electron in question becomes excited to a positive energy state, usually far above the Fermi level. The kinetic energy of the photoelectron is then measured as discussed in Section 3.2 and the energy of the atomic level is determined. Note, however, that the photoelectron method has the advantage of not being bothered with complications near "edges." The peak energy position is that to be measured.

Just as in the X-ray absorption case, we should expect to detect electron binding energy variations with atomic valence state by measuring the photoelectron energy. This effect was first measured by Nordling, Sokolowski, and Siegbahn[83,208] in copper and its oxides. Here the *K*-shell binding energy was found to shift some +4.4(5) eV in going from the metal to the CuO oxidized form; the L_3-binding energy changed as shown in Table III. For the change of copper to the Cu_2O oxidized state, the change in *K*-shell binding energy was +0.7(0.8) eV. Also, as mentioned in Chapter 6, Section 4.2, the widths of the *K*-, L_1-, and L_3-photoelectron lines increased in going from the metallic copper converter to the semiconductor CuO. However, a difference in the Fermi levels for the metal and the semiconcdutor (the CuO or Cu_2O) was interpreted as the cause of the binding energy change in going from the metal to the oxide (see Section 3.2.1). That is, the valence band was to have become more tightly bound by an amount equal to the increase in the K-level binding energy; the Fermi level was then defined as lying midway between the valence and conduction bands. This definition of the Fermi level position is conventional,[74,75] but, according to Section 3.2.1, it is not known to be valid.

Gilberg and Kern[209] measured the copper $K\alpha_1$ X-ray energy shift from the metal to the oxide and, using two different methods, found shifts toward increasing values (+0.11(4) and +0.15(3) eV). Considering the previously cited changes in K- and L_3-level energies, we would expect roughly an increase in X-ray energy of ΔK–ΔL_3 = 1.9(1.0) eV for Cu → CuO. Again, here, as in the case of Al → Al_2O_3 in Section 4.1.1, the relative change in K- and L_3-binding energies is much larger than the change in X-ray energy resulting from a transition involving the two levels. This would tend to indicate not only that the K- and L-shell electron binding energies are different in the two cases, but also that the position of the "Fermi level" of the converter is altered with respect to that of the metal components of the instrument. A correlation between these quantities is unclear (see Section 4.1 and Table III). However, in general chemical shifts of photoelectron and X-ray-absorption binding energy determinations lie in the lower end of the 1 to 10 eV energy range. The K- and L_3-shift differences also appear to be of this order. However, X-ray-emission energies shift 10^{-2} to 10^{-1} eV.

This photoelectron investigation of copper was followed up shortly by a similar investigation of the L-levels in tin and cadmium.[210] Shifts were observed between the metals and their oxides, ~ +1 eV for tin and ~ −0.7 eV for cadmium. The alloy SnAu was also investigated, but no change in energy of the L_3-level in tin compared with the metal was found.

It was not until 1964 that interest was renewed in chemical effects on atomic binding energies, with the discovery by Hagström et al.[84] that the K-photoelectron line for sulfur in sodium thiosulfate ($Na_2S_2O_4$) was a doublet. Here was clear evidence that the K-shell binding energies for the two sulfur atoms, of valences −2 and +6, differed by about 7 eV as a result of different degrees of nuclear charge screening by valence electrons. (A third component corresponding to that in Linde's X-ray-absorption measurement was not observed.) The investigation of the K-, L_1-, and $L_{2,3}$-binding energies in sulfur as a function of oxidation number, or valence, was soon performed and later reported (e.g., Ref. 211). In addition a difference in energy was found of ~4 eV of the sulfur KL_2L_3 ($^1D_2$) Auger line for the two sulfur atoms in sodium thiosulfate.[144,211]

The applications of the alteration of atomic electron binding energies with valence state of the atom were immediately obvious. Work immediately focused on investigating the feasibility of basing a chemical analysis method on this effect.

Such an application has been made by Fadley et al.[51,52] in the study of bonding of iodine in a number of compounds and of europium in the +2 and +3 oxide forms. The iodine compounds (and oxidation numbers) were KI(1−), KIO_3(5+), KIO_4(7+), and the potassium salt of p-iodobenzoic acid (uncertain). Photoelectrons, expelled by copper $K\alpha$ X-radiation, were analyzed with a $\pi\sqrt{2}$ iron-free spectrometer.

For a given element, the magnitudes of energy shifts between two oxidation states were essentially the same for all core levels investigated [including, in iodine, levels from $L_1(2s_{1/2})$ to $N_5(4d_{5/2})$]. Hence reference can be made to an "average core shift" between compounds of differing oxidation number. For iodine this average shift was 0.8 eV per unit change in oxidation number; that for europium was 9.6 eV.

Analysis of the experimental data was made considering two models:

1. For the europium case, the semiempirical model was used, in which the electronic charge participating in the bonding is removed from the valence shell to a spherical charge shell of larger radius r. The energy shift of a core level is then due to the resulting reduced shielding energy minus the repulsive potential of this spherical charge shell. From calculations of these quantities, a value of r was determined from the experimentally measured shifts; $r \cong 1.3$ Å, about half the europium–oxygen interatomic distance in Eu_2O_3.
2. The binding energy shift for iodine in the compounds investigated was analyzed using the method described at the end of Section 2. Solid state effects on binding energies—that is, the classical Coulomb interaction energy of an electron in the given ion with all other lattice ions—were determined from its relation to the Madelung constant. The iodine charges in KIO_3 and KIO_4 were calculated and compared with values from Mössbauer measurements.

Since photoelectron line widths contain an appreciable contribution from the irradiating X-radiation energy width, it is advantageous to use the lightest possible elements (instead of the copper $K\alpha$ or chromium $K\alpha$ radiation employed previously). Aluminum $K\alpha_{1,2}$ radiation lying at 1486.6 eV has a total width of about 1.0 eV, and that of magnesium is even smaller. The photoelectron spectra obtained employing these radiations will provide strong enhancements of spectral component resolution. Also, photoelectron lines of such low energies will stand free of discrete energy loss structure for high instrumental resolutions.

Krause and Wuilleumier[211a] have recently shown that the M_5N_3 (i.e. $M\zeta$) emission X-ray from ytterbium is intense, has a favorably low energy (132.29 eV), and is relatively narrow (<0.6 eV). The corresponding X-rays from zirconium, niobium, and molybdenum are similarly favored for future photoelectron work.

A recent investigation by Nordberg et al.[85,212] involved the change of the K-photoelectron line energy from nitrogen in a wide variety of chemical compounds. Aluminum and magnesium $K\alpha$ X-radiation was used, and the converter compound was either pressed into a copper mesh or stuck in powdered form onto a piece of cellulose adhesive tape. A plot was made of the K-shell binding energy as a function of total charge q (equation 30). The experimental binding energy values showed a smooth variation (over a range of about 10 eV) with charge q, except for some four out of about 30 points. This smooth variation seems to indicate that the spectrometer Fermi level is a suitable energy reference level for such measurements. The extension of this study to that of the K-shell binding energy for carbon in a large number of organic compounds was made soon afterward.[213]

A similar study has been carried out on sulfur in a large variety of organic and inorganic compounds. Again a smooth relationship is found between the binding energies of the $2p$-shell of sulfur and the associated charge q for the investigated compounds. The results of the nitrogen and sulfur investigations, as well as those of a number of other elements,[214,215] have been reported in an extensive and interesting publication by Siegbahn et al.[39] The investigation of free molecules is reported in Ref. 215a.

Similar K-shell binding energy measurements for nitrogen in a variety of compounds have been reported by Hollander et al.[216] An extensive study, using electron spectrometry, of the analytical chemistry of quaternary nitrogen has recently been reported by Jack.[216a] Several photoelectron spectrum studies for a variety of chemical compounds have also been reported by the Oak Ridge group.[216b]

We might expect that the cooperative investigation of chemical compounds as measured by the photoelectron, X-ray-emission and absorption, and other methods, would result in a thorough analysis of these compounds with regard to atomic and molecular orbital binding energies and characteristics with changing chemical state. Such an investigation (by X-ray and photoelectron methods) of several elements and their carbides for the $Z \cong 22$ region has been reported.[217] Considerations of what multitechnique investigations involving still other methods might reveal about the atomic and molecular electron orbitals in solids and core energy levels have been presented by Fadley and Shirley.[218]

The application of the relation between the level energies and the chemical state of an element has been investigated to establish a method of chemical analysis. The method has been given the name electron spectroscopy for chemical analysis (ESCA).

Both photo-ESCA and Auger-ESCA (the latter somewhat less suitably) may be employed in chemical investigations and analysis. Other methods are X-ray emission and absorption, which together with photo-ESCA are compared in Appendix D with regard to the nature of the physical processes involved and to their relative sensitivities as chemical analysis methods.

Finally, mention should be made of the X-ray-emission spectral investigation of the valence and conduction bands of semiconductor elements and their compounds in the solid state. Weich compared the $L_{2,3}$- and $K\beta$-emission bands for elemental silicon, the carbide, and the dioxide,[219] and the results were matched with calculations of the electronic band structure. Experiments were performed for phosphorous and several phosphide and phosphate compounds.[220] Selection rules governing the transitions require that $L_{2,3}$-emission bands involve s- and d-electrons in the valence band, whereas $K\beta$-bands must involve p-electrons. The comparison of the spectra will give the energy dependent level densities of the s- and p- and the d-electrons in the valence band. (See these articles for references to other work.)

Similar research has been done on sulfur and several of its compounds by Meisel and Szargan,[221] and on copper by Blokhin and Volkov.[222] (See also Ref. 223.)

The photoelectron investigation of the spin-polarization energy distribution in the electronic band structure in ferromagnetic compounds has been reported by Busch et al.[224] The sample was subjected to a strong magnetic field (33 kG), perpendicular to the converter film. Photoelectrons, ejected from EuS and EuSe samples at liquid helium temperatures, by 3 to 5 eV photons were found to have degrees of spin polarization between ~ 10 and 34%, the spin direction of the electrons having been conserved between emission and detection. A prominent peak in the spectra was attributed to transitions from the pure spin-up band, $4f^7$. These preliminary measurements indicate that such photoelectron sources may become important polarized electron sources in the future, if the degree of polarization can be improved.

4.2. Other Electron Binding Energy Alteration Effects

Aside from the previously discussed perturbation effects on atomic electron binding energies due to changes in the chemical and physical states of a compound or element, there are a number of other phenomena that affect binding energy values and influence their experimental determination. One of these was mentioned in passing in Section 4.1.1, namely, the splitting of the $K\alpha_1$ X-ray line from manganese in the fluoride salt. This effect is considered further in Section 4.2.1.

Alterations of X-ray energies due to other effects, such as isotope and isomer shifts and Doppler shifts caused by thermal motion, lie in the range of 0.1 eV and less. Only in recent years have X-radiation investigation techniques been improved sufficiently to allow a more quantitative consideration of these (see Refs. 199, 200, and other work cited there).

Second, it has been found that certain internal conversion electron transitions involving the inner atomic shells of the heavier atoms become changed in energy, reflecting an alteration in binding energy, when the radioactive atoms are placed in a strong electric field. The effect is not yet explained, but it has been studied, and several of these investigations are reviewed in Section 4.2.2.

4.2.1. Complex Structure of Kα K-ray Lines due to Excitation States in Solids

In 1965 Meisel[206] presented evidence for the splitting of the manganese $K\alpha_1$ X-ray line in manganese fluoride. This effect was observed in connection with the investigation of $K\alpha_{1,2}$ X-radiation line shape alterations with chemical binding. It was found that the manganese $K\alpha_1$ line not only became broad but actually consisted of two components. The energy difference between these was approximately 0.8 eV.

Later a similar splitting was observed by Nigavekar and Bergwall[225] for the chromium $K\alpha_1$ X-ray line in some of its chromous and chromic halides. The energy splitting of the line was measured to be 0.21(3) and 0.29(3) eV, respectively, for the measured compounds. That the effect is due to an exchange interaction with 3*d*-electrons (which is affected by chemical bonding) is supported by considerations by Nefedov[226] and by Isutsumi and Nakamori.[227] This interpretation is preferred over Parratt's postulation of excitation states in the solid[228] resulting in satellite structure in X-ray emission spectra. (See also Chapter 6, Section 4.2.)

Support of Parratt's modified level diagram[228] was presented earlier by Johansson[229] when he found that the $K\alpha_1$ X-ray profile and position for both metallic chromium and Cr_2O_3 varied as a function of the excitation voltage. The shifts were close to 0.3 eV for excitation voltages up to 1000 V over threshold. One is perhaps lead to wonder, on the basis of these results, if the aforementioned $K\alpha_1$ splitting structure for the halides of chromium would show an analogous variation in X-ray peak position and shape with excitation energy near threshold.

4.2.2. External Electric Field Effects on Inner Electron Shells

Certain internal conversion electron lines may shift in energy when the decaying atoms are placed in a strong electrostatic field. This effect was first found for internal conversion in ^{233}U following the decay of ^{233}Pa as measured by Novakov and Hollander.[230]

The activity was deposited onto the edge of a razor blade (radius of curvature $\sim 10^{-4}$ cm), which was grounded. At a distance of 5 mm and parallel to the blade was placed a grid, and a positive or negative voltage of 5 to 12 kV was applied to the grid. A second grid was placed parallel to the first one, again at a distance of 5 mm, and grounded. The electric field strength at the blade edge, for a potential of 10 kV, for the previously estimated radius of curvature of the edge and for the same geometry, was estimated to be 6×10^6 V/cm, if the edge is considered to be a wire of the same radius. Since the acceleration and deceleration potentials were identical, there was no net change in electron energy.

However, we expect changes in the focusing properties of the spectrometer due to the application of the potential. When the applied voltage is of the same order of magnitude as the electron energy, the intensity and half-width of the conversion line was increased for a negative voltage applied at the first grid. With the application of a positive voltage, the effects are reversed.

The 28 M_1 and 40 L_3 conversion lines from transitions in ^{233}U were measured for zero and negative potentials up to 10 kV. Since the energy difference between the line positions was only ~ 150 eV, instrumental effects should influence the lines in the same way. The M_1-line did not appear to change in shape on application of the potential, except for the expected increase in intensity; the L_3-line, however, developed a shoulder at about 45 eV below the peak energy. Similar low energy shoulders were found on the M_2- and M_3-conversion lines of the 28-keV transition, separated by about 50 and 100 eV, respectively, from the main peak. On the other hand, the 40-L_2-line showed no changes in shape aside from those caused by instrumental effects. Furthermore, the shoulder was unaffected by reversal of the electric field.

Observed properties of the shoulders are:

1. The conversion line energy becomes reduced upon application of the (positive or negative) potential (i.e., the binding energy of the associated atomic level increases).
2. The magnitude of shift increases from the $p_{1/2}$- to the $p_{3/2}$-subshells and from the L- to the M-shells. No shift has been observed for the $s_{1/2}$ subshell.
3. The direction of shift is not dependent on the sign of the applied voltage.
4. The shift magnitude increases more rapidly than linearly with applied voltage.
5. For a given subshell, a given field strength will cause a stronger effect in a heavy atom (e.g., uranium) than in a lighter atom (e.g., hafnium).

The investigation was soon after extended by Novakov and Janičijevič[231] to include the changes in the binding energy of the L_3-subshell electrons in ^{239}Pu caused by applying an external electrostatic field. Here the energies of the L_2- and L_3-subshell conversion lines of the 57.2-keV transition in plutonium following the beta-decay of ^{239}Np were measured. An electrostatic field of $\sim 10^7$ V/cm was applied and the shifts were measured for the oxide, hydroxide, and chloride forms of the element. The measurements showed that the satellite line accompanies the 57.2-L_3-conversion line, but not the 57.2-L_2-line. The satellite depends on the chemical state of the electron source; it is visible with the oxide and chloride, but not the hydroxide source. Figure 7 compares the L_2- and L_3-conversion line shapes for a chloride sample exposed to the electric field.

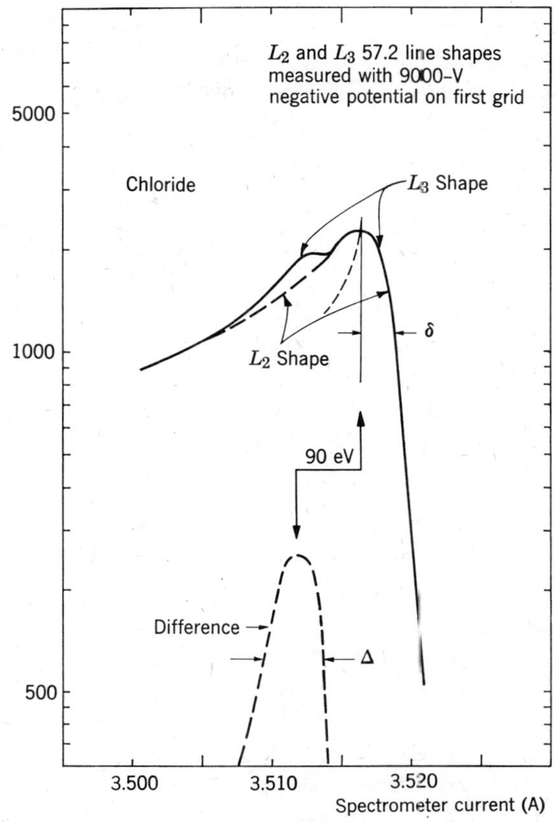

Figure 7. A comparison of the L_2- and L_3-57.2 line shapes of a plutonium chloride sample exposed to an electric field. The difference, representing the satellite, has a width corresponding to the instrumental width 2δ expected for a thin source. (From Ref. 231.)

Erman has recently reinvestigated effects of an imposed electrostatic field on ^{233}U atoms (following the decay of ^{233}Pa atoms) imbedded in aluminum oxide.[232] Beta decay of ^{233}Pa leads mainly to excited nuclear levels in the uranium isotope. The K-shell vacancies in uranium may be caused by internal conversion decay. The K X-rays following the K vacancy creation were measured by Erman. Exposure to a field of 7.5×10^6 V/cm showed no K X-ray energy shifts greater than ~ 1 eV (L_2), ~ 4 eV (L_3), ~ 15 eV (M_2, M_3), and ~ 5 eV (N_3). This is to be compared with the internal conversion results for uranium cited previously.

More recent results show that there exists also an atomic energy level splitting without an imposed electrostatic field. The photoelectron investigation of the $N_6(4f_{5/2})$, $N_7(4f_{7/2})$, $O_3(5p_{3/2})$, $O_4(5d_{3/2})$, and $O_5(5d_{5/2})$ levels in metallic converters of tantalum, platinum, gold, thorium, uranium, and plutonium, and some salts of thorium and uranium, has been reported by Novakov and Hollander.[233] This time the converters were not exposed to an electrostatic field gradient; the radiation used was aluminum $K\alpha$ or magnesium $K\alpha$ X-rays. Even here energy splittings were found, but only of the $O_3(5p_{3/2})$ subshell photoelectron line. The remaining photoelectron lines remained single. The magnitude of the splitting was obviously dependent on the chemical state of the atom: for uranium the splitting was largest for uranyl acetate (10 eV) and smallest for UO_2 (3 eV). Splitting of the $O_3(5p_{3/2})$ state was also found for thorium and plutonium. For the other three metals investigated, no evidence appeared of complex O_3-photoelectron line of structure.

According to the experimental results just cited, splitting or energy shift in the $p_{3/2}$-levels may arise under the combined influence of external and internal fields, or solely the latter. No accountable mechanism for these observations is certain now, but it is probable that core $p_{3/2}$-level splitting results from an effective electric field gradient in the interior of the atom. This in turn may result from alterations in symmetries of outer orbitals due to chemical bonding.

It appears that the cause of this splitting may be similar to that of the $K\alpha_1$ X-ray energy splitting treated in Section 4.2.1.

REFERENCES

1. A. E. Sandström, *Handbuch der Physik* (*Ed. S. Flügge*) **30** (1957) 78.
2. D. H. Tomboulian, *Handbuch der Physik* (*Ed. S. Flügge*) **30** (1957) 246.
3. W. Finkelnburg and T. Peters, *Handbuch der Physik* (*Ed. S. Flügge*) **28** (1957) 79.
4. B. Edlén, *Handbuch der Physik* (*Ed. S. Flügge*) **27** (1957) 80.
5. R. Jenkins and J. L. de Vries, *Practical X-Ray Spectrometry*, Philips Technical Library, Eindhoven, Holland, 1967.
6. J. A. R. Samson, *Techniques of Vacuum Ultraviolet Spectroscopy*, John Wiley & Sons, Inc., New York, 1967.
7. *X-Ray and Electron Methods of Analysis*, H. van Olphen and W. Parish (Eds.), *Progress in Analytical Chemistry*, Vol. 1, Plenum Press, New York, 1968.
8. P. A. M. Dirac, *The Principles of Quantum Mechanics*, 3rd ed., Oxford, Clarendon Press, 1947.
9. D. R. Hartree, *The Calculation of Atomic Structures*, John Wiley & Sons, Inc., New York, 1957.
10. J. C. Slater, *Quantum Theory of Atomic Structure*, McGraw-Hill Book Company, New York, 1960.
11. I. Lindgren, *Phys. Letters* **19** (1965) 382.
12. I. Lindgren, *Ark. Fys.* **31** (1966) 59.
13. H. E. Bethe and E. E. Salpeter, *Handbuch der Physik* (*Ed. S. Flügge*) **35** (1957) 88.
14. H. G. J. Moseley, *Phil Mag.* **26** (1913) 1024; *Phil. Mag.* **27** (1914) 703.
15. T. V. Kishnan and A. N. Nigam, *Current Sci.* **36** (1967) 260.
16. L. Pauling and S. Goudsmit, *The Structure of Line Spectra*, McGraw-Hill Book Company, New York, 1930.
17. A. Sommerfeld, *Naturwiss.* **28** (1940) 417.
18. S. Idei, *Rept. Tohuku Univ.* (*Japan*) **191** (1930) 641.
19. S. Hagström, *Z. Phys.* **178** (1964) 82.
20. A. Fahlman and S. Hagström, *Ark. Fys.* **27** (1964) 69.
21. P. Bergvall and S. Hagström, *Ark. Fys.* **17** (1960) 61.
22. D. Shalitin, *Phys. Rev.* **140** (1965) A1857.
23. D. Shalitin, *Phys. Rev.* **155** (1967) A20.
24. K. Yonei, *J. Phys. Soc. Japan* **22** (1967) 1127.
25. J. C. Slater, *Phys. Rev.* **81** (1951) 385.
26. W. Kohn and L. J. Sham, *Phys. Rev.* **140** (1965) A1133.
27. B. Y. Tong and L. J. Sham, *Phys. Rev.* **144** (1966) 1.
28. R. D. Cowan, A. C. Larson, D. Liberman, J. B. Mann, and J. Waber, *Phys. Rev.* **144** (1966) 5.
29. F. Herman and S. Skillman, *Atomic Structure Calculations*, Prentice-Hall, Inc., Englewood Cliffs, N.J., 1963.
30. D. Liberman, J. T. Waber, and D. T. Cromer, *Phys. Rev.* **137** (1965) A27.
31. C. W. Nestor, T. C. Tucker, T. A. Carlson, L. D. Roberts, F. B. Malik, and C. Froese, Oak Ridge Nat. Lab. Rept.-4027 (UC-32) (December 1966).
32. C. L. Tolliver, C. P. Bhalla, and W. R. Garrett, *Bull. Amer. Phys. Soc.* **11** (1966) 512.
33. J. L. Schonfelder, *Proc. Phys. Soc.* (*London*) **87** (1966) 163.
34. U. Gelius and L.-A. Nordqvist, University of Uppsala, Institute of Physics Report. UUIP-495 (August 1966).
35. F. C. Smith and W. R. Johnson, *Phys. Rev.* **160** (1967) 136.
36. C. Froese-Fischer, *Can. J. Phys.* **46** (1968) 2336.
37. T. C. Tucker, L. D. Roberts, C. W. Nestor, Jr., T. A. Carlson, and F. B. Malik, *Phys. Rev.* **174** (1968) 118.
38. P. S. Bagus, *Phys. Rev.* **139** (1965) A619.
39. K. Siegbahn, C. Nordling, A. Fahlman, R. Nordberg, K. Hamrin, J. Hedman, G. Johansson, T. Bergmark, S.-E. Karlsson, I. Lindgren, and B. Lindberg, *Nova Acta Reg. Soc. Sci. Upsal.*, *Ser. IV*, **20** (1967) Uppsala; Tech. Rept. AFML-TR-68-189, Wright-Patterson A.F.B., Dayton, Ohio (October 1968).
40. T. Koopmans, *Physica* **1** (1933) 104.
41. T. A. Carlson and B. P. Pullen, Oak Ridge Nat. Lab. Rept.-4393, (UC-34) 1969.

41a. T. A. Carlson, C. C. Lu, T. C. Tucker, C. W. Nestor, and F. B. Malik, Oak Ridge Nat Lab. Rept.-4614, (UC-34). 1970.

42. M. A. Coulthard, *Proc. Phys. Soc.* (*London*) **91** (1967) 44; **90** (1967) 615.

43. R. S. Mulliken, *J. Chim. Phys.* **46** (1949) 497.

44. R. Nordberg, K. Hamrin, A. Fahlman, C. Nordling, and K. Siegbahn, *Z. Phys.* **192** (1966) 462.

45. W. Mehlhorn, *Z. Phys.* **208** (1968) 1.

46. W. Mehlhorn, Habilitationschrift, University of Münster, February 1967; W. Mehlhorn and D. Stalherm, *Z. Phys.* **217** (1968) 294.

47. Private communication from M. A. Coulthard in Ref. 45.

48. S.-E. Karlsson, private communication, 1968, on work carried out in Berkeley.

49. W. Lotz, *J. Opt. Soc. Amer.* **58** (1968) 915.

50. W. Lotz, *J. Opt. Soc. Amer.* **58** (1968) 236.

51. C. S. Fadley, S. B. M. Hagström, J. M. Hollander, M. P. Klein, and D. A. Shirley, *Science* **157** (1967) 1571.

52. C. S. Fadley, S. B. M. Hagström, M. P. Klein, and D. A. Shirley, *J. Chem. Phys.* **48** (1968) 3779.

53. P. W. Anderson, *Phys. Rev. Letters* **20** (1468) 413; R. F. W. Bader and A. D. Bandrauk, *J. Chem. Phys.* **49** (1968) 1666; R. F. W. Bader, I. Keaveny, and P. E. Cade, *J. Chem. Phys.* **47** (1967) 3381; R. F. W. Bader and A. D. Bandrauk, *J. Chem. Phys.* **49** (1968) 1653; R. F. W. Bader, W. H. Henneker, and P. E. Cade, *J. Chem. Phys.* **46** (1967) 3341.

54. E. Gilberg, thesis, University of Munich, 1969, R. D. Deslattes and R. E. LaVilla, *Appl. Opt.* **6** (1967) 39.

55. M. Siegbahn, *Spektroskopie der Röntgenstrahlen*, Julius Springer-Verlag, Berlin, 1931.

56. J. C. Slater, *Phys. Rev.* **98** (1955) 1039.

57. S. Hagström, C. Nordling, and K. Siegbahn, in *Alpha-, Beta-, and Gamma-Ray Spectroscopy*, K. Siegbahn (Ed.), North-Holland Publishing Company, Vol. 1, 1965, p. 845.

58. J. A. Bearden and A. F. Burr, "Atomic Energy Levels," NYO 2543-1, 1965 (Fed. Sci. Tech. Inform., U.S. Dept. of Commerce, Springfield, Va. 22151).

59. J. A. Bearden, *Rev. Mod. Phys.* **39** (1967) 78; *Rev. Mod. Phys.* **39** (1967) 125.

60. E. R. Cohen and J. W. M. Du Mond, *Rev. Mod. Phys.* **37** (1965) 537.

61. E. Noreland, *Ark. Fys.* **26** (1964) 341.

62. K. Codling and R. P. Madden, *Appl. Opt.* **4** (1965) 1431; R. P. Madden, D. L. Ederer, and K. Codling, *Appl. Opt.* **6** (1967) 31.

62a. R. P. Godwin, *Springer Tracts Mod. Phys.* **51** (1969) 1.

63. K. Codling and R. P. Madden, *Phys. Rev.* **167** (1968) 587.

64. R. J. Liefield, *Appl. Phys. Letters* **7** (1965) 276.

65. R. D. Deslattes, *Phys. Rev. Letters* **20** (1968) 483.

66. H. F. Zandy, *Phys. Rev.* **162** (1967) 1; B. Feuerbacher, M. Skibowski, R. P. Godwin, and T. Sasaki, *J. Opt. Soc. Amer.* **58** (1968) 1434.

67. K. C. Williams, *Proc. Phys. Soc.* (*London*) **87** (1966) 983.

68. T. M. Zimkina, V. A. Fomichev, S. A. Gribovskii, and I. I. Zhukova, *Fiz. Tverd. Tela* **9** (1967) 1447 [*Sov. Phys.—Solid State* **9** (1967) 1128].

69. U. Fano and J. W. Cooper, *Rev. Mod. Phys.* **40** (1968) 441.

70. F. H. Combley, E. A. Stewardson, and J. E. Wilson, *J. Phys.* [*Proc. Phys. Soc.* (*London*)] **1B** (1968) 120.

71. G. L. Glen and C. G. Dodd, *J. Appl. Phys.* **39** (1968) 5372, 5377; W. Seka and H. P. Hanson, *J. Chem. Phys.* **50** (1968) 344.

72. C. E. Moore, "Atomic Energy Levels," Vol. II, *Circ. 467*, *Nat. Bur. Std.* (*U.S.*) Washington D.C. (1952).

73. C. Kittel, *Elementary Solid State Physics*, John Wiley & Sons, Inc., New York, 1962.

74. W. Wang, *Solid State Electronics*, McGraw-Hill Book Company, New York, 1966.

75. W. C. Dunlop, Jr., *An Introduction to Semiconductors*, John Wiley & Sons, Inc., New York, 1957.

76. F. Seitz, *Modern Theory of Solids*, McGraw-Hill Book Company, New York, 1940.

77. G. Nadjakov, V. Vassilev, and S. Balabanov, *C. R. Acad. Bulg. Sci.* **11** (1958) 461.

78. A. Eberhagen, *Forsch. Phys.* **8** (1960) 245; M. P. Hill and B. A. Pethica, *J. Chem. Phys.* **38** (1963) 567.

79. V. H. Sanner, thesis, University of Uppsala, 1941.

80. S. Bergwall, *Z. Phys.* **187** (1965) 495; D. Redfield, *Appl. Phys. Letters* **1** (1962) 9.

81. C. Mande, R. N. Patil, and A. S. Nigavekar, *Nature* **211** (1966) 518.

82. A. S. Nigavekar, thesis, Poona University, Poona, India.

83. C. Nordling, E. Sokolowski, and K. Siegbahn, *Ark. Fys.* **13** (1958) 483.

84. S. Hagström, C. Nordling, and K. Siegbahn, *Z. Phys.* **178** (1964) 439.

85. R. Nordberg, R. G. Albridge, T. Bergmark, U. Ericson, J. Hedman, C. Nordling, K. Siegbahn, and B. J. Lindberg, *Ark. Kemi* **28** (1968) 257.

86. C. de Vries, *Nucl. Phys.* **18** (1960) 428.

87. E. Sokolowski, *Ark. Fys.* **15** (1959) 1.

88. C. Nordling, *Ark. Fys.* **15** (1959) 397.

89. C. Nordling and S. Hagström, *Ark. Fys.* **15** (1959) 431.

90. S. Hagström, and S.-E. Karlsson, *Ark. Fys.* **26** (1964) 451.

91. K. Siegbahn, *Ark. Mat. Astr. Fys.* **28B** No. 6 (1941).

92. T. A. Carlson, P. Erman, and K. Fransson, *Nucl. Phys.* **A111** (1968) 371.

93. W. E. Spicer, *Phys. Rev.* **125** (1962) 1297; J. L. Shay and W. E. Spicer, *Phys. Rev.* **161** (1967) 799; *Phys. Rev.* **169** (1968) 650; *Phys. Rev.* **175** (1968) 741; W. E. Spicer and R. C. Eden, *9th Int. Conf. Physics of Semiconductors*, Leningrad, 1968.

94. R. C. Jaklevič and D. W. Juenker, *J. Appl. Phys.* **33** (1962) 562.

95. J. W. Dewdney, *Phys. Rev.* **125** (1962) 399.

96. N. G. Nakhodkin and P. V. Mel'nik, *Sov. Phys.—Solid State* 9 (1967) 567.

97. J. Berkowitz and H. Ehrhardt, *Phys. Letters* **21** (1966) 531.

98. D. W. Turner, M. I. Al-Joboury, *J. Chem. Phys.* **37** (1962) 3007; M. I. Al-Joboury and D. W. Turner, *J. Chem. Soc.* (1963) 5141; *J. Chem. Soc.* (1964) 4434; M. I. Al-Joboury, D. P. May, D. W. Turner, *J. Chem. Soc.* (1965) 616; D. W. Turner and D. P. May, *J. Chem. Phys.* **45** (1966) 471; *J. Chem. Phys.* **46** (1967) 1156; M. I. Al-Joboury, and D. W. Turner, *J. Chem. Soc.* B (1967) 373; C. R. Brundle and D. W. Turner, *Chem. Commun.* (1967) 314; D. W. Turner, *Nature* **213** (1967) 795.

98a. D. W. Turner, A. D. Baker, C. Baker, and C. R. Brundle, *Molecular Photoelectron Spectroscopy*, Wiley, New York, 1970.

99. D. C. Frost, C. A. McDowell, and D. A. Vroom, *Phys. Rev. Letters* **15** (1965) 612.

100. R. Spohr and E. von Puttkamer, *Z. Naturforsch.* **22a** (1967) 705.

101. D. C. Frost, C. A. McDowell, and D. A. Vroom, *Can. J. Chem.* **45** (1967) 1343; *Chem. Phys. Letters* **1** (1967) 93.

102. D. C. Frost, C. A. McDowell, and D. A. Vroom, *J. Chem. Phys.* **46** (1967) 4255; *Proc. Roy. Soc. (London)* **A296** (1967) 566; D. C. Frost, C. A. McDowell, J. S. Sandhu, and D. A. Vroom, *J. Chem. Phys.* **46** (1967) 2008; W. C. Price, *Endeavor*, May 1967; R. I. Schoen, *J. Chem. Phys.* **40** (1964) 1830.

103. J. Berkowitz, H. Ehrhardt, and T. Tekaat, *Z. Phys.* **200** (1967) 69.

104. G. T. Ewan, J. S. Geiger, R. L. Graham, and D. R. MacKenzie, *Can. J. Phys.* **37** (1959) 174.

105. J. M. Hollander, M. D. Holtz, T. Novakov, and R. L. Graham, *Ark. Fys.* **28** (1965) 375.

106. P. Jahn, W. Todt, and H. Daniel, *Phys. Letters* **20** (1966) 395.

107. S.-E. Karlsson, S. Högberg, and Ö. Nilsson, *Phys. Letters* **24B** (1967) 148.

108. S.-E. Karlsson, C.-H. Norberg, Ö. Nilsson, S. Högberg, A. H. El-Farrash, C. Nordling, and K. Siegbahn, University of Uppsala, Physics Institute Rept. UUIP-553 (July 1967).

109. H. Körber and W. Mehlhorn, *Z. Phys.* **191** (1966) 217.

110. W. Mehlhorn, *Z. Phys.* **187** (1965) 21.

111. K. Codling and R. P. Madden, *Phys. Rev. Letters* **12** (1964) 106.

112. G. Ruthemann, *Naturwiss.* **30** (1942) 145.

113. G. Ruthemann, *Ann. Phys. (Leipzig)* **6** (1948) 135.

114. J. Hillier and R. F. Baker, *J. Appl. Phys.* **15** (1944) 663.

115. J. L. Robins and J. B. Swan, *Proc. Phys. Soc. (London)* **76** (1960) 857.

116. J. L. Robins, *Proc. Phys. Soc. (London)* **79** (1962) 119.

117. J. L. Robins, *Proc. Phys. Soc. (London)* **78** (1961) 1177.

118. J. L. Robins and P. E. Best, *Proc. Phys. Soc. (London)* **79** (1962) 110.

119. M. J. Lynch and J. B. Swan, *Austral. J. Phys.* **20** (1967) 345.

120. B. M. Hartley, *Phys. status Solidi* **31** (1969) 259.

121. N. Swanson and C. J. Powell, *Phys. Rev.* **167** (1968) 592.

122. H. Boersch, J. Geiger, and H.-J. Reich, *Z. Phys.* **161** (1961) 296; H. Boersch and S. Schweda, *Z. Phys.* **167** (1962) 1; N. R. Daly and R. E. Powell, *Phys. Rev. Letters* **19** (1967) 1165.

123. G. J. Schulz, *Phys. Rev.* **125** (1962) 229.

124. E. N. Lassettre, M. E. Krasnow, and S. Silverman, *J. Chem. Phys.* **40** (1964) 1242; H. Ehrhardt and K. Willmann, *Z. Phys.* **203** (1967) 1; P. S. Hoeper, W. Franzen, and R. Gupta, *Phys. Rev.* **168** (1968) 50.

125. J. P. Doering and A. J. Williams, III, *J. Chem. Phys.* **47** (1967) 4180.

126. E. N. Lassettre and S. A. Francis, *J. Chem. Phys.* **40** (1964) 1208.

127. J. Geiger, *Z. Phys.* **177** (1964) 138; R. A. Bonham and H. L. Cox, Jr., *J. Chem. Phys.* **47** (1967) 3508.

128. R. J. Anderson, E. T. P. Lee, and C. C. Lin, *Phys. Rev.* **157** (1967) 31.

129. J. A. Simpson and S. R. Mielczarek, *J. Chem. Phys.* **39** (1963) 1606.

130. H. Boersch, J. Geiger, and H.-J. Reich, *Naturwiss.* **21** (1959) 596; H. Boersch, J. Geiger, and M. Topschowsky, *Phys. Letters* **17** (1965) 266; J. Geiger, *Z. Phys.* **181** (1964) 413; H.-J. Reich and H. Schmoranzer, *Phys. Letters* **9** (1964) 127; S. P. Khare, *Phys. Rev.* **157** (1967) 107; R. L. Wilkins and H. S. Taylor, *J. Phys. Chem.* **47** (1967) 3532.

131. L. S. Frost and A. V. Phelps, *Phys. Rev.* **127** (1962) 1621.

132. E. N. Lassettre and M. E. Krasnow, *J. Chem. Phys.* **40** (1964) 1248; J. Geiger and W. Stickel, *J. Chem. Phys* **43** (1965) 4535; H. Ehrhardt and K. Willmann, *Z. Phys.* **204** (1967) 462.

133. E. N. Lassettre, A. Skerbele, and M. A. Dillon, *J. Chem. Phys.* **46** (1967) 4536.

134. J. D. Carette, *Can. J. Phys.* **45** (1967) 2931.

135. E. N. Lassettre and S. M. Silverman, *J. Chem. Phys.* **40** (1964) 1256.

136. R. Spohr, T. Bergmark, N. Magnusson, and K. Siegbahn, Uppsala University, Institue of Physics Rept. UUIP-560 (October 1, 1967).

137. H. Boersch, J. Geiger, and W. Stickel, *Phys. Letters* **10** (1964) 285; H. Boersch, *6th Int. Congr. Electron Microscopy, Kyoto*, 1966, p. 85.

138. J. Geiger and K. Wittmaack, *Z. Phys.* **187** (1965) 433.

139. J. C. Y. Chen, *Phys. Rev.* **146** (1966) 61.

140. W. Bühring, *Z. Phys.* **208** (1968) 286.

141. K. Takayanagi and S. Geltman, *Phys. Letters* **13** (1964) 135; *Phys. Rev.* **138** (1965) A1003; K. Takayanagi, *J. Phys. Soc. Japan* **21** (1966) 507.

142. O. H. Crawford, *J. Chem. Phys.* **47** (1967) 1100.

143. B. van Zyl and G. H. Dunn, *Phys. Rev.* **163** (1967) 43.

144. A. Fahlman, K. Hamrin, R. Nordberg, C. Nordling, and K. Siegbahn, *Phys. Letters* **20** (1966) 159.

145. P. E. Best, *Proc. Phys. Soc. (London)* **80** (1962) 1308; *Proc. Phys. Soc. (London)* **79** (1962) 133.

146. L. B. Leder and L. Marton, *Phys. Rev.* **95** (1954) 1345.

147. F. A. Gianturco, *J. Phys. [Proc. Phys. Soc. (London)]* **1B** (1968) 614; C. A. Coulson and F. A. Gianturco, *J. Phys. [Proc. Phys. Soc. (London)]* **1B** (1968) 605.

148. J. C. Slater, *Phys. Rev.* **36** (1930) 57.

149. E. Clementi and D. L. Raimondi, *J. Chem. Phys.* **38** (1963) 2686.

150. G. Burns, *J. Chem. Phys.* **41** (1964) 1521.

151. S. Fraga and G. Malli, Dept. of Chemistry, University of Alberta, Tech. Rept. TC-6601, TC-6602, TC-6603 (1966).

152. R. H. Garstang and D. F. Mayers, *Proc. Cambridge Phil. Soc.* **62** (1966) 777.

153. G. Malli and S. Fraga, *Theoret. Chim. Acta* **7** (1967) 80; J. Thorhallsson, C. Fisk, and S. Fraga, *Theoret. Chim. Acta* **11** (1968) 455.

154. C. Valdemoro and S. Fraga, *Can. J. Phys.* **45** (1967) 2483.

155. G. Malli, *Int. J. Quant. Chem.* **1s** (1967) 115; G. Malli and S. Fraga, *Theoret. Chim. Acta* **7** (1967) 75.

156. G. Malli and C. Froese, *Int. J. Quant. Chem.* **1s** (1967) 103.

157. G. Malli and S. Fraga, *Ann. Span. Soc. Phys. Chem.* **58(A)** (1967) 329.

158. G. Malli and C. Froese, *Int. J. Quant. Chem.* **1s** (1967) 99.

159. G. Malli and C. Froese, *Int. J. Quant. Chem.* **1s** (1967) 95.

160. S. Fraga and J. Thorhallsson, *Phys. Rev.* **178** (1969) 70.

161. G. Malli, *J. Chem. Phys.* **48** (1968) 1092.

162. G. Malli, *J. Chem. Phys.* **48** (1968) 1088.

163. G. Malli and S. Fraga, *Can. J. Phys.* **45** (1967) 219.

164. G. Malli, *J. Chem. Phys.* **47** (1967) 2217.

165. S. Fraga and G. Malli, *J. Chem. Phys.* **46** (1967) 4754.

166. R. F. O'Connell and C. O. Carroll, in *Internal Conversion Processes*, J. Hamilton (Ed.), Academic Press, New York, 1966.

167. C. Mande and P. S. Damle, *Proc. Phys. Soc. (London)* **87** (1966) 1005.

168. Y. Y. Chu and M. L. Perlman, *Phys. Rev.* **135** (1964) B319.

169. A. E. Lindh, *Z. Phys.* **6** (1921) 303; thesis, University of Lund, Sweden, 1923.

170. A. E. Lindh, *Z. Phys.* **63** (1930) 106.

171. O. Stelling, thesis, University of Lund, Sweden, 1923.

172. E. Bäcklin, *Z. Phys.* **38** (1926) 215.

173. O. Lundquist, *Z. Phys.* **60** (1930) 642.

174. A. Faessler, *Röntgenspektren und Chemische Bindung*, International Symposium, Leipzig (September 1965), p. 103.

175. A. E. Lindh, *Hand. Exp. Phys.* (Leipzig) **24** pt. 2 (1930) 314.

176. L. G. Parratt, *Phys. Rev.* **49** (1936) 14.

177. J. Tandberg, *Ark. Mat. Astr. Fys.* **18**, No. 14 (1921).

178. A. E. Lindh and O. Lundquist, *Ark. Mat. Astr. Fys.* **18**, Nos. 14, 34, 35 (1924).

179. B. Cioffari, *Phys. Rev.* **51** (1937) 630.

180. See Ref. 1.

181. Landoit-Börnstein, *Zahlenwerte und Funktionen*, 6th ed., **1** (1955), 858.

182. R. Faessler and E. D. Schmid, *Z. Phys.* **138** (1964) 71.

183. R. Faessler and M. Goehring, *Z. Phys.* **142** (1965) 558.

184. W. Nefedow, *Phys. Status Solidus* **2** (1962) 904.

185. H.-U. Chun and D. Hendel, *Z. Naturforsch.* **22a** (1967) 1401.

186. E. W. White and G. V. Gibbs, *Amer. Mineral.* **52** (1967) 985.

187. C. A. Coulson and C. Zauli, *Mol. Phys.* **6** (1963) 525.

188. M. E. Crawford and A. L. Schawlow, *Phys. Rev.* **76** (1949) 1310.

189. E. V. Petrovich, O. I. Sumbaev, V. S. Zykov, Yu. P. Smirnov, A. I. Egorov, and A. I. Grushko, *Zh. Eksp. i Teor. Fiz.* **53** (1967) 796 [*Russ. J. Exp. Theor. Phys.* **26** (1968) 489].

190. I. Lindgren, *Röntgenspektren und Chemische Bindung*, International Symposium, Leipzig (September 1965), p. 182.

191. F. A. Gianturco and C. A. Coulson, *Mol. Phys.* **14** (1968) 223.

192. G. B. Deodhar, *Röntgenspektren und Chemische Bindung*, International Symposium, Leipzig (September 1965) p. 65.

193. E. E. Vainshtein, M. N. Bril, I. B. Staryi, and E. I. Yarembash, *Izv. Akad. Nauk SSSR, Neorg. Mat.* **3** (1967) 1685; T. M. T. M. Zimkina and A. P. Lukerskii, *Fiz. Tver. Tela* **7** (1965) 1462 [*Sov. Phys.—Solid State* **7** (1965) 1175].

194. R. Deslattes, private communication in Ref. 63.

195. B. Nordfors, *Ark. Fys.* **10** (1956) 279.

196. N. Swanson and K. Codling, *J. Opt. Soc. Amer.* **58** (1968) 1192.

197. W. L. Baun and D. W. Fischer, *J. Appl. Phys.* **38** (1967) 2092.

198. O. I. Sumbaev, A. F. Mezentsev, V. I. Marushenko, E. V. Petrovich, and A. S. Ryl'nikov, *Zh. Eksp. i Teor. Fiz. SSSR*, **50** (1966) 861 [*J. Exp. Theor. Phys.* **23** (1966) 572].

199. O. I. Sumbaev and A. F. Mezentsev, *Zh. Eksp. i Teor. Fiz.* **48** (1965) 445 [*Russ. J. Exp. Theor. Phys.* **21** (1965) 295]; O. I. Sumbaev, E. V. Petrovich, V. S. Sykov, A. S. Ryl'nikov, and A. I. Grushko, *J. Nucl. Phys.* (*SSSR*) **5** (1967) 544. [*Sov. J. Nucl. Phys.* **5** (1967) 387].

200. R. B. Chesler, F. Boehm, and R. T. Brockmeier, *Phys. Rev. Letters* **18** (1967) 953; R. B. Chesler and F. Boehm, *Phys. Rev.* **166** (1968) 1206.

201. L. Pauling, *The Nature of the Chemical Bond and the Structure of Molecules and Crystals*, 2nd ed., Cornell University Press, Ithaca, 1945; 3rd ed., 1960.

202. B. G. Gokhale, R. B. Chesler, and F. Boehm, *Phys. Rev. Letters* **18** (1967) 957.

203. O. I. Sumbaev, E. V. Petrovich, Yu. P. Smirnov, A. I. Egorov, V. S. Zykov, and A. I. Gushko, *Zh. Eksp. i Teor. Fiz.* **53** (1964) 1545 [*Russ. J. Exp. Theor. Phys.* **26** (1968) 891].

204. A. Meisel, M. Köstler, and A. Merkel, *J. Prakt. Chem.* **34** (1966) 112.

205. A. Meisel and D. Köstler, *Z. Phys. Chem.* **231** (1966) 99.

206. A. Meisel, *Röntgenspektren und Chemische Bindung*, International Symposium, Leipzig (September 1965), p. 212.

207. A. Meisel, *Phys. Status Solidus* **10** (1965) 365.

208. E. Sokolowski, C. Nordling, and K. Siegbahn, *Phys. Rev.* **110** (1958) 776.

209. E. Gilberg and B. Kern, *Z. Phys.* **174** (1963) 372.

210. C. Nordling, *Ark. Fys.* **15** (1959) 241.

211. A. Fahlman, K. Hamrin, J. Hedman, R. Nordberg, C. Nordling, and K. Siegbahn, *Nature* **210** (1966) 4.

211a. M. O. Krause, submitted to *Chem. Phys. Lett.*; M. O. Krause and F. Wuilleumier, submitted to *Phys. Lett.*; private communication from M. O. Krause, 1971.

212. R. Nordberg, R. G. Albridge, T. Bergmark, U. Ericson, J. Hedman, C. Nordling, and K. Siegbahn, *Nature* **214** (1967) 481.

213. R. Nordberg, U. Gelius, P. F. Hedén, J. Hedman, C. Nordling, K. Siegbahn, and B. J. Lindberg, University of Uppsala, Institute of Physics Rept. UUIP-581 (March 1968).

214. G. Axelson, U. Ericson, A. Fahlman, K. Harmin, J. Hedman, R. Nordberg, C. Nordling, and K. Siegbahn, *Nature* **212** (1967) 70.

215. C. Nordling and K. Siegbahn, *Rev. Roum. Phys.* **11** (1966) 797.

215a. K. Siegbahn, C. Nordling, G. Johansson, J. Hedman, P. F. Hedén, K. Hamrin, U. Gelius, T. Bergmark, L. O. Werme R. Manne, Y. Baer, *ESCA Applied to Free Molecules*, North-Holland Publishing Co., Amsterdam, 1969.

216. J. M. Hollander, D. N. Hendrickson, and W. L. Jolly, *J. Chem. Phys.* **49** (1968) 3315.

216a. J. J. Jack, thesis, Department of Chemistry, Massachusetts Institute of Technology, September 1970.

216b. L. D. Hulett and T. A. Carlson, *Clinical Chem.* **16** (1970) 677; B. P. Pullen, T. A. Carlson, W. E. Moddeman, G. K. Schweitzer, W. E. Bull, and F. A. Grimm, *J. Chem. Phys.* **53** (1970) 768; W. E. Bull, B. P. Pullen, F. A. Grimm, W. E. Moddeman, G. K. Schweitzer, and T. A. Carlson, *Inorg. Chem.* **9** (1970) 2474.

217. L. Ramqvist, K. Hamrin, G. Johansson, A. Fahlman, C. Nording, *J. Phys. Chem. Solids* **30** (1969) 1835; L. Ramqvist, B. Ekstig, E. Källne, E. Noreland, and R. Manne, *J. Phys. Chem. Solids* **30** (1969) 1849; L. Ramqvist, thesis, Uppsala University, 1969.

218. C. S. Fadley and D. A. Shirley, *Phys. Rev. Letters* **21** (1968) 980.

219. G. Weich, *Z. Phys.* **207** (1967) 428.

220. G. Weich, *Z. Phys.* **216** (1968) 472.

221. A. Meisel and R. Szargan, contribution to the Kiev Conference, 1968.

222. M. A. Blokhin and V. F. Volkov, *Dokl. Akad. Nauk SSSR* **183** (1968) 64.

223. R. D. Deslattes, *Phys. Rev.* **172** (1968) 625.

224. G. Busch, M. Campagna, and H. C. Siegmann, *Solid State Commun.* **7** (1969) 775.

225. A. S. Nigavekar and S. Bergwall, *J. Phys.* (Proc. Phys. Soc.) **2B** (1969) 507.

226. V. I. Nefedov, *Izv. Akad. Nauk SSSR, Fiz. Ser.* **28** (1964) 816 [*Bull. Acad. Sci. USSR, Phys. Ser.* **28** (1964) 724].

227. K. Tsutsumi and H. Nakamori, *J. Phys. Soc. Japan* **25** (1968) 1418.

228. L. G. Parratt, *Rev. Mod. Phys.* **31** (1959) 616.

229. P. Johansson, *Ark. Fys.* **18** (1960) 289.

230. T. Novakov and J. M. Hollander, *Phys. Letters.* **13** (1964) 301.

231. T. Novakov and P. Janičijevič, *Z. Phys.* **205** (1967) 359.

232. P. Erman, *Ark. Fys.* **39** (1969) 451.

233. T. Novakov and J. M. Hollander, *Phys. Rev. Letters* **21** (1968) 1133.

Chapter Eight

Electron Energy Losses by Scattering In Solids

1. INTRODUCTION

Energy loss interactions of electrons with matter were first investigated as early as 1930 (e.g., Refs. 1, 2). When low energy bombarding electrons (20–600 eV) were used and their energy spectra after reflection from the surface of a material of interest were studied, it was observed that energy losses had been suffered in multiples or combinations of certain discrete amounts.

With time and with the improvement of experimental apparatus and techniques, research on discrete electron energy losses was extended to increasingly detailed investigations of the quantitative relations between the many parameters pertinent to the process.

Simultaneously, the theoretical investigation of electron plasma excitation modes in a solid provided a base for the interpretation of experimental results and for the subsequently stimulated and expanded research on the subject.

Following either the execution of an important step in the theoretical understanding of the process or the observation of a new effect, the heightened interest in testing or explaining that result by the experimentalists or the theoreticians, respectively, has shown a most striking and productive interplay between these two groups. The rapid progress in the field can be followed by consulting reviews on the subject in the order of their appearance, beginning some 15 years ago.[3–6]

Since the last review in 1965, advances in the field, both experimental and theoretical, have been realized with impressive rapidity. For this reason the present treatment is in effect a review anew, which attempts to bring the subject up to date. This survey lays more weight on energy losses due to interband transitions than have previous efforts. This emphasis is inspired mainly by recent experimental work carried out on this aspect of the subject. See Section 3, which includes a review and bibliography of experimental work performed on each of the specific subjects treated therein (Sections 3.2.1–3.2.5). Also discussed are the small energy loss effects ($0.05\ \text{eV} \leq \Delta E \leq 4\ \text{eV}$) due to the interaction of electrons with a crystalline lattice or its individual atomic or ionic components (Section 5).

Section 2 covers volume and surface plasmon excitation and reviews recent theoretical and experimental work in which consideration of retardation effects has been emphasized. Section 4 treats the radiative decay of plasma excitations. Section 6 deals briefly with multiple energy losses, and the effects of characteristic energy losses on electron spectral line positions and shapes are the subject of Section 7.

2. PLASMA EXCITATIONS

2.1. Volume Excitations

The electrons in the conduction band of a metal closely approximate a degenerate gas or plasma of free electrons in a general homogeneous positive potential. Excitations of this plasma can be made to arise in the form of electron density oscillations of quantized energy values that are dependent on the free electron density and dielectric properties of the metal. For individual theoretical treatments of the subject, see Refs. 7 through 14, as well as works referred to in the chapter text.

2.1.1. Dispersion Relation

Assume that a density fluctuation of angular frequency ω is traveling through an electron plasma of average electron density n with a wave number $\mathbf{k}$, where $|\mathbf{k}| = 2\pi/\lambda$. By considering the perturbation interaction of the wave with the plasma electrons, Ferrell[15] derived the following condition for the self-sustaining wave:

$$\omega^2(k) = \omega_p^{\ 2} + k^2\langle v^2 \rangle + \omega_p^{\ -2} k^4 (\Delta v^2)^2 + \frac{\hbar k^4}{4m_o^{\ 2}} + \cdots, \tag{1a}$$

where

$$\omega_p^{\ 2} = \frac{4\pi n e^2}{m_o} \tag{1b}$$

is the plasma frequency,[14] m_o and e are the free electron mass and charge, n is the number of free electrons per unit volume, and

$$\Delta v^2 \equiv (\langle v^4 \rangle - \langle v^2 \rangle^2)^{1/2},$$

v being the velocity of the plasma electrons. For a Fermi gas, $\langle v^2 \rangle = 3v_F^{\ 2}/5$ and $\Delta v^2 = \frac{12}{175} v_F^{\ 2}$, where v_F is the electron velocity associated with the Fermi energy $E_F = \frac{1}{2} m_o v_F^{\ 2} = \hbar^2 k_F^{\ 2}/2m_o$. Expression 1a is identical with that first obtained by Bohm and Pines;[16] it expresses a dispersion relation, the dependence of ω on k.

The first two terms of equation 1a can be obtained directly from classical considerations of the conservation of the energy and momentum of plasma oscillations.[5]

Taking the square root of equation 1a and making simplifying substitutions, we obtain

$$\omega(k) = \omega_p \left[1 + \alpha \frac{\hbar}{m_o \omega_p} k^2 + \frac{1}{\hbar \omega_p} \left(\frac{\hbar}{2m_o} \right)^2 \left(\frac{1}{2} - \frac{6}{35} \frac{E_F}{\hbar^2 \omega_p^{\ 2}} \right) k^4 + \cdots \right] \tag{2}$$

with the dispersion coefficient $\alpha = \frac{3}{5} E_F/\hbar\omega_p$.

In several treatments of the oscillation excitation of a plasma (e.g., Refs. 17–19) it has been concluded that a correction term due to exchange must be added to the dispersion relation (equation 2), a term that is proportional to k^2. Several workers[20–23] have treated this problem, using various approaches which indicate that the exchange correction of the second term in equation 2 consists[24] of the multiplication of the dispersion constant α by the term $(1 - \hbar^2 \omega_p^{\ 2}/16E_F)$.

A second refinement of equation 2 has been treated by Ferrell[25] and by Pines.[13] The effect of atomic polarization on the plasma oscillation value ω_p should be considered. This effect alters the plasma oscillation energy to a new value $\hbar\tilde{\omega}_p = \hbar\omega_p(1 + 4\pi\chi)^{-1/2}$, where $4\pi\chi = 4\pi\alpha_p/(1 - \frac{4}{3}\pi N \alpha_p)$, where N is the number of atoms per cubic centimeter and α_p is the ionic polarizability. Kunz[26,27] was able to improve the agreement between the theoretical and his experimental values for the volume plasma oscillation energies for lithium, sodium, potassium, rubidium, and cesium by applying this correction. In the following sections, however, neither of the two corrections just set forth is considered further.

An impinging electron may cause the creation of a plasmon of momentum $\hbar\mathbf{k}$, thereby becoming scattered through an angle ϑ.

Assume that the incoming electron is of initial energy E_0 and momentum $p_0 = \hbar k_0$. Through conservation of momentum, the momentum change for the electron will equal the momentum of the plasmon, or $(\Delta \mathbf{p}_0) = \hbar(\mathbf{k}_0 - \mathbf{k}') = \hbar\mathbf{k}$, where $\hbar\mathbf{k}_0$ and $\hbar\mathbf{k}'$ are the initial and final momenta of the electron. The momentum of the plasmon will be then $\hbar k \cong p_0 \vartheta$, assuming that ϑ, the angle of the electron scattering, is small.

Considering equation 2, we see that the expression $\omega(k)$ for a plasmon created through electron scattering will become

$$\omega(\vartheta) = \omega_p \left[1 + \frac{\alpha \cdot 2E_0 \vartheta^2}{\hbar\omega_p} + \frac{E_0^{\ 2}}{\hbar\omega_p} \times \left(\frac{1}{2} - \frac{6}{35} \frac{E_F^{\ 2}}{\hbar^2 \omega_p^{\ 2}} \right) \vartheta^4 + \cdots \right]. \tag{3}$$

Expressing equation 3 in terms of energy, we have

$$\Delta E(\vartheta) = \hbar\omega(\vartheta) = \Delta E(0) + \alpha \cdot 2E_0 \vartheta^2 + \left(\frac{1}{2} - \frac{6}{35} \frac{E_F^{\ 2}}{\hbar^2 \omega_p^{\ 2}} \right) E_0^{\ 2} \vartheta^4 + \cdots, \tag{4}$$

where $\Delta E\,(0) = \hbar\omega_p$.

The dispersion of the 15-eV volume plasma excitation in aluminum has been measured by Kunz.[28] The energy loss spectrum is shown in Figure 1*a*, and the dispersion relationship appears in Figure 1*b*. The experimental results are compared in Figure 1*b* with the theoretical values obtained from the first two terms in equation 4. Clearly, the coefficient of the ϑ^4 term must be positive if better agreement is to be obtained. This agrees with the dispersion measurement result for antimony and its comparison with theory by Geiger.[24] Earlier experiments on aluminum by Watanabe[29] and Meyer,[30] however, indicated that this coefficient is negative.

The peak in Figure 1*a*, which lies at 15 eV for the larger angles, is interpreted to be a result of a two-step process. First, the electron is elastically scattered through the given angle, and second, the creation of a volume plasmon takes place through zero-angle scattering.[28] The same effect is seen in the comparable situation for measurements on magnesium.[31]

2.1.2. *The Cut-off Wave Number*

A plasma oscillation can become damped either through the emission of electromagnetic radiation

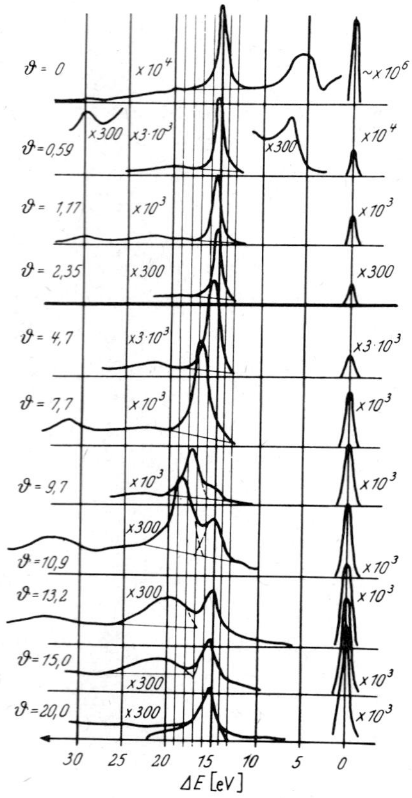

(*a*)

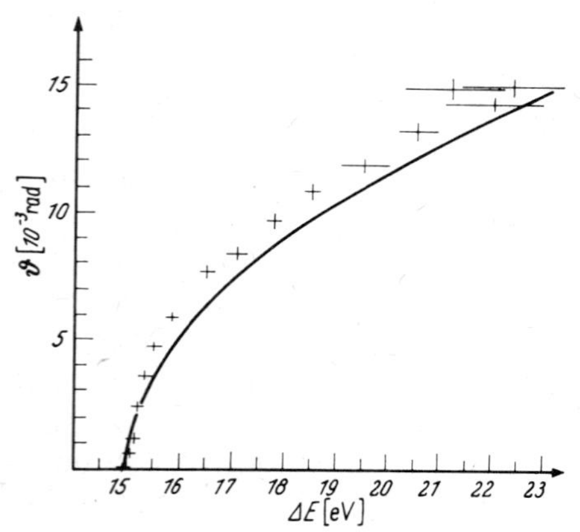

(*b*)

Figure 1. (*a*) The energy loss spectrum for 39-keV electrons bombarding a 300-Å aluminum foil for a variety of scattering angles, as reported by Kunz[28] The angles are given in milliradians. The shift and broadening of the 15-eV volume plasma peak due to dispersion increases with increasing scattering angles. The peak at 15 eV for large scattering angles is due to single large-angle elastic scattering followed by a volume plasma energy loss. (*b*) The dispersion relation for aluminum. The solid line represents values from equation 3 without the ϑ^4 term. The coefficient of this term must be positive if agreement is to be improved. The crosses give the measured values with error limits.

(Section 4), through the creation of an exciton,[32] through plasmon-phonon interactions (Section 2.2.1), or through energy transfer to a single plasma electron. The last type of damping dominates at $k \approx k_c$, the cut-off wave number (see below). For $k < k_c$, plasmon-phonon interactions are most prevalent.

Electron plasma oscillations of wavelengths that are large compared to the mean interelectron distance are collective in nature. However, as the wavelength decreases (and k increases) to approach this mean distance, the significance of collective motion becomes obscure. In addition, as the wavelength shortens, the probability for the transfer of the plasmon energy to an individual electron increases. In this way energy is absorbed from the plasma oscillation by single electrons, and the wave is damped out. The damping probability increases as the phase velocity of plasma wave propagation approaches the velocity of electrons in the plasma, and at this value damping is complete. This is the cut-off velocity or wave number for plasma oscillations. (See also Ref. 15.)

In electron bombardment excitation of plasma oscillations, the cut-off wave number k_c is associated with a cut-off scattering angle ϑ_c through equation 4. For $k \to k_c$ and $\vartheta \to \vartheta_c$, damping effects cause the probability of excitation of a plasmon to fall to zero. Here

$$\frac{\omega_p}{k_c} \cong v_F \qquad \text{or} \qquad \vartheta_c = \frac{\omega_p}{k_F v_F} = \frac{\hbar\omega_p}{2E_F}.$$

2.1.3. *Angular Dependence of Loss Probability*

Ferrell[25] has derived the scattering-angle-dependent differential cross section for the excitation of a volume plasmon as:

$$\sigma(\vartheta) = \frac{1}{2\pi a_0\, n} \frac{\vartheta_E}{\vartheta_E^{\,2} + \vartheta^2} \qquad \vartheta < \vartheta_c. \tag{5}$$

The quantity is equal to zero for other values of ϑ. Here a_0 is the Bohr radius, n is the free electron density, and $\vartheta_E = \Delta E/2E_0 = \hbar\omega_p/2E_0$. Equation 5 is valid for values of $\vartheta < \vartheta_c$, where $\sigma(\vartheta)$ decreases as ϑ^{-2}; but in the vicinity of ϑ_c a correction factor[15] $G^{-1}(\vartheta)$ must be multiplied to equation 5. This factor contributes steadily to the decrease of $\sigma(\vartheta)$ in the near vicinity of ϑ_c. Figure 2 illustrates the variation of σ with ϑ for aluminum. In Figure 2*a*, the solid line is the theoretical curve normalized to $\sigma = 1$ at $\vartheta = 0$. The experimental points follow a ϑ^{-2}-dependence until about $\vartheta = 6 \times 10^{-3}$ rad, as in Figure 2*b*. Here the $G^{-1}(\vartheta)$ correction fits the points well at first but finally decreases too rapidly. The experimental cut-off value obtained in the figure, 1.5×10^{-2} rad, agrees well with previous measured values.

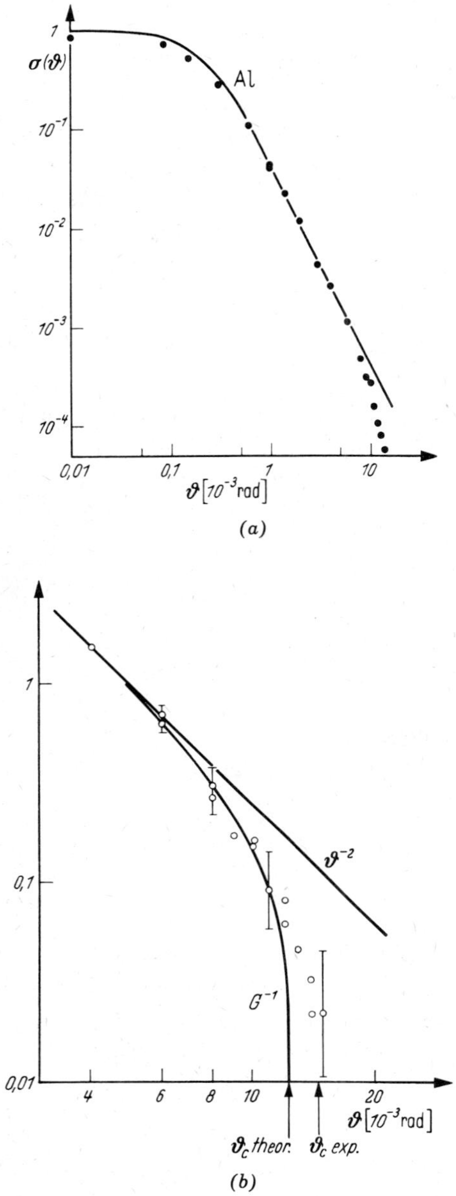

Figure 2. (*a*) The angular dependence of the volume plasma loss intensity $\sigma(\vartheta)$ for aluminum as measured by Schmüser.[31] (*b*) The experimental volume loss intensity for aluminum in the vicinity of the cut-off angle ϑ_c, where G^{-1} is the cut-off curve according to Ferrell[15]. (From Ref. 31).

2.1.4. *Free Path Length*

Integration of the differential excitation cross section (equation 5) for all angles up to ϑ_c multiplied by the mean electron density of the plasma n will give the absolute probability for a volume plasma energy loss per unit path length. This free path between inelastic interactions can be expressed as

$$\lambda_i = \frac{a_0}{\vartheta_E \ln \vartheta_c/\vartheta_E}. \tag{6a}$$

(See Refs. 33, 28, and Section 6.) It follows then that for a foil thickness a the probability P_n for n losses having taken place will be expressed by the Poisson distribution

$$P_n = \frac{1}{n!}\left(\frac{a}{\lambda_i}\right)^m e^{-a/\lambda_i}. \tag{6b}$$

The probability of multiple-volume plasma losses will increase with film thickness. These relations are considered further in Section 6.

2.1.5. *Volume Plasma Loss Half-Width*

As we see in Figure 1*b*, the width of the plasma energy loss increases with scattering angle ϑ. This plasma loss half-width, $\Delta E_{1/2}$, was estimated by Nozières and Pines[10] to be proportional to the square of the momentum transfer. Du Bois[19] calculated the plasma level width, assuming the combined contribution of two modes of plasma decay. Ninham et al.[34] later recalculated the level width and obtained a result of the same form as that of Du Bois, $\Delta E_{1/2} = A + B\vartheta^2 + C\vartheta^4$, but with coefficients B, C differing from those of the former author. (The constant A does not follow from the calculations; it is the plasma loss distribution width for $\vartheta = 0$.) The expression was tested on a measurement of the angular dependence of the plasma loss half-width for 20-keV electrons passing through a 1000 Å-thick aluminum foil. The experimental values B and C obtained by fitting this general expression agreed with the calculated values.

Figure 3*a* shows the behavior of the plasma loss width with changing scattering angle. Included in the graph are results from workers using the metal in the form of large crystals[35] and in the form of the bulk material.[28,34] The values agree well with one another and with the theoretical curve of Ninham et al.[34]

It has been shown experimentally that the half-width is dependent on the crystal size of the metal being investigated.[25] v. Festenberg[35] has measured a width of about 0.53 eV for the plasma loss in aluminum in the form of crystals 250 Å large. This value agrees well with that measured by Hartl.[36] From this we can calculate the relaxation time of 1.25×10^{-15} sec for the plasmons.

For smaller crystals, about 75 Å, the half-width of the plasma energy loss for zero scattering angle is much larger.[35] The crystal size dependence of the plasma loss half-width displays a damping of a different origin.

Figure 3*b* shows the scattering-angle dependence of the energy loss width for small angles for both large and small crystal sizes. Associated with the small

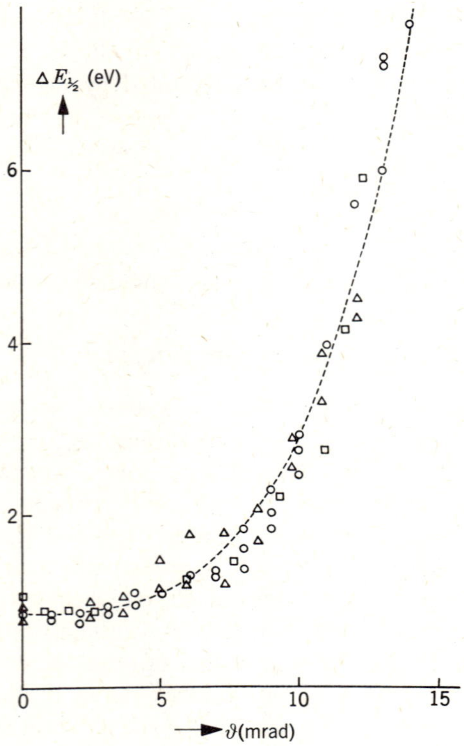

(*a*)

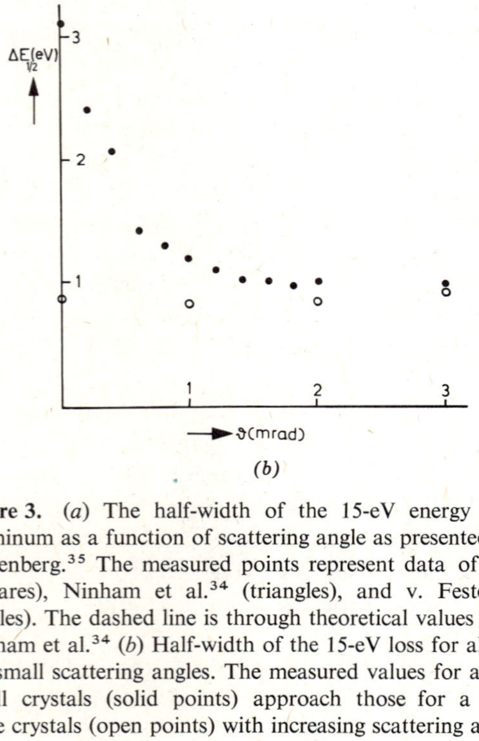

(*b*)

Figure 3. (*a*) The half-width of the 15-eV energy loss for aluminum as a function of scattering angle as presented by von Festenberg.[35] The measured points represent data of Kunz[28] (squares), Ninham et al.[34] (triangles), and v. Festenburg[35] (circles). The dashed line is through theoretical values given by Ninham et al.[34] (*b*) Half-width of the 15-eV loss for aluminum for small scattering angles. The measured values for a layer of small crystals (solid points) approach those for a layer of large crystals (open points) with increasing scattering angle.

crystals is a width which at $\vartheta = 0$ is a few times larger than that for the large crystals. This value falls rapidly with increasing scattering angle to meet the value for the large crystals.

To interpret this effect, consider the plasma wavelength Λ, which is defined by $\Lambda = \Lambda_0(\vartheta^2 + \vartheta_E^2)^{-1/2}$, where again $\vartheta_E = \Delta E/2E_0$. Hence for small scattering angles ϑ, Λ is large. v. Festenberg[35] showed that just at $\vartheta \approx 2$ to 3×10^{-3} rad, where the damping ceases, Λ is nearly equal to the size of the crystal grain. For smaller values of ϑ, the plasma wavelength is larger than the crystal size. Thus we may conclude that the damping of the plasma wave is strong at the grain surface.

2.2. The Dielectric Theory Treatment of Characteristic Energy Losses

An alternative point of view, which has had great success with regard to the treatment of characteristic energy losses, is that represented by the dielectric theory approach. The dielectric constant for a solid is a measure of the interaction between impinging electromagnetic radiation and the electrons of the solid, either in a plasma or in bound atomic states. This electrostatic interaction has a frequency (energy) dependency that varies in nature from solid to solid.

The absorption, reflection, and transmission of electromagnetic radiation by matter are known to depend on the complex dielectric constant

$$\varepsilon = \varepsilon_1 + i\varepsilon_2 . \qquad (7)$$

The radiation energy at which transmission, ε_1 positive, changes to reflection, ε_1 negative, is just at the plasma oscillation energy value.

In the optical absorption process, the transverse components of the electromagnetic field interact with matter. The probability for absorption is proportional to the product of the frequency and the imaginary part of ε, that is, to ω Im (ε). Therefore, the transmission threshold is at the ω value where Re $(\varepsilon) = \varepsilon_1 = 0$; but if absorption at this point is still strong, the transmission may remain small. Thus results of measurements of transmissivity will be dependent on both the quantities Re (ε) and Im (ε), which can be determined from optical data (Section 2.2.1).

Electron characteristic energy losses in metals are brought about by Coulomb interaction of the electron with the electron plasma. This longitudinal interaction of the electromagnetic field of the electron with matter is related to the imaginary part of the inverse of ε,[4, 37–39] that is,

$$-\mathrm{Im}\left(\frac{1}{\varepsilon}\right) = \frac{\varepsilon_2}{\varepsilon_1^2 + \varepsilon_2^2} . \qquad (8)$$

The correlation between optical absorption and characteristic electron energy losses is valid also for the several alternative types of induced excitation, treated in the following sections: surface plasmon excitation (Section 2.3), interband transitions (Section 3.1.2), inner shell ionization (Section 3.2.5), polarization wave losses (Section 2.4), and crystal lattice interactions with the impact electron (Section 5).

2.2.1. *Determination of* Re (ε) *and* Im (ε) *from Optical Data*

The determination of the real and imaginary parts of the dielectric constant ε may be carried out in a number of ways. Beaglehole[40] measured the reflectivity of light for nonnormal incidence angles and then found the corresponding values of the quantities n and κ using standard theoretical relationships. Here n and κ are the real and imaginary parts of the complex refraction index of the material n_c, which is related to the complex dielectric constant as follows:

$$n_c^2 = (n + i\kappa)^2 = \varepsilon \qquad (9)$$

(whether a minus or plus sign is written in equation 9 is determined by the sign choice in the expression for the electromagnetic field time dependence, $e^{\pm i\omega t}$). Then we see that

$$\text{Re}\,(\varepsilon) = \varepsilon_1 = n^2 - \kappa^2 \qquad (10a)$$

and

$$\text{Im}\,(\varepsilon) = \varepsilon_2 = 2n\kappa. \qquad (10b)$$

Alternatively, normal incidence reflectance data can give the quantities n and κ.[41] This method has been employed with success by Philipp and Taft.[42] Expressing r as

$$r = (n - i\kappa - 1)/(n - i\kappa + 1) = |r|e^{i\theta}, \qquad (11)$$

where $|r|^2$ is the reflectance, the phase θ for any frequency may be computed from the measured reflectance using the Kramers–Kronig (see Ref. 43) expression relating the real and imaginary parts of the function

$$\ln r = \ln |r| + i\theta, \qquad (12a)$$

$$\theta(\omega) = \frac{1}{\pi}\int_0^\infty \frac{d}{d\omega'} \ln |r(\omega')| \cdot \ln \left|\frac{\omega' + \omega}{\omega' - \omega}\right| d\omega'. \qquad (12b)$$

Equation 11 may then be used to obtain n and κ as functions of angular frequency from r and θ. In an analogous manner, the Kramers–Kronig expression relating ε_1 and ε_2 is[40,44]

$$\varepsilon_1(\omega) = 1 + \frac{2}{\pi}\int_0^\infty \frac{\omega'\varepsilon_2(\omega')}{\omega'^2 - \omega^2}\, d\omega'. \qquad (12c)$$

The unity term is the value of the real part of the dielectric constant for plasma oscillations at infinite oscillation frequency, that is,[45] $\varepsilon_1(\infty) = 1$.

The quantities Re $\varepsilon(\omega)$ and Im $\varepsilon(\omega)$ (i.e., ε_1 and ε_2) may also be derived under special conditions using the Drude theory. By this free-electron gas theory the parts of the dielectric constant may be expressed as[37,6]

$$\varepsilon(\omega) = 1 - \frac{\omega_p^2\tau}{\omega^2\tau + i\omega}, \qquad (13a)$$

or

$$\text{Re}\,\varepsilon(\omega) = 1 - \frac{\omega_p^2\tau^2}{1 + \omega^2\tau^2}, \qquad (13b)$$

and

$$\text{Im}\,\varepsilon(\omega) = \frac{\omega_p^2\tau}{\omega(1 + \omega^2\tau^2)}, \qquad (13c)$$

where τ is the relaxation time corresponding to the interaction of the plasma electrons with the lattice, and ω_p is defined in equation 1b. For large $\omega\tau$ (with ω still small enough that no interband transitions are excited), the condition of a free gas with no interaction is

$$\text{Re}\,\varepsilon(\omega) = 1 - \frac{\omega_p^2}{\omega^2}, \qquad (14a)$$

$$\text{Im}\,\varepsilon(\omega) = \frac{\omega_p^2}{\omega^3\tau}. \qquad (14b)$$

Rudberg and Slater[46] demonstrated that the maxima in the quantity $-\text{Im}\,(1/\varepsilon)$ or $n\kappa$ in terms of the optical properties of a material may also indicate interband transitions; however, such maxima can alternatively be caused by the creation of polarization waves (Section 2.4).

The integration over $\varepsilon_2(\omega)$, expressing the sum rule for the function,[47] is related to the frequency of the experimentally observable plasma oscillation ω_p:

$$\int_0^\infty \omega\varepsilon_2(\omega)\, d\omega = \frac{2\pi^2 ne^2}{m_o} = \frac{\pi}{2}\,\omega_p^2. \qquad (15a)$$

The quantity $-\text{Im}\,(1/\varepsilon)$, which may be called the electron energy loss function, is equal to $\varepsilon_1/(\varepsilon_1^2 + \varepsilon_2^2)$ or, in terms of optical constants, to $2n\kappa/(n^2 + \kappa^2)^2$. In the vicinity of interband transitions, this function should be small. At plasma oscillation excitation energies, ε_1 goes to zero from minus values (i.e., $n^2 = \kappa^2$), and $\varepsilon_2 = 2n\kappa$ is small. This is to say that $|\varepsilon| = n^2 + \kappa^2$ passes through a minimum.

In analogy to equation 15a, the sum rule for $-\text{Im}\,(1/\varepsilon)$ gives the following expression:[48,47,38]

$$-\int_0^\infty \omega\,\text{Im}\left(\frac{1}{\varepsilon}\right) d\omega = \frac{2\pi^2 ne^2}{m_o} = \frac{\pi}{2}\,\omega_p^2. \qquad (15b)$$

2.2.2. *Volume Plasma Loss Energy Dependence on the Target Density*

Following the Bohm and Pines[16] theory of electron plasma oscillations, we arrive at the expression given in equation 1b for the volume plasma energy: $\hbar\omega_p = \hbar(4\pi ne^2/m_o)^{1/2}$. The number of free electrons per unit volume n is dependent on the number of atoms in that volume and therefore on the density of the material under investigation. Richter and Ruckwied[49] have reported the measurement of the volume plasma excitation energy for the same material in different forms (i.e., crystalline and amorphous). They found that the volume plasma loss for the crystalline form of arsenic was broader and shifted ~ 2 eV toward higher energies with respect to the result with the amorphous form. This shift was attributed to the greater density of the crystalline form. There was also measured a temperature dependence of the plasmon energy resulting from the temperature-density interrelationship. A similar investigation of germanium was also performed, but the effect was of lesser magnitude.

Comparable measurements of the volume plasmon energy dependency on temperature have also been reported by Meyer.[30]

2.2.3. *Differential Free Path for Inelastic Electron Scattering*

The probability that an electron will interact with a free electron gas having a dielectric constant ε, through inelastic scattering processes can be expressed in terms of the dielectric theory. From this we can derive the differential free path length $d^2(1/\lambda_i)$ for inelastic scattering. Theory gives[37,6]

$$d^2\left(\frac{1}{\lambda_i}\right) = \frac{1}{2\pi^2 a_0 E_0} \cdot \frac{1}{\vartheta^2 + \vartheta_E{}^2}\left(-\operatorname{Im}\left(\frac{1}{\varepsilon}\right)\right) dE\, d\Omega, \quad (16)$$

where a_0 is the Bohr radius, E_0 the electron energy, ϑ the scattering angle, $\vartheta_E = \Delta E/2E_0$, ΔE the loss energy, and dE and $d\Omega$ are the energy loss and solid angle increments. This expression can be altered to express the differential cross section for plasma excitation by setting $n\sigma = 1/\lambda_i$, where n is the density of electrons participating and σ is the scattering cross section. This expression (equation 16) is valid only for a single loss of any type (e.g., intra- and interband transitions, excitation of bound electrons, etc.) in the volume of the material. The surface plasma excitation cross section is therefore not included (see Section 2.3).

For zero scattering angle

$$d^2\left(\frac{1}{\lambda_i}\right) = nd^2\sigma \propto -\operatorname{Im}\left(\frac{1}{\varepsilon}\right)(\Delta E)^{-2}\, dE\, d\Omega$$

for a given electron impact energy. Therefore the differential cross section for electron energy loss is proportional to equation 8, which can be determined by procedures given in Sections 2.2 and 2.2.1. The $(\Delta E)^{-2}$ dependence makes measurements of Im $(1/\varepsilon)$ increasingly difficult with increasing loss energy ΔE.

2.3. Surface Excitations

In the foregoing sections the phenomenon of plasma oscillations in the bulk of a material has been treated. However, charge density oscillations may also occur in the surface of the material. This effect was first discussed by Ritchie,[50] who showed that the energy of such surface plasma oscillations differs from those occurring in the bulk.

In the bulk of a material the real part of the dielectric constant has a continuous value ε_1 throughout. At the bulk surface this quantity changes abruptly to that of the adjoining medium ε_1'. Assuming that there is a plasma oscillation in the first medium, the necessary continuity of the normal component of the electrical displacement **D** across the boundary requires that $\varepsilon = -\varepsilon'$. From expression 14a we can set $1 - \omega_p{}^2/\omega = -\varepsilon_1'$, which gives the oscillation frequency for the surface plasmon

$$\omega_s = \omega = \frac{\omega_p}{\sqrt{1+\varepsilon_1'}}. \quad (17)$$

If the adjoining medium is vacuum, then $\varepsilon_1' = 1$ and

$$\omega_s = \frac{\omega_p}{\sqrt{2}}. \quad (18)$$

Since ε' always equals $-\varepsilon$ at the materials' interface, at the surface plasma excitation energy, $\varepsilon = -\varepsilon_1$. For conductor-vacuum boundaries, then, $\varepsilon = -1$.

From the foregoing it follows that, if the second medium is also a free-electron medium having a volume plasma frequency of ω_p', then $\varepsilon' = 1 - \omega_p'^2/\omega^2$ and

$$\omega_s = \sqrt{(\omega_p{}^2 + \omega_p'^2)/2}. \quad (19)$$

For the last three equations it was assumed that the interface between the two media is smooth. Variations in this surface uniformity will alter the value of ω_s. Jensen[51] has shown that the plasma oscillation frequency for a small sphere in a vacuum is a factor $1/\sqrt{3}$ less than the value for the infinite plasma. Therefore a surface made up of an agglomeration of small spheres will have a surface oscillation frequency of $\omega = \omega_p/\sqrt{3}$.

Dispersion relations for surface plasma excitations have been performed and applied using methods similar to those for volume excitations[52,53] (Section 2.1.1) and have been used in the analysis of the experimental investigation of such.[27] However, the derivation of the dispersion relations of surface excitations treated in the following sections are based on the dielectric theory of matter.

The experimental investigation of surface oscillations is usually achieved by measuring energy losses of electrons passing through a thin film of the material of interest. For very thin films (thickness $\gtrsim$200–300 Å and depending on the instrumental resolution), coupling between surface excitations on the two film surfaces will give rise to a splitting in the excitation energy, as we learn in Section 2.3.1. As the film thickness is increased, the fine structure of the surface plasmon energy loss diminishes. Of course, as the film thickness increases, the probability for surface excitations remains effectively unchanged. With increasing film thickness, the surface loss structure by transmission electron loss measurements often becomes progressively more obscured by volume losses.

Low energy electron reflection by a surface, however, will permit the investigation of the energy spectrum of surface excitations without undue volume excitations. The excitation probability for surface oscillations increases with respect to that for volume excitations with decreasing reflection angle.[54]

The results of studies of surface excitations in thin films of various thicknesses can, on the other hand, be extrapolated to infinite thickness, as can formulas describing the various aspects of surface excitations.

2.3.1. Total Electron Scattering Probability by Thin Films

2.3.1.1. NORMAL INCIDENCE. There have been several calculations of the total inelastic scattering cross section in thin films for nonrelativistic electrons. Ritchie[50] gave the first treatment for electrons of energy E_0 and momentum p_0 striking the film surface normally (see Figure 4a). Here it was assumed that the velocity of light c is infinite (i.e., retardation is neglected). The bombarding electrons are assumed to lose an amount of energy lying between $\hbar\omega$ and $\hbar\omega + \hbar\, d\omega$ and to become scattered through an angle ϑ into a solid angle element $d\Omega$.

Ritche expressed the scattering function in the form of equation 1 of Appendix H. This general expression has since been reexpressed in two alternative forms, which are useful in treating problems of electron scattering, particularly in foils.

The Ritchie formula has been expressed by Hattori and Yamada[55] in the form (see also Appendix H equation 4):

$$W(\omega, \vartheta)\, d\omega\, d\Omega$$

$$= -\frac{e^2}{\pi^2\hbar^2 v^2}\left\{\operatorname{Im}\left(\frac{1}{\varepsilon(\omega)}\right)\frac{a}{\vartheta^2+\vartheta_E{}^2} - \frac{2\vartheta}{k_0(\vartheta^2+\vartheta_E{}^2)^2}\right.$$

$$\times \operatorname{Im}\left[\frac{(\varepsilon(\omega)-1)^2}{\varepsilon(\omega)}\left(\frac{\sin^2\left(\dfrac{ak_0\vartheta_E}{2}\right)}{\varepsilon(\omega)+\tanh\left(\dfrac{ak_0\vartheta}{2}\right)}\right.\right.$$

$$\left.\left.\left.+\frac{\cos^2\left(\dfrac{ak_0\vartheta_E}{2}\right)}{\varepsilon(\omega)+\coth\left(\dfrac{ak_0\vartheta}{2}\right)}\right)\right]\right\} d\omega\, d\Omega,$$

$$\vartheta_E = \frac{\Delta E}{2E_o} = \frac{\hbar\omega}{m_o v^2}, \qquad \vartheta = \frac{\hbar k}{p_0}. \tag{20}$$

This form has been of use extensively by Otto, Swan, and Fellenzer,[56–59]

Setting $\frac{1}{2}m_o v^2 = E_0$ and $\hbar^2/m_o e^2 = a_0$, we see that the first term is identical with the thickness a times the volume inelastic scattering cross-section, (i.e., the differential free path length) given in equation 16. The magnitude of this term is proportional to the film thickness. The second term in equation 20 is the cross section for the creation of a surface plasma oscillation, which in the first-order approximation is not dependent on the film thickness.

The angular dependence of electron scattering differs in the two cases, through volume or surface excitations: the volume excitation probability approaches a ϑ^{-2} dependence with increasing scattering angle, as seen earlier in Sections 2.1.3 and 2.2.2, whereas the surface excitation probability approaches a ϑ^{-3} dependence.

A surface plasma excitation is associated with an electric field extending both above and below the surface position. The intensity of this field varies exponentially with distance z from the surface plane, that is, as $e^{-k|z|}$, where k is the plasmon wave number. The electric field strength acting on the opposite surface of the foil will then have a strength proportional to e^{-ka}, where a is the foil thickness.[6,56] (Actually k here has different values for inside and outside the foil when retardation is included.) The action of this field on the opposite side of the foil will stimulate there one of two possible surface oscillation modes, symmetric ($-$) or antisymmetric ($+$). In the first case the plasma charge density in the first surface is in phase with that on the second surface; in the second case these two are out of phase by $\pi/2$.

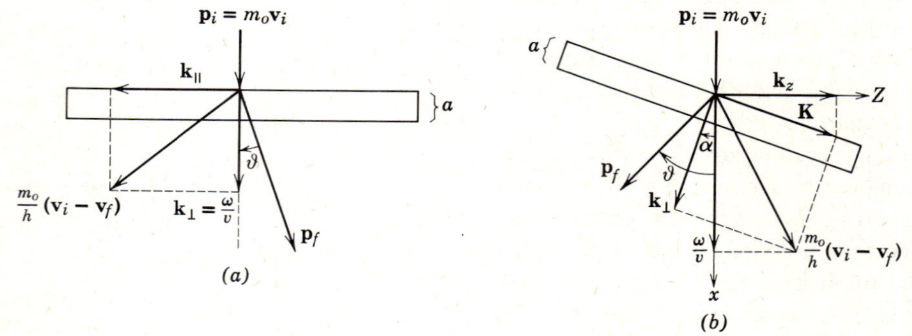

Figure 4. (*a*) Relation between quantities involved in the scattering of an electron by a foil of thickness a. The initial electron momentum is $\mathbf{p}_i$ and the final $\mathbf{p}_f$ (after scattering through an angle ϑ). A volume plasmon of momentum $m(\mathbf{v}_i - \mathbf{v}_f)/\hbar$ is created whose wave vector components perpendicular and parallel to the foil surface are $\mathbf{k}_\perp$ and $\mathbf{k}_\parallel$, respectively (vectors not drawn to scale). (*b*) Relations between quantities involved for an incidence angle α. Only the xz-plane is shown. The final wave vector component parallel to the foil surface is $\mathbf{k}_u = \sqrt{|\mathbf{k}_y|^2 + |\mathbf{K}|^2}$, where $\mathbf{K}$ is that surface parallel component which is perpendicular to the foil surface–yz-plane line of intersection.

An alternative form of the Ritchie differential scattering probability has been presented by Geiger and Wittmaack.[60] This form, a slight variation of the Romanov form, is obtained from the Hattori–Yamada expression (equation 20) by reexpressing the second term in that equation.

The differential scattering probability now takes the form (see Appendix H):

$$W(\omega, \vartheta)\, d\omega\, d\Omega$$

$$= -\frac{e}{\pi^2 \hbar^2 v^2} \operatorname{Im} \left\{ \left[\frac{a}{\vartheta^2 + \vartheta_E{}^2} - \frac{2\vartheta}{k_0(\vartheta^2 + \vartheta_E{}^2)^2} \cdot \frac{\cosh(ak_0\vartheta) - \cos(ak_0\vartheta_E)}{\sinh(ak_0\vartheta)} \right] \frac{1}{\varepsilon(\omega)} + \frac{8\vartheta}{k_0(\vartheta^2 + \vartheta_E{}^2)^2(1 - e^{-2ak_0\vartheta})} \times \left[\frac{\sin^2(ak_0\vartheta_E/2)}{\varepsilon(\omega) + \tanh(ak_0\vartheta/2)} + \frac{\cos^2(ak_0\vartheta_E/2)}{\varepsilon(\omega) + \coth(ak_0\vartheta/2)} \right] \right\} d\omega\, d\Omega. \qquad (21a)$$

Each of the terms in equation 21a has one of two different angular dependences P and Q, so that we may represent equation 21a as

$$W = P - Q_p + Q_{(+)} + Q_{(-)}, \qquad (21b)$$

each term standing for its respective counterpart in equation 21a.

The first term is the same as that in equation 20, and the layer thickness dependence of the medium's differential inelastic scattering results in volume plasma excitation. The terms Q_p, $Q_{(+)}$, $Q_{(-)}$ express the influence of the surface plasma excitation probability on W, and the sum of these terms shows a ϑ^{-3} dependence for large angles and disappears for $\vartheta = 0$.

The characteristics of equation 21a can be made clearer by setting $\varepsilon(\omega)$ equal to the Drude dielectric constant for a free electron gas (equation 13a). We obtain

$$W(\omega, \vartheta)\, d\omega\, d\Omega$$

$$= \frac{e^2}{\pi^2 \hbar^2 v^2} \left\{ \left[\frac{a}{\vartheta^2 + \vartheta_E{}^2} - \frac{2\vartheta}{k_0(\vartheta^2 + \vartheta_E{}^2)^2} \times \frac{\cosh(ak_0\vartheta) - \cos(ak_0\vartheta_E)}{\sinh(ak_0\vartheta)} \right] \times \frac{\tau\omega\,\omega_p{}^2}{\tau^2(\omega^2 - \omega_p{}^2)^2 + \omega^2} + \frac{\vartheta}{k_0(\vartheta^2 + \vartheta_E{}^2)} \times \frac{\tau\omega\,\omega_p{}^2(1 + e^{-ak_0\vartheta})}{[\omega^2 - (\omega_p{}^2/2)(1 + e^{-ak_0\vartheta})]^2\tau^2 + \omega^2} \times \frac{1 - \cos(ak_0\vartheta_E)}{1 - e^{-ak_0\vartheta}} + \frac{\tau\omega\,\omega_p{}^2(1 - e^{-ak_0\vartheta})}{[\omega^2 - (\omega_p{}^2/2)(1 - e^{-ak_0\vartheta})]^2\tau^2 + \omega^2} \times \frac{1 + \cos(ak_0\vartheta_E)}{1 + e^{-ak_0\vartheta}} \right] \right\} d\omega\, d\Omega. \qquad (22)$$

The order of terms in equation 22 is as in equations

21. Expression 22 shows resonances at three different frequencies. The term P, associated with the volume plasmon, has its resonance at the plasma frequency $\omega = \omega_p$ as does the second term Q_p. The contribution of P is reduced by that of Q_p because of a surface effect contribution. The third and fourth terms $Q_{(+)}$ and $Q_{(-)}$ express the contributions due to surface excitations at the frequencies

$$\omega_{(\pm)} = \frac{\omega_p}{\sqrt{2}} \sqrt{1 \pm e^{-ak_0\vartheta}}, \tag{23}$$

which reduces to equation 18 for $a \to \infty$ or $\vartheta \to$ large.

From expression 22 it is seen that for small scattering angles, $\hbar\omega_{(+)}$ approaches $\hbar\omega_{(-)}$ in energy and contributes, with P and Q_p, to the differential probability for the volume plasma loss. Effects of retardation alter this, however, so that the $\hbar\omega_{(+)}$ contribution to $W(\omega, \vartheta)$ approaches zero along with $\hbar\omega_{(-)}$ for vanishing by small scattering angles.

On the other hand, $Q_{(-)}$ at small scattering angles is of much greater intensity than the volume loss $(P - Q_p)$ for low energy interband losses where ϑ_E is small, that is, when the energy loss is small (see Section 2.3.1.2.1).

2.3.1.1.1. *Dispersion Relation.* The expression relating ω and k for a surface plasma excitation can be obtained from equation 20 or 21a. Again neglecting for the moment retardation, the sum of the last two terms (pertaining to surface oscillation excitation) becomes maximum when the denominators are zero. This leads to the expressions

$$\varepsilon(\omega) = -\tanh\left(\frac{ka}{2}\right) \quad \text{and} \quad \varepsilon(\omega) = -\coth\left(\frac{ka}{2}\right), \tag{24}$$

where $\vartheta = k/k_0$ has been used. Setting the dielectric constant equal to that for a free electron gas (equation 14a), we obtain the two oscillation values expressed in (equation 23), where $(-)$ and $(+)$ give the symmetric and antisymmetric modes, respectively. These two values are shown in Figure 5*a* as a function of k for a given value of a.

Experimental separation of the two surface excitation modes was first accomplished for thin aluminum foils by Schmüser.[31] However, although the dispersion of the $\hbar\omega_{(-)}$ plasmon agreed with theory, no dispersion of the $\hbar\omega_{(+)}$ plasmon could be observed. Boersch et al.,[61] using a higher resolving power, reported shifts of both the $\hbar\omega_{(+)}$ and $\hbar\omega_{(-)}$ losses with foil thickness for the same metal. These measured results agreed with values given by equation 23 down to about 60-Å thickness (see Section 2.3.3).

As in the case of volume excitations, Schmüser obtained the surface plasmon excitation energy dependence on the wave vector of the oscillations, $\omega(k)$ (equations 23 and 24). There the velocity of light was considered to be infinite and retardation effects were not included.

In the relativistic approximation, the surface loss energies for small scattering angles are altered from the values given by equation 24. Here the velocity of light is no longer assumed infinite and retardation must thereby be taken into account. Ritchie and Eldridge[63] first considered this problem from the relativistic standpoint by solving the representative relativistic Dirac wave equation (see also Ref. 64). Later Otto[56] obtained the following expression for the dispersion relation by electric field continuity considerations at the surface of the foil of thickness a:

$$\varepsilon(\omega) = -\frac{v'}{v}\coth\left(v'\frac{a}{2}\right)$$
$$v^2 = k^2 - \left(\frac{\omega}{c}\right)^2 \qquad v'^2 = k^2 - \varepsilon\left(\frac{\omega}{c}\right)^2, \tag{25a}$$

for the symmetric solution, where the waves on the two surfaces are in phase with each other, and

$$\varepsilon(\omega) = -\frac{v'}{v}\tanh\left(v'\frac{a}{2}\right) \tag{25b}$$

for the antisymmetric solution, the two waves being out of phase by $\pi/2$. Setting $\text{Im}(\varepsilon) = 0$ and $\varepsilon = \text{Re}(\varepsilon) = 1 - \omega_p{}^2/\omega^2$, the solution of these equations for ω will give the dispersion relations between $\omega_{(\pm)}$ and k. Since the surface frequencies are always of a value $\omega_{(\pm)} \leq kc$, they cannot decay through the emission of electromagnetic radiation.

Neglecting retardation, equations 25a and 25b reduce to equations 24. Figure 5*a* compares values of $\omega_{(\pm)}$ as obtained from equations 24 (without retardation) and equations 25a and 25b (with retardation) for a 50-Å-thick aluminum foil for normal incidence $(\alpha = 0°)$.

2.3.1.1.2. *Inclusion of Retardation.* The calculation of the energy loss cross section for fast electrons in thin foils, considering retardation effects, has been carried out by Kröger,[65] whose article provides a thorough treatment of the problem. It is of interest here to point out the alteration of the foil-thickness-dependent first term, the volume excitation term. This term, having its counterpart in the term P in equations 21, is expressed as[82]

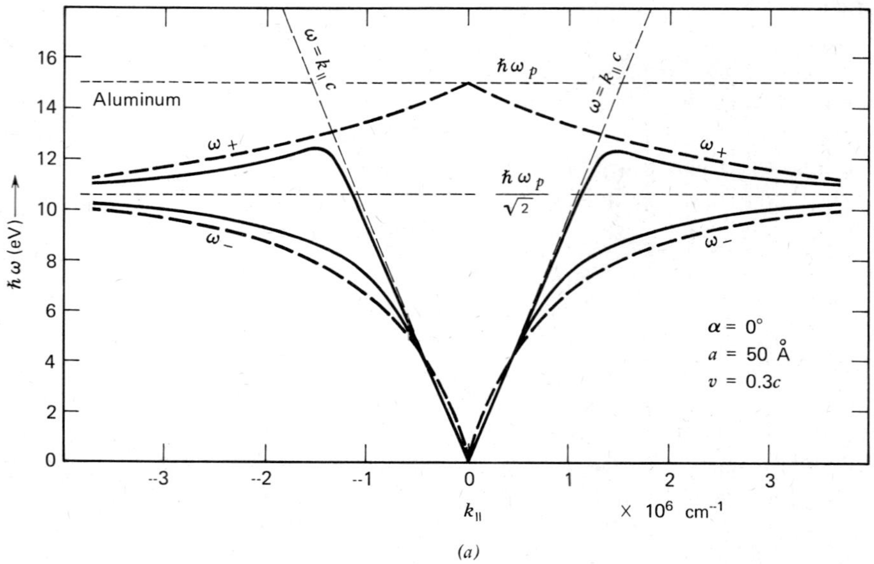

(a)

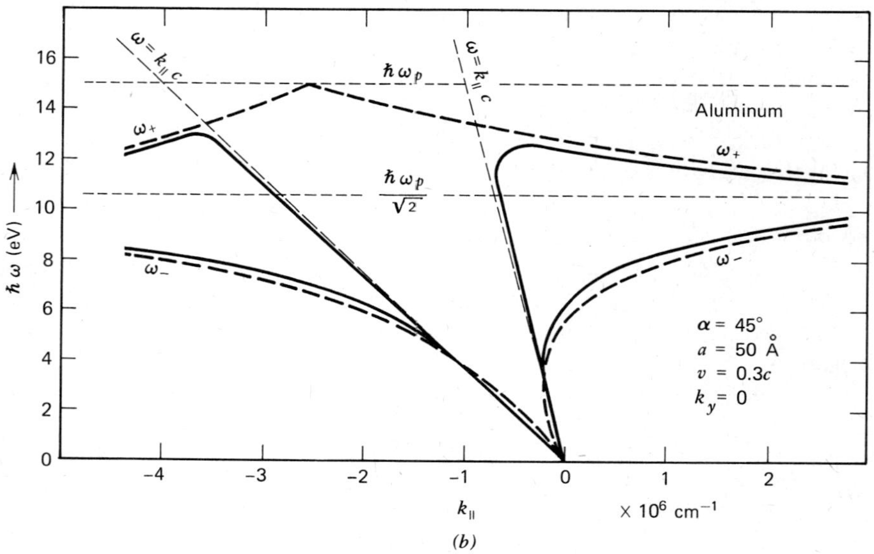

(b)

Figure 5. (*a*) The dispersion relations for electrons of velocity 0.3 *c* (~25 keV) striking perpendicularly an aluminum foil 50 Å thick. Results for omission (dashed curves) and inclusion (solid curves) of retardation effects are indicated. (*b*) Same as (*a*), but with the electron direction incidence 45° to the foil normal.

$$W_{\text{vol}}(\omega, \vartheta)\, d\omega\, d\Omega$$

$$= -a\,\frac{e^2}{\pi^2\hbar^2v^2}\cdot \operatorname{Im}\left(\frac{1}{\varepsilon}\right)$$

$$\cdot\frac{\vartheta^2+\vartheta_E{}^2[(1-\varepsilon_1\beta^2)^2+(\varepsilon_2\beta^2)^2]}{[\vartheta^2+\vartheta_E{}^2(1-\varepsilon_1\beta^2)]^2+(\vartheta_E{}^2\varepsilon_2\beta^2)^2}\,d\omega\, d\Omega, \tag{26}$$

where $\beta = v/c$ and the other symbols are as before. Here the final term includes the Čerenkov radiation probability for a fast electron transversing a medium, and the real part of the dielectric constant of this medium is large enough so that the $\varepsilon_1\beta^2 > 1$ condition for such radiation is fulfilled. Clearly in this case equation 26 is very different from that in which retardation is neglected. Here there is an extremely strong energy loss peak at a certain scattering angle ϑ_{max} where the denominator in equation 26 is at a minimum. According to Kröger,[65] a value ϑ_{max} can exist if $\varepsilon_1\beta^2 > \frac{1}{2}$. This means that for silicon and GaP, ϑ_{max} is of the order of 10^{-5} rad and is dependent upon $\varepsilon(E)$ and ϑ_E, and is therefore a function of the energy loss ΔE.

Measurements of the energy loss spectra for silicon

and GaP for zero scattering angle, using an angular resolution of 3×10^{-4} rad, showed energy loss peaks at ~ 3.0 and 3.3 eV[66] which were due to energy losses resulting from the Čerenkov radiation, which peaks out at these energies.

2.3.1.2. INCLINED INCIDENCE. The dielectric theory can again be applied for scattering of electrons falling on a thin foil at oblique, or inclined, incidence. The problem was initially treated by Stern and Ferrell[67] for a degenerate electron gas in bulk material. Subsequently, the reports of several theoretical treatments[55,58–60,68,69] and experimental investigations[50,70,58,68] of oblique incidence scattering have been made in recent years.

The scattering probability for inclined incidence as a function of scattering angle may be treated as follows (Figure 4*b* illustrates the relations among the several parameters involved). An incident electron of initial momentum $\mathbf{p}_i$ is scattered through an angle ϑ so that the momentum transfer $m_o(\mathbf{v}_i - \mathbf{v}_f)$ (see figure) takes place. (Note that neither $\mathbf{p}_i$ nor $\mathbf{p}_f$ is drawn to scale.) Only scattering in the xz-plane is shown; $\mathbf{k}_y$ can be considered later. A surface plasmon in the xz-plane will have a momentum of $\hbar\mathbf{K}$, and the energy loss $\hbar\omega$ will be dependent on this quantity.

Considering an additional momentum component in the y-direction, $\hbar\mathbf{k}_y$, we obtain for the scattering cross section,[59] in analogy to that expressed in equation 20

$$\begin{aligned} W(\omega, k_y, k_z) &= \frac{-e^2}{\pi^2 v^2 \cos\alpha}\left\{\frac{a}{k_y^2 + k_z^2 + \omega^2/v^2}\right. \\ &\quad \times \operatorname{Im}\left(\frac{1}{\varepsilon}\right) - \frac{2k_\parallel}{(k_y^2 + k_z^2 + \omega^2/v^2)^2} \\ &\quad \times \operatorname{Im}\left[\frac{(\varepsilon - 1)^2}{\varepsilon}\left(\frac{\cos^2(k_\perp a/2)}{\varepsilon + \coth(k_\parallel a/2)}\right.\right. \\ &\quad \left.\left.\left. + \frac{\sin^2(k_\perp a/2)}{\varepsilon + \tanh(k_\parallel a/2)}\right)\right]\right\}, \end{aligned} \tag{27a}$$

where $k_\parallel = \sqrt{K^2 + k_y^2}$, $k_\perp = (\omega/v)\cos\alpha - k_z \sin\alpha$, $K = (\omega/v)\sin\alpha + k_z\cos\alpha$, $\varepsilon = \varepsilon(\omega)$. Effects of retardation have not been taken into account, and the formula is not valid for $k_\parallel \gtrsim \omega/c$, as was the case for equation 20. Equation 27a may be reexpressed in quantities comparable to those used in equation 20 by setting

$$\vartheta_y = -\frac{\hbar k_y}{m_o v} \quad \text{and} \quad \vartheta_z = -\frac{\hbar k_z}{m_o v}. \tag{27b}$$

Restatement of equation 27a, assuming a free electron gas $\varepsilon = 1 - \omega_p^2/\omega^2$, has been calculated by Otto.[59] The expression corresponding to equation 27a based on the Romanov form of equation 21a is given by Geiger.[68]

2.3.1.2.1. *Dispersion Relation.* From Figure 4*b* we can see that a rotation of the xz-plane by an angle of α about the y-axis (thus obtaining the new x' and z' axes) will make it possible to restate the initial dispersion relation

$$\omega(\alpha, k_y, k_z) = \omega(0, k_y, K) \tag{28a}$$

where

$$\mathbf{k} = (0, k_y, k_z) \quad \text{and} \quad K = \frac{\omega}{v}\sin\alpha + k_z\cos\alpha.$$

Using equation 27b, we can obtain the dependence of the loss energy $\hbar\omega$ on the scattering angle $\boldsymbol{\vartheta} = (\vartheta_y, \vartheta_z)$

$$\hbar\omega(\alpha, \vartheta_y, \vartheta_z) = \hbar\omega(0, \vartheta_y, \vartheta_z\cos\alpha - \vartheta_E\sin\alpha),$$

$$\vartheta_E = \frac{\hbar\omega}{2E_0}. \tag{28b}$$

Using expression 28a, we can plot the surface plasma energy loss for an inclined angle α using the expression 24, without retardation, and equations 25 with retardation. For the case of $k_y = 0$, and $\alpha = 45°$, we obtain the curves in Figure 5*b* for aluminum (neglecting retardation). The inclusion of retardation will alter the curves for small $k_\parallel$ as shown diagramatically in the same figure.

The dispersion relation as a function of both k_y and k_z has the general form

$$\begin{aligned} \omega_{(\pm)}(k_y, k_z) &= \frac{\omega_p}{\sqrt{2}}(1 \pm e^{-k_\parallel a}) \\ &= \frac{\omega_p}{\sqrt{2}}(1 \pm e^{-a\sqrt{k_y^2 + (\omega_{(\pm)}\sin\alpha/v + k_z\cos\alpha)^2}}). \end{aligned} \tag{29}$$

Clearly, the expression is symmetric in k_y.

In the limit of an infinitely thick foil, we have[59] for the surface excitation contribution to equation 27a

$$W_s(\omega, k_y, k_z)\,(a \to \infty) = \frac{e^2}{\pi^2 v^2}\frac{1}{(k_y^2 + k_z^2 + \omega^2/v^2)^2}\frac{2k_\parallel}{\cos\alpha}\operatorname{Im}\left[\frac{(\varepsilon - 1)^2}{\varepsilon(\varepsilon + 1)}\right]. \tag{30}$$

2.3.1.3. LIMIT OF $\alpha \to 90°$. Expression 27a is inadequate in the limit of $\alpha = 90°$ since $1/\cos\alpha$ tends to infinity. Also, the excitation of a surface plasmon does not require a momentum transfer $\hbar k_z$ to the foil for $\alpha = 90°$.

The probability of excitation of a surface plasmon will be dependent on the distance b of the electron path from the foil surface. We obtain[59] for this probability per unit electron path length

$$\frac{dW(\omega, k_y)}{dx} = \frac{e^2}{v^2 \pi k_{\|}} e^{-2k_{\|} b} \times \mathrm{Im}\left\{(\varepsilon - 1)\left[\frac{\frac{1}{2}(1 + e^{-k_{\|} a})}{\varepsilon + \coth(k_{\|} a/2)} + \frac{\frac{1}{2}(1 - e^{-k_{\|} a})}{\varepsilon + \tanh(k_{\|} a/2)}\right]\right\},$$

$$k_{\|} = \sqrt{k_y{}^2 + \omega^2/v^2}. \tag{31}$$

Interaction between the passing electron and the bulk involves only the surface of the latter; volume plasmons are therefore not excited.

Reference 59 gives an extension of the treatment of this incidence limit case by the inclusion of retardation.

2.3.2. *Surface Excitations in Multilayer Films*

If a bulk material is bounded at its surface by a second bulk material of dielectric constant $\varepsilon' \neq 1$, oscillations in the interface will have a frequency given by equation 17. For the case of a thin foil, of dielectric constant ε, bounded on both surfaces by a material of infinite thickness having a dielectric constant ε', if we neglect retardation effects for the two surface plasma oscillation modes, we will have[31]

$$\omega_{(\pm)} = \frac{\omega_p}{\sqrt{1+\varepsilon'}}\left(\frac{1 \pm e^{-ka}}{1 \mp \gamma e^{-ka}}\right)^{1/2} \qquad \gamma = \frac{\varepsilon' - 1}{\varepsilon' + 1}, \tag{32}$$

where a is the film thickness. For $a \to \infty$, we find expression 17.

For a film of thickness a bound on both sides by a layer of dielectric material of thickness t, we must modify equation 32 to[31]

$$\omega_{(\pm)} = \omega_p \left\{\frac{1 \pm e^{-ka}}{(1+\varepsilon')(1+\gamma e^{-ka})} \cdot \frac{1 + \gamma e^{-2kt}}{1 - \gamma e^{-2kt}}\right\}^{1/2}. \tag{33}$$

(This equation differs from that given by Schmüser,[31] which does not provide the correct limits for $k \to 0$.) It is seen that equation 33 goes to $\omega_{(+)} = \omega_p$ and $\omega_{(-)} = 0$ for $k \to 0$ and to $\omega_{(\pm)} \to \omega_p/\sqrt{1+\varepsilon'}$ for $k \to \infty$, regardless of the value of t. This implies that, for sufficiently small values of $t > 0$, the dispersion relation 33 has a maximum at some value of k in the region $0 < k < \infty$.

Swan et al.[57] have calculated the dispersion relations with retardation for a three-layered foil, the inner material having a dielectric constant $\varepsilon(\omega)$, the outer two layers both having $\varepsilon'(\omega)$. For the surface plasma oscillations, $\omega_{(+)}$ and $\omega_{(-)}$, corresponding to tanh and coth, respectively, in the equation, they found

$$\varepsilon = -\varepsilon' \frac{\sqrt{k^2 - \varepsilon\omega^2/c^2}}{\sqrt{k^2 - \varepsilon'\omega^2/c^2}} \cdot \frac{1 - \gamma' e^{-2t\sqrt{k^2 - \varepsilon'\omega^2/c^2}}}{1 + \gamma' e^{-2t\sqrt{k^2-\varepsilon'\omega^2/c^2}}} \times \begin{Bmatrix}\tanh \\ \coth\end{Bmatrix}\left(\sqrt{k^2 - \varepsilon\omega^2/c^2}\right)\frac{a}{2} \tag{34}$$

$$\gamma' = \frac{\varepsilon'\sqrt{k^2 - \omega^2/c^2} - \sqrt{k^2 - \varepsilon'\omega^2/c^2}}{\varepsilon'\sqrt{k^2 - \omega^2/c^2} + \sqrt{k^2 - \varepsilon'\omega^2/c^2}}.$$

Swan and co-workers applied this theory to the investigation of surface excitation for an aluminum foil coated on both sides with thin layers of alumina. They observed that the energy $\hbar\omega_{(+)}$ at $k = 0$ and $\alpha = 0°$ does not increase to $\hbar\omega_p$ but decreases in accordance with the dispersion relation with retardation included.

The main effect of retardation is to limit the value of both $\hbar\omega_{(+)}$ and $\hbar\omega_{(-)}$ at low k-values. However, for a simple thin film the values of $\hbar\omega_{(\pm)}$ for low values of k approach $\hbar\omega = \hbar kc$ (see Figure 5*a*), but a metal film covered on both sides by a material of dielectric constant ε' will have for low k-values the energy asymptotic limit value of $\hbar\omega = \hbar kc/\sqrt{\varepsilon'}$.

The expression for the scattering probability for such a three-layer foil, without retardation and for $\alpha = 0$, may be written

$$W(\omega, \vartheta)\, d\omega\, d\Omega = -\frac{e^2}{\pi^2\hbar^2 v^2}\left\{\frac{a}{\vartheta^2 + \vartheta_E}\,\mathrm{Im}\left(\frac{1}{\varepsilon}\right) - \frac{2\vartheta}{k_0(\vartheta^2 + \vartheta_E{}^2)^2}\,\mathrm{Im}\left[\frac{1}{\varepsilon}\left(\frac{\varepsilon}{\varepsilon'} - 1\right)^2 \times \left(\frac{\sin^2(ak_0\vartheta_E/2)}{\varepsilon/\varepsilon_1 + \tanh(ak_0\vartheta/2)} + \frac{\cos^2(ak_0\vartheta_E/2)}{\varepsilon/\varepsilon_1 + \coth(ak_0\vartheta/2)}\right)\right]\right\} d\omega\, d\Omega, \tag{35a}$$

where ε and ε' are the dielectric constants for the central foil and the two surface layers, respectively. As before, the dispersion relations can be obtained by setting the denominators of the last two enclosed terms equal to zero, which gives

$$\frac{\varepsilon}{\varepsilon'} = -\tanh\left(\frac{ak_0\vartheta}{2}\right) \quad \text{and} \quad \frac{\varepsilon}{\varepsilon'} = -\coth\left(\frac{ak_0\vartheta}{2}\right) \tag{35b}$$

in analogy with equations 25. Also, expression 34 reduces to equation 35b for $c \to \infty$ and $\tau \to \infty$. On the other hand, for large a and setting $c \to \infty$ in expression 34, we obtain the relation given by Stern and Ferrell[67] in their treatment of surface oscillations.

Otto and Swan[58] investigated the dispersion relation of an oxide-covered aluminum foil for an inclined incidence angle of $\alpha = 45°$. The aluminum foils had thicknesses in the range of 100 Å and were covered on both sides with alumina layers about 40 Å thick. The measured spectra for scattering angles of $-1.86 \times 10^{-4} \leq \vartheta \leq 1.86 \times 10^{-4}$ in the xz-plane. The alumina surface layers had a dielectric constant of $\varepsilon' \cong 4$, so that the surface oscillation energy should have reached an asymptotic value of $\hbar\omega_p/\sqrt{1+\varepsilon'} =$ 6.7 eV. The asymmetry of the scattering probability for angles about the central angle $\vartheta = 0$ was clear.

An alternative possibility would be that of a thin conducting film covered on its two sides with materials of two different dielectric constants ε' and ε''. For large k and $a \to \infty$ the oscillation energies for the two interfaces will approach different asymptotic values, namely, $\omega_p/\sqrt{1+\varepsilon'}$ and $\omega_p/\sqrt{1+\varepsilon''}$. If one of the surfaces is bounded by vacuum, the corresponding asymptotic value for that surface will become $\omega_p/\sqrt{2}$. The dispersion relation, without retardation, will again give the two values ω_p and 0 for the limits of these two branches for $k \to 0$. Therefore, for such a foil bounded by materials of two different dielectric constants, the $\omega_{(\pm)}$ branches remain separated even in the limit $k \to \infty$.

2.3.3. *Dependence of the Dispersion Relation on Film Thickness*

The dispersion relations of equation 35b can be restated in exponential form for the general case of a thin foil covered on both sides by a layer of material of dielectric constant ε'. Replacing the expression $ak_0\vartheta/2$ by $ka/2$ and multiplying both numerator and denominator by $e^{-ka/2}$, we obtain the general expression

$$\frac{\varepsilon}{\varepsilon'} = -\frac{1 \mp e^{-ka}}{1 \pm e^{-ka}}$$

which, when solved for a, yields

$$a = \frac{1}{k_0\vartheta} \ln \left| \frac{\varepsilon/\varepsilon' - 1}{\varepsilon/\varepsilon' + 1} \right|, \tag{36}$$

after k has been set equal to $k_0\vartheta$; k_0 is the wave number for the incoming electron. Setting $\varepsilon = \text{Re}(\omega)$ in equation 14a for the free-electron gas and solving for $\hbar\omega$, we have in the limit of $a = 0$, $\hbar\omega_{(+)} = \hbar\omega_p$ and $\hbar\omega_{(-)} = 0$.

Including retardation effects, Imbusch[71] obtains in a like manner from the dispersion relationship

$$a = \frac{1}{((k_0\vartheta)^2 - \varepsilon(\omega/c)^2)^{1/2}} \ln \left| \frac{f-1}{f+1} \right| \tag{37}$$

where

$$f = \frac{\varepsilon}{\eta} \left(\frac{(k_0\vartheta)^2 - \eta(\omega/c)^2}{(k_0\vartheta)^2 - \varepsilon(\omega/c)^2} \right)^{1/2},$$

and c is the speed of light.

In the limit of $a = 0$, expression 37 gives the following for the $\hbar\omega_{(-)}$ and $\hbar\omega_{(+)}$ surface plasma excitations

$$\hbar\omega_{(-)}(a=0) = 0$$

$$\hbar\omega_{(+)}(a=0) = \frac{c}{\sqrt{\varepsilon'}} \hbar k_0 \vartheta. \tag{38}$$

Placing the $\omega_{(-)}$ value into the Drude formula 14a for a free-electron gas, we find $\varepsilon_1 = \infty$ for the associated dielectric constant. Associated with the $\hbar\omega_{(+)}$ branch we have $(k_0\vartheta)^2 - \eta(\omega/c)^2 = 0$. From equation 38, $\hbar\omega_{(+)}$ has a limit for each scattering angle ϑ, which can be less than the volume plasma loss $\hbar\omega_p$.

This means that, in the case of a single thin foil (Figure 5*a*) and for a given value of $k < \omega_p/c$, the maximum value of $\hbar\omega_{(+)}$ in the limit of $a \to 0$, will be just $\hbar\omega_{(+)}(a=0;\ k<\omega_p/c) = \hbar kc$. When the foil is covered on both sides by a layer of material of dielectric constant ε', then for a value of $k < \omega\sqrt{\varepsilon'}/c$ the value of $\hbar\omega_{(+)}$ for $a \to 0$ will be $\hbar\omega_{(+)}(a=0;\ k < \omega\sqrt{\varepsilon'}/c) = \hbar kc/\sqrt{\varepsilon'}$ as in equation 38.

In the limit of $a = \infty$, a single energy value, $\hbar\omega_{(\pm)}$, is obtained from equation 37. This value can be obtained[71] by solving

$$\varepsilon = \varepsilon' \left(\frac{(k_0\vartheta)^2 - \varepsilon(\omega/c)^2}{(k_0\vartheta)^2 - \varepsilon'(\omega/c)^2} \right)^{1/2}. \tag{39}$$

2.4. Polarization Wave Excitations

Electron energy loss experiments carried out in the early 1960s by Robins, Swan, and Best[72–74] yielded information on plasma excitation energy losses and bound shell ionization losses. Also observed in the loss spectra for a large number of the elements were broad, intense peaks in the 20 to 60 eV region, called "*G*" loss peaks, whose origin could not be explained. That the loss peaks were not due to atomic absorption processes involving single-electron excitation was shown by comparing the loss spectra for potassium and calcium with measurements of electron energy loss measurements for the metallic vapors.[75]

The elements potassium and calcium[74] and titanium, vanadium, and chromium[72] show the "*G*" loss peak, but it disappears for manganese. The second long period of the periodic table starting with rubidium[76] and strontium[77] showed the same type of structure, at least to molybdenum[78] (i.e., from $Z = 37$ to 42). The third long period, beginning with cesium[79] also shows this peak, and it appears certain that the structure is present for all members of the rare earth group elements up to rhenium (i.e., from $Z = 55$ to 75), although only a few of the elements have been investigated (see Ref. 80). A separate study of the second and third series was reported by Lynch and Swan.[81] It was noted that the nature of the energy loss function alters progressively in going from the first to the third long period. Also, energy loss structure appears to arise in elements having the *d*- or *f*-shell near the Fermi surface: the 3*d* for the first long period, 4*d* for the second long period, 4*f* and 5*d* for the third long period.

Hartley suggested a cause for this effect in connection with his study of electron energy losses in strontium.[77] A broad peak was observed at 29.5-eV energy loss and was attributed to the excitation of a polarization wave in the $N_{2,3}$-band. In other words, the oscillator strength of the $N_{2,3}$-band is sufficient to depress the value of ε_1 to negative values. As ε_1 becomes positive again at a higher energy, the possibility of a plasmon or "polarization wave" excitation arises. Therefore, the "polarization wave" is actually a volume plasma excitation that arises at the energy value at which the real part of the total dielectric constant passes through zero from minus to plus.

Subsequently, Hartley[80] reported calculations on this effect. The analysis was made by considering the complex dielectric constant of the solid.

Calculations were performed on an assigned Gaussian shape of the polarization wave contribution to the total imaginary part of the dielectric constant

$$\delta\varepsilon_2\,(\Omega) = \exp -\frac{(\Omega - B)^2}{A}, \tag{40}$$

where B determines the energy position and A the width of the function. Calculations of $\delta\varepsilon_1\,(\Omega_0)$ were carried out for various values of A and B using the Kramers–Kronig relationship

$$\delta\varepsilon_1\,(\Omega_0) = \frac{2}{\pi}\int_0^\infty \frac{\Omega\,\delta\varepsilon_2\,(\Omega)\,d\Omega}{\Omega^2 - {\Omega_0}^2}, \tag{41}$$

where $\delta\varepsilon_1\,(\infty) = 0$; see also equation 12c and related text. The Ω can be made dimensionless by setting $\Omega = \omega/\omega_c$; ω_c is an arbitrary constant. Equation 41 was solved for several values of Ω with normalization being effected using equation 15a:

$$\int_0^\infty \Omega\,\delta\varepsilon_2\,(\Omega)\,d\Omega = \frac{2\pi^2 ne^2}{m_o {\omega_c}^2} = \text{constant},$$

which shows that $\delta\varepsilon_2(\Omega)$ is dependent on the number of electrons involved in the absorption process. When changed, these electrons will cause an alteration in height of $\delta\varepsilon_2\,(\Omega)$ and $\delta\varepsilon_1\,(\Omega_0)$.

Calculations of the energy loss function $-\text{Im}\,(1/\varepsilon)$ were made assuming $\varepsilon_1 = 1 + (\pi/2)F\,\delta\varepsilon_1$ and $\varepsilon_2 = C + (\pi/2)F\,\delta\varepsilon_2$ for various values of F. In the first equation $\varepsilon_1(\infty) = 1$ gave the first term; C is a damping constant that controls the energy loss function peak height. Figure 6 shows $-\text{Im}\,(1/\varepsilon)$ for several values of F. The peak of the function is seen at $\Omega = 3$, marked by the longer vertical line, and has a width $2\Omega/3$. For $1.2 \leq F \leq 2.8$, $\varepsilon_1(\Omega)$ is negative in some region, which leads to extra structure in the loss function. The peak does not occur at the $\varepsilon_2(\Omega)$ peak position regardless of the sign of $\varepsilon_1(\Omega)$, but it is displaced to higher values of Ω. The most constantly positioned feature for the different F-values is the minimum preceding the ionization peak (marked by the short vertical line); this position corresponds to the absorption edge in the function for $\varepsilon_2(\Omega)$.

As explained earlier, a maximum in $-\text{Im}\,(1/\varepsilon)$, where $\varepsilon_1(\Omega) = 0$ and $\varepsilon_2(\omega)$ is small, is an approximation of the criterion for volume plasma oscillation creation. For $1.2 \leq F \leq 2.8$, induced collective oscillation is a possible mode of energy loss. However, for $F = 0.4$ and 0.8, collective behavior is not possible, although individual electron excitation contributes weakly to $-\text{Im}\,(1/\varepsilon)$.

The energy loss spectrum for graphite has been attributed to collective oscillations (at $\sim$7 and $\sim$25 eV).[44] Hartley,[80] however, suggested that the peak at $\sim$25 eV is due to the excitation of a polarization wave (see Section 3.2.1).

The binding energies of $L_{2,3}$, $M_{2,3}$, $N_{2,3}$, $O_{2,3}$ levels in some 24 elements have been reevaluated by taking polarization wave effects into account.[80] (See also Chapter 7, Section 3.6.)

3. SINGLE-ELECTRON EXCITATIONS

Electron energy losses due to the creation of a single plasma excitation, volume or surface, lie in the energy region from about 1 to 20 eV. This excludes the special case of "polarization wave" excitation, which generally lies at higher energies. Other energy absorption processes lying in this same energy region and capable of arising in conductors and semiconductors alike include those producing interband transitions (Section 3.1). In addition, at lower energies we can

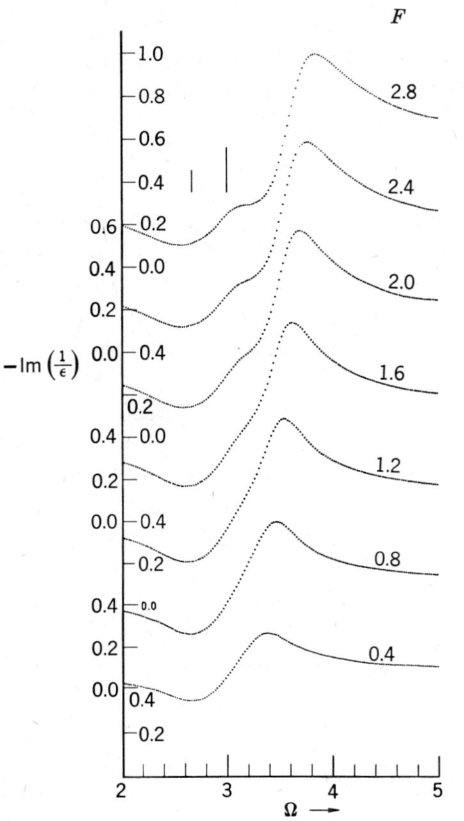

Figure 6. The function $-\mathrm{Im}\,(1/\varepsilon) = \varepsilon_2/(\varepsilon_1{}^2 + \varepsilon_2{}^2)$ for various values of F, as calculated by Hartley.[81]

expect intraband transitions in conductors where a transition takes place from a lower filled to a higher vacant state in the conduction band (Section 3.1).

However, since in inter- and intraband transitions we do not expect the electron to be removed far from the related newly formed vacancy, the interaction between the excited electron and the hole can lead to a type of coupling interaction that is called an exciton state. Interband transitions may create final excited states in the form of excitons both in semiconductors (Section 3.2.2) and in crystals of solidified rare gases (Section 3.2.4).

Toward higher loss energies ($E \gtrsim 20$ eV), energy absorption processes due to ionization of inner shell electrons may be investigated. Even higher binding energies can be measured by this method, although the cross section for inner shell electron excitation through electron impact decreases strongly with increasing binding energy (Section 3.2.5).

In the energy region of approximately 20 to 60 eV, however, and in special cases at lower energies, collective mode excitation of an energy lying close to the binding energy value of interest can influence the measured value of the latter (treated in Section 2.4).

Investigations made into the foregoing phenomena by measuring electron energy loss spectra should be carried out for small scattering angles, if comparisons with optical investigation results are desired. This is because the direct comparison of electron energy loss spectra with optical data is justified only in the limit of zero scattering angle where dispersion effects do not arise.

Recent computations on energy loss cross sections for high energy ($\gtrsim 50$ keV) impact electrons have shown that, when retardation effects are included, energy losses can be experienced by the electrons through Čerenkov radiation emission.[65] For this to happen, ε_1 for the scatterer material must be high. Losses occur in silicon and in GaP[82,66] at energies of ~3 eV, which indicates that low energy losses in other semiconductors or insulators of like energies are to be expected (see Section 3.2.2). It is possible that some of the low energy losses heretofore identified in like materials as arising from interband transition excitation are actually owing to Čerenkov radiation energy losses.

3.1. Exciton Creation and Intra- and Interband Transitions

Electron transitions in conductors and semiconductors from the valence or a more tightly bound band to a free state in the valence or conduction band can be induced by photon absorption or electron bombardment, among other methods. Transitions to the Fermi level usually dominate in conductors, although transitions to other higher lying bands are also possible.

In photoabsorption, the induced interband transitions, plasma excitations, and so on appear as fluctuations in the relative reflectivity and transmissivity of the material. From such optical measurements the complex dielectric constant can be determined. In electron energy loss measurements, the occurence of such transitions again appears as structure in the scattered electron energy spectrum. This structure can be directly correlated with the $-\mathrm{Im}\,(1/\varepsilon)$ function, which can be calculated from the optical measurements.

Intraband transitions may also be investigated by both optical and electron scattering methods, but since these lie at energies that are in general much lower than the interband transition energy region, they have evinced little interest. This process represents the transfer of an electron from its original state to an unoccupied one in the same band, and therefore we expect it is to be the strongest in conductors in which the valence band is partially filled.

An electron may be transferred by some process of

excitation from an atomic state or band in a solid material to a vacant state in a band of higher energy. It may then be confined to the vicinity of that vacancy through its coupling with the created positive hole. The electron-hole pair configuration thus created is called an exciton.

The exciton state resembles a hydrogen atom in that the excited electron is in an orbital state about the positive hole, as determined by the electron-hole Coulomb interaction. For materials having small atomic separation and strong interatomic interaction and for a large exciton orbital radius, the electron and hole are effectively separated by a dielectric medium (as, e.g., in solidified argon and xenon, Section 3.2.4). However, this situation is not allowed by the electronic structures of most solids, and exciton states usually exhibit tight binding and small radii compared to the interatomic distance.

There is nevertheless a certain tendency for the hole and the excited electron of an exciton to wander independently of each other in the solid. With greater excitation energies (i.e., larger electron-hole distances), coupling between the two components diminishes and the system is more and more strongly perturbed by the independent interactions of the excited electron and of the hole with their immediate environments. As close binding is thereby diminished, it is increasingly appropriate to treat the situation in terms of separately moving holes and electrons, a topic of central interest to the field of solid state physics.

The possibility of exciton creation in metals has been a point of controversy for some time. This is because the associated electron-hole interaction might be expected to be screened out by the sea of conduction electrons. However, in the light of more recent considerations, it appears that metallic excitons are indeed possible.[83]

Recently the prediction of a new type of excitation state in metals was published. This state, called a plasmaron, arises through the interaction between a plasmon and a hole in the electron plasma.[84] Evidence of the creation of such an excitation has been sought by Cuthill et al.[85] in the $L_{2,3}$ X-ray-emission spectrum for aluminum. A weak structure was found at the predicted position.

The expected narrower width of the plasmaron edge for the alkali metals, compared with that for aluminum, makes it desirable to extend the investigation to these elements in search of definite proof of the excitation's existence (see Ref. 85).

For ionic crystals, besides the multicomponent energy absorption spectra due to exciton creation, we see a spectral contribution due to plasma oscillations in the valence band (Section 3.2.3).

Except for a few special cases, the topic of interband transitions is now considered and treated in terms of the dielectric theory. For further treatments of exciton creation in semiconductors see the work by Dexter and Knox,[86] Kohn,[87] Phillips,[88] and Evans and Young.[89]

3.1.1. *Intraband Transitions*

We can expect transitions of low energy and low intensity to take place between filled and unfilled states of the same band. The probability of such transitions will be roughly dependent on the product of the numbers of initially filled and vacant states. Therefore such transition intensities are greatest for conductors in which the conduction band is about half-filled.

Intraband transitions will be of energies less than approximately the bandwidth energy value, which can have an upper limit of a few electron volts. Weak transitions of this type might also occur in semiconductors at temperatures far above absolute zero, however, if the forbidden gap is small enough.

Structure in optical dielectric constant functions and in electron energy loss functions due to interband transitions should resemble each other because of the phenomenological similarities between these functions. For this reason it will be difficult to distinguish between the two origins on the basis of spectral structure. Therefore the treatment of interband transitions in the next section also pertains by analogy to intraband transitions.

3.1.2. *Interband Transitions*

Assume that an interband transition may take place from a state g in the band a to a state f in band b. The energy of such a transition will be $\hbar\omega(a_g, b_f) = E(a_g) - E(b_f)$, where the E's are the energies of the states a_g and b_f as indicated. Such a resonance transition adds in effect a negative term to the expression for the dielectric constant[90,91,6,4,37,92,93] equation 13a:

$$\begin{aligned}\varepsilon(\omega) &= 1 - \frac{\omega_p^{\,2}}{\omega}\left(\omega + \frac{i}{\tau}\right)^{-1} + \delta\varepsilon^{(b)}(\omega)\\ &= \varepsilon^{(f)}(\omega) + \delta\varepsilon^{(b)}(\omega)\\ &= \varepsilon_1(\omega) + i\varepsilon_2(\omega), \end{aligned} \tag{42}$$

where $\varepsilon^{(f)}(\omega)$ is that contribution which is owing to the excitation of the free electron plasma and which is close to zero for semiconductors. The last expression in equation 42 is given in terms of the real and imaginary parts of the dielectric constant as obtainable from optical measurements (see equation 44). The term $\delta\varepsilon^{(b)}(\omega)$ is the contribution to the dielectric constant owing to induced transitions of bound electrons. It consists essentially of the sum over the individual states g and f in the bands themselves. This term in equation 42 has the form

$$\delta\varepsilon^{(b)}(\omega) = -\frac{e^2}{m_o\pi^2}\int\sum_{a,b}{}' f_a(g) \times \frac{f^{\mu}_{a,b}}{(\omega + i/\tau_{a,b})^2 - \omega^2_{a,b}}\,dg. \quad (43a)$$

The factors are: $f_a(g)$, the Fermi distribution function for the state g in the band a; $f^{\mu}_{a,b}$, the oscillator strength;[90] and $\tau_{a,b}$, the interband relaxation time.

It is useful to note that interband transitions do not set in until a lower energy limit of about 0.75 eV (i.e., $\varepsilon_2{}^{(b)} = 0$ and $\varepsilon_1{}^{(b)} \neq 0$, but is small). This makes it possible to fit experimental data for low energies to

$$\varepsilon(\omega) = \varepsilon^{(f)}(\omega) = 1 - \frac{f_0\,\omega_p{}^2}{\omega^2 + i\omega/\tau}, \quad (43b)$$

where, neglecting the interband term, f_0 is the oscillator strength for the free-electron contribution to $\varepsilon(\omega)$. For some metals interband transitions may not set in until higher values (e.g., for silver at about 4 eV); however, the finite value of $\delta\varepsilon_1{}^{(b)}$ for $\hbar\omega < 4$ eV is essential in explaining the loss at 3.78 eV.

In Drude-type formulation of plasma excitations of a free-electron gas, it is not necessary to consider effects due to interband transitions. However, even for conductors in regions of predominating interband transitions, the dielectric constant is expected to show structure resembling that for semiconductors. Thus the Drude theory is inadequate here.

From optical measurements we may determine the real and imaginary parts of the dielectric constant to which both electron plasma excitations and interband transitions contribute. With the help of equation 14a and neglecting $\tau_{a,b}$ in equation 42, we may express the real part of the dielectric constant as

$$\varepsilon_1(\omega) = 1 - \frac{\omega_p{}^2}{\omega^2} = 1 - \frac{\omega_p{}^{*2}}{\omega^2} + \delta\varepsilon_1{}^{(b)}(\omega). \quad (44)$$

The value ω_p is the volume plasma oscillation frequency measured experimentally; $\omega_p{}^*$ is the free plasma oscillation value. This frequency value coincides with the point at which the sum of the contributions to $\varepsilon_1(\omega)$ from both interband transitions and plasma oscillation excitations passes through zero from negative values. That is,

$$\omega^2(\varepsilon_1 = 0) = \omega_p{}^2 = \frac{\omega_p{}^{*2}}{1 + \delta\varepsilon^{(b)}(\omega_p)}. \quad (45)$$

Therefore the experimental plasma oscillation lies at an energy that is less or greater than the free electron plasma value depending on whether the interband transition function $\delta\varepsilon_1{}^{(b)}(\omega)$, is situated at an energy above or below, respectively, the free plasma oscillation energy $\hbar\omega_p{}^*$. If there are interband transitions lying at other ω-values, they are included in the expression $\delta\varepsilon^{(b)}(\omega)$, equation 42. Each will be expected to influence individually and to a small extent the value ω_p, depending on how close their energy values lie to that of $\hbar\omega_p$.

The amplitude of the function $\delta\varepsilon^{(b)}(\omega)$ is proportional to the oscillator strength $f^{\mu}_{a,b}$, which in turn is roughly equal to the electron density in the valence band. We would therefore expect transitions from filled s- and p-bands to contribute more weakly than those from filled bands. This is borne out by experiments on certain metals. The results are treated briefly in Section 3.2.1 for illustration.

3.2. Electron Energy Losses in Solids

The characteristics of energy losses experienced by electrons in passing through a solid are closely dependent on both the atomic electron orbital structure of the individual atoms comprising the sample and the electron band structure of the material. Energy loss structure for conductors in the low energy region will be comprised of intra- and interband transitions, as well as possible low-lying plasma excitations.

Several examples of the general characteristics of energy loss spectra obtained for the various types of materials follow. The materials are divided under headings bearing on their main unifying characteristics: for example, conductors refers in general to all metallic materials, and alloys also come under this heading. The semiconductors are divided into several groups under the headings of intrinsic semiconductors, insulator compounds, alkali halides, and solidified rare gases.

3.2.1. Conductors

The noble metals, copper, silver, and gold, are of particular interest because their just-filled $3d$, $4d$, and $5d$ levels, respectively, form bands having very low binding energies. The outer atomic levels of these metals have been subjected to optical and photoelectron measurement by several investigators.

3.2.1.1. COPPER. According to photoelectron measurements, the $3d$ and the $4s$ electrons in copper lie at a binding energy of about 2 to 6 eV.[94] Interband transitions from this $3d$-band are expected. However, according to Figure 7a the sum of $\varepsilon_1{}^{(f)}$ and $\delta\varepsilon_1{}^{(b)}$ at no energy below 6 eV gives $\varepsilon_1 = 0$. Therefore interband excitation processes from the $3d$ band will not cause a high cross section at any unique energy value in the

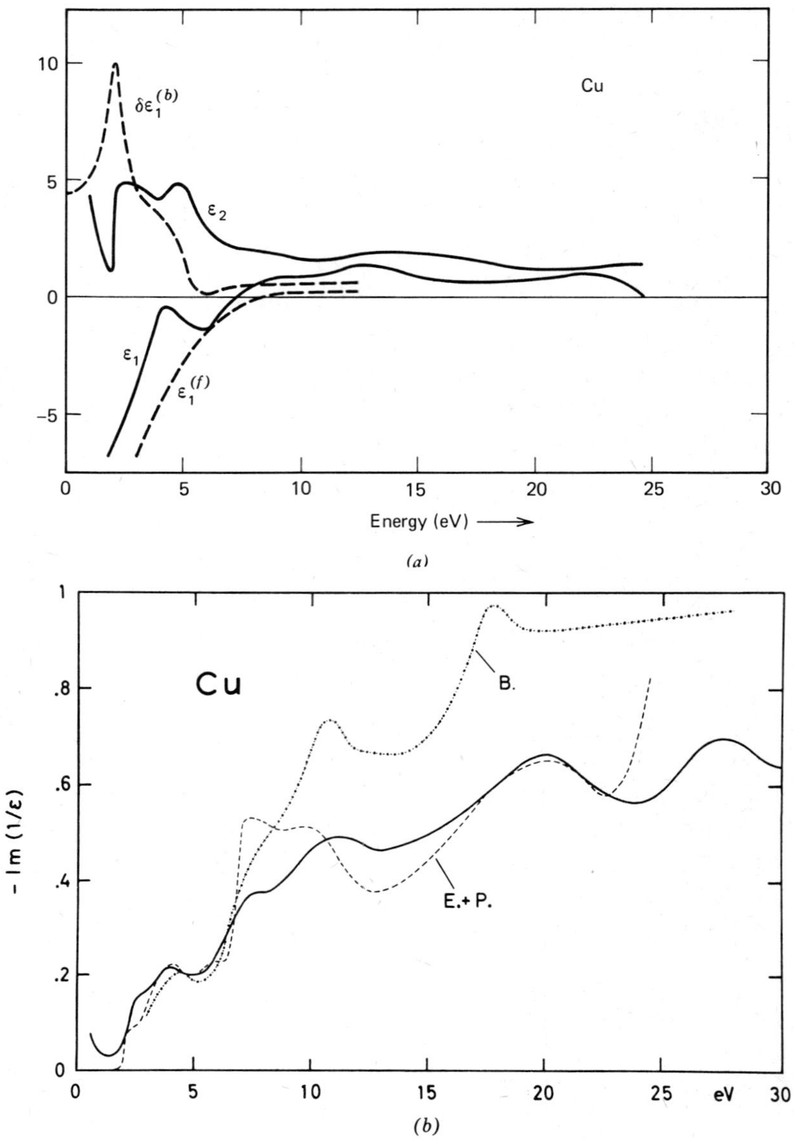

Figure 7. (*a*) The real and imaginary parts ε_1 and ε_2 of the dielectric constant for copper as determined from optical measurements by Ehrenreich and Philipp[91]. The function ε_1 is analyzed into the free $\varepsilon^{(f)}$ and bound, $\delta\varepsilon_1{}^{(b)}$, contributions. (*b*) Energy loss function, $-\mathrm{Im}\,(\delta/\varepsilon)$ for copper as obtained from the electron energy loss spectrum by Creuzburg[99] (solid curve) compared with those calculated from optical measurement determination of the dielectric constant: B, from Ref. 40; E + P, from Ref. 91.

energy loss spectrum. This is demonstrated in Figure 7*b*, where the function $-\mathrm{Im}\,(1/\varepsilon)$ obtained from optical measurements is compared with electron scattering measurements, with fair agreement. (For other work see also Refs. 4, 95–98.)

3.2.1.2. SILVER. Figure 8*a* reveals an intense peak at about 3.9 eV in the function $-\mathrm{Im}\,(1/\varepsilon)$ as calculated from optical absorption data for silver. The electron energy loss data clearly agree with this. The values of ε_1 and ε_2 appear in Figure 8*b*, where the ε_1 is analyzed into its $\varepsilon_1{}^{(f)}$ and $\delta\varepsilon_1{}^{(b)}$ component functions. The sum of these functions gives a value of $\varepsilon_1 = 0$ at the strong energy loss and optical absorption peak positions. The total real part of the dielectric constant can be compared with the energy loss spectrum as in Figure 8*c*. In the high resolution measurement in Figure 8*c*, the doublet loss structure, resolved into close-lying surface and volume plasma excitations, is visible.

That the 3.63- and 3.78-eV losses are surface and volume losses, respectively, has been tested by observing the change in the loss spectrum when a thin foil is covered on both sides with layers of carbon.[101] The peak positions agree with the $\varepsilon_1 = -1$ and 0 positions. It is found that the 3.78-eV peak remains unaltered,

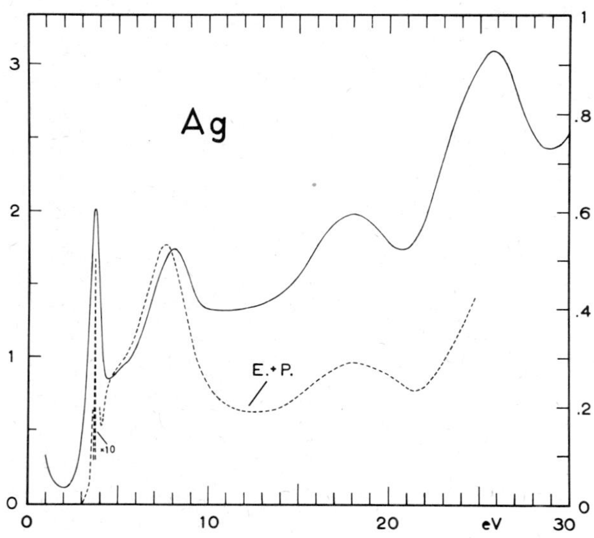

(a)

Figure 8. (*a*) The energy loss function $-\text{Im}(1/\varepsilon)$ for silver as obtained from the electron energy loss spectrum measured by Creuzburg[99] (solid curve, left-hand scale). Comparison is made with the loss function calculated from optical measurement determination of the dielectric constant in Ref. [91] (right-hand scale).

but the 3.63-eV peak is shifted and weakened with increasing carbon-layer thickness. (See also Refs. 56, 69.)

Figure 8*d* shows the photoelectron spectrum for silver obtained[102] with monochromatic vacuum ultraviolet light of the energy 10.1 eV. In this spectrum the binding energy distribution of the outermost electron states in silver is visible. The 4*d*-band lies in the binding energy region 3.5 to 5.5 eV, and the flatter structure arises from the 5*s*-conduction band. The position of the Fermi level is also seen.

Figure 8*a* displays a loss peak at about 7.5 eV for both the $-\text{Im}(1/\varepsilon)$ function and the electron energy loss spectrum. We are led to suspect that this loss is owing to a collective oscillation in the conduction band, since the ε_1 and ε_2 functions vary to give a maximum in the $-\text{Im}(1/\varepsilon)$ value at this point and $-\text{Im}(1/\varepsilon)$ shows a free-electron energy dependence. However, an interband transition also exists near this energy, as discussed below.

It is interesting to note further that for silver in vapor form electron energy losses have been found[103] at the energies 3.7, 7.3, and 8.0 eV. Here collective excitation is not possible and we interpret the reso-

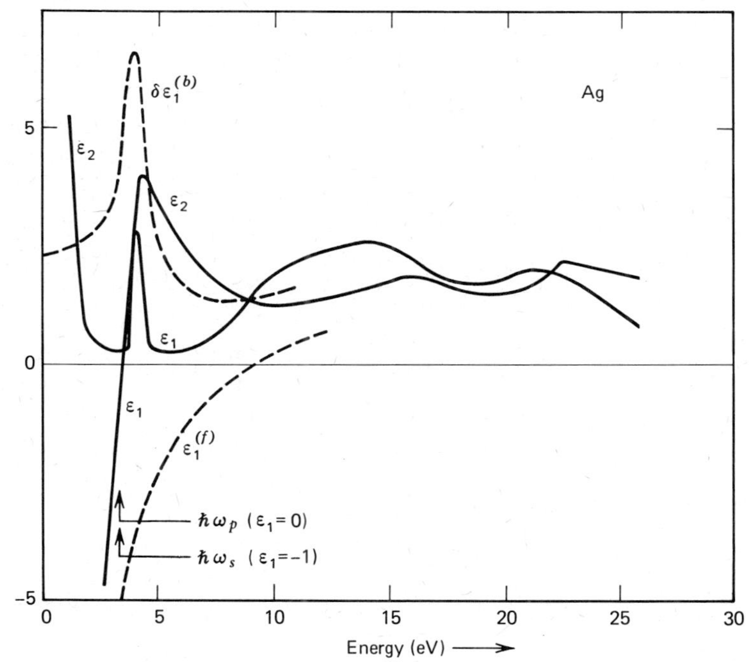

(b)

Figure 8. (*b*) The real and imaginary parts of ε for Silver.[91] The function ε is analyzed into the free $\varepsilon_1^{(f)}$ and bound $\delta\varepsilon_1^{(b)}$ contributions.

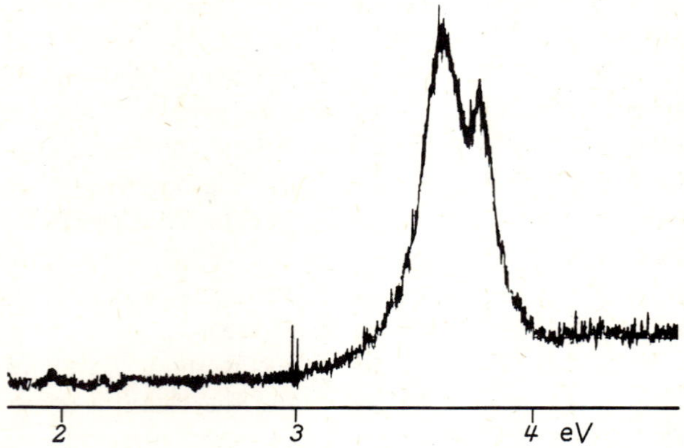

(c)

Figure 8. (*c*) High resolution measurement of the 3 to 4 eV electron energy loss region for Silver as measured by Daniels[100] A 640-Å thick foil was used and the impact electrons were of 50-keV initial energy. It is possible to distinguish between the volume and surface plasmon excitation contributions.

nances as electron excitation from the 5*s*- to 5*p*-states (~3.7 eV) and from the 4*d* to 5*p* states. This interpretation is consistent with optical data. (For other work see also Refs. 28, 60, 95, 104–107.)

3.2.1.3. GOLD. Gold, having the same basic electron band structure as silver and copper, might be expected to exhibit a low energy interband transition contribution in an electron scattering experiment or in optical

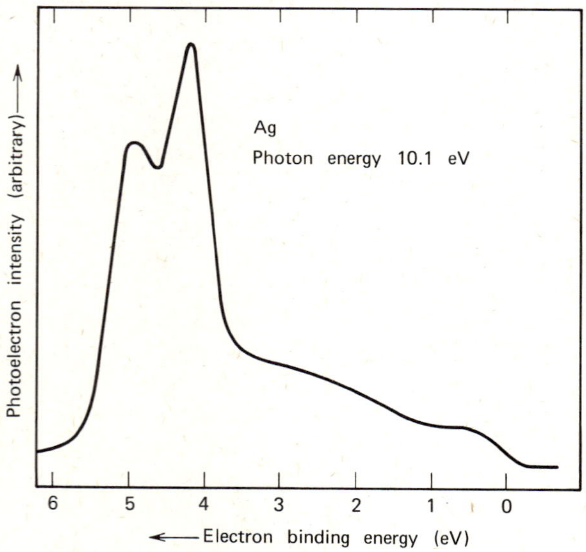

(d)

Figure 8. (*d*) The photoelectron spectrum for Silver using 10.1-eV photons. The strong contribution from the 4*d*-photoelectrons at 3.8 to 5.5 eV binding energies overlaps with the 5s-contribution extending to zero binding energy. The 4*d*-peak has two components, associated with the $4d_{3/2,5/2}$ subshell splitting. (From Ref. 102.)

absorption data. Several workers have indeed reported a low energy loss lying in the 2.5-eV region.[97,108]

The real and imaginary parts of the dielectric constant for gold measured by Joos and Klopfer[109] between about 2 and 5 eV show that $\varepsilon_1 = 0$ at about 3 eV, at a point where ε_2 is increasing strongly. Therefore we would expect a maximum in $-\text{Im}(1/\varepsilon)$ to arise at an energy just below this energy value (see Ref. 110).

The loss at this energy is less outstanding in the electron scattering spectrum than the 3.7 eV peak is in the spectrum for silver. Furthermore, Carillon[111] has shown that its intensity relative to that of the large broad peak at 24 eV decreases by a factor 2 as the gold foil thickness goes from 153 to 635 Å. The energy of the loss alters from 2.4 to 3.0 eV for the same foil thickness change, as well. This is in essence the behavior of the 3.7-eV peak in silver, and it suggests that the peak may actually be an unresolved combination of a surface and volume plasma loss as in the silver case. The gold loss peak at 2.5 eV then may occur because of an interband transition at about that energy that causes ε_1 to take values of 0 and -1 associated with volume and surface losses, in analogy to the situation for silver. (For other work see Ref. 95.)

The effect of temperature changes of the band structure in metals will be altered through a volume dilation effect, shifting the volume and surface plasma oscillation energies together with other temperature dependent effects (i.e., mainly interband transition probabilities). Such temperature dependence of free and bound electron excitation contributions to the dielectric constant may be investigated by a number of optical and other methods (e.g., Ref. 112).

Two other conductors are of special interest with regard to interband transitions: aluminum and carbon (graphite). In the first case, an interband transition exists at 1.48 eV. This transition is of such low energy that we are confined to optical methods for measuring its characteristics in any great detail. Graphite is special because of the uncomplicated atomic electron state distribution and the unique interpretation of its electron energy loss spectrum. Carbon in diamond form exhibits similar unique qualities,[113,114] which are not treated here.

3.2.1.4. ALUMINUM. The observation of the 1.48-eV interband transition excitation in aluminum has been reported a number of times in past years. The investigation of this transition with highest resolution and in greatest detail, however, has been provided only by optical techniques. Ehrenreich et al.[47] thoroughly investigated and analyzed the transition. They used their measurement of the dielectric constant to calculate the function $-\text{Im}(1/\varepsilon)$ and thereby obtained an electron energy loss function. The dielectric constant could be separated into $\varepsilon^{(f)}(\omega)$ and $\delta\varepsilon^{(b)}(\omega)$ (equation 42.) For low energies ($\gtrsim 0.75$ eV) $\varepsilon^{(f)}(\omega)$ is the dominating term in $\varepsilon(\omega)$, and $\delta\varepsilon^{(b)}(\omega)$ is negligible. Fitting the experimental values of ε_1 and ε_2 at 0.4 eV with the corresponding components in equation 42, the values of $\hbar\omega_p = \hbar\omega_p{}^*$ (if $\delta\varepsilon_1{}^{(b)} = 0$) and τ are 12.7 eV and 5.12×10^{-15} sec, respectively. This means that the negative value of $\delta\varepsilon_1{}^{(b)}$ for energies $\lesssim 2$ eV (see Figure 9) causes the free plasma oscillation energy to shift from $\hbar\omega_p{}^* = 12.7$ eV, as determined previously, to the measured value of $\hbar\omega_p = 15$ eV. (See also Ref. 35.)

The energy loss function of aluminum has also been calculated in the 1.5-eV energy region by Shklyarevskii and Yarovaya[116] from very accurate optical measurements. These measurements showed that the loss probability was dependent on the thickness of the foil. The intensity of this loss relative to the loss of the volume

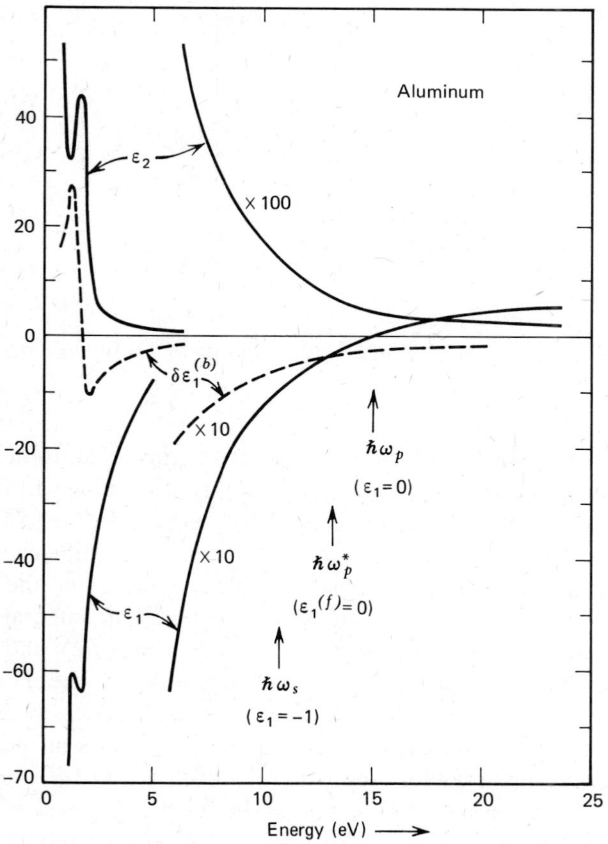

Figure 9. The real and imaginary parts of the dielectric constant for aluminum.[47] The $\delta\varepsilon_1^{(b)}$ contribution to ε_1 is drawn in, but not the $\varepsilon_1^{(f)}$ component. Clearly the interband contribution to ε_1 shifts the volume plasma energy from $\hbar\omega_p{}^*$ for $\varepsilon_1^{(f)} = 0$ to $\hbar\omega_p$ for $\varepsilon_1 = 0$.

plasma oscillation decreased by a factor 100 in going from 60 to 500 Å in film thickness.

Imbusch[71] was able to interpret this behavior by applying the Geiger–Wittmaack form of the Romanov expression[61] for the inelastic electron scattering formula altered to include effects of infinitely thick surface layers of oxide. The experimental intensity ratio values of I_{IB}/I_P versus film thickness a were fitted very effectively by

$$\frac{(P - Q_P + Q_{(-)})_{1.5\text{ eV}}}{(P - Q_P)_{15\text{ eV}}},$$

where the other terms are negligible and IB (in the first ratio) ≡ interband. Also, for the 1.48-eV interband transition it was found that $Q_{(-)}/(P - Q_P)$ increased from about 2 at 500 Å to about 1000 at 80 Å, and thereby has the essential characteristics of a surface loss.

3.2.1.5. CARBON. Carbon in graphite form is a case of special interest in regard to optical absorption and electron scattering measurements. The real and imaginary parts of the dielectric constant, as calculated from optical measurements by Taft and Philipp[44] appear in Figure 10*a*. It can be seen that ε_1 is equal to zero at two energies, and ε_2 is of a moderately low value at both. Peaks are indeed observed at the corresponding energy values, 7 and 25 eV, in electron energy loss measurements,[80,117] as in Figure 10*b*. Taft and Philipp thought that these structures resulted from plasma resonances.

Carbon atoms in graphite are bonded between the carbon-layer planes making up the crystal by $2p_z$-orbitals extending above and below the planes. These orbitals contain one electron per atom and create π-bands. Absorption peaks caused by intra- and interband transitions lying below about 9 eV involve these electrons. A broad optical absorption peak near 15 eV originates from interband transitions involving the three σ-electrons per atom. These electrons join each carbon atom with three others in coplanar bonding.

If we calculate the energy loss function in Figure 10*b* from the components of the dielectric constant given in Figure 10*a*, we see that the measured loss spectrum agrees with the loss function.

In order to analyze the energy-dependent components of the dielectric constant, assume that there are two main contributions, those due to the π- and to the σ-electrons, separately:

$$\varepsilon(\omega) = \varepsilon^{\pi}(\omega) + \delta\varepsilon^{\sigma}(\omega), \tag{46}$$

where we state that $\delta\varepsilon_2{}^{\sigma} = 0$ below 9 eV and that $\varepsilon_2{}^{\pi}$ approaches zero above 9 eV. Hence the spectrum

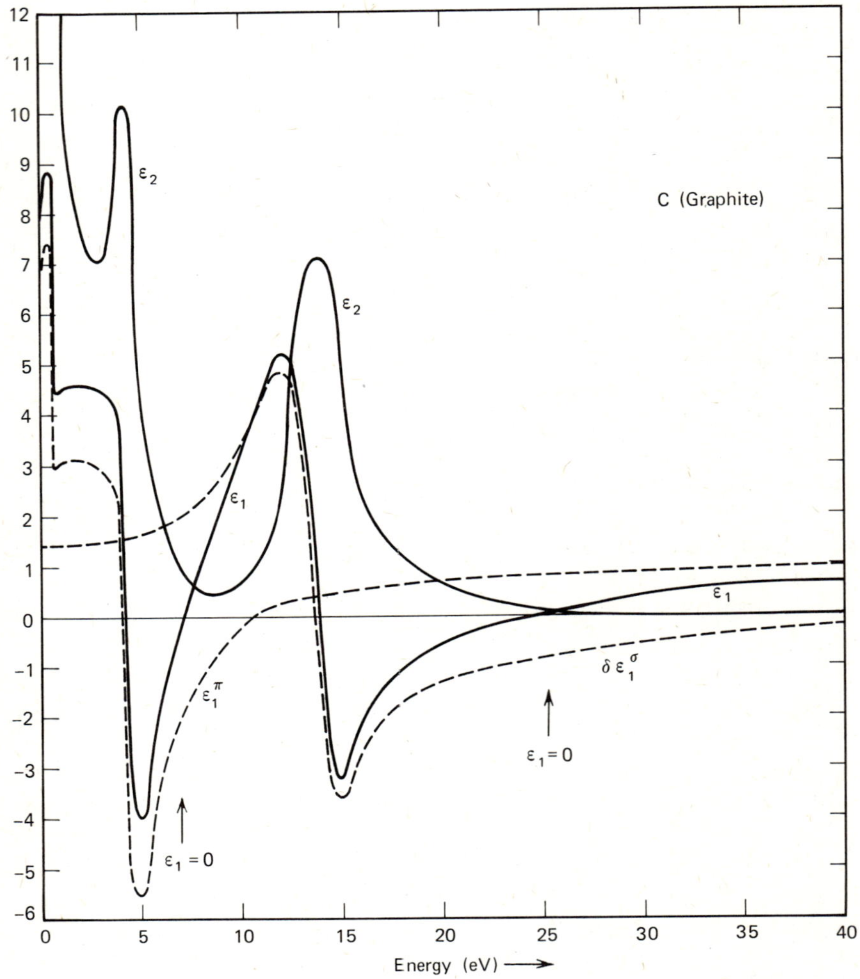

(a)

Figure 10*a*. The components of the dielectric constant of graphite. The real part ε_1 is analyzed into the π-electron ε_1^{π} and the σ electron ($\delta\varepsilon_1^{\sigma}$) contributions.[44]

is considered as arising from two plasma oscillations, each related to a separate interband transition. The value of $\delta\varepsilon_1^{\sigma}$ can then be calculated using the Kramers–Kronig relation. Both $\delta\varepsilon_1^{\sigma}$ and $\varepsilon_1^{\pi} = \varepsilon_1 - \delta\varepsilon_1^{\sigma}$ are included in Figure 10*a*.

By applying either equation 15a or 15b we can calculate the effective number of electrons $n_{\text{eff}} = n/N$ participating in the excitation. (N is the atomic density of the crystal.) We find that the integral involving ε_2 (equation 15a) shows that n_{eff} increases from zero, approaches a saturation value of unity near an energy of 10 eV, and then increases again to an asymptotic value close to 4 for higher energies. The first saturation indicates that the single π-electron band oscillation strength is exhausted; the second shows that the oscillation strengths of all four π- and σ-electrons are saturated.

Since only the quantity ε_2 is directly involved in the integration (equation 15a), the contributions owing to the π- and σ-electrons become distinguishable. Such is not the case for the integral involving $-\text{Im}(1/\varepsilon)$ (equation 15b), which also includes ε_1, whose π- and σ-components contribute at all energies. This integral shows a sharp, weak increase in n_{eff} at about 7 eV and another to a saturation value of $n_{\text{eff}} = 4$. The second increase has an inflection point at $\sim$25 eV. These two values are in agreement with the peak energies in the electron energy loss spectrum.

Considering the behavior of the functions ε_1 and ε_2 in the vicinity of the loss peaks, we see that both losses have characteristics indicating that they are plasma losses. The two plasma losses at $\sim$7 and $\sim$25 eV are shifted from the one position where $\varepsilon^{\pi} = 0$ at $\sim$11 eV by the presence of the $\delta\varepsilon_1^{\sigma}$ function.

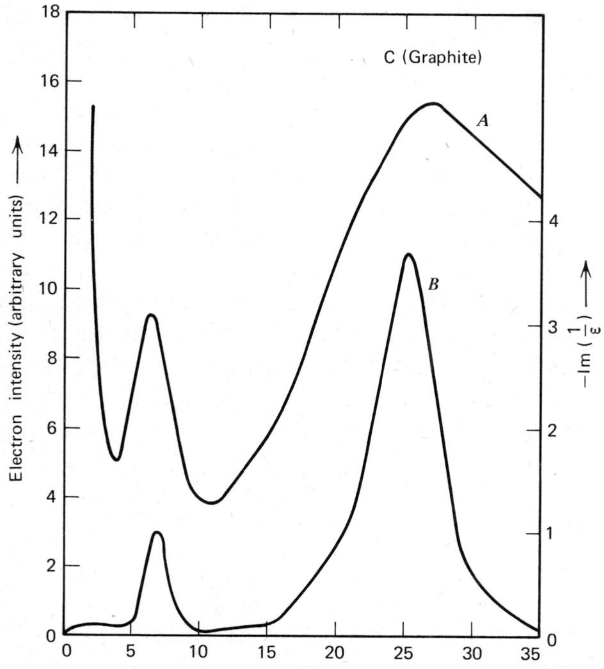

(*b*)

Figure 10*b*. The electron energy loss spectrum for graphite[80] (curve A) using electrons of bombardment energy close to 1 keV. Comparison the $-\mathrm{Im}\,(1/\varepsilon)$ loss function (curve *B*) calculated from optical dielectric constant data in Figure 10*a*.[44]

By this interpretation, the interband contribution to ε has in effect shifted the single plasma oscillation of π-electrons to the two experimental energy values. Thus the ~25-eV peak is attributed to plasma oscillations involving both π- and σ-electrons. However, Hartley[80] interprets this peak in a different manner. The effect of the π-electrons on the energy of this resonance is small, and since the negative value of ε_1 above 14 eV is caused by the peak in the ε_2 curve, the ~25-eV loss peak is interpreted as a polarization wave in the σ-electrons alone (see Section 2.4). Polarization waves are identified as due to structure in the optical constant similar to that observed for carbon.

Zeppenfeld[119] has shown that measurements of energy losses in graphite for $k_{\parallel} = 0$, giving a measure of $\varepsilon_{\perp}$, and $k_{\perp} = 0$, giving a measure of $\varepsilon_{\parallel}$, are different functions of loss energy. It was possible to determine each of these functions uniquely by carefully adjusting the angles α and ϑ with energy loss.[121]

Finally, the investigation of the electron energy loss spectrum for anthracene by Kunstreich and Otto[122] showed structure similar to that for graphite. Here there are only three condensed benzene rings per molecule instead of the infinite hexagonal planes of graphite.

The alkali metals are of special interest due to their conduction bands closely resembling a free electron gas. Several investigations of these metals have been presented over the years[26,74,120,130]. Two more recent articles are noteworthy here. That of Swan[123] reports the measurements of the surface and volume plasmon energies for sodium, 3.82(10), 5.85(5) and potassium, 2.75(10), 4.05(15) eV. These values agree well with those reported by Kunz[27] for sodium (3.85 and 5.72 eV) but somewhat less well for potassium (2.63 and 3.72 eV). The energy loss spectrum for these metals has also been obtained for sodium and potassium by exposing their halide salts to electron impact long enough for small metal agglomerates to form through decomposition of the compounds. Figure 11 shows the electron energy loss spectrum for a sodium bromide foil for bombardment of a fresh foil and after a 70-min continuous bombardment. The characteristic spectrum for metallic sodium is visible in the latter spectrum. The two plasma oscillation peaks are of energies 2.2 and 5.7 eV, the former value being less than those cited previously. The surface loss is much broader than the volume loss. Similar results were obtained for sodium iodide. The values of these two losses for metallic potassium in partially decomposed potassium fluoride and potassium iodide, obtained in the same manner, give the values 2.3 and 3.8 eV, in better agreement with the cited values by Kunz. Speculations concerning the metallic alkali formation in the crystalline compounds were presented.

Kunz[27] also presented the results of energy loss measurements on the other alkali metals from lithium to cesium and of the alkali earth metals calcium, strontium, and barium, together with a comparison of his results with those obtained earlier. The energies and half-widths of the losses were measured and compared with theory. The large discrepancy between the measured volume plasma loss energy and the value obtained from equation 1b was satisfactorily explained by atomic electron core polarization effects (see Section 2.1.1).

The structure in the energy loss spectrum for sodium owing to interband transitions of the energy of ~33 eV involving $2p$-electrons was investigated for both the metal and its oxide.[27] The position of the main structural component (a double peak for the metal) shifted by ~1 eV to higher energies for the oxide (where only a single peak was observed). The latter result compared favorably with published values of $2p$-excitation energies in isolated Na^+ ions and in crystalline sodium halides.

Energy loss spectra were measured and analyzed for multimetallic layer foils using the metals aluminum, magnesium, and silver supported by a carbon foil substrate.

The electron energy loss spectra for a large number of conductors have been investigated over the years; in the following list, references to optical work are in parentheses. For further references, turn to the articles of most recent date, as well as to the reviews by Klemperer and Shepherd,[5] Raether,[6] and Robins.[124]

3.2.1.6. ALLOYS. Energy loss spectra for alloys are composed mainly of surface and volume plasma excitations and intra- and interband transitions. Interband transitions unique to each metal in an alloy contribute to the electron energy loss spectrum in proportion to the relative amounts of the alloyed metals. Since plasma excitation energies are directly related to the electron density in the conduction band, we expect the plasma excitation energy to be to the first approximation dependent on this density in the alloy. Alteration of this value will be caused by the contribution of intra- and interband transitions to the loss cross section.

Electron energy loss investigations have indicated, therefore, that the obtained spectra are unique for the alloy or intermetallic compound and not simply superpositions of the loss spectra for the constituent metals. Probably losses associated with the constituent phases rather than with the constituent elements are the main contribution to the loss spectra.[170] The superposition of these contributions may then cause the usual measured excessive widths of the losses.

Powell[171] investigated electron energy losses in liquid indium–aluminum and indium–bismuth alloys. By use of equations similar to equations 43, he attempted to deduce the energy and oscillator strength of an interband transition in indium using a procedure previously suggested by Pines.[14]

Reports of electron loss measurements carried out on alloys follow. Also entered, in parentheses, are references to investigations of the indicated alloy using optical methods. See the listed articles for further references.

Al–Mg[172]
Al–Au (liquid, solid)[150]
Cu–Au[(173)]
Cu–Pd, Ag–Pd, Au–Pd, Cu–Mn, Ag–Mn[(174)]
In–Al, In–Bi (liquid)[171]
Ag–Au[175,110]
Ag–Al[176]

3.2.2. *Semiconductors*

The category of semiconductors is divided into four separate subcategories.

Li[27,105,120]
Be[96,31,125–132]
C[107,133,134,113,114,129,44]
Na[27,74,105,123,130]
Mg[27,31,129,135,136,130,137]
Al[27,71,57,60,58,138,139,98,105,140,141,31,61,142,35,28,97, 143–155,134,126–131,34,36,(72),(120),(156)]
K[27,141,74,105,123,130,123]
Ca[27,96,74,157,132]
Ti[74,158–160,130]
V[72]
Cr[72,127,154,130]
Mn[72,154,130]
Fe[72,130]
Co[72,130]
Ni[72,125,134,152,127,154,130]
Cu[95–99,72,125,161,157,1,154,127,152,130,132]
Zn[129,73]
W[81,165]
Re[81]
Ir[81]
Pt[81,96,125,154,152,132]
Au[81,111,95,99,97,166,148,150,125,157,1,134,167,127,73,152,168, 130,(108)]
Rb[27,76]
Sr[77,27]
Y[81]
Zr[81,73]
Nb[81]
Mo[81,129]
Rh[81]
Pd[81,162,125,73,130]
Ag[81,103–107,60,138,139,95,162,143,99,28,163,100,148,125,157,1, 126,154,161,127,57,73,130,143,101]
Cd[96,73,152,131,132]
In[148,129]
Sn[96,148,127,130,132]
Sb[24,96,158,159,163,152,127,128,136,130,132,(160)]
Cs[27,79,]
Ba[27,74,157]
Gd, Tb, Dy, Ho, Er, Tm, Yb, Lu[164]
Hf[81]
Ta[81]
Hg[103]
Tl[124]
Pb[96,154,136,132]
Bi[96,148,128,154,130,132] (liquid, solid)[169]
U[154]

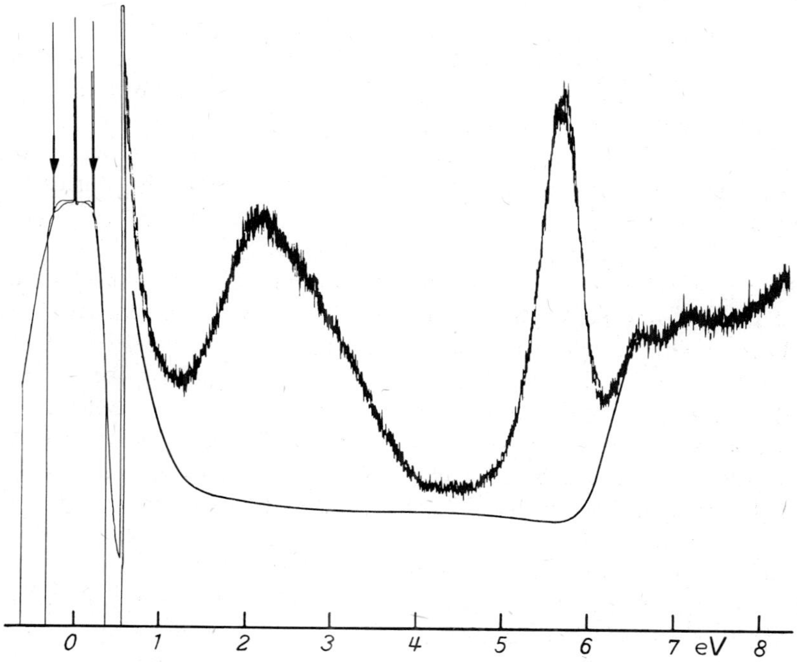

Figure 11. Portion of the energy loss spectrum for sodium bromide (continuous curve) and the spectrum for the same sodium bromide foil after 70-min irradiation at 2.5×10^{-5} amp cm^2. The surface plasma loss is at 2.2 eV and the volume plasma loss is at 5.7 eV for sodium metal. (By Creuzburg[199].) Similar dissociation effects have been reported by Palmberg and Rhodin.[200]

1. The intrinsic semiconductors are those elements or compounds which have a forbidden gap (usually of small energy, $\gtrsim$1–2 eV) between the valence and conduction bands. In some cases the gap is so small that at room temperature there is a good probability that electrons will become thermally excited from the lower to the higher of these (e.g., germanium).

Semiconducting compounds are usually of the groups 3 to 5 (e.g., GaP) or groups 2 to 6 (e.g., ZnTe) types. These also show the characteristic small forbidden gaps.

2. Compounds in which the forbidden gap is larger than for those cited above ($\sim$4–8 eV) are often classified under the heading of insulators. It is interesting to compare optical and inelastic electron scattering measurements for a metal with the same measurements for the oxide or sulfide compounds. This comparison can give information about the effect of chemical bonding on alterations in plasma oscillation and intra- and interband transition contributions to energy loss spectra. Since as yet such comparisons have been performed in only a few cases, this section contains primarily a reference list of the work.

3. The alkali halides form a unique group because of their ionic bonding. Electron energy losses of the interband type, forming excitonic states, can often be compared to excitation energies in the corresponding singly positive alkali ion.

4. Solidified rare gases show the creation of exciton states in their inelastic electron scattering spectra. Spectra obtained from such frozen gases may be very detailed, and they can be analyzed with high accuracy.

3.2.2.1. INTRINSIC SEMICONDUCTORS. Electron energy loss spectra for semiconductors, primarily silicon and germanium and such compounds as GaP, GaAs, and InAs, are expected to be composed solely of plasma oscillation and interband transition structure; intraband transition will not occur when the valence and conduction bands are completely filled and empty, respectively. The electrons in the valence band, being filled, will oscillate collectively at a frequency lying close to that expected for a free-electron gas. As in the case of the conductors, the energy of the plasma loss will be shifted by the possible close-lying presence of interband transitions.

According to Philipp and Ehrenreich,[177] silicon gives a plasmon peak at 16.9 eV that is very close to the value from equation 1b, assuming four valence electrons and from a $-\text{Im}\ (1/\varepsilon)$ plot using dielectric constants from optical measurements. Also observed in dielectric constant functions is structure indicating interband transitions at energies below 5 to 10 eV. In this region the function $-\text{Im}\ (1/\varepsilon)$ is rather monotonous and structure in the electron energy loss spectrum is expected to be weak.

Kuntz's investigation[28] of the change in energy of the 17-eV volume plasma loss in silicon revealed a dependence on scattering angle similar to that observed for aluminum (see Sections 2.1.3 and 3.2.1.4). The result was compatible with the theoretical analysis of the effect.

Electron energy losses experienced by 50-keV electrons because of interband transitions in silicon and germanium were observed by Zeppenfeld and Raether.[178] The energy loss region below 9 eV was investigated, and there were fairly strong peaks for silicon at ~3 and ~5 eV. These positions agreed well with peak positions in the $-\text{Im}\,(1/\varepsilon)$ function determined from optical measurements. As pointed out in Section 2.3.1.1.2, the lower of these two loss peaks is owing to Čerenkov radiation emission experienced by the electron in passing through the dielectric material.[66,82] Germanium exhibited a strong peak in the loss spectrum at ~2.5 eV and another at ~6.5 eV. Again agreement with the $-\text{Im}\,(1/\varepsilon)$ function was good. Furthermore, a comparison of the widths and positions of the volume plasma losses for the two elements was made between the amorphous and crystalline structures of the solid elements. An observed shift toward higher energies, visible in going from the amorphous to the crystalline state, had been expected because of an increase in the sample density. Similar measurements were carried out earlier by Richter and Ruckwied.[49]

A study of the electron scattering angle dependence of the loss energies for silicon and germanium was made by Creuzburg,[70] who measured the skewing of the distribution in scattering angle caused by inclining the foil to the electron beam.

Semiconductor compounds combining elements from the third and fourth periods of the periodic table are of especial interest in that the several *d*-electrons are lightly bound and can affect both optical absorption and electron energy loss spectra.

Interband transition contributions to the optical reflectance functions for germanium, ZnSe, GaP, GaAs, InAs, and InSb in the 1 to 12 eV energy region have been reported by Ehrenreich et al.[179] The spectra were interpreted in terms of exciton excitation in the solids and the results were compared with those published earlier by other workers (see references in Ref. 179).

von Festenburg and Kröger[82,66] have investigated the energy loss spectrum for GaP, with 50-keV electrons and have analyzed it in terms of Kröger's[65] inelastic scattering formula with retardation included. A strong loss peak at 3.3 eV was observed, which altered in intensity with scattering angle as expected for a Čerenkov radiation energy loss.

The postulated collective polarization wave excitation in the *d*-electron band and its influence on the dielectric properties of the material is treated in Section 2.4. Note, however, that interband transitions from the *d*-level to a free state will also cause structure in the electron energy loss spectrum. Furthermore, the excessive width of the plasma loss has been attributed to a strong interaction between the valence and *d*-bands.[180,181,18]

The influence of *d*-bands on optical absorption and reflection measurements and on the dielectric constants obtained therefrom in group 4, and group 3-5 semiconductors has been treated by Philipp and Ehrenreich.[177] The dielectric constant for silicon, having no *d*-electrons, was compared with those for germanium and the compounds GaP, GaAs, and InSb, to estimate the contribution owing to the *d*-band presence; in GaP the *d*-band contribution would be owing entirely to the gallium. For these elements and compounds the *d*-electron binding energies range from ~20 eV for gallium to ~55 eV for arsenic. Energy losses due to interband transitions from the *d*-shell in GaP were observed by von Festenburg.[66]

The temperature dependence of the photon absorption and emission peaks in InP at ~1.4 eV associated with exciton creation and decay have been investigated by Turner et al.[182] Analysis of the data provided a determination of the variation of energy gap and exciton peak energy with temperature, which was shown to be linear above 100°K.

A list of measurements on intrinsic semiconductors follows, with references to optical investigations enclosed in parentheses.

Si[70,183,28,178,82(177,45)]
Ge[96,97,149,70,184,125,178,127,130,49,132(177,74,131,185)]
InP[(182)]
GaS, GaSe, GaTe[(186)]
GaP, GaAs, InAs, InSb[82,85(177)]

3.2.2.2. INSULATOR COMPOUNDS. The metal oxides, sulfides, and other compounds have been under inelastic electron scattering investigation since the early 1930s.[1,157] As expected, such spectra are composed mainly of interband transition and plasma excitation losses.

In metal compounds an interband transition from a given inner level in the metallic ion will be of an energy closely corresponding to the binding energy in the metallic ion. This is also true for the anion. For example, of the three main peaks found for magnesium oxide by Watanabe,[187] the one lying at 11.45 eV agreed well in energy value with the calculated ionization energy of O^{2-}.

Leder and Marton,[136] who investigated the energy loss spectra for a number of metals and their oxides or sulfides, observed that the spectrum for the compound was the same as that for the metal but shifted to higher energies. The lower lying peaks in the loss spectrum for the metals, at ~4 to 10 eV and probably owing to interband transitions, shifted some 2 to 4 eV to higher energies for the compounds. It was suggested that this shift was of the type in which atomic electron binding energies are dependent on the chemical state of the atom (see Chapter 7). The plasma oscillation peak also shifted by an amount of the same order. This shift was surely caused in part by the change in the energy position where $\varepsilon_1 = 0$ due to the alteration of the interband transition energy and in part by a change in the effective number of electrons free to participate in the plasma excitation.

Following is a compilation of references to work done on the investigation of insulator compounds. Those in parentheses refer to optical measurements; those in italics also give results for the investigation of the associated elemental metal. The article by Best[118] gives measurements on some 28 compounds, as well as further discussion.

BeO[*127*]
MgO[187,*136*,*137*]
Al_2O_3[*152*,*61*,*155*,141,*149*,126,*127*]
SiO_2[*130*,*136*]
$Ca(OH)_2$[*127*]
CaO[*157*]
ZnO, ZnS[(188)]
CdS[(180,189,188)]
$AgCl$[(190)]
SnO_2[*127*]
Sb_2S_3[*136*]
BaO[*157*]
PbS[*136*(191)]
$PbSe$, $PbTe$[(191,45)]

3.2.3. *Alkali Halides*

In the manner stated earlier, optical measurements on the alkali halides can be used to calculate the electron energy loss functions $-\text{Im}(1/\varepsilon)$ for these compounds. Here again, interband transitions and exciton creation predominate in contribution to the loss function, along with a single broad volume plasma excitation peak.

Broad loss peaks associated with plasmon excitation in the valence band agree to within about 1 eV in energy with values obtained from equation 1b and assuming eight participating electrons per atom.[192,115] However, the estimated energy shift in oscillation energy of only the six *p*-electrons in the valence band in potassium bromide due to the separation of about 7.3 eV between this and the conduction band gave a theoretical value agreeing well with experiment.[115] The two *s*-electrons of the same shell are much more tightly bound and were not taken to participate in the excitation. The agreement results from accounting for the gap energy when calculating $\hbar\omega_p$ (see Section 3.2.4).

Of the other two contributions to the loss spectra, excitonic structure dominates. Recent high resolution measurements have made it possible to analyze and interpret these spectra with great accuracy.[193,194,99] Because of the energy gap of usually 5 to 10 eV between the valence and conduction bands, structure in the energy loss spectra does not begin until that energy loss value is reached (see Figure 11). Toward higher loss energies, one sharp energy peak of special interest is that in which an electron from the *p*-shell in the alkali ion is excited to an excitonic state corresponding to a free higher lying *s*- or *d*-level in the free cation. These transitions were seen by Creuzburg[99] for lithium, potassium, rubidium, and cesium in all the halide compounds. Energy agreement with the corresponding transitions in the free ions was good. Keil[194] was able to show that for potassium bromide the doublet lying at ~20 eV had an energy splitting of 0.23 eV, in agreement with the $p_{3/2,1/2}$ splitting for the 3*p*-state in the free K^- ion.

Following are references to articles reporting work on the optical (in parentheses) or electron energy loss investigations of the alkali halides. For additional references, consult the more recent works listed. The article by Best[192] provides further discussion. Geiger et al.[115] and Creuzburg[99] have studied all the compounds listed.

	F	Cl	Br	I
Li	192,195 (196,197)	192	—	—
Na	192(197)	192,127,198 (197)	192,199	192,199 (197)
K	192,199(197)	192,198,200 (197,201)	192–194,198 202(201)	192,199 (197,201)
Rb	199	—	—	(197)
Cs	—	—	(197)	(197)

3.2.4. *Solidified Rare Gases*

Electron energy loss measurements have been performed on argon[203–205] and on xenon[202,206] in the solid state.

In the case of argon, the strongest peak lay at 12.46 eV and was assigned to the first of the series of excitonic states whose energies are related by the Rydberg relation $E_n = E_0 - G/n^2$ $(n = 1, 2, 3, \ldots)$ with $E_0 = 13.85$ eV and $G = 1.4$ eV. The states corresponding to $n = 2$ and 3 lay at 13.5 and 13.7 eV, respectively.

Although a comparison of the spectra from solid and gaseous argon displayed no visually obvious correlation, the same comparison for xenon does so.[206] Figure 12 shows the energy loss spectrum for both gaseous and solidified xenon. A comparison of the two spectra indicates that the $\Gamma_{1s}^{3/2}$- and the $L^{3/2}$-maxima are analogous to the 6*s*- and 5*d*-maxima. The two pair 5*d*, 5*d′* and 6*d*, 6*d′* maxima are spin-orbit doublets due to $5p \rightarrow 5d$ and $6d$ transitions, their splittings reflecting that of the 5*p*-level.

A broad hump was observed for the solid centered at 15.2 eV, which agrees with that value obtained using $\hbar\omega = ((\hbar\omega_p)^2 + E_G{}^2)^{1/2}$; here ω_p is given by equation 1b where n corresponds to six *p*-electrons and the *s*-level does not participate because of its much higher binding energy; E_G is the gap energy.

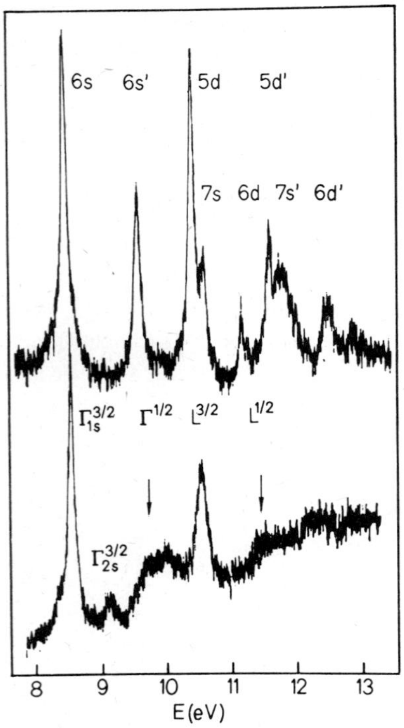

Figure 12. The energy loss spectrum for 51-keV electrons impacting on gaseous (above) and solid (below) xenon, as measured by Keil[206].

3.2.5. *Ionization of Atomic Levels by Electron Impact*

It was shown in Section 3.1.2. that interband transitions induced through electron impact contribute structure to the energy loss spectrum. Also, the energy loss spectrum may be quite accurately determined from the energy-dependent dielectric constant of the material (Section 2.2).

In some cases the interband transition contributes to both ε_1 and ε_2, but without sufficient strength to cause ε_1 to go through zero, and thus produce a peak in the electron energy loss spectrum (e.g., the 1.48-eV interband transition in aluminum, Section 3.2.1.4).

On the other hand, there is the case of the oscillator strength for a given interband transition being so strong that the contribution to ε_1 depresses this quantity to below zero. When ε_1 increases again through zero with increasing energy, a plasma oscillation, or polarization wave, is created. As shown in Section 2.4, the ionization edge for the level or band in question lies in the loss spectrum at the minimum preceding the polarization wave maximum.

The procedure of ionization energy determination from electron energy loss spectra is dependent on the oscillator strengths for the transitions in question for at least two reasons.

1. The oscillator strength for the ionization transition is not sufficient to suppress ε_1 to below zero. The ionization energy is in this case equal to the loss energy value of the inflection point of the ionization edge. This is analogous to X-ray absorption binding energy determination methods.[78] In earlier electron energy loss work on binding energy determinations, the values were incorrectly taken as being the energy position of the spectral maxima.[207–210]

2. If the oscillator strength of the level under investigation is sufficient to cause ε_1 to become negative, a plasma excitation occurs where ε_1 passes again through zero at a greater energy loss value (the polarization wave phenomenon). The peak thus caused in the energy loss spectrum is that referred to as the "*G*" peak by Robins, Swan, and Best (see Section 2.4).

Polarization waves are created when the oscillator strength for the related atomic shell is large enough to cause the already positive variable ε_1 to pass below zero, as is often the case for lightly bound $L_{2/3}$, $M_{2,3}$, $M_{4,5}$, and so on, shells. This oscillator strength for a given shell decreases with increasing binding energy. Excitation of a given atomic shell may fall under either points 1 or 2, depending on the binding energy for the shell and its oscillator strength.

For large energy losses, due mainly to ionization from tightly bound levels, we expect $\varepsilon_1 \approx 1$ and

$\varepsilon_2 \ll 1$. We then have the approximate expression $-\text{Im}(1/\varepsilon) \approx \text{Im}(\varepsilon) = \varepsilon_2$, and we may expect this variable to describe the electron energy loss spectrum with good accuracy. Polarization waves will not be excited, since ε_1 does not alter far from unity. Therefore the ionization energy can be taken as the loss energy of the inflection point of the absorption edge.[155]

Electron energy loss measurement binding energy determinations of the $L_{2,3}$-level in aluminum and alumina[155] and the K-shell in beryllium and beryllium oxide[211] have been reported. For aluminum metal the inflection point of the $L_{2,3}$-edge, estimated by the midpoint position of the intensity step, gave a binding energy value of 72.5(5) eV, in agreement with X-ray absorption measurements. The same edge for aluminum oxide was found to shift some 3.3 eV toward higher energies, owing to a chemical effect.

Similar results were obtained for beryllium and its oxide, the midpoint of the K-absorption step in the loss spectrum for the metal lying at 111.7(4) eV. In the case of the oxide, structure in the loss spectrum could be associated with excited energy levels of the isolated Be^{++} ion with reference to its ground state. (See also Refs. 34 and 212.)

For reports of work on atomic binding energy determinations by electron energy loss measurements, refer to the articles cited previously and in Section 2.4.

4. PHOTON EMISSION FROM ELECTRON–BOMBARDED FOILS

Ferrell[213] has predicted that plasma oscillations in metal foils may decay by radiation emission. The radiation energy will therefore coincide with that of the decaying plasmon. In general, all radiation from smooth surfaces, except "*bremsstrahlung*," are grouped under the phenomenological name of Lilienfeld radiation. This includes both volume plasmons and, for thick foils, surface plasmons. The so-called non-radiative surface plasmons, $\omega_\pm$ may decay by radiation emission only on rough surfaces.

Most of the first experiments carried out on radiative plasmon decay employed electron impact at normal incidence to the surface for initial plasmon excitation. Results for silver showed a radiation emission peak at an energy agreeing with the volume plasmon energy. Furthermore, the variation of emission intensity with angle from the foil normal direction agreed with theoretical prediction. The predicted film thickness dependence of the radiation intensity, due to surface charge fluctuation periodicity coincidence with foil transversal time of the impact electron, was also supported by experiment. (For a treatment of this work, see the review by Raether[6].)

Measurements by Boersch et al.[214] performed for normal impact incidence (2–12 keV electrons), brought out radiation emissions whose angular and wavelength distributions of intensity for various bombarded metals agreed with predictions of Lilienfeld radiation emission: namely, as an electron approaches a metal surface, a changing dipole field is created between the electron and its image charge. As the dipole "collapses," the visible Lilienfeld radiation is emitted. The intensity of this radiation for a 500-Å silver foil and 25-keV electrons for $\lambda <$ 3250 to 3350 Å is about the same intensity as radiation from plasmon decay[215] (see also Refs. 216, 217).

More recent experiments have been done with non-normal electron impact.[218] Grazing impact of electrons on thick silver films[219] has shown radiation at ~3550 Å which increased greatly in intensity for impact angles to the foil normal of 86 to 89°. The energy of the emission agreed well with the surface plasmon energy, although such radiative decay is theoretically impossible for thin films. However, it was not absolutely clear that the radiation was owing to the decay surface plasmons, whose energy differs so little from that of the volume plasmon in silver.

In a similar experiment, Cram and Arakawa[220] measured the emission radiation spectrum from aluminum bombarded by 40 to 80 keV electrons at angles of incidence from 20 to 87°. They questioned the validity of the surface plasmon emission interpretation.

Ashley et al.,[62] also working with thin aluminum foils, employed 80-keV electron bombardment and measured the emission radiation as a function of the electron incidence angle. A radiation emission peak at about 16 eV was observed at normal incidence. This peak decreased in intensity with increasing incidence angle. Just before an intensity minimum was reached at 60°, the peak energy began to decrease rapidly. As the incidence angle increased above 60°, a broader peak began to grow just below 16 eV, and it increased in energy with incidence angle. According to theory, the photon emission should continue to decrease with increasing electron incidence angle. The experimental spectral intensity variation was interpreted to consist of a plasmon decay component, which decreased in intensity with increasing incidence angle, and a second component that increases in intensity with increasing angles above ~60°. The characteristics of this second component showed agreement with those of a *bremsstrahlung* photon spectrum. Discussion and descriptions of further work on the subject are contained in the references cited previously.

Hence, at the beginning of 1969, the journalized

debate between the Berlin and Oak Ridge groups was still open. However, recent evidence tends to favor the radiative decay of surface plasmons, for example, the work of Sauerbrey and Dobberstein,[221] who experimentally separated the *bremsstrahlung* and plasmon radiation for silver, and that of Bürker and Steinmann.[221]

Further information on radiation effects is given in the review articles by Frank and by Steinmann.[222]

5. BOMBARDMENT–ELECTRON AND CRYSTAL–LATTICE INTERACTIONS

Interaction between electrons and a crystal lattice on which they impinge may result in energy losses of five primary types.

1. Elastic scattering energy losses in which the electron strikes a single atom and is scattered at some angle—the scattering is termed elastic because the atom upon which the electron scatters does not become excited in the process. The energy loss is related to the recoil of the single lattice atom upon collision. A broadening of the scattered electron's energy will be expected to depend on the thermal motion of the lattice atom. Electrons of 30-keV initial energy can lose from ~30 to 0.2 eV energy in such interactions with hydrogen to platinum atoms, respectively.
2. Excitation of vibrational states of radicals in compounds in crystalline form—induced excitation of such states in the $(NH_4)^+$ radical in NH_4Cl show electron energy losses ranging from 0.4 to 0.8 eV.
3. Excitation of crystal lattice vibrations, phonons, due to electron bomdardment induced lattice vibraions. (Here it is also possible that a quantum of vibrational energy that already exists through thermal excitation may be given up through the interaction to the transmitted electron. See point 4.) The probability of the occurrence of energy losses of this type drops sharply with increasing scattering angle. The phonon energies for the materials thus far investigated—lithium fluoride, sodium fluoride, magnesium oxide, and alumina—are less than 0.1 eV.
4. Electron energy gains.
5. Diffuse streak-patterns in electron diffraction patterns.

5.1. Elastic Energy Loss of Crystal Scattered Electrons

Elastic electron scattering is defined as scattering in which no alteration of the internal energy of the scattering atom occurs. Electrons scattered in such a process do, however, show a change in energy. In the laboratory coordinate system, this elastic energy loss is caused by the impulse imparted to the scatterer by the scattered electron, which has thereby experienced a kinetic energy change. For a scatterer of mass M, an electron of mass m_o where $m_o \ll M$ and initial energy E_o becoming scattered through an angle ϑ, will experience an elastic energy loss ΔE_1

$$\Delta E_1 = 4\,\frac{m_o}{M}\,E_o \cdot \sin^2 \frac{\vartheta}{2}. \qquad (47)$$

The energy loss is thus small compared to the initial energy.

For small scattering angles ($\vartheta < 45°$) for which electron diffraction is carried out, electron interference patterns are observed that result from electron–crystal-lattice interactions, (i.e., the atoms of the crystal lattice collectively). For the electron energy losses associated with interference patterns, M of equation 47 refers to the mass of the crystal as a whole, and the losses are immeasurably small.

For large scattering angles ($\vartheta > 45°$) electron interference is not observable; this may be partially owing to inelastic scattering, but it is mainly a result of electron scattering by individual lattice atoms. Here the mass M in equation 47 refers to the mass of the individual atoms of the scatterer.

In performing a measurement of inelastic energy losses, we encounter the difficulty of low scattered electron intensity, which varies as the Rutherford scattering factor $\sin^{-4} \vartheta/2$. For this reason, the investigation of elastic electron scattering by Boersch et al.[223] was carried out using a modified energy analyzer, of which the resolving power (~0.1 eV at 30 keV) was sacrificed in order to increase transmission.

Measurements of the elastically scattered electron spectra were made for the scatterers carbon, alumina, nickel, silver, and platinum for the scattering angles of 45 and 90°. Comparisons of energy spectra for 45 to 135° scattering were made on graphite; the influence of temperature on the carbon scattering spectrum was also measured. Figure 13*a* shows the energy distribution for primary and the elastically scattered electrons for a graphite scatterer and at a scattering angle of $\vartheta = 90°$. (The figure actually shows the total scattered electron intensity below a given energy value plotted as a function of that energy.) The energy distribution has clearly shifted to lower energies and at the same time has broadened: the higher the temperature of the scatterer, the greater the broadening. (See also Braglia's[224] recent treatment of the mean energy loss of electrons in inelastic encounters.)

Boersch et al.[223] made calculations based on a simple model of the thermal broadening of the elastic energy loss distribution. Calculated values for 30-keV electrons scattered on carbon ($T \lesssim 1600°K$), nickel,

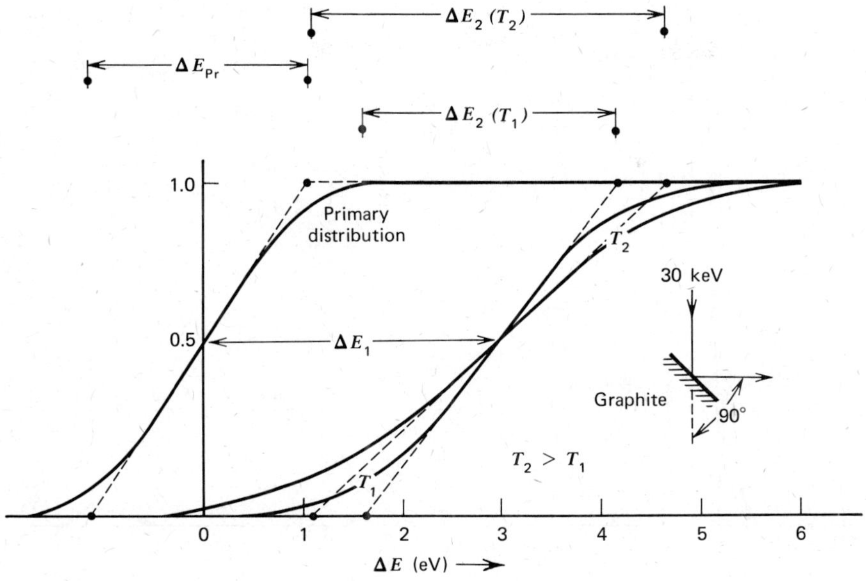

(*a*)

Figure 13. (*a*) Energy distribution for 30-keV electrons elastically scattered from a graphite surface. The measured energy loss ΔE_1 is the same independent of temperature. However, the slope of the shifted curve decreases with increasing scatterer temperature, reflecting a Doppler broadening effect. (*b*) The comparison of measured energy losses ΔE_1, for scatterers of different mass number M and for different scattering angles ϑ, with the elastic energy loss ΔE_1 (full curves) given by equation 47. (Presented by Boersch et al.[223])

and platinum ($T = 150°\text{K}$) at $\vartheta = 45°$, on carbon ($T \gtrsim 1600°\text{K}$) at $\vartheta = 90$ and $135°$, and on carbon for $\vartheta = 90°$ ($T = 150°\text{K}$) compared very well with experimental results.

Figure 13*b* displays the measured elastic energy loss values ΔE_1 as a function of atomic mass number M for several scattering angles, compared with values calculated from equation 47. The mass number for alumina (Al_2O_3) was taken as the weighted average of aluminum and oxygen.

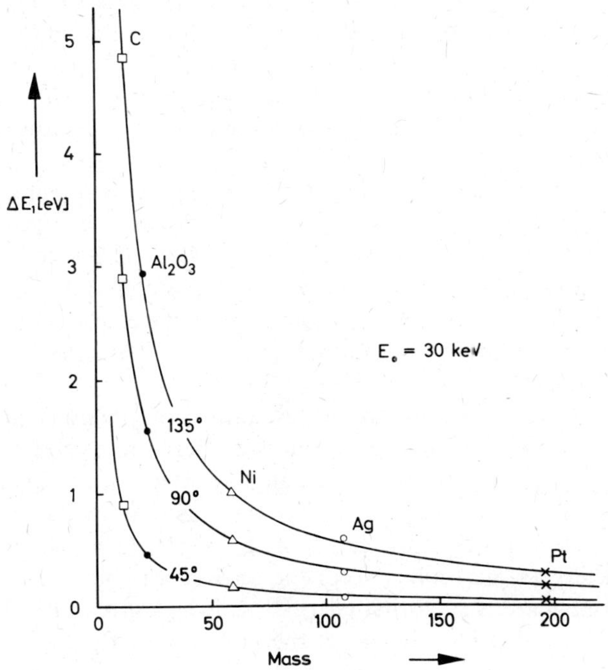

(*b*)

5.2. Vibrational Excitation of the $(NH_4)^+$ Radical in Crystalline NH_4Cl

Bohg and Boersch and co-workers[225] have measured the electron energy loss spectrum due to scattering in polycrystalline ammonium salts. Near the 20-meV energy loss and gain positions they found strong spectral components caused by excitation and de-excitation of translational lattice vibrations. (See also Section 5.4.)

Energy losses above 100 meV were also found, and these were attributed to excitations of ammonium ion fundamental vibrations and their interaction with torsional vibration and, presumably, with translational lattice vibrations (see Figure 14). The energy loss spectrum was compared with that from infrared absorption spectra. General agreement was good, but comparison of the finer details of the two spectra revealed striking deviations. This phenomenon was ascribed to a difference in the interaction mechanisms of lattice vibrations with electrons and photons.

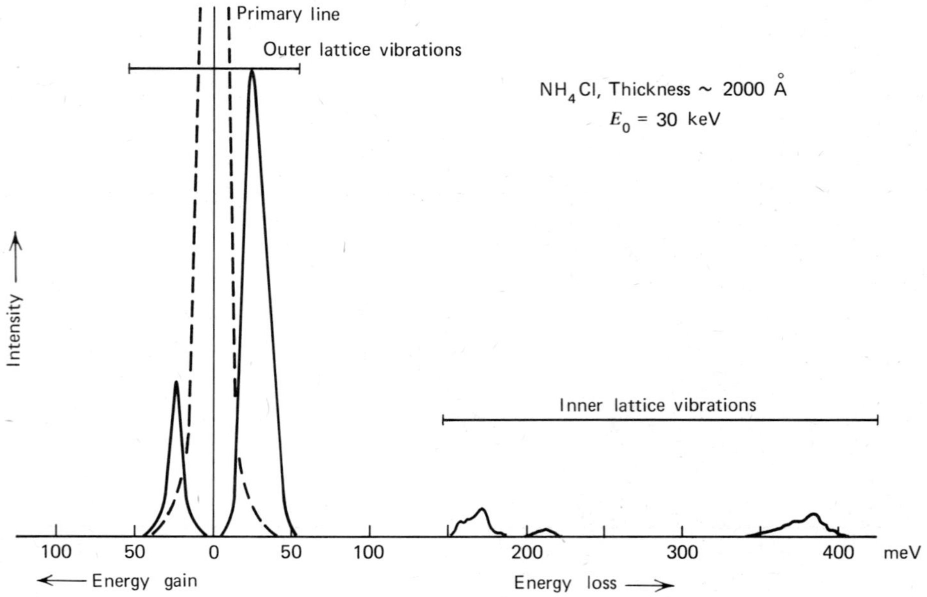

Figure 14. The electron energy loss spectrum for 30-keV electrons through a foil of NH_4Cl, as reported by Bohg and Boersch et al.[225] The loss and gain from phonon interaction lie at ~20 meV. Loss resulting from $(NH_4{}^+)^+$ radical vibrational excitation lie at the larger loss energies.

5.3. Impact Electron Interactions with a Crystal Lattice

Excitation of phonons in a crystal lattice is a topic of interest to workers dealing with neutron scattering (e.g., Refs. 228, 229, 230) as well as infrared absorption (e.g., Refs. 231, 232) in solids. It has also been shown that the interaction of bombarding electrons with a crystal lattice provides another means of lattice vibrational excitation.[139,233–235,138]

Electromagnetic radiation will excite transverse lattice vibrations, of frequency ω_T, whereas charged particles will excite longitudinal modes, of a second frequency ω_L. This excitation mode preference for the two excitation methods holds also for electron plasmon excitations, but in that case the two energies are assumed to be equal.

Structure owing to the excitation of phonons in crystalline lithium fluoride at a temperature of 300°K appears in the upper (zero-angle electron scattering) spectrum in Figure 15.[236] We also see that an energy gain peak lies at the same energy displacement *above* the peak position. (Note: the scale at the bottom of the figure is such that energy loss is a positive value.) At the cited temperature, thermally excited phonons already exist in the crystal; the bombarding electron can experience either a phonon energy quantum gain or a loss through interaction with the lattice. As the lattic temperature is increased, the intensity of the gain peak will approach that of the gain loss. Toward lower temperatures the gain peak intensity will decrease to zero as thermal phonon excitation becomes less probable (see the lower part of Figure 15). Section 5.4 explains this further.

Following the theoretical treatments by Kliewer, Fuchs, and Pardee[237–239] as applied by Boersch et al.,[236] we can, in the Born approximation, express

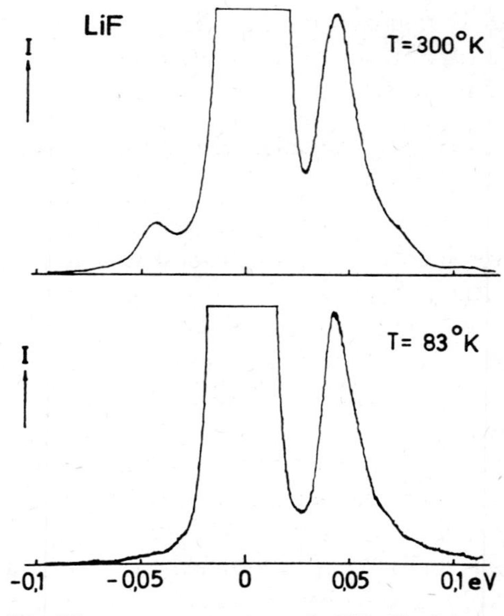

Figure 15. The energy spectra of 25-keV electrons after transmission through a lithium fluoride slab at room temperature (above) and liquid air temperature (below), as reported by Boersch et al. [236]

the differential probability W for scattering by Coulomb interaction of an electron of momentum k into a solid angle $d\Omega$ as

$$W(\omega, K)\, d\omega\, d\Omega = \frac{4aN}{a_0{}^2 k^4} S(\omega, K)\, d\omega\, d\Omega. \quad (48)$$

Here $\hbar K$ is the momentum transferred by the impact, a the crystalline slab thickness, N the scatterer density in the solid body, a_0 the Bohr radius, and $S(\omega, K)$ the dynamic form factor. For $S = 1$, equation 48 becomes the Rutherford scattering probability of a particle of charge e by the crystalline scatterer. A value $S \neq 1$ expresses the condition that the individual atoms forming the solid are not independent of one another in their capacity as scatterers.

Calculation of $S(\omega, K)$ on the basis of the dielectric theory gives an expression that is dependent on the dielectric constant of the crystal. For a thin crystalline slab, Geiger and Wittmaack's[69] expression of the function becomes

$$S(\omega, K) = \frac{a_0{}^2 k^2}{4aN}(P - Q_P + Q_{(+)} + Q_{(-)}), \quad (49)$$

where P, Q_P, $Q_{(+)}$, and $Q_{(-)}$ are given in equations 21. Therefore $W(\omega, K)$ in equation 48 is proportional to $W(\omega, \vartheta)$ in equation 21a, K being a linear function of ϑ for small ϑ. Here again, retardation effects are neglected.

The preceding comments relate to a slab that is bounded on both sides by vacuum. The scattering probability expression and dispersion relation for multilayer films would be comparatively much more complicated. In measurements by Boersch et al.[236] the lithium fluoride, sodium fluoride, and magnesium oxide slabs were supported on a carbon or collodion foil and expression 49 is only approximate. For the self-supporting alumina foil investigated, equation 49 is exact.

The theory for oscillations in thin slabs, penetratingly investigated in Refs. 237 to 239, applies for thicknesses that are small compared with the wavelength of the excited lattice vibration. In such crystalline slabs the frequency-dependent dielectric constant can be expressed by[238]

$$\varepsilon(\omega) = \varepsilon_\infty + \frac{\tau_l \omega_T{}^2(\varepsilon_0 - \varepsilon_\infty)}{\tau_l(\omega_T{}^2 - \omega^2) - i\omega}, \quad (50)$$

where τ_l is the relaxation time for lattice vibrations, $\varepsilon_0 = \varepsilon(0)$, and ε_∞ is the dielectric constant for $\omega \gg \omega_T$.

The dispersion curves for lattice vibrations in a lithium fluoride slab, without[240] and with[237,238] retardation, are given in Figure 16. The impact electron strikes the slab surface perpendicularly and in the z-coordinate direction, as shown in Figure 16*a*. Neglecting retardation, we see that two branches of surface oscillations $\omega_{(+)}$ and $\omega_{(-)}$ approach ω_L and ω_T in the $k_{\parallel} a \to 0$ limit. These two limits correspond to $\varepsilon \to -\infty$ at ω_T and $\varepsilon \to 0$ at ω_L. For $k_{\parallel} a \to \infty$, both ω_L and ω_T approach a value determined by $\varepsilon = -1$. (See 16*b*.)

The dispersion relation for the two $\omega_{(\pm)}$ phonons is again given by

$$\varepsilon(\omega_{(+)}) + \tanh\left(\frac{ka}{2}\right) = 0$$

and (51)

$$\varepsilon(\omega_{(-)}) + \coth\left(\frac{ka}{2}\right) = 0,$$

which show that $\varepsilon \to -\infty$ at ω_T and $\varepsilon \to 0$ at ω_L.

In the limit $k_{\parallel} a \to \infty$, both ω_L and ω_T approach a value ω_S determined by $\varepsilon = -1$, which for smooth-surfaced slabs is

$$\omega_S = \sqrt{(\varepsilon_0 + 1)/(\varepsilon_\infty + 1)}\,\omega_T. \quad (52)$$

For small spheres we have an expression in which the values of unity in equation 52 are replaced[232] by 2. For lithium fluoride in a thin slab $\omega_S = 1.875\ \omega_T$, and in small spheres $\omega_S = 1.695\omega_T$.

As pointed out by Boersch et al.,[236] the z-component of k will be confined to the values $k_z = m\pi/a$ ($m = 1, 2, 3, \cdots$), where a is the slab thickness. For lithium fluoride, however, $k_z = k_0 \hbar\omega_L/2E_0 = 1.4 \times 10^4\ \text{cm}^{-1}$ and the slab thickness must be at least $a = \pi/k_z = 23{,}000$ Å if volume vibrations are to be possible. The slab thickness of ~500 Å used by Boersch et al. drastically limited the probability of creation of volume excitations. Comparing P with Q_P for $a = 23{,}000$ Å, equations 49, 21a, and 21b, we see that $P > Q_P$. However, for smaller values of a (e.g., $a \gtrsim 1000$ Å), the difference between P and Q_P is small. Volume oscillation excitation is thereby not forbidden but is very improbable.

Strong coupling occurs between transverse optical phonons and the photon field when retardation of Coulomb interactions is included, and the resulting mixed excitation state is called a polariton. Polaritons occur in regions L_1 and L_1', which are non-radiative (see Figure 16*c*).

The distinction between radiative and nonradiative regions in Figure 16 depends on the decay of the electric fields associated with the excitation, outside the crystal. When the decay is exponential, the region of excitation is nonradiative; if the exponent of the

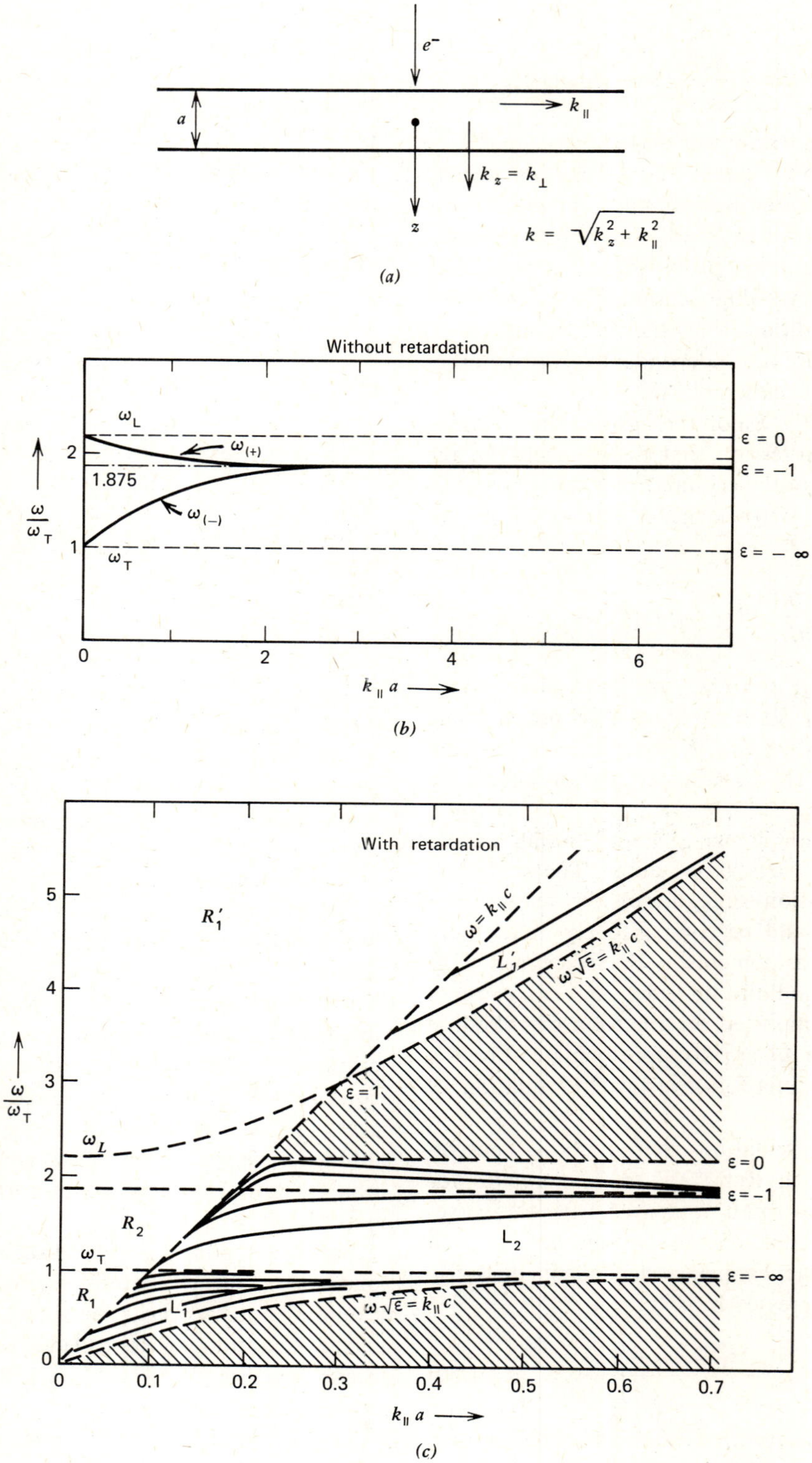

Figure 16. Dispersion relations for electron scattering in a thin (thickness a) slab of lithium fluoride: (a) quantities and geometry; (b) plots without retardation included; (c) plots with retardation. (From Ref.[236])

exponential is imaginary, the excitation is radiative, as is the case in regions R_1, R_1', and R_2 (see also Ref. 241).

The $\omega_{(+)}$ surface oscillation no longer approaches ω_T for small $k_{\parallel} a$ when retardation is included but approaches $\omega = k_{\parallel} c$ asymptotically.

Boersch et al. calculated the theoretical curve for $W(\omega, K)$ in equation 48 for a thin lithium fluoride slab by first integrating each of P, Q_P, $Q_{(+)}$, and $Q_{(-)}$ over the instrumental acceptance angle $0 \leq \vartheta \leq 1.1 \times 10^{-4}$. These contributions to W appear in Figure 17*a*. Next, the sum of these components was folded with the resolution function (Figure 17*b*), and the result was compared with the experimental spectrum for a slab thickness of 240 Å. The agreement was extremely good.

From the calculations it was seen that the main component of the scattering probability came from excitation of the lower energy surface oscillation $Q_{(-)}$. The volume oscillation contribution was almost completely nullified by the Q_P contribution, and the $Q_{(+)}$ contribution intensity was of the same order of magnitude as that of the $P - Q_P$ component.

Similar calculations and measurements were made

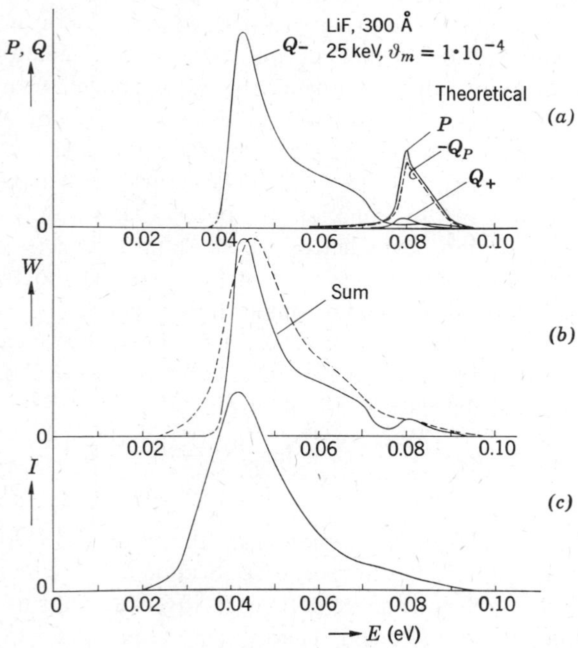

Figure 17. (*a*) Plot of $W(\omega, K)$ versus ΔE for lithium fluoride; $a = 300$ Å. The components of the total cross section were plotted separately after their respective expressions had been integrated over the acceptance angle of the energy analyzing instrument. (*b*) Sum of P, Q_P, $Q_{(+)}$, $Q_{(-)}$ (solid curve) and after folding with the window function of the spectrometer for a resolution of 0.012 eV (dashed curve). (*c*) The experimentally measured energy loss profile obtained with a slab of $a = 240$Å for the instrumental resolution of 0.021 eV. (From Ref. 236.)

for magnesium oxide, but the agreement between the experimental and theoretical spectra was not as good as for lithium fluoride—discrepancies in both maximum position and form were appreciable.

5.4. Electron Energy Gains

It was mentioned in Sections 5, 5.2, and 5.3 that interaction with lattice vibrational excitation may cause energy losses to be experienced by electrons passing through a crystalline lattice. It is also possible for a quantum of phonon energy already present in a lattice to be taken by a penetrating electron. Such an electron will then show an energy gain. The energy of these vibrational phonons must be sufficiently low that they may be created with high enough probability at experimental temperatures. As the temperature is lowered, the probability that such lattice vibrations will be present is diminished (see Figures 14 and 15).

Several such electron energy gain phenomena have been reported, not only for electron transmission through solid lattices but also through gases in which molecular rotational excitation is created through thermal impact; the latter phenomenon has been reported by Ehrhardt and Linder.[226]

Boersch et al.[227] have recently reported energy gains for electrons passed through a gas discharge tube. In the case of helium gas filling the gas discharge tube, a considerable fraction of the atoms were in excited $2^{1,3}S$ and $2^{1,3}P$ states. Energy losses were observed corresponding to transitions from these initial states to final higher lying states (energies of ~ 1–~ 4 eV). However, the result of primary interest here is that a strong energy gain peak at an energy gain of the $2^3P \rightarrow 2^3S$ transition energy was also discovered. This was the first in the series $2^3P \rightarrow n^3S$ ($n \geq 2$) transitions, of which those having $n \geq 3$ will show up as energy loss spectral components. This $n = 2$ gain transition was at the mirror position, across $\Delta E = 0$, of the $2^3S \rightarrow 2^3P$ energy loss transition. The $2^1P \rightarrow 2^1S$ transition, the first of the $2^1P \rightarrow n^1S$ ($n \geq 2$) series and at about 0.6-eV in energy gain, was expected but not found. This was because the width and intensity of the no-loss line were too large to resolve effectively the energy losses or gains of this energy.

It was pointed out that the ratio of the intensities of the energy loss $2^3S \rightarrow 2^3P$ to the energy gain $2^3P \rightarrow 2^3S$ in orthohelium will define a temperature. That is, this will give the equilibrium ratio of the helium atoms in the two excited states as a result of the discharge. The experimental data showed this temperature to be about 10,000°K, which is compatible with the expected electron temperature in the discharge.

5.5. Diffuse Streak Patterns in Electron Diffraction Patterns

Herbst and Jeschke[228] have observed diffuse streak patterns in their electron diffraction diagrams from sufficiently thick crystals. These patterns were hexagonal in form, joining the Bragg reflection spots. It has been observed[228] that the accurate recording of these patterns requires the irradiated crystal to be within narrow limits of a crucial crystal thickness. Recent measurements on tin, with a tetragonal lattice, show a square pattern of streaks.

The streaks are tentatively attributed to energy losses through excitation of transverse acoustic vibrational excitation (the L_1-region in Figure 16*c* with $k_{\parallel} a$ large).

6. MULTIPLE ENERGY LOSSES IN SOLIDS

In the foregoing sections, single energy loss interactions of electrons with a solid were considered. Relative loss probabilities for the excitation of volume and surface plasma modes depend on the method of excitation and the depth of penetration into the solid: the surface plasmon excitation probability depends only on the number of times the surface is passed by the electron, whereas the volume plasmon excitation probability is effectively dependent on the electron's path length in the solid.[54] Accordingly, the reflection technique used by the Australian group[135,146] may excite surface and volume plasmons with comparable probability when low impact energies are used. The ratio of volume to surface loss probabilities will increase with increasing impact energy, first according to equation 21a and second because of the increase of probable path length in the solid.[146] For transmission measurements, the penetration depth of the foil is equal to its thickness and the above-mentioned ratio will depend solely on its energy dependence expressed by equation 21a. Therefore, when foil thicknesses become so large that multiple volume losses can occur, the relative intensity of surface losses—whose highest possible multiplicity is 2 and whose intensity is small—may often be neglected.

For a given impact electron energy, the probability for n volume plasmon losses suffered by an electron passing through a foil of thickness a is given by the Poisson equation 6*b*.

Experimental measurements of the mean free path for inelastic scattering λ_i due to volume plasmon creation can be obtained by measuring the relative intensities of the loss peaks as a function of n for a given value of a. Two points must be considered here.

1. The inelastic scattering angle associated with a plasmon excitation loss—the variation of the scattered electron intensity with angle, given in Section 2.1.3, falls off rapidly with increasing scattering angle. The full angular width at half-maximum intensity is quite small: $\sim 2 \times 10^{-4}$ rad for aluminum for 40-keV electrons (Figure 2*b*). The angular intensity distribution for n successive plasmon excitation losses can be obtained by folding the angular-dependent initial electron beam intensity function with the window function of the analyzing instrument and the result n times with the function $\sigma(\vartheta)$ of equation 5[242] (see also Chapter 7, Section 3, for a treatment of the folding procedure).

From equation 6*b*, $\ln(n!P_n) = n \ln(a/\lambda_i) - a/\lambda_i$ and a plot of $\ln(n!P_n)$ versus n should be a straight line of slope $\ln(a/\lambda_i)$. This has been shown to be true for aluminum.[242]

2. The impact electron energy dependence of the plasmon excitation probability—the inelastic scattering angle distribution function referred to in 1 is also dependent on the initial electron energy. Using equation 5 for the volume plasma loss probability, we obtain for the angle at which the distribution intensity has the half-maximum value $\vartheta = \vartheta_E = \hbar\omega_p/2E_0$.

In addition, the mean free path of an electron in a metal will be dependent on the plasmon creation probability, which in turn is energy-dependent. From equations 21, the probability per unit length for an energy loss through the creation of a volume plasmon will be equal to $1/a(P - Q_P)$. The term P is identical to that given by equation 16, but Q_P shows only a second-order dependence on the film thickness a. This term does, however, contribute strongly to the loss probability for low bombardment energies.

Quinn[243] has calculated the energy-dependent mean free path for plasmon excitation in metals, assuming a free electron plasma. He obtains

$$\lambda_i = \frac{2a_0 E_0}{\hbar\omega_p}\left[\ln\left(\frac{(E_F + \hbar\omega_p)^{1/2} - E_F^{1/2}}{E_0^{1/2} - (E_0 - \hbar\omega_p)^{1/2}}\right)\right]^{-1}, \quad (53)$$

where a_0 is the Bohr radius, E_0 the impact electron energy, E_F the Fermi energy (see Section 2.1.1). Figure 18 shows the mean free path λ_i in Ångström units as a function of electron energy for aluminum. For $E_0 \gg E_F$, equation 53 reduces to equation 6a.

Two more effects that must be considered concern the electron scattering angle-dependent shift of volume plasmon energy (the dispersion effect) and variation of the plasma loss width.

The first is treated in Section 2.1.1, where it is shown that this effect is strong. However, since the scattering probability falls rapidly with increasing scattering

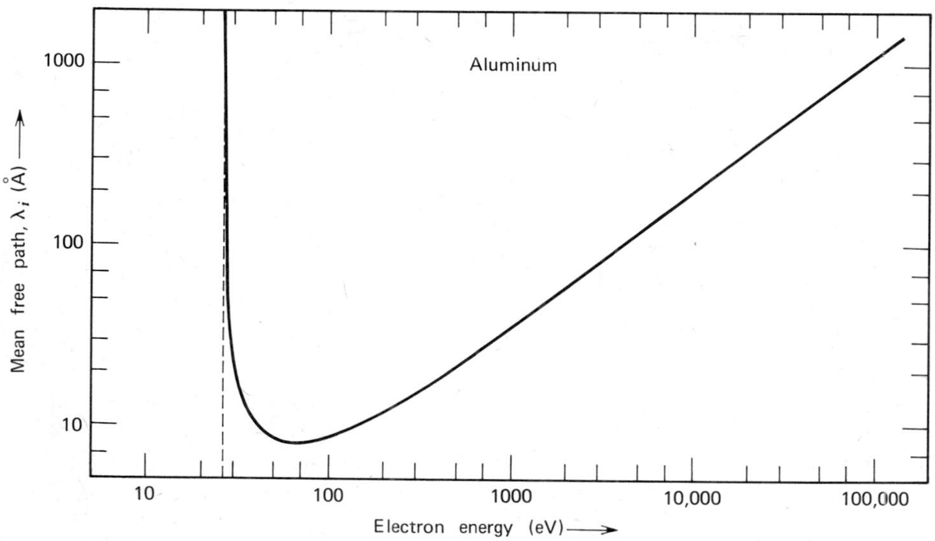

Figure 18. The energy-dependent mean free path for plasmon excitation in aluminum assuming a free electron plasma, according to Quinn.[243]

angle ϑ, one sees from Figures 1*a* and 1*b* that, for aluminum at $\vartheta = \vartheta_c \cong 5 \times 10^{-3}$ rad ($E_0 = 39$ keV), the peak position has moved from ~15.0 to ~15.6 eV, a shift of minor importance.

The second point is sometimes more significant (see Section 2.1.5). The equation for the scattering angle dependence of the plasma loss peak width $\Delta E_{1/2}$ is identical to that for the peak shift. Here the width increase for aluminum at $\vartheta_c \cong 7 \times 10^{-3}$ rad ($E_0 = 50$ keV) over that for $\vartheta = 0$ is appreciable: about 50%, or 0.5 eV. However, for small crystals of the scattering material, the width of the volume plasma loss is large for zero scattering angle; it first decreases with increasing scattering angle and then increases. Figure 3*b* shows that, for $E_0 = 50$ keV and for an aluminum scatterer, the width at $\vartheta = 0$, where the scattering cross section is greatest, can be more than three times greater than for the larger angles $10^{-3} \leq \vartheta \leq \vartheta_c$, and this effect must be considered in treating inelastic scattering in small crystals.

A final point is the contribution of elastically scattered electrons to electron intensities at large angles. The elastically scattered electron intensity falls off with approximately the Rutherford ϑ^{-4} dependence with scattering angle. This intensity falls with increasing ϑ until a point is reached (where the intensity is about 10^{-3} that at $\vartheta = 0$) from which the variation decreases more slowly with ϑ. The intensity of the first plasma loss falls as ϑ^{-2} for small scattering angles and then tends to parallel the elastic intensity function variation with ϑ at a higher intensity value.[244] Second and third plasma loss intensity distributions parallel the first at still larger intensities. Therefore, the inelastic losses at large scattering angles result from large-angle inelastic scattering followed by one or more small-angle inelastic scattering events. This must be so, since electrons cannot scatter by means of plasma interactions through angles larger than ϑ_c in a single scattering event.

Burge and Misell have carried out calculations on the electron energy loss spectra for carbon[133] and aluminum[142] films. Electron energies from 23 to 70 keV were assumed, and the volume loss for carbon and both the surface and volume losses for aluminum were considered. The spectra for losses less than 200 and 100 eV, respectively, were calculated. The agreement with experimental results was very good in both cases.

Many articles have been published dealing with single, average number of collisions <1,[245] plural, >1,[246] and multiple, $\gg 1$, scattering. The work has been carried out over a wide initial energy range. The treatments are usually formulated from the viewpoint of suitability of application in a specific field of interest (e.g., electron microscopy, electron diffraction, beta-spectroscopy). For further work on inelastic scattering, see the articles mentioned previously, Refs. 166, 247 to 259, and other references cited therein.

7. ELECTRON ENERGY LOSSES AS A SECONDARY EFFECT IN ELECTRON SPECTROMETRY

The subject of inelastic electron scattering in solids, especially the processes of plasmon and interband transition excitation, is in its own right interesting and important both experimentally and theoretically.

However, the effects of these processes are also significant in a secondary manner to experimentalists who are working in other fields.

Characteristic energy losses of electrons can give rise to structure in isochromatic X-ray spectra,[260,164,261] X-ray absorption (e.g., Ref. 262), and in electron microscopy[253,245,246] studies. They are also of importance to photoelectron, Auger, and internal conversion electron investigations, and corrections for their effects on measured quantities must at times be made (see, e.g., Chapter 3, Figure 4). Two effects of special importance to electron spectrometrists, are as follows.

1. Influence on energy determinations of internal conversion and photoelectron line positions—in both cases an electron of a specific initial kinetic energy is emitted from an atom. As the electron passes through the material bulk out into the spectrometer tank, it may lose energy primarily through excitation of plasmons or possible interband transitions. The probability for the former process is greater, and multiple plasma excitation energy loss structure will arise in the electron spectrum. Since the electron passes only one surface, the surface-to-volume loss probability ratio will be small except possibly for extremely thin converters or radioactive sources. In the copper $K\alpha$ X-ray photoelectron spectrum from the L-subshells of a thin layer of germanium, for example, two loss peaks were observed.[263] These were attributed to volume and surface losses.

As the probability of energy losses of energies less than the first intense plasma loss energy, surface or volume, is small, then if the initial electron energy distribution is sufficiently sharp, it should be possible to distinguish between the nonloss peak and the loss structure. This is the case for low energies and high instrumental resolution for both internal conversion and photoelectron spectra. Figure 19 shows the M_3-conversion line of the 4.821-keV transition in ^{151}Sm. The line peak position does not alter when the source thickness is increased a factor of 8. This is because the loss and nonloss electrons are well resolved.[264]

Assuming that the initial electron (internal conversion or photoelectron) energy distribution is much narrower than the first plasma loss energy, the instrumental resolution will contribute more and more strongly to the spectral line width for increasing electron energies. A momentum resolution of $\sim 0.05\%$ will make a 1-eV contribution to the line width at an energy of 1 keV, but a 24-eV contribution will be realized at 25 keV. The resulting agglomerate spectral lines will thereby become shifted to lower energies because of the energy loss structure contribution. This is effectively what happens for internal conversion electron lines of low energy, which are wide owing to the natural width of the atomic level in which the conversion takes place (see Chapter 7, Section 3, for a discussion). In the photoelectron situation, the initial photon energy distribution will also contribute to the line width.

2. Influence on the measurement of the natural widths of atomic levels from internal conversion or photoelectron line measurement—again, electron spectral structure due to characteristic energy losses will lie on the lower sides of the electron peaks to be analyzed. The relative enlargement of the peak width is in general of greater consequence than the shift of the peak position considered in 1. For low kinetic energies and high resolutions, however, this broadening effect may often be greatly reduced if the width of the peak produced is measured by mirroring its higher energy side across the peak energy value. In some cases unfolding procedures can be used, together with knowledge of the spectrometer window function and an assumed atomic level energy distribution function.

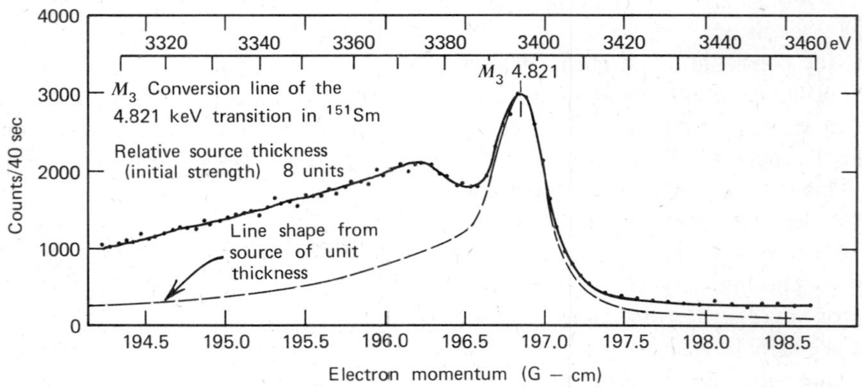

Figure 19. The M_3-internal conversion line of the 4.821-keV nuclear transition in ^{151}Sm as measured by Geiger et al.[264] The peak position does not alter when the source thickness is increased by a factor of 8 because of the negligible overlapping of characteristic energy loss structure with the zero loss peak.

REFERENCES

1. E. Rudberg, *Proc. Roy. Soc.* (*London*) **A127** (1930) 111.
2. J. Soller, *Phys. Rev.* **36** (1930) 1212.
3. D. Pines, in *Solid State Physics*, Academic Press, New York, (1955).
4. L. Marton, L. B. Leder, and H. Mendlowitz, *Advan. Electron. Electron Phys.* **7** (1955) 183.
5. O. Klemperer and J. P. G. Shepherd, *Advan. Phys.* **12** (1963) 355.
6. H. Raether, *Springer Tracts Mod. Phys.* **38** (1965) 84.
7. D. Pines and D. Bohm, *Phys. Rev.* **85** (1952) 338.
8. P. Nozières and D. Pines, *Phys. Rev.* **109** (1958) 1062.
9. P. Nozières and D. Pines, *Phys. Rev.* **109** (1958) 714.
10. P. Nozières and D. Pines, *Phys. Rev.* **113** (1959) 1254.
11. R. A. Ferrell, *Rev. Mod. Phys.* **28** (1956) 308.
12. P. Nozières and D. Pines, *Nuovo Cim.* **9** (1958) 470.
13. D. Pines, *Elementary Excitation in Solids*, W. A. Benjamin, Inc., New York, 1963.
14. D. Pines, *Physica* **26** (1960) S103.
15. R. A. Ferrell, *Phys. Rev.* **107** (1957) 450.
16. D. Bohm and D. Pines, *Phys. Rev.* **92** (1953) 609.
17. D. Pines, *Rev. Mod. Phys.* **28** (1956) 184.
18. P. Wolff, *Phys. Rev.* **92** (1953) 18.
19. D. F. Du Bois, *Ann. Phys.* **8** (1959) 24.
20. V. P. Silin, *Zh. Eksp. i Teor. Fiz.*, *SSSR*, **37** (1959) 273 [*J. Exp. Theor. Phys.* **10** (1960) 192].
21. P. Nozières and D. Pines, *Phys. Rev.* **111** (1958) 442.
22. H. Kanazawa, S. Misawa, and E. Fujita, *Proc. Theor. Phys.* (*Japan*) **23** (1960) 426.
23. O. von Roos and J. S. Zmuidzinas, *Phys. Rev.* **121** (1961) 941.
24. J. Geiger, *Z. Naturforsch.* **17a** (1962) 696.
25. R. A. Ferrell, *Phys. Rev.* **101** (1956) 554.
26. C. Kunz, *Phys. Letters* **15** (1965) 312.
27. C. Kunz, *Z. Phys.* **196** (1966) 311.
28. C. Kunz, *Z. Phys.* **167** (1962) 53.
29. H. Watanabe, *J. Phys. Soc.* (*Japan*) **11** (1956) 112.
30. G. Meyer, *Z. Phys.* **148** (1957) 61.
31. P. Schmüser, *Z. Phys.* **180** (1964) 105.
32. V. M. Agranovich and M. A. Mekhtiev, *Fiz. Tverd. Tela* (*SSSR*) **10** (1968) 2913.
33. D. Pines, *Phys. Rev.* **92** (1953) 626.
34. B. W. Ninham, C. J. Powell, and N. Swanson, *Phys. Rev.* **145** (1966) 209.
35. C. v. Festenberg, *Z. Phys.* **207** (1967) 47.
36. W. Hartl. *Z. Phys.* **191** (1966) 487.
37. H. Fröhlich and H. Pelzer, *Proc. Phys. Soc.* (*London*) **A68** (1955) 525.
38. J. Hubbard, *Proc. Phys. Soc.* (*London*) **A68** (1955) 441.
39. R. La Villa and H. Mendlowitz, *Phys. Rev. Letters* **9** (1962) 149.
40. D. Beaglehole, *Proc. Phys. Soc.* (*London*) **85** (1965) 1007.
41. T. S. Robinson, *Proc. Phys. Soc.* (*London*) **B65** (1952) 910.
42. H. R. Philipp and E. A. Taft, *Phys. Rev.* **113** (1959) 1002.
43. A. von Hipple, *Dielectrics and Waves*, John Wiley & Sons, Inc., New York, 1954.
44. E. A. Taft and H. R. Philipp, *Phys. Rev.* **A138** (1965) 197.
45. T. S. Moss, *Proc. Phys. Soc.* (*London*) **66B** (1953) 141.
46. E. Rudberg and J. C. Slater, *Phys. Rev.* **50** (1936) 150.
47. H. Ehrenreich, H. R. Philipp, and B. Segall, *Phys. Rev.* **132** (1963) 1918.
48. U. Fano, *Phys. Rev.* **103** (1956) 1202.
49. H. Richter and A. Ruckwied, *Z. Phys.* **160** (1960) 473.
50. R. H. Ritchie, *Phys. Rev.* **106** (1957) 874.
51. H. Jensen, *Z. Phys.* **106** (1937) 620.

52. H. Kanazawa, *Prog. Theor. Phys. (Japan)* **26** (1961) 851.
53. R. H. Ritchie, *Progr. Theor. Phys. (Japan)* **29** (1963) 607.
54. O. Klemperer and J. Thirlwell, *Solid State Commun.* **4** (1966) 15.
55. M. Hattori and K. Yamada, *J. Phys. Soc. (Japan)* **18** (1963) 200.
56. A. Otto, *Z. Phys.* **185** (1965) 232.
57. J. B. Swan, A. Otto, and H. Fellenzer, *Phys. Status Solidus* **23** (1967) 171.
58. A. Otto and J. B. Swan, *Z. Phys.* **206** (1967) 277.
59. A. Otto, *Phys. Status Solidus* **22** (1967) 401.
60. J. Geiger and K. Wittmaack, *Z. Phys.* **195** (1966) 44.
61. H. Boersch, J. Geiger, A. Imbusch, and H. Niedrig, *Phys. Letters* **22** (1966) 146.
62. J. C. Ashley, L. S. Cram, and E. T. Arakawa, *Phys. Rev.* **160** (1967) 313.
63. R. H. Ritchie and H. B. Eldridge, *Phys. Rev.* **126** (1962) 1935.
64. U. Fano, *J. Opt. Soc. Amer.* **31** (1941) 213.
65. E. Kröger, *Z. Phys.* **216** (1968) 115.
66. C. v. Festenberg, *Z. Phys.* **214** (1968) 464.
67. E. A. Stern and R. A. Ferrell, *Phys. Rev.* **120** (1960) 130.
68. J. Geiger, *Phys. Status Solidus* **24** (1967) 457.
69. G. M. Garibian, *Zh. Eksp. i Teor. Fiz., SSSR,* **38** (1960) 1814 [*J. Exp. Theor. Phys.* **11** (1960) 1306].
70. M. Creuzberg, *Z. Phys.* **174** (1963) 511.
71. A. Imbusch, *Z. Phys.* **216** (1968) 94.
72. J. L. Robins and J. B. Swan, *Proc. Phys. Soc. (London)* **76** (1960) 857.
73. J. L. Robins, *Proc. Phys. Soc. (London)* **78** (1961) 1177.
74. J. L. Robins and P. E. Best, *Proc. Phys. Soc. (London)* **79** (1962) 110.
75. L. B. Leder, *Phys. Rev.* **107** (1957) 1569.
76. B. M. Hartley and J. B. Swan, *Phys. Rev.* **144** (1966) 295.
77. B. M. Hartley, *Phys. Letters.* **27A** (1968) 499.
78. M. J. Lynch and J. B. Swan, *Austral. J. Phys.* **20** (1967) 345.
79. B. M. Hartley, *Phys. Letters.* **24A** (1967) 396.
80. B. M. Hartley, *Phys. Status Solidi* **31** (1969) 259. I thank J. B. Swan for the preliminary report copy he kindly sent.
81. M. J. Lynch and J. B. Swan, *Austral. J. Phys.* **21** (1968) 811.
82. C. v. Festenberg and E. Kröger, *Phys. Letters* **A26** (1968) 339.
83. G. D. Mahan, *Phys. Rev. Letters* **18** (1967) 448.
84. B. I. Lundqvist, *Phys. Kond. Mater.* **6** (1967) 193; *Phys. Kond. Mater.* **6** (1967) 206; *Phys. Kond. Mater.* **7** (1968) 117.
85. J. R. Cuthill, R. C. Dobbyn, A. J. McAlister, and M. L. Williams, *Phys. Rev.* **174** (1968) 515.
86. D. L. Dexter and R. S. Knox, "Excitons," Interscience Tracts on Physics and Astronomy, No. 25, Interscience Publishers, New York, 1965.
87. W. Kohn, *Phys. Rev. Letters* **19** (1967) 439.
88. J. C. Phillips, *Phys. Rev.* **133** (1964) A452; *Phys. Rev.* **136** (1964) A1705; *Phys. Rev.* **136** (1964) A1721.
89. B. L. Evans and P. A. Young, *Proc. Roy. Soc. (London)* **298** (1967) 74.
90. H. Ehrenreich and M. H. Cohen, *Phys. Rev.* **115** (1959) 786.
91. H. Ehrenreich and H. R. Philipp, *Phys. Rev.* **128** (1962) 1622.
92. J. Hubbard, *Proc. Phys. Soc. (London)* **A68** (1955) 976.
93. C. B. Wilson, *Proc. Phys. Soc. (London)* **76** (1960) 481.
94. K. Siegbahn, Carl Nordling, A. Fahlman, R. Nordberg, K. Hamrin, J. Hedman, G. Johansson, T. Bergmark, S.-E. Karlsson, I. Lindgren, and B. Lindberg, *Nova. Acta Reg. Soc. Sci. Upsal. Ser. IV* **20** (1967); Tech. Rept. AFML-TR-68-189, Wright-Patterson A.F.B., Dayton, Ohio (October 1968).
95. C. Gout, F. Pradal, and R. Simon, *Compt. Rend.* **254** (1962) 1233.
96. C. J. Powell, *Proc. Phys. Soc. (London)* **76** (1960) 593.
97. J. Thirlwell, *Proc. Phys. Soc. (London)* **91** (1967) 552.
98. J. Thirlwell, *J. Phys. C* [*Proc. Phys. Soc. (London)*] **1** (1968) 979.
99. M. Creuzburg, *Z. Phys.* **196** (1966) 433.
100. J. Daniels, *Z. Phys.* **203** (1967) 235.
101. J. Daniels, *Z. Naturforsch.* **21a** (1966) 662.

102. C. Norris and P. O. Nilsson, *Solid State Commun.* **6** (1968) 649.
103. H. Boersch, J. Geiger, H. Hellwig, and H. Michel, *Z. Phys.* **169** (1962) 252.
104. M. Hattori, K. Yamada, and H. Suzuki, *J. Phys. Soc.* (*Japan*) **18** (1963) 203.
105. C. Kunz, *Z. Phys.* **180** (1964) 127.
106. J. Geiger, *Z. Phys.* **161** (1961) 243.
107. J. Daniels, *Z. Phys.* **213** (1968) 227.
108. L. R. Canfield, G. Hass, and W. R. Hunter, *J. Phys.* (*Paris*) **25** (1964) 124.
109. G. Joos and A. Klopfer, *Z. Phys.* **138** (1954) 251.
110. O. Sueoka and F. Fujimoto, *J. Phys. Soc.* (*Japan*) **20** (1965) 569.
111. A. Carillon, *Compt. Rend.* **267** (1968) 332.
112. W. J. Scouler, *Phys. Rev. Letters* **18** (1967) 445.
113. L. B. Leder and J. A. Suddeth, *J. Appl. Phys.* **31** (1960) 1422.
114. K. Zeppenfeld, *Phys. Letters* **25A** (1967) 335.
115. J. Geiger, P. Odelga, and H.-C. Pfeiffer, *Phys. Status Solidus* **15** (1966) 531.
116. J. N. Shklyarevskii and R. Yarovaya, *Opt. Spectrosc. USSR*, **16** (1964) 45.
117. C. A. Williamson, M.Sc. thesis, University of London, 1966; I thank R. E. Burge for kindly sending a copy of this thesis.
118. P. E. Best, *Proc. Phys. Soc.* (*London*) **80** (1962) 1308.
119. K. Zeppenfeld, *Z. Phys.* **211** (1968) 391; see also F. Bassani and E. Tosatti, *Phys. Letters* **27** (1969) 446.
120. H. Fellenzer, *Z. Phys.* **165** (1961) 419.
121. K. Zeppenfeld, thesis, University of Hamburg, 1969.
122. S. Kunstreich and A. Otto, *Opt. Commun.* **1** (1969) 45.
123. J. B. Swan, *Phys. Rev.* A **135** (1964) 1467.
124. J. L. Robins, *Proc. Phys. Soc.* (*London*) **79** (1962) 119.
125. B. Gauthé, *Ann. Phys.* (*Paris*) **3** (1958) 29.
126. G. Ruthemann, *Ann. Phys.* (*Leipzig*) **2** (1948) 113.
127. H. Watanabe, *J. Phys. Soc.* (*Japan*) **9** (1954) 920.
128. G. Radi, *Z. Phys.* **213** (1968) 244.
129. W. Kleinn, *Optik* **11** (1954) 226.
130. L. Marton and L. B. Leder, *Phys. Rev.* **94** (1954) 203.
131. N. Swanson, *J. Opt. Soc. Amer.* **54** (1964) 1130.
132. C. J. Powell, *Proc. Phys. Soc.* (*London*) **76** (1960) 593.
133. R. E. Burge and D. L. Misell, *Phil. Mag.* **18** (1968) 251.
134. J. A. Simpson, T. F. McCraw, and L. Marton, *Phys. Rev.* **104** (1956) 64.
135. C. J. Powell and J. B. Swan, *Phys. Rev.* **116** (1959) 81.
136. L. B. Leder and L. Marton, *Phys. Rev.* **95** (1954) 1345.
137. J. R. Townsend, *Phys. Rev.* **92** (1953) 556.
138. H. Boersch, *Mikroskopie* **21** (1966) 122.
139. J. Geiger, *Physikertagung*, Munich, 1966.
140. H. Boersch, H. Miessner, W. Raith, *Z. Phys.* **168** (1962) 404.
141. T. Kloos, *Z. Phys.* **210** (1968) 303.
142. R. E. Burge and D. L. Misell, *Phil. Mag.* **18** (1968) 261.
143. K.-H. Gaukler, *Z. Phys.* **196** (1966) 85.
144. T. Kloos, *Z. Phys.* **208** (1968) 77.
145. J. Lohff, *Z. Phys.* **171** (1963) 442.
146. C. J. Powell and J. B. Swan, *Phys. Rev.* **115** (1959) 869.
147. P. Wielhold, *Z. Phys.* **208** (1968) 313.
148. W. C. Walker, O. P. Rustgi, and G. L. Weissler, *J. Opt. Soc. Amer.* **49** (1959) 471.
149. G. Haberstroh, *Z. Phys.* **145** (1956) 20.
150. C. J. Powell, *Bull. Amer. Phys. Soc.* **10** (1965) 433.
151. E. A. Stern, *Bull. Amer. Phys. Soc.* **4** (1959) 235.
152. G. Möllenstedt, *Optik* **5** (1949) 499.
153. F. N. Huffman, J. S. Cheka, B. G. Saunders, R. H. Ritchie, and R. D. Birkhoff, *Phys. Rev.* **106** (1957) 435.
154. F. Pradal and R. Saporte, *Compt. Rend.* **246** (1958) 2880.

155. N. Swanson and C. J. Powell, *Phys. Rev.* **167** (1968) 592.
156. H. Mendlowitz, *Proc. Phys. Soc.* (*London*) **75** (1960) 664.
157. E. Rudberg, *Phys. Rev.* **50** (1936) 138.
158. W. C. Walker, *J. Opt. Soc. Amer.* **48** (1958) 71.
159. H. Mendlowitz, *J. Opt. Soc. Amer.* **50** (1960) 739.
160. O. P. Rustgi, W. C. Walker, and G. L. Weissler, *J. Opt. Soc. Amer.* **51** (1961) 1357.
161. P. P. Reichertz and H. E. Farnsworth, *Phys. Rev.* **75** (1949) 1902.
162. R. C. Vehse, J. L. Stanford, and E. T. Arakawa, *Phys. Rev. Letters* **19** (1967) 1041.
163. O. P. Rustgi, W. C. Walker, and G. L. Weissler, *Bull. Amer. Phys. Soc. II* **4** (1959) 226.
164. S. Bergwall, R. K. Tyagi, and P. Ohlin, University of Uppsala Internal Rept. UUIP-460, 1966.
165. C. J. Powell, J. L. Robins, and J. B. Swan, *Phys. Rev.* **110** (1958) 657.
166. N. G. Nakhodkin and P. V. Mel'nik, *Fiz. Tver. Tela* **9** (1967) 727 [*Sov. Phys.-Solid State* **9** (1967) 567].
167. L. Marton, J. A. Simpson, and T. F. McCraw, *Phys. Rev.* **99** (1955) 495.
168. D. Gabor and G. W. Jull, *Nature* **175** (1955) 718.
169. C. J. Powell, *Phys. Rev. Letters* **15** (1965) 852.
170. O. Klemperer and J. P. G. Shepherd, *Brit. J. Appl. Phys.* **14** (1963) 89.
171. C. J. Powell, *Advan. Phys.* **16** (1967) 203.
172. D. R. Spalding and A. J. F. Metherell, *Phil. Mag.* **18** (1968) 41.
173. P.-O. Nilsson, A. Persson, and S. Hagström, *Solid State Commun.* **6** (1968) 297.
174. H. P. Myers, L. Walldén, and Å. Karlsson, *Phil. Mag.* **18** (1968) 725.
175. H. Fukutani and O. Sueoka, *J. Phys. Soc.* (*Japan*) **20** (1965) 620.
176. F. Fujimoto and O. Sueoka, *J. Phys. Soc.* (*Japan*) **19** (1964) 2069.
177. H. R. Philipp and H. Ehrenreich, *Phys. Rev.* **129** (1963) 1550.
178. K. Zeppenfeld and H. Raether, *Z. Phys.* **193** (1966) 471.
179. H. Ehrenreich, H. R. Philipp, and J. C. Phillips, *Phys. Rev. Letters* **8** (1962) 59.
180. M. Balkanski and Y. Pétroff, *J. Phys.* (*Paris*) **27** (1966) C2-84.
181. E. A. Taft and H. R. Philipp, *Phys. Rev.* **121** (1961) 1100.
182. W. J. Turner, W. E. Reese, and G. D. Pettit, *Phys. Rev.* **136** (1964) A1467.
183. M. Creuzburg and H. Raether, *Z. Phys.* **171** (1963) 436.
184. N. B. Gornyi, *Fiz. Tverd. Tela, SSSR*, **3** (1961) 698 [*Sov. Phys.—Solid State* **3** (1961) 507].
185. B. Feuerbacher, M. Skibowski, R. P. Godwin, and T. Sasaki, *J. Opt. Soc. Amer.* **58** (1968) 1434.
186. M. Shinada and S. Sugano, *J. Phys. Soc.* (*Japan*) **21** (1966) 1936.
187. H. Watanabe, *Phys. Rev.* **95** (1954) 1684.
188. W. W. Piper, *Phys. Rev.* **92** (1953) 23.
189. D. C. Reynolds, C. W. Litton, and T. C. Collins, *Phys. Rev.* **174** (1968) 8451.
190. M. Giammarinaro, M. B. Palma-Vittorelli, and M. U. Palma, *Phys. Rev. Letters* **21** (1968) 1382.
191. D. G. Avery, *Proc. Phys. Soc.* (*London*) **66B** (1953) 134.
192. P. E. Best, *Proc. Phys. Soc.* (*London*) **79** (1962) 133.
193. P. Keil, *Z. Phys.* **214** (1968) 266.
194. P. Keil, *Z. Naturforsch.* **23a** (1968) 948.
195. M. Creuzburg and H. Raether, *Solid State Commun.* **2** (1964) 175.
196. A. Milgram and M. P. Givens, *Phys. Rev.* **125** (1962) 1506.
197. D. M. Roessler, unpublished data, see Ref. 99.
198. L. B. Leder, H. Mendlowitz, and L. Marton, *Phys. Rev.* **101** (1956) 1460.
199. M. Creuzburg, *Z. Phys.* **194** (1966) 211.
200. P. W. Palmberg and T. N. Rhodin, *J. Phys. Chem. Sol.* **29** (1968) 1917.
201. H. R. Philipp and H. Ehrenreich, *Phys. Rev.* **131** (1963) 2016.
202. P. Keil, *Z. Naturforsch.* **21a** (1966) 503.
203. O. Bostanjoglo and L. Schmidt, *Phys. Letters* **22** (1966) 130.
204. H. Boersch, O. Bostanjoglo, B. Lischke, H. Niedrig, and L. Schmidt, *Sixth International Congress for Electron Microscopy*, Kyoto, 1966, p.167.
205. K. Moorjani, *Phys. Letters* **23** (1966) 652.
206. P. Keil, *Z. Phys.* **214** (1968) 251.

207. G. Ruthemann, *Naturwiss.* **30** (1942) 145.
208. G. Ruthemann, *Ann. Phys.* (*Leipzig*) **2** (1948) 135.
209. J. Hillier and R. F. Baker, *J. Appl. Phys.* **15** (1944) 663.
210. A. Fahlman, K. Hamrin, G. Axelson, C. Nordling, and K. Siegbahn, *Z. Phys.* **192** (1966) 484.
211. N. Swanson and K. Codling, *J. Opt. Soc. Amer.* **58** (1968) 1192.
212. N. Swanson and C. J. Powell, *Phys. Rev.* **145** (1966) 195.
213. R. A. Ferrell, *Phys. Rev.* **111** (1958) 1214.
214. H. Boersch, C. Radeloff, and G. Sauerbrey, *Phys. Rev. Letters* **7** (1961) 52.
215. G. Sauerbrey, *Fifth International Congress for Electron Microscopy*, Academic Press, New York, 1966, AA-13.
216. H. Boersch, P. Dobberstein, D. Fritzsche, and G. Sauerbrey, *Z. Phys.* **187** (1965) 97.
217. W. F. Hanson, R. J. Herickhoff, E. T. Arakawa, and R. D. Birkhoff, *Bull. Amer. Phys. Soc.* **10** (1965) 433; J. Bösenberg, *Z. Phys.* **218** (1969) 282.
218. E. T. Arakawa, N. O. Davis, L. C. Emerson, and R. D. Birkhoff, *J. Phys. Radium* **25** (1964) 129.
219. G. E. Jones, L. S. Cram, and E. T. Arakawa, *Phys. Rev.* **147** (1966) 515.
220. L. S. Cram and E. T. Arakawa, *Phys. Rev.* **153** (1967) 455.
221. G. Sauerbrey and P. Dobberstein, contribution to the *Second International Conference on Vacuum Ultraviolet Physics*, Gatlinburg, 1968; U. Bürker and W. Steinmann, *Phys. Rev. Letters* **21** (1968) 143.
222. I. M. Frank, *Usp. Fiz. Nauk* **8** (1966) 729: W. Steinmann, *Phys. Status Solidus* **28** (1968) 437.
223. H. Boersch, R. Wolter, and H. Schoenbeck, *Z. Phys.* **199** (1967) 124.
224. G. L. Braglia, *Phys. Letters* **28A** (1968) 48.
225. A. Bohg, thesis, I. Physikalische Institut der Technische Universität Berlin, 1969; H. Boersch, J. Geiger, and A. Bohg, *Z. Phys*). **227** (1969) 141.
226. H. Ehrhardt and F. Linder, *Phys. Rev. Letters* **21** (1968) 419.
227. H. Boersch, J. Geiger, and M. Topschowsky, Internal report, I. Physikal. Inst., Techn. Univ. Berlin, 1969.
228. H. P. Herbst and G. Jeschke, *Fourth European Regional Conference on Electron Microscopy*, Rome 1968, p. 293; private communication from G. Jeschke, 1969.
229. M. Nelkin, *Inelastic Scattering of Neutrons in Solids and Liquids*, International Atomic Energy Agency, Vienna, 1961, p. 3.
230. Y. Kaskiwase, *J. Phys. Soc.* (*Japan*) **21** (1966) 1263.
231. M. Klier, *Z. Phys.* **150** (1958) 49.
232. G. Berthold, *Z. Phys.* **181** (1964) 333.
233. H. Boersch, J. Geiger, and W. Stickel, *Phys. Rev. Letters* **17** (1966) 379.
234. H. Boersch, *Sixth International Congress for Electron Microscopy*, Kyoto, 1966, p. 9.
235. H. Boersch, *J. Electron Microsc.* **16** (1967) 39.
236. H. Boersch, J. Geiger, and W. Stickel, *Z. Phys.* **212** (1968) 130.
237. K. L. Kliewer and R. Fuchs, *Phys. Rev.* **144** (1966) 495.
238. K. L. Kliewer and R. Fuchs, *Phys. Rev.* **150** (1966) 575.
239. R. Fuchs, K. L. Kliewer, and W. J. Pardee, *Phys. Rev.* **150** (1966) 589.
240. R. Fuchs and K. L. Kliewer, *Phys. Rev.* **140** (1965) A2076.
241. K. Hisano, *J. Phys. Soc.* (*Japan*) **25** (1968) 1091.
242. L. Marton, J. A. Simpson, H. A. Fowler, and N. Swanson, *Phys. Rev.* **126** (1962) 182.
243. J. J. Quinn, *Phys. Rev.* **126** (1962) 1453.
244. M. Creuzburg and H. Dimigen, *Z. Phys.* **174** (1963) 24.
245. R. E. Burge and G. H. Smith, *Proc. Phys. Soc.* (*London*) **79** (1962) 673.
246. G. H. Smith and R. E. Burge, *Proc. Phys. Soc.* (*London*) **81** (1963) 612.
247. W. Hilgner and J. Kessler, *Z. Phys.* **187** (1965) 119.
248. A. F. Akkerman, Yu. A. Andreev, and Yu. G. Lgutov, *Fiz. Tverd. Tela* **9** (1967) 766. [*Sov. Phys—Solid State* **9** (1967) 599].
249. H. Frank, *Z. Naturforsch.* **14a** (1959) 247.
250. W. Dietrich and H. Sciler, *Z. Phys.* **157** (1960) 576.
251. O. Klemperer, A. Thetford, and F. Lenz, *Proc. Phys. Soc.* (*London*) **76** (1960) 705.
252. N. Itoh, *Phys. Status Solidus* **30** (1968) 199.
253. A. J. F. Metherell, *J. Appl. Phys. Japan* **7** (1968) 99.
254. H. Kanter, *Ann. Phys.* (*Leipzig*) **20** (1957) 9; *Phys. Rev.* **121** (1961) 461.
255. J. K. Bienlein and G. Schlosser, *Z. Phys.* **174** (1963) 91.

256. C. Feldman, *Phys. Rev.* **117** (1960) 455.

257. R. D. Birkhoff, *Handbuch der Physik*, Vol 34, Springer-Verlag, Berlin, 1958, p. 71.

258. R. M. Sternheimer, in *Methods of Experimental Physics*, L. C. L. Yuan and C. S. Wu (Eds.), Vol. 5, pt. A, Academic Press, New York, 1961.

259. G. Knop and W. Paul, in *Alpha-, Beta-, and Gamma-Ray Spectroscopy*, K. Siegbahn (Ed.), North-Holland Publishing Company, 1965, Vol. I, p. 1.

260. H. Claus and K. Ulmer, *Z. Phys.* **185** (1965) 139.

261. S. Bergwall, *Z. Phys.* **187** (1965) 495.

262. W. W. Beeman and H. Friedmann, *Phys. Rev.* **56** (1939) 392; G. A. Rooke, *Phys. Letters* **3** (1963) 234; F. Brouers, *Phys. Status Solidus*, **11** (1965) K25; *Phys. Status Solidus*, **22** (1967) 213.

263. R. Carlson, A. Fahlman, and K. Siegbahn, *Ark. Phys.* **32** (1966) 103.

264. J. S. Geiger, R. L. Graham, and J. S. Merritt, *Nucl. Phys.* **48** (1963) 97.

Appendix A

Physical Constants and Conversion Factors

Table 1.

Fundamental constants

The following information and values, as well as those of Tables 2 and 3, have been taken primarily from E. R. Cohen and J. W. M. DuMond, Rev. Mod. Phys. 37 (1965) 537 and from the "Table of Isotopes", C.M Lederer, J. M. Hollander, and I. Perlman, 6th edition, John Wiley & Sons, Inc., New York (1967). The symbols denoting the constants are arranged in alphabetical order. Units are given in cgs and mks systems, the following abbreviations being used:

C = Coulomb	m = meter
cm = centimeter	N = Newton
deg = degree Kelvin	rad = radian
emu = electromagnetic unit	s = second
esu = electrostatic unit	T = Tesla
G = Gauss	amu = atomic mass unit (unified scale)
g = gram	W = Watt
J = Joule	Wb = Weber
kg = kilogram	

Symbol	Constant	Value	Unit cgs	Unit mks
$a_0 = \hbar^2/m_o e^2$	First Bohr radius	5.29167(2)	10^{-9} cm	10^{-11} m
$1/\alpha = \hbar c/e^2$	Fine structure constant	137.0388(6)		
c	Speed of light in vacuum	2.997925(1)	10^{10} cm s^{-1}	10^8 m s^{-1}
$c_1 = 2\pi hc^2$	First radiation constant	3.74150(9)	10^{-5} erg cm^2 s^{-1}	10^{-16} W m^2
$c_2 = hc/k$	Second radiation constant	1.43879(6)	cm deg	10^{-2} m deg
e	Base of natural logarithms	2.71828183		
e	Elementary charge	4.80298(7)	10^{-10} esu	
e/c		1.60210(2)	10^{-20} emu	10^{-19} C
e/m_o	Charge-to-mass ratio of the electron	5.27274(2)	10^{17} esu g^{-1}	
$e/m_o c$		1.758796(6)	10^7 emu g^{-1}	10^{11} C kg^{-1}
$F = N_A e/c$	Faraday constant (unified mass scale)	9.64870(5)	10^3 emu $mole^{-1}$	10^4 C $mole^{-1}$
γ_p	Gyromagnetic ratio of the proton	2.675192(7)	10^4 rad s^{-1} G^{-1}	10^8 rad s^{-1} T^{-1}
G	Gravitational constant	6.670(5)	10^{-8} dyne cm^2 g^{-2}	10^{-11} N m^2 kg^{-2}
h	Planck's constant	6.62559(16)	10^{-27} erg s	10^{-34} J s
$\hbar = h/2\pi$		1.054494(25)	" " "	" " "
h/e	Quantum of magnetic flux	1.379474(13)	10^{-17} erg s esu^{-1}	

Table I (cont.)

hc/e		4.13556(4)	10^{-7} G cm^2	10^{-11} Wb
$k = R_o/N_A$	Boltzmann constant	1.38054(6)	10^{-16} erg deg^{-1}	10^{-23} J deg^{-1}
$\lambda_{C,E} = h/m_o c$	Compton wavelength of the electron	2.42621(2)	10^{-10} cm	10^{-12} m
$\lambda_{C,p} = h/m_p c$	Compton wavelength of the proton	1.321398(13)	10^{-13} cm	10^{-15} m
$\lambda_{C,n} = h/m_n c$	Compton wavelength of the neutron	1.319589(13)	10^{-13} cm	10^{-15} m
λ_g/λ_s	Conversion factor from Siegbahn X-units to milli-ångströms	1.002080(6)		
$M_e = N_A m_o$	Atomic mass of the electron (unified mass scale)	5.48597(3)	10^{-4} amu	10^{-4} amu
$M_H = N_A m_H$	Atomic mass of hydrogen (unified mass scale)	1.00782519(8)	amu	amu
$M_n = N_A m_n$	Atomic mass of the neutron (unified mass scale)	1.00866520(10)	amu	amu
$M_p = N_A m_p$	Atomic mass of the proton (unified mass scale)	1.00727662(8)	amu	amu
$\mu_B = \hbar e/2m_o c$	Bohr magneton	9.2732(2)	10^{-21} erg G^{-1}	10^{-24} J T^{-1}
μ_e/μ_B	Electron magnetic moment in Bohr magnetons	1.001159622(27)		
$\mu_N = \hbar e/2m_p c$	Nuclear magneton	5.05050(13)	10^{-24} erg G^{-1}	10^{-27} J T^{-1}
N_A	Avogadro's constant (unified mass scale)	6.02252(9)	10^{23} $mole^{-1}$	
π	Ratio of the circumference of a circle to its diameter	3.14159265		
R_o	Ideal gas constant (unified mass scale)	8.31434(35)	10^7 erg $mole^{-1}$ deg^{-1}	J $mole^{-1}$ deg^{-1}
$R_\infty = m_o e^4/4\pi c\hbar^3$	Rydberg constant for infinite mass	1.0973731(1)	10^5 cm^{-1}	10^7 m^{-1}
r	Ratio of chemical to unified mass scales $M(O = 16)/M(C^{12} = 12)$	1.000043(5)		
$r_e = e^2/m_o c^2$	Classical electron radius	2.81777(4)	10^{-13} cm	10^{-15} m
σ	Stefan-Boltzmann constant	5.6697(10)	10^{-5} erg cm^{-2} s^{-1} deg^{-4}	10^{-3} W m^{-2} deg^{-4}

Table I. (cont.)

$u = 1/N_A$	Mass unit (unified mass scale)	1.66043(2)	10^{-24} g	10^{-27} kg
v_e (1 eV)	Velocity of an electron of 1 eV energy	5.9319	10^{7} cm sec^{-1}	10^{5} m sec^{-1}
Å	Ångström unit of length	1	10^{-8} cm	10^{-10} m

Table II.

Atomic units

1 au of energy $= m_o e^2/\hbar = 2E_H = 2Ry_\infty = 27.212$ eV

1 au of length $= a_o = \hbar^2/m_o e^2 = 0.52917$ Å

1 au of velocity $= e^2/\hbar = 2.1876\times10^{8}$ cm sec^{-1}

1 au of time $= \hbar^3/m_o e^4 = 2.4189\times10^{-17}$ sec

1 au of frequency $= m_o e^4/\hbar^3 = 4.134\times10^{16}$ sec^{-1}

Hence, 1 eV corresponds to 1.519×10^{15} sec^{-1}

Table III.

Energy conversion factors

erg	eV	s^{-1}	cm^{-1}	deg K	g	atomic mass units amu	cal
1 erg	6.2418×10^{11}	1.50929×10^{26}	5.0345×10^{15}	7.2436×10^{15}	1.112649×10^{-21}	6.7010×10^{2}	2.3901×10^{-8}
1.60210×10^{-12}	1 eV	2.41804×10^{14}	8.0657×10^{3}	1.16049×10^{4}	1.78258×10^{-33}	1.07356×10^{-9}	3.8291×10^{-20}
6.6256×10^{-27}	4.13558×10^{-15}	1 s^{-1}	3.335640×10^{-11}	4.7992×10^{-48}	7.3720×10^{-11}	4.4398×10^{-24}	1.5836×10^{-34}
1.9863×10^{-16}	1.23981×10^{-4}	2.997925×10^{10}	1 cm^{-1}	1.4388	2.2101×10^{-37}	1.33101×10^{-13}	4.7474×10^{-24}
1.3805×10^{-16}	8.6171×10^{-5}	2.0836×10^{10}	6.9503×10^{-1}	1 deg K	1.5361×10^{-37}	9.2509×10^{-14}	3.2996×10^{-24}
8.987554×10^{20}	5.6098×10^{32}	1.35649×10^{47}	4.5248×10^{36}	6.5102×10^{36}	1 g	6.0225×10^{23}	2.1481×10^{13}
1.49232×10^{-3}	9.31478×10^{8}	2.2524×10^{23}	7.5131×10^{12}	1.0810×10^{13}	1.66043×10^{-24}	1 amu	3.5667×10^{-11}
4.1840×10^{7}	2.6116×10^{19}	6.3149×10^{33}	2.1064×10^{23}	3.0307×10^{23}	4.6554×10^{-14}	2.8037×10^{10}	1 cal

Appendix B

Energy Versus Velocity and Mass for Electrons

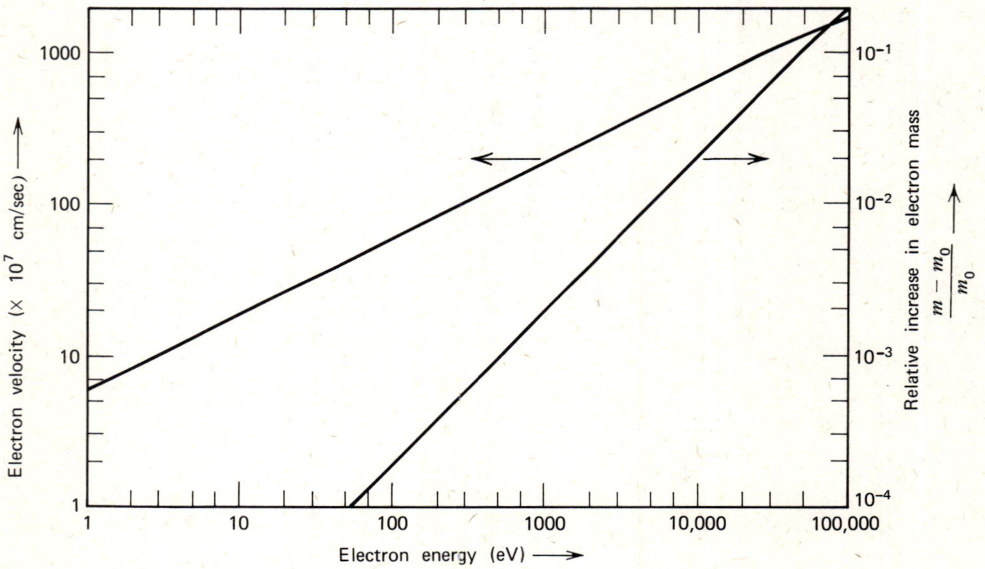

Appendix C

Electron Energy Versus Electron Momentum

The relativistic equation that relates an electron's kinetic energy E to its momentum expressed in Gauss-cm, or $B\rho$-value, may be written

$$E = \left[(m_o c^2)^2 \left(\frac{e}{m_o c}\right)^2 (B\rho)^2 + (m_o c^2)^2\right]^{1/2} - m_e c^2 \quad (1)$$

or

$$E = m_o c^2 \left\{\left[\left(\frac{e}{m_o c}\right)^2 (B\rho)^2 + 1\right]^{1/2} - 1\right\}, \quad (2)$$

where m_o and e are the electron rest mass and charge, respectively, and c is the velocity of light. This equation, together with values of the 1965 fundamental constant values,[1] can be used to obtain a table of kinetic energy values as a function of $B\rho$.* The following values are used in the calculation:

$$m_o c^2 = 0.5510066061 \times 10^3 \text{ keV},$$

$$(m_o c^2)^2 = 0.2611277515 \times 10^6 \text{ keV}^2,$$

$$\left(\frac{e}{m_o c}\right)^2 = 0.89875785 \times 10^{-1} \text{ (G-cm)}^{-2}.$$

The first constant should be compared with that given in Ref. 2, $m_o c^2 = 0.551006(2) \times 10^3$ keV, which has an accuracy of 3.5 ppm. By equation 2, then, the accuracy of the calculated energy values will be limited by the accuracy of the last of the constants, or to 15 ppm; this, then, will be the limit of accuracy of the tabulated values. The agreement between the energy values in this appendix and those in other compilations[3] is better than 2 ppm, and therefore much better than the accuracy of the values themselves.

The format of the table is the same as that used by Gibbins and Kaye,[3] where the mean difference between two adjacent energy values in each row is given on the right-hand side under DIFF.

The $B\rho$-values in this table range from 0 to 799.9 G-cm and are listed in steps of 0.1 G-cm. This corresponds to energies up to just above 50 keV, the energy region of primary interest in this book.

* I am indebted to C. Sébille, Institut de Physique Nucléaire, Orsay, France, for carrying out these computations.

REFERENCES

1. E. R. Cohen and J. W. M. Du Mond, *Rev. Mod. Phys.* **37** (1965) 537; see also Appendix A.
2. *Table of Isotopes*, C. M. Lederer, J. M. Hollander, and I. Perlman, 6th ed., John Wiley & Sons, Inc., New York. 1968, Appendix II, Table 4.
3. C. K. Gibbins and G. Kaye, The Chadwick Laboratory, University of Liverpool, England; J. Hedman, Internal Rept., Institute of Physics, University of Uppsala, Uppsala, Sweden, 1967.

	.0	.1	.2	.3	.4	.5	.6	.7	.8	.9	DIFF	
0	.000000	.000001	.000004	.000008	.000014	.000022	.000032	.000043	.000056	.000071	7	0
1	.000088	.000106	.000127	.000149	.000172	.000198	.000225	.000254	.000285	.000317	25	1
2	.000352	.000388	.000426	.000465	.000507	.000550	.000594	.000641	.000689	.000740	43	2
3	.000791	.000845	.000901	.000958	.001017	.001077	.001140	.001204	.001270	.001338	60	3
4	.001407	.001478	.001551	.001626	.001703	.001781	.001861	.001943	.002026	.002111	78	4
5	.002198	.002287	.002378	.002470	.002564	.002660	.002758	.002857	.002958	.003061	95	5
6	.003166	.003272	.003380	.003490	.003602	.003715	.003831	.003948	.004066	.004187	113	6
7	.004309	.004433	.004559	.004686	.004816	.004947	.005079	.005214	.005350	.005488	131	7
8	.005628	.005770	.005913	.006058	.006205	.006354	.006504	.006656	.006810	.006966	148	8
9	.007123	.007282	.007443	.007606	.007770	.007937	.008104	.008274	.008446	.008619	166	9
10	.008794	.008971	.009149	.009329	.009511	.009695	.009881	.010068	.010257	.010448	183	10
11	.010641	.010835	.011031	.011229	.011429	.011630	.011833	.012038	.012245	.012453	201	11
12	.012663	.012875	.013089	.013304	.013521	.013740	.013961	.014184	.014408	.014634	218	12
13	.014862	.015091	.015322	.015555	.015790	.016027	.016265	.016505	.016747	.016991	236	13
14	.017236	.017483	.017732	.017983	.018235	.018489	.018745	.019003	.019262	.019523	254	14
15	.019786	.020051	.020317	.020585	.020855	.021127	.021401	.021676	.021953	.022232	271	15
16	.022512	.022794	.023078	.023364	.023652	.023941	.024232	.024525	.024820	.025116	289	16
17	.025414	.025714	.026015	.026319	.026624	.026931	.027240	.027550	.027862	.028176	306	17
18	.028492	.028809	.029128	.029449	.029772	.030097	.030423	.030751	.031081	.031412	324	18
19	.031745	.032080	.032417	.032756	.033096	.033438	.033782	.034127	.034475	.034824	342	19
20	.035175	.035527	.035882	.036238	.036596	.036955	.037317	.037680	.038045	.038412	359	20
21	.038780	.039150	.039522	.039896	.040271	.040649	.041028	.041408	.041791	.042175	377	21
22	.042561	.042949	.043338	.043730	.044123	.044518	.044914	.045313	.045713	.046115	394	22
23	.046518	.046923	.047331	.047739	.048150	.048563	.048977	.049393	.049810	.050230	412	23
24	.050651	.051074	.051499	.051925	.052353	.052783	.053215	.053648	.054084	.054521	429	24
25	.054960	.055400	.055842	.056286	.056732	.057180	.057629	.058080	.058533	.058988	447	25
26	.059444	.059902	.060362	.060824	.061287	.061752	.062219	.062688	.063158	.063630	465	26
27	.064104	.064580	.065057	.065537	.066018	.066500	.066985	.067471	.067959	.068449	482	27
28	.068940	.069434	.069929	.070425	.070924	.071424	.071926	.072430	.072936	.073443	500	28
29	.073952	.074463	.074976	.075490	.076006	.076524	.077044	.077565	.078088	.078613	517	29
30	.079140	.079668	.080198	.080730	.081264	.081800	.082337	.082876	.083417	.083959	535	30
31	.084503	.085049	.085597	.086147	.086698	.087251	.087806	.088362	.088921	.089481	553	31
32	.090043	.090606	.091172	.091739	.092307	.092878	.093451	.094025	.094601	.095178	570	32
33	.095758	.096339	.096922	.097506	.098093	.098681	.099271	.099863	.100456	.101051	588	33
34	.101648	.102247	.102848	.103450	.104054	.104660	.105267	.105877	.106488	.107101	605	34
35	.107715	.108331	.108949	.109569	.110191	.110814	.111439	.112066	.112695	.113325	623	35
36	.113957	.114591	.115227	.115864	.116504	.117145	.117787	.118432	.119078	.119726	640	36
37	.120376	.121027	.121680	.122335	.122992	.123651	.124311	.124973	.125637	.126302	658	37
38	.126970	.127639	.128309	.128982	.129656	.130332	.131010	.131690	.132371	.133054	676	38
39	.133739	.134426	.135114	.135804	.136496	.137190	.137885	.138582	.139281	.139982	693	39
40	.140685	.141389	.142095	.142802	.143512	.144223	.144936	.145651	.146367	.147086	711	40
41	.147806	.148527	.149251	.149976	.150703	.151432	.152163	.152895	.153629	.154365	728	41
42	.155103	.155842	.156583	.157326	.158070	.158817	.159565	.160315	.161067	.161820	746	42
43	.162575	.163332	.164091	.164851	.165613	.166377	.167143	.167910	.168680	.169451	763	43
44	.170223	.170998	.171774	.172552	.173332	.174113	.174897	.175682	.176469	.177257	781	44
45	.178047	.178839	.179633	.180429	.181226	.182025	.182826	.183629	.184433	.185239	799	45
46	.186047	.186857	.187668	.188481	.189296	.190113	.190931	.191751	.192573	.193397	816	46
47	.194222	.195050	.195879	.196709	.197542	.198376	.199212	.200050	.200889	.201730	834	47
48	.202573	.203418	.204265	.205113	.205963	.206815	.207668	.208524	.209381	.210240	851	48
49	.211100	.211963	.212827	.213692	.214560	.215429	.216301	.217173	.218048	.218924	869	49

	.0	.1	.2	.3	.4	.5	.6	.7	.8	.9	DIFF	
50	.219803	.220682	.221564	.222448	.223333	.224220	.225108	.225999	.226891	.227785	886	50
51	.228681	.229578	.230477	.231378	.232281	.233186	.234092	.235000	.235910	.236821	904	51
52	.237734	.238649	.239566	.240485	.241405	.242327	.243251	.244176	.245104	.246033	922	52
53	.246964	.247896	.248831	.249767	.250705	.251644	.252586	.253529	.254474	.255420	939	53
54	.256369	.257319	.258271	.259224	.260180	.261137	.262096	.263057	.264019	.264983	957	54
55	.265949	.266917	.267886	.268858	.269831	.270805	.271782	.272760	.273740	.274722	974	55
56	.275705	.276691	.277678	.278666	.279657	.280649	.281643	.282639	.283637	.284636	992	56
57	.285637	.286640	.287644	.288651	.289659	.290669	.291680	.292694	.293709	.294726	1009	57
58	.295744	.296765	.297787	.298811	.299836	.300864	.301893	.302924	.303957	.304991	1027	58
59	.306027	.307065	.308105	.309146	.310190	.311235	.312281	.313330	.314380	.315432	1044	59
60	.316486	.317541	.318599	.319658	.320718	.321781	.322845	.323911	.324979	.326049	1062	60
61	.327120	.328193	.329268	.330344	.331423	.332503	.333585	.334668	.335753	.336841	1080	61
62	.337929	.339020	.340112	.341207	.342302	.343400	.344499	.345601	.346703	.347808	1097	62
63	.348915	.350023	.351133	.352244	.353358	.354473	.355590	.356709	.357829	.358951	1115	63
64	.360075	.361201	.362328	.363458	.364589	.365721	.366856	.367992	.369130	.370270	1132	64
65	.371411	.372555	.373700	.374846	.375995	.377145	.378297	.379451	.380607	.381764	1150	65
66	.382923	.384084	.385246	.386411	.387577	.388745	.389914	.391085	.392259	.393433	1167	66
67	.394610	.395788	.396969	.398150	.399334	.400519	.401707	.402895	.404086	.405278	1185	67
68	.406473	.407669	.408866	.410066	.411267	.412470	.413674	.414881	.416089	.417299	1202	68
69	.418511	.419724	.420939	.422156	.423375	.424596	.425818	.427042	.428267	.429495	1220	69
70	.430724	.431955	.433188	.434422	.435659	.436897	.438137	.439378	.440621	.441866	1238	70
71	.443113	.444362	.445612	.446864	.448118	.449373	.450631	.451890	.453151	.454413	1255	71
72	.455678	.456944	.458211	.459481	.460752	.462025	.463300	.464577	.465855	.467135	1273	72
73	.468417	.469701	.470986	.472273	.473562	.474853	.476145	.477439	.478735	.480033	1290	73
74	.481332	.482634	.483937	.485241	.486548	.487856	.489166	.490477	.491791	.493106	1308	74
75	.494423	.495742	.497062	.498384	.499708	.501034	.502361	.503691	.505022	.506354	1325	75
76	.507689	.509025	.510363	.511703	.513044	.514388	.515733	.517079	.518428	.519778	1343	76
77	.521130	.522484	.523839	.525197	.526556	.527917	.529279	.530643	.532009	.533377	1360	77
78	.534747	.536118	.537491	.538866	.540243	.541621	.543001	.544383	.545766	.547152	1378	78
79	.548539	.549928	.551318	.552711	.554105	.555500	.556898	.558297	.559699	.561101	1395	79
80	.562506	.563912	.565321	.566730	.568142	.569555	.570971	.572387	.573806	.575226	1413	80
81	.576649	.578072	.579498	.580926	.582355	.583786	.585218	.586653	.588089	.589527	1430	81
82	.590966	.592408	.593851	.595296	.596743	.598191	.599641	.601093	.602547	.604002	1448	82
83	.605460	.606919	.608379	.609842	.611306	.612772	.614240	.615709	.617180	.618653	1465	83
84	.620128	.621604	.623083	.624563	.626044	.627528	.629013	.630500	.631989	.633479	1483	84
85	.634972	.636466	.637961	.639459	.640958	.642459	.643962	.645466	.646973	.648481	1501	85
86	.649990	.651502	.653015	.654530	.656047	.657566	.659086	.660608	.662132	.663657	1518	86
87	.665184	.666714	.668244	.669777	.671311	.672847	.674385	.675925	.677466	.679009	1536	87
88	.680554	.682100	.683649	.685199	.686750	.688304	.689859	.691416	.692975	.694536	1553	88
89	.696098	.697662	.699228	.700796	.702365	.703936	.705509	.707084	.708660	.710238	1571	89
90	.711818	.713399	.714983	.716568	.718155	.719743	.721334	.722926	.724520	.726115	1588	90
91	.727713	.729312	.730912	.732515	.734119	.735726	.737333	.738943	.740554	.742168	1606	91
92	.743782	.745399	.747017	.748637	.750259	.751883	.753508	.755135	.756764	.758395	1623	92
93	.760027	.761661	.763297	.764935	.766574	.768216	.769858	.771503	.773149	.774798	1641	93
94	.776447	.778099	.779752	.781408	.783065	.784723	.786384	.788046	.789710	.791375	1658	94
95	.793043	.794712	.796383	.798055	.799730	.801406	.803084	.804763	.806445	.808128	1676	95
96	.809813	.811499	.813188	.814878	.816570	.818264	.819959	.821656	.823355	.825056	1693	96
97	.826758	.828462	.830168	.831876	.833585	.835296	.837009	.838724	.840440	.842159	1711	97
98	.843878	.845600	.847323	.849049	.850776	.852504	.854235	.855967	.857701	.859436	1728	98
99	.861174	.862913	.864654	.866396	.868141	.869887	.871635	.873385	.875136	.876889	1746	99

	.0	.1	.2	.3	.4	.5	.6	.7	.8	.9	DIFF	
100	.878644	.880401	.882159	.883919	.885681	.887445	.889210	.890977	.892746	.894517	1763	100
101	.896289	.898063	.899839	.901617	.903396	.905178	.906960	.908745	.910532	.912320	1781	101
102	.914110	.915901	.917695	.919490	.921287	.923085	.924886	.926688	.928492	.930297	1798	102
103	.932105	.933914	.935725	.937537	.939352	.941168	.942986	.944805	.946627	.948450	1816	103
104	.950275	.952101	.953930	.955760	.957592	.959425	.961261	.963098	.964937	.966777	1833	104
105	.968620	.970464	.972310	.974157	.976007	.977858	.979711	.981565	.983422	.985280	1851	105
106	.987140	.989001	.990865	.992730	.994597	.996465	.998336	1.000208	1.002082	1.003957	1868	106
107	1.005835	1.007714	1.009594	1.011477	1.013361	1.015247	1.017135	1.019025	1.020916	1.022809	1886	107
108	1.024704	1.026601	1.028499	1.030399	1.032301	1.034204	1.036110	1.038017	1.039926	1.041836	1903	108
109	1.043749	1.045663	1.047578	1.049496	1.051415	1.053336	1.055259	1.057184	1.059110	1.061038	1921	109
110	1.062968	1.064899	1.066833	1.068768	1.070704	1.072643	1.074583	1.076525	1.078469	1.080415	1938	110
111	1.082362	1.084311	1.086262	1.088214	1.090168	1.092124	1.094082	1.096042	1.098003	1.099966	1956	111
112	1.101931	1.103897	1.105865	1.107835	1.109807	1.111780	1.113756	1.115733	1.117711	1.119692	1973	112
113	1.121674	1.123658	1.125644	1.127631	1.129620	1.131611	1.133604	1.135599	1.137595	1.139593	1990	113
114	1.141592	1.143594	1.145597	1.147602	1.149609	1.151617	1.153627	1.155639	1.157653	1.159668	2008	114
115	1.161685	1.163704	1.165725	1.167747	1.169771	1.171797	1.173825	1.175854	1.177885	1.179918	2025	115
116	1.181953	1.183989	1.186027	1.188067	1.190109	1.192152	1.194197	1.196244	1.198293	1.200343	2043	116
117	1.202395	1.204449	1.206505	1.208562	1.210621	1.212682	1.214744	1.216809	1.218875	1.220943	2060	117
118	1.223012	1.225084	1.227157	1.229231	1.231308	1.233386	1.235466	1.237548	1.239632	1.241717	2078	118
119	1.243804	1.245893	1.247983	1.250075	1.252169	1.254265	1.256363	1.258462	1.260563	1.262666	2095	119
120	1.264770	1.266876	1.268984	1.271094	1.273205	1.275319	1.277433	1.279550	1.281669	1.283789	2113	120
121	1.285911	1.288034	1.290160	1.292287	1.294416	1.296547	1.298679	1.300813	1.302949	1.305087	2130	121
122	1.307226	1.309367	1.311510	1.313655	1.315801	1.317949	1.320099	1.322251	1.324404	1.326559	2148	122
123	1.328716	1.330875	1.333035	1.335197	1.337361	1.339526	1.341694	1.343863	1.346033	1.348206	2165	123
124	1.350380	1.352556	1.354734	1.356914	1.359095	1.361278	1.363463	1.365649	1.367837	1.370027	2183	124
125	1.372219	1.374413	1.376608	1.378805	1.381004	1.383204	1.385406	1.387610	1.389816	1.392023	2200	125
126	1.394233	1.396443	1.398656	1.400871	1.403087	1.405305	1.407524	1.409746	1.411969	1.414194	2217	126
127	1.416420	1.418649	1.420879	1.423111	1.425344	1.427580	1.429817	1.432056	1.434296	1.436538	2235	127
128	1.438783	1.441028	1.443276	1.445525	1.447776	1.450029	1.452284	1.454540	1.456798	1.459058	2252	128
129	1.461319	1.463582	1.465847	1.468114	1.470383	1.472653	1.474925	1.477199	1.479474	1.481751	2270	129
130	1.484030	1.486311	1.488593	1.490877	1.493163	1.495451	1.497740	1.500032	1.502324	1.504619	2287	130
131	1.506915	1.509214	1.511513	1.513815	1.516118	1.518423	1.520730	1.523039	1.525349	1.527661	2305	131
132	1.529975	1.532291	1.534608	1.536927	1.539248	1.541570	1.543895	1.546221	1.548548	1.550878	2322	132
133	1.553209	1.555542	1.557877	1.560213	1.562552	1.564891	1.567233	1.569577	1.571922	1.574269	2339	133
134	1.576617	1.578968	1.581320	1.583674	1.586030	1.588387	1.590746	1.593107	1.595470	1.597834	2357	134
135	1.600200	1.602568	1.604937	1.607309	1.609682	1.612057	1.614433	1.616811	1.619191	1.621573	2374	135
136	1.623957	1.626342	1.628729	1.631118	1.633508	1.635900	1.638294	1.640690	1.643088	1.645487	2392	136
137	1.647888	1.650290	1.652695	1.655101	1.657509	1.659919	1.662330	1.664743	1.667158	1.669575	2409	137
138	1.671993	1.674413	1.676835	1.679258	1.681684	1.684111	1.686540	1.688970	1.691402	1.693837	2427	138
139	1.696272	1.698710	1.701149	1.703590	1.706033	1.708477	1.710923	1.713371	1.715821	1.718273	2444	139
140	1.720726	1.723181	1.725637	1.728096	1.730556	1.733018	1.735481	1.737947	1.740414	1.742883	2461	140
141	1.745353	1.747826	1.750300	1.752776	1.755253	1.757732	1.760213	1.762696	1.765181	1.767667	2479	141
142	1.770155	1.772645	1.775136	1.777630	1.780124	1.782621	1.785120	1.787620	1.790122	1.792625	2496	142
143	1.795131	1.797638	1.800147	1.802657	1.805170	1.807684	1.810200	1.812717	1.815237	1.817758	2514	143
144	1.820281	1.822805	1.825331	1.827859	1.830389	1.832921	1.835454	1.837989	1.840526	1.843064	2531	144
145	1.845604	1.848146	1.850690	1.853235	1.855783	1.858332	1.860882	1.863435	1.865989	1.868545	2548	145
146	1.871102	1.873662	1.876223	1.878785	1.881350	1.883916	1.886484	1.889054	1.891626	1.894199	2566	146
147	1.896774	1.899351	1.901929	1.904509	1.907091	1.909675	1.912261	1.914848	1.917437	1.920027	2583	147
148	1.922620	1.925214	1.927810	1.930407	1.933007	1.935608	1.938211	1.940815	1.943421	1.946029	2601	148
149	1.948639	1.951251	1.953864	1.956479	1.959096	1.961714	1.964334	1.966956	1.969580	1.972206	2618	149

	.0	.1	.2	.3	.4	.5	.6	.7	.8	.9	DIFF	
150	1.974833	1.977462	1.980092	1.982725	1.985359	1.987995	1.990632	1.993272	1.995913	1.998556	2635	150
151	2.001200	2.003846	2.006494	2.009144	2.011796	2.014449	2.017104	2.019761	2.022419	2.025079	2653	151
152	2.027741	2.030405	2.033070	2.035738	2.038406	2.041077	2.043749	2.046423	2.049099	2.051777	2670	152
153	2.054456	2.057137	2.059820	2.062505	2.065191	2.067879	2.070569	2.073260	2.075953	2.078648	2688	153
154	2.081345	2.084043	2.086744	2.089446	2.092149	2.094855	2.097562	2.100270	2.102981	2.105693	2705	154
155	2.108408	2.111123	2.113841	2.116560	2.119281	2.122004	2.124728	2.127455	2.130183	2.132912	2722	155
156	2.135644	2.138377	2.141112	2.143848	2.146587	2.149327	2.152069	2.154812	2.157558	2.160305	2740	156
157	2.163054	2.165804	2.168556	2.171310	2.174066	2.176824	2.179583	2.182344	2.185107	2.187871	2757	157
158	2.190637	2.193405	2.196175	2.198946	2.201719	2.204494	2.207271	2.210049	2.212829	2.215611	2774	158
159	2.218394	2.221180	2.223967	2.226755	2.229546	2.232338	2.235132	2.237928	2.240725	2.243524	2792	159
160	2.246325	2.249128	2.251932	2.254738	2.257546	2.260356	2.263167	2.265980	2.268795	2.271611	2809	160
161	2.274430	2.277250	2.280071	2.282895	2.285720	2.288547	2.291376	2.294206	2.297038	2.299872	2826	161
162	2.302708	2.305545	2.308384	2.311225	2.314067	2.316912	2.319758	2.322605	2.325455	2.328306	2844	162
163	2.331159	2.334014	2.336870	2.339728	2.342588	2.345450	2.348313	2.351178	2.354045	2.356914	2861	163
164	2.359784	2.362656	2.365530	2.368405	2.371283	2.374162	2.377042	2.379925	2.382809	2.385695	2878	164
165	2.388583	2.391472	2.394363	2.397256	2.400150	2.403047	2.405945	2.408845	2.411746	2.414649	2896	165
166	2.417554	2.420461	2.423370	2.426280	2.429192	2.432105	2.435021	2.437938	2.440857	2.443777	2913	166
167	2.446700	2.449624	2.452550	2.455477	2.458406	2.461337	2.464270	2.467205	2.470141	2.473079	2930	167
168	2.476018	2.478960	2.481903	2.484848	2.487794	2.490743	2.493693	2.496645	2.499598	2.502553	2948	168
169	2.505510	2.508469	2.511430	2.514392	2.517356	2.520321	2.523289	2.526258	2.529229	2.532201	2965	169
170	2.535176	2.538152	2.541130	2.544109	2.547090	2.550073	2.553058	2.556044	2.559033	2.562023	2982	170
171	2.565014	2.568008	2.571003	2.574000	2.576998	2.579999	2.583001	2.586004	2.589010	2.592017	3000	171
172	2.595026	2.598037	2.601049	2.604064	2.607079	2.610097	2.613116	2.616138	2.619160	2.622185	3017	172
173	2.625211	2.628239	2.631269	2.634301	2.637334	2.640369	2.643405	2.646444	2.649484	2.652526	3034	173
174	2.655570	2.658615	2.661662	2.664711	2.667761	2.670814	2.673868	2.676923	2.679981	2.683040	3052	174
175	2.686101	2.689164	2.692228	2.695294	2.698362	2.701432	2.704503	2.707576	2.710651	2.713727	3069	175
176	2.716806	2.719886	2.722967	2.726051	2.729136	2.732223	2.735311	2.738402	2.741494	2.744588	3086	176
177	2.747683	2.750781	2.753880	2.756980	2.760083	2.763187	2.766293	2.769401	2.772510	2.775621	3104	177
178	2.778734	2.781849	2.784965	2.788083	2.791203	2.794324	2.797448	2.800573	2.803699	2.806828	3121	178
179	2.809958	2.813090	2.816223	2.819359	2.822496	2.825635	2.828775	2.831918	2.835062	2.838207	3138	179
180	2.841355	2.844504	2.847655	2.850808	2.853962	2.857118	2.860276	2.863435	2.866597	2.869760	3156	180
181	2.872925	2.876091	2.879259	2.882429	2.885601	2.888774	2.891949	2.895126	2.898305	2.901485	3173	181
182	2.904667	2.907851	2.911037	2.914224	2.917413	2.920603	2.923796	2.926990	2.930186	2.933384	3190	182
183	2.936583	2.939784	2.942987	2.946191	2.949398	2.952606	2.955815	2.959027	2.962240	2.965455	3207	183
184	2.968672	2.971890	2.975110	2.978332	2.981555	2.984781	2.988008	2.991236	2.994467	2.997699	3225	184
185	3.000933	3.004169	3.007406	3.010645	3.013886	3.017128	3.020373	3.023619	3.026866	3.030116	3242	185
186	3.033367	3.036620	3.039875	3.043131	3.046389	3.049649	3.052911	3.056174	3.059439	3.062706	3259	186
187	3.065974	3.069244	3.072516	3.075790	3.079065	3.082342	3.085621	3.088902	3.092184	3.095468	3277	187
188	3.098754	3.102041	3.105331	3.108622	3.111914	3.115209	3.118505	3.121802	3.125102	3.128403	3294	188
189	3.131706	3.135011	3.138318	3.141626	3.144936	3.148247	3.151561	3.154876	3.158193	3.161511	3311	189
190	3.164832	3.168154	3.171477	3.174803	3.178130	3.181459	3.184789	3.188122	3.191456	3.194792	3328	190
191	3.198129	3.201469	3.204810	3.208152	3.211497	3.214843	3.218191	3.221541	3.224892	3.228245	3346	191
192	3.231600	3.234956	3.238315	3.241675	3.245036	3.248400	3.251765	3.255132	3.258500	3.261871	3363	192
193	3.265243	3.268617	3.271992	3.275369	3.278748	3.282129	3.285511	3.288896	3.292281	3.295669	3380	193
194	3.299058	3.302449	3.305842	3.309237	3.312633	3.316031	3.319430	3.322832	3.326235	3.329640	3397	194
195	3.333046	3.336455	3.339865	3.343277	3.346690	3.350105	3.353522	3.356941	3.360361	3.363783	3415	195
196	3.367207	3.370632	3.374060	3.377489	3.380920	3.384352	3.387786	3.391222	3.394660	3.398099	3432	196
197	3.401540	3.404983	3.408427	3.411874	3.415321	3.418771	3.422223	3.425676	3.429131	3.432587	3449	197
198	3.436045	3.439505	3.442967	3.446431	3.449896	3.453363	3.456831	3.460302	3.463774	3.467248	3466	198
199	3.470723	3.474200	3.477679	3.481160	3.484643	3.488127	3.491613	3.495100	3.498590	3.502081	3484	199

	.0	.1	.2	.3	.4	.5	.6	.7	.8	.9	DIFF	
200	3.505573	3.509068	3.512564	3.516062	3.519562	3.523063	3.526566	3.530071	3.533578	3.537086	3501	200
201	3.540596	3.544108	3.547621	3.551136	3.554653	3.558172	3.561692	3.565214	3.568738	3.572263	3518	201
202	3.575791	3.579320	3.582850	3.586383	3.589917	3.593452	3.596990	3.600529	3.604070	3.607613	3535	202
203	3.611158	3.614704	3.618252	3.621801	3.625353	3.628906	3.632460	3.636017	3.639575	3.643135	3553	203
204	3.646697	3.650260	3.653825	3.657392	3.660961	3.664531	3.668103	3.671677	3.675252	3.678829	3570	204
205	3.682408	3.685989	3.689571	3.693155	3.696741	3.700328	3.703917	3.707508	3.711101	3.714695	3587	205
206	3.718292	3.721889	3.725489	3.729090	3.732693	3.736298	3.739904	3.743512	3.747122	3.750734	3604	206
207	3.754347	3.757962	3.761579	3.765197	3.768818	3.772439	3.776063	3.779688	3.783315	3.786944	3621	207
208	3.790575	3.794207	3.797841	3.801477	3.805114	3.808753	3.812394	3.816036	3.819681	3.823327	3639	208
209	3.826974	3.830624	3.834275	3.837928	3.841582	3.845239	3.848897	3.852556	3.856218	3.859881	3656	209
210	3.863546	3.867213	3.870881	3.874551	3.878223	3.881896	3.885572	3.889248	3.892927	3.896608	3673	210
211	3.900290	3.903973	3.907659	3.911346	3.915035	3.918726	3.922418	3.926112	3.929808	3.933506	3690	211
212	3.937205	3.940906	3.944609	3.948313	3.952019	3.955727	3.959437	3.963148	3.966861	3.970576	3707	212
213	3.974293	3.978011	3.981731	3.985452	3.989176	3.992901	3.996627	4.000356	4.004086	4.007818	3725	213
214	4.011552	4.015287	4.019024	4.022763	4.026504	4.030246	4.033990	4.037735	4.041483	4.045232	3742	214
215	4.048983	4.052735	4.056490	4.060246	4.064003	4.067763	4.071524	4.075287	4.079051	4.082817	3759	215
216	4.086586	4.090355	4.094127	4.097900	4.101675	4.105451	4.109230	4.113010	4.116791	4.120575	3776	216
217	4.124360	4.128147	4.131936	4.135726	4.139518	4.143312	4.147107	4.150904	4.154703	4.158504	3793	217
218	4.162306	4.166110	4.169916	4.173724	4.177533	4.181344	4.185156	4.188971	4.192787	4.196605	3810	218
219	4.200424	4.204245	4.208068	4.211893	4.215719	4.219547	4.223377	4.227209	4.231042	4.234877	3828	219
220	4.238714	4.242552	4.246392	4.250234	4.254077	4.257923	4.261770	4.265618	4.269469	4.273321	3845	220
221	4.277175	4.281030	4.284887	4.288746	4.292607	4.296469	4.300333	4.304199	4.308067	4.311936	3862	221
222	4.315807	4.319680	4.323554	4.327430	4.331308	4.335188	4.339069	4.342952	4.346837	4.350723	3879	222
223	4.354611	4.358501	4.362393	4.366286	4.370181	4.374077	4.377976	4.381876	4.385778	4.389682	3896	223
224	4.393587	4.397494	4.401402	4.405313	4.409225	4.413139	4.417054	4.420972	4.424891	4.428811	3913	224
225	4.432734	4.436658	4.440584	4.444511	4.448440	4.452371	4.456304	4.460239	4.464175	4.468113	3930	225
226	4.472052	4.475993	4.479936	4.483881	4.487827	4.491775	4.495725	4.499677	4.503630	4.507585	3948	226
227	4.511542	4.515500	4.519460	4.523422	4.527386	4.531351	4.535318	4.539286	4.543257	4.547229	3965	227
228	4.551203	4.555178	4.559155	4.563134	4.567115	4.571097	4.575081	4.579067	4.583055	4.587044	3982	228
229	4.591035	4.595028	4.599022	4.603018	4.607016	4.611015	4.615016	4.619019	4.623024	4.627030	3999	229
230	4.631038	4.635048	4.639060	4.643073	4.647088	4.651104	4.655123	4.659143	4.663164	4.667188	4016	230
231	4.671213	4.675240	4.679269	4.683299	4.687331	4.691365	4.695400	4.699437	4.703476	4.707517	4033	231
232	4.711559	4.715603	4.719649	4.723696	4.727745	4.731796	4.735848	4.739903	4.743959	4.748016	4050	232
233	4.752076	4.756137	4.760200	4.764264	4.768330	4.772398	4.776468	4.780539	4.784612	4.788687	4067	233
234	4.792764	4.796842	4.800922	4.805003	4.809087	4.813172	4.817259	4.821347	4.825437	4.829529	4085	234
235	4.833623	4.837718	4.841815	4.845914	4.850014	4.854116	4.858220	4.862326	4.866433	4.870542	4102	235
236	4.874653	4.878765	4.882879	4.886995	4.891113	4.895232	4.899353	4.903475	4.907600	4.911726	4119	236
237	4.915854	4.919983	4.924114	4.928247	4.932382	4.936518	4.940656	4.944796	4.948937	4.953081	4136	237
238	4.957225	4.961372	4.965520	4.969670	4.973822	4.977975	4.982130	4.986287	4.990446	4.994606	4153	238
239	4.998768	5.002932	5.007097	5.011264	5.015433	5.019603	5.023776	5.027950	5.032125	5.036303	4170	239
240	5.040482	5.044662	5.048845	5.053029	5.057215	5.061402	5.065592	5.069783	5.073975	5.078170	4187	240
241	5.082366	5.086564	5.090763	5.094964	5.099167	5.103372	5.107578	5.111786	5.115996	5.120208	4204	241
242	5.124421	5.128636	5.132852	5.137071	5.141291	5.145512	5.149736	5.153961	5.158188	5.162416	4221	242
243	5.166646	5.170878	5.175112	5.179347	5.183585	5.187823	5.192064	5.196306	5.200550	5.204795	4238	243
244	5.209043	5.213292	5.217543	5.221795	5.226049	5.230305	5.234562	5.238822	5.243083	5.247345	4255	244
245	5.251610	5.255876	5.260144	5.264413	5.268684	5.272957	5.277232	5.281508	5.285786	5.290066	4272	245
246	5.294347	5.298630	5.302915	5.3 7202	5.311490	5.315780	5.320072	5.324365	5.328660	5.332957	4289	246
247	5.337255	5.341555	5.345857	5.350161	5.354466	5.358773	5.363082	5.367392	5.371704	5.376018	4306	247
248	5.380334	5.384651	5.388970	5.393291	5.397613	5.401937	5.406263	5.410590	5.414919	5.419250	4324	248
249	5.423583	5.427917	5.432253	5.436591	5.440930	5.445271	5.449614	5.453958	5.458305	5.462653	4341	249

	.0	.1	.2	.3	.4	.5	.6	.7	.8	.9	DIFF	
250	5.467002	5.471353	5.475706	5.480061	5.484418	5.488776	5.493135	5.497497	5.501860	5.506225	4358	250
251	5.510592	5.514960	5.519330	5.523702	5.528075	5.532450	5.536827	5.541206	5.545586	5.549968	4375	251
252	5.554352	5.558737	5.563124	5.567513	5.571904	5.576296	5.580690	5.585085	5.589482	5.593881	4392	252
253	5.598282	5.602684	5.607089	5.611494	5.615902	5.620311	5.624722	5.629135	5.633549	5.637965	4409	253
254	5.642383	5.646802	5.651223	5.655646	5.660070	5.664497	5.668925	5.673354	5.677786	5.682219	4426	254
255	5.686653	5.691090	5.695528	5.699968	5.704409	5.708852	5.713297	5.717744	5.722192	5.726642	4443	255
256	5.731094	5.735548	5.740003	5.744460	5.748918	5.753378	5.757840	5.762304	5.766769	5.771236	4460	256
257	5.775705	5.780176	5.784648	5.789122	5.793597	5.798074	5.802553	5.807034	5.811516	5.816000	4477	257
258	5.820486	5.824973	5.829463	5.833953	5.838446	5.842940	5.847436	5.851934	5.856433	5.860934	4494	258
259	5.865437	5.869941	5.874447	5.878955	5.883465	5.887976	5.892489	5.897004	5.901520	5.906038	4511	259
260	5.910558	5.915079	5.919602	5.924127	5.928654	5.933182	5.937712	5.942244	5.946777	5.951312	4528	260
261	5.955849	5.960387	5.964927	5.969469	5.974013	5.978558	5.983105	5.987653	5.992204	5.996756	4545	261
262	6.001310	6.005865	6.010422	6.014981	6.019541	6.024103	6.028667	6.033233	6.037800	6.042369	4562	262
263	6.046940	6.051512	6.056086	6.060662	6.065240	6.069819	6.074400	6.078982	6.083567	6.088153	4579	263
264	6.092740	6.097330	6.101921	6.106514	6.111108	6.115704	6.120302	6.124901	6.129503	6.134106	4596	264
265	6.138710	6.143317	6.147925	6.152535	6.157146	6.161759	6.166374	6.170990	6.175609	6.180229	4613	265
266	6.184850	6.189473	6.194098	6.198725	6.203354	6.207984	6.212615	6.217249	6.221884	6.226521	4630	266
267	6.231160	6.235800	6.240442	6.2 5085	6.249731	6.254378	6.259027	6.263677	6.268329	6.272983	4647	267
268	6.277639	6.282296	6.286955	6.291615	6.296278	6.300942	6.305607	6.310275	6.314944	6.319615	4663	268
269	6.324287	6.328961	6.333637	6.338315	6.342994	6.347675	6.352358	6.357042	6.361728	6.366416	4680	269
270	6.371105	6.375796	6.380489	6.385184	6.389880	6.394578	6.399277	6.403979	6.408682	6.413386	4697	270
271	6.418093	6.422801	6.427510	6.432222	6.436935	6.441650	6.446366	6.451085	6.455805	6.460526	4714	271
272	6.465250	6.469975	6.474701	6.479430	6.484160	6.488891	6.493625	6.498360	6.503097	6.507836	4731	272
273	6.512576	6.517318	6.522061	6.526807	6.531554	6.536302	6.541053	6.545805	6.550559	6.555314	4748	273
274	6.560072	6.564830	6.569591	6.574353	6.579117	6.583883	6.588650	6.593419	6.598190	6.602962	4765	274
275	6.607737	6.612512	6.617290	6.622069	6.626850	6.631632	6.636417	6.641203	6.645990	6.650780	4782	275
276	6.655571	6.660363	6.665158	6.669954	6.674752	6.679551	6.684352	6.689155	6.693960	6.698766	4799	276
277	6.703574	6.708384	6.713195	6.718008	6.722823	6.727639	6.732457	6.737277	6.742098	6.746921	4816	277
278	6.751746	6.756573	6.761401	6.766231	6.771063	6.775896	6.780731	6.785568	6.790406	6.795246	4833	278
279	6.800088	6.804931	6.809776	6.814623	6.819472	6.824322	6.829174	6.834027	6.838883	6.843740	4850	279
280	6.848599	6.853459	6.858321	6.863185	6.868050	6.872917	6.877786	6.882656	6.887529	6.892402	4867	280
281	6.897278	6.902155	6.907034	6.911915	6.916797	6.921681	6.926567	6.931454	6.936343	6.941234	4883	281
282	6.946127	6.951021	6.955916	6.960814	6.965713	6.970614	6.975517	6.980421	6.985327	6.990235	4900	282
283	6.995144	7.000055	7.004968	7.009882	7.014798	7.019716	7.024635	7.029557	7.034480	7.039404	4917	283
284	7.044330	7.049258	7.054188	7.059119	7.064052	7.068987	7.073923	7.078861	7.083801	7.088742	4934	284
285	7.093685	7.098630	7.103577	7.108525	7.113475	7.118426	7.123379	7.128334	7.133291	7.138249	4951	285
286	7.143209	7.148171	7.153134	7.158099	7.163066	7.168034	7.173004	7.177976	7.182950	7.187925	4968	286
287	7.192902	7.197880	7.202860	7.207842	7.212826	7.217811	7.222798	7.227787	7.232777	7.237769	4985	287
288	7.242763	7.247758	7.252755	7.257754	7.262755	7.267757	7.272761	7.277766	7.282773	7.287782	5002	288
289	7.292793	7.297805	7.302819	7.307835	7.312852	7.317871	7.322891	7.327914	7.332938	7.337964	5018	289
290	7.342991	7.348020	7.353051	7.358083	7.363118	7.368153	7.373191	7.378230	7.383271	7.388314	5035	290
291	7.393358	7.398404	7.403451	7.408501	7.413552	7.418604	7.423659	7.428715	7.433773	7.438832	5052	291
292	7.443893	7.448956	7.454021	7.459087	7.464155	7.469224	7.474295	7.479368	7.484443	7.489519	5069	292
293	7.494597	7.499677	7.504758	7.509841	7.514926	7.520012	7.525100	7.530190	7.535281	7.540374	5086	293
294	7.545469	7.550566	7.555664	7.560764	7.565865	7.570968	7.576073	7.581180	7.586288	7.591398	5103	294
295	7.596510	7.601623	7.606738	7.611854	7.616973	7.622093	7.627215	7.632338	7.637463	7.642590	5120	295
296	7.647718	7.652848	7.657980	7.663114	7.668249	7.673386	7.678524	7.683664	7.688806	7.693950	5136	296
297	7.699095	7.704242	7.709391	7.714541	7.719693	7.724847	7.730002	7.735159	7.740318	7.745478	5153	297
298	7.750640	7.755804	7.760969	7.766137	7.771305	7.776476	7.781648	7.786822	7.791997	7.797174	5170	298
299	7.802353	7.807534	7.812716	7.817900	7.823086	7.828273	7.833462	7.838652	7.843845	7.849039	5187	299

	.0	.1	.2	.3	.4	.5	.6	.7	.8	.9	DIFF	
300	7.854235	7.859432	7.864631	7.869832	7.875034	7.880238	7.885444	7.890651	7.895861	7.901071	5204	300
301	7.906284	7.911498	7.916714	7.921931	7.927151	7.932371	7.937594	7.942818	7.948044	7.953272	5220	301
302	7.958501	7.963732	7.968965	7.974199	7.979435	7.984673	7.989912	7.995153	8.000396	8.005640	5237	302
303	8.010886	8.016134	8.021383	8.026634	8.031887	8.037142	8.042398	8.047656	8.052915	8.058176	5254	303
304	8.063439	8.068704	8.073970	8.079238	8.084507	8.089778	8.095052	8.100326	8.105602	8.110880	5271	304
305	8.116160	8.121441	8.126724	8.132009	8.137295	8.142583	8.147873	8.153164	8.158457	8.163752	5288	305
306	8.169049	8.174347	8.179646	8.184948	8.190251	8.195556	8.200862	8.206170	8.211480	8.216792	5304	306
307	8.222105	8.227420	8.232736	8.238054	8.243374	8.248696	8.254019	8.259344	8.264670	8.269999	5321	307
308	8.275329	8.280660	8.285993	8.291329	8.296665	8.302003	8.307343	8.312685	8.318029	8.323373	5338	308
309	8.328720	8.334069	8.339419	8.344770	8.350124	8.355479	8.360836	8.366194	8.371554	8.376916	5355	309
310	8.382279	8.387644	8.393011	8.398380	8.403750	8.409122	8.414495	8.419870	8.425247	8.430626	5371	310
311	8.436006	8.441388	8.446771	8.452157	8.457543	8.462932	8.468322	8.473714	8.479108	8.484503	5388	311
312	8.489900	8.495299	8.500699	8.506101	8.511504	8.516910	8.522317	8.527725	8.533136	8.538548	5405	312
313	8.543961	8.549377	8.554794	8.560212	8.565633	8.571055	8.576478	8.581904	8.587331	8.592760	5422	313
314	8.598190	8.603622	8.609056	8.614491	8.619928	8.625367	8.630808	8.636250	8.641694	8.647139	5438	314
315	8.652586	8.658035	8.663485	8.668938	8.674391	8.679847	8.685304	8.690763	8.696223	8.701686	5455	315
316	8.707150	8.712615	8.718082	8.723551	8.729022	8.734494	8.739968	8.745443	8.750921	8.756399	5472	316
317	8.761880	8.767362	8.772846	8.778332	8.783819	8.789308	8.794798	8.800291	8.805785	8.811280	5488	317
318	8.816778	8.822276	8.827777	8.833279	8.838783	8.844289	8.849796	8.855305	8.860816	8.866328	5505	318
319	8.871842	8.877358	8.882875	8.888394	8.893915	8.899437	8.904961	8.910487	8.916014	8.921543	5522	319
320	8.927074	8.932606	8.938140	8.943676	8.949213	8.954752	8.960293	8.965835	8.971380	8.976925	5539	320
321	8.982473	8.988022	8.993572	8.999125	9.004679	9.010235	9.015792	9.021351	9.026912	9.032474	5555	321
322	9.038038	9.043604	9.049171	9.054741	9.060311	9.065884	9.071458	9.077034	9.082611	9.088190	5572	322
323	9.093771	9.099353	9.104937	9.110523	9.116111	9.121700	9.127290	9.132883	9.138477	9.144073	5589	323
324	9.149670	9.155269	9.160870	9.166473	9.172077	9.177682	9.183290	9.188899	9.194510	9.200122	5605	324
325	9.205736	9.211352	9.216969	9.222589	9.228209	9.233832	9.239456	9.245082	9.250709	9.256338	5622	325
326	9.261969	9.267601	9.273235	9.278871	9.284509	9.290148	9.295789	9.301431	9.307075	9.312721	5639	326
327	9.318369	9.324018	9.329668	9.335321	9.340975	9.346631	9.352288	9.357947	9.363608	9.369270	5655	327
328	9.374935	9.380600	9.386268	9.391937	9.397607	9.403280	9.408954	9.414630	9.420307	9.425986	5672	328
329	9.431667	9.437349	9.443033	9.448719	9.454407	9.460096	9.465787	9.471479	9.477173	9.482869	5689	329
330	9.488566	9.494265	9.499966	9.505668	9.511372	9.517078	9.522785	9.528494	9.534205	9.539918	5705	330
331	9.545632	9.551347	9.557065	9.562784	9.568504	9.574227	9.579951	9.585676	9.591404	9.597133	5722	331
332	9.602864	9.608596	9.614330	9.620066	9.625803	9.631542	9.637282	9.643025	9.648769	9.654514	5738	332
333	9.660262	9.666011	9.671761	9.677514	9.683267	9.689023	9.694780	9.700539	9.706300	9.712062	5755	333
334	9.717826	9.723592	9.729359	9.735128	9.740898	9.746671	9.752445	9.758220	9.763998	9.769776	5772	334
335	9.775557	9.781339	9.787123	9.792909	9.798696	9.804484	9.810275	9.816067	9.821861	9.827657	5788	335
336	9.833454	9.839253	9.845053	9.850855	9.856659	9.862464	9.868272	9.874080	9.879891	9.885703	5805	336
337	9.891517	9.897332	9.903149	9.908968	9.914788	9.920610	9.926434	9.932259	9.938087	9.943915	5822	337
338	9.949746	9.955578	9.961411	9.967247	9.973084	9.978922	9.984763	9.990605	9.996449	10.002294	5838	338
339	10.008141	10.013989	10.019840	10.025692	10.031545	10.037400	10.043257	10.049116	10.054976	10.060838	5855	339
340	10.066702	10.072567	10.078434	10.084302	10.090172	10.096044	10.101918	10.107793	10.113670	10.119548	5871	340
341	10.125429	10.131310	10.137194	10.143079	10.148966	10.154854	10.160744	10.166636	10.172529	10.178424	5888	341
342	10.184321	10.190220	10.196119	10.202021	10.207925	10.213830	10.219736	10.225645	10.231555	10.237466	5904	342
343	10.243380	10.249295	10.255211	10.261129	10.267049	10.272971	10.278894	10.284819	10.290746	10.296674	5921	343
344	10.302604	10.308535	10.314468	10.320403	10.326340	10.332278	10.338218	10.344159	10.350102	10.356047	5938	344
345	10.361994	10.367942	10.373891	10.379843	10.385796	10.391750	10.397707	10.403665	10.409625	10.415586	5954	345
346	10.421549	10.427513	10.433480	10.439448	10.445417	10.451388	10.457362	10.463336	10.469312	10.475290	5971	346
347	10.481270	10.487251	10.493234	10.499218	10.505204	10.511192	10.517182	10.523173	10.529166	10.535160	5987	347
348	10.541156	10.547154	10.553153	10.559155	10.565157	10.571161	10.577168	10.583175	10.589185	10.595195	6004	348
349	10.601208	10.607222	10.613238	10.619256	10.625275	10.631296	10.637319	10.643343	10.649369	10.655396	6020	349

	.0	.1	.2	.3	.4	.5	.6	.7	.8	.9	DIFF	
350	10.661425	10.667456	10.673488	10.679523	10.685558	10.691596	10.697635	10.703676	10.709718	10.715762	6037	350
351	10.721808	10.727855	10.733904	10.739955	10.746007	10.752061	10.758117	10.764174	10.770233	10.776293	6053	351
352	10.782356	10.788419	10.794485	10.800552	10.806621	10.812691	10.818764	10.824837	10.830913	10.836990	6070	352
353	10.843069	10.849149	10.855231	10.861315	10.867400	10.873487	10.879576	10.885666	10.891758	10.897851	6086	353
354	10.903947	10.910044	10.916142	10.922242	10.928344	10.934448	10.940553	10.946660	10.952768	10.958878	6103	354
355	10.964990	10.971103	10.977218	10.983335	10.989453	10.995573	11.001695	11.007818	11.013943	11.020070	6119	355
356	11.026198	11.032328	11.038460	11.044593	11.050728	11.056864	11.063002	11.069142	11.075284	11.081427	6136	356
357	11.087572	11.093718	11.099866	11.106016	11.112167	11.118320	11.124475	11.130631	11.136789	11.142948	6152	357
358	11.149110	11.155272	11.161437	11.167603	11.173771	11.179940	11.186112	11.192284	11.198459	11.204635	6169	358
359	11.210813	11.216992	11.223173	11.229356	11.235540	11.241726	11.247913	11.254103	11.260294	11.266486	6185	359
360	11.272680	11.278876	11.285074	11.291273	11.297474	11.303676	11.309880	11.316086	11.322293	11.328502	6202	360
361	11.334713	11.340925	11.347139	11.353355	11.359572	11.365791	11.372011	11.378234	11.384457	11.390683	6218	361
362	11.396910	11.403139	11.409369	11.415601	11.421835	11.428070	11.434307	11.440546	11.446786	11.453028	6235	362
363	11.459272	11.465517	11.471764	11.478013	11.484263	11.490514	11.496768	11.503023	11.509280	11.515538	6251	363
364	11.521798	11.528060	11.534323	11.540588	11.546855	11.553123	11.559393	11.565664	11.571938	11.578212	6268	364
365	11.584489	11.590767	11.597047	11.603328	11.609611	11.615896	11.622182	11.628470	11.634760	11.641051	6284	365
366	11.647344	11.653639	11.659935	11.666233	11.672533	11.678834	11.685136	11.691441	11.697747	11.704055	6301	366
367	11.710364	11.716675	11.722988	11.729302	11.735618	11.741935	11.748255	11.754575	11.760898	11.767222	6317	367
368	11.773548	11.779876	11.786204	11.792535	11.798868	11.805202	11.811537	11.817875	11.824214	11.830554	6333	368
369	11.836896	11.843240	11.849586	11.855933	11.862282	11.868632	11.874984	11.881338	11.887693	11.894050	6350	369
370	11.900409	11.906769	11.913131	11.919495	11.925860	11.932227	11.938595	11.944965	11.951337	11.957710	6366	370
371	11.964086	11.970462	11.976840	11.983220	11.989602	11.995985	12.002370	12.008757	12.015145	12.021535	6383	371
372	12.027926	12.034319	12.040714	12.047110	12.053508	12.059908	12.066309	12.072712	12.079117	12.085523	6399	372
373	12.091931	12.098340	12.104751	12.111164	12.117579	12.123995	12.130412	12.136832	12.143253	12.149675	6416	373
374	12.156100	12.162525	12.168953	12.175382	12.181813	12.188245	12.194679	12.201115	12.207553	12.213992	6432	374
375	12.220432	12.226874	12.233318	12.239764	12.246211	12.252660	12.259110	12.265562	12.272016	12.278472	6448	375
376	12.284929	12.291387	12.297847	12.304309	12.310773	12.317238	12.323705	12.330173	12.336644	12.343115	6465	376
377	12.349589	12.356064	12.362540	12.369019	12.375499	12.381980	12.388463	12.394948	12.401435	12.407923	6481	377
378	12.414413	12.420904	12.427397	12.433892	12.440388	12.446886	12.453386	12.459887	12.466390	12.472894	6497	378
379	12.479400	12.485908	12.492417	12.498929	12.505441	12.511955	12.518471	12.524989	12.531508	12.538029	6514	379
380	12.544552	12.551076	12.557601	12.564129	12.570658	12.577188	12.583721	12.590255	12.596790	12.603327	6530	380
381	12.609866	12.616407	12.622949	12.629493	12.636038	12.642585	12.649134	12.655684	12.662236	12.668789	6546	381
382	12.675345	12.681901	12.688460	12.695020	12.701581	12.708145	12.714710	12.721276	12.727845	12.734415	6563	382
383	12.740986	12.747559	12.754134	12.760710	12.767288	12.773868	12.780450	12.787032	12.793617	12.800203	6579	383
384	12.806791	12.813381	12.819972	12.826564	12.833159	12.839755	12.846352	12.852952	12.859553	12.866155	6595	384
385	12.872759	12.879365	12.885972	12.892582	12.899192	12.905805	12.912419	12.919034	12.925652	12.932270	6612	385
386	12.938891	12.945513	12.952137	12.958762	12.965389	12.972018	12.978648	12.985280	12.991914	12.998549	6628	386
387	13.005186	13.011824	13.018464	13.025106	13.031749	13.038394	13.045041	13.051689	13.058339	13.064990	6644	387
388	13.071643	13.078298	13.084954	13.091612	13.098272	13.104933	13.111596	13.118261	13.124927	13.131595	6661	388
389	13.138264	13.144935	13.151608	13.158282	13.164958	13.171636	13.178315	13.184996	13.191678	13.198362	6677	389
390	13.205048	13.211735	13.218424	13.225115	13.231807	13.238501	13.245196	13.251893	13.258592	13.265293	6693	390
391	13.271995	13.278698	13.285403	13.292110	13.298819	13.305529	13.312241	13.318954	13.325669	13.332386	6710	391
392	13.339104	13.345824	13.352546	13.359269	13.365994	13.372720	13.379448	13.386178	13.392909	13.399642	6726	392
393	13.406377	13.413113	13.419850	13.426590	13.433331	13.440074	13.446818	13.453564	13.460312	13.467061	6742	393
394	13.473812	13.480564	13.487318	13.494074	13.500831	13.507590	13.514351	13.521113	13.527877	13.534642	6758	394
395	13.541410	13.548178	13.554948	13.561721	13.568494	13.575269	13.582046	13.588825	13.595605	13.602386	6775	395
396	13.609170	13.615955	13.622741	13.629530	13.636320	13.643111	13.649904	13.656699	13.663495	13.670293	6791	396
397	13.677093	13.683894	13.690697	13.697501	13.704307	13.711115	13.717925	13.724735	13.731548	13.738362	6807	397
398	13.745178	13.751996	13.758815	13.765636	13.772458	13.779282	13.786107	13.792935	13.799764	13.806594	6823	398
399	13.813426	13.820260	13.827095	13.833932	13.840771	13.847611	13.854453	13.861296	13.868141	13.874988	6840	399

	.0	.1	.2	.3	.4	.5	.6	.7	.8	.9	DIFF	
400	13.881836	13.888686	13.895538	13.902391	13.909246	13.916102	13.922960	13.929820	13.936681	13.943544	6856	400
401	13.950409	13.957275	13.964143	13.971012	13.977883	13.984756	13.991630	13.998506	14.005384	14.012263	6872	401
402	14.019144	14.026026	14.032910	14.039796	14.046683	14.053571	14.060462	14.067354	14.074248	14.081143	6888	402
403	14.088040	14.094939	14.101839	14.108741	14.115644	14.122549	14.129456	14.136364	14.143275	14.150186	6905	403
404	14.157099	14.164014	14.170930	14.177849	14.184768	14.191689	14.198612	14.205537	14.212463	14.219391	6921	404
405	14.226320	14.233251	14.240184	14.247118	14.254054	14.260991	14.267931	14.274871	14.281814	14.288758	6937	405
406	14.295703	14.302650	14.309599	14.316550	14.323502	14.330455	14.337411	14.344368	14.351326	14.358286	6953	406
407	14.365248	14.372212	14.379176	14.386143	14.393111	14.400081	14.407053	14.414026	14.421001	14.427977	6969	407
408	14.434955	14.441934	14.448915	14.455899	14.462883	14.469869	14.476857	14.483846	14.490837	14.497829	6986	408
409	14.504823	14.511819	14.518816	14.525815	14.532816	14.539818	14.546822	14.553827	14.560835	14.567843	7002	409
410	14.574854	14.581866	14.588879	14.595894	14.602911	14.609929	14.616949	14.623971	14.630994	14.638019	7018	410
411	14.645046	14.652074	14.659103	14.666135	14.673167	14.680202	14.687238	14.694276	14.701315	14.708356	7034	411
412	14.715399	14.722443	14.729489	14.736537	14.743585	14.750636	14.757689	14.764742	14.771798	14.778855	7050	412
413	14.785914	14.792974	14.800036	14.807100	14.814165	14.821232	14.828300	14.835370	14.842442	14.849515	7066	413
414	14.856590	14.863667	14.870745	14.877825	14.884906	14.891989	14.899074	14.906160	14.913248	14.920337	7082	414
415	14.927428	14.934521	14.941615	14.948711	14.955808	14.962907	14.970008	14.977111	14.984215	14.991320	7099	415
416	14.998427	15.005536	15.012646	15.019759	15.026872	15.033987	15.041104	15.048223	15.055343	15.062464	7115	416
417	15.069588	15.076713	15.083839	15.090967	15.098097	15.105228	15.112361	15.119496	15.126632	15.133770	7131	417
418	15.140909	15.148050	15.155193	15.162337	15.169483	15.176630	15.183780	15.190930	15.198083	15.205236	7147	418
419	15.212392	15.219549	15.226708	15.233868	15.241030	15.248194	15.255359	15.262526	15.269694	15.276864	7163	419
420	15.284036	15.291209	15.298384	15.305560	15.312738	15.319918	15.327099	15.334282	15.341467	15.348653	7179	420
421	15.355841	15.363030	15.370221	15.377414	15.384608	15.391803	15.399001	15.406200	15.413400	15.420602	7195	421
422	15.427806	15.435012	15.442219	15.449427	15.456638	15.463849	15.471063	15.478278	15.485495	15.492713	7211	422
423	15.499933	15.507154	15.514377	15.521602	15.528829	15.536056	15.543286	15.550517	15.557750	15.564984	7227	423
424	15.572220	15.579458	15.586697	15.593938	15.601180	15.608424	15.615670	15.622917	15.630166	15.637416	7243	424
425	15.644669	15.651922	15.659177	15.666434	15.673693	15.680953	15.688214	15.695478	15.702743	15.710009	7260	425
426	15.717277	15.724547	15.731818	15.739091	15.746366	15.753642	15.760920	15.768199	15.775480	15.782762	7276	426
427	15.790047	15.797332	15.804620	15.811909	15.819199	15.826491	15.833785	15.841081	15.848378	15.855676	7292	427
428	15.862977	15.870278	15.877582	15.884887	15.892193	15.899501	15.906811	15.914123	15.921436	15.928751	7308	428
429	15.936067	15.943385	15.950704	15.958025	15.965348	15.972672	15.979998	15.987326	15.994655	16.001985	7324	429
430	16.009318	16.016652	16.023987	16.031324	16.038663	16.046003	16.053345	16.060688	16.068034	16.075380	7340	430
431	16.082729	16.090079	16.097430	16.104784	16.112138	16.119494	16.126852	16.134212	16.141573	16.148936	7356	431
432	16.156300	16.163666	16.171034	16.178403	16.185774	16.193146	16.200520	16.207895	16.215273	16.222651	7372	432
433	16.230032	16.237414	16.244797	16.252182	16.259569	16.266958	16.274348	16.281739	16.289132	16.296527	7388	433
434	16.303924	16.311321	16.318721	16.326122	16.333525	16.340929	16.348335	16.355743	16.363152	16.370563	7404	434
435	16.377975	16.385389	16.392805	16.400222	16.407641	16.415061	16.422483	16.429907	16.437332	16.444759	7420	435
436	16.452187	16.459617	16.467048	16.474482	16.481916	16.489353	16.496791	16.504230	16.511672	16.519114	7436	436
437	16.526559	16.534005	16.541452	16.548902	16.556352	16.563804	16.571259	16.578714	16.586171	16.593630	7452	437
438	16.601090	16.608552	16.616016	16.623481	16.630948	16.638416	16.645886	16.653358	16.660831	16.668305	7468	438
439	16.675782	16.683260	16.690739	16.698220	16.705703	16.713187	16.720673	16.728161	16.735650	16.743140	7484	439
440	16.750633	16.758127	16.765622	16.773119	16.780618	16.788118	16.795620	16.803123	16.810629	16.818135	7500	440
441	16.825644	16.833153	16.840665	16.848178	16.855692	16.863209	16.870727	16.878246	16.885767	16.893290	7516	441
442	16.900814	16.908340	16.915867	16.923396	16.930927	16.938459	16.945993	16.953528	16.961065	16.968604	7532	442
443	16.976144	16.983685	16.991229	16.998774	17.006320	17.013868	17.021418	17.028970	17.036523	17.044077	7548	443
444	17.051633	17.059191	17.066750	17.074311	17.081873	17.089437	17.097003	17.104570	17.112139	17.119710	7564	444
445	17.127282	17.134856	17.142431	17.150007	17.157586	17.165166	17.172748	17.180331	17.187916	17.195502	7579	445
446	17.203090	17.210679	17.218271	17.225863	17.233458	17.241053	17.248651	17.256250	17.263851	17.271453	7595	446
447	17.279057	17.286663	17.294270	17.301878	17.309489	17.317100	17.324714	17.332329	17.339946	17.347564	7611	447
448	17.355184	17.362805	17.370428	17.378053	17.385679	17.393306	17.400936	17.408567	17.416199	17.423833	7627	448
449	17.431469	17.439106	17.446745	17.454386	17.462028	17.469671	17.477317	17.484964	17.492612	17.500262	7643	449

	.0	.1	.2	.3	.4	.5	.6	.7	.8	.9	DIFF	
450	17.507914	17.515567	17.523222	17.530878	17.538536	17.546196	17.553857	17.561519	17.569184	17.576850	7659	450
451	17.584517	17.592186	17.599857	17.607529	17.615203	17.622879	17.630556	17.638234	17.645915	17.653596	7675	451
452	17.661280	17.668965	17.676651	17.684340	17.692029	17.699720	17.707414	17.715108	17.722804	17.730502	7691	452
453	17.738201	17.745902	17.753604	17.761309	17.769014	17.776721	17.784430	17.792140	17.799853	17.807566	7707	453
454	17.815281	17.822998	17.830716	17.838436	17.846158	17.853881	17.861605	17.869332	17.877060	17.884789	7723	454
455	17.892520	17.900253	17.907987	17.915723	17.923460	17.931199	17.938939	17.946681	17.954425	17.962171	7738	455
456	17.969918	17.977666	17.985416	17.993168	18.000921	18.008675	18.016432	18.024190	18.031950	18.039711	7754	456
457	18.047473	18.055238	18.063004	18.070771	18.078540	18.086311	18.094083	18.101857	18.109632	18.117409	7770	457
458	18.125188	18.132968	18.140750	18.148533	18.156318	18.164104	18.171893	18.179682	18.187474	18.195266	7786	458
459	18.203061	18.210857	18.218654	18.226454	18.234254	18.242056	18.249861	18.257666	18.265473	18.273282	7802	459
460	18.281092	18.288904	18.296717	18.304532	18.312349	18.320167	18.327987	18.335808	18.343631	18.351455	7818	460
461	18.359282	18.367109	18.374938	18.382769	18.390602	18.398435	18.406271	18.414108	18.421947	18.429787	7833	461
462	18.437629	18.445473	18.453318	18.461164	18.469012	18.476862	18.484714	18.492567	18.500421	18.508277	7849	462
463	18.516135	18.523994	18.531855	18.539718	18.547582	18.555447	18.563314	18.571183	18.579054	18.586926	7865	463
464	18.594799	18.602674	18.610551	18.618429	18.626309	18.634190	18.642073	18.649958	18.657844	18.665731	7881	464
465	18.673621	18.681512	18.689404	18.697298	18.705194	18.713091	18.720990	18.728890	18.736792	18.744696	7897	465
466	18.752601	18.760507	18.768415	18.776325	18.784237	18.792150	18.800064	18.807980	18.815898	18.823817	7912	466
467	18.831738	18.839661	18.847585	18.855510	18.863438	18.871366	18.879297	18.887228	18.895162	18.903097	7928	467
468	18.911034	18.918972	18.926912	18.934853	18.942796	18.950741	18.958687	18.966635	18.974584	18.982535	7944	468
469	18.990487	18.998441	19.006397	19.014354	19.022312	19.030273	19.038235	19.046198	19.054163	19.062130	7960	469
470	19.070098	19.078068	19.086039	19.094012	19.101986	19.109962	19.117940	19.125919	19.133900	19.141882	7976	470
471	19.149866	19.157852	19.165839	19.173828	19.181818	19.189809	19.197803	19.205798	19.213794	19.221792	7991	471
472	19.229792	19.237793	19.245796	19.253801	19.261807	19.269814	19.277823	19.285834	19.293846	19.301860	8007	472
473	19.309876	19.317893	19.325911	19.333931	19.341953	19.349976	19.358001	19.366027	19.374056	19.382085	8023	473
474	19.390116	19.398149	19.406183	19.414219	19.422256	19.430295	19.438336	19.446378	19.454422	19.462467	8038	474
475	19.470514	19.478563	19.486613	19.494664	19.502717	19.510772	19.518828	19.526886	19.534946	19.543007	8054	475
476	19.551069	19.559134	19.567199	19.575267	19.583335	19.591406	19.599478	19.607552	19.615627	19.623703	8070	476
477	19.631782	19.639862	19.647943	19.656026	19.664111	19.672197	19.680285	19.688374	19.696465	19.704557	8086	477
478	19.712651	19.720747	19.728844	19.736943	19.745043	19.753145	19.761248	19.769353	19.777460	19.785568	8101	478
479	19.793678	19.801789	19.809902	19.818016	19.826132	19.834249	19.842369	19.850489	19.858612	19.866735	8117	479
480	19.874861	19.882988	19.891116	19.899247	19.907378	19.915511	19.923646	19.931782	19.939921	19.948060	8133	480
481	19.956201	19.964344	19.972488	19.980634	19.988781	19.996930	20.005080	20.013232	20.021386	20.029541	8148	481
482	20.037698	20.045856	20.054016	20.062178	20.070341	20.078505	20.086672	20.094839	20.103009	20.111179	8164	482
483	20.119352	20.127526	20.135701	20.143879	20.152057	20.160237	20.168419	20.176603	20.184788	20.192974	8180	483
484	20.201162	20.209352	20.217543	20.225736	20.233930	20.242126	20.250324	20.258523	20.266723	20.274925	8195	484
485	20.283129	20.291334	20.299541	20.307750	20.315960	20.324171	20.332384	20.340599	20.348815	20.357033	8211	485
486	20.365253	20.373473	20.381696	20.389920	20.398146	20.406373	20.414602	20.422832	20.431064	20.439297	8227	486
487	20.447532	20.455769	20.464007	20.472247	20.480488	20.488731	20.496975	20.505221	20.513469	20.521718	8242	487
488	20.529969	20.538221	20.546474	20.554730	20.562987	20.571245	20.579505	20.587767	20.596030	20.604295	8258	488
489	20.612561	20.620829	20.629098	20.637369	20.645642	20.653916	20.662191	20.670469	20.678747	20.687028	8274	489
490	20.695310	20.703593	20.711878	20.720165	20.728453	20.736742	20.745034	20.753326	20.761621	20.769917	8289	490
491	20.778214	20.786514	20.794814	20.803116	20.811420	20.819725	20.828032	20.836340	20.844651	20.852962	8305	491
492	20.861275	20.869590	20.877906	20.886224	20.894543	20.902864	20.911187	20.919511	20.927836	20.936163	8320	492
493	20.944492	20.952822	20.961154	20.969488	20.977823	20.986159	20.994497	21.002836	21.011178	21.019521	8336	493
494	21.027865	21.036211	21.044558	21.052907	21.061258	21.069609	21.077963	21.086318	21.094675	21.103034	8352	494
495	21.111393	21.119755	21.128118	21.136482	21.144848	21.153216	21.161585	21.169956	21.178329	21.186702	8367	495
496	21.195078	21.203455	21.211833	21.220213	21.228595	21.236978	21.245363	21.253749	21.262137	21.270527	8383	496
497	21.278918	21.287310	21.295704	21.304100	21.312497	21.320896	21.329297	21.337698	21.346102	21.354507	8398	497
498	21.362914	21.371322	21.379731	21.388143	21.396555	21.404969	21.413386	21.421803	21.430222	21.438643	8414	498
499	21.447065	21.455488	21.463913	21.472340	21.480769	21.489198	21.497630	21.506063	21.514498	21.522934	8429	499

	.0	.1	.2	.3	.4	.5	.6	.7	.8	.9	DIFF	
500	21.531372	21.539811	21.548251	21.556694	21.565138	21.573583	21.582030	21.590478	21.598929	21.607380	8445	500
501	21.615834	21.624288	21.632744	21.641202	21.649662	21.658123	21.666585	21.675049	21.683515	21.691982	8460	501
502	21.700451	21.708921	21.717393	21.725866	21.734341	21.742818	21.751296	21.759775	21.768257	21.776739	8476	502
503	21.785224	21.793709	21.802197	21.810686	21.819176	21.827668	21.836162	21.844656	21.853153	21.861651	8491	503
504	21.870152	21.878653	21.887155	21.895660	21.904166	21.912673	21.921182	21.929693	21.938205	21.946719	8507	504
505	21.955234	21.963751	21.972269	21.980789	21.989311	21.997834	22.006358	22.014884	22.023412	22.031941	8523	505
506	22.040472	22.049005	22.057538	22.066074	22.074611	22.083149	22.091689	22.100231	22.108774	22.117319	8538	506
507	22.125865	22.134413	22.142962	22.151513	22.160066	22.168619	22.177175	22.185732	22.194291	22.202851	8553	507
508	22.211413	22.219976	22.228541	22.237108	22.245675	22.254245	22.262816	22.271388	22.279963	22.288538	8569	508
509	22.297116	22.305694	22.314275	22.322856	22.331440	22.340025	22.348611	22.357199	22.365789	22.374380	8584	509
510	22.382973	22.391567	22.400163	22.408760	22.417359	22.425959	22.434561	22.443165	22.451770	22.460377	8600	510
511	22.468985	22.477594	22.486206	22.494819	22.503433	22.512048	22.520666	22.529285	22.537906	22.546528	8615	511
512	22.555151	22.563776	22.572403	22.581031	22.589661	22.598292	22.606925	22.615560	22.624196	22.632833	8631	512
513	22.641473	22.650113	22.658755	22.667398	22.676044	22.684690	22.693339	22.701989	22.710640	22.719293	8646	513
514	22.727948	22.736604	22.745261	22.753920	22.762581	22.771243	22.779907	22.788572	22.797239	22.805907	8662	514
515	22.814578	22.823249	22.831922	22.840596	22.849272	22.857950	22.866630	22.875310	22.883992	22.892676	8677	515
516	22.901362	22.910049	22.918737	22.927427	22.936118	22.944811	22.953506	22.962202	22.970900	22.979599	8693	516
517	22.988300	22.997002	23.005706	23.014411	23.023118	23.031826	23.040537	23.049248	23.057961	23.066676	8708	517
518	23.075393	23.084110	23.092829	23.101550	23.110272	23.118996	23.127722	23.136449	23.145177	23.153907	8723	518
519	23.162639	23.171372	23.180107	23.188843	23.197580	23.206320	23.215061	23.223803	23.232547	23.241292	8739	519
520	23.250040	23.258788	23.267538	23.276289	23.285043	23.293797	23.302554	23.311311	23.320071	23.328831	8754	520
521	23.337594	23.346358	23.355123	23.363890	23.372658	23.381428	23.390200	23.398973	23.407748	23.416524	8770	521
522	23.425302	23.434082	23.442862	23.451644	23.460428	23.469213	23.478001	23.486789	23.495579	23.504371	8785	522
523	23.513164	23.521959	23.530755	23.539552	23.548352	23.557152	23.565955	23.574759	23.583564	23.592371	8800	523
524	23.601180	23.609990	23.618801	23.627614	23.636429	23.645245	23.654063	23.662882	23.671703	23.680525	8816	524
525	23.689349	23.698174	23.707001	23.715829	23.724659	23.733491	23.742324	23.751159	23.759995	23.768832	8831	525
526	23.777672	23.786513	23.795355	23.804198	23.813044	23.821890	23.830739	23.839589	23.848440	23.857293	8846	526
527	23.866148	23.875004	23.883862	23.892720	23.901581	23.910443	23.919307	23.928173	23.937039	23.945907	8862	527
528	23.954778	23.963649	23.972522	23.981396	23.990272	23.999149	24.008029	24.016909	24.025791	24.034675	8877	528
529	24.043561	24.052447	24.061335	24.070225	24.079116	24.088009	24.096904	24.105799	24.114697	24.123596	8892	529
530	24.132497	24.141399	24.150302	24.159207	24.168113	24.177022	24.185932	24.194843	24.203755	24.212670	8908	530
531	24.221586	24.230503	24.239422	24.248342	24.257264	24.266187	24.275113	24.284039	24.292967	24.301897	8923	531
532	24.310828	24.319761	24.328695	24.337630	24.346567	24.355506	24.364447	24.373389	24.382332	24.391277	8938	532
533	24.400223	24.409171	24.418121	24.427072	24.436024	24.444978	24.453934	24.462891	24.471849	24.480809	8953	533
534	24.489772	24.498735	24.507699	24.516666	24.525633	24.534602	24.543574	24.552546	24.561520	24.570495	8969	534
535	24.579473	24.588451	24.597431	24.606412	24.615396	24.624380	24.633367	24.642354	24.651343	24.660334	8984	535
536	24.669327	24.678320	24.687315	24.696312	24.705310	24.714310	24.723312	24.732315	24.741319	24.750325	8999	536
537	24.759333	24.768342	24.777352	24.786364	24.795378	24.804393	24.813410	24.822428	24.831448	24.840469	9015	537
538	24.849492	24.858516	24.867542	24.876569	24.885598	24.894628	24.903661	24.912694	24.921729	24.930765	9030	538
539	24.939804	24.948843	24.957884	24.966927	24.975971	24.985016	24.994064	25.003112	25.012162	25.021214	9045	539
540	25.030268	25.039323	25.048379	25.057436	25.066496	25.075556	25.084619	25.093683	25.102748	25.111815	9060	540
541	25.120884	25.129954	25.139026	25.148098	25.157173	25.166249	25.175327	25.184406	25.193487	25.202569	9076	541
542	25.211653	25.220738	25.229825	25.238913	25.248003	25.257094	25.266187	25.275281	25.284377	25.293474	9091	542
543	25.302574	25.311674	25.320776	25.329880	25.338984	25.348091	25.357199	25.366309	25.375420	25.384532	9106	543
544	25.393647	25.402763	25.411880	25.420998	25.430118	25.439240	25.448364	25.457489	25.466615	25.475742	9121	544
545	25.484872	25.494003	25.503135	25.512269	25.521404	25.530541	25.539680	25.548820	25.557961	25.567104	9136	545
546	25.576249	25.585395	25.594543	25.603692	25.612842	25.621994	25.631149	25.640304	25.649460	25.658618	9152	546
547	25.667778	25.676940	25.686102	25.695266	25.704432	25.713599	25.722769	25.731939	25.741111	25.750284	9167	547
548	25.759459	25.768636	25.777813	25.786993	25.796174	25.805356	25.814541	25.823726	25.832913	25.842101	9182	548
549	25.851292	25.860484	25.869677	25.878871	25.888067	25.897264	25.906464	25.915665	25.924867	25.934071	9197	549

	.0	.1	.2	.3	.4	.5	.6	.7	.8	.9	DIFF	
550	25.943276	25.952483	25.961691	25.970901	25.980112	25.989325	25.998540	26.007755	26.016973	26.026191	9212	550
551	26.035412	26.044634	26.053857	26.063082	26.072309	26.081536	26.090766	26.099997	26.109230	26.118464	9227	551
552	26.127700	26.136937	26.146175	26.155415	26.164657	26.173900	26.183145	26.192391	26.201638	26.210887	9243	552
553	26.220139	26.229391	26.238644	26.247900	26.257156	26.266414	26.275675	26.284936	26.294198	26.303463	9258	553
554	26.312729	26.321996	26.331265	26.340535	26.349807	26.359080	26.368356	26.377632	26.386910	26.396189	9273	554
555	26.405470	26.414753	26.424037	26.433322	26.442609	26.451897	26.461188	26.470479	26.479772	26.489067	9288	555
556	26.498363	26.507661	26.516960	26.526260	26.535562	26.544866	26.554171	26.563478	26.572786	26.582095	9303	556
557	26.591407	26.600720	26.610034	26.619349	26.628666	26.637985	26.647306	26.656627	26.665951	26.675275	9318	557
558	26.684602	26.693930	26.703259	26.712590	26.721922	26.731255	26.740591	26.749928	26.759266	26.768606	9333	558
559	26.777948	26.787291	26.796635	26.805981	26.815328	26.824677	26.834028	26.843379	26.852733	26.862088	9348	559
560	26.871445	26.880803	26.890162	26.899523	26.908885	26.918249	26.927615	26.936982	26.946350	26.955720	9363	560
561	26.965092	26.974465	26.983840	26.993216	27.002593	27.011972	27.021353	27.030735	27.040118	27.049504	9379	561
562	27.058891	27.068279	27.077668	27.087059	27.096452	27.105846	27.115242	27.124639	27.134037	27.143437	9394	562
563	27.152840	27.162243	27.171647	27.181053	27.190461	27.199870	27.209281	27.218693	27.228107	27.237522	9409	563
564	27.246939	27.256357	27.265777	27.275198	27.284621	27.294045	27.303471	27.312898	27.322327	27.331757	9424	564
565	27.341189	27.350622	27.360057	27.369493	27.378931	27.388370	27.397811	27.407254	27.416697	27.426142	9439	565
566	27.435590	27.445038	27.454488	27.463939	27.473392	27.482846	27.492302	27.501759	27.511218	27.520678	9454	566
567	27.530141	27.539604	27.549068	27.558535	27.568002	27.577472	27.586943	27.596415	27.605889	27.615364	9469	567
568	27.624842	27.634320	27.643800	27.653281	27.662764	27.672248	27.681734	27.691221	27.700710	27.710201	9484	568
569	27.719693	27.729186	27.738681	27.748177	27.757675	27.767174	27.776676	27.786178	27.795682	27.805187	9499	569
570	27.814694	27.824203	27.833712	27.843224	27.852736	27.862251	27.871767	27.881284	27.890803	27.900323	9514	570
571	27.909846	27.919369	27.928894	27.938420	27.947948	27.957477	27.967009	27.976541	27.986074	27.995610	9529	571
572	28.005147	28.014685	28.024225	28.033767	28.043309	28.052853	28.062400	28.071947	28.081496	28.091046	9544	572
573	28.100598	28.110152	28.119706	28.129262	28.138820	28.148380	28.157941	28.167503	28.177067	28.186632	9559	573
574	28.196199	28.205768	28.215337	28.224909	28.234481	28.244055	28.253632	28.263209	28.272788	28.282368	9574	574
575	28.291950	28.301533	28.311118	28.320704	28.330292	28.339881	28.349472	28.359064	28.368658	28.378253	9589	575
576	28.387851	28.397449	28.407048	28.416649	28.426252	28.435856	28.445462	28.455069	28.464678	28.474288	9604	576
577	28.483900	28.493513	28.503128	28.512744	28.522362	28.531981	28.541602	28.551224	28.560848	28.570472	9619	577
578	28.580100	28.589728	28.599357	28.608988	28.618621	28.628255	28.637891	28.647528	28.657166	28.666806	9634	578
579	28.676449	28.686091	28.695736	28.705382	28.715029	28.724678	28.734329	28.743981	28.753634	28.763289	9648	579
580	28.772947	28.782604	28.792264	28.801925	28.811587	28.821251	28.830917	28.840584	28.850252	28.859922	9663	580
581	28.869594	28.879267	28.888941	28.898617	28.908294	28.917973	28.927654	28.937335	28.947019	28.956703	9678	581
582	28.966390	28.976078	28.985767	28.995458	29.005150	29.014844	29.024539	29.034236	29.043934	29.053634	9693	582
583	29.063336	29.073038	29.082742	29.092448	29.102155	29.111864	29.121574	29.131286	29.140999	29.150713	9708	583
584	29.160430	29.170148	29.179867	29.189587	29.199309	29.209033	29.218758	29.228485	29.238213	29.247942	9723	584
585	29.257674	29.267406	29.277140	29.286875	29.296612	29.306350	29.316091	29.325832	29.335575	29.345319	9738	585
586	29.355066	29.364813	29.374562	29.384312	29.394064	29.403817	29.413572	29.423328	29.433086	29.442845	9753	586
587	29.452607	29.462369	29.472132	29.481898	29.491664	29.501432	29.511203	29.520973	29.530746	29.540520	9768	587
588	29.550296	29.560073	29.569852	29.579632	29.589413	29.599196	29.608981	29.618767	29.628555	29.638343	9783	588
589	29.648134	29.657926	29.667720	29.677515	29.687311	29.697109	29.706908	29.716709	29.726511	29.736315	9797	589
590	29.746121	29.755928	29.765736	29.775546	29.785357	29.795170	29.804984	29.814800	29.824617	29.834435	9812	590
591	29.844256	29.854078	29.863901	29.873725	29.883551	29.893379	29.903208	29.913039	29.922871	29.932704	9827	591
592	29.942540	29.952376	29.962214	29.972053	29.981894	29.991736	30.001581	30.011426	30.021273	30.031121	9842	592
593	30.040972	30.050823	30.060675	30.070529	30.080385	30.090242	30.100101	30.109962	30.119823	30.129686	9857	593
594	30.139551	30.149417	30.159285	30.169154	30.179024	30.188896	30.198770	30.208645	30.218522	30.228399	9871	594
595	30.238279	30.248160	30.258043	30.267926	30.277812	30.287698	30.297587	30.307477	30.317368	30.327260	9886	595
596	30.337156	30.347051	30.356948	30.366847	30.376747	30.386648	30.396552	30.406456	30.416362	30.426270	9901	596
597	30.436179	30.446090	30.456002	30.465915	30.475830	30.485746	30.495665	30.505584	30.515505	30.525427	9916	597
598	30.535351	30.545276	30.555203	30.565131	30.575061	30.584992	30.594925	30.604859	30.614795	30.624732	9931	598
599	30.634671	30.644611	30.654552	30.664495	30.674439	30.684385	30.694333	30.704282	30.714232	30.724184	9945	599

	.0	.1	.2	.3	.4	.5	.6	.7	.8	.9	DIFF	
600	30.734138	30.744093	30.754049	30.764007	30.773966	30.783926	30.793889	30.803853	30.813818	30.823784	9960	600
601	30.833753	30.843723	30.853693	30.863666	30.873640	30.883615	30.893593	30.903571	30.913551	30.923532	9975	601
602	30.933516	30.943500	30.953485	30.963472	30.973461	30.983451	30.993443	31.003437	31.013431	31.023427	9990	602
603	31.033425	31.043424	31.053425	31.063426	31.073430	31.083435	31.093442	31.103450	31.113459	31.123470	10004	603
604	31.133482	31.143496	31.153511	31.163528	31.173546	31.183566	31.193588	31.203610	31.213634	31.223659	10019	604
605	31.233687	31.243716	31.253745	31.263777	31.273809	31.283844	31.293880	31.303918	31.313956	31.323996	10034	605
606	31.334039	31.344082	31.354126	31.364172	31.374220	31.384269	31.394320	31.404372	31.414426	31.424480	10049	606
607	31.434538	31.444596	31.454655	31.464715	31.474778	31.484842	31.494907	31.504974	31.515042	31.525112	10063	607
608	31.535184	31.545256	31.555330	31.565406	31.575482	31.585561	31.595642	31.605723	31.615806	31.625890	10078	608
609	31.635976	31.646064	31.656152	31.666242	31.676334	31.686427	31.696522	31.706619	31.716716	31.726815	10093	609
610	31.736916	31.747018	31.757121	31.767226	31.777333	31.787440	31.797550	31.807661	31.817773	31.827887	10107	610
611	31.838003	31.848119	31.858237	31.868357	31.878478	31.888600	31.898725	31.908850	31.918977	31.929105	10122	611
612	31.939236	31.949367	31.959500	31.969634	31.979770	31.989907	32.000046	32.010186	32.020328	32.030470	10137	612
613	32.040616	32.050762	32.060909	32.071058	32.081208	32.091360	32.101514	32.111669	32.121825	32.131982	10151	613
614	32.142142	32.152303	32.162465	32.172628	32.182793	32.192959	32.203128	32.213297	32.223468	32.233641	10166	614
615	32.243815	32.253990	32.264167	32.274345	32.284524	32.294705	32.304889	32.315073	32.325258	32.335445	10181	615
616	32.345634	32.355824	32.366015	32.376208	32.386402	32.396598	32.406796	32.416995	32.427195	32.437396	10195	616
617	32.447600	32.457804	32.468010	32.478218	32.488426	32.498637	32.508849	32.519063	32.529277	32.539493	10210	617
618	32.549712	32.559931	32.570151	32.580373	32.590597	32.600821	32.611049	32.621276	32.631506	32.641737	10224	618
619	32.651969	32.662203	32.672438	32.682675	32.692913	32.703153	32.713394	32.723637	32.733881	32.744126	10239	619
620	32.754374	32.764622	32.774871	32.785123	32.795375	32.805629	32.815886	32.826143	32.836401	32.846661	10254	620
621	32.856924	32.867187	32.877450	32.887716	32.897984	32.908252	32.918523	32.928795	32.939068	32.949343	10268	621
622	32.959620	32.969897	32.980175	32.990456	33.000738	33.011021	33.021307	33.031593	33.041881	33.052170	10283	622
623	33.062461	33.072753	33.083046	33.093341	33.103638	33.113935	33.124236	33.134537	33.144839	33.155142	10297	623
624	33.165448	33.175755	33.186063	33.196373	33.206684	33.216996	33.227311	33.237626	33.247942	33.258261	10312	624
625	33.268581	33.278903	33.289225	33.299549	33.309875	33.320202	33.330531	33.340861	33.351192	33.361525	10327	625
626	33.371860	33.382195	33.392533	33.402872	33.413211	33.423553	33.433897	33.444241	33.454587	33.464934	10341	626
627	33.475284	33.485634	33.495986	33.506339	33.516694	33.527050	33.537408	33.547767	33.558127	33.568490	10356	627
628	33.578854	33.589218	33.599585	33.609952	33.620321	33.630692	33.641065	33.651438	33.661813	33.672190	10370	628
629	33.682568	33.692947	33.703328	33.713710	33.724094	33.734479	33.744867	33.755255	33.765644	33.776035	10385	629
630	33.786428	33.796822	33.807217	33.817614	33.828012	33.838412	33.848814	33.859216	33.869620	33.880026	10399	630
631	33.890433	33.900842	33.911252	33.921663	33.932076	33.942490	33.952906	33.963323	33.973742	33.984161	10414	631
632	33.994584	34.005006	34.015431	34.025856	34.036283	34.046712	34.057143	34.067575	34.078008	34.088442	10428	632
633	34.098879	34.109316	34.119755	34.130195	34.140637	34.151080	34.161525	34.171971	34.182419	34.192868	10443	633
634	34.203319	34.213771	34.224224	34.234679	34.245135	34.255592	34.266052	34.276513	34.286975	34.297438	10457	634
635	34.307904	34.318370	34.328838	34.339307	34.349778	34.360250	34.370724	34.381199	34.391675	34.402153	10472	635
636	34.412633	34.423114	34.433596	34.444080	34.454565	34.465052	34.475540	34.486030	34.496521	34.507013	10486	636
637	34.517508	34.528003	34.538499	34.548997	34.559497	34.569998	34.580502	34.591005	34.601511	34.612017	10501	637
638	34.622527	34.633036	34.643547	34.654060	34.664574	34.675089	34.685607	34.696125	34.706645	34.717166	10515	638
639	34.727690	34.738214	34.748739	34.759266	34.769795	34.780325	34.790857	34.801390	34.811924	34.822460	10529	639
640	34.832997	34.843536	34.854076	34.864617	34.875160	34.885705	34.896251	34.906798	34.917347	34.927897	10544	640
641	34.938449	34.949002	34.959557	34.970113	34.980670	34.991229	35.001790	35.012352	35.022914	35.033479	10558	641
642	35.044045	35.054613	35.065182	35.075752	35.086324	35.096897	35.107472	35.118049	35.128626	35.139205	10573	642
643	35.149786	35.160368	35.170951	35.181536	35.192122	35.202709	35.213299	35.223890	35.234481	35.245075	10587	643
644	35.255671	35.266267	35.276865	35.287463	35.298064	35.308666	35.319270	35.329875	35.340481	35.351089	10602	644
645	35.361699	35.372310	35.382922	35.393535	35.404150	35.414766	35.425385	35.436004	35.446625	35.457247	10616	645
646	35.467871	35.478496	35.489122	35.4 9751	35.510380	35.521010	35.531643	35.542277	35.552912	35.563549	10630	646
647	35.574187	35.584826	35.595468	35.606110	35.616753	35.627398	35.638046	35.648694	35.659343	35.669994	10645	647
648	35.680647	35.691301	35.701956	35.7 2613	35.723270	35.733930	35.744592	35.755254	35.765918	35.776583	10659	648
649	35.787251	35.797919	35.808588	35.819259	35.829932	35.840605	35.851282	35.861958	35.872636	35.883316	10673	649

	.0	.1	.2	.3	.4	.5	.6	.7	.8	.9	DIFF	
650	35.893998	35.904680	35.915364	35.926049	35.936736	35.947424	35.958115	35.968806	35.979498	35.990192	10688	650
651	36.000888	36.011585	36.022283	36.032983	36.043684	36.054386	36.065091	36.075797	36.086504	36.097211	10702	651
652	36.107922	36.118633	36.129346	36.140059	36.150775	36.161492	36.172211	36.182931	36.193652	36.204374	10716	652
653	36.215099	36.225825	36.236552	36.247280	36.258009	36.268741	36.279474	36.290208	36.300944	36.311680	10731	653
654	36.322420	36.333159	36.343901	36.354643	36.365387	36.376133	36.386880	36.397629	36.408379	36.419130	10745	654
655	36.429883	36.440637	36.451393	36.462150	36.472908	36.483668	36.494430	36.505193	36.515956	36.526722	10759	655
656	36.537490	36.548258	36.559028	36.569799	36.580572	36.591346	36.602122	36.612899	36.623678	36.634458	10774	656
657	36.645239	36.656022	36.666806	36.677592	36.688379	36.699167	36.709958	36.720749	36.731542	36.742336	10788	657
658	36.753132	36.763929	36.774727	36.785527	36.796328	36.807131	36.817936	36.828741	36.839548	36.850357	10802	658
659	36.861167	36.871978	36.882791	36.893605	36.904420	36.915237	36.926057	36.936876	36.947698	36.958520	10817	659
660	36.969345	36.980171	36.990997	37.001826	37.012656	37.023487	37.034320	37.045154	37.055990	37.066827	10831	660
661	37.077666	37.088506	37.099347	37.110189	37.121033	37.131879	37.142726	37.153574	37.164424	37.175275	10845	661
662	37.186129	37.196983	37.207839	37.218695	37.229553	37.240413	37.251275	37.262137	37.273002	37.283867	10859	662
663	37.294734	37.305603	37.316473	37.327343	37.338216	37.349090	37.359966	37.370843	37.381721	37.392601	10874	663
664	37.403482	37.414365	37.425249	37.4 6134	37.447021	37.457909	37.468800	37.479691	37.490583	37.501477	10888	664
665	37.512373	37.523270	37.534168	37.545067	37.555968	37.566871	37.577775	37.588680	37.599587	37.610495	10902	665
666	37.621405	37.632316	37.643229	37.654142	37.665058	37.675974	37.686893	37.697812	37.708733	37.719656	10916	666
667	37.730580	37.741505	37.752431	37.763360	37.774289	37.785220	37.796153	37.807086	37.818021	37.828958	10930	667
668	37.839897	37.850836	37.861777	37.872719	37.883662	37.894608	37.905555	37.916502	37.927452	37.938402	10945	668
669	37.949355	37.960309	37.971263	37.982220	37.993178	38.004137	38.015098	38.026060	38.037024	38.047988	10959	669
670	38.058956	38.069923	38.080893	38.091863	38.102835	38.113808	38.124784	38.135760	38.146738	38.157717	10973	670
671	38.168698	38.179680	38.190663	38.201648	38.212634	38.223621	38.234611	38.245602	38.256593	38.267586	10987	671
672	38.278582	38.289578	38.300575	38.311574	38.322575	38.333576	38.344580	38.355585	38.366590	38.377598	11001	672
673	38.388608	38.399618	38.410629	38.4 1642	38.432657	38.443673	38.454690	38.465709	38.476729	38.487751	11015	673
674	38.498775	38.509799	38.520824	38.531852	38.542880	38.553910	38.564942	38.575975	38.587009	38.598045	11030	674
675	38.609083	38.620121	38.631161	38.642203	38.653245	38.664289	38.675336	38.686383	38.697431	38.708481	11044	675
676	38.719533	38.730586	38.741640	38.752695	38.763752	38.774810	38.785871	38.796932	38.807994	38.819058	11058	676
677	38.830124	38.841191	38.852259	38.863328	38.874400	38.885472	38.896546	38.907621	38.918698	38.929776	11072	677
678	38.940856	38.951937	38.963019	38.974103	38.985188	38.996274	39.007363	39.018453	39.029543	39.040636	11086	678
679	39.051730	39.062825	39.073921	39.085019	39.096118	39.107219	39.118321	39.129425	39.140530	39.151636	11100	679
680	39.162745	39.173853	39.184964	39.196076	39.207189	39.218304	39.229421	39.240538	39.251657	39.262777	11114	680
681	39.273900	39.285023	39.296147	39.307273	39.318401	39.329529	39.340661	39.351792	39.362925	39.374059	11128	681
682	39.385196	39.396333	39.407472	39.418612	39.429753	39.440896	39.452041	39.463187	39.474334	39.485482	11142	682
683	39.496633	39.507784	39.518937	39.530091	39.541246	39.552403	39.563562	39.574722	39.585884	39.597046	11156	683
684	39.608211	39.619376	39.630543	39.641711	39.652881	39.664052	39.675225	39.686399	39.697574	39.708750	11171	684
685	39.719929	39.731108	39.742289	39.753471	39.764655	39.775840	39.787027	39.798215	39.809404	39.820595	11185	685
686	39.831788	39.842981	39.854176	39.865372	39.876570	39.887769	39.898970	39.910172	39.921375	39.932580	11199	686
687	39.943787	39.954994	39.966203	39.977414	39.988625	39.999838	40.011054	40.022270	40.033487	40.044705	11213	687
688	40.055926	40.067148	40.078371	40.089595	40.100821	40.112048	40.123277	40.134507	40.145739	40.156971	11227	688
689	40.168206	40.179442	40.190679	40.201917	40.213157	40.224398	40.235641	40.246885	40.258131	40.269377	11241	689
690	40.280626	40.291876	40.303127	40.314379	40.325633	40.336888	40.348145	40.359403	40.370662	40.381923	11255	690
691	40.393186	40.404449	40.415715	40.426981	40.438249	40.449518	40.460789	40.472061	40.483335	40.494609	11269	691
692	40.505886	40.517164	40.528442	40.539723	40.551005	40.562288	40.573573	40.584859	40.596147	40.607435	11283	692
693	40.618726	40.630018	40.641311	40.652605	40.663900	40.675198	40.686497	40.697797	40.709098	40.720401	11297	693
694	40.731706	40.743011	40.754318	40.765626	40.776936	40.788247	40.799561	40.810874	40.822190	40.833506	11311	694
695	40.844825	40.856145	40.867465	40.878788	40.890111	40.901436	40.912764	40.924092	40.935421	40.946752	11325	695
696	40.958084	40.969418	40.980752	40.992089	41.003427	41.014765	41.026107	41.037448	41.048791	41.060136	11339	696
697	41.071483	41.082830	41.094179	41.1 5529	41.116881	41.128234	41.139589	41.150945	41.162302	41.173660	11353	697
698	41.185021	41.196383	41.207745	41.219109	41.230474	41.241842	41.253211	41.264580	41.275951	41.287324	11367	698
699	41.298698	41.310074	41.321451	41.332829	41.344208	41.355589	41.366972	41.378356	41.389740	41.401127	11380	699

	.0	.1	.2	.3	.4	.5	.6	.7	.8	.9	DIFF	
700	41.412516	41.423904	41.435295	41.4 6687	41.458080	41.469475	41.480872	41.492270	41.503669	41.515069	11394	700
701	41.526472	41.537875	41.549279	41.560685	41.572092	41.583501	41.594912	41.606323	41.617736	41.629150	11408	701
702	41.640567	41.651984	41.663402	41.674822	41.686243	41.697666	41.709090	41.720516	41.731942	41.743371	11422	702
703	41.754801	41.766232	41.777664	41.789098	41.800533	41.811969	41.823408	41.834847	41.846288	41.857730	11436	703
704	41.869174	41.880619	41.892065	41.9 3512	41.914961	41.926412	41.937864	41.949317	41.960772	41.972228	11450	704
705	41.983686	41.995145	42.006605	42.018066	42.029529	42.040993	42.052460	42.063927	42.075395	42.086865	11464	705
706	42.098337	42.109809	42.121284	42.132759	42.144236	42.155714	42.167194	42.178675	42.190157	42.201641	11478	706
707	42.213127	42.224613	42.236101	42.247590	42.259080	42.270572	42.282066	42.293561	42.305058	42.316555	11492	707
708	42.328054	42.339555	42.351057	42.362559	42.374064	42.385570	42.397078	42.408587	42.420096	42.431608	11505	708
709	42.443121	42.454635	42.466151	42.477668	42.489186	42.500706	42.512228	42.523750	42.535274	42.546799	11519	709
710	42.558326	42.569854	42.581384	42.592914	42.604446	42.615980	42.627516	42.639052	42.650589	42.662128	11533	710
711	42.673670	42.685212	42.696755	42.708299	42.719845	42.731392	42.742942	42.754492	42.766044	42.777596	11547	711
712	42.789152	42.800707	42.812264	42.823822	42.835382	42.846943	42.858507	42.870070	42.881636	42.893202	11561	712
713	42.904771	42.916341	42.927912	42.939484	42.951057	42.962632	42.974209	42.985787	42.997366	43.008947	11575	713
714	43.020529	43.032113	43.043697	43.055283	43.066871	43.078459	43.090050	43.101642	43.113235	43.124829	11588	714
715	43.136425	43.148023	43.159621	43.171221	43.182822	43.194424	43.206029	43.217634	43.229241	43.240849	11602	715
716	43.252460	43.264070	43.275682	43.287296	43.298911	43.310527	43.322146	43.333765	43.345385	43.357007	11616	716
717	43.368631	43.380256	43.391881	43.403509	43.415138	43.426768	43.438400	43.450033	43.461667	43.473303	11630	717
718	43.484941	43.496579	43.508219	43.519860	43.531502	43.543146	43.554792	43.566439	43.578087	43.589736	11643	718
719	43.601388	43.613040	43.624693	43.636348	43.648005	43.659662	43.671322	43.682982	43.694644	43.706307	11657	719
720	43.717973	43.729639	43.741306	43.752974	43.764644	43.776316	43.787989	43.799664	43.811339	43.823016	11671	720
721	43.834695	43.846375	43.858056	43.869738	43.881422	43.893106	43.904794	43.916482	43.928171	43.939862	11685	721
722	43.951554	43.963248	43.974943	43.986639	43.998336	44.010035	44.021736	44.033438	44.045140	44.056845	11698	722
723	44.068552	44.080258	44.091967	44.103677	44.115388	44.127100	44.138815	44.150530	44.162247	44.173965	11712	723
724	44.185686	44.197406	44.209128	44.220852	44.232577	44.244303	44.256032	44.267761	44.279491	44.291223	11726	724
725	44.302957	44.314692	44.326427	44.338164	44.349903	44.361643	44.373385	44.385128	44.396872	44.408617	11740	725
726	44.420365	44.432114	44.443863	44.455614	44.467366	44.479120	44.490876	44.502632	44.514390	44.526149	11753	726
727	44.537910	44.549673	44.561436	44.573200	44.584966	44.596734	44.608503	44.620274	44.632045	44.643818	11767	727
728	44.655593	44.667368	44.679145	44.690924	44.702703	44.714484	44.726268	44.738051	44.749836	44.761623	11781	728
729	44.773412	44.785201	44.796992	44.808784	44.820577	44.832372	44.844169	44.855966	44.867765	44.879565	11794	729
730	44.891367	44.903171	44.914975	44.926780	44.938587	44.950396	44.962206	44.974018	44.985830	44.997644	11808	730
731	45.009460	45.021276	45.033094	45.044914	45.056734	45.068556	45.080380	45.092206	45.104032	45.115859	11822	731
732	45.127689	45.139519	45.151350	45.163183	45.175018	45.186853	45.198691	45.210530	45.222370	45.234210	11835	732
733	45.246054	45.257898	45.269743	45.281590	45.293437	45.305287	45.317139	45.328990	45.340844	45.352699	11849	733
734	45.364556	45.376413	45.388272	45.400132	45.411994	45.423857	45.435722	45.447587	45.459455	45.471323	11863	734
735	45.483193	45.495065	45.506937	45.518811	45.530686	45.542563	45.554441	45.566321	45.578201	45.590084	11876	735
736	45.601968	45.613853	45.625739	45.637626	45.649515	45.661405	45.673297	45.685190	45.697084	45.708980	11890	736
737	45.720878	45.732776	45.744676	45.756577	45.768479	45.780383	45.792289	45.804196	45.816103	45.828012	11903	737
738	45.839924	45.851836	45.863750	45.875664	45.887580	45.899498	45.911417	45.923337	45.935258	45.947181	11917	738
739	45.959106	45.971032	45.982959	45.994887	46.006816	46.018747	46.030681	46.042614	46.054549	46.066486	11931	739
740	46.078424	46.090364	46.102304	46.114245	46.126189	46.138133	46.150080	46.162027	46.173976	46.185926	11944	740
741	46.197878	46.209831	46.221785	46.233740	46.245697	46.257655	46.269615	46.281576	46.293538	46.305501	11958	741
742	46.317467	46.329433	46.341401	46.353370	46.365340	46.377312	46.389286	46.401260	46.413236	46.425213	11971	742
743	46.437192	46.449172	46.461153	46.473136	46.485119	46.497105	46.509092	46.521080	46.533069	46.545060	11985	743
744	46.557053	46.569046	46.581041	46.593037	46.605034	46.617033	46.629034	46.641036	46.653038	46.665042	11998	744
745	46.677049	46.689055	46.701064	46.713073	46.725084	46.737096	46.749111	46.761126	46.773142	46.785160	12012	745
746	46.797180	46.809200	46.821222	46.833245	46.845269	46.857295	46.869323	46.881352	46.893382	46.905413	12025	746
747	46.917446	46.929480	46.941515	46.953552	46.965590	46.977629	46.989671	47.001713	47.013756	47.025801	12039	747
748	47.037848	47.049895	47.061944	47.073994	47.086045	47.098098	47.110153	47.122209	47.134265	47.146324	12052	748
749	47.158384	47.170445	47.182508	47.194571	47.206636	47.218702	47.230771	47.242840	47.254910	47.266982	12066	749

	.0	.1	.2	.3	.4	.5	.6	.7	.8	.9	DIFF	
750	47.279056	47.291130	47.303206	47.315283	47.327362	47.339441	47.351523	47.363606	47.375690	47.387775	12079	750
751	47.399862	47.411950	47.424039	47.436130	47.448222	47.460315	47.472411	47.484507	47.496604	47.508702	12093	751
752	47.520803	47.532905	47.545007	47.557111	47.569217	47.581324	47.593432	47.605542	47.617653	47.629765	12106	752
753	47.641879	47.653994	47.666110	47.678228	47.690347	47.702467	47.714589	47.726712	47.738836	47.750962	12120	753
754	47.763090	47.775218	47.787348	47.799479	47.811611	47.823744	47.835880	47.848017	47.860155	47.872293	12133	754
755	47.884435	47.896576	47.908720	47.920864	47.933010	47.945157	47.957306	47.969456	47.981607	47.993760	12147	755
756	48.005914	48.018069	48.030226	48.042384	48.054543	48.066703	48.078866	48.091029	48.103194	48.115360	12160	756
757	48.127528	48.139697	48.151866	48.164038	48.176210	48.188385	48.200561	48.212737	48.224915	48.237094	12174	757
758	48.249276	48.261458	48.273642	48.285826	48.298012	48.310199	48.322389	48.334579	48.346771	48.358963	12187	758
759	48.371158	48.383354	48.395550	48.407749	48.419948	48.432149	48.444352	48.456555	48.468760	48.480966	12200	759
760	48.493175	48.505383	48.517593	48.529805	48.542018	48.554232	48.566449	48.578665	48.590884	48.603103	12214	760
761	48.615325	48.627547	48.639771	48.651996	48.664222	48.676449	48.688679	48.700910	48.713141	48.725374	12227	761
762	48.737609	48.749845	48.762082	48.774320	48.786560	48.798800	48.811044	48.823287	48.835532	48.847779	12241	762
763	48.860027	48.872276	48.884527	48.896778	48.909031	48.921286	48.933542	48.945799	48.958057	48.970317	12254	763
764	48.982579	48.994841	49.007105	49.019370	49.031636	49.043904	49.056174	49.068444	49.080716	49.092989	12267	764
765	49.105264	49.117540	49.129817	49.142096	49.154375	49.166656	49.178940	49.191223	49.203508	49.215795	12281	765
766	49.228084	49.240373	49.252663	49.264955	49.277247	49.289542	49.301839	49.314136	49.326434	49.338734	12294	766
767	49.351036	49.363338	49.375642	49.387947	49.400253	49.412561	49.424871	49.437181	49.449493	49.461806	12307	767
768	49.474122	49.486437	49.498755	49.511073	49.523393	49.535714	49.548037	49.560361	49.572686	49.585012	12321	768
769	49.597341	49.609670	49.622000	49.634332	49.646665	49.658999	49.671336	49.683673	49.696012	49.708351	12334	769
770	49.720693	49.733036	49.745379	49.757724	49.770071	49.782418	49.794768	49.807118	49.819470	49.831823	12347	770
771	49.844179	49.856534	49.868891	49.881249	49.893610	49.905970	49.918334	49.930697	49.943062	49.955429	12361	771
772	49.967797	49.980166	49.992537	50.004908	50.017281	50.029655	50.042032	50.054409	50.066787	50.079167	12374	772
773	50.091548	50.103931	50.116314	50.128699	50.141086	50.153473	50.165863	50.178253	50.190645	50.203038	12387	773
774	50.215433	50.227829	50.240225	50.252624	50.265023	50.277424	50.289827	50.302230	50.314635	50.327042	12400	774
775	50.339450	50.351859	50.364269	50.376680	50.389093	50.401507	50.413924	50.426341	50.438758	50.451178	12414	775
776	50.463600	50.476022	50.488445	50.500870	50.513296	50.525723	50.538153	50.550583	50.563015	50.575447	12427	776
777	50.587882	50.600317	50.612754	50.625192	50.637631	50.650072	50.662515	50.674958	50.687403	50.699849	12440	777
778	50.712297	50.724745	50.737195	50.749647	50.762099	50.774553	50.787009	50.799466	50.811924	50.824383	12453	778
779	50.836844	50.849306	50.861769	50.874233	50.886699	50.899166	50.911636	50.924106	50.936576	50.949049	12467	779
780	50.961524	50.973999	50.986475	50.998953	51.011432	51.023912	51.036395	51.048878	51.061362	51.073847	12480	780
781	51.086335	51.098824	51.111313	51.123804	51.136296	51.148790	51.161286	51.173782	51.186279	51.198778	12493	781
782	51.211279	51.223781	51.236284	51.248788	51.261293	51.273800	51.286309	51.298819	51.311329	51.323841	12506	782
783	51.336355	51.348870	51.361386	51.373903	51.386422	51.398942	51.411464	51.423987	51.436511	51.449036	12520	783
784	51.461564	51.474091	51.486620	51.499151	51.511683	51.524216	51.536751	51.549287	51.561824	51.574363	12533	784
785	51.586903	51.599444	51.611987	51.624530	51.637075	51.649621	51.662170	51.674719	51.687270	51.699821	12546	785
786	51.712375	51.724929	51.737485	51.750041	51.762600	51.775159	51.787721	51.800283	51.812847	51.825411	12559	786
787	51.837978	51.850546	51.863114	51.875685	51.888256	51.900828	51.913404	51.925979	51.938555	51.951133	12572	787
788	51.963714	51.976294	51.988876	52.001459	52.014043	52.026629	52.039217	52.051806	52.064395	52.076987	12585	788
789	52.089580	52.102174	52.114769	52.127365	52.139963	52.152562	52.165163	52.177764	52.190367	52.202971	12599	789
790	52.215578	52.228185	52.240793	52.253402	52.266013	52.278625	52.291240	52.303854	52.316470	52.329088	12612	790
791	52.341707	52.354327	52.366949	52.379571	52.392195	52.404820	52.417448	52.430076	52.442705	52.455335	12625	791
792	52.467968	52.480601	52.493236	52.505871	52.518508	52.531147	52.543787	52.556428	52.569070	52.581714	12638	792
793	52.594360	52.607006	52.619654	52.632302	52.644952	52.657604	52.670258	52.682912	52.695567	52.708224	12651	793
794	52.720883	52.733542	52.746202	52.758864	52.771528	52.784192	52.796859	52.809526	52.822195	52.834865	12664	794
795	52.847537	52.860209	52.872883	52.885558	52.898234	52.910912	52.923592	52.936272	52.948954	52.961636	12677	795
796	52.974321	52.987007	52.999694	53.012382	53.025071	53.037762	53.050455	53.063148	53.075843	53.088539	12690	796
797	53.101237	53.113935	53.126636	53.139337	53.152039	53.164743	53.177449	53.190156	53.202863	53.215572	12703	797
798	53.228283	53.240995	53.253708	53.266422	53.279138	53.291855	53.304574	53.317293	53.330014	53.342736	12716	798
799	53.355461	53.368185	53.380911	53.393639	53.406367	53.419097	53.431829	53.444562	53.457295	53.470031	12730	799

Appendix D

Accuracy Comparison of Chemical Analysis Techniques

It is not an uncommon development in science, as well as in other fields, for there to be established a relationship between the alteration of accurately measurable quantities of a system and a given, known change in the chemical, physical, or other state of that system. When this occurs, that effect may be investigated with thoroughness and the results tabulated so that the measurement of these quantities will identify, perhaps uniquely, an initially unknown state of the system.

The several ways in which the chemical state of an atom may be determined provide one very apt example of this application. Although these methods differ widely in nature and measure in some cases very different quantities, their results can give information on the chemical state of an atom with accuracies unique to the method employed.

Following are descriptions of several physical methods of chemical analysis. The accuracies of these are considered and the most accurate compared. This is not an attempt to tie the fields together, but rather a summary of the various approaches to the problems of chemical analysis from several investigational viewpoints.

Five classes of physical techniques are considered:

1. Nuclear magnetic resonance (NMR).
2. Photoelectron spectroscopic analysis for solids: (photo)electron spectroscopy for chemical analysis (ESCA).
3. X-ray emission spectral methods.
4. X-ray absorption spectral methods.
5. Other methods of electron spectrometry.

1. NUCLEAR MAGNETIC RESONANCE

A nucleus with a magnetic moment $\boldsymbol{\mu} = \gamma\hbar\mathbf{I}$, where γ is the magnetogyric ratio and $\mathbf{I}$ is the nuclear spin, will form one of m substates when exposed to a magnetic field $\mathbf{H}$. The projection of the nuclear angular momentum $[I(I+1)]^{1/2}$ on the main field axis is $\hbar m$ $(m = I, I-1, \cdots, -I)$. These substates will be of energies $E = \gamma\hbar\langle\mathbf{H}\cdot\mathbf{I}\rangle = \gamma\hbar H\langle I_z\rangle = \gamma\hbar mH$. Taking the simplest case—$I = \frac{1}{2}$, where the eigenvalues of I_z are $\pm\hbar/2$—then a second, oscillating field imposed on the system in a direction perpendicular to $\mathbf{H}$ will result in a finite transition probability between the two states. The frequency of the transition will be $\nu = \Delta E/\hbar = \gamma H$, where ΔE is the energy between the two substrates $m = \pm\frac{1}{2}$.

It is most suitable to use atoms having nuclei of spin $I = \frac{1}{2}$, since higher spins usually have associated quadrupole moments leading to quadrupole broadening of absorption lines. In addition, the nuclei must have large enough magnetic dipole moments and at the same time be stable enough to be employed with sufficient ease. Several such nuclei are listed here, together with their resonant frequencies in a 10,000-Gauss field and their natural abundances. The resonant frequency for the electron is also entered for comparison.[1]

	Resonant frequency (Mc/sec)	Natural Abundance (%)
e^-	28.000	—
^{1}H	42.576	99.98
^{2}H	6.536	1.56×10^{-2}
^{11}B	13.660	81.17
^{13}C	10.705	1.108
^{14}N	3.076	99.635 ($I = 1$)
^{15}N	4.315	0.365
^{19}F	40.055	100
^{31}P	17.235	100
^{33}S	3.266	0.74
^{35}Cl	4.172	75.4
^{37}Cl	3.472	24.6
^{205}Tl	24.57	70.48

The effect of a "chemical shift," the so-called Knight shift, of a resonance value was first found by Knight in 1949[2] for certain metals and metallic salts. Subsequently, Proctor and Yu[3] made measurements on ^{14}N and Dickinson[4] studied ^{19}F.

Nuclei may be magnetically different either because they are in chemically distinct groups or because they have different spacial environments. An example of the first situation is given by the two fluorine atoms in $(CF_3)_2CFI$; of the second is typified by the two hydrogen atoms in primary amides:

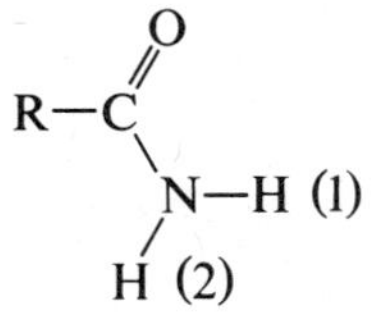

The molecular symmetry is most important in deciding how many peaks are expected in the absorption spectrum. The $B^{19}F_3$, which shows one absorption band, has the structure

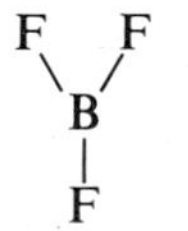

and $Cl^{19}F_3$ shows two peaks of the intensity ratio 2 : 1, as

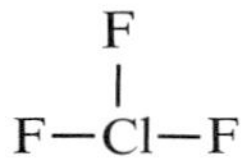

The shift in resonance frequency arises from a change in the diamagnetic shielding by the electrons about the nucleus: $H = H_0 - H' = (1 - \sigma)H_0$, where σ is a dimensionless number called the shielding coefficient or constant. Magnetic shielding of a nucleus by a spherical electron distribution was first calculated by Lamb.[5] Smaller splittings and shifts also arise through coupling with other constituents of the molecule. (See Refs. 6, 1.)

1.1. Plotting of Parameters

We have seen that resonance frequency values alter with changes in magnetic shielding of the nucleus. This is in turn related to the chemical state of the atom in question, and therefore to the valence state of the atom. In order to obtain parameters that are more accurately representative of the electronegativity of the atom, Pauling (see Chapter 7) and Hammett parameters have been postulated (Ref. 7 treats these and other parameters). Pauling parameters are employed for atoms in most molecules (or for polyatomic substituents using the Cavanaugh–Dailey treatment[8]), whereas those of Hammett are employed exclusively for chemicals including benzene rings.

Once the relative frequency shift for an element has been determined as a function of electronegativity, this knowledge may be used to determine the initially unknown chemical state of the element in another compound from its NMR spectrum. These spectral component frequency values are usually measured as a frequency displacement from a given standard value.

Alternatively, it is found that the NMR value for a particular isotope in a given group varies little for various large molecules in which that group is included. These small shifts are included among the tertiary or substitution effects.

Several specific effects cause chemical shifts. Classed as primary, secondary, and so on, in accordance with their relative magnitudes, these are listed here in the order of their magnitude.

1. Primary effect—the change in shielding caused by change in valence or oxidation state of the atom. Primary effects are not usually investigated in NMR studies.

2. Secondary effect—shielding changes due to changes in multiplicity of bonding.

3. Tertiary effect—changes in shielding due to the chemical replacement of a group that is remote from the site of the nucleus measured. For example, the NMR value of ^{19}F will shift[9] in going from CF_3–$C{\equiv}CH$ to CF_3–$C{\equiv}C$–CF_3, since the hydrogen has been replaced by the –CF_3 group. This is also called the substitution effect.

4. Isotopic substitution effect—there will occur a decrease in shielding constant of the order of 10^{-7} for the hydrogen atom in HD as compared with H_2. Such shifts for other elements have also been observed in replacing an H by a D atom, or, for example, for ^{19}F in replacing a ^{12}C atom with a ^{13}C atom, where the ^{19}F resonance may shift ≈ 0.1 ppm for $H = 10{,}000$ G. It has been suggested that this effect is probably attributable to a difference in ground state vibrational amplitudes in the molecules concerned, thus changing the shielding constant for a nucleus of interest.

Although many compounds have been measured with NMR methods, only a few have also been measured by any of the other methods considered here. Also, NMR measurements usually employ the chemical of interest in a dissolved form. Therefore, measurements may be made on radicals containing the pertinent atom. This is the case for the nitrogen compounds included in Table I. The NMR values are for NO_3^- and CN^-, relative to that for NO_2^-. Also included in the table, for comparison, are measured binding energy shifts for similar compounds using a photoelectron method in which the investigated chemicals are in the solid state. The photo-ESCA measured values of the nitrogen *K*-shell binding energies were obtained for the sodium and potassium salts (see Section 2). However, the last compound in the list, $N(CH_3)_4Cl$, was used in both cases. The table also gives two relative NMR shifts for inorganic salts of thallium in the solid state. These may be roughly compared with the relative thallium L_3-X-ray absorption energy shifts for the same salts.

Unfortunately, $K\alpha_1$ X-ray energy shifts were not available for compounds also measured with NMR. A rough relative $K\alpha_1$ energy shift determination accuracy relative to NMR shift values can be obtained via comparison with photo-ESCA measurements.

Finally, Table II and Figure 1 compare NMR measurements on ^{13}C in a variety of compounds for which photo-ESCA shifts have also been determined. See Section 2 for further discussion.

2. PHOTOELECTRON SPECTROMETRIC ANALYSIS FOR SOLIDS

It was observed experimentally in 1964 that the photoelectron peak from the *K*-shell in sulfur in $Na_2S_2O_3$ is actually a doublet.[10] Since the two atoms are in two different oxidation states in the compound, this difference in *K*-photoelectron energies was taken to be a direct result of the difference in charge of the two atoms. (The binding energy is that energy required to remove a electron from its initial atomic orbital to infinity. Different degrees of electronegativity of the photoionized atom will result in different binding enegies. See Chapter 7, Sections 4, 4.1, and 4.1.2 for further discussion.)

This method was named electron spectroscopy for chemical analysis (ESCA),[10] but here it is referred to as photo-ESCA to distinguish it from Auger-ESCA (see Section 5). Photo-ESCA has one strong advantage over NMR methods of chemical analysis; namely, it can be applied to all elements of the periodic table, whereas NMR methods require that the nucleus of the atom of interest have a magnetic moment and preferably a spin value of 1/2. Hence it has been possible to investigate sulfur in a wide variety of organic and inorganic compounds using the photoelectron method. Few measurements have been made on sulfur by NMR methods.

The final kinetic energy of a photoelectron is dependent on the charge of the atom investigated. Therefore, initial measurements of chemical shifts in photoelectron energy for sulfur, tin, chlorine, and copper were related to the valence state or degree of ionization of the atom. Later, however, the correlation between the measured binding energy and the charge on a constituent atom of interest in a compound obtained from Pauling's electronegativity values for the elements was investigated and found to be very good. Thus we see that NMR and photo-ESCA are mutually analogous: the former measures the magnetic shielding of the nucleus from an external magnetic field, and the latter measures the electrostatic shielding of a photoelectron as it is removed from the atom. Table I compares NMR and binding energy shifts. Note, however, that in the former case the measurements are carried out generally for ions in solution, and in the latter chemicals in the solid form are used.

It is possible to contrast the accuracies of the relation between the chemical state and relative respective shifts for the two methods for several organic compounds. These are listed in Table II and are plotted in Figure 1. Clearly, however, there is only a moderate trend between decreasing $\delta_{^{13}C}$ and increasing carbon *K*-binding energy, which indicates that the two effects are moderately related to each other. Table II allows us to obtain relative shifts between two chemicals for the two effects. Taking these to be the extreme cases of CCl_4 and CH_4, we see that the relative accuracies for the shifts are:

	NMR $\Delta\delta$ (ppm)	Photo-ESCA ΔE (eV)
Shift magnitude	99.8(2.2)	7.80(0.14)
Method accuracy (%)	2.2	1.8
Approximate accuracy limit (%)	<0.1	~ 1.2

These values are obtained from the corresponding values and their errors given in Table II; the accuracies may well be better for the NMR case. In fact, since isotopic substitution effects are detectable (being of the magnitude $\gtrsim 0.1$ ppm), then the NMR shift accuracies will have the approximate limit shown in the tabulation. Photoelectron measurements can be made accurate to ~ 0.1 eV, hence the limit quoted.

For investigating substitution effects, the two methods have similar sensitivities, judging by Figure 1 (i.e., for $C^*H_3CH_2X$, $X = Cl$, OH), whereas for chemical shifts NMR appears to be superior. Note also that although one method may differentiate poorly between two compounds, the other may serve to

Table I.

Comparison of NMR, photo-ESCA, $K\alpha_1$ X-ray energy, and X-ray absorption edge energy shifts.

Chemical state	NMR[1)] shifts (ppm) ± ∿2ppm for H = 10,000 Gauss	K, L_1, L_2, L_3 photo-ESCA shifts (eV) ref.	K	L_1	L_2	L_3	$K\alpha_1$ X-ray energy shifts	K, L_3 X-ray absorption energy shifts (eV) ref.	K	L_3
N in NO_2^-	0		0	--	--	--				
NO_3^-	+254(2)		+2.9(1)	(in $NaNO_3$)	--	--				
CN^-	+380(2) §	10)	-5.1(1)	(in KCN)	--	--				
$N(CH_3)_4Cl$	+581(2)		-2.8(1)	--	--	--				
Cl in NaCl			0	0	0	0	0			
$NaClO_2$		10)	+4.5(5)	+4.0(5)	+3.8(5) (L_2, L_3)		--			
$KClO_3$			+6.5(5)	+6.4(5)	+7.1(5)		+0.96 eV[23)]			
$KClO_4$			+9.6(5)	+9.3(5)	+9.5(5)		+1.4 eV[23)]			
Tl in TlCl	0								--	0
$Tl_2(SO_4)_2$	+5200							21)	--	+0.8(4)
TlBr	-4700								--	+5.4(4)
S in Na_2S			0	--	--	--	0			
S			+2.0	--	--	--	+0.39 eV[23)]			
BaS			--	--	--	--	+0.46 eV			
CuS			--	--	--	--	+0.85 eV			
Na_2SO_4		10)	+7.8	--	--	--	+1.59 eV			
Na_2SO_3			+6.5	--	--	--	+1.39 eV			
$Na_2S_2O_3$			+7.3 +0.3	-- --	-- --	-- --	+1.09 eV			

Table I. (cont.)

Cu in Cu			0	0	0	0	0		0	--
CuO		18)	+4.4(5)	+4.4(1.0)	+3.3(1.5)	+2.5(8)	+0.11(4) eV[19)] +0.15(3) eV	22)	-3.87	--
Cu_2O			+0.7(8)	--	--	--	--		+0.67	--
Sn in Sn			--	0	0	0	0			
SnO		20)	--	+1.3(3)	+1.1(3)	+1.3(2)	-93.5(1.5) meV[15)] -76(8) meV[12)]			
SnO_2			--	--	--	+1.1(3)	-192(10) meV[15)] -152(5) meV[13)] -167(5) meV[12)]			

§Obtained with a magnetic field of 10,700 Gauss.

Table II.

Relative shifts in the carbon K shell binding energy and the NMR values for ^{13}C for several different chemical compounds. The carbon atom measured in the compound is designated by †.

Compound	ΔE C(K)[24)] (eV ± 1 eV)	δ w.r.t. $CH_3C^{\dagger}OOH$[1)] (ppm ± 1-2 ppm)
CCl_4	7.8	81
$CH_3C^{\dagger}H_2OH$	1.5	119.6
CH_2Cl_2	3.0	124
$CH_3C^{\dagger}H_2Cl$	1.3	138.0
CH_2Br_2	2.3	156
$C^{\dagger}H_3CH_2OH$	0.1	159.8
$C^{\dagger}H_3CH_2Cl$	0.2	159.9
CH_4	0.00	180.8

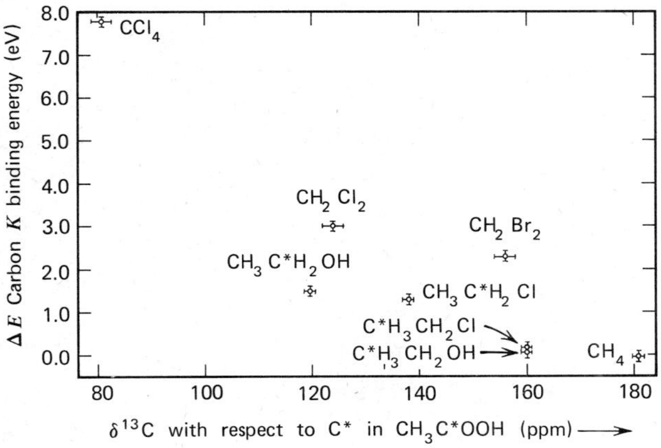

Figure 1. A plot of chemical shifts in the carbon K shell binding energy[10] against those in the ^{13}C NMR spectrum[1] for the several organic compounds as given in Table II. In the molecules containing more than one carbon atom, the atom subjected to measurement is denoted by an asterisk. This comparison provides a rough indication of the relative respective shifts and thereby the relative sensitivities of the two methods.

distinguish them quite clearly. This is not true, however, for the $C^*H_3CH_2Cl$ and $C^*H_3CH_2OH$ substitution pair.

Extending the last observation, it is possible to analyze a multigroup chemical molecule with greater precision by using the two methods: if K-photoelectron line energies and NMR spectral components are compared, for an element in a given compound, with experimentally determined plots of ΔE and δ versus chemical group plots, one may uniquely determine the simultaneous presence of several such groups.

Hercules[11] has pointed out that photo-ESCA measurement results plotted thus could be very useful in chemical analysis work. Figure 2 shows his plot for sulfur K-shell binding energies for various groups. It is seen that the K-shell binding energy for sulfur in a given group, which itself may be included in a wide variety of compounds, is confined to a rather limited range of binding energies.

3. X-RAY EMISSION ENERGY SHIFTS FOR SOLIDS

The X-ray emission method of determining atomic electron binding energies is described and treated in Chapter 7, Section 4.1.1. An atomic electron transition from an initially filled level to an inner atomic vacancy (e.g., an L_3–K transition) will be, essentially, the difference in the binding energies between the two levels. When the screenings of the nuclei for these two states are altered by a change in atomic chemical state, the shift in X-ray energy will be the difference in the two resulting binding energy shifts for the levels. Therefore, chemical X-ray energy shifts will be expected to be much smaller than photoelectron energy shifts because the contribution of two effects: (*a*) X-ray shifts are due to *differences* in the bound state energy shifts for the pertinent chemical states as a

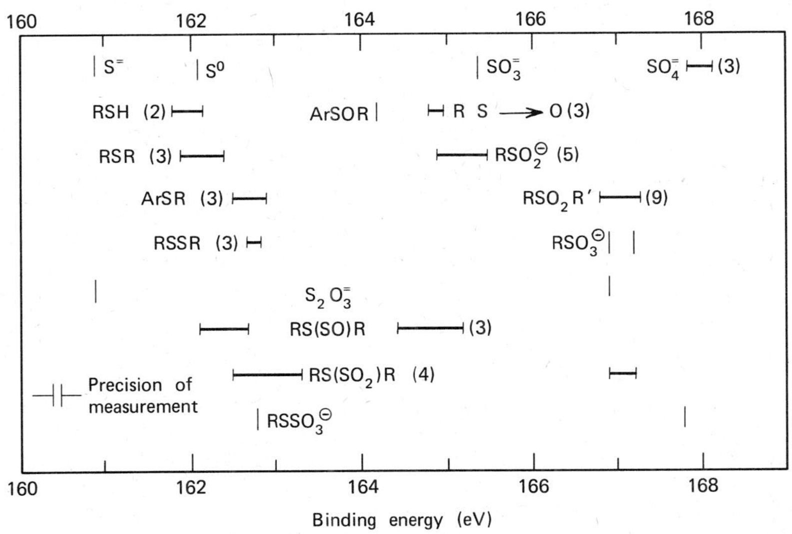

Figure 2. A scheme showing sulfur L_1-shell binding energy with respect to functional groups as constructed by Hercules[11]. Here |—| denotes the range of observed energies; the number of compounds used in correlation appears in parentheses.

result of change in their associated screenings; photoelectron energy shifts involve this bound state energy shift itself, and (*b*) X-ray shifts result from changes in the electrostatic field within the atom. Photoelectron shifts are due to changes in the electrostatic field both internal and external to the atom.

As a result of the preceding phenomenological differences, the *K* X-ray shifts are from 5 to 10 times smaller than photoelectron shifts. However, workers in the Soviet Union,[12–15] and in East Germany[16] have reported results that indicate that measurement accuracies of X-ray energy shifts relative to the shifts themselves are 2 to 10%. Comparing this with values cited in the foregoing section, the sensitivity of the measurements appear to be good. (Chapter 7, Section 4.1.1, and the references cited there provide more discussion.) Several measured $K\alpha_1$ X-ray energy shifts due to chemical binding are shown in Table I, where they may be compared with corresponding shifts for other analysis methods.

4. X-RAY ABSORPTION ENERGY SHIFTS FOR SOLIDS

Again, as explained in Chapter 7, Section 4.1.1, shifts in the energy of an X-ray absorption edge for a given element in going from one chemical form to another are due primarily to two effects: (*a*) a change in the binding energy of the transition electron in the initial state (as for Sections 2 and 3) and (*b*) alterations in the wave function characteristics and energies of the final state.

This effect has been investigated over the years since the 1920s (see Chapter 7), but has exhibited weakness as a chemical analysis technique. This is primarily because of the large number of possible alterations in the nature of the final orbitals resulting from the change in chemical state. Instead, X-ray absorption measurements have often been used to provide information about the distribution and nature of outer atomic levels in the atom in a given chemical state. Such investigational methods are inversely analogous to those that focus attention on X-ray emission spectra involving the outermost atomic levels. Table I cites two cases of shifts of X-ray absorption edges that are due to chemical binding. These values may be compared with those from other methods.

More recently, synchrotron radiation absorption techniques have been employed to gain information on chemical shifts in binding energy (see Chapter 7). Again, however, the changes in the nature and densities of final available orbitals with the chemical state of the atom limit the accuracy of the method for determining binding energy shifts.

5. OTHER ANALYSIS METHODS

The methods discussed in Sections 2 to 4 may be applied to chemicals in the gaseous state. X-ray absorption methods, then, will show alterations in the difference in binding energies between the initial and final states for the transition.

In Oxford, Turner and co-workers[17] demonstrated that the photoelectron effect for atoms in gaseous compounds is complicated by effects of the excitation of molecular vibrational and rotational states at the moment of photoionization. The helium-resonance line was used, and the spectrum of photoelectrons coming from the outermost atomic energy levels was measured. Since the vibrational structure dominates in the spectrum, and since this is dependent on the structure of the molecule in question, a correlation between the molecular structure and the photoelectron spectrum can be expected. Establishment of this relationship could result in what the authors called "molecular photoelectron spectroscopy," a useful method of chemical identification of at least simple gaseous molecules.[17a]

Auger electron spectra for atoms in the gaseous state may also provide information about the chemical state of structure of the compound. This phenomenon, as investigated by the group in Uppsala, may be referred to as Auger-ESCA, to distinguish it from the comparable photoelectron method. The carbon *KLL* and bromine *MNN* Auger spectra in $CH_{4-n}Br_n$ ($n = 0$–4) have been investigated (see Chapter 3, Section 4.1.3.4). These spectra are expected to be complicated by the combined effects of vibrational modes excited in the initial ionization and those excited by the following Auger transitions. For the resolution used, however, Mehlhorn's work on the *KLL*-Auger spectrum for N_2 (Chapter 3, Section 4.1.3.4) has not revealed identifiable vibrational structure. The spectrum was explained solely in terms of molecular orbitals. The carbon and bromine spectra, although complex, are probably primarily of the same final orbital origin.

Aside from the above-mentioned spectrometric and X-ray methods for chemical analysis, there are also the methods of ultraviolet and infrared absorption, the Raman effect, and others. However, the methods enumerated previously have been exclusively treated here due to their interrelationship and their intimate relationship to the subject interest of this book.

REFERENCES

1. J. W. Emsley, J. Feeney, and L. H. Sutcliffe, *High Resolution Nuclear Magnetic Resonance Spectroscopy*, Pergamon Press, Oxford, 1966, Vols. 1 and 2.
2. W. D. Knight, *Phys. Rev.* **76** (1949) 1259.
3. W. G. Proctor and F. C. Yu, *Phys. Rev.* **77** (1950) 717.
4. W. C. Dickinson, *Phys. Rev.* **77** (1950) 736.
5. W. E. Lamb, *Phys. Rev.* **60** (1941) 817.
6. A. Carrington and A. D. McLachlan, *Introduction to Magnetic Resonance*, Harper & Row, Publishers, New York, 1967.
7. P. R. Wells, *Linear Free Energy Relationships*, Academic Press, New York, 1968, Chapter 2.
8. J. R. Cavanaugh and B. P. Dailey, *J. Chem. Phys.* **34** (1961) 1099.
9. C. Béguin, *Bull. Soc. Chim. France*, **11** (1967) 4214; also private communication, 1969.
10. K. Siegbahn, C. Nordling, A. Fahlman, R. Nordberg, K. Hamrin, J. Hedman, G. Johansson, T. Bergmark, S.-E. Karlsson, I. Lindgren, and B. Lindberg, "ESCA, Atomic Molecular, and Solid State Structure Studied by Means of Electron Spectroscopy," *Nova Acta Reg. Soc. Sci. Upsal., Ser. IV* **20** (1967) Uppsala; Rept. AFML-TR-68-189, Wright-Patterson A.F.B., Dayton, Ohio (October 1968).
11. D. M. Hercules, *Anal. Chem.* **42** (1970) 20A.
12. E. V. Petrovich, O. I. Sumbaev, V. S. Zykov, Yu. P. Smirnov, A. I. Egorov, and A. I. Grushko, *Zh. Eksp. i Teor. Fiz.* **53** (1967) 796 [*J. Exp. Theor. Phys.* **26** (1968) 489].
13. O. I. Sumbaev, A. F. Mezentsev, V. I. Marushenko, E. V. Petrovich, and A. S. Ryl'nikov, *Zh. Eksp. i Teor. Fiz,. SSSR*, **50** (1966) 861 [*J. Exp. Theor. Phys.* **23** (1966) 572].
14. O. I. Sumbaev and A. F. Mezentsev, *Zh. Eksp. i Teor. Fiz., SSSR*, **48** (1965) 445 [*J. Exp. Theor. Phys.* **21** (1965) 295]; O. I. Sumbaev, E. V. Petrovich, V. S. Sykov, A. S. Ryl'nikov, and A. I. Grushko, *J. Nucl. Phys., SSSR*, **5** (1965) 544 [*J. Nucl. Phys. (USSR)* **5** (1967) 387].
15. O. I. Sumbaev, E. V. Petrovich, Yu. P. Smirnov, A. I. Egorov, V. S. Zykov, and A. I. Gushko, *Zh. Eksp. i Teor. Fiz.* **53** (1967) 1545 [*J. Exp. Theor. Phys.* **26** (1968) 891].
16. W. Nefedow, *Phys. Status Solidus* **2** (1962) 904.
17. D. W. Turner and D. P. May, *J. Chem. Phys.* **46** (1967) 1156; M. I. Al-Joboury and D. W. Turner, *J. Chem. Soc.* **(B)** (1967) 373; A. D. Baker, D. P. May, and D. W. Turner, *J. Chem. Soc.* **(B)** (1968) 22. See these for other references.

17a. D. W. Turner, C. Baker, A. O. Baker, and C. R. Brundle, *Molecular Photoelectron Spectroscopy*, Wiley-Interscience, New York, 1970.

18. C. Nordling, E. Sokolowski, and K. Siegbahn, *Ark. Fys.* **13** (1958) 483; E. Sokolowski, C. Nordling, and K. Siegbahn, *Phys. Rev.* **110** (1958) 776.
19. E. Gilberg and B. Kern, *Z. Phys.* **174** (1963) 372.
20. C. Nordling, *Ark. Fys.* **15** (1959) 241.
21. B. K. Agarwal and L. P. Verma, *J. Phys. C.* [*Proc. Phys. Soc.* (*London*)] **2** (1969) 104.
22. L. P. Verma and B. K. Agarwal, *J. Phys. C.* [*Proc. Phys. Soc.* (*London*)] **1** (1968) 1658.
23. *Handbook of X-Rays*, E. F. Kaelble (Ed.), McGraw-Hill, Inc., New York, 1967.
24. R. Nordberg, U. Gelius, P. F. Hedén, J. Hedman, C. Nordling, K. Siegbahn, and B. J. Lindberg, University of Uppsala, Institute of Physics Rept. UUIP-581 (1968).

Appendix E

$K\alpha_{1,2}$ X-Ray Energies and Ultraviolet Resonance Line Energies

1. $K\alpha_{1,2}$ X-RAY ENERGIES

The following $K\alpha_{1,2}$ X-ray energies are taken from the table by Bearden.[1] Values are obtained from wavelength measurements with respect to the single "X-Ray Wavelength Standard," defined as $\lambda W K\alpha_1 \equiv$ (0.2090100 Å*), where 1 Å* = 1 Å ±5 ppm; the error here expressed is that of the original $WK\alpha_1$ value. The wavelengths of the silver $K\alpha_1$, molybdenum $K\alpha_1$, copper $K\alpha_1$, and chromium $K\alpha_2$ were established as secondary standards with errors of about 1 ppm.

In this table the X-ray energies, with their approximate probable errors, are expressed in kiloelectron volts. The asterisk denotes values obtained from interpolation on Moseley-type diagrams.

Z	Element	$K\alpha_1$ (keV)	$K\alpha_2$ (keV)	Z	Element	$K\alpha_1$ (keV)	$K\alpha_2$ (keV)
3	Li	—0.0543(2)—		30	Zn	8.63886(4)	8.61578(4)
4	Be	—0.1085(9)—		31	Ga	9.25174(7)	9.22482(7)
5	B	—0.1833(9)—		32	Ge	9.88642(8)	9.85532(8)
6	C	—0.277(2)—		33	As	10.54372(9)	10.50799(9)
7	N	—0.3924(52)—		34	Se	11.2224(2)	11.1814(2)
8	O	—0.5249(7)—		35	Br	11.9242(2)	11.8776(2)
9	F	—0.6768(7)—		36	Kr	12.649(1)	12.598(1)
10	Ne	—0.8486(2)—		37	Rb	13.3953(1)	13.3358(1)
11	Na	—1.0410(1)—		38	Sr	14.1650(2)	14.0979(2)
12	Mg	—1.25360(2)—		39	Y	14.9584(2)	14.8829(2)
13	Al	1.48670(2)	1.48627(2)	40	Zr	15.7751(2)	15.6909(2)
14	Si	1.73998(1)	1.73938(1)	41	Nb	16.6151(2)	16.5210(2)
15	P	2.0137(3)*	2.0127(3)*	42	Mo	17.47934(3)	17.3743(1)
16	S	2.30784(3)	2.30664(3)	43	Tc	18.3671(9)*	18.2508(9)*
17	Cl	2.62239(5)	2.62078(5)	44	Ru	19.2792(1)	19.1504(2)
18	Ar	2.95770(4)	2.95563(4)	45	Rh	20.2161(2)	20.0737(2)
19	K	3.3138(2)	3.3111(2)	46	Pd	21.1771(1)	21.0201(1)
20	Ca	3.69168(3)	3.68809(3)	47	Ag	22.16292(2)	21.9903(2)
21	Sc	4.0906(1)*	4.0861(1)*	48	Cd	23.1736(1)	22.9841(1)
22	Ti	4.51084(3)	4.50486(3)	49	In	24.2097(2)	24.0020(2)
23	V	4.95220(4)	4.94464(4)	50	Sn	25.2713(2)	25.0440(2)
24	Cr	5.41472(5)	5.405510(6)	51	Sb	26.3591(2)	26.1108(2)
25	Mn	5.89875(3)	5.88765(6)	52	Te	27.4723(2)	27.2017(2)
26	Fe	6.40384(3)	6.39084(3)	53	I	28.6120(3)	28.3172(5)
27	Co	6.93032(3)	6.91530(3)	54	Xe	29.779(2)*	29.458(2)*
28	Ni	7.47815(3)	7.46089(3)	55	Cs	30.9728(3)	30.6251(3)
29	Cu	8.04778(1)	8.02783(1)	56	Ba	32.1936(3)	31.8171(4)

57	La	33.4418(2)	33.0341(2)
58	Ce	34.7197(2)	34.2789(2)
59	Pr	36.0263(2)	35.5502(2)
60	Nd	37.3610(2)	36.8474(2)
61	Pm	38.7247(5)	38.1712(5)
62	Sm	40.1181(3)	39.5224(3)
63	Eu	41.5422(3)	40.9019(3)
64	Gd	42.9962(3)	42.3089(3)
65	Tb	44.4816(3)	43.7441(3)
66	Dy	45.9984(3)	45.2078(3)
67	Ho	47.5467(4)	46.6997(4)
68	Er	49.1277(4)	48.2211(4)
69	Tm	50.7416(4)	49.7726(4)
70	Yb	52.3889(4)	51.3540(4)
71	Lu	54.0698(5)	52.9650(6)
72	Hf	55.7902(7)	54.6114(7)
73	Ta	57.532(1)	56.277(2)
74	W	59.31824 Std.	57.9817(6)
75	Re	61.1403(6)	59.7179(3)
76	Os	63.0005(6)	61.4867(6)
77	Ir	64.8956(6)	63.2867(6)
78	Pt	66.832(1)	65.122(1)
79	Au	68.8037(7)	66.9895(7)
80	Hg	70.819(1)	68.895(1)
81	Tl	72.8715(8)	70.8319(8)
82	Pb	74.9694(9)	72.8042(9)
83	Bi	77.1079(9)	74.8148(9)
84	Po	79.290(5)*	76.862(5)*
85	At	81.52(1)*	78.95(1)*
86	Rn	83.78(2)*	81.07(2)*
87	Fr	86.10(2)*	83.23(2)*
88	Ra	88.47(1)*	85.43(1)*
89	Ac	90.884(5)*	87.67(1)*
90	Th	93.350(1)	89.953(1)
91	Pa	95.868(2)	92.287(6)
92	U	98.439(3)	94.665(3)

2. ULTRAVIOLET RESONANCE LINE ENERGIES

The following are energies of ultraviolet resonance emission lines commonly employed in measurements of low energy photoelectrons, radiation absorption, and so on. Energies tabulated are those between the ground and first excited states of the neutral atom (except for He^+). For the noble gases, the two energies given are for transitions to the final $^2P_{3/2}$ and $^2P_{5/2}$ states.

	Resonance Line Energies (eV)	Reference
H	10.199	2
He	21.218	2
He^+	40.812	2
Ne	16.848	3
	16.671	3
Ar	11.828	3
	11.623	3
Kr	10.643	2
	10.132	2
Xe	9.570	2
	8.436	2

REFERENCES

1. J. A. Bearden, *Rev. Mod. Phys.* **39** (1967) 78.
2. C. E. Moore, "Atomic Energy Levels," Vol. II, *Nat. Std. (U.S.), Circ.* 467, 1952.
3. "Atomic Emission Lines Below 2000 Å," Naval Rese. Lab. Rept. 6648.

Appendix F

Table of Atomic Electron Binding Energies

The following table of atomic electron binding energies was compiled from published values obtained by a variety of measurement methods, as discussed in Chapter 7.

The value for each atomic level considered here to be most accurate among those listed is underlined. All energies are in electron volts.

Many of the values cited are taken directly from the table of Bearden and Burr,[3] and they are based on weighted adjustments of values obtained from X-ray and ultraviolet emission and absorption measurements[4] and photoelectron measurements.

In all cases where the preferred (underlined) value is the average of several experimental values, those values contributing to the average are listed in parenthesis directly under.

For the rare gases—neon, argon, krypton, xenon, radon—the values cited refer to binding energy values for the gaseous state of the element. These values are expected to differ slightly from values that would be obtained, were the atoms situated in a solid state environment or frozen into the solid phase. The *K*-level binding energy for sodium has been measured for isolated atoms, for the solid metallic state, and for the solid oxide state: none of the values agree. (See Chapter 7, Section 2.)

For the remaining elements, the binding energies obtained by the various experimental methods are expected to be dependent on the oxidation state of the atom. For most of the metallic elements, the pure metal was used in the binding energy determination. However, alterations in binding energies with atomic valence state (Chapter 7, Section 4) will be expected. Binding energy values obtained for atoms in chemical compound forms will then differ from those for the atoms in the elemental form by amounts depending on the nature of the associated chemical bond. No attempts have been made to correct for such effects. Such binding energy alterations are expected to seldom exceed ±1 to 2 eV per unit change in oxidation state, although they can become as high as 10 eV (Chapter 7, Section 4.1.2).

Where experimental binding energy values are not available, they are estimated by interpolation or extrapolation from neighboring values or by theoretical calculations.

In the table are included source references for the data as well as information on the methods used in the determinations. These methods, described briefly in Chapter 7, are as follows:

- X value given by Bearden and Burr,[3] often a weighted least-mean-square adjustment of values from available X-ray and ultraviolet absorption and emission as well as photoelectron data
- x other X-ray and ultraviolet emission and absorption data
- pe photoelectron data
- ag binding energies determined from Auger electron energy data
- ic binding energies determined from internal conversion electron energy data for nuclear transitions
- ai binding energies from autoionization level series limit determinations
- dl binding energies as determined from discrete (characteristic) electron energy loss data
- av average over available binding energy values of various origins (other than found in Ref. 3)
- in binding energy values obtained by interpolation in a modified Moseley diagram
- ex binding energy values obtained by extrapolation in a modified Moseley diagram (expected to be rough values)
- sd binding energy values from a relativistic self-consistent Slater–Dirac energy level calculation (accurate to roughly 100 eV).

Level	Ref.	Meth.	1 H (gas)	Ref.	Meth.	2 He (gas)	Ref.	Meth.	3 Li	Ref.	Meth.	4 Be
K	1,4	X	13.59811(8)	2,4	X	24.5868(2)	3	X	54.75(2)	4a	x	112.1(2)
							4	x	(54.75)	3	X	111.0(1.0)
										4	x	111.0
			5 B			6 C			7 N			8 O
K	3	X	188.0(4)	3	X	283.8(4)	3	X	401.6(4)	3	X	532.0(4)
	5	pe	(188.0)	5	pe	(283.8,	5	pe	(401.6)	5	pe	(532.0)
				4	x	283.8)	6	pe	399.4(4)			
L_1			--			--			--	3	X	23.7(4)
										7	pe	(23.7)
$L_{2,3}$	3	X	4.7(9)	3	X	6.4(1.9)	3	X	9.2(6)	3	X	7.1(8)
				30	dl	10.3(1)						
			9 F			10 Ne (gas)			11 Na			12 Mg
K	3	X	685.4(4)	8	av	870.1(2)	3	X	1072.1(4)	3	X	1305.0(4)
	5	pe	(685.4)	8	ag	(870.0(4))	5	pe	(1072.1,	5	pe	(1305.0,
							4	x	1072.)	4	x	1303.)
L_1	7	pe	34(1)	9	x	48.47(10)	3	X	63.3(4)	3	X	89.4(4)
				10	x	48.47	7	pe	(63.3)	7	pe	(89.4)
				11	x	48.5						
L_2 } L_3	3	X	8.6(8)	12	x	21.66	3	X	31.1(4)	3	X	51.4(5)
				12	x	21.56						
M_1			--			--	4	x	0.7(4)	4	x	2.1(4)
			13 Al			14 Si			15 P			16 S
K	3	X	1559.6(4)	3	X	1838.9(4)	3	X	2145.5(4)	3	X	2472.0(4)
	5	pe	(1559.6,	5	pe	(1838.9)	7	pe	(2145.5)	5	pe	(2472.0,
	4	x	1559.8)							4	x	2470.)
L_1	3	X	117.7(4)	3	X	148.7(4)	3	X	189.3(4)	3	X	229.2(4)
	7	pe	(117.7)	7	pe	(148.7)	7	pe	(189.3)	7	pe	(229.2)

Level	Ref.	Meth.	13 Al	Ref.	Meth.	14 Si	Ref.	Meth.	15 P	Ref.	Meth.	16 S
L_2	13	x	73.15(2)	4	x	99.5(4)	4	x	136.2(4)	4	x	165.4(4)
L_3	13	x	72.72(2)	4	x	98.9(4)	4	x	135.3(4)	4	x	164.2(4)
M_1	4	x	0.7(4)	4	x	7.6(5)	4	x	16.2(6)	4	x	15.8(6)
$M_{2,3}$			--	4	x	3.0(4)	4	x	9.9(4)	4	x	8.0(4)
			17 Cl			18 Ar(gas)			19 K			20 Ca
K	3 5	X pe	2822.4(3) (2822.4)	3 4 10	X x x	3202.9(3) (3202.9) 3202.3	3 5 4	X pe x	3607.4(4) (3607.4, 3607.8)	3 5 4	X pe x	4038.1(4) (4038.1, 4038.1)
L_1	3 7	X pe	270.2(4) (270.2)	14 18	ag pe	326.5(5) 325(1)	3 7	X pe	377.1(4) (377.1)	3 7	X pe	437.8(4) (437.8)
L_2	3	X	201.6(3)	15 16	ag X	250.53(25) 250.55(5)	3	X	296.3(4)	3	X	350.0(4)
L_3	3	X	200.0(3)	15 16	ag X	248.39(25) 248.52(5)	3	X	293.6(4)	3	X	346.4(4)
M_1	3	X	17.5(4)	12	x	29.24	3	X	33.9(4)	3	X	43.7(4)
M_2 M_3	3	X	6.8(4)	12 12	x x	15.93 15.76	3 30	X dl	17.8(4) 18(1)	3 30	X dl	25.4(4) 26(1)
N_1			--			--	4	x	-0.1(5)	4	x	1.8(6)
			21 Sc			22 Ti			23 V			24 Cr
K	3 5	X pe	4492.8(4) (4492.8) --	3 5 4	X pe x	4966.4(4) (4966.4, 4964.5)	3 5 4	X pe x	5465.1(3) (5465.1, 5464.)	3 5 4	X pe x	5989.2(3) (5989.2, 5989.)
L_1	3 7	X pe	500.4(4) (500.4)	3 7	X pe	563.7(4) (563.7)	3 7	X pe	628.2(4) (628.2)	3 7	X pe	694.6(4) (694.6)
L_2	3	X	406.7(4)	3	X	461.5(4)	3	X	520.5(3)	3	X	583.7(3)
L_3	3	X	402.2(4)	3	X	455.5(4)	3	X	512.9(3)	3	X	574.5(3)
M_1	3	X	53.8(4)	3	X	60.3(4)	3	X	66.5(4)	3	X	74.1(4)
$M_{2,3}$	3	X	32.3(5) --	3 30	X dl	34.6(4) 30(1)	3 30	X dl	37.8(3) 33(1)	3 30	X dl	42.5(3) 40(1)
$M_{4,5}$	3	X	6.6(5)	3,4	X,x	3.7(4)	3	X	2.2(3)	3	X	2.3(4)
			25 Mn			26 Fe			27 Co			28 Ni
K	3 5 4	X pe x	6539.0(4) (6539.0, 6538.) --	3 5 4 19	X pe x pe	7112.0(9) (7111.3, 7111.2) 7113.6(5)	3 5 4	X pe x	7708.9(3) (7708.9, 7709.5) --	3 5 4	X pe x	8332.8(4) (8332.8, 8331.6) --
L_1	3 7	X pe	769.0(4) (769.0)	3 7	X pe	846.1(4) (846.1)	3 7	X pe	925.6(4) (925.6)	3 7	X pe	1008.1(4) (1008.1)
L_2	3	X	651.4(4) --	3 4	X x	721.1(9) (720.8)	3 4	X x	793.6(3) (793.8)	3 4	X x	871.9(4) (870.6)
L_3	3	X	640.3(4) --	3 4	X x	708.1(9) (707.4)	3 4	X x	778.6(3) (779.0)	3 4	X x	854.7(4) (853.6)

Level	Ref.	Meth.	25 Mn	Ref.	Meth.	26 Fe	Ref.	Meth.	27 Co	Ref.	Meth.	28 Ni
M_1	3	X	83.9(5)	3	X	92.9(9)	3	X	100.7(4)	3	X	111.8(6)
$M_{2,3}$	3	X	48.6(4) --	3 4	X x	54.0(9) (54.)	3 4	X x	59.5(3) (61.)	3 4	X x	68.1(4) (66.)
$M_{4,5}$	3	X	3.3(5)	3	X	3.6(9)	3	X	2.9(3)	3	X	3.6(4)
			29 Cu			30 Zn			31 Ga			32 Ge
K	3 21,22 4	X pe x	8978.9(4) (8978.9, 8980.3)	3 21 4	X pe x	9658.6(6) (9658.6, 9660.7)	3 21 4	X pe x	10367.1(5) (10367.1, 10368.2)	3 21 4	X pe x	11103.1(7) (11103.8, 11103.6)
L_1	3 7	X pe	1096.1(4) (1096.0)	3	X	1193.6(9) --	3	X	1297.7(1.1) --	3 21	X pe	1414.3(7) (1413.6)
L_2	3 22 4	X pe x	951.0(4) (950.0, 953.)	3 4	X x	1042.8(6) (1045.) --	3	X	1142.3(5) -- --	3 4	X x	1247.8(7) (1249.) --
L_3	3 22 4	X pe x	931.1(4) (931.4, 933.)	3 4	X x	1019.7(6) (1022.) --	3 4	X x	1115.4(5) (1117.) --	3 4	X x	1216.7(7) (1217.0) --
M_1	3	X	119.8(6)	3	X	135.9(1.1)	3	X	158.1(5)	3	X	180.0(8)
M_2	3 4	X x	73.6(4) (75.)	3 4	X x	86.6(6) (86.)	3	X	106.8(7)	3	X	127.9(9)
M_3							3	X	102.9(5)	3	X	120.8(7)
$M_{4,5}$	3	X	1.6(4)	3	X	8.1(6)	3	X	17.4(5)	3	X	28.7(7)
N_1			--			--			--			--
$N_{2,3}$			--			--	4	x	0.8(5)	4	x	2.3(8)
			33 As			34 Se			35 Br			36 Kr (gas)
K	3 23 4	X pe x	11866.7(7) (11866.7, 11865.)	3 21 4	X pe x	12657.8(7) (12657.8, 12654.5)	3 4	X x	13473.7(4) (13470.) --	3 4	X x	14325.6(8) (14324.4) --
L_1	3 4	X x	1526.5(8) (1529.)	3 4	X x	1653.9(3.5) (1652.5)	3 24	X pe	1782.0(4) (1782.0)	3 25	X pe	1921.0(6) (1921.2)
L_2	3 4	X x	1358.6(7) (1358.7) --	3 4	X x	1476.2(7) (1474.7) --	3 24	X pe	1596.0(4) (1596.2) --	3 25 4	X pe x	1727.2(5) (1727.2, 1730.)
L_3	3 4	X x	1323.1(7) (1323.5) --	3 4	X x	1435.8(7) (1434.0) --	3 24	X pe	1549.9(4) (1549.7) --	3 25 4	X pe x	1674.9(5) (1674.8, 1677.)
M_1	3	X	203.5(7)	3	X	231.5(7)	3	X	256.5(4)	26	ag	292.1(1.0)
M_2	3	X	146.4(1.2) -- --	3	X	168.2(1.3) -- --	3	X	189.3(4) -- --	27 26 3	ag ag X	221.8(4) 222.1(6) 222.7(1.1)
M_3	3	X	140.5(8) -- --	3	X	161.9(1.0) -- --	3	X	181.5(4) -- --	27 26 3	ag ag X	214.5(4) 214.6(6) 213.8(1.1)

Level	Ref.	Meth.	33 As	Ref.	Meth.	34 Se	Ref.	Meth.	35 Br	Ref.	Meth.	36 Kr(gas)
M_4	3	X	41.2(7)	3	X	56.7(8)	3	X	70.1(4)	28	x	95.04
M_5							3	X	69.0(4)	28	x	93.82
N_1			--			--	3	X	27.3(5)	12	x	27.52
N_2	3	X	2.5(1.0)	3	X	5.6(1.3)	3	X	5.2(4)	12	x	14.66
N_3							3	X	4.6(4)	12	x	14.00
			37 Rb			38 Sr			39 Y			40 Zr
K	3 4	X x	15199.7(3) (15202.)	3 4	X x	16104.6(3) (16107.)	3 4	X x	17038.4(3) (17038.)	3 4	X x	17997.6(4) 17999.)
L_1	3 24	X pe	2065.1(3) (2065.4)	3 29	X pe	2216.3(3) (2216.2)	3 29	X pe	2372.5(3) (2372.7)	3 29	X pe	2531.6(3) (2531.6)
L_2	3 24	X pe	1863.9(3) (1863.4) --	3 29 4	X pe x	2006.8(3) (2006.6, 2008.5)	3 29 4	X pe x	2155.5(3) (2155.0, 2154.0)	3 29 4	X pe x	2306.7(3) (2306.5, 2305.3)
L_3	3 24	X pe	1804.4(3) (1804.6) --	3 29 4	X pe x	1939.6(3) (1939.9, 1941.)	3 29 4	X pe x	2080.0(3) (2080.2, 2079.4)	3 29 4	X pe x	2222.3(3) (2222.5, 2222.5)
M_1	3	X	322.1(3)	3	X	357.5(3)	3	X	393.6(3)	3	X	430.3(3)
M_2	3	X	247.4(3)	3	X	279.8(3)	3	X	312.4(4)	3	X	344.2(4)
M_3	3	X	238.5(3)	3	X	269.1(3)	3	X	300.3(4)	3	X	330.5(4)
M_4	3	X	111.8(3)	3	X	135.0(3)	3	X	159.6(3)	3	X	182.4(3)
M_5	3	X	110.3(3)	3	X	133.1(3)	3	X	157.4(3)	3	X	180.0(3)
N_1	3	X	29.3(3)	3	X	37.7(3)	3	X	45.4(3)	3	X	51.3(3)
N_2	3	X	14.8(4)	3	X	19.9(3)	3	X	25.6(4)	3	X	28.7(4)
N_3	3	X	14.0(3)	30	dl	20.0(1)	30	dl	21.1(5)	30	dl	24.6(5)
$N_{4,5}$			--			--	3	X	2.4(1.2)	3	X	3.0(3)
			41 Nb			42 Mo			43 Tc			44 Ru
K	3 4	X x	18985.6(4) (18987.)	3 4	X x	19999.5(3) (20004.)	3	X	21044.0(7) --	3 4	X x	22117.2(3) (22119.)
L_1	3 29	X pe	2697.7(3) (2697.7)	3 29	X pe	2865.5(3) (2866.0)	3 29	X pe	3042.5(4) (3042.5)	3 29	X pe	3224.0(3) (3224.3)
L_2	3 29	X pe	2464.7(3) (2464.7) --	3 29 4	X pe x	2625.1(3) (2624.5, 2627.)	3 29	X pe	2793.2(4) (2793.2) --	3 29 4	X pe x	2966.9(3) (2966.8, 2966.3)
L_3	3 29	X pe	2370.5(3) (2370.6) --	3 29 4	X pe x	2520.2(3) (2520.2, 2523.2)	3 29	X pe	2676.9(4) (2676.9) --	3 29 4	X pe x	2837.9(3) (2837.7, 2837.7)
M_1	3	X	468.4(3)	3	X	504.6(3)	31	in	544	3	X	585.0(3)
M_2	3	X	378.4(4)	3	X	409.7(4)	3	X	444.9(1.5)	3	X	482.8(3)
M_3	3	X	363.0(4)	3	X	392.3(3)	3	X	425.0(1.5)	3	X	460.6(3)

Level	Ref.	Meth.	41 Nb	Ref.	Meth.	42 Mo	Ref.	Meth.	43 Tc	Ref.	Meth.	44 Ru
M_4	3	X	207.4(3)	3	X	230.3(3)	3	X	256.4(5)	3	X	283.6(3)
M_5	3	X	204.6(3)	3	X	227.0(3)	3	X	252.9(4)	3	X	279.4(3)
N_1	3	X	58.1(3)	3	X	61.8(3)	31	in	68	3	X	74.9(3)
$N_{2,3}$	3 30	X dl	33.9(4) 28.7(5)	3 30	X dl	34.8(4) 33.6(5)	3	X	38.9(1.9) --	3	X	43.1(4) --
$N_{4,5}$	3	X	3.2(3)	3	X	1.8(3)	31	in	2	3	X	2.0(3)

Level	Ref.	Meth.	45 Rh	Ref.	Meth.	46 Pd	Ref.	Meth.	47 Ag	Ref.	Meth.	48 Cd
K	3 4	X x	23219.9(3) (23219.8)	3 4	X x	24350.3(3) (24348.)	3 4	X x	25514.0(3) (25516.)	3 4	X x	26711.2(3) (26716.)
L_1	3 29 4	X pe x	3411.9(3) (3412.0, 3417.)	3 29 4	X pe x	3604.3(3) (3604.6, 3607.)	3 32 4	X pe x	3805.8(3) (3806.2, 3807.)	3 32 4	X pe x	4018.0(3) (4018.1, 4019.)
L_2	3 29 4	X pe x	3146.1(3) (3146.3, 3145.)	3 29 4	X pe x	3330.3(3) (3330.3, 3330.3)	3 21,32 4	X pe x	3523.7(3) (3523.6, 3526.)	3 32 4	X pe x	3727.0(3) (3727.1, 3728.)
L_3	3 29 4	X pe x	3003.8(3) (3003.5, 3002.)	3 21,29 4	X pe x	3173.3(3) (3173.0, 3173.0)	3 21 4	X pe x	3351.1(3) (3350.8, 3351.0)	3 21 4	X pe x	3537.5(3) (3537.3, 3537.6)
M_1	3	X	627.1(3)	3	X	669.9(3)	3	X	717.5(3)	3	X	770.2(3)
M_2	3	X	521.0(3)	3	X	559.1(3)	3	X	602.4(3)	3	X	650.7(3)
M_3	3	X	496.2(3)	3	X	531.5(3)	3	X	571.4(3)	3	X	616.5(3)
M_4	3	X	311.7(3)	3	X	340.0(3)	3	X	372.8(3)	3	X	410.5(3)
M_5	3	X	307.0(3)	3	X	334.7(3)	3	X	366.7(3)	3	X	403.7(3)
N_1	3	X	81.0(3)	3	X	86.4(3)	3	X	95.2(3)	3	X	107.6(3)
N_2	3 30	X dl	47.9(4) 48.5	3 30	X dl	51.1(4) 52.0	3	X	62.6(3)	3	X	66.9(4)
N_3							3	X	55.9(3)			
$N_{4,5}$	3	X	2.5(4)	3	X	1.5(3)	3	X	3.3(3)	3	X	9.3(3)
O_1			--			--			--			--
$O_{2,3}$			--			--			--	4	x	2.2(1.4)

Level	Ref.	Meth.	49 In	Ref.	Meth.	50 Pd	Ref.	Meth.	51 Sb	Ref.	Meth.	52 Te
K	3	X	27939.9(3) --	3 4	X x	29200.1(4) (29195.)	3 4	X x	30491.2(3) (30486.)	3 4	X x	31813.8(3) (31811.)
L_1	3 32 4	X pe x	4237.5(3) (4237.7, 4237.3)	3 21 4	X pe x	4464.7(3) (4464.5, 4464.8)	3 32 4	X pe x	4698.3(3) (4698.3, 4698.4)	3 32 4	X pe x	4939.2(3) (4939.3, 4939.7)
L_2	3 32 4	X pe x	3938.0(3) (3937.8, 3939.3)	3 21 4	X pe x	4156.1(3) (4156.2, 4157.)	3 32 4	X pe x	4380.4(3) (4380.6, 4382.)	3 32 4	X pe x	4612.0(3) (4612.0, 4612.6)
L_3	3 32 4	X pe x	3730.1(3) (3730.0, 3730.2)	3 21 4	X pe x	3928.8(3) (3928.8, 3928.8)	3 32 4	X pe x	4132.2(3) (4132.2, 4132.3)	3 32 4	X pe x	4341.4(3) (4341.2, 4341.8)

Level	Ref.	Meth.	49 In	Ref.	Meth.	50 Pd	Ref.	Meth.	51 Sb	Ref.	Meth.	52 Te
M_1	3	X	825.6(3)	3	X	883.8(3)	3	X	943.7(3)	3	X	1006.0(3)
M_2	3	X	702.2(3)	3	X	756.4(4)	3	X	811.9(3)	3	X	869.7(3)
M_3	3	X	664.3(3)	3	X	714.4(3)	3	X	765.6(3)	3	X	818.7(3)
M_4	3	X	450.8(3)	3	X	493.3(3)	3	X	536.9(3)	3	X	582.5(3)
M_5	3	X	443.1(3)	3	X	484.8(3)	3	X	527.5(3)	3	X	572.1(3)
N_1	3	X	121.9(3)	3	X	136.5(4)	3	X	152.0(3)	3	X	168.3(3)
$N_{2,3}$	3	X	77.4(4)	3	X	88.6(4)	3	X	98.4(5)	3	X	110.2(5)
$N_{4,5}$	3	X	16.2(3)	3	X	23.9(3)	3	X	31.4(3)	3	X	39.8(3)
O_1	3	X	0.1(4.5)	3	X	0.9(5)	3	X	6.7(5)	3	X	11.6(6)
$O_{2,3}$	3	X	0.8(4)	3	X	1.1(5)	3	X	2.1(4)	3	X	2.3(5)
			53 I			54 Xe(gas)			55 Cs			56 Ba
K	3 4	X x	33169.4(4) (33167.)	3 4	X x	34561.4(1.1) (34590.)	3 4	X x	35984.6(4) (35987.)	3 4	X x	37440.6(4) (37452.)
L_1	3 24	X pe	5188.1(3) (5188.1) --	3 4	X x	5452.8(4) (5452.8) --	3 24 4	X pe x	5714.3(4) (5712.7, 5721.)	3 24 4	X pe x	5988.8(4) (5986.8, 5996.)
L_2	3 24	X pe	4852.1(3) (4852.0) --	3 4	X x	5103.7(4) (5103.7) --	3 24 4	X pe x	5359.4(3) (5359.5, 5358.)	3 24 4	X pe x	5623.6(3) (5623.6, 5623.3)
L_3	3 24	X pe	4557.1(3) (4557.1) --	3 4	X x	4782.2(4) (4782.2) --	3 24 4	X pe x	5011.9(3) (5012.0, 5011.3)	3 24 4	X pe x	5247.0(3) (5247.3, 5247.0)
M_1	3	X	1072.1(3)	31	in	1145	3	X	1217.1(4)	3	X	1292.8(4)
M_2	3	X	930.5(3)	3	X	999.0(2.1)	3	X	1065.0(5)	3	X	1136.7(5)
M_3	3	X	874.6(3)	3	X	937.0(2.1)	3	X	997.6(5)	3	X	1062.2(5)
M_4	3	X	631.3(3)	18 (17)	pe	690.6(7)	3	X	739.5(4)	3	X	796.1(3)
M_5	3	X	619.4(3) --	18 3 (17)	pe X	677.6(6) 672.3(5)	3	X	725.5(5) --	3	X	780.7(3) --
N_1	3	X	186.4(3)	10	x	207.4	3	X	230.8(4)	3	X	253.0(5)
N_2 }	3	X	122.7(5)	3	X	146.7(3.1)	3	X	172.3(6)	3	X	191.8(7)
N_3 }							3	X	161.6(6)	3	X	179.7(6)
N_4 }	3	X	49.6(3)	28 10	x x	69.52 69.52	3	X	78.8(5) --	3	X	92.5(5) --
N_5 }				28 10	x x	67.55 67.54	3	X	76.5(5) --	3	X	89.9(5) --
O_1	3	X	13.6(6) --	12 10	x x	23.39 23.40	3	X	22.7(5) --	3	X	29.1(6)[a)] --
O_2 }	3	X	3.3(5)	12 10	x x	13.43 13.44	3	X	13.1(5) --	3	X	16.6(5) --
O_3 }				12 10	x x	12.13 12.13	3	X	11.4(5) --	3	X	14.6(5) --

Level	Ref.	Meth.	57 La	Ref.	Meth.	58 Ce	Ref.	Meth.	59 Pr	Ref.	Meth.	60 Nd
K	3 4	X x	38924.6(4) (38934.)	3 4	X x	40443.0(4) (40453.)	3 4	X x	41990.6(5) (42002.)	3 4	X x	43568.9(4) (43574.)
L_1	3 33	X pe	6266.3(5) (6266.3) --	3 33	X pe	6548.8(5) (6548.5) --	3 33	X pe	6834.8(5) (6834.9) --	3 33 4	X pe x	7126.0(4) (7125.8, 7129.)
L_2	3 33	X pe	5890.6(4) (5890.7) --	3 33	X pe	6164.2(4) (6164.3) --	3 33	X pe	6440.4(5) (6440.2) --	3 33 4	X pe x	6721.5(4) (6721.8, 6723.)
L_3	3 33	X pe	5482.7(4) (5482.6) --	3 33	X pe	5723.4(4) (5723.6) --	3 33	X pe	5964.3(4) (5964.3) --	3 33 4	X pe x	6207.9(4) (6208.0, 6209.)
M_1	3	X	1361.3(3)	3	X	1434.6(6)	3	X	1511.0(8)	3	X	1575.3(7)
M_2	3	X	1204.4(6)	3	X	1272.8(6)	3	X	1337.4(7)	3	X	1402.8(6)
M_3	3	X	1123.4(5)	3	X	1185.4(5)	3	X	1242.2(6)	3	X	1297.4(5)
M_4	3	X	848.5(4)	3	X	901.3(6)	3	X	951.1(6)	3	X	999.9(6)
M_5	3	X	831.7(4)	3	X	883.3(5)	3	X	931.0(6)	3	X	977.7(6)
N_1	3	X	270.4(8)	3	X	289.6(7)	3	X	304.5(9)	3	X	315.2(8)
N_2	3	X	205.8(1.2)	3	X	223.3(1.1)	3	X	236.3(1.5)	3	X	243.3(1.6)
N_3	3	X	191.4(9)	3	X	207.2(9)	3	X	217.6(1.1)	3	X	224.6(1.3)
$N_{4,5}$	3	X	98.9(8)	3	X	110.0(6)	3	X	113.2(7)	3	X	117.5(7)
$N_{6,7}$			--	3	X	0.1(1.2)	3	X	2.0(6)	3	X	1.5(9)
O_1	3	X	32.3(7.2)	3	X	37.8(1.3)	3	X	37.4(1.0)	3	X	37.5(9)
$O_{2,3}$	3	X	14.4(1.2)	3	X	19.8(1.2)	3	X	22.3(7)	3	X	21.1(8)

Level	Ref.	Meth.	61 Pm	Ref.	Meth.	62 Sm	Ref.	Meth.	63 Eu	Ref.	Meth.	64 Gd
K	3 4	X x	45184.0(7) (45198.)	3 4	X x	46834.2(5) (46849.)	3 4	X x	48519.0(4) (48519.)	3 4	X x	50239.1(5) (50233.)
L_1	3 34	X pe	7427.9(8) (7427.9) --	3 33 4	X pe x	7736.8(5) (7736.2, 7748.)	3 33 4	X pe x	8052.0(4) (8051.7, 8061.)	3 33 4	X pe x	8375.6(5) (8375.4, 8386.)
L_2	3 34	X pe	7012.8(6) (7012.8) --	3 33 4	X pe x	7311.8(4) (7312.0, 7313.)	3 33 4	X pe x	7617.1(4) (7617.6, 7620.)	3 33 4	X pe x	7930.3(4) (7930.5, 7931.)
L_3	3 34	X pe	6459.3(6) (6459.4) --	3 33 4	X pe x	6716.2(5) (6716.8, 6717.)	3 33 4	X pe X	6976.9(4) (6976.7, 6981.)	3 33 4	X pe x	7242.8(4) (7242.8, 7243.)
M_1	31	in	1650	3	X	1722.8(8)	3	X	1800.0(5)	3	X	1880.8(5)
M_2	3	X	1471.4(6.2)	3	X	1540.7(1.2)	3	X	1613.9(7)	3	X	1688.3(7)
M_3	3	X	1356.9(1.4)	3	X	1419.8(1.1)	3	X	1480.6(6)	3	X	1544.0(8)
M_4	3	X	1051.5(9)	3	X	1106.0(8)	3	X	1160.6(6)	3	X	1217.2(6)
M_5	3	X	1026.9(1.0)	3	X	1080.2(6)	3	X	1130.9(6)	3	X	1185.2(6)

Level	Ref.	Meth.	61 Pm	Ref.	Meth.	62 Sm	Ref.	Meth.	63 Eu	Ref.	Meth.	64 Gd
N_1	31	in	331	3	X	345.7(9)	3	X	360.2(7)	3	X	375.8(7)
N_2 }	3	X	242(16)	3	X	265.6(1.9)	3	X	283.9(1.0)	3	X	288.5(1.2)
N_3 }				3	X	247.4(1.5)	3	X	256.6(8)	3	X	270.9(9)
$N_{4,5}$	3	X	120.4(2.0)	3	X	129.0(1.2)	3	X	133.2(6)	3	X	140.5(8)
$N_{6,7}$	31	in	4	3	X	5.5(1.1)	3	X	0.0(3.2)	3	X	0.1(3.5)
O_1	31	in	38	3	X	37.4(1.5)	3	X	31.8(7)	3	X	36.1(8)
$O_{2,3}$	31	in	22 --	3 30	X dl	21.3(1.5) 20(1)	3	X	22.0(6) --	3	X	20.3(1.2) --

Level	Ref.	Meth.	65 Tb	Ref.	Meth.	66 Dy	Ref.	Meth.	67 Ho	Ref.	Meth.	68 Er
K	3 4	X x	51995.7(5) (52002.)	3 4	X x	53788.5(5) (53793.)	3 4	X x	55617.7(5) (55619.)	3 4	X x	57485.5(5) (57487.)
L_1	3 33 4	X pe x	8708.0(5) (8707.6, 8717.)	3 33	X pe	9045.8(5) (9046.5) --	3 33 4	X pe x	9394.2(4) (9394.3, 9399.)	3 33 4	X pe x	9751.3(4) (9751.5, 9757.)
L_2	3 33 4	X pe x	8251.6(4) (8251.8, 8253.)	3 33 4	X pe x	8580.6(4) (8580.4, 8583.)	3 33 4	X pe x	8917.8(4) (8918.2, 8916.)	3 33 4	X pe x	9264.3(4) (9264.3, 9262.)
L_3	3 33 4	X pe x	7514.0(4) (7514.2, 7515.)	3 33 4	X pe x	7790.1(4) (7789.6, 7789.7)	3 33 4	X pe x	8071.1(4) (8070.6, 8068.)	3 33 4	X pe x	8357.9(4) (8357.6, 8357.5)
M_1	3	X	1967.5(6)	3	X	2046.8(4)	3	X	2128.3(6)	3	X	2206.5(6)
M_2	3	X	1767.7(9)	3	X	1841.8(5)	3	X	1922.8(1.0)	3	X	2005.8(6)
M_3	3	X	1611.3(8)	3	X	1675.6(9)	3	X	1741.2(9)	3	X	1811.8(6)
M_4	3	X	1275.0(6)	3	X	1332.5(4)	3	X	1391.5(7)	3	X	1453.3(5)
M_5	3	X	1241.2(7)	3	X	1294.9(4)	3	X	1351.4(8)	3	X	1409.3(5)
N_1	3	X	397.9(8)	3	X	416.3(5)	3	X	435.7(8)	3	X	449.1(1.0)
N_2	3	X	310.2(1.2)	3	X	331.8(6)	3	X	343.5(1.4)	3	X	366.2(1.5)
N_3	3	X	285.0(1.0)[a)]	3	X	292.9(6)	3	X	306.6(9)	3	X	320.0(7)
N_4 }	3	X	147.0(8)	3	X	154.2(5)	3	X	161.0(1.0)	3	X	176.7(1.2)
N_5 }										3	X	167.6(1.5)
$N_{6,7}$	3	X	2.6(1.5)	3	X	4.2(1.6)	3	X	3.7(3.0)	3	X	4.3(1.4)
O_1	3	X	39.0(8)	3	X	62.9(5)	3	X	51.2(1.3)	3	X	59.8(1.7)
$O_{2,3}$	3 30	X dl	25.4(8) 20(1)	3 30	X dl	26.3(6) 22(1)	3 30	X dl	20.3(1.5) 22(1)	3 30	X dl	29.4(1.6) 21(1)

Level	Ref.	Meth.	69 Tm	Ref.	Meth.	70 Yb	Ref.	Meth.	71 Lu	Ref.	Meth.	72 Hf
K	3	X	59389.6(5) --	3 4	X x	61332.3(5) (61300.)	3 4	X x	63313.8(5) (63310.)	3 4	X x	65350.8(6) (65310)
L_1	3 33 4	X pe x	10115.7(4) (10115.6, 10121.)	3 33 4	X pe x	10486.4(4) (10487.3, 10490.)	3 33 4	X pe x	10870.4(4) (10870.1, 10874.)	3 34 4	X pe x	11270.7(4) (11271.6, 11274.)

Level	Ref.	Meth.	69 Tm	Ref.	Meth.	70 Yb	Ref.	Meth.	71 Lu	Ref.	Meth.	72 Hf
L_2	3 33 4	X pe x	9616.9(4) (9617.1, 9617.1)	3 33 4	X pe x	9978.2(4) (9977.9, 9976.)	3 33 4	X pe x	10348.6(4) (10349.0, 10345.)	3 34 4	X pe x	10739.4(4) (10738.9, 10736.)
L_3	3 33 4	X pe x	8648.0(4) (8647.8, 8649.6)	3 33 4	X pe x	8943.6(4) (8942.6, 8944.1)	3 33	X pe	9244.1(4) (9243.8) --	3 34 4	X pe x	9560.7(4) (9560.4, 9558.)
M_1	3	X	2306.8(7)	3	X	2398.1(4)	3	X	2491.2(5)	3	X	2600.9(4)
M_2	3	X	2089.8(7)	3	X	2173.0(4)	3	X	2263.5(4)	3	X	2365.4(4)
M_3	3	X	1884.5(1.1)	3	X	1949.8(5)	3	X	2023.6(5)	3	X	2107.6(4)
M_4	3	X	1514.6(7)	3	X	1576.3(4)	3	X	1639.4(4)	3	X	1716.4(4)
M_5	3	X	1467.7(9)	3	X	1527.8(4)	3	X	1588.5(4)	3	X	1661.7(4)
N_1	3	X	471.7(9)	3	X	487.2(6)	3	X	506.2(6)	3	X	538.1(4)
N_2	3	X	385.9(1.6)	3	X	396.7(7)	3	X	410.1(1.8)	3	X	437.0(5)
N_3	3	X	336.6(1.6)	3	X	343.5(5)	3	X	359.3(5)	3	X	380.4(5)
N_4 }	3	X	179.6(1.2)	3	X	198.1(5)	3	X	204.8(5)	3	X	223.8(4)
N_5 }				3	X	184.9(1.3)	3	X	195.0(4)	3	X	213.7(5)
$N_{6,7}$	3	X	5.3(1.9)	3	X	6.3(1.0)	3	X	6.9(5)	3	X	17.1(5)
O_1	3	X	53.2(3.0)	3	X	54.1(5)	3	X	56.8(5)	3	X	64.9(4)
O_2 }	3	X	32.3(1.6)	3	X	23.4(6)	3	X	28.0(6)	3 30	X dl	38.1(6) 38.3(5)
O_3 }										3 30	X dl	30.6(6) 27.7(5)
$O_{4,5}$			--			--	4	X	4.6(6)	4	X	6.6(4)

Level	Ref.	Meth.	73 Ta	Ref.	Meth.	74 W	Ref.	Meth.	75 Re	Ref.	Meth.	76 Os
K	3 4	X x	67416.4(6) (67403.)	3 4	X x	69525.0(3) (69508.)	3 4	X x	71676.4(4) (71658.)	3	X	73870.8(5) --
L_1	3 35 4	X pe x	11681.5(3) (11680.2, 11682.)	3 35 4	X pe x	12099.8(3) (12098.2, 12099.6)	3 4	X x	12526.7(4) (12530.) --	3 4	X x	12968.0(4) (12972.) --
L_2	3 35 4	X pe x	11136.1(3) (11136.1, 11132.)	3 35 4	X pe x	11544.0(3) (11541.4, 11538.)	3 35 4	X pe x	11958.7(3) (11956.9, 11954.)	3 4	X x	12385.0(4) (12381.) --
L_3	3 35 4	X pe x	9881.1(3) (9880.3, 9877.7)	3 35 4	X pe x	10206.8(3) (10204.2, 10200.)	3 35 4	X pe x	10535.3(3) (10534.2, 10531.)	3 35 4	X pe x	10870.9(3) (10870.7, 10868.)
M_1	3	X	2708.0(4)	3	X	2819.6(4)	3	X	2931.7(4)	3	X	3048.5(4)
M_2	3 35	X pe	2468.7(3) (2468.6)	3 35	X pe	2574.9(3) (2575.0)	3	X	2681.6(4) --	3 35	X pe	2792.2(3) (2791.9)
M_3	3 35	X pe	2194.0(3) (2194.1)	3 35	X pe	2281.0(3) (2281.0)	3 35	X pe	2367.3(3) (2367.3)	3 35	X pe	2457.2(4) (2457.4)

Level	Ref.	Meth.	73 Ta	Ref.	Meth.	74 W	Ref.	Meth.	75 Re	Ref.	Meth.	76 Os
M_4	3 35	X pe	1793.2(3) (1793.1)	3 35	X pe	1871.6(3) (1871.4)	3 35	X pe	1948.9(3) (1948.9)	3 35	X pe	2030.8(3) (2031.0)
M_5	3 35	X pe	1735.1(3) (1735.2)	3 35	X pe	1809.2(3) (1809.3)	3 35	X pe	1882.9(3) (1882.9)	3 35	X pe	1960.1(3) (1960.2)
N_1	3	X	565.5(5)	3	X	595.0(4)	3	X	625.0(4)	3	X	654.3(5)
N_2	3	X	464.8(5)	3	X	491.6(4)	3	X	517.9(5)	3	X	546.5(5)
N_3	3	X	404.5(4)	3	X	425.3(5)	3	X	444.4(5)	3	X	468.2(6)
N_4	3	X	241.3(4)	3	X	258.8(4)	3	X	273.7(5)	3	X	289.4(5)
N_5	3	X	229.3(3)	3	X	245.4(4)	3	X	260.2(4)	3	X	272.8(6)
N_6	36	pe	27.5(2) --	36 3	pe X	37.9(2) 36.5(4)	36	pe	48.1(2) --	36	pe	53.8(2) --
N_7	36	pe	25.6(2) --	36 3	pe X	35.8(2) 33.6(4)	36	pe	45.7(2) --	36	pe	51.0(2) --
O_1	3	X	71.1(5)	3	X	77.1(4)	3	X	82.8(5)	3	X	83.7(6)
O_2	3 30	X dl	44.9(4) 42.1(5)	3 30	X dl	46.8(5) 46.5(5)	3 30	X dl	45.6(7) 50.4(5)	3	X	58.0(1.1) --
O_3	3 36 30	X pe dl	36.4(4) 37(1) 31.4(5)	3 36 30	X pe dl	35.6(5) 42(1) 34.7(5)	3 30	X dl	34.6(6) -- 38.7(5)	3	X	45.4(1.0) -- --
$O_{4,5}$	3	X	5.7(4)	3	X	6.1(4)	3	X	3.5(5)	4	x	0.0(7)

Level	Ref.	Meth.	77 Ir	Ref.	Meth.	78 Pt	Ref.	Meth.	79 Au	Ref.	Meth.	80 Hg
K	3	X	76111.0(5) --	3 4	X x	78394.8(7) (78381.)	3 4	X x	80724.9(5) (80720.)	3	X	83102.3(8) --
L_1	3 4	X x	13418.5(3) (13423.) --	37 3 4	pe X x	13880.5(6) 13879.9(4) 13883.	3 4	X x	14352.8(4) (14353.7) --	3 4	X x	14839.3(1.0) (14842.) --
L_2	3 20,35 4	X pe x	12824.1(3) (12824.0, 12820.) --	37 3 20,35 4	pe X pe x	13272.6(7) 13272.6(3) 13272.6 13272.3	3 20,35 4	X pe x	13733.6(3) (13733.5, 13736.) --	3 4	X x	14208.7(7) (14215.) -- --
L_3	3 20,35 4	X pe x	11215.2(3) (11215.1, 11212.) --	37 3 20,35 4	pe X pe x	11563.8(6) 11563.7(3) 11563.7 11562.	3 20,35 4	X pe x	11918.7(3) (11918.2, 11921.) --	3 20,35 4	X pe x	12283.9(4) (12284.0, 12286.) --
M_1	3	X	3173.7(1.7) --	37 3	pe X	3297.6(6) 3296.0(9)	3 35	X pe	3424.9(3) (3424.8)	3	X	3561.6(1.1) --
M_2	3 35	X pe	2908.7(3) (2909.1) -- --	37 3 35 4	pe X pe x	3027.0(6) 3026.5(4) 3026.5 3029.	3 35	X pe	3147.8(4) (3149.5) -- --	3	X	3278.5(1.3) -- -- --
M_3	3 35 4	X pe x	2550.7(3) (2550.5, 2550.5) --	37 3 35 4	pe X pe x	2645.3(6) 2645.4(4) 2645.5 2645.9	3 35 4	X pe x	2743.0(3) (2743.1, 2744.0) --	3 35	X pe	2847.1(4) (2847.1) -- --

Level	Ref.	Meth.	77 Ir	Ref.	Meth.	78 Pt	Ref.	Meth.	79 Au	Ref.	Meth.	80 Hg
M_4	3	X	2116.1(3)	37	pe	2201.5(4)	3	X	2291.1(3)	3	X	2384.9(3)
	35	pe	(2116.1)	3	X	2201.9(3)	35	pe	(2291.2,	35	pe	(2384.9)
			--	35	pe	2201.9	4	x	2307.)			--
M_5	3	X	2040.4(3)	37	pe	2121.1(4)	3	X	2205.7(3)	3	X	2294.9(3)
	35	pe	(2040.5)	3	X	2121.6(3)	35	pe	(2206.1,	35	pe	(2294.9)
			--	35	pe	2121.6	4	x	2220.)			--
N_1	3	X	690.1(4)	37	pe	724.0(1.0)	3	X	758.8(4)	3	X	800.3(1.0)
			--	3	X	722.0(6)			--			--
N_2	3	X	577.1(4)	37	pe	607.6(6)	3	X	643.7(5)	3	X	676.9(2.4)
			--	3	X	609.2(6)			--			--
N_3	3	X	494.3(6)	37	pe	519.1(6)	3	X	545.4(5)	3	X	571.0(1.4)
			--	3	X	519.0(6)			--			--
N_4	3	X	311.4(4)	37	pe	330.7(6)	3	X	352.0(4)	3	X	378.3(1.0)
			--	3	X	330.8(5)			--			--
N_5	3	X	294.9(4)	37	pe	313.8(6)	3	X	333.9(4)	3	X	359.8(1.2)
			--	3	X	313.3(4)			--			--
N_6	36	pe	64.0(2)	36	pe	74.5(2)	36	pe	87.8(2)	36	pe	103.8(2)
	3	X	63.4(4)	37	pe	73.6(6)	3	X	86.4(4)	3	X	102.2(5)
			--	3	X	74.3(4)			--			--
N_7	36	pe	61.0(2)	36	pe	71.1(2)	36	pe	84.1(2)	36	pe	99.8(2)
	3	X	60.5(4)	37	pe	70.3(6)	3	X	82.8(5)	3	X	98.5(5)
			--	3	X	71.1(5)			--			--
O_1	3	X	95.2(4)	3	X	101.7(4)	3	X	107.8(7)	3	X	120.3(1.3)
O_2	3	X	63.0(6)	3	X	65.3(7)	3	X	71.7(7)	3	X	80.5(1.3)
O_3	3	X	50.5(6)	37	pe	51.0(7)	30	dl	58.7	36	pe	65(1)
	30	dl	48.8	36	pe	52(1)	36	pe	58(1)	3	X	57.6(1.3)
			--	30	dl	51.7	3	X	53.7(7)			--
			--	3	X	51.7(7)			--			--
$O_{4,5}$	3	X	3.8(4)	37	pe	2.1(7)	3	X	2.5(5)	3	X	6.4(1.4)
			--	3	X	2.2(1.3)			--			--

Level	Ref.	Meth.	81 Tl	Ref.	Meth.	82 Pb	Ref.	Meth.	83 Bi	Ref.	Meth.	84 Po
K	3	X	85530.4(6)	3	X	88004.5(7)	3	X	90525.9(7)	3	X	93105.0(3.8)
			--	4	x	(88005.)	4	x	(90534.)			--
L_1	3	X	15346.7(4)	3	X	15860.8(5)	3	X	16387.5(4)	3	X	16939.3(9.8)
	4	x	(15343.)	4	x	(15855.)	4	x	(16376.)			--
L_2	3	X	14697.9(3)	3	X	15200.0(4)	3	X	15711.1(3)	3	X	16244.3(2.4)
	35	pe	(14697.3,	4	x	(15205.)	35	pe	(15708.4,			--
	4	x	14699.)			--	4	x	15719.)			--
L_3	3	X	12657.5(3)	3	X	13035.2(3)	3	X	13418.6(3)	3	X	13813.8(1.0)
	20,35	pe	(12656.3,	20,35	pe	(13034.9,	20,35	pe	(13818.3,	3	in	(13813.8)
	4	x	12660.)	4	x	13041.)	4	x	13426.)			--
M_1	3	X	3704.1(4)	3	X	3850.7(5)	3	X	3999.1(3)	3	X	4149.4(3.9)
			--			--	35	pe	(3999.1)			--
M_2	3	X	3415.7(3)	3	X	3554.2(3)	3	X	3696.3(3)	3	X	3854.1(9.8)
	35	pe	(3415.7)	35	pe	(3554.2)	35	pe	(3696.4)			--

Level	Ref.	Meth.	81 Tl	Ref.	Meth.	82 Pb	Ref.	Meth.	83 Bi	Ref.	Meth.	84 Po
M_3	3 35	X pe	2956.6(3) (2956.5)	3 35	X pe	3066.4(4) (3066.3)	3 35	X pe	3176.9(3) (3176.8)	3	X	3301.9(9.9) --
M_4	3 35	X pe	2485.1(3) (2485.2) --	3 35 4	X pe x	2585.6(3) (2585.5, 2606.)	3 35	X pe	2687.6(3) (2687.4) --	3	X	2798.0(1.2) -- --
M_5	3 35	X pe	2389.3(3) (2389.4) --	3 35 4	X pe x	2484.0(3) (2484.2, 2502.)	3 35	X pe	2579.6(3) (2579.5) --	3	X	2683.0(1.1) -- --
N_1	3	X	845.5(5) --	3	X	893.6(7) --	3 35	X pe	938.2(3) (938.7)	3	X	995.3(2.9) --
N_2	3	X	721.3(8) --	3	X	763.9(8) --	3 35	X pe	805.3(3) (805.3)	3	X	851(12) --
N_3	3	X	609.0(5) --	3	X	644.5(6) --	3 35	X pe	678.9(3) (678.9)	3	X	705(14) --
N_4	3	X	406.6(4) --	3	X	435.2(5) --	3 35	X pe	463.6(3) (463.6)	3	X	500.2(2.4) --
N_5	3	X	386.2(5) --	3	X	412.9(6) --	3 35	X pe	440.0(3) (440.1)	3	X	473.4(1.3) --
N_6	36 3	pe X	123.1(2) 122.8(4)	36 3	pe X	141.2(2) 142.9(4)	36 3	pe X	162.4(2) 161.9(5)	31	in	184
N_7	36 3	pe X	118.8(2) 118.5(4)	36 3	pe X	136.3(2) 138.1(4)	36 3	pe X	157.1(2) 157.4(6)			
O_1	3	X	136.3(7)	3	X	147.3(8)	3	X	159.3(7)	31	in	177
O_2	3	X	99.6(6)	3	X	104.8(1.0)	3	X	116.8(7)	31	in	132
O_3	36 3	pe X	73(1) 75.4(6)	36 3	pe X	83(1) 86.0(1.0)	36 3	pe X	93(1) 92.8(6)	31	in	104 --
O_4	3	X	15.3(4)	3	X	21.8(4)	3	X	26.5(5)	3	X	31.4(3.2)
O_5	3	X	13.1(4)	3	X	19.2(4)	3	X	24.4(6)			
P_1			--	3	X	3.1(1.0)	31	in	8	31	in	12
$P_{2,3}$			--	3	X	0.7(1.0)	3	X	2.7(7)	31	in	5

			85 At			86 Rn(gas)			87 Fr			88 Ra
K	3	X	95729.9(7.7)	3	X	98404(12)	3	X	101137(13)	3	X	103921.9(7.2)
L_1	3	X	17493(29) --	3	X	18049(38) --	3	X	18639(40) --	3 4	X x	19236.7(1.5) (19236.0)
L_2	3	X	16784.7(2.5) --	3	X	17337.1(3.4) --	3	X	17906.5(3.5) --	3 4	X x	18484.3(1.5) (18486.0)
L_3	3 3	X in	14213.5(2.0) (14213.5)	3 3	X in	14619.4(3.0) (14619.4)	3 3	X in	15031.2(3.0) (15031.2)	3 4	X x	15444.4(1.5) (15444.0)
M_1	3	in	4317	3	in	4482	3	in	4652	3	X	4822.0(1.5)
M_2	3	X	4008(28)	3	X	4159(38)	3	X	4327(40)	3	X	4489.5(1.8)
M_3	3	X	3426(29)	3	X	3538(38)	3	X	3663(40)	3	X	3791.8(1.7)

Level	Ref.	Meth.	85 At	Ref.	Meth.	86 Rn(gas)	Ref.	Meth.	87 Fr	Ref.	Meth.	88 Ra
M_4	3	X	2908.7(2.1)	3	X	3021.5(3.1)	3	X	3136.2(3.1)	3	X	3248.4(1.6)
M_5	3	X	2786.7(2.1)	3	X	2892.4(3.1)	3	X	2999.9(3.1)	3	X	3104.9(1.6)
N_1	3	in	1042	3	in	1097	3	in	1153	3	X	1208.4(1.6)
N_2	3	X	886(30)	3	X	929(40)	3	X	980(42)	3	X	1057.6(1.8)
N_3	3	X	740(30)	3	X	768(40)	3	X	810(43)	3	X	879.1(1.8)
N_4	3	X	533.2(3.2)	3	X	566.6(4.0)	3	X	603.3(4.1)	3	X	635.9(1.6)
N_5	31	in	507	31	in	541	3	X	577(34)	3	X	602.7(1.7)
$N_{6,7}$	31	in	210	31	in	238	31	in	268	3	X	298.9(2.4)
O_1	31	in	195	31	in	214	31	in	234	3	X	254.4(2.1)
O_2	31	in	148	31	in	164	31	in	182	3	X	200.4(2.0)
O_3	31	in	115	31	in	127	31	in	140	3	X	152.8(2.0)
$O_{4,5}$	31	in	40	31	in	48	31	in	58	3	X	67.2(1.7)
P_1	31	in	18	31	in	26	31	in	34	3	X	43.5(2.2)
$P_{2,3}$	31	in	8	31	in	11	31	in	15	3	X	18.8(1.8)
			89 Ac			90 Th			91 Pa			92 U
K	3	X	106755.3(5.3)	3	X	109650.9(9)	3	X	112601.4(2.4)	3	X	115606.1(1.6)
L_1	3	X	19840(18) --	3 4	X x	20472.1(5) (20464.)	3 4	X x	21104.6(1.8) (21128.)	3 4	X x	21757.4(3) (21771.)
L_2	3	X	19083.2(2.8) --	3 4	X x	19693.2(4) (19683.)	3 4	X x	20313.7(1.5) (20319.)	3 4	X x	20947.6(3) (20945.)
L_3	3 3	X in	15871.0(2.0) (15871.0) --	3 38 4	X pe x	16300.3(3) (16299.6, 16299.)	3 4	X x	16733.1(1.4) (16733.) --	3 39 4	X pe x	17166.3(3) (17168.5, 17165.)
M_1	3	in	5002 --	3 38	X pe	5182.3(3) (5182.3)	3	X	5366.9(1.6) --	3	X	5548.0(4) --
M_2	3	X	4656(18) --	3 38	X pe	4830.4(4) (4830.6)	3	X	5000.9(2.3) --	3 39	X pe	5182.2(4) (5180.9)
M_3	3	X	3909(18) -- --	3 38 4	X pe x	4046.1(4) (4046.1, 4041.)	3	X	4173.8(1.8) -- --	3 39 4	X pe x	4303.4(3) (4303.6, 4299.)
M_4	3	X	3370.2(2.1) -- --	3 38 4	X pe x	3490.8(3) (3490.7, 3485.)	3 4	X x	3611.2(1.4) (3608.) --	3 39 4	X pe x	3727.6(3) (3728.1, 3720.)
M_5	3	X	3219.0(2.1) -- --	3 38 4	X pe x	3332.0(3) (3332.1, 3325.)	3 4	X x	3441.8(1.4) (3436.) --	3 39 4	X pe x	3551.7(3) (3551.7, 3545.)
N_1	3	in	1269 --	3 38	X pe	1329.5(4) (1329.8)	3	X	1387.1(1.9) --	3 39	X pe	1440.8(4) (1441.3)
N_2	3	X	1080(19) --	3 38	X pe	1168.2(4) (1168.3)	3	X	1224.3(1.6) --	3 39	X pe	1272.6(3) (1272.5)
N_3	3	X	890(19) --	3 38	X pe	967.3(4) (967.6)	3	X	1006.7(1.7) --	3 39	X pe	1044.9(3) (1044.9)

<table>
<tr><th>Level</th><th>Ref.</th><th>Meth.</th><th>89 Ac</th><th>Ref.</th><th>Meth.</th><th>90 Th</th><th>Ref.</th><th>Meth.</th><th>91 Pa</th><th>Ref.</th><th>Meth.</th><th>92 U</th></tr>
<tr><td>N_4</td><td>3</td><td>X</td><td>674.9(3.7)
--</td><td>3
38</td><td>X
pe</td><td>714.1(4)
(714.4)</td><td>3</td><td>X</td><td>743.4(2.1)
--</td><td>3
39</td><td>X
pe</td><td>780.4(3)
(779.7)</td></tr>
<tr><td>N_5</td><td>31</td><td>in</td><td>639
--</td><td>3
38</td><td>X
pe</td><td>676.4(4)
(676.4)</td><td>3</td><td>X</td><td>708.2(1.8)
--</td><td>3
39</td><td>X
pe</td><td>737.7(3)
(737.6)</td></tr>
<tr><td>N_6</td><td rowspan="2">31</td><td rowspan="2">in</td><td rowspan="2">319</td><td>3
38</td><td>X
pe</td><td>344.4(3)
(344.2)</td><td>3</td><td>X</td><td>371.2(1.6)
--</td><td>3</td><td>X</td><td>391.3(6)
--</td></tr>
<tr><td>N_7</td><td>3
38</td><td>X
pe</td><td>335.2(4)
(335.0)</td><td>3</td><td>X</td><td>359.5(1.6)
--</td><td>3</td><td>X</td><td>380.9(9)
--</td></tr>
<tr><td>O_1</td><td>31</td><td>in</td><td>272</td><td>3</td><td>X</td><td>290.2(8)</td><td>3</td><td>X</td><td>309.6(4.3)</td><td>3</td><td>X</td><td>323.7(1.1)</td></tr>
<tr><td>O_2</td><td>31</td><td>in</td><td>215</td><td>3</td><td>X</td><td>229.4(1.1)</td><td rowspan="2">3</td><td rowspan="2">X</td><td rowspan="2">222.9(3.9)</td><td>3</td><td>X</td><td>259.3(5)</td></tr>
<tr><td>O_3</td><td>31</td><td>in</td><td>167
--</td><td>3
38</td><td>X
pe</td><td>181.8(4)
(181.8)</td><td>3</td><td>X</td><td>195.1(1.3)
--</td></tr>
<tr><td>O_4</td><td rowspan="2">31</td><td rowspan="2">in</td><td rowspan="2">80</td><td>3
38</td><td>X
pe</td><td>94.3(4)
(94.4)</td><td rowspan="2">3</td><td rowspan="2">X</td><td rowspan="2">94.1(2.8)</td><td>3</td><td>X</td><td>105.0(5)
--</td></tr>
<tr><td>O_5</td><td>3
38</td><td>X
pe</td><td>87.9(3)
(88.1)</td><td>3</td><td>X</td><td>96.3(1.4)
--</td></tr>
<tr><td>P_1</td><td></td><td></td><td>--</td><td>3</td><td>X</td><td>59.5(1.1)</td><td></td><td></td><td>--</td><td>3</td><td>X</td><td>70.7(1.2)</td></tr>
<tr><td>P_2</td><td></td><td></td><td>--</td><td>3</td><td>X</td><td>49.0(2.5)</td><td></td><td></td><td>--</td><td>3</td><td>X</td><td>42.3(9.0)</td></tr>
<tr><td>P_3</td><td></td><td></td><td>--</td><td>3</td><td>X</td><td>43.0(2.5)</td><td></td><td></td><td>--</td><td>3</td><td>X</td><td>32.3(9.0)</td></tr>
</table>

Level	Ref.	Meth.	93 Np	Ref.	Meth.	94 Pu	Ref.	Meth.	95 Am	Ref.	Meth.	96 Cm
K	3 44	X pe	118678(33) 118663(31)	3	X	121818(44) --	3	X	125027(55) --	42	sd	128220 --
L_1	3 44	X pe	22426.8(9) 22431(13)	3 4	X x	23097.2(1.6) (23109.)	3 3	X in	23772.9(2.0) (23772.9)	42	sd	24460 --
L_2	3 44	X pe	21600.5(4) 21612(14)	3 4	X x	22266.2(7) (22253.)	3	X	22944.0(1.0) --	42	sd	23779 --
L_3	3 4 44	X x pe	17610.0(4) (17606.2) 17617(15)	3 4	X x	18056.8(6) (18053.1) --	3 4	X x	18504.1(9) (18504.1) --	42	sd	18930 -- --
M_1	3 44	X pe	5723.2(3.6) 5748(15)	3	X	5932.9(1.4) --	3	X	6120.5(7.5) --	42	sd	6288 --
M_2	3 40 44	X pe pe	5366.2(7) (5366.4) 5370(16)	3	X	5541.2(1.7) -- --	3	X	5710.2(2.1) -- --	42	sd	5895 -- --
M_3	3 40 44	X pe pe	4432.7(5) (4434.6) 4444(15)	3	X	4556.6(1.5) --	3	X	4667.0(2.1) --	42	sd	4797 --
M_4	3 40	X pe	3850.3(4) (3849.8)	3 41	X pe	3972.6(6) (3972.7)	3	X	4092.1(1.0) --	42	sd	4227 --
M_5	3 40 44	X pe pe	3665.8(4) (3664.2) 3674(24)	3 41	X pe	3778.1(6) (3778.0) --	3	X	3886.9(1.0) -- --	42	sd	3971 -- --

Level	Ref.	Meth.	93 Np	Ref.	Meth.	94 Pu	Ref.	Meth.	95 Am	Ref.	Meth.	96 Cm
N_1	3	X	1500.7(8)	3	X	1558.6(8)	3	X	1617.1(1.1)	42	sd	1643
	40	pe	(1500.7)			--			--			--
	44	pe	1509(16)			--			--			--
N_2	3	X	1327.7(8)	3	X	1372.1(1.8)	3	X	1411.8(8.3)	42	sd	1440
	40	pe	(1327.7)			--			--			--
	44	pe	1331(16)			--			--			--
N_3	3	X	1086.8(7)	3	X	1114.8(1.6)	3	in	1135.7	42	sd	1154
	40	pe	(1086.8)			--			--			--
	44	pe	1094(16)			--			--			--
N_4	3	X	815.9(5)	3	X	848.9(6)	3	X	878.7(1.0)			--
	40	pe	(817.1)	41	pe	(848.9)			--			--
N_5	3	X	770.3(4)	3	X	801.4(6)	3	X	827.6(1.0)			--
	40	pe	(773.2)	41	pe	(801.4)			--			--
N_6	3	X	415.0(8)	3	X	445.8(1.7)			--			--
	40	pe	(415.0)			--			--			--
N_7	3	X	404.4(5)	3	X	432.4(2.1)			--			--
	40	pe	(404.4)			--			--			--
O_1	31	in	338	3	X	351.9(2.4)	31	in	367	42	sd	385
O_2	3	X	283.4(8)	3	X	274.1(4.7)	31	ex	290			--
	40	pe	(283.4)			--			--			--
O_3	3	X	206.1(7)	3	X	206.5(4.7)	31	ex	220			--
	40	pe	(206.1)			--			--			--
O_4	3	X	109.3(7)	3	X	116.0(1.2)	3	X	115.8(1.3)			--
	40	pe	(108.8)			--			--			--
O_5	3	X	101.3(5)	3	X	105.4(1.0)	3	X	103.3(1.1)			--
	40	pe	(101.4)			--			--			--
			97 Bk			98 Cf			99 Es			100 Fm
K	43	ic	131590(60)	42	sd	135960	42	sd	139490	42	sd	143090
L_1	43	ic	25275(26)	42	sd	26110	42	sd	26900	42	sd	27700
L_2	43	ic	24385(26)	42	sd	25250	42	sd	26020	42	sd	26810
L_3	43	ic	19452(30)	42	sd	19930	42	sd	20410	42	sd	20900
M_1	43	ic	6556(31)	42	sd	6754	42	sd	6977	42	sd	7205
M_2	43	ic	6147(46)	42	sd	6359	42	sd	6574	42	sd	6793
M_3	43	ic	4977(46)	42	sd	5109	42	sd	5252	42	sd	5397
M_4	42	sd	4366	42	sd	4497	42	sd	4630	42	sd	4766
M_5	42	sd	4132	42	sd	4253	42	sd	4374	42	sd	4498
N_1	43	ic	1755(33)	42	sd	1799	42	sd	1868	42	sd	1937
N_2	42	sd	1554	42	sd	1616	42	sd	1680	42	sd	1747
N_3	42	sd	1235	42	sd	1279	42	sd	1321	42	sd	1366
O_1	43	ic	398(33)	42	sd	419	42	sd	435	42	sd	454

Level	Ref.	Meth.	101 Md	Ref.	Meth.	102 No	Ref.	Meth.	103 Lw
K	42	sd	146780	42	sd	150540	42	sd	154380
L_1	42	sd	28530	42	sd	29380	42	sd	30240
L_2	42	sd	27610	42	sd	28440	42	sd	29280
L_3	42	sd	21390	42	sd	21880	42	sd	22360
M_1	42	sd	7441	42	sd	7675	42	sd	7900
M_2	42	sd	7019	42	sd	7245	42	sd	7460
M_3	42	sd	5546	42	sd	5688	42	sd	5710
M_4	42	sd	4903	42	sd	5037	42	sd	5150
M_5	42	sd	4622	42	sd	4741	42	sd	4860
N_1	42	sd	2010	42	sd	2078	42	sd	2140
N_2	42	sd	1814	42	sd	1876	42	sd	1930
N_3	42	sd	1410	42	sd	1448	42	sd	1480
O_1	42	sd	472	42	sd	484	42	sd	490

a) Error in ref.[3] (private communication from M. O. Krause).

References:

1. J. E. Mack, 1949, as given in ref.[12] and revised in accordance with the 1963 atomic constant values by Bearden and Burr[3)4)].
2. G. Herzberg, 1957, as given in ref.[12] and revised in accordance with the 1963 atomic constant values by Bearden and Burr[3)4)].
3. J. A. Bearden and A. F. Burr, Rev. Mod. Phys. 39(1967)125.
4. J. A. Bearden and A. F. Burr, "Atomic Energy Levels", NYO 2543-1 (Fed. Sci. and Tech. Inf., U. S. Dept. of Commerce, Springfield, Virginia 22151).

4a. N. Swanson and K. Codling, J. Opt. Soc. Am. 58(1968)1192.

5. S. Hagström and S.-E. Karlsson, Ark. f. Fys. 26(1964)451; S. Hagström, Zeits. f. Phys. 178(1964)82.
6. R. Nordberg, R. G. Albridge, T. Bergmark, U. Ericson, J. Hedman, C. Nordling, K. Siegbahn, and B. J. Lindberg, Ark. f. Fys. 28(1968)257.
7. A. Fahlman, K. Hamrin, R. Nordberg, C. Nordling, and K. Siegbahn, Phys. Rev. Lett. 14(1965)127; R. Nordberg, K. Hamrin, A. Fahlman, C. Nordling, and K. Siegbahn, Zeits. f. Phys. 192(1968)462.
8. H. Körber and W. Mehlhorn, Zeits. f. Phys. 191(1966)217.
9. W. Lotz, J. Opt. Soc. Amer. 58(1968)236.
10. J. A. R. Samson and F. L. Kelley, "Photoionization Cross-Section of the Rare Gases", (Geographic Corp. of Amer. Tech. Report 64-3-N, 1964), binding energies cited in C. W. Nestor, T. C. Tucker, T. A. Carlson, L. D. Roberts, F. B. Malik, and C. Froese, Oak Ridge Nat. Lab. Report-4027(UC-32) Dec. 1966.
11. P. S. Bagus, Phys. Rev. 139(1965)A619.
12. C. E. Moore, "Atomic Energy Levels" (U. S. National Bureau of Standards, Washington, D. C., 1949-1958), Vol. 1, 2, 3.
13. K. Codling and R. P. Madden, Phys. Rev. 167(1968)587.

14. W. Mehlhorn, Zeits. f. Phys. 208(1968)1.

15. W. Mehlhorn, Habilitationschrift, University of Münster (Feb. 1967); W. Mehlhorn and D. Stalherm, Zeits. f. Phys. 217 (1968) 294.

16. T. Sagawa, quoted in ref.[17]; M. Nakamura, M. Sasanuma, S. Sato, M. Watanabe, H. Yamashita, Y. Iguchi, A. Ejiri, S. Nakai, S. Yamaguchi, T. Sagawa, Y. Nakai, and T. Oshio, Phys. Rev. Lett. 21(1958)1303.

17. R. D. Deslattes, Phys. Rev. Lett. 20(1968)483.

18. Private communication from M. O. Krause.

19. A. Fahlman, S. Hagström, K. Hamrin, R. Nordberg, C. Nordling, and K. Siegbahn, Ark. f. Fys. 31(1966)479.

20. E. Sokolowski, Ark. f. Fys. 15(1959)1.

21. C. Nordling, Ark. f. Fys. 15(1959)397.

22. E. Sokolowski, C. Nordling, and K. Siegbahn, Ark. f. Fys. 12(1957)301.

23. C. Nordling and S. Hagström, Ark. f. Fys. 16(1960)515.

24. I. Andersson and S. Hagström, Ark. f. Fys. 27(1964)161.

25. M. O. Krause, Phys. Rev. 140(1965)A1845.

26. W. Mehlhorn, Zeits. f. Phys. 187(1965)21.

27. R. Spohr, T. Bergmark, N. Magnusson, and K. Siegbahn, University of Uppsala (Prelim. rep., Oct. 1967).

28. K. Codling and R. P. Madden, Phys. Rev. Lett. 12(1964)106.

29. A. Fahlman, O. Hörnfeldt, and C. Nordling, Ark. f. Fys. 23(1962)75.

30. B. M. Hartley, Phys. Stat. Sol. 31(1969)259.

31. K. Siegbahn, C. Nordling, A. Fahlman, R. Nordberg, K. Hamrin, J. Hedman, G. Johansson, T. Bergmark, S.-E. Karlsson, I. Lindgren, and B. Lindberg, Nova Acta Reg. Soc. Upsaliensis, Ser. IV, Vol. 20 Uppsala (1968).

32. P. Bergvall, O. Hörnfeldt, and C. Nordling, Ark. f. Fys. 17(1960)113.

33. P. Bergvall and S. Hagström, Ark. f. Fys. 17(1960)61.

34. S. Hagström, Zeits. f. Phys. 178(1964)82.

35. A. Fahlman and S. Hagström, Ark. f. Fys. 27(1964)69.

36. J. Hedman and R. Nordberg, University of Uppsala, Internal Report UUIP-580; R. Nordberg, J. Hedman, C. Nordling, and K. Siegbahn (to be published).

37. S.-E. Karlsson, C.-H. Norberg, Ö. Nilsson, S. Högberg, A. H. El-Farrash, C. Nordling, and K. Siegbahn, Ark. f. Phys. (to be published).

38. C. Nordling and S. Hagström, Zeits. f. Phys. 178(1964)418.

39. C. Nordling and S. Hagström, Ark. f. Fys. 15(1959)431.

40. S. Hagström, Bull. Am. Phys. Soc. 11(1966)389.

41. A Fahlman, K. Hamrin, R. Nordberg, C. Nordling, K. Siegbahn, and L. W. Holm, Phys. Lett. 19(1966)643.

42. Theoretical values from relativistic self-consistent Slater-Dirac energy level calculations as presented in ref.[3][4]

43. J. M. Hollander, M. D. Holtz, T. Novakov, and R. L. Graham, Ark. f. Fys.28 (1965)375.

44. T. Yamazaki and J. M. Hollander, Nucl. Phys. 84(1966)505; see also J. L. Wolfson and J. J. H. Park, Canad. J. Phys. 42(1964)1387.

Appendix G

Gamma-Ray Energies

The energies of nuclear gamma-ray transitions as determined by several methods to an accuracy of $\gtrsim 10$ eV are tabulated here; the listed energies are given in kiloelectron volts. The parent level is a metastable state in the daughter isotope (column 5) or a ground or metastable state in another isotope which decays to that daughter. The half-life of this parent level is given in column 4. In column 6 the multipolarities of the gamma-transitions appear.

Three general methods of measurement have been used; these are abbreviated as follows:

B.C.S. bent crystal spectrometer—used for precision measurements of gamma-rays.

C.L. conversion line measurement (internal) —$B\rho$-values and kinetic energies for the various internal conversion lines are determined by comparison with standard, accurately known lines. Gamma-ray energies E_γ are calculated from

$$E_\gamma = \text{K.E.} + \text{B.E.} + \phi_{sp}, \quad (1)$$

where the symbols on the right-hand side of the equation stand for kinetic energy, binding energy, and spectrometer work function, respectively.

G.R.S. Gamma-ray scintillation counter—gamma-energies are determined; the instrument is first calibrated using standard sources.

The last column names the calibration standard employed in determining the named gamma-ray energy. Usually previously well-known conversion electron lines, Auger lines for unknown conversion line energy determinations, or gamma-rays for gamma-energies are used. The momentum ratio method is treated in Chapter 7, Section 3.2.3.

The gamma-energies in this table may be used to calculate highly accurate conversion electron energies that can be used as calibration standards in electron spectrometry work. The conversion electron kinetic energy, for a given level in the atom, can be obtained from equation 1. Binding energy values can be taken from Appendix F.

If the conversion electron source is a solid, the spectrometer work function value must be included. If it is in the gaseous state, this quantity is dropped from the expression. (See Chapter 7, Section 3.2.1.)

Gamma Energy (keV)	Ref.	Parent Isotope or Level	Half-Life of Parent Level	Daughter Isotope	Multipolarity	Method of Measurement	Calibration and Comments
14.389(6)	1	^{57}Co	270 d.	^{57}Fe	$M1$ (+0.0006% $E2$)	C.L.	Momentum ratio
14.4125(1)	2	^{57}Co	270 d.	^{57}Fe	$M1$ (+0.0006% $E2$)	B.C.S.	—
40.88(1)	3	^{62}Zn	9.2 hr	^{62}Cu	$M1$	C.L.	Th(B + C + C″) and ^{137}Cs C.L.
9.0(2)	4	^{72}Zn	46 hr	^{72}Ga	—	G.R.S.	Other gamma-rays in decay
16.4(3)	4	^{72}Zn	46 hr	^{72}Ga	$E1$ or $M1$	G.R.S.	Other gamma-rays in decay
2.38(36)	5	^{90}Mo	5.7 hr	^{90}Nb	$M2$	G.R.S.	Gamma-ray energy standards

Gamma Energy (keV)	Ref.	Parent Isotope or Level	Half-Life of Parent Level	Daughter Isotope	Multipolarity	Method of Measurement	Calibration and Comments
2.15(3)	6	^{99m}Tc	6 hr	^{99}Tc	*E*3	C.L.	(Investigate the chemical effects on lifetime.)
39.748(8)	7	^{103}Pd	17 d.	^{103}Rh	*E*3	C.L.	*KLL* Auger
3.18(15)	8	^{110m}Ag (excited)	253 d.	^{110}Ag	*E*1	G.R.S.	Gamma-ray energy differences [Neutron capture ^{109}Ag(n, γ)^{110}Ag]
88.033(9)	9	^{109}Cd	463 d.	^{109}Ag	*E*3	B.C.S.	(see also Ref. 10)
23.875(10)	11,	^{119m}Sn	250 d.	^{119}Sn	*M*1 (+0.11% *E*2)	C.L.	Momentum ratio
65.660(10)	12	^{119m}Sn	250 d.	^{119}Sn	*M*4	C.L.	Momentum ratio
80.1640(88)	13	^{131}I	8 d.	^{131}Xe	*M*1	B.C.S.	—
661.595(76)	14	^{137}Cs	30 yr	^{137}Ba	*M*4	C.L.	^{198}Au C.L.
80.997(6)	15	^{133}Ba	7.2 yr	^{133}Cs	*M*1 (+2.6% *E*2)	C.L.	1. Th(B + C + C″) 2. Momentum ratio
24.595(4)		^{140}La	40 hr	^{140}Ce	*E*2 (<4.2% *M*1)	C.L.	^{140}Ce C.L.
64.130(7)	16,	^{140}La	40 hr	^{140}Ce	*M*1 (<3.7% *E*2)	C.L.	^{140}Ce C.L.
68.916(6)	17	^{140}La	40 hr	^{140}Ce	*M*1 (<0.7% *E*2)	C.L.	^{140}Ce C.L.
109.417(6)		^{140}La	40 hr	^{140}Ce	*M*1 + *E*2	C.L.	^{140}Ce C.L.
91.1050(16)	18	^{147}Nd	11 d.	^{147}Pd	*M*1 (+*E*2)	B.C.S.	—
4.821(3)	19, 20	^{151}Pm	28 hr	^{151}Sm	*M*1 (+0.01% *E*2)	C.L.	^{198}Hg 412-keV transition (see Fig. 2, Chap. 6)
19.85(5)	21	^{153}Sm	47 hr	^{153}Eu	*E*2	C.L.	103-keV transition, C.L.
69.672(6)	22	^{153}Sm	47 hr	^{153}Eu	*M*1 (1.9% *E*2)	C.L.	103-keV transition, C.L.
103.175(4)	23	^{153}Sm	47 hr	^{153}Eu	*M*1 (1.7% *E*2)	B.C.S.	—
121.779(6)	24	^{152m}Eu	9.3 hr	^{152}Sm	*E*2	B.C.S.	—
94.694(5)	24	^{165}Dy	139 min	^{165}Ho	*M*1 (2% *E*2)	B.C.S.	—
28.227(5)	25	^{166}Dy	82 hr	^{166}Ho	*M*1	C.L.	^{166}Er, 80.6 keV C.L.
54.226(10)	25	^{166}Dy	82 hr	^{166}Ho	*E*2	C.L.	^{166}Er, 80.6 keV C.L.
82.454(9)	25	^{166}Dy	82 hr	^{166}Ho	*M*1	C.L.	^{166}Er, 80.6 keV C.L.
80.557(4)	24	^{166}Ho	27 hr	^{166}Er	*E*2	B.C.S.	—
8.401(8)	26	^{169}Er	9.3 d.	^{169}Tm	*M*1 (0.108% *E*2)	C.L.	Momentum ratio
84.262(4)	24	^{170}Tm	130 d.	^{170}Yb	*E*2	B.C.S.	—
9.150(3)		^{171}Lu	8.3 d.	^{171}Yb	*M*1 (+*E*2)	C.L.	^{172}Yb, 181-*K* line
19.384(3)		^{171}Lu	8.3 d	^{171}Yb	*E*1	C.L.	^{172}Yb, 181-*K* line
27.126(4)*		^{171}Lu	8.3 d.	^{171}Yb	*E*1	C.L.	^{172}Yb, 181-*K* line
46.516(5)*	27, 28	^{171}Lu	8.3 d.	^{171}Yb	*M*1 (1.6% *E*2)	C.L.	^{172}Yb, 181-*K* line
55.679(5)*		^{171}Lu	8.3 d.	^{171}Yb	*M*1 (0.25% *E*2)	C.L.	^{172}Yb, 181-*K* line
66.718(5)*		^{171}Lu	8.3 d.	^{171}Yb	*M*1 (66% *E*2)	C.L.	^{172}Yb, 181-*K* line
72.365(5)		^{171}Lu	8.3 d.	^{171}Yb	*M*1 (7% *E*2)	C.L.	^{172}Yb, 181-*K* line
181.519(6)	27	^{172}Lu	6.7 d.	^{172}Yb	*E*2	C.L.	^{198}Hg, 412-*K* line[29]
112.952(3)		^{177m}Lu	155 d.	^{177}Lu	—	B.C.S.	—
121.620(3)	30	^{177m}Lu	155 d.	^{177}Lu	*M*1 (19% *E*2)	B.C.S.	—
147.165(5)		^{177m}Lu	155 d.	^{177}Lu	*M*1 (35% *E*2)	B.C.S.	—
268.801(14)		^{177m}Lu	155 d.	^{177}Lu	*E*2	B.C.S.	—
4.64(2)	31	^{173}Hf	24 hr	^{173}Lu	*E*2	C.L.	^{175}Hf conversion lines
3.9	32, 33	^{181}Hf	45 d.	^{181}Ta	(*M*1)	C.L.	*L*-Auger
6.25(2)	34	^{181}W	145 d.	^{181}Ta	*E*1	C.L.	*L*-Auger (see Fig. 17, Chap. 3)
31.731(4)	35	^{182}Ta	115 d.	^{182}W	*E*1	C.L.	Gamma-energy values in ref. 36)
42.713(7)		^{182}Ta	115 d.	^{182}W	*E*1	C.L.	Gamma-energy values in ref. 36)

65.723(1)	36	^{182}Ta	115 d.	^{182}W	*M*1 (0.8% *E*2)	B.C.S.	—
67.751(1)		^{182}Ta	115 d.	^{182}W	*E*1 (0.02% *E*2)	B.C.S.	—
84.693(1)		^{182}Ta	115 d.	^{182}W	*M*1 (12% *E*2)	B.C.S.	—
100.107(1)		^{182}Ta	115 d.	^{182}W	*E*2	B.C.S.	—
52.598(1)	36	^{183}Ta	5.1 d.	^{183}W	*M*1 (1.4% *E*2)	B.C.S.	—
107.933(1)		^{183}Ta	5.1 d.	^{183}W	*M*1 (4% *E*2)	B.C.S.	—
71.995(4)	24	^{187}W	24 hr	^{187}Re	*E*1 (<0.3% *M*2)	B.C.S.	—
134.237(9)		^{187}W	24 hr	^{187}Re	*M*1 (3% *E*2)	B.C.S.	—
137.157(8)	24	^{186}Re	90 hr	^{186}Os	*E*2	B.C.S.	—
155.032(12)	24	^{188}Re	17 hr	^{188}Os	*E*2	B.C.S.	—
24.39(1)	37	^{191}Au	3.2 hr	^{191}Pt	*E*1 (0.02% *M*2)	C.L.	316-*K* C.L. in ^{192}Pt
30.40(1)		^{191}Au	3.2 hr	^{191}Pt	*M*1 (0.08% *E*2)	C.L.	316-*K* C.L. in ^{192}Pt
48.37(1)		^{191}Au	3.2 hr	^{191}Pt	*M*2	C.L.	316-*K* C.L. in ^{192}Pt
87.74(2)		^{191}Au	3.2 hr	^{191}Pt	*M*1 (6% *E*2)	C.L.	316-*K* C.L. in ^{192}Pt
1.642(2)	38, 39	^{193}Au	17 hr	^{193}Pt	(*M*1)	C.L.	C.L. in ^{192}Pt and Th(B + C + C″) and *KLL*-Auger lines See Fig. 14, Chap. 3
12.634(8)		^{193}Au	17 hr	^{193}Pt	*M*1 (~0.02% *E*2)	C.L.	
112.515(10)		^{193}Au	17 hr	^{193}Pt	*M*1 + *E*2	C.L.	
114.155(13)		^{193}Au	17 hr	^{193}Pt	*M*1 (~18% *E*2)	C.L.	
147.773(16)	40	^{196m}Au	9.7 hr	^{196}Au	*E*2	C.L.	Momentum ratio
174.867(21)		^{196m}Au	9.7 hr	^{196}Au	*E*2	C.L.	Momentum ratio
411.799(7)	29	^{198}Au	2.7 d.	^{198}Hg	*E*2	L_3-external C.L. from U converter	External C.L. of annihilation radiation
49.828(4)	41	^{199}Au	3.2 d.	^{199}Hg	*M*1	C.L.	C.L. of ^{198}Hg 412-keV trans.[29] (see Ref. 42)
158.370(6)		^{199}Au	3.2 d.	^{199}Hg	*E*2	C.L.	
77.345(8)	24	^{197}Hg	65 hr	^{197}Au	*M*1 (10% *E*2)	B.C.S.	—
46.503(15)	43	^{210}Pb (RaD)	21 yr	^{210}Bi (RaE)	*M*1	C.L.	Th(B + C + C″) C.L.
115.166(14)	44	^{212}Pb	10.6 hr	^{212}Bi	*M*1	C.L.	39.85 L_1-line in ^{208}Tl
238.624(9)	45	^{212}Pb	10.6 hr	^{212}Bi	*M*1	C.L.	^{198}Hg 412-keV trans.
26.22(1)	46	^{205}Bi	15.3 d.	^{205}Pb	*M*2	C.L.	1. Other trans. in ^{205}Pb. 2. *L*-Auger lines
39.85(1)	45	^{212}Bi (α-decay)	1 hr	^{208}Tl	*M*1 (+*E*2)	C.L.	^{198}Hg 412-keV trans. C.L. (See also Refs. 47, 48)
17.26(1)*	38	^{233}Pa	27 d.	^{233}U	—	C.L.	C.L. standards
28.54(2)*		^{233}Pa	27 d.	^{233}U	*M*1 (3%*E*2)	C.L.	C.L. standards
40.29(2)*		^{233}Pa	27 d.	^{233}U	*M*1 (30% *E*2)	C.L.	C.L. standards
(40.35(1))		^{233}Pa	27 d.	^{233}U	*M*1 (30% *E*2)	B.C.S.	—
75.13(3)*		^{233}Pa	27 d.	^{233}U	*M*1 (1% *E*2)	C.L.	C.L. standards
(75.28(1))		^{233}Pa	27 d.	^{233}U	*M*1 (1% *E*2)	B.C.S.	—
0.030(3)	50	^{235m}U	26 min	^{235}U	*E*3	C.L.	Retarding potential
13.81(2)	51	^{237}U	6.8 d.	^{237}Np	*M*1 (0.15% *E*2)	C.L.	1. ^{198}Hg C.L. 2. Momentum ratio (see also Refs. 52, 53).
26.348(10)		^{237}U	6.8 d.	^{237}Np	*E*1	C.L.	
33.195(11)		^{237}U	6.8 d.	^{237}Np	*M*1 (1.4% *E*2)	C.L.	
43.423(20)		^{237}U	6.8 d.	^{237}Np	*M*1 (14% *E*2)	C.L.	
59.543(15)		^{237}U	6.8 d.	^{237}Np	*E*1	C.L.	

* Estimated error.

REFERENCES

1. W. Mehlhorn and R. G. Albridge, *Z. Phys.* **175** (1963) 506.
2. J. A. Bearden, *Rev. Mod. Phys.* **37** (1967) 78.
3. S. Antman, H. Pettersson, and A. Suarez, *Nucl. Phys.* **A94** (1967) 289.
4. A. Kjelberg, E. Hagebø, and R. Nordhagen, *Nucl. Phys.* **A111** (1968) 193.

5. J. A. Cooper, J. M. Hollander, and J. O. Rasmussen, *Nucl. Phys.* **A109** (1967) 603.
6. M. S. Freedman, F. T. Porter, F. Wagner, Jr., and P. P. Day, *Phys. Rev.* **108** (1957) 836.
7. Y. Grunditz, S. Antman, H. Pettersson, and M. Saraceno, *Nucl. Phys.* **A113** (1969) 369.
8. Th. W. Elze, T. v. Egidy, H. U. Freund, B. C. Dutta, and W. Kaiser, *Z. Phys.* **209** (1968) 497.
9. W. R. Pierson and R. H. Marsch, *Nucl. Phys.* **A104** (1967) 511.
10. C. Foin, A. Gizon, and J. Oms, *Nucl. Phys.* **A113** (1968) 241.
11. J.-P. Bocquet, Y. Y. Chu, O. C. Kistner, M. L. Perlman, and G. T. Emery, *Phys. Rev. Letters* **17** (1966) 809.
12. J.-P. Bocquet, Y. Y. Chu, G. T. Emery, and M. L. Perlman, *Phys. Rev.* **167** (1968) 1117.
13. H. C. Hoyt and J. W. M. Du Mond, *Phys. Rev.* **91** (1953) 1027.
14. R. L. Graham, G. T. Ewan, and J. S. Geiger, *Nucl. Instr. Meth.* **9** (1960) 245.
15. K. Siegbahn, C. Nordling, S.-E. Karlsson, S. Hagström, A. Fahlman, and I. Andersson, *Nucl. Instr. Meth.* **27** (1964) 173.
16. S.-E. Karlsson, B. Svahn, H. Pettersson, and G. Malmsten, *Nucl. Phys.* **A100** (1967) 113.
17. H. W. Baer, J. J. Reidy, and M. L. Wiedenbeck, *Nucl. Phys.* **A113** (1968) 33.
18. J. C. Hill and M. L. Wiedenbeck, *Nucl. Phys.* **A98** (1967) 599.
19. J. S. Geiger, R. L. Graham, and J. S. Merritt, *Nucl. Phys.* **48** (1963) 97.
20. J. S. Geiger, R. L. Graham, and J. S. Merritt, *Proc. Int. Conf. Role of Atomic Electrons in Nuclear Transformations*, Warsaw, 1963, p. 454.
21. E. Monnand and A. Moussa, *Nucl. Phys.* **25** (1961) 292.
22. T. Suter, P. Reyes-Suter, S. Gustafsson, and I. Marklund, *Nucl. Phys.* **29** (1962) 33.
23. P. Bergvall, *Ark. Fys.* **17** (1960) 125.
24. I. Marklund and B. Lindström, *Nucl. Phys.* **40** (1963) 329.
25. V. Brabec, O. Bergman, Y. Grunditz, E. Aasa, and S.-E. Karlsson, *Ark. Fys.* **26** (1964) 511.
26. T. A. Carlson, P. Erman, and K. Fransson, *Nucl. Phys.* **A111** (1968) 371.
27. G. Kaye, *Nucl. Phys.* **68** (1965) 529.
28. G. Kaye, *Nucl. Phys.* **86** (1966) 241.
29. G. Murray, R. L. Graham, and J. S. Geiger, *Nucl. Phys.* **45** (1963) 177.
30. B. P. K. Maier, *Z. Phys.* **184** (1965) 153.
31. J. Valentin, D. J. Horen, and J. M. Hollander, *Nucl. Phys.* **31** (1962) 353.
32. A. H. Muir, Jr., *Nucl. Phys.* **68** (1965) 305.
33. A. H. Muir, Jr., and F. Boehm, *Phys. Rev.* **122** (1961) 1564.
34. P. Alexander, H. Ryde, and E. Seltzer, *Nucl. Phys.* **76** (1966) 167.
35. Ö. Nilsson, S. Högberg, S.-E. Karlsson, and G. M. El-Sayad, *Nucl. Phys.* **A100** (1967) 351.
36. E. J. Seppi, H. Henrikson, F. Boehm, and J. W. M. Du Mond, *Nucl. Instr. Meth.* **16** (1962) 17.
37. A. Johansson, B. Nyman, G. Malmsten, and S.-E. Karlsson, *Nucl. Phys.* **98** (1967) 278.
38. B. Svahn, A. Johansson, B. Nyman, G. Malmsten, and H. Pettersson, *Z. Phys.* **210** (1968) 466.
39. A. Johansson, G. Malmsten, A. Marelius, B. Nyman, H. Pettersson, and B. Svahn, *Phys. Letters* **26B** (1967) 83.
40. C. de Vries and J. H. Dijkstra, *Nucl. Phys.* **18** (1960) 446.
41. G. Kaye and R. L. Graham, *Can. J. Phys.* **42** (1964) 1865.
42. G. Bäckström, O. Bergman, and J. Burde, *Nucl. Phys.* **7** (1958) 263.
43. M. Filley and M. Valadares, *J. Phys. Radium* **18** (1957) 468.
44. K. D. Sevier, *Ark. Fys.* **25** (1963) 87.
45. R. L. Graham, G. Murray, and J. S. Geiger, *Can. J. Phys.* **43** (1965) 171.
46. R. Stockendal, *Ark. Fys.* **17** (1960) 553.
47. J. S. Dionisio, *J. Rech. C.N.R.S.* **42** (1958) 61.
48. J. S. Dionisio, *J. Rech. C.N.R.S.* **48** (1959) 231.
49. R. G. Albridge and J. M. Hollander, *Nucl. Phys.* **27** (1961) 529.
50. H. Mazaki and S. Shimizer, *Phys. Letters* **23** (1966) 137.
51. T. Yamazaki and J. M. Hollander, *Nucl. Phys.* **84** (1966) 505.
52. J. L. Wolfson and J. J. H. Park, *Can. J. Phys.* **42** (1964) 1387.
53. P. P. Day, *Phys. Rev.* **97** (1955) 689.

Appendix H

The Differential Cross section for Inelastic Electron Scattering in Solids

The expression for electron scattering in thin foils, originally worked out by Ritchie,[1] is of importance in the study and analysis of characteristic electron energy losses (Chapter 8). Ritchie's expression has been rephrased in two main forms, that by Hattori and Yamada[2] and the modified Romanov[3] form by Geiger and Wittmaack.[4] Independently, each of these two expressions of the formula has been of extensive utility, as is seen in Chapter 8. However, because of errors due to misprints in the original articles and in view of the significance of these equations, it appears desirable to present here the derivation of the Hattori–Yamada and Romanov formulas from the original Ritchie formula.*

Ritchie's formula[1] states that

$$W(\omega, k_\perp)\, dk_\perp\, d\omega = -\frac{e^2}{\pi^2 \hbar v^2}\, dk_\perp\, d\omega \times \left[\frac{a}{k_\perp{}^2 + \omega^2/v^2} \operatorname{Im}\left(\frac{1}{\varepsilon}\right) + \frac{2k_\perp}{(k_\perp{}^2 + \omega^2/v^2)^2} \operatorname{Im} A\right] \tag{1a}$$

where

$$A = \left(\frac{1-\varepsilon}{\varepsilon}\right) \times \frac{2(\varepsilon - 1)\cos \omega a/v + (\varepsilon - 1)^2 e^{-k_\perp a} + (1 - \varepsilon^2) e^{k_\perp a}}{(\varepsilon - 1)^2 e^{-k_\perp a} - (\varepsilon + 1)^2 e^{k_\perp a}}. \tag{1b}$$

Equation 1a differs from Ritchie's by a factor (-1); also in equation 1b $(\varepsilon - 1)$ in the second term in the numerator is replaced[4] by $(\varepsilon - 1)^2$, $W(\omega, k_\perp)$ expresses the probability of absorption of an amount of energy $\hbar\omega$ and momentum $\hbar k_\perp$ in the direction perpendicular to $\hbar \mathbf{k}_0$, the initial momentum vector of the electron. The $\varepsilon = \varepsilon_1 + i\varepsilon_2$ is the dielectric constant for the material, a its thickness, and v the bombarding electron velocity.

Now $k_\perp = (mv/\hbar)\vartheta = k_0 \vartheta$ for small scattering angles ϑ, and $k_\perp\, dk_\perp = (mv/\hbar)^2 \vartheta\, d\vartheta$, where $\hbar \mathbf{k}_0$ is the initial electron momentum. Note that $W(\omega, \Omega)$ can be replaced by $W(\omega, \vartheta)$, the expression used in Chapter 8. Setting $2\pi\vartheta\, d\vartheta = d\Omega$, the solid angle into which the electron becomes scattering, and $\vartheta_E = \Delta E/2E_0 = \hbar\omega/mv^2$, we have

$$\begin{aligned} W(\omega, \Omega)\, d\omega\, d\Omega &= W(\omega, k_\perp) 2\pi k_\perp\, dk_\perp\, d\omega \\ &= -\frac{e^2}{\pi^2 \hbar v^2}\left[\frac{a}{k_\perp{}^2 + \omega^2/v^2} \operatorname{Im}\left(\frac{1}{\varepsilon}\right) + \frac{2k_\perp}{(k_\perp{}^2 + \omega^2/v^2)^2} \operatorname{Im} A\right] 2\pi k_\perp\, dk_\perp\, d\omega \\ &= -\frac{e^2}{\pi^2 \hbar v^2}\left[\frac{a}{k_0{}^2(\vartheta^2 + \vartheta_E{}^2)} \operatorname{Im}\left(\frac{1}{\varepsilon}\right) + \frac{2\vartheta}{k_0{}^3(\vartheta^2 + \vartheta_E{}^2)^2} \operatorname{Im} A\right] \frac{1}{\hbar}\left(\frac{mv}{\hbar}\right)^2 d\omega\, d\Omega \\ &= -\frac{e^2}{\pi^2 \hbar^2 v^2}\left[\frac{a}{\vartheta^2 + \vartheta_E{}^2} \operatorname{Im}\left(\frac{1}{\varepsilon}\right) + \frac{2\vartheta}{k_0(\vartheta^2 + \vartheta_E{}^2)^2} \operatorname{Im} A\right] d\omega\, d\Omega. \end{aligned} \tag{2}$$

The expression A in equation 1b may be altered as follows:

* I thank Prof. J. Geiger of Berlin, now of Kaiserslautern, for providing these derivations.

$$A = \left[\frac{(1-\varepsilon)^2}{\varepsilon}\right] \frac{(\varepsilon+1)e^{ka} - (\varepsilon-1)e^{-ka} - e^{i(\omega/v)a} - e^{-i(\omega/v)a}}{(\varepsilon-1)^2 e^{-ka} - (\varepsilon+1)^2 e^{ka}}$$

$$\left[\frac{(1-\varepsilon)^2}{\varepsilon}\right]\left\{\frac{\begin{array}{r}-\frac{1}{4}(e^{i(\omega/v)a} + e^{-i(\omega/v)a} - 2)[(\varepsilon+1)e^{ka} + 2 - (\varepsilon-1)e^{-ka}] \\ +\frac{1}{4}(e^{i(\omega/v)a} + e^{-i(\omega/v)a} + 2)[(\varepsilon+1)e^{ka} - 2 - (\varepsilon-1)e^{-ka}]\end{array}}{(\varepsilon-1)^2 e^{-ka} - (\varepsilon+1)^2 e^{ka}}\right\}$$

$$= (-1)\left[\frac{(1-\varepsilon)^2}{\varepsilon}\right]\left\{\frac{-\frac{1}{4}(e^{i(\omega/v)a} + e^{-i(\omega/v)a} - 2)(e^{ka/2} + e^{-ka/2})}{(\varepsilon+1)e^{ka/2} + (\varepsilon-1)e^{-ka/2}} + \frac{\frac{1}{4}(e^{i(\omega/v)a} + e^{-i(\omega/v)a} + 2)(e^{ka/2} - e^{-ka/2})}{(\varepsilon+1)e^{ka/2} - (\varepsilon-1)e^{-ka/2}}\right\}$$

$$= (-1)\left[\frac{(1-\varepsilon)^2}{\varepsilon}\right]\left\{\frac{\left[\frac{1}{2i}(e^{i(\omega/2v)a} - e^{-i(\omega/2v)a})\right]^2}{\varepsilon + \dfrac{e^{ka/2} - e^{-ka/2}}{e^{ka/2} + e^{-ka/2}}} + \frac{[\frac{1}{2}(e^{i(\omega/2v)a} + e^{-i(\omega/2v)a})]^2}{\varepsilon + \dfrac{e^{ka/2} + e^{-ka/2}}{e^{ka/2} - e^{-ka/2}}}\right\}$$

$$= (-1)\frac{(1-\varepsilon)^2}{\varepsilon}\left(\frac{\sin^2 \omega a/2v}{\varepsilon + \tanh ka/2} + \frac{\cos^2 \omega a/2v}{\varepsilon + \coth ka/2}\right). \tag{3}$$

Consequently, we obtain the Hattori–Yamada form of the scattering cross-section expression:†

$$W(\omega, \Omega)\, d\omega\, d\Omega = -\frac{e^2}{\pi^2\hbar^2 v^2}\left[\frac{a}{\vartheta^2 + \vartheta_E{}^2}\,\mathrm{Im}\left(\frac{1}{\varepsilon}\right) - \frac{2\vartheta}{k_0(\vartheta^2 + \vartheta_E{}^2)^2} \times \mathrm{Im}\,\frac{(1-\varepsilon)^2}{\varepsilon}\left(\frac{\sin^2 \omega a/2v}{\varepsilon + \tanh ka/2} + \frac{\cos^2 \omega a/2v}{\varepsilon + \coth ka/2}\right)\right] d\omega\, d\Omega. \tag{4}$$

The following alternative forms of some of the included expressions are sometimes used:

$$\frac{\omega a}{2v} = \frac{ak_0\vartheta_E}{2} \quad \text{and} \quad \frac{e^2}{\pi^2\hbar^2 v^2} = \frac{1}{2\pi^2 E_0 a_0},$$

where a_0 is the first Bohr radius.

Going on to the Romanov expression, we set

$$\mathrm{Im}\, A = \mathrm{Im}\,(1-\varepsilon)^2\left[\frac{\sin^2 \omega a/2v}{\varepsilon \tanh ka/2} + \frac{\cos^2 \omega a/2v}{\varepsilon \coth ka/2} - \frac{\sin^2 \omega a/2v}{(\varepsilon + \tanh ka/2)\tanh ka/2} - \frac{\cos^2 \omega a/2v}{(\varepsilon + \coth ka/2)\coth ka/2}\right]$$

$$= \left(\frac{\sin^2 \omega a/2v}{\tanh ka/2} + \frac{\cos^2 \omega a/2v}{\coth ka/2}\right)\mathrm{Im}\,\frac{(1-\varepsilon)^2}{\varepsilon} - \frac{\sin^2 \omega a/2v}{\tanh ka/2}\,\mathrm{Im}\,\frac{(1-\varepsilon)^2}{(\varepsilon + \tanh ka/2)} - \frac{\cos^2 \omega a/2v}{\coth ka/2}\,\mathrm{Im}\,\frac{(1-\varepsilon)^2}{(\varepsilon + \coth ka/2)}.$$

† In the expression by Hattori and Yamada the term $(1-\varepsilon)^2/\varepsilon$ was omitted by misprint.

The various terms may be reexpressed by use of

$$\mathrm{Im}\,\frac{(1-\varepsilon)^2}{\varepsilon} = \mathrm{Im}\left(\frac{1}{\varepsilon} - 2 + \varepsilon\right) = -\frac{\varepsilon_2}{\varepsilon_1{}^2 + \varepsilon_2{}^2} + \varepsilon_2 = \mathrm{Im}\left(\frac{1}{\varepsilon}\right) + \mathrm{Im}\,(\varepsilon),$$

$$\mathrm{Im}\,\frac{(1-\varepsilon)^2}{(\varepsilon + \tanh ka/2)} = \mathrm{Im}\left[\frac{1}{\varepsilon + \tanh ka/2} - \frac{2\varepsilon}{\varepsilon + \tan ka/2} + \frac{\varepsilon^2}{\varepsilon + \tanh ka/2}\right]$$

$$= \varepsilon_2 - \frac{\varepsilon_2(1 + \tanh ka/2)^2}{(\varepsilon_1 + \tanh ka/2)^2 - \varepsilon_2{}^2}$$

$$= \mathrm{Im}\,(\varepsilon) + \left(1 + \tan\frac{ka}{2}\right)^2 \mathrm{Im}\left(\frac{1}{\varepsilon + \tanh ka/2}\right),$$

$$\mathrm{Im}\,\frac{(1-\varepsilon)^2}{(\varepsilon + \coth ka/2)} = \varepsilon_2 - \frac{\varepsilon_2(1 + \coth ka/2)^2}{(\varepsilon_1 + \coth ka/2)^2 + \varepsilon_2{}^2}$$

$$= \mathrm{Im}\,(\varepsilon) + \left(1 + \coth\frac{ka}{2}\right)^2 \times \mathrm{Im}\left(\frac{1}{\varepsilon + \coth ka/2}\right).$$

Collecting terms,

$$\operatorname{Im} A = \left(\frac{\sin^2 \omega a/2v}{\tanh ka/2} + \frac{\cos^2 \omega a/2v}{\coth ka/2}\right) \operatorname{Im}\left(\frac{1}{\varepsilon}\right) - \frac{(\sinh ka/2 + \cosh ka/2)^2}{\sinh ka/2 \cosh ka/2} \times \operatorname{Im}\left(\frac{\sin^2 \omega a/2v}{\varepsilon + \tanh ka/2} + \frac{\cos^2 \omega a/2v}{\varepsilon + \coth ka/2}\right),$$

the coefficients of Im (ε) having vanished. Simplifying,

$$\coth\frac{ka}{2}\sin^{-2}\frac{\omega a}{2v} + \tanh\frac{ka}{2}\cos^2\frac{\omega a}{2v} = \frac{1+e^{-ka}}{1-e^{-ka}}\cdot\frac{1}{2}\left(1-\cos\frac{\omega a}{v}\right) + \frac{1-e^{-ka}}{1+e^{-ka}}\cdot\frac{1}{2}\left(1+\cos\frac{\omega a}{v}\right) = \frac{\cosh ka - \cos \omega a/2}{\sinh ka},$$

and

$$\frac{(\sinh ka/2 + \cosh ka/2)^2}{\sinh ka/2 \cosh ka/2} = 2\,\frac{\sinh ka + \cosh ka}{\sinh ka} = -\frac{4}{1-e^{-2ka}}.$$

The Romanov expression, as expressed by Geiger and Wittmaack, is now

$$W(\omega, \Omega)\, d\omega\, d\Omega = -\frac{e^2}{\pi^2\hbar^2 v^2}\left\{\left[\frac{a}{\vartheta^2+\vartheta_E{}^2} - \frac{2\vartheta}{k_0(\vartheta^2+\vartheta_E{}^2)^2}\,\frac{\cosh ak - \cos \omega a/v}{\sinh ak}\right]\operatorname{Im}\left(\frac{1}{\varepsilon}\right) + \frac{8\vartheta}{k_0(\vartheta^2+\vartheta_E{}^2)^2(1-e^{-2ak})}\operatorname{Im}\left[\frac{\sin^2 \omega a/2v}{\varepsilon + \tanh ak/2} + \frac{\cos^2 \omega a/2v}{\varepsilon + \coth ak/2}\right]\right\} d\omega\, d\Omega. \tag{5}$$

Expression 5 differs from that by Romanov in the form of the second term and by the omission here of a factor $(1 - e^{-\omega\beta})^{-1}$ multiplying the entire expression. Since $\beta = \hbar/k_B T$ ($k_B \equiv$ Boltzmann's constant), the factor is very close to unity.

REFERENCES

1. R. H. Ritchie, *Phys. Rev.* **106** (1957) 874.
2. M. Hattori and K. Yamada, *J. Phys. Soc.* (*Japan*) **18** (1963) 200.
3. Yu. A. Romanov, *J. Exp. Theor. Phys., SSSR*, **47** (1964) 2119 [*Sov. Phys. J.E.T.P.* **20** (1965) 1424].
4. J. Geiger and K. Wittmaack, *Z. Phys.* **195** (1966) 44.

Author Index

Numbers in parentheses indicate the number of times an author's work appears on a given page. Letters in italics refer to appendix so indicated.

Subject Index